Antidiabetic Medicinal Plants and Herbal Treatments

Diabetes is a chronic condition associated with metabolic disorder. Persons suffering from diabetes have shown accelerated levels of blood sugar which often harms the heart, blood vessels, eyes, kidneys, and nerves. Over the past few decades, the prevalence of diabetes has been progressively increasing. Synthetic drugs are used to treat diabetic patients to help control the disorder, but it is shown that numerous medicinal plants and herbal drugs are widely used in several traditional systems of medicine to prevent and treat diabetes. They are reported to produce beneficial effects in combating diabetes and alleviating diabetes-related complications. These plants contain phytonutrients and phytoconstituents demonstrating protective or disease preventive properties. In many developing countries, herbal drugs are recommended by traditional practitioners for diabetes treatment because the use of synthetic drugs is not affordable.

KEY FEATURES

- Provides botanical descriptions, distribution, and pharmacological investigations of notable medicinal and herbal plants used to prevent or treat diabetes.
- Discusses phytochemical and polyherbal formulations for the management of diabetes and other related complications.
- Contains reports on antidiabetic plants and their potential uses in drug discovery based on their bioactive molecules.

This volume in the *Exploring Medicinal Plants* series provides an overview of natural healing treatments in selected antidiabetic plants. The book presents valuable information to scientists, researchers, and students working with medicinal plants or for those specializing in areas of ethnobotany, natural products, pharmacognosy, and other areas of allied healthcare. It is also useful to pharmaceutical companies, industrialists, and health policy makers.

Exploring Medicinal Plants

Series Editor
Azamal Husen
Wolaita Sodo University, Ethiopia

Medicinal plants render a rich source of bioactive compounds used in drug formulation and development; they play a key role in traditional or indigenous health systems. As the demand for herbal medicines increases worldwide, supply is declining as most of the harvest is derived from naturally growing vegetation. Considering global interests and covering several important aspects associated with medicinal plants, the *Exploring Medicinal Plants* series comprises volumes valuable to academia, practitioners, and researchers interested in medicinal plants. Topics provide information on a range of subjects including diversity, conservation, propagation, cultivation, physiology, molecular biology, growth response under extreme environment, handling, storage, bioactive compounds, secondary metabolites, extraction, therapeutics, mode of action, and healthcare practices.

Led by Azamal Husen, PhD, this series is directed to a broad range of researchers and professionals consisting of topical books exploring information related to medicinal plants. It includes edited volumes, references, and textbooks available for individual print and electronic purchases.

Plants as Medicine and Aromatics
Conservation, Ecology, and Pharmacognosy
Mohd Kafeel Ahmad Ansari, Bengu Turkyilmaz Unal, Munir Ozturk, and Gary Owens

Sustainable Uses of Medicinal Plants
Learnmore Kambizi and Callistus Bvenura

Medicinal Plant Responses to Stressful Conditions
Arafat Abdel Hamed Abdel Latef

Aromatic and Medicinal Plants of Drylands and Deserts
Ecology, Ethnobiology and Potential Uses
David Ramiro Aguillón Gutiérrez, Cristian Torres León, and Jorge Alejandro Aguirre Joya

Secondary Metabolites from Medicinal Plants
Nanoparticles Synthesis and Their Applications
Rakesh Kumar Bachheti and Archana Bachheti

Aquatic Medicinal Plants
Archana Bachheti, Rakesh Kumar Bachheti, and Azamal Husen

Antidiabetic Medicinal Plants and Herbal Treatments
Edited by Azamal Husen

For more information about this series, please visit: www.routledge.com/Exploring-Medicinal-Plants/book-series/CRCEMP

Antidiabetic Medicinal Plants and Herbal Treatments

Edited by Azamal Husen

CRC Press
Taylor & Francis Group
Boca Raton London New York

CRC Press is an imprint of the
Taylor & Francis Group, an **informa** business

First edition published 2024
by CRC Press
6000 Broken Sound Parkway NW, Suite 300, Boca Raton, FL 33487–2742

and by CRC Press
4 Park Square, Milton Park, Abingdon, Oxon, OX14 4RN

CRC Press is an imprint of Taylor & Francis Group, LLC

Library of Congress Cataloging-in-Publication Data
Names: Husen, Azamal, editor.
Title: Antidiabetic medicinal plants and herbal treatments / Edited by Azamal Husen.
Description: First edition. | Boca Raton, FL : CRC Press, 2023. | Includes bibliographical references and index.
Summary: "Diabetes is a chronic problem associated with the metabolic disorder. Persons suffering from diabetes have shown an accelerated levels of blood sugar, which often harms the heart, blood vessels, eyes, kidneys and nerves. Over the past few decades, the prevalence of diabetes has been progressively increasing. In this connection, synthetic drugs are used to treat the diabetic patients and keep the disorder under control. However, in the form of substitute numerous medicinal plants and herbal drugs are widely used in several traditional systems of medicine to prevent diabetes, and Several plant-based drugs are now available in the market. They have been reported to show multiple beneficial effects in combating diabetes and alleviating the diabetes-related complications. Numerous phytonutrients and or phytoconstituents have been identified and extracted from these plants. They have shown protective or disease preventive properties. In many developing countries, diabetes treatment with synthetic drugs looks to be an expensive process due to poverty and lack of free medical facility etc. Thus, herbal drugs are recommended by traditional practitioners for their therapeutic value and/or antidiabetic response. Taken together, the aim of the book is to provide an overview on the most important and selected antidiabetic plants and the natural products obtained from these plants. This book provide valuable information to scientists, researchers and students, working specially on medicinal plants, or in areas of ethnomedicine, natural products, economic botany, plant biochemistry, biotechnology, pharmacognosy, industrial chemistry, and other allied subjects. This book also help in testifying the various reports on antidiabetic plants and their other potential uses in new drug discovery based on their bioactive molecules, and also useful for pharmaceutical companies, industrialists and health policy makers"— Provided by publisher.
Identifiers: LCCN 2022059801 | ISBN 9781032386263 (hardback) | ISBN 9781032386270 (paperback) | ISBN 9781003345947 (ebook)
Subjects: LCSH: Medicinal plants. | Pharmacognosy. | Hypoglycemic agents. | Diabetes.
Classification: LCC RC661.H4 A678 2023 | DDC 615.3/21—dc23/eng/20230414
LC record available at https://lccn.loc.gov/2022059801

ISBN: 978-1-032-38626-3 (hbk)
ISBN: 978-1-032-38627-0 (pbk)
ISBN: 978-1-003-34594-7 (ebk)

DOI: 10.1201/b23347

Typeset in Times
by Apex CoVantage, LLC

Thank you so much to my dearest wife for providing time and extended support and making my ordinary day an extraordinary one

Contents

Preface

Diabetes is a chronic problem associated with a metabolic disorder. Persons suffering from diabetes have shown accelerated levels of blood sugar, which often harms the heart, blood vessels, eyes, kidneys, and nerves. The most frequently noticed diabetes is type 2 diabetes, usually recorded in adults. In this case, the human body becomes resistant to insulin or doesn't produce enough insulin. This form of diabetes was previously known as non-insulin-dependent diabetes mellitus or adult-onset diabetes. The occurrence of type 2 diabetes has increased intensely worldwide. However, type 1 diabetes develops due to the body's malfunctioning towards insulin production and necessitates insulin injection. In this case, the pancreas produces little or no insulin by itself. This type 1 was earlier known as insulin-dependent diabetes mellitus or juvenile diabetes. Overall, over the past few decades, the prevalence of diabetes has been progressively increasing.

In this connection, synthetic drugs are used to treat the diabetic patients and keep the disorder under control. However, in the form of a substitute, numerous medicinal plants and herbal drugs are widely used in several traditional systems of medicine to prevent diabetes, and several plant-based drugs are now available in the market. They have been reported to show multiple beneficial effects in combating diabetes and alleviating diabetes-related complications. Numerous phytonutrients and/or phytoconstituents have been identified and extracted from these plants. They have shown protective or disease preventive properties. Mechanisms of action of medicinal plants and herbal drugs in terms of antidiabetic response have also been indicated, such as inhibition of renal glucose reabsorption, stimulation of insulin secretion from beta cells of islets, inhibition of insulin degradative processes, reduction in insulin resistance generation, repair of pancreatic beta cells, and so on. Many plants also produce positive effects on carbohydrate and lipid metabolisms. They improve glucose tolerance in healthy humans as well as in diabetic patients and cause significant reduction in blood glucose, glycosylated haemoglobin, and glycosylated plasma proteins. These antidiabetic plants reportedly scavenge free radicals, quench electronically excited compounds, reduce hydroperoxide formation, and attenuate production of reactive oxygen species through modulation of several enzymes including xanthine oxidase, cyclooxygenase, lipoxygenase, microsomal monooxygenase, NADH oxidase, and mitochondrial succinoxidase. Additionally, a great many phytonutrients and/or phytoconstituents, for instance polyphenols, are known to enhance the endogenous antioxidant system, improve oxidant-antioxidant balance, prevent oxidative damage, decrease lipid peroxidation, increase total antioxidant capacity of plasma, and induce several antioxidant enzymes, including superoxide dismutase, catalase, and glutathione peroxidase. Further, in many developing countries, diabetes treatment with synthetic drugs looks to be an expensive process due, for example, to poverty and lack of free medical facilities. Thus, herbal drugs are recommended by traditional practitioners for their therapeutic value and/or antidiabetic response.

Taken together, the aim of the book is to provide an overview of the most important and selected antidiabetic plants and the natural products obtained from these plants. This book will provide valuable information to scientists, researchers, and students, working specially on medicinal plants or in areas of ethnomedicine, natural products, economic botany, plant biochemistry, biotechnology, pharmacognosy, industrial chemistry, and other allied subjects. This book would also help validate the various reports on antidiabetic plants and their other potential uses in new drug discovery based on their bioactive molecules, and it may also attract the attention of pharmaceutical companies, industrialists, and health policy makers.

I am grateful to all contributors for readily accepting my invitation, sharing their knowledge in specialized areas of research, and readily adjusting the suggestions for improving the shape of their contributions. With great pleasure, I also extend my sincere thanks to Randy Brehm, Tom Connelly, and all associates at Taylor & Francis Group, LLC/CRC Press for their active cooperation. Finally, my special thanks go to Shagufta, Zaara, Mehwish, and Huzaifa for providing their time and overall support to put everything together.

Azamal Husen

The Editor

Azamal Husen has served as Professor and Head, Department of Biology, University of Gondar, Ethiopia, and is a foreign delegate at Wolaita Sodo University, Wolaita, Ethiopia. Earlier, he was a visiting faculty member of the Forest Research Institute and the Doon College of Agriculture and Forest at Dehra Dun, India. His research and teaching experience of 20 years involves studies of biogenic nanomaterial fabrication and application, plant responses to environmental stresses, and nanomaterials at the physiological, biochemical, and molecular levels, herbal medicine, and clonal propagation for improvement of tree species.

He has conducted several research projects sponsored by various funding agencies, including the World Bank (Forestry Research Education and Extension Project), the National Agricultural Technology Project (NATP), the Indian Council of Agriculture Research (ICAR), the Indian Council of Forest Research Education (ICFRE); and the Japan Bank for International Cooperation (JBIC). He received four fellowships from India and a recognition award from the University of Gondar, Ethiopia, for excellent teaching, research, and community service. Husen has been on the editorial board and the panel of reviewers of several reputed journals published by Elsevier, Frontiers Media, Taylor & Francis, Springer Nature, RSC, Oxford University Press, Sciendo, The Royal Society, CSIRO, PLOS, MDPI, John Wiley & Sons, and UPM Journals. He is on the advisory board of Cambridge Scholars Publishing, UK. He is a fellow of the Plantae group of the American Society of Plant Biologists and a member of the International Society of Root Research, Asian Council of Science Editors, and INPST. He has over 200 publications to his credit, and he is Editor-in-Chief of the *American Journal of Plant Physiology*.

He is also Series Editor of 'Exploring Medicinal Plants,' published by Taylor & Francis Group, USA; 'Plant Biology, Sustainability, and Climate Change,' published by Elsevier, USA; and 'Smart Nanomaterials Technology,' published by Springer Nature Singapore Pte Ltd. Singapore.

Contributors

Balahbib Abdelaali
Laboratory of Biodiversity, Ecology, and Genome
Faculty of Sciences
Mohammed V University in Rabat
Rabat, Morocco

Jamal Akhtar
Central Council for Research in Unani Medicine
Ministry of AYUSH
New Delhi, India

Khalid Mashay Al-Anazi
Department of Zoology
College of Science
King Saud University
Riyadh, Saudi Arabia

Raya Algonaiman
Department of Food Science and Human Nutrition
College of Agriculture and Veterinary Medicine
Qassim University
Buraydah, Saudi Arabia

Ahmad Ali
Department of Life Sciences
University of Mumbai, Vidyanagari
Mumbai, Maharashtra, India

Abdel Rahman Al-Tawaha
Department of Biological Sciences
Al-Hussein Bin Talal University
Maan, Jordan

Phaniendra Alugoju
Natural Products for Neuroprotection and Anti-Ageing Research Unit
Chulalongkorn University
Bangkok, Thailand
And
Department of Clinical Chemistry
Faculty of Allied Health Sciences
Chulalongkorn University
Bangkok, Thailand

Nighat Anjum
Central Council for Research in Unani Medicine
Ministry of AYUSH
New Delhi, India

Atul Arya
Medicinal Plant Research Laboratory
Department of Botany
Ramjas College
University of Delhi
New Delhi, India

Anoja Priyadarshani Attanayake
Department of Biochemistry
Faculty of Medicine
University of Ruhuna
Galle, Sri Lanka

El Hamri Ayoub
Department of Chemistry
Faculty of Sciences
Mohammed V University in Rabat
Rabat, Morocco

Saad Bakrim
Geo-Bio-Environment Engineering and Innovation Laboratory
Molecular Engineering
Biotechnology, and Innovation Team
Polydisciplinary Faculty of Taroudant
Ibn Zohr University
Agadir, Morocco

Hassan Barakat
Department of Food Science and Human Nutrition
College of Agriculture and Veterinary Medicine
Qassim University Buraydah
Saudi Arabia
And
Department of Food Technology
Faculty of Agriculture
Benha University
Moshtohor, Egypt

Fouzia Bashir
Central Council for Research in Unani Medicine
Ministry of Ayush
Government of India, 61–65, Institutional Area
D Block Janakpuri
New Delhi, India

Ankita Bhardwaj
Department of Silviculture and Agroforestry
Kerala Agricultural University
Thrissur, Kerala, India

Protha Biswas
Department of Life Sciences
Presidency University, 86/1 College Street
Kolkata, India

Abdelhakim Bouyahya
Laboratory of Human Pathologies Biology
Department of Biology
Faculty of Sciences
Mohammed V University
Rabat, Morocco

Kolagani Chandramohan
Forest Survey of India
Central Zone, Seminary Hills
Nagpur, Maharashtra, India

Anuj Choudhary
Department of Botany
Punjab Agricultural University
Ludhiana, Punjab, India

Andrea Copetta
CREA-Research Centre for Vegetable and
Ornamental Crops, Corso Inglesi 508
Sanremo, Italy

Tuyelee Das
Department of Life Sciences
Presidency University, 86/1 College Street
Kolkata, India

Umakanta Dash
Department of Silviculture and Agroforestry
Kerala Agricultural University
Thrissur, Kerala, India

Manik Devgan
Department of Plant Breeding and Genetics
Punjab Agricultural University
Ludhiana, Punjab, India

Abhijit Dey
Department of Life Sciences
Presidency University, 86/1 College Street
Kolkata, India

Vita Di Stefano
Department of Biological
Chemical and Pharmaceutical Science and
Technology (STEBICEF)
University of Palermo, Via Archirafi
Palermo, Italy

Alessandra Durazzo
CREA-Research Centre for Food and Nutrition
Via Ardeatina 546
Rome, Italy

Naoual El Menyiy
Laboratory of Pharmacology
National Agency of Medicinal and Aromatic
Plants
Taounate, Morocco

Mohammad Abul Farah
Department of Zoology
College of Science
King Saud University
Riyadh, Saudi Arabia

Paolo Gabrielli
CREA-Research Centre for Food and
Nutrition
Via Ardeatina 546
Rome, Italy

Manoj Gadewar
Department of Pharmacology
School of Medical and Allied Sciences
K.R. Mangalam University
Sohna Rural, Gurugram, Haryana, India

Rohini Ganorkar
Forest Survey of India
Central Zone, Seminary Hills
Nagpur, Maharashtra, India

Arun Garg
Department of Pharmacology
School of Medical and Allied Sciences
K.R. Mangalam University
Sohna Rural, Gurugram, Haryana, India

Satyakam Guha
Department of Botany
Hans Raj College
University of Delhi
New Delhi, India

Pankaj Gupta
School of Medical and Allied Sciences
K.R. Mangalam University
Sohna Rural, Gurugram, Haryana, India

Zerkani Hannou
Research Team of Chemistry of Bioactive Molecules and the Environment
Laboratory of Innovative Materials and Biotechnology of Natural Resources
Faculty of Sciences
Moulay Ismaïl University
Zitoune, Meknes, Morocco

Ikbal Hasan
Department of Life Sciences
Presidency University
86/1 College Street
Kolkata, West Bengal, India

Rajib Hossain
Department of Pharmacology
School of Medicine
Chungnam National University
Daejeon, South Korea

Azamal Husen
Wolaita Sodo University
Wolaita, Ethiopia

Muhammad Torequl Islam
Department of Pharmacy
Life Science Faculty
Bangabandhu Sheikh Mujibur Rahman Science and Technology University
Gopalganj, Bangladesh

Divya Jain
Department of Bioscience and Biotechnology
Banasthali Vidyapith
Rajasthan, India

Dolly Kain
Department of Botany
Deen Dayal Upadhyaya College
University of Delhi
New Delhi, India

Arshdeep Kaur
Department of Plant Breeding and Genetics
Punjab Agricultural University
Ludhiana, Punjab, India

Harmanjot Kaur
Department of Botany
Punjab Agricultural University
Ludhiana, Punjab, India

Jasbir Kaur
Department of Ocular Biochemistry
Dr. R.P. Centre for Ophthalmic Sciences
All India Institute of Medical Sciences
New Delhi, India

Ranjeet Kaur
Department of Ocular Biochemistry
Dr. R.P. Centre for Ophthalmic Sciences
All India Institute of Medical Sciences
New Delhi, India

Rasel Ahmed Khan
Pharmacy Discipline
Life Science School
Khulna University
Khulna, Bangladesh

Salman Khan
Forest Survey of India
Central Zone, Seminary Hills
Nagpur, Maharashtra, India

Antul Kumar
Department of Botany
Punjab Agricultural University
Ludhiana, Punjab, India

Vijay Kumar
Department of Silviculture and Agroforestry
Kerala Agricultural University
Thrissur, Kerala, India

Kishoree Krishna Kumaree
Natural Products for Neuroprotection and Anti-Ageing Research Unit
Chulalongkorn University
Bangkok, Thailand
And
College of Public Health Sciences
Chulalongkorn University
Bangkok, Thailand

Massimo Lucarini
CREA-Research Centre for Food and Nutrition
Via Ardeatina 546
Rome, Italy

Ashwath M N
Department of Forest Biology and Tree Improvement
Kerala Agricultural University
Thrissur, Kerala, India

Rohan M S
Department of Natural Resources and Management
Kerala Agricultural University
Thrissur, Kerala, India

H. D. T. Madhuranga
Department of Biomedical Science
Faculty of Health Science
KAATSU International University Sri Lanka
Sri Lanka

Sujata Mandal
Department of Life Sciences
Presidency University
86/1 College Street
Kolkata, West Bengal, India

Sahil Mehta
Department of Botany
Hans Raj College
University of Delhi
New Delhi, India

Shatakshi Mitra
Department of Life Sciences
Presidency University
86/1 College Street
Kolkata, West Bengal, India

Shuvadip Mondal
Department of Botany
Fakir Chand College
Diamond Harbour
West Bengal, India

Rania Moubachir
Bioactives
Health and Environmental Laboratory
Epigenetics Team
Faculty of Sciences
Moulay Ismail University
Meknes, Morocco

Mohammad S. Mubarak
Department of Chemistry
The University of Jordan
Amma, Jordan

Nabarun Mukhopadhyay
Department of Pharmacognosy
Bapuji Pharmacy College
Davangere, India

Sana Nafees
Department of Ocular Biochemistry
Dr. R.P. Centre for Ophthalmic Sciences
All India Institute of Medical Sciences
New Delhi, India

Arghya Naskar
Molecular and Applied Mycology and Plant Pathology Laboratory
Department of Botany
University of Calcutta
Kolkata, India

C.I. Che Noh
Borneo Research for Algesia
Inflammation and Neurosciences (BRAIN) Unit
Department of Biomedical Sciences
Faculty of Medicine and Health Sciences
Universiti Malaysia Sabah
Kota Kinabalu, Sabah, Malaysia

P. Palasuberniam
Borneo Research for Algesia
Inflammation and Neurosciences (BRAIN) Unit
Department of Biomedical Sciences
Faculty of Medicine and Health Sciences
Universiti Malaysia Sabah
Kota Kinabalu, Sabah, Malaysia

Abhijoy Paul
Department of Life Sciences
Presidency University
86/1 College Street
Kolkata, West Bengal, India

Yashashree Pradhan
Department of Life Sciences
University of Mumbai
Vidyanagari
Mumbai, Maharashtra, India

Anchalee Prasansuklab
Natural Products for Neuroprotection and Anti-ageing Research Unit
Chulalongkorn University
Bangkok, Thailand
And
College of Public Health Sciences
Chulalongkorn University
Bangkok, Thailand

M.S. Rahman
Borneo Research for Algesia
Inflammation and Neurosciences (BRAIN) Unit
Department of Biomedical Sciences
Faculty of Medicine and Health Sciences
Universiti Malaysia Sabah
Kota Kinabalu, Sabah, Malaysia

Shakeelur Rahman
Prakriti Bachao Foundation
3-C, Madina Apartment
Hill View Road Bariatu
Ranchi, Jharkhand, India

Manjula Rai
Department of Botany
St. Joseph's College
Darjeeling
West Bengal, India

Kajal Samantara
Department of Genetics and Plant Breeding
Institute of Agricultural Sciences
Siksha 'O' Anusandhan (Deemed to be University)
Bhubaneswar, Odisha, India

D. N. A. W. Samarakoon
Department of Biomedical Science
Faculty of Health Science
KAATSU International University Sri Lanka
Sri Lanka

Katayoon Samiakalantari
Department of Agriculture and Food Science
Islamic Azad University
Science and Research Branch
Tehran, Iran

M.H. Mohd. Sani
Borneo Research for Algesia
Inflammation and Neurosciences (BRAIN) Unit
Department of Biomedical Sciences
Faculty of Medicine and Health Sciences
Universiti Malaysia Sabah
Kota Kinabalu
Sabah, Malaysia

Antonello Santini
Department of Pharmacy
University of Napoli Federico II
Via D. Montesano 49
Napoli, Italy

Rupa Sanyal
Department of Botany
Bhairab Ganguly College
Kolkata, India

Surjit Sen
Department of Botany
Fakir Chand College
Diamond Harbour
West Bengal, India

Kuldipika Sharma
Department of Forest Products Utilization
Dr. Yashwant Singh Parmar University of Horticulture and Forestry
Nauni, Solan, Himachal Pradesh, India

Radhika Sharma
Department of Soil Science
Punjab Agricultural University
Ludhiana, Punjab, India

Susmita Shil
Department of Silviculture and Agroforestry
Kerala Agricultural University
Thrissur, Kerala, India.

Gurjot Singh
Department of Industrial and Systems Engineering
University of Wisconsin-Madison
Madison, United States

Anand Sonkar
Department of Botany
Hans Raj College
University of Delhi
New Delhi, India

Sidhartha T N
Department of Forest Products and Utilization
Kerala Agricultural University
Thrissur, Kerala, India

Douae Taha
Laboratory of Spectroscopy
Molecular Modelling Materials
Nanomaterials Water and Environment – CERNE2D
Faculty of Sciences, Mohammed V University
Rabat, Morocco

Maryam Tahmasebi
Department of R&D and Quality Assurance
Nazchin Company
Tehran, Iran

Tewin Tencomnao
Natural Products for Neuroprotection and Anti-Ageing Research Unit
Chulalongkorn University
Bangkok, Thailand
And
Department of Clinical Chemistry
Faculty of Allied Health Sciences
Chulalongkorn University
Bangkok, Thailand

Keddagoda Gamage Piyumi Wasana
Department of Biochemistry
Faculty of Medicine
University of Ruhuna
Galle, Sri Lanka

Weria Weisany
Department of Agriculture and Food Science
Islamic Azad University
Science and Research Branch
Tehran, Iran

Shima Yousefi
Department of Agriculture and Food Science
Islamic Azad University
Science and Research Branch
Tehran, Iran

Z.A. Zakaria
Borneo Research for Algesia
Inflammation and Neurosciences (BRAIN) Unit
Department of Biomedical Sciences
Faculty of Medicine and Health Sciences
Universiti Malaysia Sabah
Kota Kinabalu
Sabah, Malaysia

1 Introduction to Diabetes
An Overview

Radhika Sharma, Anuj Choudhary, Harmanjot Kaur, Gurjot Singh, Kuldipika Sharma, Antul Kumar, Arun Garg, Manoj Gadewar, Satyakam Guha, Anand Sonkar, and Sahil Mehta

CONTENTS

DOI: 10.1201/b23347-1

1.1 INTRODUCTION

According to the International Diabetes Federation, there are currently 463 million international people with diabetes mellitus between the ages of 20 and 79. It represents 9.3% of the world's adult population, and by 2030, it is expected that this percentage will increase to 10.2% (Dugalic et al. 2022). Diabetes is a multifunctional long-term metabolic disorder that results in high glucose levels (hyperglycemia) and poor glycemic control due to inadequate insulin synthesis or activity. Diabetes is a condition marked by the insufficient or improper activity of insulin (Demir et al. 2021). Type 1 diabetes (T1D) and type 2 diabetes (T2D) are the two most prevalent types of diabetes.

Early-onset T1D is caused by an autoimmune condition in which the immune system's cells assault the cells in the pancreas that make insulin. However, T2D typically appears later in life as a result of systemic disturbances in metabolic balance. People are more likely to develop T2D because of genetic susceptibility, which would be exacerbated by bad dietary patterns and unhealthy lifestyles (Zheng et al. 2018; Udler et al. 2019). In persons with type 2 diabetes mellitus (T2DM), microvascular and macrovascular problems are the primary causes of mortality and disability in people (Mohammedi et al. 2017; Cole and Florez 2020).

T2D, in contrast to T1D, is considerably more varied and complicated, including an excessive number of pathophysiological pathways that not only impact the pancreas but also the metabolic organs. This makes successful treatment extremely difficult (Peppa et al. 2003; Mohammedi et al. 2017; Cole and Florez 2020). Many health issues that diabetes patients deal with either exist at the time of diagnosis, such as diabetic retinopathy, or develop over time in the course of infection (Spijkerman et al. 2003; Khalid et al. 2022). Such issues include dysfunctions in numerous essential systems in the body, primarily the kidney, cardiovascular system, retina, and nervous system. Fibrosis of both the kidney and liver and cognitive impairment are other new diseases associated with diabetes that are also becoming more prevalent (Selvin et al. 2011; Tomic et al. 2022). Studies reveal that, contrary to the past, vascular disease is no longer the primary cause of mortality among those with diabetes mellitus (Harding et al. 2016). In certain nations or areas, cancer has surpassed dementia as the primary cause of death for those with diabetes mellitus, and these deaths have increased since the turn of the century (Gregg et al. 2018; Pearson-Stuttard et al. 2021).

1.2 CLASSIFICATION OF DIABETES

Since the 1950s, there was hardly any change in the diagnosis of diabetes. A patient is found to have T1D whenever an analysis has been conducted that indicates virtually a complete lack of endogenous glucose synthesis because, in islets of Langerhans, autoimmune demolition in the β-cells can be observed. All patients with this condition will ultimately require insulin treatment to maintain normal glucose tolerance. T2D refers to a deficit in the responsiveness of insulin inside the organs that regulate metabolic activity, which is the next categorization. Phenotypic traits serve as a basis for classification and include age, familial history, body mass index, islet lipid levels, C-peptide levels, and autoantibodies. Individuals with this disease have hypertension, obesity, high cholesterol, and hormonal imbalance (Balasubramanyam 2021; Ahlqvist et al. 2022). These traditional diagnostic standards fall short of taking into consideration the variety seen in the current global diabetes epidemic. In many ethnic and racial communities, doctors and clinical researchers have seen an increase in the variety and complexity of diseases (American Diabetes Association 2021).

Expert groups have recently added more explanatory categories to the diagnosis of diabetes. This would include steroid-induced hyperglycemia, which is connected to exocrine pancreatic diseases, subsequent diabetes caused by high hormonal changes that are unable to control blood sugar, and single-gene hyperglycemia including neonatal hyperglycemia and maturity-onset diabetes of the young (MODY) syndrome (Pearson-Stuttard et al. 2021). Latent autoimmune diabetes in adults (LADA), which manifests in the middle and late years of adulthood due to autoimmune destruction of antibody levels, a healthier way of life, and increased efficiency of glucose supply compared to people with traditional T2D, is another form of diabetes that results from T1D and T2D (Pieralice et al. 2018; Andersen and Hansen 2019). Demographic studies reveal heterogeneity in the way physiological variables such as sensitivity to insulin, functions of β-cells, and typical determinants including body mass index, hypertension, fat distribution, sex, and family history of diabetes affect the risk of developing T2D and the rate at which it progresses. The pathogenetic categorization of diabetes can be established using monogenic diabetes as a template, and its therapeutic applicability can be tested using that classification (Balasubramanyam 2021).

An illustration of the appropriate classification of diabetes based on phenotypic parameters is ketosis-prone diabetes. Ketoacidosis is uncommon in T2D, but when it does occur, it typically does so in conjunction with the stress of another condition, such as an infection (Adler et al. 2021). To characterize the variety of T2D, certain models nowadays have been put forth. In one of them, T2D is viewed as a collection of several subtypes that are brought about by various causes, which results in a homogenous phenotype in all of its sufferers. Another theory emphasizes the importance of quantitative changes in metabolic pathways and postulates that people acquire diabetes as a result of a variety of minor flaws in various pathways (Pearson-Stuttard et al. 2021; Ahlqvist et al. 2022). An intermediary approach that uses a bigger canvas and hues that indicate significant medical aspects was utilized by Ahlqvist et al. (2018) instead of employing different biological mechanisms. The concept remains to assume that hyperglycemia is caused by numerous intersecting routes, but it now presupposes that every patient has a different genotypic and phenotypic cause of disease. Clusters including severe insulin-resistant diabetes, severe insulin-dependent diabetes, and severe autoimmune diabetes are used in predictive modeling for the diagnosis of diabetes (Ahlqvist et al. 2018, 2022; Balasubramanyam 2021).

1.2.1 Pathogenesis and Physiopathology of T1D

Antigen presentation cells (APCs) expose the body's immune system to β-cell antigens, which leads to chronic immunological responses that culminate in β-cell death. Mortality of β-cells caused by a virus or by natural phenomena stimulates the production of antigens and the start of immunological reactions against additional β-cells. Normally, these antigens are taken up by dendritic cells, which then deliver them to T lymphocytes. Only when autoreactive T cells have evaded thymic negative selection is an autoimmune response conceivable (Wallberg and Cooke 2013). T1D is characterized by several features that point to autoimmune cell damage, and a mechanism is presented in Figure 1.1 (Al Homsi and Lukic 1993; Baynes 2015): (i) Infiltrated pancreatic islets include auxiliary and immune-competent cells; (ii) autoantibodies are detected in islet-specific cells; (iii) Class II genes that respond to the immune system are associated with illness vulnerability; (iv) changes in CD4+ T cell compartment immune-regulation; (v) interleukins, which are generated by TH1 cytokines and monokines, cause the sickness to manifest; and (vi) additional organ-specific autoimmune illnesses that often afflict affected people or members of their families.

An inability to produce enough insulin, an overproduction of glucagons, unregulated lipolysis, and a rise in plasma concentrations of fatty acids are all consequences of the autoimmune death of

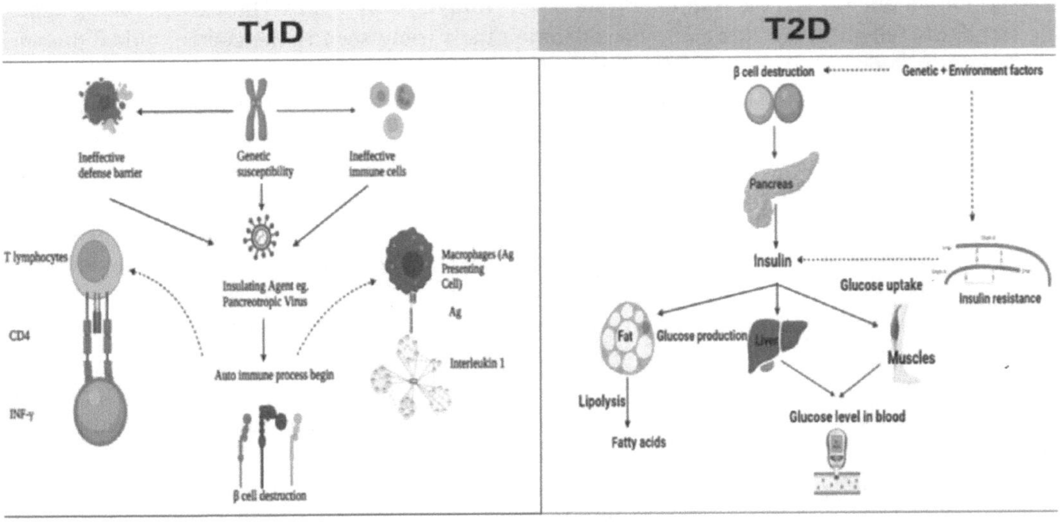

FIGURE 1.1 Pathogenesis and pathophysiology of T1D and T2D.

pancreatic β-cells. Glucokinase inside the liver as well as the GLUT4 family of transport proteins in adipose tissue represents two examples of proteins whose expression is reduced when there is a deficit of insulin in addition to impairing glucose absorption (Holt 2004; Baynes 2015).

1.2.2 Pathogenesis and Physiopathology of T2D

The two key mechanisms in T2D are reduced insulin action brought on by insulin sensitivity and pancreatic β-cell dysfunction. This type of diabetes has a greater association of genes with T1D. When insulin resistance is prevalent, the mass of cells goes through a change that can increase the insulin supply and make up for the excessive and abnormal demand. Hyperglycemia and physical and metabolic reactions in T2D are closely related. When diabetes occurs, the brain recognizes this and transmits a signal through nerve cells to the pancreas and other organs to minimize its effects (Ohiagu et al. 2021). Resistivity to insulin and hyperglycemia both contribute to uncontrolled diabetes, and their roles in the etiology and pathogenesis of T1D and T2D are unaffected (Baynes 2015). Apart from maturity-onset hyperglycemia of the young, which is caused by abnormalities in the glucokinase gene on chromosome 7p, the mode of inheritance for T2D remains unclear. MODY is characterized as diabetes discovered even before 25 years and is treatable for some more than 50 years without insulin when islet cell antibodies (ICA) are absent (Sekikawa et al. 1993; Ozougwu et al. 2013). The mechanism for T2D is presented in Figure 1.1.

1.3 COMPLICATIONS OF DIABETES

The macrovascular and microvascular systems, which pertain to the major and tiny vascular systems, accordingly, all across the body, are both linked to long-term damage in diabetes (Morrish et al. 2001). Although the main cause of death in persons with T2D is definite impairment due to glucose levels in the capillary structure, such as the heart and cerebral arteries, damage to the microvascular networks in the kidneys, retina, and neurons is more prevalent and also significantly influences death as summarized in Figure 1.2 (Cole et al. 2020). The microvascular complications are stated later.

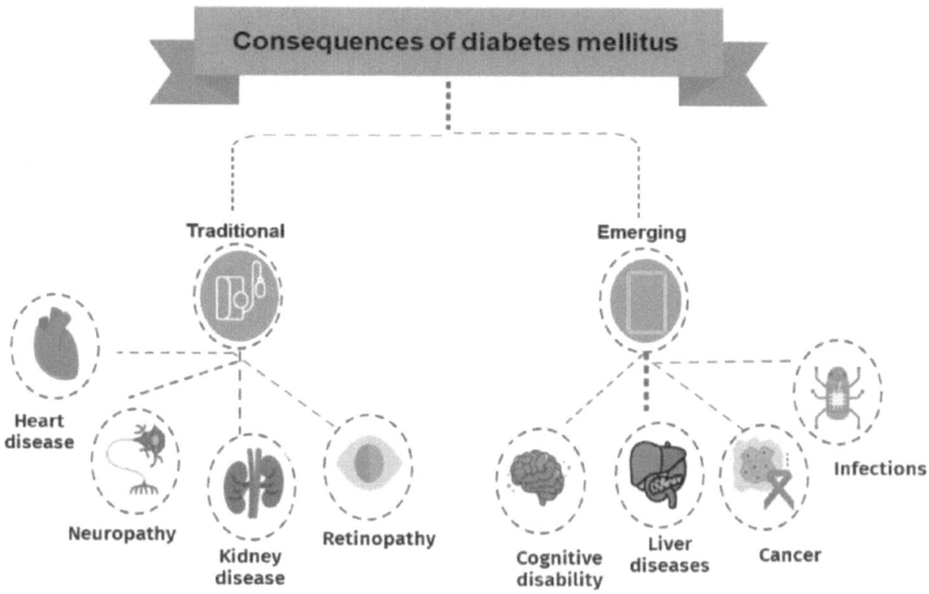

FIGURE 1.2 Diagrammatic representation of consequences of diabetes mellitus.

1.3.1 Diabetes-Related Nephropathy

It is a chronic condition marked by acute renal failure brought on by hyperglycemia and partial correlation with albuminuria (Martínez-Castelao et al. 2015). Reduced renal function is also caused by factors other than diabetes, such as obesity, dyslipidemia, and hypertension, and patients with diabetes may also develop kidney dysfunction (Cole et al. 2020). The two pathological changes that are most frequently utilized to make a diagnosis are elevated albuminuria and a reduced estimated rate of glomerular filtration, the timing of the initiation of kidney failure and the prognosis of hyperglycemia could perhaps help distinguish between specific and nonspecific diabetes (Cho et al. 2018). Diabetes-related nephropathy has no known cure; but at late stages, when dialysis or kidney transplantation are frequently required for life, therapy focuses on controlling blood sugar levels, proteinuria, and gradual kidney destruction. End-stage kidney disease is linked to higher mortality (Martínez-Castelao et al. 2015).

1.3.2 Diabetic Retinal Disorder

Retina blood vessels may gradually degenerate due to hyperglycemia, resulting in hemorrhaging, retinopathy, and loss of sight. Retinopathy at the initial and late stage is more serious and is defined by the development of the new, fragile vascular system that spills and spread throughout the retina and into the vitreous. Clinically severe macular edema, a specific kind of diabetic retinopathy, includes direct macula damage (Chu and Ali 2008). Furthermore, the most common reason for adult blindness is diabetic retinopathy. The length of diabetes is determined by the intensity of blood pressure, HbA1c levels, proteinuria insulin usage, and age at which the disease is diagnosed. According to research, the intensity of retinopathy is more likely to be influenced by inherited factors than the inclusion or exclusion of external factors (Hallman et al. 2005; Cole et al. 2020).

1.3.3 Diabetes-Related Neuropathy

Diabetes is a substantial contributor to nerve degeneration, especially again for lengthier peripheral nerves that largely impact the abdominal muscles (Johannsen et al. 2001). Distal symmetric polyneuropathy, a kind of peripheral nerve injury, has been the most popular type of diabetic neuropathies. Autonomic, atypical, and nondiabetic neuropathies are additional diabetic neuropathy subtypes. Diabetic patients have a 15-fold increased likelihood of foot bleeding ulcers, which can range in severity from 15% to 25%, combined with the growing pain and decline in quality of life caused by diabetic neuropathy (Most and Sinnock 1983; Singh et al. 2005). Although having the greatest mortality rate, diabetic neuropathy is among the least researched diabetes as it is challenging to evaluate directly and precisely, since there is no cure and treatment only focuses on prevention through glycemic control and symptom monitoring. For diabetic neuropathy, no extensive sequencing work has been done, perhaps because of phenotypic restrictions.

1.3.4 Molecular Mechanism Associated with Diabetes-Related Complications

Hyperglycemia is the primary step in the generation of diabetic complications, and it triggers the activation of the sorbitol–aldose reductase pathway, hexosamine biosynthetic pathway, mitogen-activated protein kinases, and protein kinase C (Grau and Pericas 2022). It also induces the enhancement of the growth factors and cytokines expressions such as tumor necrosis factor-α, insulin-like growth factor, platelet-derived growth factor, vascular endothelial growth factor, and transforming growth factor-β. Reactive oxygen species (ROS) are the mediator factors in most of the events to initiate the intracellular signaling pathway and further transcriptional cascades regulated principally by MAPKs and nuclear factor kappa B (Andreadi et al. 2022). The increase in oxidative stress and reduction in the potential of the antioxidant system is associated with complications that are well established. The overproduction of ROS and hyperglycemia resulted in enhanced NADPH oxidase activation is an integral component of metabolic syndrome. Additionally, insulin resistance is positively correlated with systemic oxidative stress (Barkabi-Zanjani et al. 2020). The contribution of oxidative stress is tissue-specific chiefly in microvascular diseases in diabetic complications. A clear picture of molecular events for diabetic complications is shown in Figure 1.3.

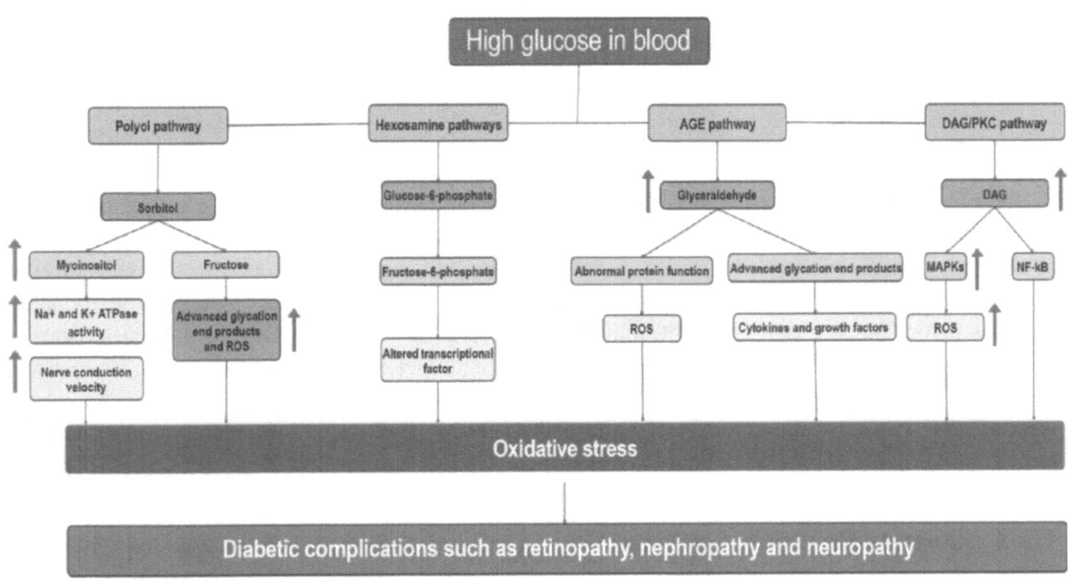

FIGURE 1.3 Molecular mechanisms associated with the diabetes complications.

1.4 DIAGNOSIS OF DIABETES

If type 2 diabetes is not diagnosed early enough, it can cause several significant, sometimes fatal consequences. Thus, machine learning-based decision support systems is used to predict chronic illness. It had drawn a lot of interest in providing healthcare professionals and the general public with improved prognosis/diagnosis help (Ismail et al. 2022). Based on various possible causes, numerous projects have now been suggested in the literature for applying machine learning classification algorithms to predict the incidence of type 2 diabetes (EMC 2015). Such algorithms can be classified as either tree-based, which builds a classification tree using the dataset features as the nodes and the class labels as the leaves, or probability-based, which employs a probability distribution function over the class labels for a specific observation. Different algorithm approaches, such as decision trees, random forest methods, etc., provides relatively good accuracy, are simple to read, mimic physician criteria, and reduced chances of variance as compared to other statistical methods (Committee ADAPP 2022; Ismail et al. 2022).

1.4.1 Methods of Clinical Diagnosis of Diabetes

Plasma glucose criteria, such as the fasting plasma glucose (FPG) value or the 2-hour plasma glucose (2-h PG) value during a 75-gram oral glucose tolerance test (OGTT) or A1C criteria may be used to diagnose diabetes. Fasting plasma glucose is affordable, widely used, and requires fasting for at least 8 hours; FPG ≥126 mg/dL (Committee ADAPP 2022). The most accurate test and the first indicator of glucose dysregulation is 2-h PG. The test should be carried out following World Health Organization (WHO) guidelines, using a glucose load that is equivalent to 75 g of anhydrous glucose dissolved in water. In glycated hemoglobin, fasting is not necessary, there is little biological variability, the results are closely related to problems, the sample remains stable in the vial throughout acute sickness, and there is worldwide uniformity. The A1C test needs to be carried out in a lab using a technique that is NGSP-certified and standardized to the DCCT assay (Committee ADAPP 2022).

1.4.2 Algorithms Used for Diabetes Management

Various algorithms used for diabetes management are presented in Table 1.1.

TABLE 1.1
Algorithms Used for Diabetes Management

Algorithm	Advantage	Method used for prediction	References
K Nearest Neighbors	Datasets with outliers	Saves the dataset and assigns new observations to a category	Aha et al. (1991)
Stacking	Eliminates prediction error	Ensemble meta-estimator	Wolpert (1992)
One Rule	Simple to use	Frequency distribution	Holte (1993)
Decision Table	Easy to use	IF-THEN rules	Kohavi (1995)
Artificial Neural Networks	Manage null values, high-dimensional datasets with many samples	Multilayer perception model	Hassoun (1995)
JRip	Appropriate for nonlinear data	Pruning often and incrementally to reduce errors	Cohen (1995)
Support Vector Machine	Nonlinear data	Hyperplane method	Cortes and Vapnik (1995)

(Continued)

TABLE 1.1
(Continued)

Algorithm	Advantage	Method used for prediction	References
Bagging	Tolerates missing values, uses high-dimensional datasets, and reduces the overfitting of data	Ensemble meta-estimator	Breiman (1996)
K-means	Appropriate for huge data	Clustering technique	Alsabti et al. (1997)
Zero Rule	Simple to comprehend	Frequency distribution	Hall and Frank (2008)
Bayesian Network	Appropriate with incomplete data	Graphically represents probability relationship	Muralidharan and Sugumaran (2012)
Logistic Regression	Appropriates for huge datasets	Probability distribution	Hosmer et al. (2013)
Random Forest	Tolerates missing values, uses high-dimensional datasets	Decision trees	Liaw and Wiener (2002)
Naïve Bayes	Simple to use	Based on Bayes' theorem	EMC (2015)
Decision Tree	Simple to use and understand	Constructs a tree structure	EMC (2015)

1.5 MANAGEMENT OF DIABETES

1.5.1 Drugs

In this heading, we have represented the antidiabetic group. Various drugs used in diabetes treatment and their underlying processes are stated in Table 1.2.

1.5.2 Management through Lifestyle

1.5.2.1 Altering One's Way of Life

Diabetic patients typically need to modify their dietary habits and level of fitness in terms of avoiding or controlling the condition, in addition to having to make significant life decisions regarding their treatment (Piepoli et al. 2016). Persons with chronic diseases need to change their self-care behaviors since they need to gain information about their health problems to promote healthcare, manage the disease, and prevent consequences (Chrvala et al. 2016).

1.5.2.2 Obesity Management

Bodyweight control is a crucial component of lifestyle adjustment, especially for patients who are overweight or obese (Piepoli et al. 2016). As soon as practical following diagnosis, effective diet treatment can help to improve DM glycemic control. Additionally, obese individuals with DM who maintained a 5% weight loss from their starting point appeared to have better blood pressure, lipids, and glucose regulation. In research including food interventions, time is an important factor.

The fact is that if patients' dedication to healthy diet modifications is increased over time and widely acknowledged, diabetes typically can be controlled more significantly. Weight can also be managed by a moderate-carbohydrate, low-fat, high-protein, and low-glycemic index diet (Andrews et al. 2011). Besides the apparent concerns of binge drinking, people with DM also run the danger of late hypoglycemia or hyperglycemia, as well as excess weight (Andrews et al. 2011; Lambrinou et al. 2019).

TABLE 1.2
Drugs Used in Diabetes Treatment and Their Underlying Processes

Categories	Familiar drugs	Mode of action	Treatment	Advantages	Effect on weight	Hypoglycemia risk	Other concerns	Route	References
2nd generation Sulfonylureas	Pramlintide, Glipizide, Glibenclamide, Gliclazide, Glimepiride	Inactivation of the K_{ATP} channel stimulates pancreatic insulin secretion	1st/2nd line	Cost effective	Gain	Yes	Cardiovascular	Oral	Prasad-Reddy and Isaacs (2015)
α-Glucosidase inhibitors	Acarbose, Vocarbose, Miglitol	Disturbance of the digestive system's ability to digest and absorb glucose	-	Cardiovascular benefits	No	No	-	Oral	Lamos et al. (2016)
Thiazolidinediones	Rosiglitazone Pioglitazone	Improvement of insulin action with PPAR activation γ	2nd/3rd line	Improved lipid profile	Gain	No	Edema, heart failure	Oral	Sanchez-Rangel E and Inzucchi (2017)
Biguanides	Metformin, Phenformin	Increasing sensitivity to insulin raises hepatic glucose uptake	1st line	Cost effective	None	No	Lactic acidosis	Oral	Sanchez-Rangel E and Inzucchi (2017)
DPP4 inhibitors	Saxagliptin, Vildagliptin, Alogliptin, Sitagliptin, Linagliptin	Increase the half-life of GLP-1 to lower glucagon and blood glucose levels	2nd/3rd line	Improved mass of β-cell	Neutral	No	-	Oral	He et al. (2019)
GLP-1 agonists	Exenatide, Liraglutide, Albiglutide, Exenatide	Enhance insulin signaling and reduce excessive glucagon levels	1st/2nd line	Improved mass of β-cell	Loss	No	Risk of pancreatitis	Injection	He et al. (2019)
SGLT2 inhibitors	Dapagliflozin Canagliflozin	Reduce glucose secretion in the kidney	1st/2nd/3rd line	Decrease in cardiovascular and renal diseases	Loss	No	-	Oral	He et al. (2019)

1.5.2.3 Physical Activity

Exercises are crucial for the control of hyperglycemia and the avoidance of problems in patients with DM. They have been found to help with weight reduction, lower cardiovascular diseases, and improve blood sugar management (Tikkanen-Dolenc et al. 2017; Lambrinou et al. 2019). Even when there is no weight reduction, an exercise intervention lasting at least eight weeks can lower the HbA1c level in persons with DM by 0.66% (Boulé et al. 2001). People with DM should be encouraged to do brief periods of standing, walking, or other mild exercises to break up the time spent sitting down.

1.5.2.4 Integrated Lifestyle Changes

Recent studies on healthy eating, losing weight, or just physical exercise can be beneficial for curing diabetes. A study on lifestyle modification in people with diabetes found that diastolic and systolic blood pressure, body mass index, and HbA1c all improved significantly after making adjustments to one's lifestyle. However, these combination therapies have not consistently demonstrated a decreased risk of death up to this point (Andrews et al. 2011; Lambrinou et al. 2019).

1.5.2.5 Quitting Smoking

Recent epidemiologic, case-control, and cohort studies have demonstrated the connection between smoking and health risk. It is remarkable that even after a diagnosis, smoking rates are greater in those with chronic conditions like diabetes mellitus (Chrvala et al. 2016). According to recent statistics, patients with diabetes who actively smoke have much higher chances of cardiovascular events. Most specifically, the chances of acquiring cardiovascular disease, passing away early, and getting micro- and macrovascular complications are 1.44 times greater in smokers with DM. In addition, the analysis showed that people with diabetes who smoke have far increased risks of dying altogether and having a cardiac disease (Lambrinou et al. 2019).

1.5.3 Use of Traditional Plants

It has been demonstrated that traditional herbs have powerful antidiabetic properties with no adverse side effects. Studies show that traditional botanicals are the first-line therapy for 80% of patients in poor nations. Traditional medicine plays a crucial role in basic healthcare in many developing countries (Abate et al. 2021; Bachheti et al. 2021; Husen and Iqbal 2021; Husen 2021, 2022; Husen et al. 2021; Sonkar et al. 2022; Rahman and Husen 2021, 2022). Because medicinal plants are less expensive and more readily available than conventional pharmaceuticals, herbal therapy is increasingly routinely used to treat diabetes. When paired with other antidiabetic drugs or insulin, this therapeutic approach can be used to manage diabetes with minimal side effects (Suvarna et al. 2021; Przeor et al. 2020). Many phytoconstituents in medicinal plants, such as glycosides, terpenoids, alkaloids, saponins, flavonoids, and carotenoids, have antidiabetic properties (Chhetri et al. 2005). The diverse plant network is a carrier of multiple phytochemical compounds that characterizes the particular relationships of these compounds, despite being challenging to imitate and offering health advantages. Terpenes, saponins, and polyphenols are thought to be the most important natural inhibitors since they are found in many antidiabetic plants (Xiao et al. 2013). Many medicinal plants having antidiabetic or antihyperglycemic properties, utilized in various parts of the world, are presented in the literature (Przeor et al. 2020). Various herbs used in the management of diabetes are presented in Table 1.3.

TABLE 1.3
Herbs That Are Being Used in the Treatment of Diabetes

Scientific name	Part used	Bioactive compounds	Antidiabetic effects	Reference
Tinospora crispa	Stem	Flavones and flavone glycosides	Stimulate insulin secretion through the modulation of β-cell Ca^{2+} concentration	Datey et al. (2018)
Momordica charantia	Pulp, leaves, seeds	Triterpenoids, saponins, and polypeptides	Scavenge the oxygen radicals	Raveendran et al. (2018)
Gymnema sylvestre	Leaves	Oleananesaponin, triterpenoid, saponins, and gurmarin	β-cell regeneration	Jacob and Narendhirakannan (2019)
Aloe barbadensis	Flower and leaf	Alkaloids, flavonoids, tannins, phenols, and saponins	Reduce blood sugar levels and potential hypoglycemic effects.	Jacob and Narendhirakannan (2019)
Euphorbia hirta	Leaf, flower, and stem	Flavonoids and terpenoids	Inhibit the α-glucosidase activity	Jacob and Narendhirakannan (2019)
Ocimum sanctum	Leaves	Eugenol, methyleugenol, and p-caryophylen	Boost the hepatic glycogen synthesis, lower blood glucose, and improve oral glucose tolerance	Kumar et al. (2020)
Zingiber officinale	Rhizome, roots	Terpenoids and phenolics	Increase the glucose utilization	Salleh et al. (2021)
Momordica charantia	Fruit	Polypeptide-p, momordicosides, saponins, and momordin	Act as insulin-like protein, decrease blood glucose level, release intestinal GLP-1	Tran et al. (2020)
Panax ginseng	Root	Ginsenosides	Reduces the fasting blood glucose or postprandial glucose in type 2 diabetic patients	Salleh et al. (2021)
Tinospora cordifolia	Stem	Alkaloids, terpenoids, essential oils, glycosides, steroids, phenolic constituents, aliphatic compounds, and polysaccharides	β-cells regeneration	Kumar et al. (2021)
Allium cepa	Bulb	Quercetin, allicin, alliin, and diallyl disulfide	Strong antioxidant properties to reduce blood sugar levels and their potential hypoglycemic effects	Rahman et al. (2022)
Pterocarpus marsupium	Heartwood, leaves, flowers	Marsupsin, pterosupin and pterostilbene and flavonoidspteroside, pteroisoauroside, carsupin and marsupol	Increase the glucose absorption and induce insulin production	Shahin et al. (2022)
Cyamposis tertragonoloba	Beans	Quercetin, daidizein, kaemferol, and 3-arabinosides	Reduction in the absorption of glucose from the gastrointestinal tract	Rahman et al. (2022)
Costus pictus	Leaves	Alkaloids, saccharides, terpenoids, glycosides, steroids, tannins, saponins, phenols, and flavonoids	Inhibit the α-glucosidase and α-amylase activities	Kumari et al. (2022)

1.5.3.1 Mode of Action of Medicinal Plants to Treat Diabetes Mellitus

- By impeding the α-glucosidase secreted from the small intestine: The inhibition delays the carbohydrates cleavage and attenuated the postprandial level of glucose. It was reported that the methanolic extract of root bark and fruit pulp inhibited the activity in *Annona muricata* (Agu et al. 2019).
- By inhibiting the activity of the DPP4 enzyme: The inhibitors of DPP4 allow for sustaining the activity of glucagon-like peptide-1 and glucose-dependent insulinotropic polypeptide, trigger the release of insulin, and inhibit glucagon secretion to regulate the blood glucose level. The extracts of *Camelia sinensis*, *Pometia pinnata*, *Syzygium polyanthum*, *Artocarpus heterophyllous*, *Lagerstroemia speciosa*, and *Persea americana* were showing the inhibition of DPP4 (Elya et al. 2015).
- By inhibiting the activity of α-amylase secreted from the salivary gland: The hydrolysis of glycogen and starch is carried out with the help of α-amylase mainly in the salivary and pancreatic juice. Therefore, the inhibition of this enzyme helps to prevent the levels of postprandial glucose in the blood. The hydro-ethanolic extract of *Withania frutescens* was showing antihyperglycemic and antidiabetic activities (Mechchate et al. 2021).
- By enhancing insulin secretion: The rise in the calcium level is tightly associated with the release of insulin from the β-cell vesicles of the pancreas. The membrane depolarization is caused by the ATP-sensitive K^+ channels closure belonging to insulin-secreting β-cells. It leads to activation of the voltage-dependent calcium channels and enhanced insulin secretion. The cinnamic acid and its related compounds were observed to enhance the insulin-secreting activity in the β-cells of perfused rat pancreas and increase calcium levels *in vitro* (Adisakwattana et al. 2008; Zhao et al. 2019).
- By enhancing insulin sensitivity and enhancing glucose uptake by muscle cells and adipose tissue: Several phytochemicals help to strengthen the sensitivity of nonpancreatic cells to insulin for refining glycemic control (Schenk et al. 2008).
- By nourishing the β-cells of the pancreas: The herbal formulation of survival, restoration, and repairing of pancreatic β-cells hinders the complications of diabetes mellitus. *Panax ginseng*, *Momordica charantia*, *Gymnema sylvestre*, *Carica papaya*, and *Bidens pilosa* were reported to improve the activity of pancreatic β-cells (Oh 2015).
- By reducing the HbA1c and the glycated plasma protein level: The nonenzymatic reaction between the proteins in blood and monosaccharides, and adheres to amino acids to produce a modified protein complex through glycation. The reaction will produce glycated plasma proteins and glycated hemoglobin (HbA1c). These products undergo intramolecular modifications by irreversible reactions such as dehydration, cyclization, cross-linking, condensation, and glycoxidation to generate advanced glycation end products (AGEs). They accumulate and pose deleterious effects on vascular health and metabolism to generate diabetic complications. Therefore, the inhibitors of glycation impede it by competitive binding to the amino group of the respective protein, cutting the structure of monosaccharides, binding at the site of glycation, and adhering to the intermediate products during the glycation reaction (Welsh et al. 2016; Ghazanfari-Sarabi et al. 2019).
- By enhancing the level of glucagon-like peptide-1 (GLP-1): GLP-1 is secreted by the L cells of the colon and distal ileum of the gastrointestinal portion. GLP-1 subsequently binds to the G-protein-coupled receptor of the β-cell of the pancreas to raise the glucose-induced insulin secretion and reduce glucagon secretion (Shaefer et al. 2015).
- By regulating the concentration of GLUT4: A sugar transporter protein of 12-transmembrane domain type allows the insulin to trigger the blood glucose influx into fat cells and skeletal muscle through facilitated diffusion and therefore maintains the glucose homeostasis in the body. Phytochemicals are helped in improving the insulin-induced signaling

pathways by stimulating GLUT-4. D-limomine in *Aegle marmelos* and piceatannol in *Passiflora edulis* helped in rebuilding the insulin-mediated signaling pathways (Minakawa et al. 2012; Tan et al. 2016).

1.5.4 Ayurveda and Yoga

Prameha (diabetes mellitus), a group of complicated clinical illnesses with frequent irregular urination, is described in Ayurveda and correlates with diabetes, metabolic syndrome, and obesity. Sthula pramehi, also known as Apathyanimittaja or obese diabetes, is comparable to T2D. Prameha, also known as acquired diabetes, is brought on by making bad lifestyle decisions. The best treatment for Dosha removal is Sam-shodhana (purification). According to Vagbhata, Doshas should be expelled by the most direct route (Sharma et al. 2011). Yoga has a favorable effect on the management of type 2 diabetes, according to several kinds of research.

Various poses, including Surya Namaskara, strengthen the digestive system and produce more insulin (Kumari et al. 2022). Both Apatarpana (healthy diet with calorie reduction) and Santarpana (nutritious, high-calorie diet meant to raise weight) are recommended as Ayurvedic therapy for patients with T2D and T1D (Raveendran et al. 2018; Kumari et al. 2022). The experiment conducted by Datey et al. (2018) reveals that a combination of Ayurveda and yoga, together with stringent dietary and behavioral guidelines, such as meal and bedtime restrictions and exercise, may be effective in treating newly identified increased blood sugar levels. It also implies that supplementing yoga programs with Ayurvedic herbal juice treatment may improve their effectiveness.

1.6 NANOTECHNOLOGY IN DIABETES

The use of nanotechnology in the healthcare field has paved the way for the development of a new branch of nano-science called nano-medicine. Three broad categories may be used to summarize the advances made by nanotechnology in medicine.

1.6.1 Drug Delivery Methods (DDM)

The creation of new nanomaterial-based carrier systems attempts to regulate the release and biodistribution of a medicinal component in a targeted manner. Therapeutic DDM based on nanocarriers is drawing greater interest than traditional DDM because of its promising uses. Because nanocarriers have a higher surface-to-volume ratio, at the same drug concentration, more of the drug surface can touch the body (Simos et al. 2021; Choudhury and Patra 2022). Therefore, the DDM can be more effective at the same dose, and the harmful effects of medications can be lessened at a lower level. The external properties of nanocarriers for different therapeutic medicines and targeting modalities are also adjustable (Simos et al. 2021; Li et al. 2022).

The most popular nano-based drug delivery methods used in DM management include liposomes, polymer-based nanoparticles (NPs), and inorganic NPs. Diverse polymer-based NPs, such as nanocapsules, dendrimers, micelles, and nanospheres, are among them and have been proven to be effective drug carriers. These nanocarriers have the potential to be beneficial in several ways, including boosting bioavailability, preventing enzymatic degradation of medicines, and improving stability by conquering various biological obstacles *in vivo*. Additionally, they can act as an automated adaptive system to simulate endogenous insulin supply and show a dynamic behavior to an outside signal, reducing the risk of hypoglycemia and boosting patient compliance (Mishra et al. 2021; Simos et al. 2021). NPs might be used to increase the uptake of drugs in paracellular tissues. For epithelial endocytosis, NPs' hydrophobic surface is advantageous. However, the negatively charged mucus layer interacts with cationic NPs, preventing their absorption (Mishra et al. 2021). Table 1.4 presents several findings based on the use of nanocarriers in the treatment of diabetes and related problems.

TABLE 1.4
Nanoparticle-Associated Drugs Used in the Treatment of Diabetes

Nanoparticle	Drug	Tested on	Effects on glucose level	Effect on insulin level	Delivery mechanism	Intake mechanism	References
Dendrimers	Insulin	Mice	Decrease	–	Nano-drug carriers	Subcutaneous	Nowacka et al. (2016)
Liposomes	GLP-1	Rat	Decrease	Increase	(GLP-1)-based nanomaterials	Oral	Zhang et al. (2017)
Peptide-conjugated selenium	VPAC2 agonist	Mice	Decrease	Decrease	Pituitary adenylate cyclase-activating peptide (PACAP)	Oral	Dai et al. (2019)
HTCA-pDNA complex	GLP-1	Mice	Decrease	Increase	(GLP-1)-based nanomaterials	Oral	Reichelt-Wurm et al. (2019)
Niosomes	Repaglinide	Rat	Decrease	–	Nano-drug carriers	Oral	Singhal et al. (2019)
Nanospheres	Viglagliptin	Mice	Decrease	–	Nano-drug carriers	Oral	Singhal et al. (2019)
Polymeric	Glipizide	Rat	Decrease	–	Nano-drug carriers	Oral	Singhal et al. (2019)
Chitosan, porous silicon	GLP-1/i DPP4	Rat	Decrease	Increase	(GLP-1)-based nanomaterials/ nano-DPP4 inhibitors	Oral	Pouran et al. (2020)
Polymeric	Liraglutide	Rat	Decrease	Increase	(GLP-1)-based nanomaterials/ nano-DPP4 inhibitors	Oral	Simos et al. (2021)
Self-assembled peptides	GLP-1	Mice	Decrease	Increase	(GLP-1)-based nanomaterials	Oral	Simos et al. (2021)

1.6.2 Diagnosis/Imaging

By employing specific NPs as labels for diagnostic tests, nano-medicine can enhance early illness detection, diagnosis, and prevention. Substrates for high-resolution imaging or tools are generally used for the construction of biosensors (Simos et al. 2021). Eventually, the use of nanosensors will result in the creation of extremely sensitive biological instruments for the quick and greater detection of infectious biomarkers (Strakosas et al. 2019). Nanomaterials' physical, chemical, biological, optical, and magnetic characteristics make them excellent for diagnostic tumor and atherosclerotic plaque imaging (Ventola 2012). Compared to larger imaging materials, the NPs are more flexible to optical and magnetic features due to variations in quantum systems at the nanoscale. Since the size of the NP influences the color created, marking items with varied color coding may be quite helpful during diagnostic procedures. Additionally, the use of nanoscale magnetic materials results in improved MRI pictures with even more details (Gul et al. 2019).

1.6.3 Biomaterials

Nanomaterials are utilized to create synthetic cells and implants that can be used to replace or heal damaged tissues. Nanotechnology enables the creation of biodegradable substrates that imitate both the functioning and the diversity of extracellular matrix (ECM) and are utilized for tissue regeneration (Dahman et al. 2017; Gul et al. 2019). Additionally, nano-featured scaffolds are intended to contain medications and regulate their spatiotemporal release (e.g. growth factors). In orthopedics, for instance, nanotechnology-based biomaterials (nano-coatings or nano-structured surfaces) are utilized to address several implant material drawbacks, such as bacterial adhesion or corrosion resistance (Kumar et al. 2020).

For management, the three major kinds of NPs are used: organic, inorganic, and hybrid. Simple metallic NPs make up inorganic NPs and in the shapes of nanospheres, nanocapsules, and nanocages. Organic NPs, on the other hand, are exceptionally soft and flexible organic matrixes, such as liposomes, dendrimers, liposomes, and micelles (Mishra et al. 2021; Simos et al. 2021). Hybrid nanoparticles are standardized mixtures of inorganic-organic, inorganic-inorganic, and organic-organic that have been developed to meet certain requirements. The use of all three major classes is presented in Table 1.4.

1.6.4 Nanotechnological Advances in Diabetes

1.6.4.1 Nano-Medicine

Mucus-penetrating nanoparticles or covered insulin nanoparticles can enhance the mobility of insulin. The bioadhesion covering of a nanoparticle promotes effective drug delivery close to the site of uptake in the gut. Insulin is shielded from the stomach's acidic condition and biochemical breakdown by being enclosed within nanoparticle carriers. Encasing insulin in a nanoparticle increases its stability because it is not readily soluble (Bahman et al. 2019)

1.6.4.2 Pancreatic Islet Cell Transplantation

One of the most hopeful therapy modalities for people with severe T1DM is pancreatic islet transplantation. The formation of continuous neo-islet tissues is enabled by the transplantation of tailored islet cell sheets, which offers remarkable management potential for T1DM. However, a substantial number of islet cells are needed during a single injection, which aims to treat hyperglycemia in diabetic animals (Mishra et al. 2021).

1.6.4.3 Artificial Pancreas

A lengthy treatment of diabetes mellitus may be made possible by the development of an artificial pancreatic system that combines a continuous glucose level tracking device, a glucose sensor, and an insulin injection unit for monitoring validation. A closed-loop artificial pancreas has shown to be an effective diabetic therapy. The artificial pancreas involves checking blood sugar levels and gives T1DM patients combinations of insulin. The artificial pancreas has the potential to transform the lives of several T1DM sufferers and is creative (Kumar et al. 2020; Simos et al. 2021).

1.6.4.4 Tissue Engineering (TE)

Three elements used in the TE method to repair damaged tissues are scaffolds, growth factors, and stem cells. These treatment methods produce respectable results while being secure and comfy for patients. The unhealed diabetic damage was successfully treated using TE-based treatment using cord blood mononuclear stem cells, peripheral blood mononuclear stem cells, and plasma-rich platelets. This approach nurtures insulin-producing cells using ECM elements in three-dimensional structures and may increase the effectiveness of cell-based diabetes treatments (Amer et al. 2014).

1.6.4.5 Gene Therapy

The genetic engineering method for treating diabetes has been established just on the genetic co-expression system from both insulin and glucokinase in skeletal muscles utilizing adeno-associated viral vectors. The long-term survival of gene therapy can achieve normal blood glucose levels, except for the exogenous insulin supply. Without integrating the host cell genome, these vectors infect both partitioning and lethargic cells, resulting in a moderate resistance response (Tudurí et al. 2012).

1.7 RECENT ADVANCES IN DIABETES MANAGEMENT

1.7.1 Sensors

1.7.1.1 Glucose Monitoring in Blood

Diabetic patients need to frequently check their blood glucose levels throughout the day, especially following meals and physical exercise. This uses two different sorts of sensors, a continuous glucose monitor and a blood glucose meter (BGM; Zhang et al. 2021). The "finger-pricking" approach based on BGM is still often used today to control the health of DM patients. A finger is pricked to acquire a sample of blood, which is then examined using a test strip and a glucose meter. The method's drawbacks include discomfort, inconvenience, and time restraints. It must use "finger-pricking" at various times throughout the day, especially for measuring glucose levels. Exercise, glucose testing after meals, and insulin administration are common monitoring techniques (Zhang et al. 2021). No doubt the latest continuous glucose monitor systems are available in markets, but real-time glucose monitoring data for a practical application cannot be achieved using this sensor (Bratlie et al. 2012; Nathan et al. 2014).

1.7.1.2 Glucose Monitoring in Sweat

Sweat contains biological information that has been investigated using the right sensors for diagnostic purposes. It is possible to use the illness indicators found in sweat, such as K^+, Ca^{2+}, Na^+, and glucose (Coyle et al. 2014). Another benefit of utilizing sweat is that it may be accessible continually. For perspiration diabetes management, a Bluetooth-enabled bracelet is also paired with another sensor instrument (Gao et al. 2016). Under this idea, many sensing units are utilized to overcome problems that occur whenever a sole sensor unit is being used to analyze the intricate components of sweat.

1.7.1.3 Glucose Monitoring of Interstitial Fluid

The interstitial fluid contains a variety of biomarkers that can be exploited for clinical diagnosis (Corrie et al. 2015). Research initiatives have been put into place in the field of glucose monitoring, which can give important information about the blood glucose levels of people with diabetes. A thorough analysis of the associated techniques for implementing glucose monitoring through the skin may be found (Bandodkar and Wang 2014). These methods include biosensor technology, thermal emission, heat, ultrasonic, optical, and others. Monitoring glucose levels in interstitial fluid is frequently done using a sensor built around a micro-needle array. An amperometric sensor is a three-electrode and a part of the sensing system that may directly detect the hydrogen peroxide product of the enzyme conversion of glucose. The micro-needle array consists of about two hundred silicon micro-needles with a 6 mm × 6 mm surface area. Each micro-needle has a lumen that is 50 μm × 50 μm and measures around 300 μm in length. As a result, this particular glucose detecting unit is rather tiny (Zhang et al. 2021).

1.7.1.4 Glucose Monitoring in Ocular Fluid

A standard medical checkup includes checking the fundus of the eyes. The ocular fluid, or aqueous humor, contains biological components such as glucose, proteins, lipids, and hormones. It uses a digital contact lens that is driven by a battery and has an embedded enzymatic glucose sensor (Senior 2014). As seen in Figure 1.4, the related microchip and the enzyme GOx are employed as

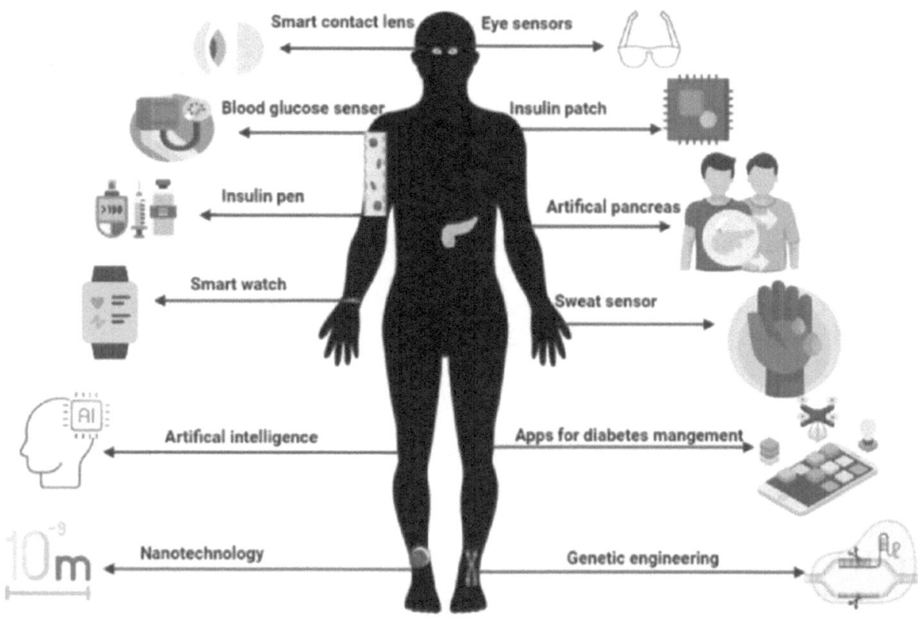

FIGURE 1.4 Schematic representation of trends in diabetes management.

the sensor components and thus are formed between the two layers of a smart contact lens. In this sensor, a redox reaction caused by hydrogen peroxide just at the electrode boosts the glucose concentration inside the ocular fluid (Watt et al. 2004). The sensor relay technology is utilized for data transfer for glucose monitoring.

1.7.2 Insulin Delivery Devices

1.7.2.1 Insulin Pens

Pens make administering insulin more straightforward, accurate, and practical than syringes. A disposable needle, an insulin cartridge, and a single-click per unit dose method make up an insulin pen. The device can be used again or thrown away. Greater patient autonomy, choice, and long-term cost efficiency all encourage adherence to the prescribed course of therapy. Because of this, insulin pens offer improved blood glucose management and are spreading in popularity (Guerci et al. 2019; Rai et al. 2021). Insulin pens with smart connectivity are the newest development in this class of devices, offering features much beyond memory. The InPen System, a Bluetooth-enabled cordless pen with such a device input and bolus advisor, was launched by Companion Medical in 2017. Pendiq 2.0, new InPen™, Humalog Junior KwikPen, FlexTouch (Levemir®, Fiasp®), and All-Star PRO are some models of insulin pens (Domingo-Lopez et al. 2022).

1.7.2.2 Patches for Diabetic

These are the insulin patches that are wearable and effective for conventional injectors. By delivering bolus or basal insulin injection by touch directly, they do away with the difficulty of the long infusion tubes of traditional pumps and the related risk of line disconnections (Rai et al. 2021). The lifespan of an insulin patch is up to three days and it can provide up to 200 U for each infusion. However, because of their reduced size, their ability to use insulin and skin responses are the major factors limiting their lifetime (Domingo-Lopez et al. 2022).

1.7.2.3 Personal Diabetes Managers (PDMs)

When employing nonautomated delivery methods, PDMs aid diabetic patients in making decisions and better controlling their blood sugar levels. PDMs include software programs, internet monitoring systems, and portable cell phone gadgets (typically sensor-specific). The majority of PDMs employ Bluetooth and other wireless technologies to transmit data between the sensors, pump, and display device (Domingo-Lopez et al. 2022). PDMs have the following capabilities: (i) information storage; (ii) long-term recording of BGLs; (iii) regulation of glucose bolus delivery in systems; (iv) production of graphs, diagrams, and analyses; (v) sharing of data with medical practitioners; and (vi) more data extraction from other programs and gadgets, such smartwatches, armbands, and activity trackers (Rai et al. 2021; Domingo-Lopez et al. 2022). The aforementioned advances in diabetes cures are also represented in Figure 1.4.

1.8 COVID-19 AND DIABETES

Diabetes is one of the most prevalent comorbidities among COVID-19 patients. When compared to nondiabetic participants who had access to more practical methods of glucose monitoring, diabetics infected with COVID-19 had greater rates of hospital admission, severe pneumonia, and death (Pranata et al. 2021). Following are several molecular pathomechanisms that may make diabetic people more susceptible to COVID-19. First, diabetes has been associated with impaired immunity, neutrophil chemotaxis, decreased phagocytic activity, and decline in functioning of T cells. Therefore, diabetes should be identified and managed as part of post-acute care regimens for individuals with COVID-19 (Muniyappa and Gubbi 2020). Second, ACE2 (angiotensin-converting enzyme-2) functions as an entry receptor and is abundantly expressed in vascular endothelium, cardiomyocytes, humans, and other distinct sites because of its significant affinities for the SARS-CoV-2 (Zhang et al. 2020). Third, due to immune system dysregulation and anti-inflammatory response, high glucose levels directly promote SARS-CoV-2 replication and may have fatal consequences (Lim et al. 2021). Previous SARS-CoV outbreaks have demonstrated the potential for long-lasting metabolic changes that might predispose individuals to cardiovascular problems (Pranata et al. 2021).

1.9 CONCLUSION

Diabetes is a serious chronic disorder, and today more and more options for glucose monitoring in individuals with DM are becoming accessible because of advancements in biosensor technology and algorithms. These techniques offer workable solutions for making logical recommendations on how to carry out the treatment. In conjunction with the blood-sensing approach, other body fluids are being used to measure and forecast blood glucose levels. The explosive growth of algorithms for determining the glucose content of diverse fluids has brought practical usage closer. In addition, innovative technologies and wearable electronics have provided other useful strategies for glucose monitoring. Nevertheless, a significant gap still exists between the practical applicability of the current findings.

REFERENCES

Abate, L., A. Bachheti, R.K. Bachheti & A. Husen. 2021. Antibacterial properties of medicinal plants: Recent trends, progress, and challenges. In: *Traditional Herbal Therapy for the Human Immune System* (Ed. Husen A.). CRC Press, Boca Raton, FL, pp. 13–54.

Adisakwattana, S., P. Moonsan & S. Yibchok-Anun. 2008. Insulin-releasing properties of a series of cinnamic acid derivatives in vitro and in vivo. *Journal of Agricultural and Food Chemistry* 56(17): 7838–7844.

Adler, A., P. Bennett, S.C. Chair, E. Gregg, K.V. Narayan, M.I. Schmidt, E. Sobngwi, N. Tajima, N. Tandon, N. Unwin & S. Wild. 2021. Reprint of: Classification of diabetes mellitus. *Diabetes Research and Clinical Practice*: 108972.

Agu, K.C., N. Eluehike, R.O. Ofeimun, D. Abile, G. Ideho, M.O. Ogedengbe & O.O. Elekofehinti. 2019. Possible anti-diabetic potentials of *Annona muricata* (soursop): Inhibition of α-amylase and α-glucosidase activities. *Clinical Phytoscience* 5(1):1–13.

Aha, D.W., D. Kibler & M.K. Albert. 1991. Instance-based learning algorithms. *Machine Learning* 6(1): 37–66.

Ahlqvist, E., R.B. Prasad & L. Groop. 2022. 100 years of insulin: Towards improved precision and a new classification of diabetes mellitus. *Journal of Endocrinology* 252(3): R59–R70.

Ahlqvist, E., P. Storm, A. Käräjämäki, M. Martinell, M. Dorkhan, A. Carlsson, P. Vikman, R.B. Prasad, D.M. Aly, P. Almgren & Y. Wessman. 2018. Novel subgroups of adult-onset diabetes and their association with outcomes: A data-driven cluster analysis of six variables. *The Lancet Diabetes & Endocrinology* 6(5): 361–369.

Al Homsi, M.F. & M.L. Lukic. 1993. An update on the pathogenesis of diabetes mellitus. *Dubai Diabetes and Endocrinology Journal* 1: 1–12.

Alsabti, K., S. Ranka & V. Singh. 1997. An efficient k-means clustering algorithm. In: *Proceedings of the First Workshop High Performance Data Mining*. https://surface.syr.edu/eecs/43

Amer, L.D., M.J. Mahoney & S.J. Bryant. 2014. Tissue engineering approaches to cell-based type 1 diabetes therapy. *Tissue Engineering Part B: Reviews* 20(5): 455–467.

American Diabetes Association. 2021. Classification and diagnosis of diabetes: Standards of medical care in diabetes. *Diabetes Care* 44: S15–S33. https://doi.org/10.2337/dc21-S002

Andersen, M.K. & T. Hansen. 2019. Genetic aspects of latent autoimmune diabetes in adults: A mini-review. *Current Diabetes Reviews* 15(3): 194–198.

Andreadi, A., A. Bellia, N. Di Daniele, M. Meloni, R. Lauro, D. Della-Morte & D. Lauro. 2022. The molecular link between oxidative stress, insulin resistance, and type 2 diabetes: A target for new therapies against cardiovascular diseases. *Current Opinion in Pharmacology* 62: 85–96.

Andrews, R.C., A.R. Cooper, A.A. Montgomery, A.J. Norcross, T.J. Peters, D.J. Sharp, N. Jackson, K. Fitzsimons, J. Bright, K. Coulman & C.Y. England. 2011. Diet or diet plus physical activity versus usual care in patients with newly diagnosed type 2 diabetes: The Early ACTID randomised controlled trial. *The Lancet* 378(9786): 129–139.

Bachheti, A., Deepti, R.K. Bachheti & A. Husen. 2021. Medicinal plants and their pharmaceutical properties under adverse environmental conditions. In: *Harsh Environment and Plant Resilience* (Ed. Husen A.) Springer International Publishing, Cham, pp. 457–502. https://doi.org/10.1007/978-3-030-65912-7_19

Bahman, F., K. Greish & S. Taurin. 2019. Nanotechnology in insulin delivery for management of diabetes. *Pharmaceutical Nanotechnology* 7(2): 113–128.

Balasubramanyam, A. 2021. Defining and classifying new subgroups of diabetes. *Annual Review of Medicine* 72: 63–74.

Bandodkar, A.J. & J. Wang. 2014. Non-invasive wearable electrochemical sensors: A review. *Trends in Biotechnology* 32(7): 363–371.

Barkabi-Zanjani, S., V. Ghorbanzadeh, M. Aslani, A. Ghalibafsabbaghi & L. Chodari. 2020. Diabetes mellitus and the impairment of male reproductive function: Possible signaling pathways. *Diabetes & Metabolic Syndrome: Clinical Research & Reviews* 14: 1307–1314.

Baynes, H.W. 2015. Classification, pathophysiology, diagnosis and management of diabetes mellitus. *Journal of Diabetes and Metabolism* 6(5): 1–9.

Boulé, N.G., E. Haddad, G.P. Kenny, G.A. Wells & R.J. Sigal. 2001. Effects of exercise on glycemic control and body mass in type 2 diabetes mellitus: A meta-analysis of controlled clinical trials. *Jama* 286(10): 1218–1227.

Bratlie, K.M., R.L. York, M.A. Invernale, R. Langer & D.G. Anderson. 2012. Materials for diabetes therapeutics. *Advanced Healthcare Materials* 1(3): 267–284.

Breiman, L. 1996. Bagging predictors. *MachineLearning* 24: 123–140.

Chhetri, D.R., P. Parajuli & G.C. Subba. 2005. Antidiabetic plants used by Sikkim and Darjeeling Himalayan tribes, India. *Journal of Ethnopharmacology* 99(2): 199–202.

Cho, N.H., J.E. Shaw, S. Karuranga, Y. Huang, J.D. da Rocha Fernandes, A.W. Ohlrogge & B. Malanda. 2018. IDF Diabetes Atlas: Global estimates of diabetes prevalence for 2017 and projections for 2045. *Diabetes Research and Clinical Practice* 138: 271–281.

Choudhury, S. & P. Patra. 2022. Recent developments of nano-formulations against Diabetes: A systematic review. *Recent Patents on Nanotechnology* 16. https://doi.org/10.2174/1872210516666220622114505.

Chrvala, C.A., D. Sherr & R.D. Lipman. 2016. Diabetes self-management education for adults with type 2 diabetes mellitus: A systematic review of the effect on glycemic control. *Patient Education and Counseling* 99(6): 926–943.

Chu, J. & Y. Ali. 2008. Diabetic retinopathy: A review. *Drug Development Research* 69(1): 1–14.

Cohen, W.W. 1995. Fast effective rule induction. In: *Machine Learning Proceedings*, 115–123. www.sciencedirect.com/science/article/pii/B9781558603776500232

Cole, J.B. & J.C. Florez. 2020. Genetics of diabetes mellitus and diabetes complications. *Nature Reviews Nephrology* 16(7): 377–390.

Committee, A.D.A.P.P. 2022. Classification and diagnosis of diabetes: Standards of medical care in Diabetes–2022. *Diabetes Care* 45: 17–38. https://doi.org/10.2337/dc22-S002

Corrie, S.R., J.W. Coffey, J. Islam, K.A. Markey & M.A.F. Kendall. 2015. Blood, sweat, and tears: Developing clinically relevant protein biosensors for integrated body fluid analysis. *Analyst* 140(13): 4350–4364.

Cortes, C. & V. Vapnik. 1995. Support-vector networks. *Machine Learning* 20: 273–297.

Coyle, S., V.F. Curto, F. Benito-Lopez, L. Florea & D. Diamond. 2014. Wearable bio and chemical sensors. *Wearable Sensors* 2: 65–83.

Dahman, Y. 2017. Nanotechnology and functional materials for engineers. *Nanomedicine*: 229–249.

Dai, Z., A. Yang, X. Bao & R. Yang. 2019. Facile non-enzymatic electrochemical sensing for glucose based on Cu2O–BSA nanoparticles modified GCE. *Sensors* 19(12): 2824.

Datey, P., A. Hankey & H.R. Nagendra. 2018. Combined Ayurveda and yoga practices for newly diagnosed type 2 diabetes mellitus: A controlled trial. *Complementary Medicine Research* 25(1): 16–23.

Demir, T., S. Turan, K. Unluhizarci, O. Topaloglu, T. Tukek & D. Gogas Yavuz. 2021. Use of insulin degludec/insulin aspart in the management of diabetes mellitus: Expert panel recommendations on appropriate practice patterns. *Frontiers in Endocrinology* 12: 616514.

Domingo-Lopez, D.A., G. Lattanzi, L.H. Schreiber, E.J. Wallace, R. Wylie, J. O'Sullivan, E.B. Dolan & G.P. Duffy. 2022. Medical devices, smart drug delivery, wearables and technology for the treatment of diabetes mellitus. *Advanced Drug Delivery Reviews*: 114280.

Dugalic, S., M. Petronijevic, B. Vasiljevic, J. Todorovic, D. Stanisavljevic, A. Jotic, L. Lukic, T. Milicic, N. Lalić, K. Lalic & M. Stoiljkovic. 2022. Trends of the prevalence of pre-gestational Diabetes in 2030 and 2050 in Belgrade cohort. *International Journal of Environmental Research and Public Health* 19(11): 6517.

Elya, B., R. Handayani, R. Sauriasari, U.S. Hasyyati, I.T. Permana & Y.I. Permatasari. 2015. Antidiabetic activity and phytochemical screening of extracts from Indonesian plants by inhibition of alpha amylase, alpha glucosidase and dipeptidyl peptidase IV. *Pakistan Journal of Biological Sciences* 18(6): 279–287.

EMC Education Services. 2015. *Data Science and Big Data Analytics: Discovering, Analyzing, Visualizing and Presenting Data*. Wiley, New York.

Gao, W., S. Emaminejad, H.Y.Y. Nyein, S. Challa, K. Chen, A. Peck, H.M. Fahad, H. Ota, H. Shiraki, D. Kiriya & D.H. Lien. 2016. Fully integrated wearable sensor arrays for multiplexed in situ perspiration analysis. *Nature* 529(7587): 509–514.

Ghazanfari-Sarabi, S., M. Habibi-Rezaei, R. Eshraghi-Naeeni & A.A. Moosavi-Movahedi. 2019. Prevention of haemoglobin glycation by acetylsalicylic acid (ASA): A new view on old mechanism. *PLoS ONE* 14(4): e0214725.

Grau, M. and C. Pericas. 2022. Diabetes: A multifaceted disorder. *Biomedicines* 10: 1698.

Gregg, E.W., Y.J. Cheng, M. Srinivasan, J. Lin, L.S. Geiss, A.L. Albright & G. Imperatore. 2018. Trends in cause-specific mortality among adults with and without diagnosed diabetes in the USA: An epidemiological analysis of linked national survey and vital statistics data. *The Lancet* 391(10138): 2430–2440.

Guerci, B., N. Chanan, S. Kaur, J.G. Jasso-Mosqueda & E. Lew. 2019. Lack of treatment persistence and treatment nonadherence as barriers to glycaemic control in patients with type 2 diabetes. *Diabetes Therapy* 10(2): 437–449.

Gul, S., S.B. Khan, I.U. Rehman, M.A. Khan & M.I. Khan. 2019. A comprehensive review of magnetic nanomaterials modern day theranostics. *Frontiers in Materials* 6: 179.

Hall, M. & E. Frank. 2008. Combining naive Bayes and decision tables. In: *Proc 21st Florida Artificial Intelligence Research Society Conference*. AAAI Press, Miami, FL.

Hallman, D.M., J.C. Huber Jr, V.H. Gonzalez, B.E. Klein, R. Klein & C.L Hanis. 2005. Familial aggregation of severity of diabetic retinopathy in Mexican Americans from Starr County, Texas. *Diabetes Care* 28(5): 1163–1168.

Harding, J.L., J.E. Shaw, A. Peeters, S. Davidson & D.J. Magliano. 2016. Age-specific trends from 2000–2011 in all-cause and cause-specific mortality in type 1 and type 2 diabetes: A cohort study of more than one million people. *Diabetes Care* 39(6): 1018–1026.

Hassoun, M.H. 1995. *Fundamentals of Artificial Neural Networks*. MIT Press, Cambridge, MA.

He, J.H., L.X. Chen & H. Li. 2019. Progress in the discovery of naturally occurring anti-diabetic drugs and in the identification of their molecular targets. *Fitoterapia* 134: 270–289.

Holt, R.I. 2004. Diagnosis, epidemiology and pathogenesis of diabetes mellitus: An update for psychiatrists. *The British Journal of Psychiatry* 184(S47): s55–s63.

Holte, R.C. 1993. Very simple classification rules perform well on most commonly used datasets. *Machine learning* 11(1): 63–90.

Hosmer Jr, D.W., S. Lemeshow & R.X. Sturdivant. 2013. *Applied Logistic Regression*. John Wiley & Sons. ISBN 978-0-470-58247-3.

Husen, A. 2021. *Traditional Herbal Therapy for the Human Immune System*. CRC Press, Boca Raton, FL. https://doi.org/10.1201/9781003137955

Husen, A. 2022. *Herbs, Shrubs and Trees of Potential Medicinal Benefits*. Taylor & Francis, Boca Raton, FL. https://doi.org/10.1201/9781003205067

Husen, A., R.K. Bachheti & A. Bachheti. 2021. *Non-Timber Forest Products: Food, Healthcare and Industrial Applications*. Springer, Cham, Switzerland. https://doi.org/10.1007/978-3-030-73077-2

Husen, A., & M. Iqbal. 2021. Plant-based potential nutraceuticals for improving the human immune system. In: *Traditional Herbal Therapy for the Human Immune System* (Ed. Husen A.). CRC Press, Boca Raton, FL, pp. 1–11.

Ismail, L., H. Materwala, M. Tayefi, P. Ngo & A.P. Karduck. 2022. Type 2 diabetes with artificial intelligence machine learning: Methods and evaluation. *Archives of Computational Methods in Engineering* 29(1): 313–333.

Jacob, B. & R.T. Narendhirakannan. 2019. Role of medicinal plants in the management of diabetes mellitus: A review. *3 Biotech* 9(1): 1–17.

Johannsen, L., T. Smith, A.M. Havsager, C. Madsen, M.J. Kjeldsen, N.J. Dalsgaard, D. Gaist, H.D. Schrøder & S.H. Sindrup. 2001. Evaluation of patients with symptoms suggestive of chronic polyneuropathy. *Journal of Clinical Neuromuscular Disease* 3(2): 47–52.

Khalid, M., G. Petroianu & A. Adem. 2022. Advanced glycation end products and diabetes mellitus: Mechanisms and perspectives. *Biomolecules* 12(4): 542.

Kohavi, R. 1995. Feature subset selection as search with probabilistic estimates. In: *Proceedings AAAI Fall Symposium on Relevance*. New Orleans, LA, pp. 122–126.

Kumar, S., A. Mittal, D. Babu & A. Mittal. 2021. Herbal medicines for diabetes management and its secondary complications. *Current Diabetes Reviews* 17(4): 437–456.

Kumar, S., M. Nehra, D. Kedia, N. Dilbaghi, K. Tankeshwar & K.H. Kim. 2020. Nanotechnology-based biomaterials for orthopaedic applications: Recent advances and future prospects. *Materials Science and Engineering* 106: 110154.

Kumari, S., S.D. Laxmikant, B. Sonika & S. Khanal. 2022. Efficacy of integrated Ayurveda treatment protocol in type 2 Diabetes Mellitus – A case report. *Journal of Ayurveda and Integrative Medicine* 13(1): 100512.

Lambrinou, E., T.B. Hansen & J.W. Beulens. 2019. Lifestyle factors, self-management and patient empowerment in diabetes care. *European Journal of Preventive Cardiology* 26(2): 55–63.

Lamos, E.M., D.L. Levitt & K.M. Munir. 2016. A review of dopamine agonist therapy in type 2 diabetes and effects on cardio-metabolic parameters. *Primary Care Diabetes* 10(1): 60–65.

Li, D., Y. Liu & N. Wu. 2022. Application progress of nanotechnology in regenerative medicine of diabetes mellitus. *Diabetes Research and Clinical Practice*: 109966.

Liaw, A. & M. Wiener. 2002. Classification and regression by random Forest. *R News* 2(3): 18–22.

Lim, S., J.H. Bae, H.S. Kwon & M.A. Nauck. 2021. COVID-19 and diabetes mellitus: From pathophysiology to clinical management. *Nature Reviews Endocrinology* 17(1): 11–30.

Martínez-Castelao, A., J.F. Navarro-González, J.L. Górriz & F. De Alvaro. 2015. The concept and the epidemiology of diabetic nephropathy have changed in recent years. *Journal of Clinical Medicine* 4(6): 1207–1216.

Mechchate, H., I. Es-Safi, A. Louba, A.S. Alqahtani, F.A. Nasr, O.M. Noman, M. Farooq, M.S. Alharbi, A. Alqahtani, A. Bari & H. Bekkari. 2021. In vitro alpha-amylase and alpha-glucosidase inhibitory activity and in vivo antidiabetic activity of Withania frutescens L. Foliar extract. *Molecules* 26(2): 293.

Minakawa, M., Y. Miura & K. Yagasaki. 2012. Piceatannol, a resveratrol derivative, promotes glucose uptake through glucose transporter 4 translocation to plasma membrane in L6 myocytes and suppresses blood glucose levels in type 2 diabetic model db/db mice. *Biochemical and Biophysical Research Communications* 422(3): 469–475.

Mishra, V., P. Nayak, M. Sharma, A. Albutti, A.S. Alwashmi, M.A. Aljasir, N. Alsowayeh & M.M. Tambuwala. 2021. Emerging treatment strategies for diabetes mellitus and associated complications: An update. *Pharmaceutics* 13(10): 1568.

Mohammedi, K., M. Woodward, M. Marre, S. Colagiuri, M. Cooper, S. Harrap, G. Mancia, N. Poulter, B. Williams, S. Zoungas & J. Chalmers. 2017. Comparative effects of microvascular and macrovascular disease on the risk of major outcomes in patients with type 2 diabetes. *Cardiovascular Diabetology* 16(1): 1–9.

Morrish, N.J., S.L. Wang, L.K. Stevens, J.H. Fuller & H. Keen. 2001. Mortality and causes of death in the WHO multinational study of vascular disease in diabetes. *Diabetologia* 44(2): 14–21.

Most, R.S. & P. Sinnock. 1983. The epidemiology of lower extremity amputations in diabetic individuals. *Diabetes Care* 6(1): 87–91.

Muniyappa, R. & S. Gubbi. 2020. COVID-19 pandemic, coronaviruses, and diabetes mellitus. *American Journal of Physiology-Endocrinology and Metabolism* 318(5): E736–E741.

Muralidharan, V. & V. Sugumaran. 2012. A comparative study of Naïve Bayes classifier and Bayes net classifier for fault diagnosis of monoblock centrifugal pump using wavelet analysis. *Applied Soft Computing* 12: 2023–2029.

Nathan, D.M., P. McGee, M.W. Steffes & J.M. Lachin. 2014. Group DER. Relationship of glycated albumin to blood glucose and HbA1c values and to retinopathy, nephropathy, and cardiovascular outcomes in the DCCT/EDIC study. *Diabetes* 63(1): 282–290.

Nowacka, O., K. Milowska, S. Belica-Pacha, B. Palecz, K. Šipošová, Z. Gazova & M. Bryszewska. 2016. Generation-dependent effect of PAMAM dendrimers on human insulin fibrillation and thermal stability. *International Journal of Biological Macromolecules* 82: 54–60.

Oh, Y.S. 2015. Plant-derived compounds targeting pancreatic beta cells for the treatment of diabetes. *Evidence-Based Complementary and Alternative Medicine* 2015: 629863.

Ohiagu, F.O., P.C. Chikezie & C.M. Chikezie. 2021. Pathophysiology of diabetes mellitus complications: Metabolic events and control. *Biomedical Research and Therapy* 8(3): 4243–4257.

Ozougwu, J.C., K.C. Obimba, C.D. Belonwu & C.B. Unakalamba. 2013. The pathogenesis and pathophysiology of type 1 and type 2 diabetes mellitus. *Journal of Physiology and Pathophysiology* 4(4): 46–57.

Pearson-Stuttard, J., J. Buckley, M. Cicek & E.W. Gregg. 2021. The changing nature of mortality and morbidity in patients with diabetes. *Endocrinology and Metabolism Clinics* 50(3): 357–368.

Peppa, M., J. Uribarri & H. Vlassara. 2003. Glucose, advanced glycation end products, and diabetes complications: What is new and what works. *Clinical Diabetes* 21(4): 186–187.

Piepoli, M.F., A.W. Hoes, S. Agewall, C. Albus, C. Brotons, A.L. Catapano, M.T. Cooney, U. Corrà, B. Cosyns, C. Deaton, I. Graham, M.S. Hall, F.D.R. Hobbs, M.L. Løchen, H. Löllgen, P. Marques-Vidal, J. Perk, E. Prescott, J. Redon, D.J. Richter, N. Sattar, Y. Smulders, M. Tiberi, H.B. van der Worp, I. van Dis, W.M.M. Verschuren, S. Binno & ESC Scientific Document Group. 2016. European Guidelines on cardiovascular disease prevention in clinical practice: The sixth joint task force of the European society of cardiology and other societies on cardiovascular disease prevention in clinical practice (constituted by representatives of 10 societies and by invited experts) developed with the special contribution of the European association for cardiovascular prevention & rehabilitation (EACPR). *European Heart Journal* 37(29): 2315–2381.

Pieralice, S. & P. Pozzilli. 2018. Latent autoimmune diabetes in adults: A review on clinical implications and management. *Diabetes & Metabolism Journal* 42(6): 451–464.

Pouran, S.R., A. Bayrami, F.M. Arvanag, A. Habibi-Yangjeh, R.D.C. Soltani, R. Singh, A.A.A. Raman, K.H. Chae, A. Khataee & H.K. Kang. 2020. Biogenic integrated ZnO/Ag nanocomposite: Surface analysis and in vivo practices for the management of type 1 diabetes complications. *Colloids and Surfaces B: Biointerfaces* 189: 110878.

Pranata, R., J. Henrina, W.M. Raffaello, S. Lawrensia & I. Huang. 2021. Diabetes and COVID-19: The past, the present, and the future. *Metabolism* 121: 154814.

Prasad-Reddy, L. and D. Isaacs. 2015. A clinical review of GLP-1 receptor agonists: Efficacy and safety in diabetes and beyond. *Drugs in Context* 4: 212283.

Przeor, M., E. Flaczyk, D. Kmiecik, M.S. Buchowski, H. Staniek, A. Tomczak-Graczyk, J. Kobus-Cisowska, A. Gramza-Michałowska & J. Foksowicz-Flaczyk. 2020. Functional properties and antioxidant activity of Morus alba L. leaves var. *Zolwinska Wielkolistna* (WML-P) – The effect of controlled conditioning process. *Antioxidants* 9(8): 668.

Rahman, M.M., M.R. Islam, S. Shohag, M.E. Hossain, M.S. Rahaman, F. Islam, M. Ahmed, S. Mitra, M.U. Khandaker, A.M. Idris & K. Chidambaram. 2022. The multifunctional role of herbal products in the management of diabetes and obesity: A comprehensive review. *Molecules* 27(5): 1713.

Rahman, S. & A. Husen. 2021. Potential role of medicinal plants in the cure of liver and kidney diseases. In: *Non-Timber Forest Products* (Eds. Husen A., Bachheti R.K., Bachheti A.). Springer International Publishing, Cham, pp. 229–225. https://doi.org/10.1007/978-3-030-73077-2_10

Rahman, S. & A. Husen. 2022. *Asparagus racemosus* (Shatavari) and *Dioscorea villosa* (Wild yams). In: *Herbs, Shrubs, and Trees of Potential Medicinal Benefits* (Ed. Husen A.). CRC Press, Boca Raton, FL, pp. 95–112.

Rai, I., A. Wanjari & S. Acharya. 2021. Recent advances in insulin delivery devices and modes of insulin therapy. *Journal of Pharmaceutical Research International* 33: 358–367.

Raveendran, A.V., A. Deshpandae & S.R. Joshi. 2018. Therapeutic role of yoga in type 2 diabetes. *Endocrinology and Metabolism* 33(3): 307–317.

Reichelt-Wurm, S., T. Wirtz, D. Chittka, M. Lindenmeyer, R.M. Reichelt, S. Beck, P. Politis, A. Charonis, M. Kretz, T. B. Huber & S. Liu. 2019. Glomerular expression pattern of long non-coding RNAs in the type 2 diabetes mellitus BTBR mouse model. *Scientific Reports* 9(1): 1–13.

Salleh, N.H., I.N. Zulkipli, H. Mohd Yasin, F. Ja'afar, N. Ahmad, W.A.N. Wan Ahmad & S.R. Ahmad. 2021. Systematic review of medicinal plants used for treatment of diabetes in human clinical trials: An ASEAN perspective. *Evidence-Based Complementary and Alternative Medicine*: Article ID 5570939.

Sanchez-Rangel, E. & S.E. Inzucchi. 2017. Metformin: Clinical use in type 2 diabetes. *Diabetologia* 60(9): 1586–1593.

Schenk, S., M. Saberi & J.M. Olefsky. 2008. Insulin sensitivity: Modulation by nutrients and inflammation. *Journal of Clinical Investment* 118(9): 2992–3002.

Sekikawa, A., M. Tominaga, K. Takahashi, H. Eguchi, M. Igarashi, H. Ohnuma, K. Sugiyama, H. Manaka, H. Sasaki, H. Fukuyama & K. Miyazawa. 1993. Prevalence of diabetes and impaired glucose tolerance in Funagata area, Japan. *Diabetes Care* 16(4): 570–574.

Selvin, E., Y. Ning, M.W. Steffes, L.D. Bash, R. Klein, T.Y. Wong, B.C. Astor, A.R. Sharrett, F.L. Brancati & J. Coresh. 2011. Glycated hemoglobin and the risk of kidney disease and retinopathy in adults with and without diabetes. *Diabetes* 60(1): 298–305.

Senior, M. 2014. Novartis signs up for Google smart lens. *Nature Biotechnology* 32(9): 856–857.

Shaefer, J.C.F., P. Kushner & R. Aguilar. 2015. User's guide to mechanism of action and clinical use of GLP-1 receptor agonists. *Postgraduate Medicine* 127(8): 818–826.

Shahin D.H.H., R. Sultana, J. Farooq, T. Taj, U.F. Khaiser, N.S.A Alanazi, M.K. Alshammari, M.N. Alshammari, F.H. Alsubaie, S.M.B. Asdaq & A.A. Alotaibi. 2022. Insights into the uses of traditional plants for diabetes nephropathy: A review. *Current Issues in Molecular Biology* 44(7): 2887–2902.

Sharma, H. & H.M. Chandola. 2011. Prameha in ayurveda: Correlation with obesity, metabolic syndrome, and diabetes mellitus. Part 1 – etiology, classification, and pathogenesis. *The Journal of Alternative and Complementary Medicine* 17(6): 491–496.

Simos, Y.V., K. Spyrou, M. Patila, N. Karouta, H. Stamatis, D. Gournis, E. Dounousi & D. Peschos. 2021. Trends of nanotechnology in type 2 diabetes mellitus treatment. *Asian Journal of Pharmaceutical Sciences* 16(1): 62–76.

Singh, N., D.G. Armstron & B.A. Lipsky. 2005. Preventing foot ulcers in patients with diabetes. *Jama* 293(2): 217–228.

Singhal, T., M. Mujeeb, A. Ahad, M. Aqil, S.O. Rahman, A.K. Najmi & W.A. Siddiqui. 2019. Preparation, optimization and biological evaluation of gymnemic acid loaded niosomes against streptozotocin-nicotinamide induced diabetic-nephropathy in Wistar rats. *Journal of Drug Delivery Science and Technology* 54: 101328.

Sonkar, S., A.K. Singh & A. Husen. 2022. *Convolvulus pluricaulis* (Shankhpushpi) and *Erythroxylum coca* (Coca plant). In: *Herbs, Shrubs, and Trees of Potential Medicinal Benefits* (Ed. Husen A.). CRC Press, Boca Raton, FL, pp. 83–93.

Spijkerman, A.M., J.M. Dekker, G. Nijpels, M.C. Adriaanse, P.J. Kostense, D. Ruwaard, C.D. Stehouwer, L.M. Bouter & R.J. Heine. 2003. Microvascular complications at time of diagnosis of type 2 diabetes are similar among diabetic patients detected by targeted screening and patients newly diagnosed in general practice: The hoorn screening study. *Diabetes Care* 26(9): 2604–2608.

Strakosas, X., J. Selberg, P. Pansodtee, N. Yonas, P. Manapongpun, M. Teodorescu & M. Rolandi. 2019. A non-enzymatic glucose sensor enabled by bioelectronic pH control. *Scientific Reports* 9(1): 1–7.

Suvarna, R., R.P. Shenoy, B.S. Hadapad & A.V. Nayak. 2021. Effectiveness of polyherbal formulations for the treatment of type 2 Diabetes mellitus-A systematic review and meta-analysis. *Journal of Ayurveda and Integrative Medicine* 12(1): 213–222.

Tan, X.C., K.H. Chua, R.M. Ravishankar & U.R. Kuppuswamy. 2016. Monoterpenes: Novel insights into their biological effects and roles on glucose uptake and lipid metabolism in $3T_3L_1$ adipocytes. *Food Chemistry* 196: 242–250.

Tikkanen-Dolenc, H., J. Wadén, C. Forsblom, V. Harjutsalo, L.M. Thorn, M. Saraheimo, N. Elonen, M. Rosengård-Bärlund, D. Gordin, H.O. Tikkanen & P.H. Groop. 2017. Frequent and intensive physical activity reduces risk of cardiovascular events in type 1 diabetes. *Diabetologia* 60(3): 574–580.

Tomic, D., J.E. Shaw & D.J. Magliano. 2022. The burden and risks of emerging complications of diabetes mellitus. *Nature Reviews Endocrinology*18: 525–539.

Tran, N., B. Pham & L. Le. 2020. Bioactive compounds in anti-diabetic plants: From herbal medicine to modern drug discovery. *Biology* 9(9): 252.

Tudurí, E., J.E. Bruin & T.J. Kieffer. 2012. Restoring insulin production for type 1 diabetes. *Journal of Diabetes* 4: 319–331.

Udler, M.S., M.I. McCarthy, J.C. Florez & A. Mahajan. 2019. Genetic risk scores for diabetes diagnosis and precision medicine. *Endocrine Reviews* 40(6): 1500–1520.

Ventola, C.L. 2012. The nanomedicine revolution: Part 1: Emerging concepts. *Pharmacy and Therapeutics* 37(9): 512.

Wallberg, M. & A. Cooke. 2013. Immune mechanisms in type 1 diabetes. *Trends in Immunology* 34(12): 583–591.

Watt, B.E., A.T. Proudfoot & J.A. Vale. 2004. Hydrogen peroxide poisoning. *Toxicological Reviews* 23(1): 51–57.

Welsh, K.J., M.S. Kirkman & D.B. Sacks. 2016. Role of glycated proteins in the diagnosis and management of diabetes: Research gaps and future directions. *Diabetes Care* 39(8): 1299–1306.

Wolpert, D.H. 1992. Stacked generalization. *Neural Networks* 5(2): 241–259.

Xiao, J., X. Ni, G. Kai & X. Chen. 2013. A review on structure–activity relationship of dietary polyphenols inhibiting α-amylase. *Critical Reviews in Food Science and Nutrition* 53(5): 497–506.

Zhang, H., J.M. Penninger, Y. Li, N. Zhong & A.S. Slutsky. 2020. Angiotensin-converting enzyme 2 (ACE2) as a SARS-CoV-2 receptor: Molecular mechanisms and potential therapeutic target. *Intensive Care Medicine* 46(4): 586–590.

Zhang, M., Y.Z. Zhao, W.C. Ma, J.L. Xu, J.L. Wang, M.J. Chen, L. Yu & Y.N. Chen. 2017. Therapeutic effect of combined use of FGF1-loaded nano-liposomes and ultrasound-targeted microbubble destruction technique on treating rats with experimental diabetic cardiomyopathy. *Zhonghua Xin Xue Guan Bing Za Zhi* 45(5): 427–433.

Zhang, Y., J. Sun, L. Liu & H. Qiao. 2021. A review of biosensor technology and algorithms for glucose monitoring. *Journal of Diabetes and its Complications* 35(8): 107929.

Zhao, C., C. Yang, S.T.C. Wai, Y.P. Zhang, M. Portillo, P. Paoli, Y. Wu, W. San Cheang, B. Liu, C. Carpéné & J. Xiao. 2019. Regulation of glucose metabolism by bioactive phytochemicals for the management of type 2 diabetes mellitus. *Critical Reviews in Food Science and Nutrition* 59(6): 830–847.

Zheng, Y., S.H. Ley & F.B. Hu. 2018. Global aetiology and epidemiology of type 2 diabetes mellitus and its complications. *Nature Reviews Endocrinology* 14(2): 88–98.

2 Medicinal and Herbal Plant-Based Phytochemicals

God's Souvenir for Antidiabetic Response

Kuldipika Sharma, Manik Devgan, Arshdeep Kaur, Radhika Sharma, Antul Kumar, Anuj Choudhary, Pankaj Gupta, Satyakam Guha, Anand Sonkar, and Sahil Mehta

CONTENTS

DOI: 10.1201/b23347-2

2.1 INTRODUCTION

The human body is the most intricate organization of neuropeptides and hormones, which are released from the endocrine glands (the endocrine cells of the pancreas, brain, ovaries/testes, parathyroid gland, and thyroid gland), sebaceous glands, sweat glands, muscle cells, and Brunner's glands. The pancreas is the glucose sensor to modulate the secretion and synthesis of insulin and helps to regulate blood sugar levels (Wickramasinghe et al. 2021; Mudau et al. 2022). The apoptosis following necrosis leads to the disintegration of beta cells and then the progression of diabetes. Type 1 diabetes mellitus (T1DM) and type 2 diabetes mellitus (T2DM) is a single disorder differentiated by β-cell degeneration and accelerators to trigger the β-cell loss. The function of these accelerators is to differentially regulate the metabolic function of the body. During apoptosis, antigens are released by the β-cells to trigger an autoimmune attack in the β-cells of the pancreas in susceptible individuals (Shahin et al. 2022). Hyperglycemia and elevated inflammation may result in a secondary kind of issues, including gastrointestinal and urological complications. DM is the assemblage of metabolic disorders signaled by complex interactions of environmental and genetic factors (Volpe et al. 2018). Diabetes can cause multiple organ failures, heart diseases, nerve damage, hypertension, stroke, adult-onset blindness, and kidney failure (Abejew et al. 2015). Prolonged hyperglycemia results in impairment in growth, and the body becomes more prone to infectious diseases. The prolonged complications involve nephropathy (failure of the kidney), retinopathy (vision loss), foot ulcers, gastrointestinal complications, cardiovascular complications, and a high risk of sexual dysfunctions (Rahman et al. 2022), as shown in Figure 2.1.

The diagnosis of DM requires medical management with an altered lifestyle. Several synthetic drugs are prescribed for the symptomatic cure of diabetes (Ojo et al. 2022). These include injecting

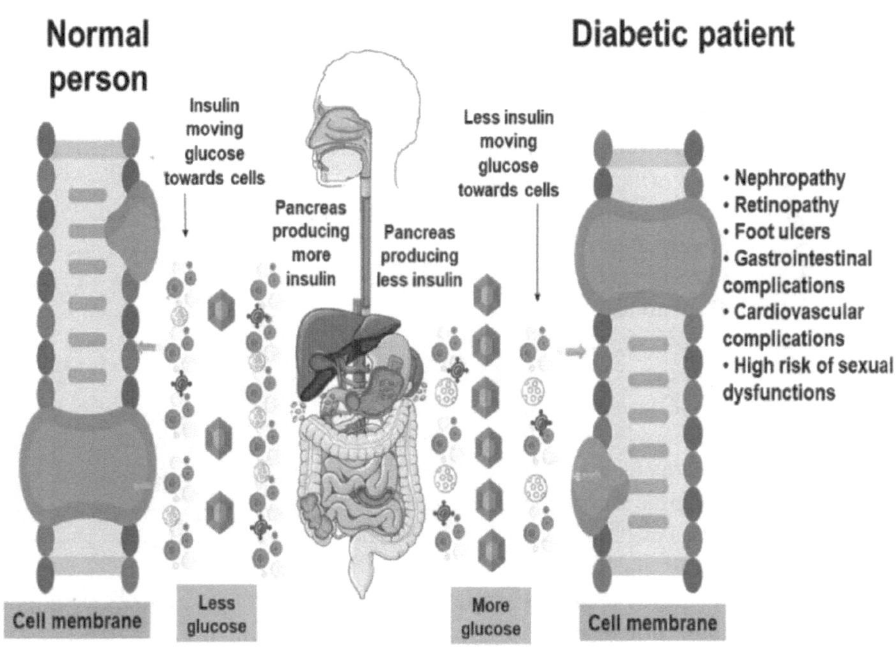

FIGURE 2.1 Difference between normal and diabetic patients.

insulin or taking antidiabetic medications orally (α-glucosidase inhibitors, such as sulfonylureas, biguanides, dipeptidyl peptidase-4 (DPP4) inhibitors, thiazolidinediones, and sodium-glucose co-transporter-2 inhibitors) (Gourgari et al. 2017). Therefore, optional treatments are the primary need for more efficient strategies for the disease. Alternative and complementary medicine is gradually used as a management approach for DM. In this view, the conventional herbal system gains considerable importance for disease treatment all over the world (Benchoula et al. 2022).

Over the last five decades, changes in lifestyle and globalization have had a profound impact on society, politics, the environment, and human behavior; as a result, diabetes has emerged as a global epidemic. In both developed and developing nations, the number of diabetes patients has significantly increased, with indigenous peoples and socially disadvantaged groups typically bearing the greater burden.

2.2 DIABETES

Hyperglycemia caused by insulin synthesis, insulin action, or both characterizes a group of metabolic diseases recognized as diabetes mellitus (Patle et al. 2021). Chronic hyperglycemia caused by diabetes is associated with long-term damage, dysfunction, and failure of several organs, including the heart, blood vessels, nerves, eyes, and kidneys (ADA 2004). One of the first populations to describe diabetes, which is marked by weight loss and frequent urination, was the Egyptians. The term diabetes mellitus, on the other hand, was created by the Greek physician Aertaeus. "Diabetes" is a Greek word that means "to pass through," and "mellitus" is a Latin word that means "honey" (Kaul et al. 2013).

2.2.1 Type I Diabetes and Type 2 Diabetes

T1DM, also known as insulin-dependent diabetes or juvenile diabetes, is an autoimmune disease in which activated $C0^{4+}$ and $C0^{8+}$ T cells and macrophages penetrate the pancreatic islets and damage the cells (Phillips et al. 2013). Although type 1 diabetes can occur at any age, the risk is higher in the majority of populations between the years of birth and 14 than between the ages of 15 and 29 (Konieczynski et al. 2022). People with type 1 diabetes that slowly progresses may have enough beta cells to avoid serious issues for a long time, but when the body is under stress from an illness, trauma, or surgery, their diabetic control can quickly worsen. Insulin dependence develops in people with T1DM that is slowly progressing (Balbaa et al. 2021). Although there seem to be several significant variables, the exact causes of T1DM are still not entirely known.

T2DM is characterized by low insulin secretion as a result of insulin resistance. It is reported to increase in incidence and prevalence with age and accounts for roughly 90% of all cases of diabetes worldwide. (Masharani 2001). It has been linked to several metabolic changes, including obesity, hypertension, and dyslipidemia, and the patients are at a higher risk of developing atherosclerosis (Kumar et al. 2021b).

2.2.2 Causes and Consequences

2.2.2.1 Insulin Resistance

The pancreatic islets of Langerhans cells release the peptide hormone insulin, which modulates protein, lipid, and carbohydrate metabolism, aids in cellular glucose uptake, and, through its mitogenic effects, encourages cell growth and proliferation (Wilcox 2005). Overweight individuals have worse insulin resistance. It is also connected to a variety of other disorders, including alterations in blood coagulation, high cholesterol, central obesity, high blood pressure, and high insulin levels (hyperinsulinemia). The risk of strokes and heart attacks is raised by these anomalies, which are sometimes referred to as insulin resistance syndrome (Nazarko 2009; Al-Ishaq et al. 2019).

2.2.2.2 Hyperglycemia

Hyperglycemia refers to an excess of sugar in the blood, which is a sign of diabetes. It leads to an energy deficit. This is due to an issue with glycolysis, the process by which sugar is converted into adenosine triphosphate (ATP) (Al-Ishaq et al. 2019). Cells and tissues can get energy from ATP. Hyperglycemia can also elevate the chances of tissue injury because of a rise in the amounts of urea and amino acids in the blood. In some conditions, it may cause diabetic coma (Ardalani et al. 2021).

2.2.3 Treatment Approaches

The management of co-occurring risk factors, glycemic control, and lifestyle modification must all be objectives of diabetes treatment. Exercise and diet are essential components of the management of diabetes and metabolic syndrome. Increased physical activity and weight loss will improve hypertension, dyslipidemia, insulin resistance, and high blood sugar (Miranda 2021). Diabetes patients with hypertension have been demonstrated to benefit most from managing their blood pressure. Similar to this, several clinical studies demonstrate that lowering cholesterol is more essential for diabetes control than in the normal population for lowering the possibility of cardiovascular diseases (Chauhan et al. 2020).

Numerous studies have shown that treating this illness effectively depends on maintaining strict glycemic control. Metformin, alpha-glucosidase inhibitors, insulin secretagogues, insulin, and thiazolidinediones are a few of the pharmacological medicines that are readily accessible (Rosenberg et al. 2004). The pharmacokinetic characteristics, subsequent failure rates, and concomitant adverse effects of these treatments limit their use. Even insulin treatment has a higher risk of atherogenesis and hypoglycemia and does not permanently restore a normal pattern of glucose homeostasis (Kalra et al. 2018). Medicinal plants offer the benefit of having no or few negative effects. Many cultures throughout the world have used some of these for hundreds of years in their traditional medical systems. In the Indian system of medicine and many other traditional medical systems across the world, a variety of plants have been utilized to treat diabetes mellitus. Only a small number of them have been assessed using the standards of contemporary medicine (Padhi et al. 2020).

Currently, extracts of medicinal plants and herbs are utilized for their antidiabetic properties. The antidiabetic and pancreatic beta-cell action of medicinal plant extract has been demonstrated in several clinical investigations (Verma et al. 2018). For the treatment of T2DM, a variety of treatments have been developed. These fall under the subsequent sub-headings:

2.2.3.1 Insulin Secretagogues

This class of medications, particularly sulfonylureas and metiglinides, works by boosting insulin production by interacting with the receptors of sulfonylureas located on the pancreatic cells, which is an ATP sensitive to potassium channels (Padhi et al. 2020). Acetohexamide, chlorpropamide tolbutamide, and tolazamide are examples of first-generation sulfonylurea, whereas glimepiride and glipizide are examples of second-generation sulfonylurea (Kalra et al. 2018; Seino et al. 2017). Increased potency, quicker start of the activity, short plasma half-lives, and a longer period of action led to the development of second-generation sulfonylurea. Sulfonylurea's side effects may include symptoms of low blood sugar such as perspiration, dark urine, disorientation, hunger, weight gain, stomach discomfort, and skin response (Kalra et al. 2018).

2.2.3.2 Biguanides

They limit glucose uptake from the gut and hepatic production of glucose by enhancing the body's reaction to natural insulin. By reducing the phenomena of gluconeogenesis and promoting the process of glycolysis, biguanides lower hepatic glucose production. By boosting the activation of insulin receptors, they enhance the signaling of insulin (Quillen et al. 1999; Rubino et al. 2019). These compounds do not even directly affect the production of insulin like insulin secretagogues do.

These include the different compounds metformin and phenformin. However, biguanides frequently have gastrointestinal side effects, like nausea, vomiting, diarrhea, and cramps. Its prolonged usage is linked to a decline in vitamin B1 absorption (Padhi et al. 2020).

2.2.3.3 Insulin Sensitizers

These substances are sometimes referred to as agonists of the peroxisome proliferator-activated receptor (PPARs). PPARs control protein and carbohydrate metabolism and keep blood sugar levels stable. These ligand-activated transcription factors are members of the nuclear hormone receptor superfamily. They lower hemoglobin levels and increase the risk of bone fractures when used with other antidiabetic medications (Lebovitz 2019; Balakumar et al. 2019). These have a history of causing widespread adverse effects such as heart failure, edema, and weight gain.

2.2.3.4 Alpha-Glucosidase Inhibitors (AGIs)

The two main enzymes responsible for the metabolism of carbohydrates are alpha-amylase and alpha-glucosidase (Hieronymus & Griffin 2015). Oral antidiabetic medications called alpha-glucosidase inhibitors (AGIs) are typically used to treat T2DM. AGIs move the undigested glucose into the small intestine, delaying the process of carbohydrate absorption there. When used in conjunction with other diabetes medications, these medications offer advantages in lowering post-meal blood sugars (Godbout & Chiasson 2007). They aid in increasing GLP-1 levels after meals, which aids in slowing down digestion and reducing appetite. Gastrointestinal discomfort, bloating, and flatulence are common side effects of AGIs and may subside in a few weeks (Shabab et al. 2021).

2.2.3.5 Amylin Substitutes

The hormone amylin is composed of a single 37 amino acid chain. It is secreted with insulin by pancreatic beta cells (Rang et al. 2003). It keeps the blood sugar levels in the fasting and postprandial states by postponing stomach evacuation and hindering glucagon secretion. By altering the brain's appetite center, it controls the amount of food consumed. Amylin cannot be used as a medication because it collects and is not soluble in solution; thus, chemical substitutes that can mimic the effects of amylin were created. The parenteral administration of amylin analogs is utilized to treat both T1DM and T2DM. Amylin analogs' most frequent adverse effects when used with insulin include hypoglycemia, headache, nausea, and vomiting. When the patient becomes used to the drug, these adverse effects disappear (DeFronzo 2009).

2.2.3.6 Incretin Mimetics (GLP-1 Agonists and DPP-IV Inhibitors)

The new class of injectable medications for the treatment of type 2 diabetes is GLP-1 agonists or analogs. GLP-1 is metabolized by DPP-IV as a result of an alanine residue at the N terminus. So, by replacing the alanine group with other amino acids, including serine, threonine, and glycine, novel analogs of GLP-1 were created (Sun et al. 2017; MacDonald et al. 2002). Blood sugar levels can be lowered by incretin mimics that aid in decreasing HbA1c levels by boosting the production of insulin and reducing the release of glucagon (Holst 2019). Nausea, diarrhea, constipation, vomiting, indigestion, headaches, lack of appetite, dizziness, and increased perspiration are a few of the negative effects of incretin mimics.

2.2.3.7 Sodium-Glucose Antagonists/Inhibitors

Facilitative glucose transporter (GLUT), an active co-transporter, and sodium-glucose co-transporter (SGLT), a passive transporter, work together to reabsorb glucose in the proximal convoluted tubule (PCT) (Hsia et al. 2017; Padhi et al. 2020). The SGLT2 enzyme in PCT is inhibited by

SGLT2 inhibitors, which increase the excretion of glucose in urine and inhibits glucose from reabsorption. The blood glucose level and other glycemic indicators are maintained while glucose is eliminated in urine (Padhi et al. 2020).

2.3 ANTIDIABETIC POTENTIAL OF PHYTOCONSTITUENTS

Continuous use of synthetic medications to treat diabetes has led to the development of potentially fatal side effects, which already require the use of synthetic medications for treatment (Godbout & Chiasson 2007). This continuous use of chemically synthesized drugs degenerates human health and well-being (Tiwari et al. 2014). The adverse effects of several synthetic medicines have rekindled interest in using ancient medical methods to manage diabetes. Plant-derived medicines prevent the onset of diabetes problems and regulate metabolic irregularities through several mechanisms (Mukherjee et al. 2006). The phytochemicals are highlighted in Table 2.1.

TABLE 2.1
List of Different Phytochemicals Having Antidiabetic Potentials

Category	Phytochemicals	Sources	References
Alkaloids	Aegelin, marmesin, and marmelosin	*Aegle marmelos*	Ponnachan et al. (1993)
	Trigocoumarin, nicotinic acid	*Trigonella foenum graecum*	Abou EL-Soud et al. (2007)
	Ginkgolides	*Ginkgo biloba*	Pinto et al. (2009)
	Vindoline	*Catharanthus roseus*	Tiong et al. (2013)
	Piperine	*Piper longum*, *Piper nigrum*	Naidu et al. (2014)
	Koenidine	*Murraya koenigii*	Patel et al. (2015)
	Berberine	*Berberis aristata*, *Berberis vulgaris*	Ma et al. (2016)
Flavonoids	Naringenin	*Camellia sinensis*	Waltner-Law et al. (2002)
	Diosmin	Citrus plants, *Scrophularia nodusa*	Saravanan and Pari (2005)
	α-Cephalin and myricetin-3'-glucoside	*Abelmoschus moschatus*	Ngueyem et al. (2009)
	Quercetin, quercetrin, apigenin, rutin, and apigenin-7-O-glucoside	*Bauhinia varigata*, *Ginkgo biloba*	Naseer et al. (2017)
	Chrysin	*Oroxylum indicum*, *Passiflora caerulea*	Sirovina et al. (2013)
	Rutin	*Prunus persica*	Niture et al. (2014)
	Kaempferol	*Ginkgo biloba*	Al-Numair et al. (2015)
	Luteolin-7-glucoside	*Vitex negundo*	Huang et al. (2015)
Terpenoids	Trans-dehydrocrotonin	*Croton cajucara*	Silva et al. (2005)
	Thymol	*Thymus vulgaris*, *Trachyspermum ammi*	Saravanan and Pari (2005)
	Carvacrol	*Origanum vulgare*, *Marsilea minuta*	Ezhumalai et al. (2015)
	Betulinic acid	*Ziziphus mauritiana*, *Betula pubescens*	Castro et al. (2015)
	Ursolic acid and oleanolic acid	*Origanum majorana*	Sun et al. (2015)
	Zerumbone	*Zingiber zerumbet*	Liu et al. (2016)
Saponins	Diogenin	*Diosccorea rotundata*	Hemalatha et al. (2010)
	Arjunolic acid	*Terminalia arjuna*	Hemalatha et al. (2010)
	Geinoside	*Panax notoginseng*	Lee et al. (2011)
	Platyconic acid	*Platycodi radix*	Pinto et al. (2009)
Glycosides	Gymnenic acid	*Gymnema sylvestre*	Verma et al. (2008)
	Quercetin and quercetin 3-o-glycoside	*Vaccinium vitis*	Eid et al. (2014)

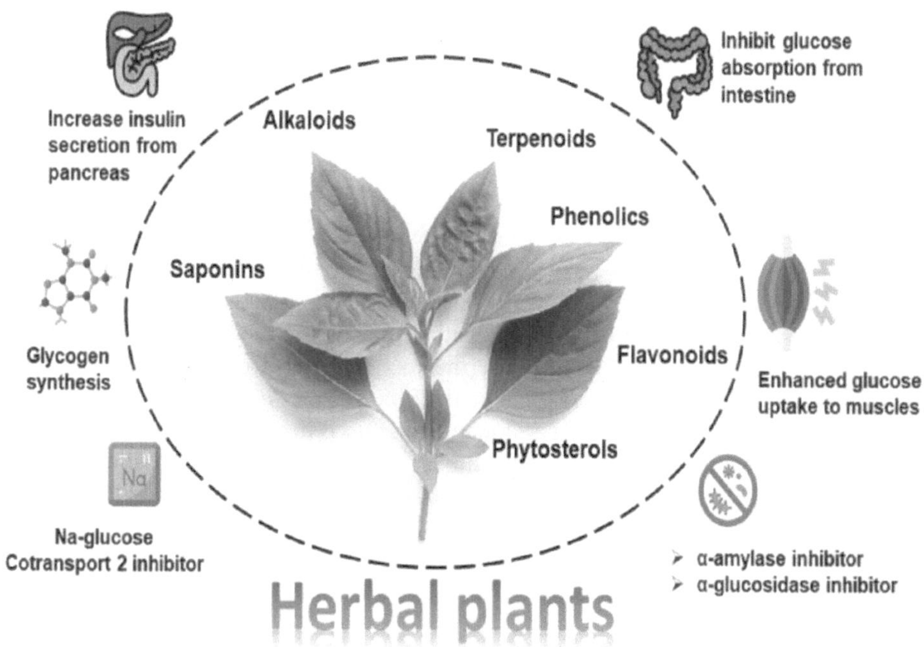

FIGURE 2.2 Advantages of phytochemicals for diabetes management.

2.3.1 Classification of Antidiabetic Phytoconstituents

Antidiabetic phytochemicals have been classified into certain groups based on their chemical structures and characteristics as shown in Figure 2.2 (Ardalani et al. 2021).

2.3.1.1 Alkaloids

Among the main components of plant-derived medications are alkaloids. These compounds are quite prevalent in nature and are classified into true, pseudo-, and proto-alkaloids. About 80% of alkaloids are synthesized from amino acids (Ziegler & Facchini 2008). These compounds have rapid distribution and moderate absorption in the human system (Rasouli et al. 2020). Alkaloids derived from different plant species and trees belonging to such plant families as Moraceae, Euphorbiaceae, Rutaceae, Fabaceae, Apocynaceae, Ranunculaceae, and Campanulaceae have significant diabetes-resisting potential (Rasouli et al. 2020). Alkaloids reduce diabetes mellitus by controlling the activities of two main enzymes, α-glucosidase and α-amylase. Both of these enzymes are involved in catalyzing polysaccharide to monosaccharide production reactions (Rasouli et al. 2017). For instance, the plant *Morus atropurpurea* possesses six different alkaloids with an α-glucosidase inhibition effect (Tang et al. 2013). Similarly, pyrrole derivatives derived from the plant *Chrozophora plicata* are effective in down-regulating carbohydrate metabolic enzymes (Tabussum et al. 2013).

Over 2000 species of the Rutaceae family feature antidiabetic alkaloids, and the majority of these species have an inhibitory action on the α-amylase enzyme (Kumar et al. 2010). These investigations have illustrated the potency and medical applicability of plant alkaloids in diabetes management. Mahanimbine (carbazole alkaloid) was isolated from the *Murraya koenigii* leaves having 83.72 lg/mL of IC_{50} value showed an inhibitory effect against α-amylase (Kumar et al. 2010). Similarly, vindogentianine was obtained from the *Catharanthus roseus* leaves having fainted activity with IC_{50} 74.43 lg/mL, and acarbose inhibits the α-amylase with 28.35 lg/mL of IC_{50} value (Tiong et al. 2015). Nigelladine A, B, and C were suggested to increase the parameters related to glucose metabolism with the 12.5 μM levels for hexokinase activity and glycogen synthesis (Tang et al. 2017). Norephedrine, pseudoephedrine, norpseudoephedrine, N-methylpseudoephedrine, and N-methylephedrine were reported in *Ephedra alata* and *E. sinica* to dipeptidyl peptidase-4 inhibiting activity 124–2800 μM of IC_{50} where ephedrine has the highest activity (Ojeda-Montes et al. 2017).

Berberine is recognized to act on diabetes by following different mechanisms such as mimicking insulin activity, triggering 5′ adenosine monophosphate-activated protein kinase activity to enhance insulin action, decreasing insulin resistance by the expression of protein kinase C insulin receptor, increasing glucagon-like peptide 1 secretion to cause glycolysis, and reducing dipeptidyl peptidase-4 activity (Potdar et al. 2012). Berberine was extracted from *Berberis aristata* at 0.5 g three times a day against type 2 diabetes (Yin et al. 2008). Twice daily intake of 588/105 mg of the *Berberis aristata/Silybum marianum* in combination was used in a six-month trial on type 1 diabetes in 46 females and 39 males to decrease insulin use (Derosa et al. 2016). Additionally, berberine showed promising efficiency by decreasing the hemoglobin A1c and postprandial glucose in the blood (Sarker et al. 2020).

2.3.1.2 Flavonoids and Phenols

In vitro and animal model studies on phenols and flavonoids have demonstrated their ability to prevent diabetes and its complication (Al-Ishaq et al. 2019). By regulating the activities of hepatic enzymes, regulating glucose metabolism, and altering lipid profiles, flavonoids reduce the development of diabetes and its consequences. Based on the structure, there are five antidiabetic classifications for flavonoids: flavonol, flavanones, flavones, isoflavones, and anthocyanins. Quercetin, one of the most prevalent flavonoids in the plant kingdom, is known for its effect on positive diabetes control by regulating intestinal glucose absorption and increasing insulin insensitivity (Eid et al. 2017). Similarly, rutin, which is also a flavonol, controls insulin release from pancreatic beta cells and inhibits tissue gluconeogenesis (Ola et al. 2015). 6-shogol (monomethoxybenzene) was reported to suppress the development of diabetic complications and arrest methylglyoxal for the development of glycation end products (Zhu et al. 2015).

Catechins were observed to trigger the erythroid-derived 2 transcriptional factor and related protein to strengthen the anti-oxidative effect (Fu et al. 2017). Hesperidin, a naturally abundant flavanone observed in citrus, berries, and so on has an upregulation effect on gene glucose transporter 4 (GLUT4) translocation that decreases the concentration of glucose-6-phosphate in the blood (Agrawal et al. 2014). Likewise, other naturally occurring flavonoids, such as naringin, apigenin, tangeretin, and delphinidin, have antidiabetic effects (Al-Ishaq et al. 2019). However, understanding the mechanism and side effects of these compounds needs more precise clarification, which can be gained by more research.

2.3.1.3 Terpenoids

Terpenoids have a promising effect on reducing the development of insulin resistance in humans as well as normalizing cell glucose levels (Nazaruk & Borzym-Kluczyk 2015). Moreover, terpenoids have been found to counter diabetic complications such as obesity and hyperglycemia. The most common categories of terpenoids possessing antidiabetic effects are sesquiterpenoids, triterpenes, and monoterpenes. These compounds have an inhibitory effect on enzymes α-glucosidase and α-amylase (Nazaruk & Borzym-Kluczyk 2015; Tan et al. 2016). Terpenes also combat diabetic microvascular complications by regulating enzyme aldose reductase as well as ameliorating insulin resistance by inhibiting enzymes such as protein tyrosine phosphatases (Thareja et al. 2012).

Piceatannol and scirpusin B were isolated from the extract of stem bark of *Callistemon rigidus* to suppress the α-amylase activity (Kobayashi et al. 2006). Bassic acid (unsaturated triterpene acid) isolated from the root bark ethanol extract of *Bumelia surtorum* exhibited considerable hypoglycemic properties in the alloxan-triggered diabetic rats. It enhanced glycogen synthesis and glucose uptake by increasing the insulin level in plasma. Bacosine (triterpene) was isolated from the *Bacopa monnieri* in the ethyl acetate fraction and showed a reduction in the glucose level of blood in the diabetic rat. This compound has anti-hyperglycemic activities similar to insulin to reduce oxidative damage due to alloxan-induced damage. It increases the uptake of glucose to enhance the glycogen content in the liver of diabetic rats (Ghosh et al. 2011). Bixin is the natural carotenoid pigment showing the hypoglycemic effect observed from the *Bixa orellana*. Annatto is the commercialized name for bixin, a highly promising medicine as a nutraceutical (Vilar et al. 2014).

Similarly, *B. orellana* induces a reduction in blood sugar levels in dogs with streptozotocin-triggered diabetes. It enhanced the peripheral utilization of glucose to reduce the blood glucose level and increase the affinity of insulin receptors to insulin and the insulin content in plasma (Rivera-Madrid et al. 2016). Cucurbitacin B is the tetracyclic triterpenoid isolated from *Trichosanthes dioica* to improve glucose utilization and insulin tolerance through GLUT4 translocation for the activation of the PI3K/Akt signaling pathway (Ramachandran & Saravanan 2015). Guavenoic acid at 0.3–30 nmol/L had substantially improved insulin resistance and was related to the upregulation expression of peroxisome proliferators-activated receptor gene and the downregulation expression of protein tyrosine phosphatase 1B gene. The ginsenoside Rb1 triggered the translocation of GLUT4 by signaling the leptin receptors and activation of PI3K to improve the insulin sensitivity and metabolism of glucose (Tabandeh et al. 2017).

2.3.1.4 Saponins

In the treatment of diabetes mellitus, saponin is a highly effective antioxidant due to its capacity to lower blood glucose levels. Antidiabetic properties of saponins include renewing beta cells in the islets and activating the glucose-utilizing enzymes in the human body (Abdel-Zaher et al. 2005). Several studies have demonstrated the effect of saponins in managing diabetes. Momordicine II, asaponin extracted from the plant *Mimordica charantia*, has hypoglycemic activity in animals (Fuangchana et al. 2011). Saponins in the root bark of *Berberis* sp. also have been found to have similar effects (Meliani et al. 2011). More study is required to assess and comprehend the effects of saponins on diabetes management.

2.3.1.5 Tannins

Tannins, hydrolyzable and non-hydrolyzable, have hyperglycemic effects on the animal body; furthermore, they are involved in slowing down starch digestion leading to the inhibition of α-glucosidase activity (Li et al. 2018). Tannins extracted in condensed form from raw food grains such as finger millet, field beans, and drumsticks have illustrated positive antidiabetic effects. Tannins can also help improve a person's harmful oxidative condition in diabetes (Kumari et al. 2012).

2.3.1.6 Phytosteroids

Plant steroids are secondary metabolites that can control diabetes-related complications such as cholesterol levels in diabetic patients and enhance the activity of antidiabetic drugs (Baker et al. 2009; Semova et al. 2019). Researchers reported that feeding 5-campestenone (0.3%) to mice prominently lowers their blood glucose levels (Suzuki et al. 2002). The oral administration of 5-campestenone (0.6%) to diabetic fatty rats leads to a limited rise in blood glucose levels after heavy glucose oral administration (Konno et al. 2005). Similarly, 20 μM of ß-sitosterol allows the improvement of glucose intake by enhancing the translocation of GLUT4 to the plasma membrane (Hwang et al. 2008).

2.3.2 Herbal Drugs and Formulations for Diabetes

Phytochemical drugs conversing antidiabetic and hypoglycemic effects offer promising solutions to synthetic diabetic drugs by reducing the complications due to the latter. Even so, the introduction of herbal antidiabetic drugs will not completely replace the synthetic drugs that have been previously used to treat diabetes, as it will take a significant amount of research to create the best combination of antidiabetic plants that will have the finest therapeutic benefit and minimum side effects (Ghorbani 2014). Integrating multiple herbal compounds as specific formulations has proven synergistic effects in combating diabetes (Kaur & Valecha 2014). Polyherbal formulations include the incorporation of two or more phytochemicals in a single formulation. To describe the current status of available antidiabetic polyherbal formulations and potential phytochemicals that can be equipped as antidiabetic drugs, this section is elaborated in two sections. A list of plants involving plant parts usage in the treatment of diabetes are summarized in Table 2.2.

2.3.2.1 Phytochemical-Based Antidiabetic Drugs

Currently, different polyherbal formulations are available as commercial antidiabetic medicines (Kaur & Valecha 2014). These formulations are normally consumed as decoctions, infusions, tinctures, and powders. As herbal formulations comprise several species and phytoconstituents with varied activity, a balanced formulation must be manufactured with minimum side effects. Common polyherbal medicines available in the market currently are further explained in Table 2.3.

2.3.2.2 Potential Antidiabetic Phytochemicals

Even if there are currently numerous phyto-extracted diabetic medications on the market, the research for further antidiabetic phytochemicals cannot be restricted to these available polyherbal compositions. Furthermore, phytochemically derived therapeutic diabetic medicines have shown decreased *in vivo* activity due to insufficient absorption (Parveen et al. 2018). Around the world, new formulations are being developed that address diabetes with more specificity and precision and fewer side effects (Ardalani et al. 2021). For instance, phospholipid encapsulated polyherbal formulations have been developed from plant extracts of *Momordica* sp., *foenum-graceum*, and *Withania somnifera* (2:2:1), which have better solubility in the human body as compared to present herbal formulations (Gauttam & Kalia 2013).

Similarly, traditional or Ayurvedic plants primitively used in diabetes treatment by people have been investigated to estimate their antidiabetic activities. ADJ6, a polyherbal formulation developed from traditionally used medicinal plants *Syzygium cumini*, *Psidium guajava*, *Phyllanthus emblica*, *Momordica charantia*, and so on was found to have potent antidiabetic activity almost similar to the synthetic drug acarbose (Duraiswamy et al. 2015). Moreover, inhibitory action was observed on enzymes, α-glucosidase, and α-amylase. Similarly, naringenin, a flavonoid present abundantly across *Citrus* sp., was observed with antidiabetic, anti-inflammatory, and anti-obesity activity with significantly higher *in vivo* absorption (Tsuhako et al. 2020). Resveratrol, a non-flavonoid polyphenol, is another potential phytochemical that encodes anti-cancer, antidiabetic, and anti-inflammatory actions and is abundant in grapes (Kong et al. 2021). These findings point to the potential that prospective antidiabetic phytochemicals may have and their use in the development of new antidiabetic medicines with greater efficacies.

TABLE 2.2
Tabulated Representation of Medicinal Parts Used in the Treatment of Diabetes

S. No.	Scientific name	Family	Common name	Part(s) used	Activities	Active constituent	References
1.	*Juniperus communis*	Cupressaceae	Common juniper	Berries	Hypoglycemic	Tannins and phenolic compounds	Bais et al. (2014)
2.	*Urtica dioica*	Urticaceae	Stinging nettle	Aerial parts	Anti-hyperglycemic	Polyphenols, triterpenes, sterols	Nazarian-Samani et al. (2018)
3.	*Salacia reticulata*	Celastraceae	Marking nut tree	Root, stem, leaves	Antidiabetic, anti-hyperlipidemic	Naringenin	Nazarian-Samani et al. (2018)
4.	*Nelumbo nucifera*	Nelumbonacea	Indian lotus	Seeds, leaves	Hypoglycemic, antihyperlipidemic	Flavonoids	Tungmunnithum et al. (2018)
5.	*Berberis vulgaris*	Berberidaceae	Barberry	Root, fruit	Hypoglycemic, antidiabetic	Phenols	Bindu and Narendhirakannan (2018)
6.	*Capparis deciduas*	Capparidaceae	Karira	Fruit, flowers, bark	Antidiabetic, hypolipidemic	Glycosides, phenols, saponins,	Dahiya et al. (2019)
7.	*Dillenia indica*	Dilleniaceae	Elephant apple	Leaves	Antidiabetic, hypolipidemic	Methanolic extract	Kamboj et al. (2019)
8.	*Eugenia uniflora*	Myrtaceae	Brazilian Cherry	Fruit	Antihyperglycemic, antidyslipidemic	Gallic acid	Sobeh et al. (2019)
9.	*Phyla nodiflora*	Verbenaceae	Capeweed	Whole plant	Antihyperlipidemic, anti-hyperglycemic	Phenol, steroids, essential oils	Suky et al. (2019)
10.	*Opuntia streptacantha*	Cactaceae	Prickly pear cactus	Leaves	Anti-hyperglycemic	Phenols	Gouws et al. (2019)
11.	*Phyllanthus niruri*	Phyllanthaceae	Stonebreaker	Aerial parts	Antidiabetic	Flavonoid, tannin, and quinine	Sutrisna et al. (2019)
12.	*Ipomoea batatas*	Convolvulaceae	Sweet potato	Leaves, root	Hypoglycemic	Flavonoids	Li et al. (2019)
13.	*Punica granatum*	Punicaceae	Pomegranate	Leaves, fruit	Antidiabetic, anti-hyperlipidemic	Quercetin, ellagic acid, punicalagi	Puneeth et al. (2020)
14.	*Tetraena alba*	Zygophyllaceae	White bean caper	Whole plant	Antihyperglycemic	Pyrroline	Unuofin and Lebelo (2020)
15.	*Hibiscus rosa-sinensi*	Malvaceae	Chinese hibiscus	Flower, leaves	Hypoglycemic, anti-dyslipidemic	Polyphenol extract from leaves	Unuofin and Lebelo (2020)
16.	*Bombax ceiba*	Bombacaceae	Cotton tree	Leaves	Antihyperglycemic, anti-hyperlipidemic	Methanol and acetone	Komati et al. (2020)

(Continued)

TABLE 2.2
(Continued)

S. No.	Scientific name	Family	Common name	Part(s) used	Activities	Active constituent	References
17.	*Azadirachta indica*	Meliaceae	Neem tree	Leaf, seeds	Hypoglycemic, β-cell regeneration	Polyphenolic flavonoids	Srivastava et al. (2020)
18.	*Annona squamosa*	Annonaceae	Sugar apple	Leaf, fruit pulp	Hypoglycemic, anti-lipidemic	Polyphenolics	Pathak et al. (2021)
19.	*Boerhavia diffusa*	Nyctaginaceae	Spreading hogweed	Leaves	Antidiabetic	Methanol	Sinan et al. (2021)
20.	*Eucalyptus globules*	Myrtaceae	Blue-gum	Leaves	Anti-hyperglycemic	Catechin, ergosterol, pinitol	Bello et al. (2021)
21.	*Allium cepa*	Amaryllidaceae	Onion	Bulbs	Anti-hyperlipidemic, antidiabetic and antioxidant	Flavonoids, S-propyl cysteine sulfoxide	Galavi et al. (2021)
22.	*Ficus religiosa*	Moraceae	Sacred fig	Root	Antidiabetic	Lupenyl acetate, lanosterol, and piperine	Yueniwati et al. (2021)
23.	*Lantana camara*	Verbenaceae	Bigsage, wild sage	Fruits	Anti-hyperglycemic	Ethanolic extract from leaves	Balti et al. (2022)
24.	*Lithocarpuspolystachyus*	Fagaceae	Sweet tea	Leaves	Hypoglycemic	Flavonoids	Fang et al. (2022)
25.	*Piper nigrum*	Piperaceae	Black pepper	Fruit	Anti-hyperglycemic	Phenol	Kondapalli et al. (2022)

TABLE 2.3
Herbal Medicines, Their Constituents, and Their Properties

Product	Constituents	Properties	Manufacturer	Dose (mg/kg)	Reference
Karmin plus	*Momordica charantia, Ocimum sanctum, Azadirachta indica*	Antidiabetic	Government of India	200–400	Kaur and Valecha (2014)
Diasol	*Terminilia chebula, Foenum graecum, Emblica officinalis*	Hypoglycemic	Ipca Laboratories Pvt. Ltd.	250	Ghorbani (2014)
Diakure	*Vetiveria zizanioides, Cassia auriculata, Acacia catechu*	Hypoglycemic, anti-obesity	Deep Ayurveda	500	Tomy et al. (2016)
Glyoherb	*Momordica charantia, Phyllanthus emblica, Gymnema sylvestre, Azadirachta indica*	Anti-hyperlipidemic, hypoglycemic, antioxidant	Himalaya herbal healthcare	200–600	Choudhury et al. (2017)
Diabecon	*Sphaeranthus indicus, Curcuma longa, Aloe vera, Tribulus terrestris*	β-cell repair, anti-hyperlipidemic	Himalaya herbal healthcare	100 for 56 days	Choudhury et al. (2017)
SR10	*Cortex lycii, Radix astragali, Radix codonopsis*	Hypoglycemic, improves pancreatic β-cell function	-	927 with 0.2 mL water	Tan and Vinanzaca (2018)
Diabeta plus	*Curcuma longa, Gymnema sylvestre, Syzygium cumuni, Tribulus terrestis*	Glucose tolerance	Himalaya herbal healthcare	500	Chauhan et al. (2020)
Pancreatic tonic 180cp	*Aegle marmelose, Pterocarpus marsupium, Momordica charantia, Syzygium cumuni*	Hypoglycemic	Ayurvedic herbal supplements	-	Prabhakar and Banerjee (2020)
Syndrex	*Trigonella foenumgraecum*	Antidiabetic, β-cell repair	Plethico Laboratory	-	Prabhakar and Banerjee (2020)

2.4 ANTIDIABETIC ACTIVITIES OF PHYTOCHEMICALS

2.4.1 Phytochemicals and Their Response Action in the Body

The medical sector is increasingly interested in the development of herbal diabetic medications as a result of scientific validation of phytochemicals' beneficial benefits for diabetes management in people. Drug-to-drug validation for assessing the actual impacts of available herbal formulations is still under process. Certain synergistic responses of phytochemical-derived diabetic drugs are presented as follows.

2.4.1.1 Increased Insulin Secretion

DPP4 inhibition causes a reduction in glucagon levels and increased insulin secretion in the liver. This overcasts hyperglycemic effects on the body (Prabhakar & Banerjee 2020). Polyphenols such as stilbenes are natural DPP4 inhibitors and present abundantly across species (Huang et al. 2018). Diabetes type 2 harshly affects insulin production and causes necrosis in pancreatic β-cells. In the presence of phytochemicals, this diabetic impact was reduced *in vitro* in streptozotocin-induced diabetic rats (Wu & Yan 2015). D-limonene present in *Aegle marmelos* (Tan et al. 2016) is very effective in mitigating and repairing pancreatic β-cells in mammals. Increased insulin resistance precedes T2D and, therefore, leads to its development.

Certain phytochemicals such as betulinic (Ko et al. 2016) and glycyrrhizic acid (Hou et al. 2014) help accelerate insulin activity and, therefore, decrease diabetes symptoms. GLUT4 mediates insulin secretion, which further reduces glucose in the bloodstream. A mutation in this receptor causes T2D (Sayem et al. 2018). Phytochemicals help in improving insulin signaling pathways through GLUT4 stimulation. Metabolites such as D-limonene present in *Aegle marmelos* (Tan et al. 2016) and piceatannol observed in *Passiflora edulis* (Minakawa et al. 2012) help in rebuilding insulin signaling pathways in cells.

2.4.1.2 Inhibition of Hepatic Glucose Production

Trigonelline (Hamden et al. 2013), oleanolic acid (Mukundwa et al. 2016), and so on are a few of the active plant metabolites that help in downregulating gluconeogenesis and glycogenolysis in liver cells by inhibiting the activity of enzyme glucose-6-phosphate translocase that catalyzes the transformation of glucose to glucose-6-phosphate.

2.4.1.3 Inhibition of Glucose Absorption

Certain plant-extracted metabolites inhibit the action of the enzyme α-glucoside, which is responsible for the catabolic breakdown of carbohydrates into glucose. This reduces the glucose level to a great extent and, therefore, reduces hyperglycemia (Kumar et al. 2011). Quercetin abundant in *Ficus racemosa* (Keshari et al. 2016) is a natural α-glucoside inhibitor. Similarly, kaempferol extracted from *Hyophorbela genicaulis* shows α-glucoside inhibition (William et al. 2019). Nordihydroguaiaretic acid is present in *Larrea tridentate* (Zhang et al. 2015) and is an active α-amylase inhibiting compound that can replace synthetic drugs such as acarbose.

2.5 CONCLUSION

Diabetes is a chronic, heterogeneous health problem with multiple side effects that increase pathogenicity as a person's age progresses. Drugs synthesized by chemical processing have been mainly used to treat this disorder all across the world, but continued usage of these synthetic medications has resulted in serious consequences and unpredictable side effects. Numerous new compounds from numerous unique plants have been discovered to possess antidiabetic activity because of fast advancements in modern analytical techniques and an increase in studies on natural products with antidiabetic bioactivity.

Numerous slivers of scientific data have demonstrated that these phytochemicals have antihyperglycemic potentials and can be used to control diabetic and metabolic problems without the negative side effects of synthetic medications. The ideal combination of several antidiabetic plants must be developed through significant research to offer the most therapeutic efficacy with minimal negative effects. Through this chapter, we comprehensively provide in-depth knowledge of prospective plants and their potential antidiabetic phytoconstituents isolated from them with their mechanism of action.

REFERENCES

Abdel-Zaher, A.O., S.Y. Salim & M.H. Assaf. 2005. Antidiabetic activity and toxicity of *Zizyphus spina-christi* leaves. *Journal of Ethnopharmacology* 101(1–3): 129–138.

Abejew, A.A., A.Z. Belay & M.W. Kerie. 2015. Diabetic complications among adult diabetic patients of a tertiary hospital in Northeast Ethiopia. *Advances in Public Health*: 290920.

Abou El-Soud, K.M.Y., F.S.H. Hussein & F.A.R. Hussein. 2007. Antidiabetic effects of Fenugreek alkaloid extract in streptozotocin induced hyperglycemic rat. *Journal of Applied Sciences Research* 3(10): 1073–1083.

Agrawal, Y.O., P.K. Sharma, B. Shrivastava, S. Ojha, H.M. Upadhya, D.S. Arya & S.N. Goyal. 2014. Hesperidin produces cardioprotective activity via PPAR-gamma pathway in ischemic heart disease model in diabetic rats. *PLoS ONE* 9: 111212.

Al-Ishaq, R.K., M. Abotaleb, P. Kubatka, K. Kajo & D. Büsselberg. 2019. Flavonoids and their anti-diabetic effects: Cellular mechanisms and effects to improve blood sugar levels. *Biomolecules* 9(9): 430.

Al-Numair, K.S., G. Chandramohan, C. Veeramani & M.A. Alsaif. 2015. Ameliorative effect of kaempferol, a flavonoid, on oxidative stress in streptozotocin-induced diabetic rats. *Redox Report* 20(5): 198–209.

American Diabetes Association. 2004. Dyslipidemia management in adults with diabetes. *Diabetes Care* 27: 68–71.

Ardalani, H., F. HejaziAmiri & A. Hadipanah. 2021. Potential antidiabetic phytochemicals in plant roots: A review of in vivo studies. *Journal of Diabetes & Metabolic Disorders* 20: 1837–1854.

Bais, S., N.S. Gill, N. Rana & S. Shandil. 2014. A phytopharmacological review on a medicinal plant: *Juniperus communis. International Scholarly Research Notices*: 634723.

Baker, W.L., E.L. Baker & C.I. Coleman. 2009. The effect of plant sterols or stanols on lipid parameters in patients with type 2 diabetes: A meta-analysis. *Diabetes Research and Clinical Practice* 84(2): 33–37.

Balakumar, P., N. Mahadevan & R. Sambathkumar. 2019. A contemporary overview of PPARα/γ dual agonists for the management of diabetic dyslipidemia. *Current Molecular Pharmacology* 12: 195–201.

Balbaa, M., M. El-Zeftawy & S.A. Abdulmalek. 2021. Therapeutic screening of herbal remedies for the management of diabetes. *Molecules* 26(22): 6836.

Balti, T., K. Charradi, M. Mahmoudi, N. Oueslati, F. Limam & E. Aouani. 2022. Paradoxical anti-diabetic effect of *Lantana camara* leaf extract and pancreatic oxidative stress relieved by grape seed and skin extract. *Pharmaceutical Chemistry Journal* 55(11): 1219–1228.

Bello, M., B. Jiddah-Kazeem, T.H. Fatoki, E.O. Ibukun & A.C. Akinmoladun. 2021. Antioxidant property of *Eucalyptus globulus* Labill. Extracts and inhibitory activities on carbohydrate metabolizing enzymes related to type-2 diabetes. *Biocatalysis and Agricultural Biotechnology* 36: 102111.

Benchoula, K., M.S. Vohra, I.S. Parhar & W.E. Hwa. 2022. Metabolomics based biomarker identification of anti-diabetes and anti-obesity properties of Malaysian herbs. *Metabolomics* 18(2): 1–12.

Bindu, J. & R.T. Narendhirakannan. 2018. Role of medicinal plants in the management of diabetes mellitus: A review. *3 Biotech* 9(1): 4.

Castro, A.J., M.J. Frederico, L.H. Cazarolli, C.P. Mendes, L.C. Bretanha, E.C. Schmidt, Z.L. Bouzon, V.A. de Medeiros Pinto, C. da Fonte Ramos, M.G. Pizzolatti & F.R. Silva. 2015. The mechanism of action of ursolic acid as insulin secretagogue and insulinomimetic is mediated by cross-talk between calcium and kinases to regulate glucose balance. *Biochim et Biophysica Acta* 1850(1): 51–61.

Chauhan, D.S., P. Gupta, F.H. Pottoo & M. Amir. 2020. Secondary metabolites in the treatment of diabetes mellitus: A paradigm shift. *Current Drug Metabolism* 21(7): 493–511.

Choudhury, H., M. Pandey, C. Kui Hua, C. Shi Mun, J.K. Jing, L. Kong, L.Y. Ern, N.A. Ashraf, S.W. Kit, T.S. Yee, M.R. Pichika, B. Gorain & P. Kesharwania. 2017. An update on natural compounds in the remedy of diabetes mellitus: A systematic review. *Journal of Traditional and Complementary Medicine* 8(3): 361–376.

Dahiya, R., J.S. Vaghela & R.K. Singh. 2019. A brief review on pharmacological and phytochemical studies of *Capparis decidua. Advance Pharmaceutical Journal* 4(6): 133–139.

DeFronzo, R. 2009. From the triumvirate to the ominous octet: A new paradigm for the treatment of type 2 diabetes mellitus. *Diabetes* 58: 773–795.

Derosa, G., A. D'Angelo & P. Maffioli. 2016. The role of a fixed *Berberis aristata/Silybum marianum* combination in the treatment of type 1 diabetes mellitus. *Clinical Nutrition (Edinburgh Scotland)* 35:1091–1095.

Duraiswamy, A., D. Shanmugasundaram, C.S. Sasikumar, S.M. Cherian & K.M. Cherian. 2015. Development of an antidiabetic formulation (ADJ6) and its inhibitory activity against a-amylase and a-glucosidase. *Journal of Traditional and Complementary Medicine* 6(3): 204–208.

Eid, M.H. & P.S. Haddad. 2017. The antidiabetic potential of quercetin: Underlying mechanisms. *Current Medicinal Chemistry* 24(4): 355–364.

Eid, H.M., M. Ouchfoun, A. Brault, D. Vallerand, L. Musallam, J.T. Arnason & P.S. Haddad. 2014. Lingonberry (*Vaccinium vitis-idaea* L.) exhibits antidiabetic activities in a mouse model of diet-induced obesity. *Evidence-Based Complementary and Alternative Medicine* 10.

Ezhumalai, M., N. Ashokkumar & K.V. Pugalendi. 2015. Combination of carvacrol and rosiglitazone ameliorates high fat diet induced changes in lipids and inflammatory markers in C57BL/6J mice. *Biochimie* 110: 129–136.

Fang, H.L., M.L. Liu, S.Y. Li, W.Q. Song, H. Ouyang, Z.P. Xiao & H.L. Zhu. 2022. Identification, potency evaluation, and mechanism clarification of α-glucosidase inhibitors from tender leaves of *Lithocarpus polystachyus* Rehd. *Food Chemistry* 371: 131128.

Fu, Q.Y., Q.S. Li, X.M. Lin, R.Y. Qiao, R. Yang, X.M. Li, Z.B. Dong, L.P. Xiang, X.Q. Zheng, J.L. Lu & C.B. Yuan. 2017. Antidiabetic effects of tea. *Molecules* 22(5): 849.

Fuangchana, A., P. Sonthisombata, T. Seubnukarnb, R. Chanouan, P. Chotchaisuwat, V. Sirigulsatien, K. Ingkaninan, P. Plianbangchang & S.T. Haines. 2011. Hypoglycemic effect of bitter melon compared with metformin in newly diagnosed type 2 diabetes patients. *Journal of Ethnopharmacology* 134(2): 422–428.

Galavi, A., H. Hosseinzadeh & B.M. Razavi. 2021. The effects of *Allium cepa* L.(onion) and its active constituents on metabolic syndrome: A review. *Iranian Journal of Basic Medical Sciences* 24(1): 3.

Gauttam, V.K. & A.N. Kalia. 2013. Development of polyherbal antidiabetic formulation encapsulated in the phospholipids vesicle system. *Journal of Advanced Pharmaceutical Technology & Research* 4(2): 108–117.

Ghorbani, A. 2014. Clinical and experimental studies on polyherbal formulations for diabetes: Current status and future prospective. *Journal of Integrative Medicine* 12(4): 336–345.

Ghosh, T., T.K. Maity & J. Singh. 2011. Antihyperglycemic activity of bacosine, a triterpene from *Bacopa monnieri*, in alloxan-induced diabetic rats. *Planta Medica* 77: 804–808.

Godbout, A. & J.L. Chiasson. 2007. Who should benefit from the use of alpha-glucosidase inhibitors? *Current Diabetes Reports* 7(5): 333–339.

Gourgari, E., E.E Wilhem, H. Hassanzadeh, V.R. Aroda & I. Shoulson. 2017. A comprehensive review of the FDA-approved labels of diabetes drugs: Indications, safety, and emerging cardiovascular safety data. *Journal of Diabetes Complications* 31: 1719–1727.

Gouws, C.A., E.N. Georgousopoulou, D.D. Mellor, A. McKune & N. Naumovski. 2019. Effects of the consumption of prickly pear cacti (*Opuntia* spp.) and its products on blood glucose levels and insulin: A systematic review. *Medicina* 55(5): 138.

Hamden, K., A. Bengara, Z. Amri & A. Elfeki. 2013. Experimental diabetes treated with trigonelline: Effect on key enzymes related to diabetes and hypertension, b-cell and liver function. *Molecular and Cellular Biochemistry* 381(1–2): 85–94.

Hemalatha, T., S. Pulavendran, C. Balachandran, B.M. Manohar & R. Puvanakrishnan. 2010. Arjunolic acid: A novel phytomedicine with multifunctional therapeutic applications. *Indian Journal of Experimental Biology* 48(3): 238–247.

Hieronymus, L. & S. Griffin. 2015. Role of amylin in type 1 and type 2 diabetes. *Diabetes Education* 41: 47S–56S.

Holst, J. 2019. From the Incretin concept and the discovery of GLP-1 to today's diabetes therapy. *Frontiers in Endocrinology* 10: 260.

Hou, S., F. Zheng, Y. Li, L. Gao & J. Zhang. 2014. The protective effect of glycyrrhizic acid on renal tubular epithelial cell injury induced by high glucose. *International Journal of Molecular Sciences* 15(9): 15026–15043.

Hsia, D., O. Grove & W. Cefalu. 2017. An update on sodium-glucose co-transporter-2 inhibitors for the treatment of diabetes mellitus. *Current Opinion in Endocrinology, Diabetes and Obesity* 24: 73–79.

Huang, M., Y. Zhang, S. Xu, W. Xu, K. Chu, W. Xu, H. Zhao & J. Lu. 2015. Identification and quantification of phenolic compounds in *Vitex negundo* L. var. cannabifolia (Siebold et Zucc.) Hand.-Mazz. using liquid chromatography combined with quadrupole time-of-flight and triple quadrupole mass spectrometers. *Journal of Pharmaceutical Biomedical Analysis* 108: 11–20.

Huang, Y., J. Hao, D. Tian, Y. Wen, P. Zhao, H. Chen & Y. Lv. 2018. Antidiabetic activity of a flavonoid-rich extract from *Sophora davidii* (Franch.) skeels in KK-Ay Mice via activation of AMP-activated protein kinase. *Frontiers in Pharmacology* 9: 1–15.

Hwang, S.L., H.N. Kim, H.H. Jung, J.E. Kim, D.K. Choi, J.M. Hur, J.Y. Lee, H. Song, K.S. Song & T.L. Huh. 2008. Beneficial effects of beta-sitosterol on glucose and lipid metabolism in L6 myotube cells are mediated by AMP-activated protein kinase. *Biochemical and Biophysical Research Communications* 377(4): 1253–1258.

Kalra, S., S. Bahendeka, R. Sahay, S. Ghosh, F. Md & A. Orabi. 2018. Consensus recommendations on sulfonylurea and sulfonylurea combinations in the management of Type2 diabetes mellitus – International Task Force. *Indian Journal of Endocrinology and Metabolism* 22: 132.

Kamboj, P., N.C. Talukdar & S.K. Banerjee. 2019. Therapeutic benefit of *Dillenia indica* in diabetes and its associated complications. *Journal of Diabetes Research*: 4632491.

Kaul, K., J.M. Tarr, S.I. Ahmad, E.M. Kohner & R. Chibber. 2013. Introduction to diabetes mellitus. *Diabetes*: 1–11.

Kaur, M. & V. Valecha. 2014. Diabetes and antidiabetic herbal formulations: An alternative to allopathy. *European Journal of Medicinal Chemistry* 6(4): 226–240.

Keshari, A.K., G. Kumar, P.S. Kushwaha, M. Bhardwaj, P. Kumar, A. Rawat & S. Saha. 2016. Isolated flavonoids from *Ficus racemosa* stem bark possess antidiabetic, hypolipidemic and protective effects in albino Wistar rats. *Journal of Ethnopharmacology* 181: 252–262.

Ko, B.S., S. Kang, B.R. Moon, J.A. Ryuk & S. Park. 2016. A 70% ethanol extract of mistletoe rich in betulin, betulinic acid, and oleanolic acid potentiated b-cell function and mass and enhanced hepatic insulin sensitivity. *Evidence-Based Complementary and Alternative Medicine*: 1–13.

Kobayashi, K., T. Ishihara, E. Khono, T. Miyase & F. Yoshizaki. 2006. Constituents of stem bark of *Callistemon rigidus* showing inhibitory effects on mouse alpha-amylase activity. *Biological and Pharmaceutical Bulletin* 29: 1275–1277.

Komati, A., A. Anand, H. Shaik, M.K.R. Mudiam, K.S. Babu & A.K. Tiwari. 2020. *Bombax ceiba* (Linn.) calyxes ameliorate methylglyoxal-induced oxidative stress via modulation of RAGE expression: Identification of active phytometabolites by GC-MS analysis. *Food & Function* 11(6): 5486–5497.

Kondapalli, N.B., R. Hemalatha, S. Uppala, S.R. Yathapu, S. Mohammed, M. Venkata Surekha, A. Rajendran & D.K. Bharadwaj. 2022. *Ocimum sanctum*, *Zingiber officinale*, and *Piper nigrum* extracts and their effects on gut microbiota modulations (prebiotic potential), basal inflammatory markers and lipid levels: Oral supplementation study in healthy rats. *Pharmaceutical Biology* 60(1): 437–450.

Kong, M., K. Xie, M. Lv, L. Jufei, Y. Jianyu, Y. Kaixuan, W. Xiaoqin, X. Ying & Y. Dewei. 2021. Anti-inflammatory phytochemicals for the treatment of diabetes and its complications: Lessons learned and future promise. *Biomedicine & Pharmacotherapy* 133: 110975.

Konieczynski, P., M. Gappa, M. Wesolowski, E. Pinto & A. Almeida. 2022. Trace elements in medicinal plants traditionally used in the treatment of diabetes – Do they have a role in the claimed therapeutic effect? *Foods* 11(5): 667.

Konno, R., Y. Kaneko, K. Suzuki & Y. Matsui. 2005. Effect of 5-campestenone (24-methylcholest-5-en-3-one) on Zucker diabetic fatty rats as a type 2 diabetes mellitus model. *Hormone and Metabolic Research* 37: 79–83.

Kumar, B., A. Mitra & M. Mahadevappa. 2010. Antidiabetic and hypolipidemic effects of mahanimbine (carbazole alkaloid) from *Murraya koenigii* (Rutaceae) leaves. *International Journal of Phytomedicine* 2(1): 22–30.

Kumar, M., V. Saurabh, M. Tomar, M. Hasan, S. Changan, M. Sasi, C. Maheshwari, U. Prajapati, S. Singh, R.K. Prajapat & S. Dhumal. 2021a. Mango (*Mangifera indica* L.) leaves: Nutritional composition, phytochemical profile, and health-promoting bioactivities. *Antioxidants*10(2): 299.

Kumar, S., A. Mittal, D. Babu & A. Mittal. 2021b. Herbal medicines for diabetes management and its secondary complications. *Current Diabetes Reviews* 17(4): 437–456.

Kumar, S., S. Narwal, V. Kumar & O. Prakash. 2011. α-glucosidase inhibitors from plants: A natural approach to treat diabetes. *Pharmacognosy Reviews* 5(9): 19–29.

Kumari, M. & S. Jain. 2012. Tannins: An antinutrient with positive effect to manage diabetes. *Research Journal of Recent Sciences* 1(12): 1–8.

Lebovitz, H. 2019. Thiazolidinediones: The forgotten diabetes medications. *Current Diabetes Reports* 19: 151.

Lee, K.T., T.W. Jung, H.J. Lee, S.G. Kim & Y.S. Shin. 2011. The anti-diabetic effect of genosenoside Rb2 via activation of AMPK. *Archives of Pharmacal Research* 34: 1201–1208.

Li, G., Z. Lin, H. Zhang, Z. Liu, Y. Xu, G. Xu, H. Li, R. Ji, W. Luo, Y. Qiu, S. Qiu & H. Tang. 2019. Anthocyanin accumulation in the leaves of the purple sweet potato (*Ipomoea batatas* L.) cultivars. *Molecules* 24(20): 3743. https://doi.org/10.3390/molecules24203743

Li, K., F. Yao, J. Du, X. Deng & C. Li. 2018. Persimmon tannin decreased the glycemic response through decreasing the digestibility of starch and inhibiting α-amylase, α glucosidase, and intestinal glucose uptake. *Journal of Agricultural and Food Chemistry* 66(7): 1629–1637.

Liu, W.Y., T.F. Tzeng & I.M. Liu. 2016. Zerumbone, a bioactive sesquiterpene, ameliorates diabetes-induced retinal microvascular damage through inhibition of Phospho-p38 mitogen-activated protein kinase and nuclear factor-kappaB pathways. *Molecules* 21: 1708.

Ma, Y.G., Y.B. Zhang, Y.G. Bai, Z.J. Dai, L. Liang, M. Liu, M.J. Xie & H.T. Guan. 2016. Berberine alleviates the cerebrovascular contractility in streptozotocin-induced diabetic rats through modulation of intracellular Ca handling in smooth muscle cells. *Cardiovascular Diabetology* 15: 63.

MacDonald, P., W. El-kholy, M. Riedel, A. Salapatek, P. Light & M. Wheeler. 2002. The multiple actions of GLP-1 on the process of glucose-stimulated insulin secretion. *Diabetes* 51: S434–S442.

Masharani, U.K.J.H. 2001. Diabetes mellitus and hypo glycemia. In *Current Medical Diagnosis and Treatment*, ed. Tierney, J.L.M., S.J. McPhee & M. Papadaki. New York: McGraw-Hill.

Meliani, N., M.A. Dib, H. Allali & B. Tabti. 2011. Hypoglycaemic effect of *Berberis vulgaris* L. in normal and streptozotocin-induced diabetic rats. *Asian Pacific Journal of Tropical Biomedicine* 1(6): 468–471.

Minakawa, M., Y. Miura & K. Yagasaki. 2012. Piceatannol, a resveratrol derivative, promotes glucose uptake through glucose transporter 4 translocation to plasma membrane in L6 myocytes and suppresses blood glucose levels in type 2 diabetic model db/db mice. *Biochemical and Biophysical Research Communications* 422(3): 469–475.

Miranda, J.J.M. 2021. Medicinal plants and their traditional uses in different locations. In *Phytomedicine*. Cambridge, MA: Academic Press, pp. 207–223.

Mudau, T.E., J.O. Olowoyo & S.O. Amoo. 2022. Ethnobotanical assessment of medicinal plants used traditionally for treating diabetes in Vhembe district, Limpopo Province, South Africa. *South African Journal of Botany* 146: 304–324.

Mukherjee, P.K., K. Mukherjee & K. Maiti. 2006. Leads from Indian medicinal plants with hypoglycemic potentials. *Journal of Ethnopharmacology* 106: 1–28.

Mukundwa, A., S.O. Langa, S. Mukaratirwa & B. Masola. 2016. In vivo effects of diabetes, insulin and oleanolic acid on enzymes of glycogen metabolism in the skin of streptozotocin-induced diabetic male Sprague-Dawley rats. *Biochemical and Biophysical Research Communications* 471: 315–319.

Naidu, B.P., H. Nemani, B. Meriga, S.K. Mehar, S. Potana & S. Ramgopalrao. 2014. Mitigating efficacy of piperine in the physiological derangements of high fat diet induced obesity in Sprague Dawley rats. *Chemico-Biological Interactions* 221: 42–51.

Naseer, Z., E. Ahmad & E.T. Epikmen. 2017. Quercetin supplemented diet improves follicular development, oocyte quality, and reduces ovarian apoptosis in rabbits during summer heat stress. *Theriogenology* 96: 136–141. https://doi.org/10.1016/j.theriogenology.2017.03.029

Nazarian-Samani, Z., R.D.E. Sewell, Z. Lorigooini & M. Rafieian-Kopaei. 2018. Medicinal plants with multiple effects on diabetes mellitus and its complications: A systematic review. *Current Diabetes Reports* 18(10):72.

Nazarko, L. 2009. Causes and consequences of diabetes. *British Journal of Healthcare Assistants* 3(11): 534–538.

Nazaruk, J. & M. Borzym-Kluczyk. 2015. The role of triterpenes in the management of diabetes mellitus and its complications. *Phytochemistry Reviews* 14(4): 675–690.

Ngueyem, T.A., G. Brusotti, G. Caccialanza & P.V. Finzi. 2009. The genus *Bridelia*: A phytochemical and ethnopharmacological review. *Journal of Ethnopharmacology* 124(3): 339–349.

Niture, S.K., R. Khatri & A.K. Jaiswal. 2014. Regulation of Nrf2 – an update. *Free Radical Biology and Medicine* 66: 36–44.

Ojeda-Montes, M.J., A. Ardid-Ruiz, S. Tomas-Hernandez, A. Gimeno, A. Cereto-Massague, R. Beltrán-Debon, M. Mulero, S. Garcia-Vallvé, G. Pujadas & C. Valls. 2017. Ephedrine as a lead compound for the development of new DPP-IV inhibitors. *Future Medicinal Chemistry* 9(18): 2129–2146.

Ojo, O.A., J.C. Amanze, A.I. Oni, S. Grant, M. Iyobhebhe, T.C. Elebiyo, D. Rotimi, N.T. Asogwa, B.E. Oyinloye, B.O. Ajiboye & A.B. Ojo. 2022. Antidiabetic activity of avocado seeds (*Persea americana* Mill.) in diabetic rats via activation of PI3K/AKT signaling pathway. *Scientific Reports* 12(1): 1–17.

Ola, M.S., M.M. Ahmed, R. Ahmad, H.M. Abuohashish, S.S. Al-Rejaie & A.S. Alhomida. 2015. Neuroprotective effects of rutin in streptozotocin-induced diabetic rat retina. *Journal of Molecular Neuroscience* 56: 440–448.

Padhi, S., A.K. Nayak & A. Behera. 2020. Type II diabetes mellitus: A review on recent drug based therapeutics. *Biomedicine & Pharmacotherapy* 131: 110708.

Parveen, A., M. Jin & S.Y. Kim. 2018. Bioactive phytochemicals that regulate the cellular processes involved in diabetic nephropathy. *Phytomedicine* 39: 146–159.

Patel, D.K., T.K. Gajaria, R.V. Devkar & A.V. Ramachandran. 2015. Flavonoid rich extract of *Murraya koenigii* alleviates in-vitro LDL oxidation and oxidized LDL induced apoptosis in raw 264.7 Murine macrophage cells. *Journal of Food Science and Technology* 52(6): 3367–3375.

Pathak, K., A. Das, A. Shakya, R. Saikia & H. Sarma. 2021. Evaluation of antidiabetic and anti-hyperlipidemic activity of isolated bioactive compounds of leaves of *Annona reticulata* linn. *The Natural Products Journal* 11(3): 414–421.

Patle, D., M. Vyas & G.L. Khatik. 2021. A review on natural products and herbs used in the management of diabetes. *Current Diabetes Reviews* 17(2): 186–197.

Phillips, J.M., N.M. Parish & T. Raine. 2013. Type I diabetes development requires both C D4+ and CD8+ T-cell s and can be reversed by non-depleting antibodies targeting both T-Cell populations. *Review of Diabetic Studies* 6(2): 97–103.

Pinto, M.D.S., Y.I. Kwon & E. Apostolidis E. 2009. Potential of *Ginkgo biloba* L. leaves in the management of hyperglycemia and hypertension using in vitro models. *Bioresource Technology* 100: 6599–6609.

Ponnachan, P.T., C.S. Paulose & K.R. Panikkar.1993. Effect of leaf extract of *Aegle marmelose* in diabetic rats. *Indian Journal of Experimental Biology* 31: 345–347.

Potdar, D., R.R. Hirwani & S. Dhulap. 2012. Phyto-chemical and pharmacological applications of *Berberis aristata*. *Fitoterapia* 83: 817–830.

Prabhakar, P. & M. Banerjee. 2020. Antidiabetic phytochemicals: A comprehensive review on opportunities and challenges in targeted therapy for herbal drug development. *International Journal of Pharmaceutical Research* 12: 1673–1696.

Puneeth, H.R. & S.S.P. Chandra. 2020. A review on potential therapeutic properties of pomegranate (*Punica granatum* L.). *Plant Science Today* 7(1): 9–16.

Quillen, D.M., G. Samraj & L. Kuritzky. 1999. Improving management of type 2 diabetes mellitus: 2. Biguanides. *Hospital Practice* 34: 41–44.

Rahman, M.M., M.R. Islam, S. Shohag, M.E. Hossain, M.S. Rahaman, F. Islam, M. Ahmed, S. Mitra, M.U. Khandaker, A.M. Idris & K. Chidambaram. 2022. The multifunctional role of herbal products in the management of diabetes and obesity: A comprehensive review. *Molecules* 27(5): 1713.

Ramachandran, V. & R. Saravanan. 2015. Glucose uptake through translocation and activation of GLUT4 in PI3K/Akt signaling pathway by asiatic acid in diabetic rats. *Human and Experimental Toxicology* 34: 884–893.

Rang, H., M. Dale, J. Ritter & P. Moore. 2003. *Pharmacology*, Ninth ed. Edinburg: Churchill Livingstone, pp. 408–419.

Rasouli, H., G. Hosseini, H. Adibi, R. Khodarahmi & M. Baghir. 2017. Differential αamylase/α-glucosidase inhibitory activities of plant-derived phenolic compounds: A virtual screening perspective for the treatment of obesity and diabetes. *Food & Function* 8(5).

Rasouli, H., R. Yarani, F. Pociot & J. Popović-Djordjević. 2020. Anti-diabetic potential of plant alkaloids: Revisiting current findings and future perspectives. *Pharmacological Research* 155: 104723.

Rivera-Madrid, R., M. Aguilar-Espinosa, Y. Cárdenas-Conejo & L.E. Garza-Caligaris. 2016. Carotenoid derivates in Achiote (*Bixa orellana*) seeds: Synthesis and health promoting properties. *Frontiers in Plant Science* 7: 1406.

Rosenberg, L., M. Lipsett, J.W. Yoon, M. Prentki, R. Wang, H.S. Jun, G.L. Pittenger, D. Taylor-Fishwick & A.I. Vinik. 2004. A pentadecapeptide fragment of islet neogenesis-associated protein increases beta-cell mass and reverses diabetes in C57BL/6J mice. *Annals of Surgery* 240(5): 875.

Rubino, E., E. Carrillo, G. Alcala, A. Domínguez-Martín, J. Marchal & H. Boulaiz. 2019. Phenformin as an anticancer agent: Challenges and prospects. *International Journal of Molecular Sciences* 20: 3316.

Saravanan, R. & L. Pari. 2005. Antihyperlipidemic and antiperoxidative effect of Diasulin, a polyherbal formulation in alloxan induced hyperglycemic rats. *BMC Complementary and Alternative Medicine* 5: 14.

Sarker, M.M.R. & M.A. Soma. 2020. Updates on clinical study reports of phytotherapy in the management of type 2 diabetes mellitus. *Phytotherapy in the Management of Diabetes and Hypertension* 4: 1–60.

Sayem, A.S.M., A. Arya, H. Karimian, N. Krishnasamy, A. Hasamnis & C.F. Hossain. 2018. Action of phytochemicals on insulin signaling pathways accelerating glucose transporter (GLUT4) protein translocation. *Molecules* 23: 258.

Seino, S., K. Sugawara, N. Yokoi & H. Takahashi. 2017. β-Cell signalling and insulin secretagogues: A path for improved diabetes therapy. *Diabetes, Obesity and Metabolism* 19: 22–29.

Semova, I., A.E. Levenson, J. Krawczyk, K. Bullock, K.A. Williams, R.P. Wadwa, P.R. Khoury, T.R. Kimball, E.M. Urbina, S.D. Ferranti, D.M. Maahs, L.M. Dolan, A.S. Shah, C.B. Clish & S.B. Biddinger. 2019. Type-1 diabetes is associated with an increase in cholesterol absorption markers but a decrease in cholesterol synthesis markers in a young adult population. *Journal of Clinical Lipidology* 13(6): 940–946.

Shabab, S., Z. Gholamnezhad & M. Mahmoudabady. 2021. Protective effects of medicinal plant against diabetes induced cardiac disorder: A review. *Journal of Ethnopharmacology* 265: 113328.

Shahin, D.H., H. Sultana, J. Farooq, T. Taj, U.F. Khaiser, N.S.A. Alanazi, M.K. Alshammari, M.N. Alshammari, F.H. Alsubaie, S.M.B. Asdaq & A.A. Alotaibi. 2022. Insights into the uses of traditional plants for diabetes nephropathy: A review. *Current Issues in Molecular Biology* 44(7): 2887–2902.

Silva, R.M., F.A. Oliveira, K.M. Cunha, J.L. Maia, M.A.M. Maciel, A.C. Pinto, N.R. Nascimento, F.A. Santos & V.S. Rao. 2005. Cardiovascular effects of trans-dehydrocrotonin, a diterpene from *Croton cajucara* in rats. *Vascular Pharmacology* 43(1): 11–18.

Sinan, K.I., U. Akpulat, A.A. Aldahish, Y. CelikAltunoglu, M.C. Baloğlu, D. Zheleva Dimitrova, R. Gevrenova, D. Lobine, M.F. Mahomoodally, O.K. Etienne & G. Zengin. 2021. LC-MS/HRMS analysis, anti-cancer, anti-enzymatic and anti-oxidant effects of *Boerhavia diffusa* extracts: A potential raw material for functional applications. *Antioxidants* 10(12): 2003.

Sirovina, D., N. Oršolić, M.Z. Končić, G. Kovačević, V. Benković & G. Gregorović. 2013. Quercetin vs chrysin: Effect on liver histopathology in diabetic mice. *Human & Experimental Toxicology* 32(10): 1058–1066.

Sobeh, M., M. El-Raey, S. Rezq, M.A. Abdelfattah, G. Petruk, S. Osman, A.M. El-Shazly, H.A. El-Beshbishy, M.F. Mahmoud & M. Wink. 2019. Chemical profiling of secondary metabolites of *Eugenia uniflora* and their antioxidant, anti-inflammatory, pain killing and antidiabetic activities: A comprehensive approach. *Journal of Ethnopharmacology* 240: 111939.

Srivastava, S.K., B. Agrawal, A. Kumar & A. Pandey. 2020. Phytochemicals of *Azadirachta indica* source of active medicinal constituent used for cure of various diseases: A review. *Journal of Scientific Research* 64(1): 385–390.

Suky, M.G., T.P. Shanthi, R.G.D. Ruby, S. Sowmiya & M. Yasodha. 2019. Isolation and identification of bioactive compounds in chloroform extract of *Phyla nodiflora* (L.) Greene. *Journal of Emerging Technologies and Innovative Research* 6(1): 275–286.

Sun, E., D. de-Fontgalland, P. Rabbitt, P. Hollington, L. Sposato & S. Due. 2017. Mechanisms controlling glucose-induced GLP-1 secretion in human small intestine. *Diabetes* 66: 2144–2149.

Sun, Y., X. Sun, F. Wang & S. Liu. 2015. Inhibitory effects of ursolic acid on diabetic retinopathy in mice. *Zhonghua Yi Xue Za Zhi* 95: 2589–2593.

Sutrisna, E., M. Maryati, I. Trisharyanti, S. Suprapto, D. Usdiana, T. Husna, R. Rinaldy & I. Malik. 2019. Antidiabetic potencies of *Phyllanthus niruri* Linn and thin-layer chromatography profile. *Drug Invention Today* 11(10): 2339–2341.

Suzuki, K., M. Tanaka, R. Konno & Y. Kaneko. 2002. Effects of 5-campestenone (24-methylcholest-5-en-3-one) on the type 2 diabetes mellitus model animal C57BL/KsJ-db/db mice. *Hormone and Metabolic Research* 34: 121–126.

Tabandeh, M.R., S.A. Hosseini & M. Hosseini. 2017. Ginsenoside Rbl exerts antidiabetic action on C2C12 muscle cells by leptin receptor signaling pathway. *Journal of Receptors and Signal Transduction* 37: 370–378.

Tabussum, A., N. Riaz, M. Saleem, M. Ashraf, M. Ahmad, U. Alam, B. Jabeen, A. Malik & A. Jabbar. 2013. α-Glucosidase inhibitory constituents from *Chrozophora plicata*. *Phytochemistry Letters* 6(4): 614–619.

Tan, H. & C. Vinanzaca. 2018. The effect of Cortex Lycii on rattusnorvegicus' blood glucose level using a high glucose diet. *Journal of Diabetes Mellitus* 8: 160–170.

Tan, X.C., K.H. Chua, R.M. Ravishankar & U.R. Kuppuswamy. 2016. Monoterpenes: Novel insights into their biological effects and roles on glucose uptake and lipid metabolism in $3T_3L_1$ adipocytes. *Food Chemistry* 196: 242–250.

Tang, D., Q.B. Chen, X.L. Xin & H.A. Aisa. 2017. Anti-diabetic effect of three new norditerpenoid alkaloids in vitro and potential mechanism via PI3K/Akt signaling pathway. *Biomedicine and Pharmacotherapy* 87:145–152.

Tang, B., T. Yang, W. Yang, W. Wang, X. Zhang & W. Ye. 2013. Chemical constituents in leaves of *Morus atropurpurea* and their α-glucosidase activity. *Chinese Traditional and Herbal Drugs* 44: 3109–3113.

Thareja, S., S. Aggarwal, T.R. Bhardwaj & M. Kumar. 2012. Protein tyrosine phosphatase 1B inhibitors: A molecular level legitimate approach for the management of diabetes mellitus. *Medicinal Research Reviews* 32: 459–517.

Tiong, S.H., C.Y. Looi, A. Arya, W.F. Wong, H. Hazni, M.R. Mustafa & K. Awang. 2015. Vindogentianine, a hypoglycemic alkaloid from *Catharanthus roseus* (L.) G. Don (Apocynaceae). *Fitoterapia* 102:182–188.

Tiong, S.H., C.Y. Looi, H. Hazni, A. Arya, M. Paydar, W.F. Wong, S.C. Cheah, M.R. Mustafa & K. Awang. 2013. Antidiabetic and antioxidant properties of alkaloids from *Catharanthus roseus* (L.) G. Don. *Molecules* 18(8): 9770–9784.

Tiwari, N., A. Thakur, V. Kumar, A. Dey & V. Kumar. 2014. Therapeutic targets for diabetes mellitus: An update. *Clinical Pharmacology & Biopharmaceutics* 3: 117.

Tomy, S., T.K. Ujwala, S. Celine & C.S. Kumar. 2016. Toxicological evaluation of Diakure, an antidiabetic polyherbal formulation. *International Journal of Phytomedicine* 8(1): 127–137.

Tsuhako, R., H. Yoshida, C. Sugita & M. Kurokawa. 2020. Naringenin suppresses neutrophil infiltration into adipose tissue in high-fat diet-induced obese mice. *Journal of Natural Medicines* 74: 229–237.

Tungmunnithum, D., D. Pinthong & C. Hano. 2018. Flavonoids from *Nelumbo nucifera* Gaertn., a medicinal plant: Uses in traditional medicine, phytochemistry and pharmacological activities. *Medicines* 5(4): 127.

Unuofin, J.O. & S.L. Lebelo. 2020. Antioxidant effects and mechanisms of medicinal plants and their bioactive compounds for the prevention and treatment of type 2 diabetes: An updated review. *Oxidative Medicine and Cellular Longevity*: 1356893.

Verma, N., V.K. Shakya & R.C. Saxena. 2008. Antidiabetic activity of glycoside isolated from Gymnema sylvestre in streptozotocin induced diabetic rats. *Asian Journal of Chemistry* 20(7): 5033.

Verma, S., S. Rawat, K.L. Ho, C.S. Wagg, L. Zhang, H. Teoh, J.E. Dyck, G.M. Uddin, G.Y. Oudit, E. Mayoux & M. Lehrke. 2018. Empagliflozin increases cardiac energy production in diabetes: Novel translational insights into the heart failure benefits of SGLT2 inhibitors. *JACC: Basic to Translational Science* 3(5): 575–587.

Vilar, D.A., M.S.dA. Vilar, T.F.A.dL. Moura, F.N. Raffin, M.R.d. Oliveira, C.F.dO. Franco, P.F.dA. Filho, M.dF.F.M. Melo & J.M.B Filho. 2014. Traditional uses, chemical constituents, and biological activities of *Bixa orellana* L.: A review. *The Scientific World Journal*: 857292.

Volpe, C.M.O., P.H. Villar-Delfino, P.M.F. Dos Anjos & J.A. Nogueira-Machado. 2018. Cellular death, reactive oxygen species (ROS) and diabetic complications. *Cell Death and Disease* 9: 1–9.

Waltner-Law, M.E., X.L. Wang & B.K. Law. 2002. Epigallocatechin gallate: A constituent of green tea, represses hepatic glucose production. *Journal of Biological Chemistry* 277: 34933–34940.

Wickramasinghe, A.S.D., P. Kalansuriya & A.P. Attanayake. 2021. Herbal medicines targeting the improved β-cell functions and β-cell regeneration for the management of diabetes mellitus. *Evidence-Based Complementary and Alternative Medicine*: 2920530.

Wilcox, G. 2005. Insulin and insulin resistance. *Clinical biochemist reviews* 26(2): 19.

William, J., P. John, M.W. Mumtaz, A.R. Ch, A. Adnan, H. Mukhtar, S. Sharif, S.A. Raza & M.T. Akhtar. 2019. Antioxidant activity, α-glucosidase inhibition and phytochemical profiling of *Hyophorbe lagenicaulis* leaf extracts. PeerJ 7: e7022.

Wu, J. & L.J. Yan. 2015. Streptozotocin-induced type 1 diabetes in rodents as a model for studying mitochondrial mechanisms of diabetic β cell glucotoxicity. *Diabetes Metabolic Syndrome and Obesity* 8: 181–188.

Yin, J., H. Xing & J. Ye. 2008. Efficacy of berberine in patients with type 2 diabetes mellitus. *Metabolism* 57: 712–717.

Yueniwati, Y., M.F.R. Syaban, N.E. Erwan, G.F.A. Putra & A.D. Krisnayana. 2021. Molecular docking analysis of *Ficus religiosa* active compound with anti-inflammatory activity by targeting tumour necrosis factor alpha and vascular endothelial growth factor receptor in diabetic wound healing. *Open Access Macedonian Journal of Medical Sciences* 9(A): 1031–1036.

Zhang, H., Y. Li, J. Hu, W.J. Shen, M. Singh, X. Hou, A. Bittner, S. Bittner, Y. Cortez, J. Tabassum & F.B. Kraemer. 2015. Effect of creosote bush-derived NDGA on expression of genes involved in lipid metabolism in liver of high-fructose fed rats: Relevance to NDGA amelioration of hypertriglyceridemia and hepatic steatosis. *PLoS One* 10(9): e0138203. https://doi.org/10.1371/journal.pone.0138203

Zhu, Y., Y. Zhao, P. Wang, M. Ahmedna & S. Sang. 2015. Bioactive ginger constituents alleviate protein glycation by trapping Methylglyoxal. *Chemical Research in Toxicology* 28: 1842–1849.

Ziegler, J. & P.J. Facchini. 2008. Alkaloid biosynthesis: Metabolism and trafficking. *Annual Review of Plant Biology* 59: 735–769.

3 Antidiabetic Features of *Allium cepa* L. (Onion)

Recent Trends, Progress, and Challenges

Weria Weisany, Katayoon Samiakalantari, Shima Yousefi, and Maryam Tahmasebi

CONTENTS

ABBREVIATIONS

ABTS: 2,2'-azino-bis-(3-ethylbenzothiazoline-6-sulfonic acid)
CAT: catalase
DM: diabetes mellitus
DPPH: 1,1-diphenyl-2-picrilhydrazyl
FRAP: ferric reducing antioxidant power
GPx: glutathione peroxidase
HO-1: hemeoxygenase-1
IDDM: insulin-dependent diabetes mellitus
MDA: malondialdehyde
NIDDM: noninsulin-dependent diabetes mellitus
ORAC: oxygen radical absorbance capacity
SOD: superoxide dismutase
TAC: total antioxidant capacity
TEAC: trolox equivalent antioxidant capacity
WHO: World Health Organization

DOI: 10.1201/b23347-3

3.1 INTRODUCTION

Diabetes mellitus (DM), a severe metabolic disorder, has affected the lives of millions of people worldwide. Insufficient insulin production and incapacity of peripheral tissues to absorb insulin are considered two primary causes of chronic hyperglycemia, according to Geberemeskel, Debebe et al. (2019). The International Diabetes Federation announced in 2017 that more than 352 million people were in danger of type 2 diabetes, which verifies WHO's estimation of having more than 439 million adults with diabetes by 2030 (Shaw, Sicree et al. 2010). In addition to the high rate of spread, diabetes is one of the main reasons for adult death. Of all deaths in 2010, 15.7% in North America and 6% in Africa belonged to diabetic patients dying from this disease (Roglic and Unwin 2010). The overlap of many diseases can have irreversible health effects, and diabetes is no exception. Diabetic patients are more prone to being affected by other diseases, as a recent meta-analysis conducted during the COVID-19 pandemic in 2021 showed in the mortality rate of COVID-19 patients who have diabetes (Wu, Tang et al. 2021). This complication on patients' health caused by diabetes shows the importance of slowing down the development of this disease to maintain society's homeostasis, especially with the expected emergence of new pathogen diseases in the future. Several historical reports have shown the use of specific plants by ancient cultures and traditions as an alternative to pharmaceutical solutions to cure, assist in treatment, or prevent some diseases, including diabetes. Scientists are investigating the potential health benefits of these medicinal plants through their bioactive compounds, including antioxidants (Szczepaniak, Ligaj et al. 2020; Tajner-Czopek, Gertchen et al. 2020). There have been efforts to include these plants in patients' diets. Table 3.1 shows two different approaches to adding these plants to patients' daily food. It is critical to consider the purity of these plants of their toxic compounds before introducing them into the diet (Sowa, Marcinčáková et al. 2020).

These plants can be consumed directly in their basic shape, their modified shape, or slightly altered. In the modified shape case, we can modify the matrix of raw plants with antidiabetic properties. This modification can be done by producing functional food as dietary supplements

TABLE 3.1
Different Possible Ways of Introducing Medicinal Plants into the Diet

	How can the medicinal plant be introduced into the diet?	
	First way	
Eat on your own	Basic form:	Whole leaves
		Whole seeds
		Whole shoots
		Whole fruits
		ground
	Modified form:	crushed
		dried
		cut
		brew
	Changed state of matter:	tea
		extract
	Second way	
Change the matrix	Functional food products	
	Dietary supplements	

or designing food recipes that use these raw plants. This composition modification of plants is especially suitable for obese patients and those suffering from high blood glucose levels. These food products, which are highly in demand by consumers (Przeor, Flaczyk et al. 2018), are mainly designed by technologists (Kozlowska, Zbikowska et al. 2019). While adding these raw materials is mainly recommended to the youth as a preventive measurement, there has been a considerable emphasis on using these raw materials for maintaining a healthy lifestyle for the elderly population. So far, oral administration and recommendation have been the most common and easiest ways to advocate adding such raw materials into the human diet for an average patient. However, other administration methods of such plants are also under examination (Baset, Ali et al. 2020). Although conventional pharmaceutical solutions are available for treating diabetes, their high prices and possible side effects make them less attractive to some patients who would prefer plant-based alternatives.

3.2 THE DEFINITION OF DIABETES MELLITUS

Diabetes mellitus is a serious metabolic disorder affecting millions worldwide. As of 2004, there were more than 366 million diabetic patients worldwide. It is predicted that the number of diabetic patients will reach more than 552 million by 2030 due to a shortage of sustainable procedures (Wild, Roglic et al. 2004). Scientists believe obesity, inactive lifestyles, and high-energy diets are the primary motivations for this progressively rising rate of type 2 diabetes. By 2030, DM is expected to account for almost US$490 billion in healthcare costs worldwide and more than 11.6% of all international healthcare costs in 2010 (Colagiuri 2010). Diabetes has two main types: type I (insulin-dependent diabetes mellitus [IDDM]) and type II (noninsulin-dependent diabetes mellitus [NIDDM]). A person with IDDM suffers from insulin insufficiency, as their bodies cannot produce insulin and depend exclusively on external sources of insulin to operate. Young children and youth are more prone to IDDM, which severely damages pancreatic β-cells. Type I diabetes is categorized as autoimmune (immune-mediated) diabetes (type 1A) or idiopathic diabetes with β-cell destruction (type 1B). Unlike IDDM patients, diabetes patients with NIDDM do not react to insulin. Instead, their treatment includes exercise, dietary management, and special medication techniques. In most cases, NIDDM affects adults over 40 and has the symptoms of obesity with disturbed carbohydrate and fat metabolism. These two types of diabetes have some characteristics in common, including high blood sugar, extreme hunger and weakness, blurred vision, unusual thirst, and frequent urination (Khan, Najmi et al. 2012). Traditional and manufactured antidiabetic agents can also potentially have adverse outcomes such as weight gain, acute hypoglycemia, edema, hepatic and cardiac defects, and gastric and respiratory problems (Akash, Shen et al. 2012). These side effects have motivated researchers and scientists to investigate new approaches to treating DM (Sajid Hamid Akash, Rehman et al. 2013). Medicinal plants with hypoglycemic properties are considered as being therapeutic agents. Compared to synthetic drugs, herbal plants are only slightly harmful and have fewer side effects (Romila, Mazumder et al. 2010; Akash, Shen et al. 2012). In many countries, especially in Asian communities, herbal agents are typically used as an effective remedy for treating many disorders (Akash, Rehman et al. 2014). Using medicinal plants to treat DM is most important when resources are meager and insufficient (Anderson, Soeandy et al. 2013). The families of plants that have shown rich hypoglycemic effects include Leguminosae, Lamiaceae, Liliaceae, Cucurbitaceae, Asteraceae, Moraceae, Rosaceae, Euphorbiaceae, Araliaceae, and Ranunculaceae (Ibrahim, Farooq et al. 2013). Throughout this chapter of the book, we have summarized the general effects of *Allium cepa*. We discuss various experimental studies performed to identify the hypoglycemic properties of onion, its extracts, juices, essential oils, and constituents. This study aims to confirm that consuming onions can lower blood glucose levels and reduce the risk factors of type 2 diabetes. Onion (*A. cepa*) is widely produced and consumed worldwide (Pareek, Sagar et al. 2017).

3.3 *ALLIUM CEPA* L. (ONION)

Three color varieties of onions (cultivar groups) usually available in the food market are white, yellow, and red. There are various ways of serving onion as a cooked vegetable ingredient, such as baking, boiling, braising, grilling, frying, roasting, sautéing, or steaming. Onion can also be consumed raw in salads, made into juice, pickled in vinegar, or used as a spice. Other products can be prepared, including onion oil, onion juice, onion salt, and onion pickles. Onions are used in herbal medicine to ease or prevent several common diseases, including atherosclerosis, asthma, bronchitis, and coughs. Additionally, onions have numerous health benefits besides their various bioactive constituents like organosulfur compounds, phenolic acids, polysaccharides, and saponins (Marrelli, Amodeo et al. 2018; Teshika, Zakariyyah et al. 2019). The chemical compounds and bioactive properties of onions have remarkable health benefits (Figure 3.1), including antioxidant (Ouyang, Hou et al. 2018), antimicrobial (Loredana, Giuseppina et al. 2019), anti-inflammatory (Jakaria, Azam et al. 2019), anti-obesity (Lee, Cha et al. 2016), antidiabetic (Jini and Sharmila 2020), anticancer (Tsuboki, Fujiwara et al. 2016), cardiovascular protective (Gabriele, Frassinetti et al. 2017), neuroprotective (Salami, Tamtaji et al. 2020), hepatorenal protective (Ahmed, Al-Yousef et al. 2017), respiratory protective (Sarkar, Mondal et al. 2020), reproductive protective (Shokoohi, Madarek et al. 2018), and immunomodulatory properties (Alkhedaide, Soliman et al. 2017). Although there has been no hazard or risk reported for consuming onion, some potential health concerns need to be considered, such as pesticide residue (Malhat and Abdallah 2019), heavy metal enrichment (Cavanagh, Yi et al. 2019), microbial contamination (Mayer, Twarużek et al. 2016), and nitrate accumulation (Haftbaradaran, Khoshgoftarmanesh et al. 2018).

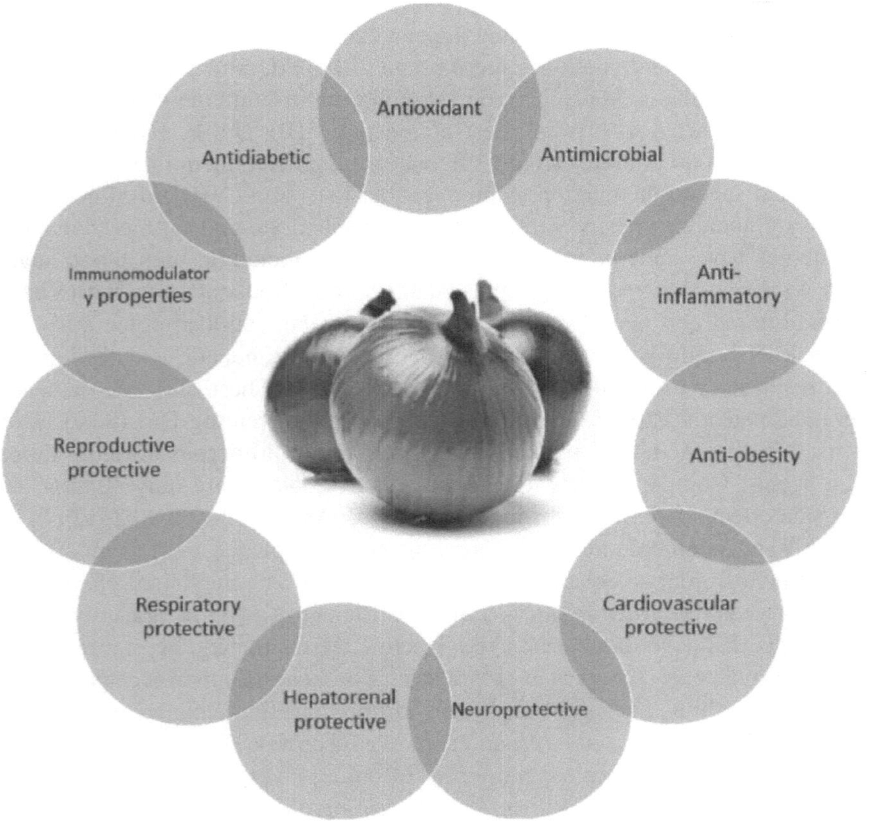

FIGURE 3.1 Bioactive compounds and health functions of onion.

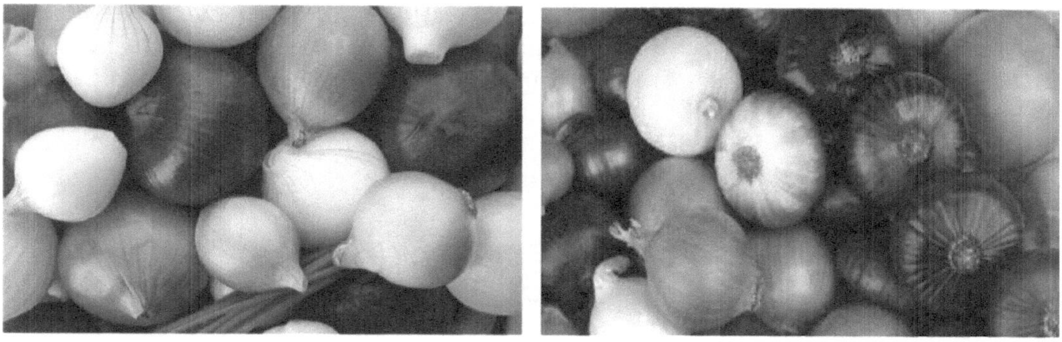

FIGURE 3.2 Different types of *Allium cepa*.

3.4 BOTANICAL OF *ALLIUM CEPA* L. (ONION)

A. cepa (onion), a culinary and medicinal herb belonging to the Amaryllidaceae botanical family, is the third most critical horticultural vegetable after potatoes and tomatoes. Its distinctive flavor has made it one of the most widely consumed plants worldwide (Teshika, Zakariyyah et al. 2019). An onion is a bulbous perennial plant with one to three depressed-globulous bulbs, varying in size depending on the cultivar or race. This plant typically has white or purple flowers with upright or spreading petals. This plant has regular leaves, fistulose blades, and persistent leaves. The fruit is withering, bearing margins with either obtuse or acute apex (Figure 3.2).

3.5 PRE-HARVEST AND POST-HARVEST TREATMENTS OF ONIONS

Onions must reach the appropriate maturity stage before harvesting. Planting season, cultivar, and climate highly influence the variations in the maturity stage of the plant. Usually, farmers pull an onion bulb from the soil before harvest and leave it on the field for a few days to prevent excessive shrinkage by removing moisture from the outer skin and neck. It is during the moisture removal stage that color development happens. Fresh onions can be sold directly to consumers or in the market in a processed form, including powders and flakes (Khan, Ansar et al. 2016). Powdered, frozen, or canned onion products are reported to have become popular in the growing international onion markets (Choi, Lee et al. 2017). There are several reasons to further process fresh onions into other forms, including reducing the product loss (20%–30%) during the storage episode. Additionally, dehydrated products contain more beneficial compounds than fresh onions do and are easier to consume. Onion powders, rich in phenolic compounds, can be used as an antioxidant and flavor enhancer to produce positive effects that combine with other foods (Arslan and Özcan 2010). Thus, the food industry offers commercial onion powder as a nutraceutical or a dietary supplement.

3.6 TRADITIONAL USES OF *ALLIUM CEPA* L.

For centuries, *A. cepa* has been used for its medicinal properties in treating various forms of illnesses. In ancient Greek, *A. cepa* was used by athletes to purify their blood; Roman gladiators used it to make firm muscles. Hippocrates, the famous Greek physician, prescribed onion for diuretic effect, healing wounds, and combating pneumonia. In the 6th century, onion was described as one of India's most crucial vegetables, spices, and medicine (Kabrah 2015). The Asian nations, Indians, and Pakistanis and in general low-developed countries are among the majority of people who use onions for various medical purposes. Most likely, this is due to a lack of medical facilities and the attainability of traditional medicine such as onions. It can be noted that *A. cepa* is commonly used for treating infectious diseases raw or as an extract. Further, it is used internally and externally to treat digestive problems, skin diseases, metabolic disorders, insect bites, and other illnesses (Ayyanar and Ignacimuthu 2011). In traditional medicine, onion has been used for a large variety of illnesses such as headache, fever, snake bites, hair loss, toothache, cough, sore throat, flu, baldness, epilepsy, rash, jaundice, constipation, flatulence, intestinal worms, low sexual power, rheumatism, body pain and muscle cramps, high blood pressure, and diabetes (Ayyanar and Ignacimuthu 2011; Ahmed, Mahmood et al. 2015; Upadhyay 2016; Teshika, Zakariyyah et al. 2019).

3.7 PHYTOCHEMISTRY OF *ALLIUM CEPA* L.

Several phytochemical studies have been conducted on *A. cepa* and have reported on the existence of several compounds that contribute to this plant's distinct flavor and medicinal value. Numerous bioactive compounds are found in onion, including sulfur-containing compounds such as onionin A, cysteine sulfoxides, and phenolic compounds such as rutin, quercetin, and quercetin glucosides (Figure 3.3). Onion is rich in phytochemicals with health-promoting artifacts, including organosulfur compounds (Moreno-Rojas, Moreno-Ortega et al. 2018; Zamri and Hamid 2019), phenolic compounds (Lee, Parks et al. 2017; Viera, Piovesan et al. 2017), polysaccharides (Ma, Zhu et al. 2018), and saponins (Lanzotti, Romano et al. 2012; Dahlawi, Nazir et al. 2020). The amount of bioactive compounds can differ between different onion varieties (Loredana, Giuseppina et al. 2019). For example, anthocyanins are shown to be highest in the red onion, followed by flavonols in the yellow onion, while flavonols were lowest in the white onion (Zhang, Deng et al. 2016). Likewise, different layers of onions contain different major compounds (Beesk, Perner et al. 2010). The main compound of the red onion's bulb is quercetin-4-glucoside, and its skin is predominantly quercetin (Park, Ryu et al. 2018). Because of processing, there can be changes in onion bioaccessibility and bioactive compound content. Although the high-pressure processing does not impact the onion flavonols' bioaccessibility, the onion matrix may enhance its bioaccessibility (Fernández-Jalao, Sánchez-Moreno et al. 2017). Research studies have reported that the onion skin quercetin aglycone is more bioavailable than pure quercetin dehydrates when taken by humans (Burak, Brüll et al. 2017). The quercetin content does not significantly change by sautéing (Zhao, Lin et al. 2021). However, the content and bioaccessibility of phenolic compounds, especially quercetin-derivatives, are increased by cooking, such as baking, grilling, and frying (Cattivelli, Conte et al. 2021). The impact of boiling, frying, steaming, and microwaving on the levels of S-alk(en)yl-L-cysteine sulfoxides (ACSOs) such as methiin, isoalliin, propiin, and cycloalliin in onions was evaluated using

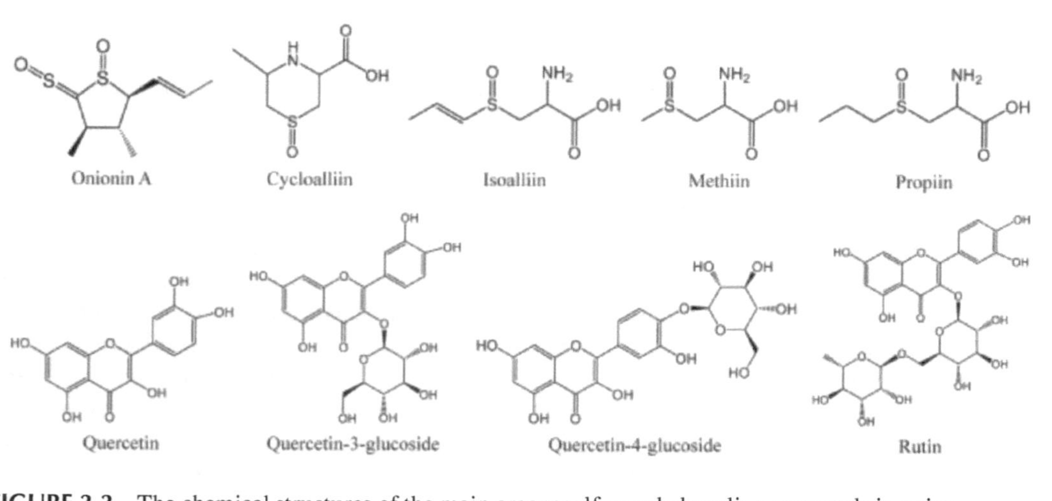

FIGURE 3.3 The chemical structures of the main organosulfur and phenolic compounds in onion.

Liquid Chromatography Electrospray Ionization-Tandem Mass Spectrometry (LC/ESI-MS/MS). The findings of the study revealed that frying, steaming, and microwaving resulted in a significant increase in ACSOs content, ranging from 34.2% to 568.0%. Conversely, boiling led to a decrease in ACSOs content, ranging from 32.6% to 69.4% (Kim, Lee et al. 2016). For example, the contents of these foods decrease if boiled, but they increase when fried, microwaved, and steamed. Additionally, black onions were dramatically reduced in flavonoid content, while their contents of isoalliin and fructose were increased (Moreno-Rojas, Moreno-Ortega et al. 2018).

3.7.1 Biological Activities of Onion Peel

The significant increases in obesity, cardiovascular disorders, diabetes, neurodegenerative disorders, and cancer are the result of the rising consumption of processed foods, lack of exercise, sedentary lifestyle, and exposure to polluted environments, while oxidative stress is one of the primary causes (Sharifi-Rad, Anil Kumar et al. 2020). Several studies have shown that phytonutrients protect humans against free radicals and oxidative stress (Sharifi-Rad, Anil Kumar et al. 2020). Numerous studies have shown that onion peel/waste and their extracts are rich sources of phytoconstituents with various therapeutic potentials and pharmacological activities. Various studies have emphasized onion peel extracts' beneficial properties, including anticancer, antimicrobial, anti-obesity, neuroprotective, cardio-protective, antidiabetic, and erectile dysfunction. By understanding these biological activities better, we can move forward with future research and develop novel uses for onion waste in biomedical and pharmaceutical applications.

3.7.2 Pharmacological Activities of *Allium cepa* L.

Flavonoids, phenolic acids, and organosulfur compounds are all part of *A. cepa*'s diverse group of phytochemicals that contribute to its bioactivity. There are numerous medicinal properties present in *A. cepa* such as antimicrobial, antioxidant, analgesic, anti-inflammatory, antidiabetic, hypolipidemic, antihypertensive, and immunoprotective properties. Recent studies (Ogunmodede, Saalu et al. 2012) have emphasized that in diabetic rabbits, *A. cepa* may be effective in reducing liver oxidative stress by preventing the decrease in antioxidant parameters such as superoxide dismutase, catalase, and glutathione peroxidase. After treatment with onion extract, alloxan-diabetic rats showed a reduction in their plasma and tissue levels of free radicals (El-Demerdash, Yousef et al. 2005), and this result was in agreement with the study of Kumari and Augusti (2002) and Campos, Diniz et al. (2003).

3.8 HEALTH FUNCTION OF ONION

A wide range of plant-based foods, such as garlic (Shang, Cao et al. 2019), ginger (Mao, Xu et al. 2019), sweet tea (Shang, Liu et al. 2020), dark tea (Lin, Wei et al. 2021), germinated edible seeds and sprouts (Gan, Lui et al. 2017), and their bioactive compounds, including resveratrol (Meng, Zhou et al. 2020), curcumin (Xu, Meng et al. 2018), rutin (Farha, Gan et al. 2022), quercetin (Ulusoy and Sanlier 2020), citrus flavonoids (Gandhi, Vasconcelos et al. 2020), and spice essential oils (Zhang, Gan et al. 2020), have been reported to offer a range of health benefits based on *in vitro*, *in vivo*, and human studies. Furthermore, onions have also been shown to possess multiple health benefits using similar research methods.

3.8.1 Antioxidant Activity

Onion is a good source of natural antioxidants (Sidhu, Ali et al. 2019). Many studies have been carried out to evaluate the antioxidant activities of onion, and they found that onion exhibits strong antioxidant properties by using a series of *in vitro* assays, including 2,2'-azino-bis-(3-ethylbenzothiazoline-6-sulfonic acid) (ABTS), 1,1-diphenyl-1-picrylhydrazyl (DPPH), ferric reducing antioxidant power (FRAP), lipid peroxidation, oxygen radical absorbance capacity (ORAC), total antioxidant capacity (TAC), and trolox equivalent antioxidant capacity (TEAC) assays. Several factors influence onions' antioxidant activity. Various factors must be considered, such as genetics, horticulture, storage conditions, parts, extraction method, and processing technology. Studies have found that different onion varieties exhibit different levels of antioxidant activity (Sagar, Pareek et al. 2020), probably related to their genetic background (Mitrova, Hrbek et al. 2016). In addition, organic cultivation practices (Ren, Reilly et al. 2017), sulfur bentonite-organic-based fertilizers (Muscolo, Papalia et al. 2020), and mycorrhizal fungi (Chan, Koch et al. 2017) enhance the scope of bioactive compounds with antioxidant activities in onions. Applying mycorrhizal inocula, humic acids, and elevated atmospheric CO_2 increases the content of phenolics in the onion bulb and its antioxidant property (Bettoni, Mogor et al. 2017). Planting time and density are the other two impacting factors that affect the antioxidant components of onion seeds (Amalfitano, Golubkina et al. 2019). It is also possible that the conditions under which the second-year onion resprouts were stored may affect the quality and bioaccessibility of phenolics and their antioxidant activity (Zudaire, Viñas et al. 2017). Washing the fresh-cut onions with a combination of nisin and citric acid was reported to increase the total phenolic and flavonoid contents and antioxidant capacity during storage (Chen, Hu et al. 2016). Onion sprouting also increased its antioxidant activity and contents of total phenolics and flavonoids (Majid and Nanda 2018). Similarly, the antioxidant activities in distinct parts of red onion, such as the dry skin and edible portion, were segregated based on the principal component analysis, probably due to the former being rich in quercetin while the latter rich in quercetin-4-glucosideantioxidant (Park, Ryu et al. 2018). The influences of food processing on antioxidant capacities of onions were investigated as well, including drying (Salamatullah, Uslu et al. 2021), freezing (Çubukçu, Kılıçaslan et al. 2019), heating (Çubukçu, Kılıçaslan et al. 2019), sautéing (Watanabe, Muro et al. 2018), and high-pressure processing (Fernández-Jalao, Sánchez-Moreno et al. 2017). For instance, heating and freezing reduce the antioxidant activity of onion (Çubukçu, Kılıçaslan et al. 2019), while sautéing did not significantly change it (Watanabe, Muro et al. 2018). Onion also exhibits antioxidant activity in cell and animal models. The expression of antioxidant enzymes, including catalase (CAT), NAD (P)H quinone dehydrogenase 1 (NQO1), and hemeoxygenase-1 (HO-1), was upregulated by onion extract in N27-A cells (Jakaria, Azam et al. 2019). In addition, several studies demonstrated that onion treatment could improve the antioxidant status of animals. Onion enhances the activity of antioxidant enzymes, such as superoxide dismutase (SOD), CAT, and glutathione peroxidase (GPx), which was effective for protecting against oxidative stress in hypercholesterolemic rats (Gabriele, Frassinetti et al. 2017). The oxidative stress in the liver and kidney was reduced by pre-treatment with red onion peel extract in carbon tetrachloride-challenged rats (Ahmed, Al-Yousef et al. 2017). Onion-fortified feed ameliorated the liver and kidney oxidative damages in rats administered with potassium bromate (Nwonuma, Osemwegie et al. 2021). Dietary addition of onion extract and combining onion peel powder with pawpaw seed increased antioxidant enzyme activity in broiler chicks (Omar, Al-Khalaifah et al. 2020) and African catfish (Fawole, Adeoye et al. 2020), respectively. Furthermore, a clinical trial revealed that drinking onion juice (100 mL) for eight weeks could reduce free radicals and superoxide anions levels while elevating the glutathione content and total antioxidant capacity in healthy subjects (Law, Chiu et al. 2016). Overall, onions exhibit strong antioxidant effects, and many factors could affect the antioxidant capacity of onion. Although the antioxidant activity of onion has been extensively investigated, the related antioxidant molecular mechanism has been much less explored, which should be clarified in the future.

3.8.2 Antidiabetic Activity of *Allium cepa* L.

Increasing evidence from *in vitro* and *in vivo* studies has demonstrated that onions can reduce not only diabetes but also treat different diabetic complications (Figure 3.4). Several studies indicate that onion exhibits antidiabetic potential *in vitro*. The extracts of onion skin or solid onion waste

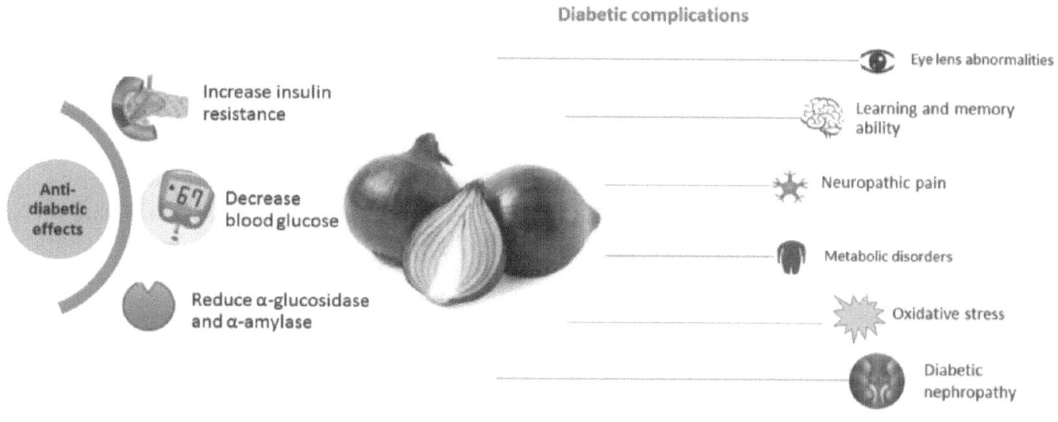

FIGURE 3.4 The effects of onion in diabetes and diabetic complications.

showed a remarkable inhibitory activity toward α-glucosidase and α-amylase, and the enzyme inhibitory activity was dose-dependent (Nile, Nile et al. 2018; Gois Ruivo da Silva, Skrt et al. 2020). Besides, onion-based green synthesized silver nanoparticles exhibit excellent α-glucosidase and α-amylase inhibitory activities (Jini and Sharmila 2020). In another study, onion fiber concentrates reduce starch digestibility and glucose production by suppressing α-amylase activity (Benítez, Mollá et al. 2017). Onion also exhibits antidiabetic potential *in vivo*. The heat-processed onion extract has a high content of Amadori rearrangement compounds, which reduces the post-prandial carbohydrate absorption and blood glucose levels by inhibiting intestinal sucrase activity in rats fed with sucrose or starch meals (Kang, Choi et al. 2018). Moreover, dietary supplementation with fenugreek seeds and onion has lessened hyperglycemia and its associated metabolic disorders in diabetic rats (Pradeep and Srinivasan 2017). Hyperglycemia-induced osmotic and oxidative stress is a primary factor in the progression of diabetic complications. It has been reported that intake of fenugreek seeds and onion could also reduce oxidative stress (Pradeep and Srinivasan 2017), relieve eye lens abnormalities (Pradeep and Srinivasan 2018a), alleviate cardiac damage (Pradeep and Srinivasan 2018b, 2018c), attenuate diabetic nephropathy (Pradeep and Srinivasan 2018c; Pradeep, Barman et al. 2019), and counter the deformity and fragility of erythrocytes (Ji, Lu et al. 2019) in STZ-induced diabetic rat (Tamtaji, Hosseinzadeh et al. 2017). The ethanolic extract of onion had a protective effect on learning and memory deficits in diabetic rats. Furthermore, the antidiabetic activities of the inedible parts of onion, including skin, seeds, and leaves, have also been investigated in STZ-induced diabetic rats. An extract of red onion scales can reduce fasting blood glucose and advanced glycation end products, increase serum insulin levels, and alleviate diabetic nephropathy (Abouzed, del Mar Contreras et al. 2018). The blood glucose and malondialdehyde (MDA) levels declined in alloxan-induced diabetic rats fed with wheat bread supplemented with onion powder or onion peel extract (Masood, Bashir et al. 2021). Onion seed extract showed protective activity against the adverse side effects of diabetes in rats (Fallah, Mahabadi et al. 2017). The leaf extract of onion reduce diabetes-induced neuropathic pain (Khan, Mohammed et al. 2020). Onion exhibits antidiabetic potential in humans. Based on a randomized placebo-controlled clinical trial, daily ingestion of fresh yellow onion in breast cancer patients receiving doxorubicin-based chemotherapy reduce hyperglycemia and insulin resistance (Jafarpour-Sadegh, Montazeri et al. 2017). In conclusion, onion can fight against diabetes by reducing oxidative stress, moderating hyperglycemia, improving insulin resistance, and ameliorating various histopathological changes.

3.9 ADVERSE EFFECTS AND TOXICITY

The negative consequences of *Allium* species in humans is categorized as most common, less frequent, and rare (Tattelman 2005). The most common side effects of taking small amounts of onion are bad breath and body odor. Less frequent undesirable effects, including gastrointestinal upsets (burning sensation and diarrhea), flatulence, and changes in the intestinal flora can happen due to consuming excessive onion on an empty stomach (Ackermann, Mulrow et al. 2001). For rare effects, dermatological problems such as allergic dermatitis, burns, and blisters have been seen in susceptible individuals. These rare side effects are produced by directly applying fresh or crushed garlic and onion to the skin (Friedman, Shalom et al. 2006). Several recent studies reported a significant reduction in the oral bioavailability of cyclosporin due to high consumption of onion with a high quercetin content, leading to organ transplant rejection (Yang, Chao et al. 2006). There are also some studies conducted to determine the impact of quercetin coadministration on plasma concentrations of the protease inhibitor saquinavir (antiviral HIV drug). However, more studies should be conducted to determine if saquinavir intracellular concentrations are altered by coadministration of quercetin due to substantial inter- and intrasubject variability (DiCenzo, Frerichs et al. 2006). The intake of onion and its derivatives should be controlled, including ingested dose, long-term medication, and safety and effectiveness of chosen preparation, to minimize the risk of adverse side effects (Davis 2005). Despite the associated risks, onion is a safe therapeutic agent. All the mentioned side effects only occur due to the high consumption of onions for a long time.

3.10 CONCLUSION AND FUTURE PERSPECTIVE

Onion, an herb rich in bioactive compounds, is planted and used in the food industry as a vegetable worldwide. The phenolic compounds (including quercetin and quercetin glucosides) and the sulfur-containing compounds (including onionin A and cysteine sulfoxide), the primary bioactive elements of onion, contribute to its multiple health benefits, including antimicrobial, anti-inflammatory, antioxidant, and immunomodulatory properties. Additionally, onion is a valuable natural resource used in functional foods or nutraceuticals for the prevention, treatment, and management of certain diseases, including obesity, diabetes, cancers, cardiovascular diseases, neurodegenerative diseases, nephropathy, respiratory disorders, colitis, and infertility. Onion powder, juice, and extracts are shown to have several health benefits. In addition to the several bioactive compounds found to be health-promoting, there is a need for further studying other bioactive compounds from onion or its by-products. Specifically, their health functions, the relevant molecular mechanisms, and whether they have synergistic effects need to be illuminated in the future and should be well studied. Since most current studies focus on raw onions' health benefits, it is necessary to investigate whether cooking and processing can influence those benefits. Moreover, more well-designed clinical trials are still required to verify the human health benefits of onion and onion-derived bioactive compounds. Finally, safety issues of onion should always be considered. These safety issues include the mentioned contaminations and other potential risks such as the overdose of bioactive compounds.

REFERENCES

Abouzed, T. K., M. del Mar Contreras, K. M. Sadek, M. Shukry, D. H. Abdelhady, W. M. Gouda, W. Abdo, N. E. Nasr, R. H. Mekky and A. Segura-Carretero (2018). "Red onion scales ameliorated streptozotocin-induced diabetes and diabetic nephropathy in Wistar rats in relation to their metabolite fingerprint." *Diabetes Research and Clinical Practice* **140**: 253–264.

Ackermann, R. T., C. D. Mulrow, G. Ramirez, C. D. Gardner, L. Morbidoni and V. A. Lawrence (2001). "Garlic shows promise for improving some cardiovascular risk factors." *Archives of Internal Medicine* **161**(6): 813–824.

Ahmed, A. F., H. M. Al-Yousef, J. H. Al-Qahtani and M. S. Al-Said (2017). "A hepatonephro-protective phenolic-rich extract from red onion (Allium cepa L.) peels." *Pakistan Journal of Pharmaceutical Sciences* **30**.

Ahmed, N., A. Mahmood, A. Ashraf, A. Bano, S. Tahir and A. Mahmood (2015). "Ethnopharmacological relevance of indigenous medicinal plants from district Bahawalnagar, Punjab, Pakistan." *Journal of Ethnopharmacology* **175**: 109–123.

Akash, M. S. H., K. Rehman and S. Chen (2014). "Effects of coffee on type 2 diabetes mellitus." *Nutrition* **30**(7–8): 755–763.

Akash, M. S. H., Q. Shen, K. Rehman and S. Chen (2012). "Interleukin-1 receptor antagonist: A new therapy for type 2 diabetes mellitus." *Journal of Pharmaceutical Sciences* **101**(5): 1647–1658.

Alkhedaide, A., M. M. Soliman and T. A. Ismail (2017). "Protective effect of onion extract against experimental immunesuppression in Wistar rats: Biological and molecular study." *African Journal of Traditional, Complementary and Alternative Medicines* **14**(5): 96–103.

Amalfitano, C., N. A. Golubkina, L. Del Vacchio, G. Russo, M. Cannoniero, S. Somma, G. Morano, A. Cuciniello and G. Caruso (2019). "Yield, antioxidant components, oil content, and composition of onion seeds are influenced by planting time and density." *Plants* **8**(8): 293.

Anderson, G. H., C. D. Soeandy and C. E. Smith (2013). "White vegetables: Glycemia and satiety." *Advances in Nutrition* **4**(3): 356S–367S.

Arslan, D. and M. M. Özcan (2010). "Study the effect of sun, oven and microwave drying on quality of onion slices." *LWT-Food Science and Technology* **43**(7): 1121–1127.

Ayyanar, M. and S. Ignacimuthu (2011). "Ethnobotanical survey of medicinal plants commonly used by Kani tribals in Tirunelveli hills of Western Ghats, India." *Journal of Ethnopharmacology* **134**(3): 851–864.

Baset, M. E., T. I. Ali, H. Elshamy, A. M. El Sadek, D. G. Sami, M. T. Badawy, S. S. Abou-Zekry, H. H. Heiba, M. K. Saadeldin and A. Abdellatif (2020). "Anti-diabetic effects of fenugreek (*Trigonella foenum-graecum*): A comparison between oral and intraperitoneal administration-an animal study." *International Journal of Functional Nutrition* **1**(1): 1–1.

Beesk, N., H. Perner, D. Schwarz, E. George, L. W. Kroh and S. Rohn (2010). "Distribution of quercetin-3, 4′-O-diglucoside, quercetin-4′-O-monoglucoside, and quercetin in different parts of the onion bulb (Allium cepa L.) influenced by genotype." *Food Chemistry* **122**(3): 566–571.

Benítez, V., E. Mollá, M. A. Martín-Cabrejas, Y. Aguilera and R. M. Esteban (2017). "Physicochemical properties and in vitro antidiabetic potential of fibre concentrates from onion by-products." *Journal of Functional Foods* **36**: 34–42.

Bettoni, M. M., Á. F. Mogor, V. Pauletti and N. Goicoechea (2017). "The interaction between mycorrhizal inoculation, humic acids supply and elevated atmospheric CO_2 increases energetic and antioxidant properties and sweetness of yellow onion." *Horticulture, Environment, and Biotechnology* **58**(5): 432–440.

Burak, C., V. Brüll, P. Langguth, B. F. Zimmermann, B. Stoffel-Wagner, U. Sausen, P. Stehle, S. Wolffram and S. Egert (2017). "Higher plasma quercetin levels following oral administration of an onion skin extract compared with pure quercetin dihydrate in humans." *European Journal of Nutrition* **56**(1): 343–353.

Campos, K., Y. Diniz, A. Cataneo, L. Faine, M. Alves and E. Novelli (2003). "Hypoglycaemic and antioxidant effects of onion, Allium cepa: Dietary onion addition, antioxidant activity and hypoglycaemic effects on diabetic rats." *International Journal of Food Sciences and Nutrition* **54**(3): 241–246.

Cattivelli, A., A. Conte, S. Martini and D. Tagliazucchi (2021). "Influence of cooking methods on onion phenolic compounds bioaccessibility." *Foods* **10**(5): 1023.

Cavanagh, J.-A. E., Z. Yi, C. Gray, K. Munir, N. Lehto and B. Robinson (2019). "Cadmium uptake by onions, lettuce and spinach in New Zealand: Implications for management to meet regulatory limits." *Science of the Total Environment* **668**: 780–789.

Chan, G. G., C. M. Koch and L. H. Connors (2017). "Blood proteomic profiling in inherited (ATTRm) and acquired (ATTRwt) forms of transthyretin-associated cardiac amyloidosis." *Journal of Proteome Research* **16**(4): 1659–1668.

Chen, C., W. Hu, R. Zhang, A. Jiang and Y. Zou (2016). "Levels of phenolic compounds, antioxidant capacity, and microbial counts of fresh-cut onions after treatment with a combination of nisin and citric acid." *Horticulture, Environment, and Biotechnology* **57**(3): 266–273.

Choi, S. M., D.-J. Lee, J.-Y. Kim and S.-T. Lim (2017). "Volatile composition and sensory characteristics of onion powders prepared by convective drying." *Food Chemistry* **231**: 386–392.

Colagiuri, R. (2010). "Diabetes: A pandemic, a development issue or both?" *Expert Review of Cardiovascular Therapy* **8**(3): 305–309.

Çubukçu, H. C., N. S. D. Kılıçaslan and İ. Durak (2019). "Different effects of heating and freezing treatments on the antioxidant properties of broccoli, cauliflower, garlic and onion. An experimental in vitro study." *Sao Paulo Medical Journal* **137**: 407–413.

Dahlawi, S. M., W. Nazir, R. Iqbal, W. Asghar and N. Khalid (2020). "Formulation and characterization of oil-in-water nanoemulsions stabilized by crude saponins isolated from onion skin waste." *RSC Advances* **10**(65): 39700–39707.

Davis, S. R. (2005). "An overview of the antifungal properties of allicin and its breakdown products – the possibility of a safe and effective antifungal prophylactic." *Mycoses* **48**(2): 95–100.

DiCenzo, R., V. Frerichs, P. Larppanichpoonphol, L. Predko, A. Chen, R. Reichman and M. Morris (2006). "Effect of quercetin on the plasma and intracellular concentrations of saquinavir in healthy adults." *Pharmacotherapy: The Journal of Human Pharmacology and Drug Therapy* **26**(9): 1255–1261.

El-Demerdash, F., M. I. Yousef and N. Abou El-Naga (2005). "Biochemical study on the hypoglycemic effects of onion and garlic in alloxan-induced diabetic rats." *Food and Chemical Toxicology* **43**(1): 57–63.

Fallah, V., J. A. Mahabadi, M. Y. Mahabadi, H. H. Kashani and H. Nikzad (2017). "Protective effect of allium cepa (Onion) seeds (AC) extract on histopathology of testis in STZ-induced male rats." *International Journal of Morphology* **35**(4).

Farha, A. K., R.-Y. Gan, H.-B. Li, D.-T. Wu, A. G. Atanasov, K. Gul, J.-R. Zhang, Q.-Q. Yang and H. Corke (2022). "The anticancer potential of the dietary polyphenol rutin: Current status, challenges, and perspectives." *Critical Reviews in Food Science and Nutrition* **62**(3): 832–859.

Fawole, F. J., A. A. Adeoye, L. O. Tiamiyu, F. C. Samuel, O. M. Omosuyi and M. T. Amusa (2020). "Dietary combination of pawpaw seed and onion peel powder: Impact on growth, haematology, biochemical and antioxidant status of Clarias gariepinus." *Aquaculture Research* **51**(7): 2903–2912.

Fernández-Jalao, I., C. Sánchez-Moreno and B. De Ancos (2017). "Influence of food matrix and high-pressure processing on onion flavonols and antioxidant activity during gastrointestinal digestion." *Journal of Food Engineering* **213**: 60–68.

Friedman, T., A. Shalom and M. Westreich (2006). "Self-inflicted garlic burns: Our experience and literature review." *International Journal of Dermatology* **45**(10): 1161–1163.

Gabriele, M., S. Frassinetti, L. Caltavuturo, L. Montero, G. Dinelli, V. Longo, D. Di Gioia and L. Pucci (2017). "Citrus bergamia powder: Antioxidant, antimicrobial and anti-inflammatory properties." *Journal of Functional Foods* **31**: 255–265.

Gan, R.-Y., W.-Y. Lui, K. Wu, C.-L. Chan, S.-H. Dai, Z.-Q. Sui and H. Corke (2017). "Bioactive compounds and bioactivities of germinated edible seeds and sprouts: An updated review." *Trends in Food Science & Technology* **59**: 1–14.

Gandhi, G. R., A. B. S. Vasconcelos, D.-T. Wu, H.-B. Li, P. J. Antony, H. Li, F. Geng, R. Q. Gurgel, N. Narain and R.-Y. Gan (2020). "Citrus flavonoids as promising phytochemicals targeting diabetes and related complications: A systematic review of in vitro and in vivo studies." *Nutrients* **12**(10): 2907.

Geberemeskel, G. A., Y. G. Debebe and N. A. Nguse (2019). "Antidiabetic effect of fenugreek seed powder solution (*Trigonella foenum-graecum* L.) on hyperlipidemia in diabetic patients." *Journal of Diabetes Research*: 8507453.

Gois Ruivo da Silva, M., M. Skrt, D. Komes, N. Poklar Ulrih and L. Pogačnik (2020). "Enhanced yield of bioactivities from onion (Allium cepa l.) skin and their antioxidant and anti-α-amylase activities." *International Journal of Molecular Sciences* **21**(8): 2909.

Haftbaradaran, S., A. Khoshgoftarmanesh and M. Malakouti (2018). "Assessment, mapping, and management of health risk from nitrate accumulation in onion for Iranian population." *Ecotoxicology and Environmental Safety* **161**: 777–784.

Ibrahim, M., T. Farooq, N. Hussain, A. Hussain, T. Gulzar, I. Hussain, M. S. H. Akash and F. S. Rehmani (2013). "Acetyl and butyryl cholinesterase inhibitory sesquiterpene lactones from Amberboa ramosa." *Chemistry Central Journal* **7**(1): 1–5.

Jafarpour-Sadegh, F., V. Montazeri, A. Adili, A. Esfehani, M.-R. Rashidi and S. Pirouzpanah (2017). "Consumption of fresh yellow onion ameliorates hyperglycemia and insulin resistance in breast cancer patients during doxorubicin-based chemotherapy: A randomized controlled clinical trial." *Integrative Cancer Therapies* **16**(3): 276–289.

Jakaria, M., S. Azam, D.-Y. Cho, M. Haque, I.-S. Kim and D.-K. Choi (2019). "The methanol extract of Allium cepa L. protects inflammatory markers in LPS-induced BV-2 microglial cells and upregulates the anti-apoptotic gene and antioxidant enzymes in N27-A cells." *Antioxidants* **8**(9): 348.

Ji, X., Y. Lu, H. Tian, X. Meng, M. Wei and W. C. Cho (2019). "Chemoresistance mechanisms of breast cancer and their countermeasures." *Biomedicine & Pharmacotherapy* **114**: 108800.

Jini, D. and S. Sharmila (2020). "Green synthesis of silver nanoparticles from Allium cepa and its in vitro antidiabetic activity." *Materials Today: Proceedings* **22**: 432–438.

Kabrah, A. (2015). *The Antibacterial Activity of Onion on MSSA and MRSA Isolates of Staphylococcus Aureus*, Quinnipiac University.

Kang, Y.-R., H.-Y. Choi, J.-Y. Lee, S. I. Jang, H. Kang, J.-B. Oh, H.-D. Jang and Y.-I. Kwon (2018). "Calorie restriction effect of heat-processed onion extract (ONI) using in vitro and in vivo animal models." *International Journal of Molecular Sciences* **19**(3): 874.

Khan, D., M. Mohammed, A. Upaganlawar, C. D. Upasani and H. D. Une (2020). "Ameliorative potential of Allium cepa Lam. leaves on diabetes induced and chronic constriction injury induced neuropathic pain in experimental rats." *Indian Journal of Pharmaceutical Education and Research* **54**: 143–149.

Khan, M. K. I., M. Ansar, A. Nazir and A. A. Maan (2016). "Sustainable dehydration of onion slices through novel microwave hydro-diffusion gravity technique." *Innovative Food Science & Emerging Technologies* **33**: 327–332.

Khan, V., A. K. Najmi, M. Akhtar, M. Aqil, M. Mujeeb and K. Pillai (2012). "A pharmacological appraisal of medicinal plants with antidiabetic potential." *Journal of Pharmacy & Bioallied Sciences* **4**(1): 27.

Kim, S., S. Lee, D. Shin and M. Yoo (2016). "Change in organosulfur compounds in onion (Allium cepa L.) during heat treatment." *Food Science and Biotechnology* **25**(1): 115–119.

Kozlowska, M., A. Zbikowska, K. Marciniak-Lukasiak and M. Kowalska (2019). "Herbal extracts incorporated into shortbread cookies: Impact on color and fat quality of the cookies." *Biomolecules* **9**(12): 858.

Kumari, K. and K. Augusti (2002). "Antidiabetic and antioxidant effects of S-methyl cysteine sulfoxide isolated from onions (Allium cepa Linn) as compared to standard drugs in alloxan diabetic rats." *Indian Journal of Experimental Biology* **40**(9): 1005–1009.

Lanzotti, V., A. Romano, S. Lanzuise, G. Bonanomi and F. Scala (2012). "Antifungal saponins from bulbs of white onion, Allium cepa L." *Phytochemistry* **74**: 133–139.

Law, Y.-Y., H.-F. Chiu, H.-H. Lee, Y.-C. Shen, K. Venkatakrishnan and C.-K. Wang (2016). "Consumption of onion juice modulates oxidative stress and attenuates the risk of bone disorders in middle-aged and post-menopausal healthy subjects." *Food & Function* **7**(2): 902–912.

Lee, J.-S., Y.-J. Cha, K.-H. Lee and J.-E. Yim (2016). "Onion peel extract reduces the percentage of body fat in overweight and obese subjects: A 12-week, randomized, double-blind, placebo-controlled study." *Nutrition Research and Practice* **10**(2): 175–181.

Lee, S. G., J. S. Parks and H. W. Kang (2017). "Quercetin, a functional compound of onion peel, remodels white adipocytes to brown-like adipocytes." *The Journal of Nutritional Biochemistry* **42**: 62–71.

Lin, F.-J., X.-L. Wei, H.-Y. Liu, H. Li, Y. Xia, D.-T. Wu, P.-Z. Zhang, G. R. Gandhi, H.-B. Li and R.-Y. Gan (2021). "State-of-the-art review of dark tea: From chemistry to health benefits." *Trends in Food Science & Technology* **109**: 126–138.

Loredana, L., A. Giuseppina, N. Filomena, F. Florinda, D. M. Marisa and A. Donatella (2019). "Biochemical, antioxidant properties and antimicrobial activity of different onion varieties in the Mediterranean area." *Journal of Food Measurement and Characterization* **13**(2): 1232–1241.

Ma, Y.-L., D.-Y. Zhu, K. Thakur, C.-H. Wang, H. Wang, Y.-F. Ren, J.-G. Zhang and Z.-J. Wei (2018). "Antioxidant and antibacterial evaluation of polysaccharides sequentially extracted from onion (Allium cepa L.)." *International Journal of Biological Macromolecules* **111**: 92–101.

Majid, I. and V. Nanda (2018). "Total phenolic content, antioxidant activity, and anthocyanin profile of sprouted onion powder." *Acta Alimentaria* **47**(1): 52–60.

Malhat, F. and O. Abdallah (2019). "Residue distribution and risk assessment of two macrocyclic lactone insecticides in green onion using micro-liquid-liquid extraction (MLLE) technique coupled with liquid chromatography tandem mass spectrometry." *Environmental Monitoring and Assessment* **191**(9): 1–10.

Mao, Q.-Q., X.-Y. Xu, S.-Y. Cao, R.-Y. Gan, H. Corke, T. Beta and H.-B. Li (2019). "Bioactive compounds and bioactivities of ginger (Zingiber officinale Roscoe)." *Foods* **8**(6): 185.

Marrelli, M., V. Amodeo, G. Statti and F. Conforti (2018). "Biological properties and bioactive components of Allium cepa L.: Focus on potential benefits in the treatment of obesity and related comorbidities." *Molecules* **24**(1): 119.

Masood, S., S. Bashir, M. El Shazly, M. Imran, P. Khalil, F. Ifthikar, H. M. Jaffar and T. Khursheed (2021). "Investigation of the anti-hyperglycemic and antioxidant effects of wheat bread supplemented with onion peel extract and onion powder in diabetic rats." *Journal of Diabetes & Metabolic Disorders* **20**(1): 485–495.

Mayer, S., M. Twarużek, A. Błajet-Kosicka and J. Grajewski (2016). "Occupational exposure to mould and microbial metabolites during onion sorting – insights into an overlooked workplace." *Environmental Monitoring and Assessment* **188**(3): 1–10.

Meng, X., J. Zhou, C.-N. Zhao, R.-Y. Gan and H.-B. Li (2020). "Health benefits and molecular mechanisms of resveratrol: A narrative review." *Foods* **9**(3): 340.

Mitrova, K., V. Hrbek, P. Svoboda, J. Hajšlová and J. Ovesna (2016). "Antioxidant activity, S-alk (en) yl-l-cysteine sulfoxide and polyphenol content in onion (Allium cepa L.) cultivars are associated with their genetic background." *Czech Journal of Food Sciences* **34**(2): 127–132.

Moreno-Rojas, J. M., A. Moreno-Ortega, J. L. Ordóñez, R. Moreno-Rojas, J. Pérez-Aparicio and G. Pereira-Caro (2018). "Development and validation of UHPLC-HRMS methodology for the determination of flavonoids, amino acids and organosulfur compounds in black onion, a novel derived product from fresh shallot onions (Allium cepa var. aggregatum)." *LWT* **97**: 376–383.

Muscolo, A., T. Papalia, G. Settineri, C. Mallamaci and M. R. Panuccio (2020). "Sulfur bentonite-organic-based fertilizers as tool for improving bio-compounds with antioxidant activities in red onion." *Journal of the Science of Food and Agriculture* **100**(2): 785–793.

Nile, A., S. H. Nile, D. H. Kim, Y. S. Keum, P. G. Seok and K. Sharma (2018). "Valorization of onion solid waste and their flavonols for assessment of cytotoxicity, enzyme inhibitory and antioxidant activities." *Food and Chemical Toxicology* **119**: 281–289.

Nwonuma, C., O. Osemwegie, O. Alejolowo, E. Irokanulo, A. Olaniran, D. Fadugba, D. Opaleke and O. Ojo (2021). "Antioxidant and the ameliorating effect of *Allium cepa* (Onion) fortified feed against potassium bromate induced oxidative damage in Wistar rats." *Toxicology Reports* **8**: 759–766.

Ogunmodede, O., L. Saalu, B. Ogunlade, G. Akunna and A. Oyewopo (2012). "An evaluation of the hypoglycemic, antioxidant and hepatoprotective potentials of onion (*Allium cepa* L.) on alloxan-induced diabetic rabbits." *International Journal of Pharmacology* **8**(1): 21–29.

Omar, A. E., H. S. Al-Khalaifah, W. A. Mohamed, H. S. Gharib, A. Osman, N. A. Al-Gabri and S. A. Amer (2020). "Effects of phenolic-rich onion (*Allium cepa* L.) extract on the growth performance, behavior, intestinal histology, amino acid digestibility, antioxidant activity, and the immune status of broiler chickens." *Frontiers in Veterinary Science* **7**: 582612.

Ouyang, H., K. Hou, W. Peng, Z. Liu and H. Deng (2018). "Antioxidant and xanthine oxidase inhibitory activities of total polyphenols from onion." *Saudi Journal of Biological Sciences* **25**(7): 1509–1513.

Pareek, S., N. Sagar, S. Sharma and V. Kumar (2017). "Onion (*Allium cepa* L.)." In *Fruit and Vegetable Phytochemicals: Chemistry and Human Health*, Wiley Blackwell, pp. 1145–1162.

Park, M. J., D. H. Ryu, J. Y. Cho, I. J. Ha, J. S. Moon and Y.-H. Kang (2018). "Comparison of the antioxidant properties and flavonols in various parts of Korean red onions by multivariate data analysis." *Horticulture, Environment, and Biotechnology* **59**(6): 919–927.

Pradeep, S. R., S. Barman and K. Srinivasan (2019). "Attenuation of diabetic nephropathy by dietary fenugreek (*Trigonella foenum-graecum*) seeds and onion (*Allium cepa*) via suppression of glucose transporters and renin-angiotensin system." *Nutrition* **67**: 110543.

Pradeep, S. R. and K. Srinivasan (2017). "Amelioration of hyperglycemia and associated metabolic abnormalities by a combination of fenugreek (*Trigonella foenum-graecum*) seeds and onion (*Allium cepa*) in experimental diabetes." *Journal of Basic and Clinical Physiology and Pharmacology* **28**(5): 493–505.

Pradeep, S. R. and K. Srinivasan (2018a). "Alleviation of cardiac damage by dietary fenugreek (*Trigonella foenum-graecum*) seeds is potentiated by Onion (*Allium cepa*) in experimental diabetic rats via blocking renin–angiotensin system." *Cardiovascular Toxicology* **18**(3): 221–231.

Pradeep, S. R. and K. Srinivasan (2018b). "Alleviation of oxidative stress-mediated nephropathy by dietary fenugreek (*Trigonella foenum-graecum*) seeds and onion (*Allium cepa*) in streptozotocin-induced diabetic rats." *Food & Function* **9**(1): 134–148.

Pradeep, S. R. and K. Srinivasan (2018c). "Ameliorative influence of dietary fenugreek (*Trigonella foenum-graecum*) seeds and onion (*Allium cepa*) on eye lens abnormalities via modulation of crystallin proteins and polyol pathway in experimental diabetes." *Current Eye Research* **43**(9): 1108–1118.

Przeor, M., E. Flaczyk, D. Kmiecik, J. Kobus-Cisowska, M. Bueschke and B. Kulczyński (2018). "Polish consumers' awareness and knowledge about functional food." *Folia Pomeranae Universitatis Technologiae Stetinensis, Agricultura, Alimentaria, Piscaria et Zootechnica* **341**(46/2): 59–68.

Ren, F., K. Reilly, M. Gaffney, J. P. Kerry, M. Hossain and D. K. Rai (2017). "Evaluation of polyphenolic content and antioxidant activity in two onion varieties grown under organic and conventional production systems." *Journal of the Science of Food and Agriculture* **97**(9): 2982–2990.

Roglic, G. and N. Unwin (2010). "Mortality attributable to diabetes: Estimates for the year 2010." *Diabetes Research and Clinical Practice* **87**(1): 15–19.

Romila, Y., P. Mazumder and M. D. Choudhury (2010). "A review on antidiabetic plants used by the people of manipur charactirized by hypoglycemic activity." *Assam University Journal of Science and Technology* **6**(1): 167–175.

Sagar, N. A., S. Pareek and G. A. Gonzalez-Aguilar (2020). "Quantification of flavonoids, total phenols and antioxidant properties of onion skin: A comparative study of fifteen Indian cultivars." *Journal of Food Science and Technology* **57**(7): 2423–2432.

Sajid Hamid Akash, M., K. Rehman and S. Chen (2013). "Goto-Kakizaki rats: Its suitability as non-obese diabetic animal model for spontaneous type 2 diabetes mellitus." *Current Diabetes Reviews* **9**(5): 387–396.

Salamatullah, A. M., N. Uslu, M. M. Özcan, M. S. Alkaltham and K. Hayat (2021). "The effect of oven drying on bioactive compounds, antioxidant activity, and phenolic compounds of white and red-skinned onion slices." *Journal of Food Processing and Preservation* **45**(2): e15173.

Salami, M., O. Tamtaji, M. Mohammadifar, S. Talaei, A. Tameh, A. Abed, S. Shirkhoda, E. Dadgostar and M. Taghizadeh (2020). "Neuroprotective effects of onion (Allium cepa) ethanolic extract on animal model of parkinson's disease induced by 6-hydroxydopamine: A behavioral, biochemical, and histological study." *Gazi Medical Journal* **31**: 25–29.

Sarkar, C., M. Mondal, M. Torequl Islam, M. Martorell, A. O. Docea, A. Maroyi, J. Sharifi-Rad and D. Calina (2020). "Potential therapeutic options for COVID-19: Current status, challenges, and future perspectives." *Frontiers in Pharmacology* **11**: 572870.

Shang, A., S.-Y. Cao, X.-Y. Xu, R.-Y. Gan, G.-Y. Tang, H. Corke, V. Mavumengwana and H.-B. Li (2019). "Bioactive compounds and biological functions of garlic (Allium sativum L.)." *Foods* **8**(7): 246.

Shang, A., H.-Y. Liu, M. Luo, Y. Xia, X. Yang, H.-Y. Li, D.-T. Wu, Q. Sun, F. Geng and H.-B. Li (2020). "Sweet tea (Lithocarpus polystachyus rehd.) as a new natural source of bioactive dihydrochalcones with multiple health benefits." *Critical Reviews in Food Science and Nutrition*: 1–18.

Sharifi-Rad, M., N. V. Anil Kumar, P. Zucca, E. M. Varoni, L. Dini, E. Panzarini, J. Rajkovic, P. V. Tsouh Fokou, E. Azzini and I. Peluso (2020). "Lifestyle, oxidative stress, and antioxidants: Back and forth in the pathophysiology of chronic diseases." *Frontiers in Physiology* **11**: 694.

Shaw, J. E., R. A. Sicree and P. Z. Zimmet (2010). "Global estimates of the prevalence of diabetes for 2010 and 2030." *Diabetes Research and Clinical Practice* **87**(1): 4–14.

Shokoohi, M., E. O. S. Madarek, A. Khaki, H. Shoorei, A. A. Khaki, M. Soltani and N. Ainehchi (2018). "Investigating the effects of onion juice on male fertility factors and pregnancy rate after testicular torsion/detorsion by intrauterine insemination method." *International Journal of Women's Health and Reproduction Sciences* **6**(4): 499–505.

Sidhu, J. S., M. Ali, A. Al-Rashdan and N. Ahmed (2019). "Onion (Allium cepa L.) is potentially a good source of important antioxidants." *Journal of Food Science and Technology* **56**(4): 1811–1819.

Sowa, P., D. Marcinčáková, M. Miłek, E. Sidor, J. Legáth and M. Dżugan (2020). "Analysis of cytotoxicity of selected Asteraceae plant extracts in real time, their antioxidant properties and polyphenolic profile." *Molecules* **25**(23): 5517.

Szczepaniak, O., M. Ligaj, J. Kobus-Cisowska, M. Tichoniuk, M. Dziedziński, M. Przeor and P. Szulc (2020). "The genoprotective role of naringin." *Biomolecules* **10**(5): 700.

Tajner-Czopek, A., M. Gertchen, E. Rytel, A. Kita, A. Z. Kucharska and A. Sokół-Łętowska (2020). "Study of antioxidant activity of some medicinal plants having high content of caffeic acid derivatives." *Antioxidants* **9**(5): 412.

Tamtaji, O. R., H. Hosseinzadeh, S. A. Talaei, M. Behnam, S. M. T. Firoozeh, M. Taghizadeh and R. Alipoor (2017). "Protective effects of red onion (Allium cepa) ethanolic extract on learning and memory impairments in animal models of diabetes." *Galen Medical Journal* **6**(3): 249–257.

Tattelman, E. (2005). "Health effects of garlic." *American Family Physician* **72**(1): 103–106.

Teshika, J. D., A. M. Zakariyyah, T. Zaynab, G. Zengin, K. R. Rengasamy, S. K. Pandian and M. M. Fawzi (2019). "Traditional and modern uses of onion bulb (Allium cepa L.): A systematic review." *Critical Reviews in Food Science and Nutrition* **59**(Supl): S39–S70.

Tsuboki, J., Y. Fujiwara, H. Horlad, D. Shiraishi, T. Nohara, S. Tayama, T. Motohara, Y. Saito, T. Ikeda and K. Takaishi (2016). "Onionin A inhibits ovarian cancer progression by suppressing cancer cell proliferation and the protumour function of macrophages." *Scientific Reports* **6**(1): 1–11.

Ulusoy, H. G. and N. Sanlier (2020). "A minireview of quercetin: From its metabolism to possible mechanisms of its biological activities." *Critical Reviews in Food Science and Nutrition* **60**(19): 3290–3303.

Upadhyay, R. K. (2016). "Nutraceutical, pharmaceutical and therapeutic uses of Allium cepa: A review." *International Journal of Green Pharmacy (IJGP)* **10**(1).

Viera, V., N. Piovesan, J. Rodrigues, R. de O Mello, R. Prestes, R. Dos Santos, R. de A Vaucher, T. Hautrive and E. Kubota (2017). "Extraction of phenolic compounds and evaluation of the antioxidant and antimicrobial capacity of red onion skin (Allium cepa L.)." *International Food Research Journal* **24**(3).

Watanabe, J., T. Muro, M. Kobori and Y. Takano-Ishikawa (2018). "Changes in quercetin content and antioxidant capacity of quercetin-rich onion cultivar'Quer-Gold'during cooking." *Nippon Shokuhin Kagaku Kogaku Kaishi= Journal of the Japanese Society for Food Science and Technology* **65**(2): 63–66.

Wild, S., G. Roglic, A. Green, R. Sicree and H. King (2004). "Global prevalence of diabetes: Estimates for the year 2000 and projections for 2030." *Diabetes Care* **27**(5): 1047–1053.

Wu, Z.-H., Y. Tang and Q. Cheng (2021). "Diabetes increases the mortality of patients with COVID-19: A meta-analysis." *Acta Diabetologica* **58**(2): 139–144.

Xu, X.-Y., X. Meng, S. Li, R.-Y. Gan, Y. Li and H.-B. Li (2018). "Bioactivity, health benefits, and related molecular mechanisms of curcumin: Current progress, challenges, and perspectives." *Nutrients* **10**(10): 1553.

Yang, C., P. Chao, Y. Hou, S. Tsai, K. Wen and S. Hsiu (2006). "Marked decrease of cyclosporin bioavailability caused by coadministration of ginkgo and onion in rats." *Food and Chemical Toxicology* **44**(9): 1572–1578.

Zamri, N. and H. A. Hamid (2019). "Comparative study of onion (Allium cepa) and leek (Allium ampeloprasum): Identification of organosulphur compounds by UPLC-QTOF/MS and anticancer effect on MCF-7 cells." *Plant Foods for Human Nutrition* **74**(4): 525–530.

Zhang, D., R. Y. Gan, J. R. Zhang, A. K. Farha, H. B. Li, F. Zhu, X. H. Wang and H. Corke (2020). "Antivirulence properties and related mechanisms of spice essential oils: A comprehensive review." *Comprehensive Reviews in Food Science and Food Safety* **19**(3): 1018–1055.

Zhang, S., P. Deng and X. Chao (2016). "Quantification and analysis of anthocyanin and flavonoids compositions, and antioxidant activities in onions with three different colors." *Journal of Integrative Agriculture* **15**: 2175–2181.

Zhao, X.-X., F.-J. Lin, H. Li, H.-B. Li, D.-T. Wu, F. Geng, W. Ma, Y. Wang, B.-H. Miao and R.-Y. Gan (2021). "Recent advances in bioactive compounds, health functions, and safety concerns of onion (Allium cepa L.)." *Frontiers in Nutrition* **8**.

Zudaire, L., I. Viñas, M. Abadias, J. Simó, G. Echeverria, L. Plaza and I. Aguiló-Aguayo (2017). "Quality and bioaccessibility of total phenols and antioxidant activity of calçots (Allium cepa L.) stored under controlled atmosphere conditions." *Postharvest Biology and Technology* **129**: 118–128.

4 Major Bioactive Compounds and Antidiabetic Activity of *Allium sativum*

Saad Bakrim, Naoual El Menyiy, Rania Moubachir, Douae Taha, Abdelhakim Bouyahya, and Mohammad S. Mubarak

CONTENTS

ABBREVIATIONS

A. sativum: *Allium sativum*
ALP: Alkaline phosphatase
ALT: Alanine aminotransferase
AST: Aspartate aminotransferase
CAT: Catalase
DM: Diabetes mellitus
GPx: Glutathione peroxidase
GSH: Glutathione
HDL-C: High-density lipoprotein cholesterol
HMG-CoA: Hydroxyl methylglutaryl CoA
HOMA-IR: Homeostasis model assessment for insulin resistance
iNOS: Inducible nitric oxide synthase
LDL-C: Low-density lipoproteins cholesterol
MCSO: S-methyl cysteine-sulfoxide
MDA: Malondialdehyde
NO: Nitric oxide
P: Phosphorus
PCSO: S-propyl-cysteine-sulfoxide
RS: Riviera spezzina
SAMC: S-ally-mercapto cysteine
SIRTs: Silent information regulators
SOD: Superoxide dismutase
TBARS: Thiobarbituric acid-reactive substances
TNF-α: Tumor necrosis factor-alpha

DOI: 10.1201/b23347-4

4.1 INTRODUCTION

For years, medicinal plants have been used for human health care in the form of dietary components, spices, and flavors (Husen, 2021, 2022; Husen et al., 2021). In this respect, *Allium sativum* (*A. sativum*) is one of the most used herbs in modern folk medicine (El-Saber Batiha et al., 2020). *A. sativum*, commonly known as garlic, is a medicinal herb belonging to the Liliaceae family and is a member of the onion family. This plant has a height of 70 to 80 cm, 13 to 15 leaves, a bulb weight of 50 g, a diameter of 5 cm, 12 to 13 teeth, a growth period of 250 days when implanted in the winter, and a latency period of 40 days (Wu et al., 2015). Garlic consists of two basic types – hardneck and softneck. Hardneck garlic, which includes *A. sativum* var. *ophioscorodon* and *A. sativum* variety *pekinense*, is distinguished by central hard, woody stems that extend to the basal plate at the bottom of the bulb (Badole et al., 2013). Although its precise origin is unknown, it is a strongly aromatic bulb crop thought to be native to northeastern Iran (Kazakhstan and Uzbekistan) and central Asia (Western China). *A. sativum* was domesticated many years ago and is recorded in ancient Indian, Chinese, Egyptian, and Greek literature. *A. sativum* is grown worldwide in tropical and temperate zones, and many different garlic cultivars have been engineered to accommodate diverse environments (Kuete, 2017b). It is propagated by separating and planting individual bulbs and grows at a pH of 6 to 7.5 (Morales-González et al., 2019). The soil condition is of great importance because it is a universal medium for plant growth, which supplies essential nutrients to the plants (Addis & Abebaw, 2017).

A. sativum is rich in several sulfur-containing phytoconstituents such as ajoenes (E-ajoene, Z-ajoene), thiosulfinates (allicin), alliin (S-allyl-L-cysteine sulfoxide), vinyldithiins (2-vinyl-(4*H*)-1,3-dithiin, 3-vinyl-(4*H*)-1,2-dithiin), sulfides (diallyl disulfide), and flavonoids such as quercetin. Allicin is garlic's predominant biologically active sulfur compound and is the source of its odor and flavor. Alliin is allicin's principal precursor, accounting for about 70% of the total thiosulfinates in crushed cloves. On the other hand, allyl mercaptan is the odorant compound that causes garlic breath and is formed by the interplay of allicin or diallyl disulfide with cysteine in the presence of S-ally-mercapto-cysteine (El-Saber Batiha et al., 2020). Extracts and isolated bioactive molecules of *A. sativum* produced by different methods have been found to possess a variety of biological and pharmacological properties, including antidiabetic, anti-inflammatory, antioxidant, antibacterial, antifungal, antiviral, antitumor, antithrombotic, anticarcinogenic, antiparasitic, and antiatherosclerotic effects, among others (Bar et al., 2022).

Garlic extracts exhibit broad-spectrum fungicidal activity against numerous fungi, including Trichophyton, Aspergillus, Candida, Cryptococcus, and Torulopsis (Pârvu et al., 2019). *A. sativum* extracts exhibited anthelmintic activity against *Hymenolepis microstoma*, *Fasciola hepatica*, *Taenia taeniaeformis*, *H. diminuta*, and *Echinostoma caproni* (Abdel-Ghaffar et al., 2011). The extract of raw garlic was found to be the most potent and specific anticancer agent among 33 raw plant extracts against several different cancer cells with no effect on non-cancerous cells (Li et al., 2018). Diabetes mellitus is a critical condition that continues to affect more individuals daily. Medications like insulin and hypoglycemic medicines are administered to manage diabetes. However, consistent physical activity and an appropriate diet are also recommended. In addition, coadjuvant therapies such as herbal remedies are routinely used to help reduce the high blood sugar levels caused by diabetes (Tadeu Volpato et al., 2017). Within this context, Ahmed et al. (2020) investigated the inhibitory effect of *A. sativum* oil against α-amylase. These researchers showed that *A. sativum* oil exhibits potent enzymatic inhibitory activity with an IC_{50} value of 3.0 ± 0.02%. Similarly, *A. sativum* was found in many reports to possess a potent antidiabetic effect *in vivo* using a diabetic animal model. In streptozotocin-induced diabetic rats, the oral administration of an aqueous extract of *A. sativum* for 14 days significantly decreased blood glucose levels (Mostofa et al., 2007; Shakya et al., 2010).

In traditional medicine, garlic is used internally to treat gastrointestinal problems, flatulence, hypertension and diabetes, menstrual cramps, leprosy, bronchitis, and respiratory problems and externally for arthritis, muscle pain, corns, sciatica, neuralgia, and warts. For example, in

Morocco, the bulb of *A. sativum* has been used against hypertension, diabetes, and cardiac diseases (Bnouham et al., 2002; Idm'hand et al., 2019). In Algeria, it has been reported that this species is used to treat diabetes mellitus (Rachid et al., 2012). In India, the bulb is used as a vermifuge, diuretic, antidiabetic expectorant, and stimulant. In the same country, the decoction of garlic with milk is used for heart disease, and the oil is used for skin rashes, as ear drops, and for atonic dyspepsia, flatulence, and colic (Goyal, 2015; Wilson et al., 2019). In Pakistan, garlic extract is traditionally taken orally to settle the stomach, treat coughs, and reduce fever (Kuete, 2017b). In Ethiopia, *A. sativum* L. was the first-ranked medicinal plant traditionally used to treat abdominal pain and prevent and treat malaria (Dagne & Belachew, 2019). Based on the previous discussion, the purpose of this chapter is to contribute a comprehensive review of *A. sativum*, such as botanical description, geographical distribution, traditional uses, culture practices, and phytochemistry. In addition, we focused on the antidiabetic properties of this plant as well as its safety considerations. We expect that this chapter will constitute a sufficient scientific basis for deeper investigations on isolated components of *A. sativum* that might eventually identify new therapeutic opportunities.

4.2 BOTANICAL DESCRIPTION

Allium sativum L. has a 1 m tall, fragrant stem divided into 6 to 12 bulblets (garlic cloves) that are linked by a thin shell to create the garlic head (Figure 4.1a). The roots of *A. sativum* grow from the basal disc part and can extend down at least 80 cm (Figure 4.1b). Near the plant base, its leaves are long, narrow, and flat; nevertheless, the tips of its leaves are cylindrical and pointed. It flowers in little, whitish-purple clusters (Figure 4.1c). Humus, heavy clay soil, and much water are necessary for growing garlic. Garlic plants have a height of 70 to 80 cm, 13 to 15 leaves, a bulb weight of 50 g, a diameter of 5 cm, 12 to 13 teeth, a growth period of 250 days when implanted in the winter, and a latency period of 40 days (Wu et al., 2015).

FIGURE 4.1 (a) *Allium sativum* tree general aspect, (b) roots, (c) flowers.

4.3 DISTRIBUTION

A. sativum, also known as garlic, is a perennial bulbous plant that may reach a height of 1.2 meters and has a strong onion-like flavor and perfume. It is also known as rocambole, allium, stinking rose, rustic treacle, the nectar of the gods, camphor of the poor, poor man treacle, and clove garlic, depending on the region where it is grown and utilized. The most popular bulb consumed after the onion is *A. sativum*, which is simple to grow and can be grown in temperate and tropical climates worldwide (Figure 4.2). Garlic comes in several varieties or subspecies, the most recognized being the hardneck and softneck varieties. Although its precise origin is unknown, it is thought to be a native of northeastern Iran and Central Asia. *A. sativum* was domesticated long ago; it is recorded in the literature from the ancient worlds of Egypt, Greece, India, and China. Currently, about 10 million tons of garlic are produced annually, with the leading producers being in China, Korea, India, the United States, Spain, Egypt, and Turkey (Acosta et al., 2008). It is also considered a significant product in Mexico, with the preponderance of its production concentrated in the states of Zacatecas, Guanajuato, and Baja California (Acosta et al., 2008).

4.4 PHYTOCHEMICAL CONSTITUENTS

There are reports that *A. sativum* contains hundreds of phytochemicals, including sulfur-containing substances such as ajoenes (E-ajoene, Z-ajoene), thiosulfinates (allicin), vinyldithiins (2-vinyl-(4*H*)-1,3-dithiin, 3-vinyl-(4*H*)-1,2-dithiin), and sulfides (diallyl disulfide) (Zeng et al., 2017). The primary aroma-producing compounds in freshly ground garlic homogenates are S-propyl-cysteine-sulfoxide (PCSO), allicin, and S-methyl cysteine sulfoxide (MCSO) (Souza et al., 2011). A mixture of MCSO,

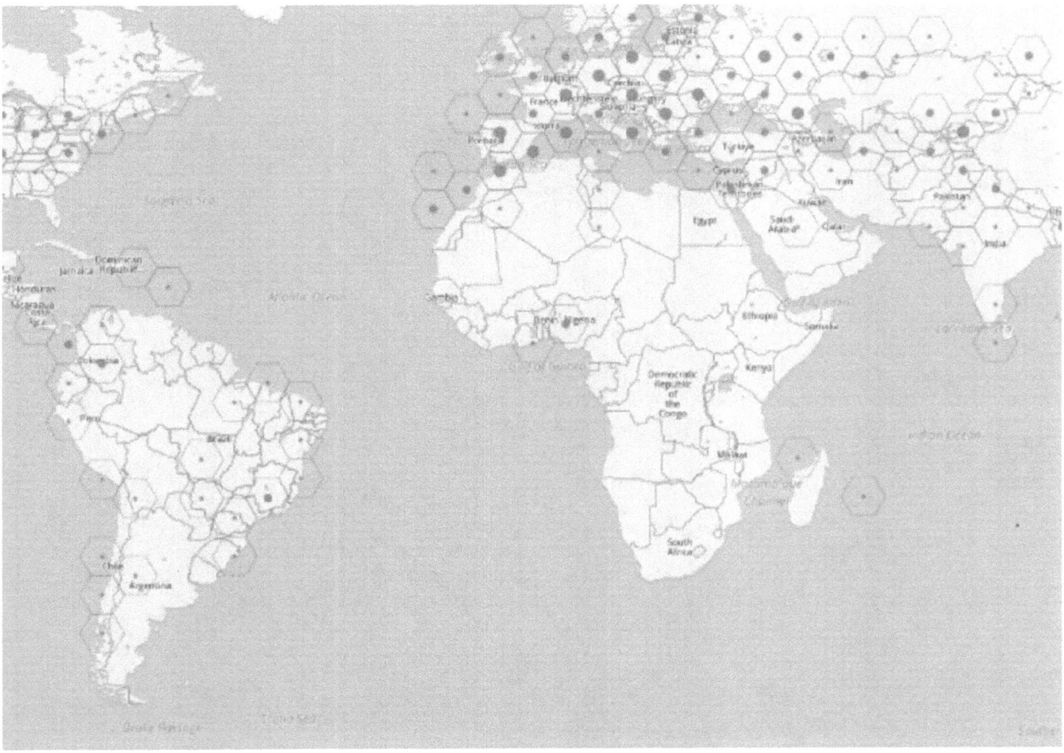

FIGURE 4.2 Geographic distribution of *Allium sativum*.

PCSO, and alliin can produce other molecules, such as allyl methane thiosulfinates, methyl methanethiosulfonate, and the corresponding thiosulfinates (R-S-S-R0), where R and R0 are allyl, propyl, and methyl groups. Depending on the water content and temperature, PCSO can produce more than 50 metabolites (Souza et al., 2011). The secondary metabolites derived from cysteine that accumulate in the Allium genus of plants are identified as S-alk(en)yl-l-cysteine sulfoxides (Souza et al., 2011) and S-ally-mercapto cysteine (SAMC), which are derived from alliin (Tran et al., 2018)

4.5 PHARMACOLOGICAL STUDIES: ANTIDIABETIC RESPONSE

Diabetes mellitus (DM) is an endocrine disorder characterized by hyperglycemia with carbohydrate, fat, and protein metabolism disturbances resulting from defects in insulin secretion or action. Poorly controlled blood glucose is believed to be the most important factor in the development of diabetic complications in both type 1 and type 2 diabetes (Londhe et al., 2011). In recent years, much research has focused on developing herbal medicines, which offer a good source for drug discovery. Based on the results of several studies, garlic can be used to help treat hyperglycemia. Recently, Ahmed et al. (2020) investigated the inhibitory effect of *A. sativum* oil against α-amylase. These researchers showed that *A. sativum* oil exhibits potent enzymatic inhibitory activity with an IC_{50} value of 3.0 ± 0.02%. Similarly, another *in vitro* study reported that an aqueous extract of bulbs of *A. sativum* exhibits an IC_{50} of 136.3 μg/mL against the α-glucosidase enzyme (Kesavanarayanan et al., 2012). Furthermore, *A. sativum* was found in many reports to also possess a potent antidiabetic effect *in vivo* using a diabetic animal model.

In streptozotocin-induced diabetic rats, the oral administration of an aqueous extract of *A. sativum* for 14 days significantly decreased blood glucose levels (Mostofa et al., 2007; Shakya et al., 2010). In addition, treatment with the *A. sativum* aqueous extract at a dose of 100, 300, and 600 mg/kg daily for eight weeks significantly increased serum insulin and lessened the increase in the level of glycated hemoglobin, cholesterol, and triglyceride. It also significantly increased the antioxidant defense systems, as indicated by increased levels of total antioxidants and antioxidant enzymes and lowered lipid peroxidation in liver and kidney tissues (Thomson et al., 2015). Using the same diabetic model, Ziamajidi et al. (2018) showed that oral administration of an aqueous extract of garlic significantly decreased glucose levels and improved histopathological changes in the liver tissues and aspartate aminotransferase (AST), alanine aminotransferase (ALT), and alkaline phosphatase (ALP) levels in the serum of garlic-treated diabetic rats, reduced oxidative stress status, and downregulated iNOS gene expression and nitric oxide (NO) level.

Similarly, Raju et al. (2008) reported that the methanolic extract of *A. sativum* attenuates the glycemia-mediated oxidative stress via restoration of electrolytes, glutathione (GSH) content, antioxidant reserves, and prevention of protein aggregation, downregulation of mRNA inducible nitric oxide synthase (iNOS) transcript, thiobarbituric acid-reactive substances (TBARS), and carbonyl content. Furthermore, Naderi et al. (2015) studied the protective effects of garlic (*A. sativum*) in the blood and heart of streptozotocin-induced diabetic rats. They showed that oral administration of raw fresh garlic homogenate at a dose of 250 mg/kg for a period of six weeks effectively modulates antioxidants status in the blood and heart of streptozotocin induced-diabetic rats by decreasing the malondialdehyde (MDA) levels and increasing the levels of superoxide dismutase (SOD), glutathione peroxidase (GPx), and catalase (CAT) in blood and heart homogenates. *A. sativum* extract also significantly affected SIRT1 and SIRT2 (silent information regulators or SIRTs) gene expressions in the livers and kidneys of diabetic rats. This effect was reported by Sadeghabadi et al. (2018), who investigated the effects of aqueous garlic extract on the expressions of SIRT1 and SIRT2 genes in the kidney and liver tissues of STZ+niacinamide-induced diabetic rats. These authors showed that after treatment with garlic extract, the gene expressions increased significantly compared with those in the diabetic rat, thus decreasing diabetic complications.

The antidiabetic activity of *A. sativum* extracts was also evaluated in alloxan-induced diabetic rats by using increasing doses of the extract. A hypoglycemic effect was observed in diabetic rats

(Belemkar et al., 2013; Ibegbulem & Chikezie, 2013; Hfaiedh et al., 2011, 2011; Johnson et al., 2012; Ozougwu et al., 2009). This effect was linked to a decrease in certain parameters such as blood sugar, total serum lipids, and cholesterol. The authors also reported that *A. sativum* extract protects from liver damage by decreasing the hepatic marker enzymes (ALT, AST, and ALP) (Johnson et al., 2012) and exhibits an antioxidative activity demonstrated by the increase of CAT, SOD, and GPx activities in the liver and pancreas and a lowering of lipid peroxidation level in these organs (Hfaiedh et al., 2011). On the other hand, the hypoglycemic effects of *A. sativum* extracts were also confirmed in the model of type 2 diabetes. Indeed, Senthil Kumar (Senthil Kumar, 2013) studied the effects of *A. sativum* aqueous extract on insulin resistance, inflammation, and oxidative stress in male Wistar rats fed with a 60% fructose-rich diet. The results showed that *A. sativum* extract decreased blood glucose, insulin, insulin resistance, and tumor necrosis factor alpha (TNF-α) and improved oxidative stress by decreasing MDA levels and increasing levels of total antioxidant status and antioxidant enzymes.

In another study, chronic consumption of garlic at the level of 5% significantly decreased fasting serum glucose and blood glycated hemoglobin levels and increased insulin level. This consumption also decreased homeostasis model assessment for insulin resistance (HOMA-IR) and total cholesterol in an animal model of type 2 diabetes (Seo et al., 2009). In diabetic patients, it was reported that consumption of garlic tablets (300 mg, twice daily) for 12 weeks significantly decreased fasting blood sugar and serum cholesterol but increased triglycerides and high-density lipoprotein cholesterol (HDL-C) (Ashraf et al., 2011). Moreover, Siddiqui et al. (2020) found that the administration of garlic at a dose of 1000 mg significantly reduced the levels of total cholesterol, triglycerides, and low-density lipoproteins cholesterol (LDL-C) and increased HDL-C. Similar anti-hyperlipidemic effects of garlic were reported previously by Hussien (2014). The improvement in blood sugar and lipid profile observed by *A. sativum* could be largely attributed to its antioxidant power and the blockage of biosynthesis of hepatic cholesterol, possibly via inhibition of hydroxyl methylglutaryl CoA (HMG-CoA) reductase by allicin, an organo-sulfur constituent of garlic (Kojuri et al., 2007).

4.6 TRADITIONAL AND OTHER POTENTIAL USES

A. sativum L. is one of the essential bulb vegetables used as a food and flavoring agent. It has become a common ingredient in the kitchen worldwide, used in preparing foods, particularly dried foods for storage and some types of soup, and it can be utilized in fresh and dehydrated states (Gudalwar et al., 2021). The garlic flower stems (scapes) are also edible and used in cooking. They are used similarly to chives but have a milder taste (Wilson et al., 2019). Traditionally, garlic has a long history of medical uses against several diseases. The therapeutic application depends on the plant parts. Along this line, the traditional Moroccan medicine uses the bulb of *A. sativum* against hypertension, diabetes, and cardiac disorders (Bnouham et al., 2002; Idm'hand et al., 2019; Jouad et al., 2001; Tahraoui et al., 2007). In Algeria, it has been reported that this species is used to treat diabetes mellitus disease (Rachid et al., 2012). On the other hand, the bulbs of this plant were used crushed in Turkey against snakebite (Ugulu, 2011) and for hypertension and rheumatism (Polat et al., 2013). The raw bulbs are used by the population of the Bartin region of Turkey like pills with one glass of milk every morning on an empty stomach for hemorrhoids (Koyuncu & BAfiER, 2005).

In India, the bulb is used as a vermifuge, diuretic, antidiabetic expectorant, and stimulant. In the same country, the decoction of garlic with milk is used for heart disease, and the oil is used for skin rashes, as ear drops, and for atonic dyspepsia, flatulence, and colic (Goyal, 2015; Wilson et al., 2019). In 2009, Cornara et al. (2009) carried out an ethnopharmacological survey in Riviera spezzina (RS), Eastern Liguria, Italy, and showed that the bulb is eaten raw or macerated in water as hypotensive, pounded in oil, and used against acne, rubbed on bee stings, wasp bites, and chilblains, and applied to wounds as hemostatic. *A. sativum* has also been used in traditional medicine in Cameroon (Emmanuel & Didier, 2012); studies reported that this plant is used against several pathologies, particularly to treat rheumatism, hypertension, nerve damage, cardiovascular diseases, asthma, cold, asthenia, mycoses, cholesterol problem, goiter, and smooth aches. In Ethiopia, *A. sativum* L. was the first-ranked medicinal plant traditionally used to treat abdominal pain and prevent and treat malaria (Dagne & Belachew, 2019). Furthermore, the fresh bulb is used as an aperitive vermifuge in Iraq, and the bulb cataplasm is used as an antirheumatic, antidandruff, and antiscabies. The people of Iraq also used bulb juice internally as a carminative, anthelmintic, anticancer, and antipyretic (Al-douri, 2000). Another ethnobotanical study indicated that fresh and steam inhalation of *A. sativum* is used to treat various diseases, including hoarseness, alopecia, hypertension, arteriosclerosis, anthelmintic, and intestinal inflammation in Jordan (Abdelhalim et al., 2017). Other reports demonstrated that bulbs of *A. sativum* are also used to treat diabetes by the people of Mauritius (Mootoosamy & Fawzi Mahomoodally, 2014), Kenya (Keter & Mutiso, 2012), Togo (Holaly et al., 2015), and Bangladesh (Kadir et al., 2012).

4.7 SAFETY ISSUES

According to a study pertaining to the acute toxicity of the aqueous extract of garlic (*A. sativum*) through the oral route at doses of 100, 1000, 2500, and 5000 mg/kg body weight, no death was recorded during the treatment period at all doses in animal models. Results showed that the animals were healthy with no sign of toxicity up to 2500 mg/kg. However, they were weak at 5000 mg/kg and had intense ethremia, tachycardia, and disorientation, but no death was recorded. Thus, the LD_{50} was more than 5000 mg/kg (Lawal et al., 2016). Another work by Mikail (2010) investigated the acute toxicity following subcutaneous administration of graded doses of the aqueous bulb extracts (300, 600, 1200, 2200, 3200, and 4200 mg/kg) in experimental rabbits. The author showed that the mortality occurred in rabbits given the extract at 3200 and 4200 mg/kg with other behavioral signs like loss of appetite and partial paralysis. The LD_{50} was found to be 3034 mg/kg, and the maximum tolerated dose was 2200 mg/kg, this high LD_{50} and absence of cyanogenic glycosides, as recorded in this study, is an indication that the extract is relatively safe. The toxic potential of aqueous extracts of *A. sativum* was also investigated by Sulaiman et al. (2014). In this study, Wistar rats were treated by oral administration of 10 mg/kg of aqueous plant extract for 30 days. The extract of *A. sativum* appears to have toxic potential and demonstrated the ability to alter biochemical indices in vital tissues.

4.8 CULTIVATION PRACTICES

A. sativum was domesticated long ago and is mentioned in ancient Egyptian, Greek, Indian, and Chinese writings. Currently, nearly 10 million tons of garlic are produced yearly, with China, Korea, India, the United States, Spain, Egypt, and Turkey as the world's largest producers. It is easy to grow and can grow in temperate and tropical regions around the globe. It is propagated by separating and planting individual bulbs and grows at a pH of 6 to 7.5 (Morales-González et al., 2019). The soil condition is of great importance because it is a universal medium for plant growth, which supplies essential nutrients to the plants (Addis & Abebaw, 2017). Garlic grows best in well-drained soil and is high in organic matter content. Sand, silt, and clay loam soils are recommended for commercial production.

Based on traditional practice, garlic growers in the central highlands of Ethiopia tend to rely on fertilizer sources that contain only nitrogen (N) and phosphorus (P), resulting in a steady decline in other nutrients in the soil. Smaller potassium and sulfur uptake relative to N uptake can predispose the crop to severe disease and insect damage (Diriba-Shiferaw et al., 2014). The majority of the highest value resulted in N, P_2O_5, and K_2O rates of 375, 90, and 30 kg/ha, respectively (Etana, 2018). To optimize and increase bioactive compound production, the plant cell and tissue culture can be an effective alternative source (Espinosa-Leal et al., 2018). In addition, according to Sachin et al. (2017), the prevalence of small-scale garlic cultivation has made it a more desirable crop production option to maximize the bulb's qualitative and quantitative yield of bulbs per unit area rather than increase the cultivated area.

4.9 CONCLUSIONS AND PERSPECTIVES

The principal purpose of this chapter was to establish a comprehensive compilation of the traditional global uses, chemical constituents of garlic, and its antidiabetic activity. As shown in this chapter, garlic is an herbaceous aromatic plant that is widely consumed throughout the world as a food and traditional remedy for different kinds of diseases. Sulfur-containing compounds, such as vinyldithiins, allicin, alliin, ajoenes, and sulfides, are the major components identified from *A. sativum* extracts. These bioactive molecules, including cysteine sulfoxide, allyl propyl disulfide, S-allyl cysteine sulfoxide, and allicin, decreased blood glucose levels by inhibiting liver-induced insulin activation, increasing secretion by pancreatic beta cells, isolating insulin from bound forms, and improving the sensitivity of cells to insulin. It would be important to continue more investigations (both preclinical and clinical) on this perennial plant to confirm its safety and to determine its antimutagenic capacity. In addition, the bioactivity of *A. sativum* should be reviewed, highlighting its mechanism of action by which it exerts its therapeutic effect, thus serving as a crucial starting point for future drug development. Therefore, studying the pharmacokinetic and pharmacodynamic impact is essential using animal models. Furthermore, addressing knowledge gaps in the methods of enhancing the stability of bioactive components in garlic extracts and identifying the optimal ways and means of delivering them to therapeutically relevant organizations can help to ensure the more efficient application of garlic extracts.

REFERENCES

Abdel-Ghaffar, F., Semmler, M., Al-Rasheid, K. A., Strassen, B., Fischer, K., Aksu, G., Klimpel, S., & Mehlhorn, H. (2011). The effects of different plant extracts on intestinal cestodes and on trematodes. *Parasitology Research*, *108*(4), 979–984.

Abdelhalim, A., Aburjai, T., Hanrahan, J., & Abdel-Halim, H. (2017). Medicinal plants used by traditional healers in Jordan, the Tafila region. *Pharmacognosy Magazine*, *13*(49), 95. https://doi.org/10.4103/0973-1296.203975

Acosta, R. G. F., Lujan, F. M., & Parra, Q. R. A. (2008). Crecimiento y rendimiento de cultivares de ajo en Delicias, Chihuahua, México. *Agricultura Técnica En México*, *34*(2), 177–188.

Addis, W., & Abebaw, A. (2017). Determination of heavy metal concentration in soils used for cultivation of *Allium sativum* L. (garlic) in East Gojjam Zone, Amhara Region, Ethiopia. *Cogent Chemistry*, *3*(1), 1419422. https://doi.org/10.1080/23312009.2017.1419422

Ahmed, M. U., Ibrahim, A., Dahiru, N. J., & Mohammed, H. S. (2020). Alpha amylase inhibitory potential and mode of inhibition of oils from *Allium sativum* (Garlic) and Allium cepa (Onion). *Clinical Medicine Insights. Endocrinology and Diabetes*, *13*, 1179551420963106. https://doi.org/10.1177/1179551420963106

Al-douri, N. A. (2000). A survey of medicinal plants and their traditional uses in Iraq. *Pharmaceutical Biology*, *38*(1), 74–79. https://doi.org/10.1076/1388-0209(200001)3811-BFT074

Arab Sadeghabadi, Z., Ziamajidi, N., Abbasalipourkabir, R., & Mohseni, R. (2018). Garlic (Allium sativum) increases SIRT1 and SIRT2 gene expressions in the kidney and liver tissues of STZ- and STZ+niacinamide-induced diabetic rats. *Journal of Basic and Clinical Physiology and Pharmacology*, *29*(5), 463–467. https://doi.org/10.1515/jbcpp-2017-0079

Ashraf, R., Khan, R. A., & Ashraf, I. (2011). Garlic (Allium sativum) supplementation with standard antidiabetic agent provides better diabetic control in type 2 diabetes patients. *Pakistan Journal of Pharmaceutical Sciences*, *24*(4). https://pubmed.ncbi.nlm.nih.gov/21959822/

Badole, S. L., Ghule, A. E., & Wagh, N. K. (2013). Chapitre 15. Antidiabetic activity of Allium sativum. In *Bioactive Food as Dietary Interventions for Diabetes* (pp. 157–161).

Bar, M., Binduga, U. E., & Szychowski, K. A. (2022). Methods of isolation of active substances from garlic (Allium sativum L.) and its impact on the composition and biological properties of garlic extracts. *Antioxidants*, *11*(7), 1345.

Belemkar, S., Dhameliya, K., & Pata, M. K. (2013). Comparative study of garlic species (Allium sativum and Allium porrum) on glucose uptake in diabetic rats. *Journal of Taibah University Medical Sciences*, *8*(2), 80–85. https://doi.org/10.1016/j.jtumed.2013.04.002

Bnouham, M., Mekhfi, H., Legssyer, A., & Ziyyat, A. (2002). Medicinal plants used in the treatment of diabetes in Morocco. *International Journal of Diabetes and Metabolism*, *10*, 33–50.

Cornara, L., La Rocca, A., Marsili, S., & Mariotti, M. G. (2009). Traditional uses of plants in the Eastern Riviera (Liguria, Italy). *Journal of Ethnopharmacology*, *125*(1), 16–30. https://doi.org/10.1016/j.jep.2009.06.021

Dagne, A., & Belachew, G. (2019). Ethnobotanical survey of plants traditionally used for malaria prevention and treatment in indigenous villages of Tepi Town South West Ethiopia. *Journal of Pharmacognosy and Phytotherapy*, *11*(1), 9–16. https://doi.org/10.5897/JPP2018.0532

Diriba-Shiferaw, G., Nigussie-Dechassa, R., Kebede, W., Sharma, J. J., & Getachew, T. (2014). Bulb quality of Garlic (Allium sativum L.) as influenced by the application of inorganic fertilizers. *African Journal of Agricultural Research*, *9*(8), 784–796. https://doi.org/10.5897/AJAR2013.7723

El-Saber Batiha, G., Magdy Beshbishy, A., Wasef, L. G., Elewa, Y. H., Al-Sagan, A. A., Abd El-Hack, M. E., Taha, A. E., Abd-Elhakim, Y. M., & Prasad Devkota, H. (2020). Chemical constituents and pharmacological activities of garlic (Allium sativum L.): A review. *Nutrients*, *12*(3), 872.

Emmanuel, M. M., & Didier, D. S. (2012). Traditional knowledge on medicinal plants use by ethnic communities in Douala, Cameroon. *European Journal of Medicinal Plants*, *2*(2), 159–176.

Espinosa-Leal, C. A., Puente-Garza, C. A., & García-Lara, S. (2018). In vitro plant tissue culture: Means for production of biological active compounds. *Planta*, *248*(1), 1–18. https://doi.org/10.1007/s00425-018-2910-1

Etana, M. B. (2018). Review on the agronomic management practices of garlic (Allium sativum L.). *Journal of Biology*, *7*.

Goyal, M. (2015). Traditional plants used for the treatment of diabetes mellitus in Sursagar constituency, Jodhpur, Rajasthan – An ethnomedicinal survey. *Journal of Ethnopharmacology*, *174*, 364–368. https://doi.org/10.1016/j.jep.2015.08.047

Gudalwar, B. R., Nimbalwar, M. G., Panchale, W. A., Wadekar, A. B., Manwar, J. V., & Bakal, R. L. (2021). Allium sativum, a potential phytopharmacological source of natural medicine for better health. *GSC Advanced Research and Reviews*, *6*(3), 220–232. https://doi.org/10.30574/gscarr.2021.6.3.0049

Hfaiedh, N., Murat, J. C., & Elfeki, A. (2011). Compared ability of garlic (*Allium sativum*) extract or α-tocopherol + magnesium association to reduce metabolic disorders and oxidative stress in diabetic rats: Stress oxidant in liver and pancreas in diabetic rats. *Phytotherapy Research*, *25*(6), 821–827. https://doi.org/10.1002/ptr.3344

Holaly, G. E., Simplice, K. D., Charlemagne, G., Kodjovi, A., Kokou, A., Tchadjobo, T., Amegnona, A., Komlan, B., & Jacques, S. (2015). Étude ethnobotanique des plantes utilisées dans le traitement du diabète dans la médecine traditionnelle de la région Maritime du Togo. *Pan African Medical Journal*, *20*. https://doi.org/10.11604/pamj.2015.20.437.5660

Husen, A. (2021). *Traditional Herbal Therapy for the Human Immune System*. CRC Press, Boca Raton, FL. https://doi.org/10.1201/9781003137955

Husen, A. (2022). *Herbs, Shrubs and Trees of Potential Medicinal Benefits*. CRC Press. https://doi.org/10.1201/9781003205067

Husen, A., Bachheti, R. K., & Bachheti, A. (2021). *Non-Timber Forest Products: Food, Healthcare and Industrial Applications*. Springer, Cham, Switzerland. https://doi.org/10.1007/978-3-030-73077-2

Hussien, Z. M. (2014). Effect of garlic on dyslipidemic patients with diabetes mellitus (type 2). *Diyala Journal of Medicine*, *6*(1), 100–104.

Ibegbulem, C. O., & Chikezie, P. C. (2013). Hypoglycemic properties of ethanolic extracts of Gongronema latifolium, Aloe perryi, Viscum album and Allium sativum administered to alloxan-induced diabetic albino rats (Rattus norvegicus). *Pharmacognosy Communications*, *3*(2), 12–16. https://doi.org/10.5530/pc.2013.2.4

Idm'hand, E., Msanda, F., & Cherifi, K. (2019). Ethnopharmacological documentation of medicinal plants used in the traditional treatment of hypertension in tarfaya province, Morocco. *International Journal of Pharmacology, Phytochemistry and Ethnomedicine*, *11*, 16–39.

Johnson, O. R., Elizabeth, M. A., Oloruntoba, A. C., & Sunday, A. I. (2012). Effects of pre-induction administration of allium sativum on some biochemical parameters in alloxan induced diabetic rats. *Journal of Biotechnology and Biochemistry*, *1*(2), 7.

Jouad, H., Haloui, M., Rhiouani, H., El Hilaly, J., & Eddouks, M. (2001). Ethnobotanical survey of medicinal plants used for the treatment of diabetes, cardiac and renal diseases in the North centre region of Morocco (Fez–Boulemane). *Journal of Ethnopharmacology*, *77*(2–3), 175–182. https://doi.org/10.1016/S0378-8741(01)00289-6

Kadir, M. F., Bin Sayeed, M. S., Shams, T., & Mia, M. M. K. (2012). Ethnobotanical survey of medicinal plants used by Bangladeshi traditional health practitioners in the management of diabetes mellitus. *Journal of Ethnopharmacology*, *144*(3), 605–611. https://doi.org/10.1016/j.jep.2012.09.050

Kesavanarayanan, K. S., Sathiya, S., Ranju, V., Sunil, A. G., Ilavarasan, R., Saravana Babu, C., Kavimani, S., & Prathiba, D. (2012). In vitro cytotoxic, antioxidative and alpha-glucosidase inhibitory potential of a herbal mixture comprised of Allium sativum and Lagerstroemia speciosa. *European Review for Medical and Pharmacological Sciences*, *16*(Suppl 3), 58–68.

Keter, L. K., & Mutiso, P. C. (2012). Ethnobotanical studies of medicinal plants used by Traditional Health Practitioners in the management of diabetes in Lower Eastern Province, Kenya. *Journal of Ethnopharmacology*, *139*(1), 74–80. https://doi.org/10.1016/j.jep.2011.10.014

Kojuri, J., Vosoughi, A. R., & Akrami, M. (2007). Effects of anethum graveolens and garlic on lipid profile in hyperlipidemic patients. *Lipids in Health and Disease*, *6*, 5. https://doi.org/10.1186/1476-511X-6-5

Koyuncu, M., & Bafier, K. H. C. (2005). Folk use of some wild and cultivated Allium species in Turkey. *Proceedings of the IVth International Congress of Ethnobotany*, Vol. 87, p. 90.

Kuete, V. (2017a). Allium cepa. In *Medicinal Spices and Vegetables from Africa* (pp. 353–361). Academic Press.

Kuete, V. (2017b). Chapter 14 - Allium sativum. In V. Kuete (Ed.), *Medicinal Spices and Vegetables from Africa* (pp. 363–377). Academic Press.

Lawal, B., Shittu, O. K., Oibiokpa, F. I., Mohammed, H., Umar, S. I., & Haruna, G. M. (2016). Antimicrobial evaluation, acute and sub-acute toxicity studies of Allium sativum. *Journal of Acute Disease*, *5*(4), 296–301. https://doi.org/10.1016/j.joad.2016.05.002

Li, Z., Le, W., & Cui, Z. (2018). A novel therapeutic anticancer property of raw garlic extract via injection but not ingestion. *Cell Death Discovery*, *4*(1), 1–10.

Londhe, V. P., Gavasane, A. T., Nipate, S. S., Bandawane, D. D., & Chaudhari, P. D. (2011). Role of garlic (allium sativum) in various diseases: An overview. *Journal of Pharmaceutical Research and Opinion*, *1*(4), 129–134.

Mikail, H. G. (2010). Phytochemical screening, elemental analysis and acute toxicity of aqueous extract of Allium sativum L. bulbs in experimental rabbits. *Journal of Medicinal Plants Research*, *4*(4), 322–326.

Mootoosamy, A., & Fawzi Mahomoodally, M. (2014). Ethnomedicinal application of native remedies used against diabetes and related complications in Mauritius. *Journal of Ethnopharmacology*, *151*(1), 413–444. https://doi.org/10.1016/j.jep.2013.10.069

Morales-González, J. A., Madrigal-Bujaidar, E., Sánchez-Gutiérrez, M., Izquierdo-Vega, J. A., Valadez-Vega, M. del C., Álvarez-González, I., Morales-González, Á., & Madrigal-Santillán, E. (2019). Garlic (Allium sativum L.): A brief review of its antigenotoxic effects. *Foods*, *8*(8), 343. https://doi.org/10.3390/foods8080343

Mostofa, M., Choudhury, M. E., Hossain, M. A., Islam, M. Z., Islam, M. S., & Sumon, M. H. (2007). Antidiabetic effects of catharanthus roseus, azadirachta indica, allium sativum and glimepride in experimentally diabetic induced rat. *Bangladesh Journal of Veterinary Medicine*, *5*(2), 99–102.

Naderi, R., Mohaddes, G., Mohammadi, M., Alihemmati, A., Badalzadeh, R., Ghaznavi, R., Ghyasi, R., & Mohammadi, S. (2015). Preventive effects of garlic (*Allium sativum*) on oxidative stress and histopathology of cardiac tissue in streptozotocin-induced diabetic rats. *Acta Physiologica Hungarica*, *102*(4), 380–390. https://doi.org/10.1556/036.102.2015.4.5

Ozougwu, J., Nwachi, U., & Eyo, J. (2009). Comparative hypolipidaemic effects of *Allium cepa, Allium sativum* and *Zingiber officinale* Aqueous extracts on alloxan-induced diabetic Rattus novergicus. *Bio-Research*, *6*(2), 384–391. https://doi.org/10.4314/br.v6i2.28672

Pârvu, M., Moţ, C. A., Pârvu, A. E., Mircea, C., Stoeber, L., Roşca-Casian, O., & Ţigu, A. B. (2019). Allium sativum extract chemical composition, antioxidant activity and antifungal effect against Meyerozyma guilliermondii and Rhodotorula mucilaginosa causing onychomycosis. *Molecules*, *24*(21), 3958.

Polat, R., Cakilcioglu, U., & Satıl, F. (2013). Traditional uses of medicinal plants in Solhan (Bingöl–Turkey). *Journal of Ethnopharmacology*, *148*(3), 951–963. https://doi.org/10.1016/j.jep.2013.05.050

Rachid, A., Rabah, D., Farid, L., Zohra, S. F., Houcine, B., & Nacera, B. (2012). Ethnopharmacological survey of medicinal plants used in the traditional treatment of diabetes mellitus in the North Western and South Western Algeria. *Journal of Medicinal Plants Research*, *6*(10), 2041–2050. https://doi.org/10.5897/JMPR11.1796

Raju, T. N., Kanth, V. R., & Lavanya, K. (2008). Effect of methanolic extract of Allium sativum (AS) in delaying cataract in STZ-induced diabetic rats. *Journal of Ocular Biology, Diseases, and Informatics*, *1*(1), 46. https://doi.org/10.1007/s12177-008-9003-5

Sachin, A., Bhalerao, P., & Patil, S. (2017). Effect of organic and inorganic sources of nitrogen on growth and yield of garlic (Allium sativum L.) var. GG-4. *International Journal of Chemical Studies*, *5*(4), 559–562.

Senthil Kumar, G. P. (2013). Attenuation of oxidative stress, inflammation and insulin resistance by *Allium sativum* in fructose-fed male rats. *Journal of Clinical and Diagnostic Research*, *7*(9), 1860. https://doi.org/10.7860/JCDR/2013/6924.3334

Seo, Y.-J., Gweon, O.-C., Im, J.-E., Lee, Y.-M., Kang, M.-J., & Kim, J.-I. (2009). Effect of garlic and aged Black garlic on hyperglycemia and dyslipidemia in animal model of type 2 diabetes mellitus. *Preventive Nutrition and Food Science*, *14*(1), 1–7. https://doi.org/10.3746/jfn.2009.14.1.001

Shakya, K., Saxena, R. C., & Shaky, A. (2010). Effect of ethanolic extract of Allium sativum bulbs on Streptozotocin induced diabetic rats. *Journal of Chemical and Pharmaceutical Research*. www.jocpr.com/abstract/effect-of-ethanolic-extract-of-allium-sativum-bulbs-on-streptozotocin-induced-diabetic-rats-464.html

Siddiqui, N. A., Taj, F. B., Ishaque, I., Shahid, A., & Ayaz, A. (2020). Role of simvastatin and allium sativum on lipid profile in diabetic dyslipidemic patients: A comparative study. *The Professional Medical Journal*, *27*(2), 353–358. https://doi.org/10.29309/TPMJ/2020.27.02.3974

Souza, G. A., Ebaid, G. X., Seiva, F. R. F., Rocha, K. H. R., Galhardi, C. M., Mani, F., & Novelli, E. L. B. (2011). N -Acetylcysteine an allium plant compound improves high-sucrose diet-induced obesity and related effects. *Evidence-Based Complementary and Alternative Medicine*, *2011*, 1–7. https://doi.org/10.1093/ecam/nen070

Sulaiman, F. A., Kazeem, M. O., Waheed, A. M., Temowo, S. O., Azeez, I. O., Zubair, F. I., Adeyemi, T. A., Nyang, A., & Adeyemi, O. S. (2014). Antimicrobial and toxic potential of aqueous extracts of Allium sativum, Hibiscus sabdariffa and Zingiber officinale in Wistar rats. *Journal of Taibah University for Science*, *8*(4), 315–322. https://doi.org/10.1016/j.jtusci.2014.05.004

Tadeu Volpato, G., Queiroz Moraes-Souza, R., Sousa Soares, T., Leal-Silva, T., & Cristina Damasceno, D. (2017). Medicinal plants for diabetes treatment during pregnancy. *Current Medicinal Chemistry*, *24*(4), 404–410.

Tahraoui, A., El-Hilaly, J., Israili, Z. H., & Lyoussi, B. (2007). Ethnopharmacological survey of plants used in the traditional treatment of hypertension and diabetes in south-eastern Morocco (Errachidia province). *Journal of Ethnopharmacology*, *110*(1), 105–117. https://doi.org/10.1016/j.jep.2006.09.011

Thomson, M., Al-Qattan, K. K., Js, D., & Ali, M. (2015). Anti-diabetic and anti-oxidant potential of aged garlic extract (AGE) in streptozotocin-induced diabetic rats. *BMC Complementary and Alternative Medicine*, *16*(1), 17. https://doi.org/10.1186/s12906-016-0992-5

Tran, G.-B., Dam, S.-M., & Le, N.-T. T. (2018). Amelioration of single clove Black garlic aqueous extract on dyslipidemia and hepatitis in chronic carbon tetrachloride intoxicated Swiss albino mice. *International Journal of Hepatology*, *2018*, 1–9. https://doi.org/10.1155/2018/9383950

Ugulu, I. (2011). Traditional ethnobotanical knowledge about medicinal plants used for external therapies in Alasehir, Turkey. *International Journal of Medicinal and Aromatic Plants*, *1*(2), 7.

Wilson, A., Pandya, D., Mankad, D. A., & Solanki, D. H. (2019). A review on allium sativum (L.) as medicinal plant. *International Journal of Research and Analytical Reviews*, *6*(1).

Wu, C., Wang, M., Dong, Y., Cheng, Z., & Meng, H. (2015). Growth, bolting and yield of garlic (Allium sativum L.) in response to clove chilling treatment. *Scientia Horticulturae*, *194*, 43–52. https://doi.org/10.1016/j.scienta.2015.07.018

Zeng, Y., Li, Y., Yang, J., Pu, X., Du, J., Yang, X., Yang, T., & Yang, S. (2017). Therapeutic role of functional components in Alliums for preventive chronic disease in human being. *Evidence-Based Complementary and Alternative Medicine*, *2017*, 1–13. https://doi.org/10.1155/2017/9402849

Ziamajidi, N., Behrouj, H., Abbasalipourkabir, R., & Lotfi, F. (2018). Ameliorative effects of Allium sativum extract on iNOS gene expression and NO production in liver of streptozotocin + nicotinamide-induced diabetic rats. *Indian Journal of Clinical Biochemistry*, *33*(2), 147–153. https://doi.org/10.1007/s12291-017-0656-3

5 *Aloe vera* (L.) Burm.f.
An Important Medicinal Herb for Diabetes Treatment

Surjit Sen, Shuvadip Mondal, Arghya Naskar, Protha Biswas, Tuyelee Das, Rupa Sanyal, Abhijit Dey, and Manjula Rai

CONTENTS

5.1 INTRODUCTION

Around the globe, numerous plants have been used for their therapeutic properties. An opinion report by WHO (World Health Organization) states, "In order to stay healthy, 80% of people worldwide depend on traditional medicines" (Bawankar et al. 2014). Of the different medicinal plants, *Aloe vera* has been utilized since ancient times for medicinal purposes to treat different human ailments and disorders and has health, beauty and skin care properties (Bawankar et al. 2014; Kumar et al. 2019; Babu and Noor 2020). *Aloe vera* is a succulent, perennial, xerophytic shrub and has been studied thoroughly and reported with various bioactive compounds like amino acids, enzymes, anthraquinones, sugars, minerals, polyphenols, and vitamin A, vitamin B, vitamin C and vitamin E (Hirata and Suga 1977; Bozzi et al. 2007; Quispe et al. 2018; Khajeeyana et al. 2019; Martínez-Sanchez et al. 2020). The richness of different bioactive compounds makes this plant a potent health promoter by performing as anti-hypercholesterolaemic, anti-ulcer, antioxidant, antiviral, antibacterial, antifungal, nutraceutical, anti-acne and humectant, providing skin protection against UV-B and UV-A, healing wounds, preventing cardiovascular disease and cancer, generating antibodies and preventing type II diabetes (Sahu et al. 2013; Chacón et al. 2019; Martínez-Sanchez et al. 2020).

DOI: 10.1201/b23347-5

Besides numerous health activities, in this context, we are going to unfurl the potential aptitude of *Aloe vera* to counter hyperglycaemic condition in diabetes mellitus.

Information states that there are main three kinds of diabetes mellitus: (a) type I, which is related to total insulin deficiency; (b) type II, which is related to progressive insulin deficiency and, lastly, (c) gestational diabetes mellitus, which generally occurs over the course of the second or third semester of pregnancy (Hutapea and Chandra 2021). Diabetes mellitus is considered a disorder of metabolism as shown by hyperglycaemia (a condition of elevated levels of blood glucose) and glucose intolerance. It results in insulin secretion problems or malfunction of glucose uptake by insulin's action (Unuofin and Sogolo 2020). According to the International Diabetes Federation (IDF) report of 2017, diabetes now affects 451 million individuals globally, and by 2045 there will likely be 693 million cases, and the "Diabetic capital of the world" has been proclaimed to be India (Choudhary et al. 2014; Unuofin and Sogolo 2020).

Controls of diabetes by using chemical drugs have reflected some setbacks (Almeida et al. 2002; Mahabady et al. 2021a). To overcome this problem, some cost-effective, natural and secure substances are required. A worldwide knowledge of traditional herbal treatment is also put forth (Almeida et al. 2002; Mahabady et al. 2021a). In this context, the glucose lowering efficacy of *Aloe vera* is commonly described (Bahrami et al. 2017; Mahabady et al. 2021a). The plant is not only good at lowering blood glucose through increasing the body tissues' sensitivity to insulin, but it also has a potent role in lowering high blood pressure (Choudhary et al. 2014). The antidiabetic effects of this plant have been documented in both clinical and experimental research (Mahabady et al. 2021a). Investigators have claimed to discover antidiabetic and antioxidant properties in *Aloe vera* in lab-induced diabetes (Ajabnoor 1990; Can et al. 2004; Abo-Youssef and Hussein 2013). The glucose-lowering effects of this plant proved to be mediated partially by its potent antioxidant effects (Rajasekaran et al. 2005; Boudreau et al. 2006; Abo-Youssef and Hussein 2013). *Aloe vera*-derived extracts have been shown in increasing amounts of research to have lipid-lowering and anti-insulin resistance impacts (Kim et al. 2009; Yagi et al. 2009; Zhang et al. 2016). Certain hormones, interleukins and cytokines are found to create the onset of depression in diabetic patients. Diabetes conjugated with depression can make severe health implications such as lower self-confidence, less energy and incidence of dementia. The availability of mood-regulating compounds in *Aloe vera* like enzymes, vitamins, minerals, salicylic acid, lignin, copper, magnesium, iron, arachidonic acid, steroids, Y-linolenic acid, mannose, L-rhamnose, glucose, aldopentose, folic acid, choline, auxins and gibberellins help to counter depression in people with diabetes (Foadoddini and Samaneh 2020). This makes the plant absolutely unique to treat diabetes in a wide-scale range. Hence, this chapter is designed to incorporate the future role of *Aloe vera* against the hyperglycaemic condition.

5.2 BOTANICAL DESCRIPTION

Aloe vera (L.) Burm.f. belonging to the family Xanthorrhoeaceae (previously under the family Liliaceae) is a light green, xerophytic, perennial, succulent, short or without stemmed shrub growing up to 60 to 100 cm in length. It grows in tropical, semi-tropical and desert areas all across the world, and commonly thrives in the wild and spreads via offsets (Surjushe et al. 2008; Haghani et al. 2022) (Figure 5.1). The leaves of the plant are fleshy and triangular, with serrated margins arranged in a rosette pattern, having an inflorescence that resembles a simple or compound spike cluster, capable of comprising 100 to 200 hermaphroditic 2–3 cm long pendulous flowers (that possess yellow-coloured tubular corolla). They are placed helically on an upright column that shoots from the plant's centre and measures around 90–100 cm in length with fruits that contain numerous seeds (Martínez-Sanchez et al. 2020; Jangra et al. 2021). These tubular yellow flowers mature gradually from bottom to top, indicating both quantitative and qualitative variations in the bioactive chemicals found in the blossoms. It is observed in the cultivation practices of farmers that they often remove the inflorescence-axis from the base to preserve the full strength of the leaves. (Martínez-Sanchez et al. 2020). Researchers have found that the antioxidant activity in flowers is four to eight times

FIGURE 5.1 *Aloe vera.*

greater than in the leaves (Quispe et al. 2018; Martínez-Sanchez et al. 2020). This plant performs crassulacean acid metabolism (CAM) pathway for conserving water within its parenchymatous tissues to prevent drought-like conditions. In a condition when water is normally available, this plant perpetuates itself as a typical CAM plant, but whenever it undergoes a water-logging condition, it shifts itself to CAM-idling (Kumar and Yadav 2014).

5.3 DISTRIBUTION

Currently, *Aloe vera* appears as a cosmopolitan plant having a worldwide distribution by the process of "plant introduction". However, it is a native to Africa's southern and eastern regions with the range exceeding to the upper Nile in Sudan (Sahu et al. 2013). From this region it was brought into northern Africa, where it became used to the climate there before spreading to other nations throughout the world (Sahu et al. 2013). The commercial cultivation has been vigorously observed in the countries like Bonaire, Aruba, Haiti, South Africa, India, Venezuela and the United States (Yeh et al. 2003; Sahu et al. 2013). It also has been described that, in the subtropical and tropical parts of the southern USA, some indigenous varieties of *Aloe vera* have been found (Reynolds and Dweck 1999; Mahor and Sharique 2016).

5.4 PHYTOCHEMICAL CONSTITUENTS

Herbal remedies have been utilized to treat chronic ailments since ancient times (Hardy 2021). In the pharmaceutical industry, about 25% of medications are of botanical origin (Choudhury et al. 2018). Several studies have looked at *Aloe vera* as an alternate kind of treatment due to its abundance of bioactive phytochemicals with considerable therapeutic potentials such as antioxidant (Kaur et al. 2022), antidiabetic (Budiastutik et al. 2022), neuroprotective (Xu et al. 2022), cytotoxic (Tong et al. 2022) and anti-inflammatory (Farid et al. 2021) properties, among others. Vitamins, proteins, sterols, anthrones, anthraquinones, pyrones, phenols, flavonoids, carbohydrates, coumarins, tannins, chromones, alkaloids and a variety of other organic substances are among these phytochemicals (Banik et al. 2019; Dey et al. 2017; Salehi et al. 2018) (Figure 5.2). The extraction method, the choice

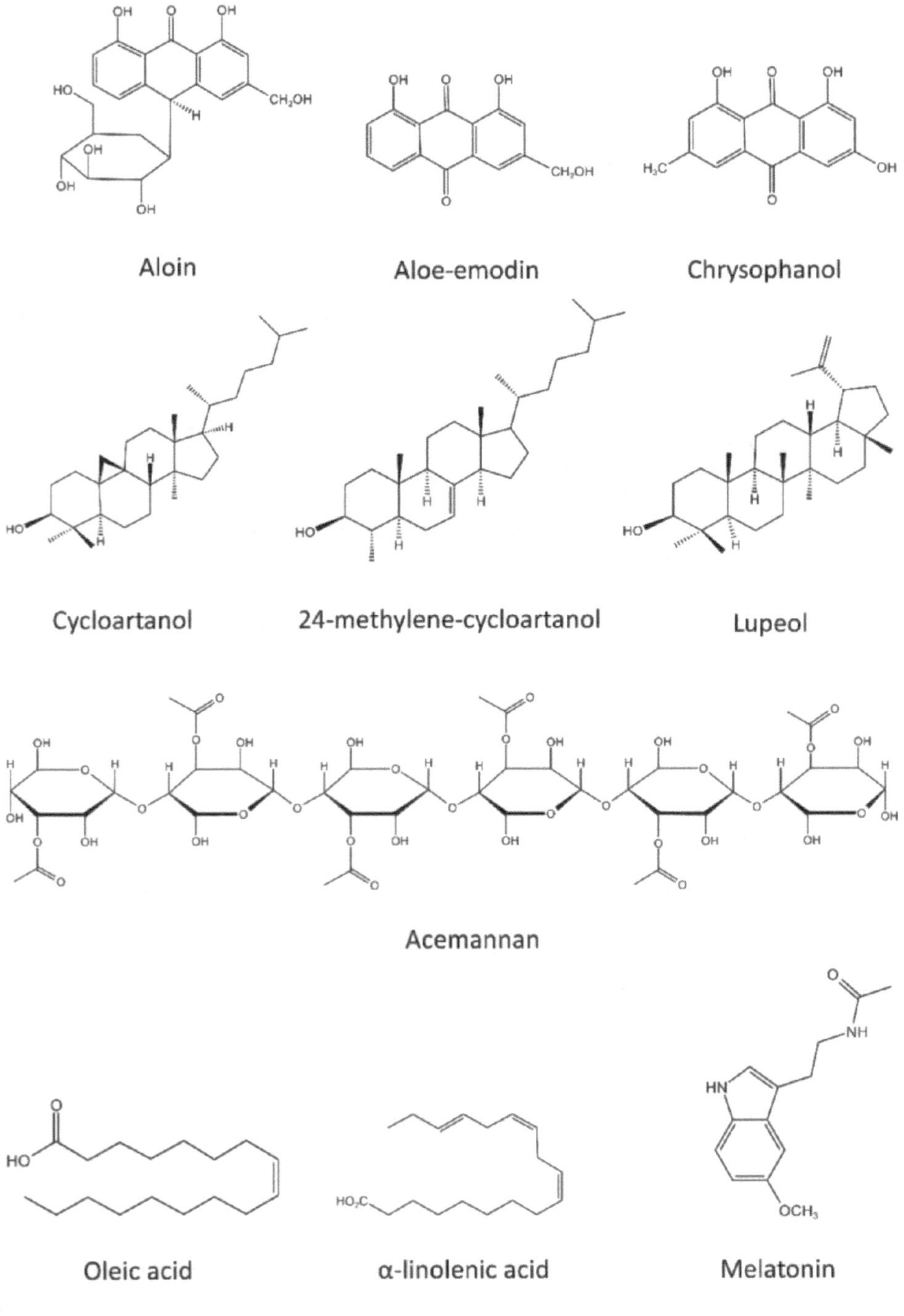

FIGURE 5.2 Some significant bioactive phytoconstituents of *Aloe vera*.

of solvent, the stage of plant development, the geographical variation, harvesting time, climatic condition and the plant component that is employed are all factors that influence the variation in concentration of these phytochemical contents. Detailed phytochemical composition is mentioned in Table 5.1.

TABLE 5.1
Phytochemical Constituents of *Aloe vera*

Class	Chemical compounds	References
Anthraquinones	Aloenin	Reynolds (1985);
	Aloe-emodin	Majumder et al. (2019);
	Helminthosporin	Mukherjee et al. (2014)
	Chrysophanol	
	Aloesaponarin-I	
	Aloesaponarin-II	
	Lactic acid D methyl ester	
	Deoxyerythrolaccin	
	Aloetic acid	Hamman (2008)
	Ester of cinnemonic acid	
	Emodin	
	Isobarbaloin	
	Anthranol	
	Chrysophanic acid	
	Ethereal oil	
	Anthracine	
	Resistannol	
	Anthranon	
	Barbaloin (Aloin A and Aloin B)	Baruah et al. (2016)
	Aloe emodin-9-anthrone	
	Aloeresin E	
	Isoaloeresin D	
	Aloeresin D	
	Chrysophanol	
	Physcion	
	Aloe emodin	
	Trihydroxy octadecanoic acid	Quispe et al. (2018)
	5,3'-Dihydroxy-6,7,4'-Trimethoxyflavone (eupatorin)	
	7-Methylether of 2'feruloylaloesin	
	Naringenin-4'-methoxy-7-O-glucuronide	
	4,5-Dimethyl ether of *Aloe* emodin	
Phenolic compounds	Chlorogenic acid	Dammak et al. (2018)
	Catechin	
	Gallic acid	
	Caffeic acid	
	Quercetin	
	Trans-3-hydroxycinnamic acid	
	Luteolin 7-O-glucoside	
	Isorhamnetin 3-O-rutinoside	

(*Continued*)

TABLE 5.1
(Continued)

Class		Chemical compounds	References
Carbohydrates	Polysaccharides	Mannan	Hamman (2008)
		Acetylated mannan	
		Acetomannan	
		Glucogalactomannan	
		Galactan	
		Galactogalactan	
		Arabinogalactan	
		Galactoglucoarabinomannan	
		Pectin	
		Xylan	
		Cellulose	
	Monosaccharides	Glucose	Baruah et al. (2016)
		Fructose	
		Mannose	
		L-rhamnose	
		Aldopentose	
Sterols		Sitosterol	Bawankar et al. (2013)
		Stigmatosterol	
		Campesterol	
		Lupeol	
		1-Dodecanol	
		1-Octadecanol	
Chromones		8-C-glucosyl-(2'-O-cinnamoyl)-7-O-methylaloediol A	Gupta and Malhotra (2012)
		8-C-glucosyl-(S)- aloesol, 8-C-glucosyl-7-O-methyl- (S)-aloesol	
		8-C-glucosyl-7-O-methylaloediol	
		8-C-glucosyl-noreugenin	
		Isorabaichromone	

Aloin (also known as barbaloin) is the most prevalent bioactive substance among the tentative 32 forms of anthraquinone and its glycosidic derivatives found in *Aloe vera* (Kahramanoğlu et al. 2019). Additionally, chrysophanol and aloe-emodin are other prevalent anthraquinones of *Aloe vera* (Misir et al. 2014). Several investigations described the anti-microbial, antidiabetic, hepatoprotective, anti-cancer, and vasodilator properties of *A. vera*-isolated anthraquinones (Hamman 2008).

Five steroids have been extracted from the gel of *A. vera* as of yet, which include 24-ethyl-lophenol (A53), 24-methylene-cycloartanol, 24-methyl-lophenol, lophenol and cycloartanol. It was found that lophenol and cycloartanol taken orally can help avoid metabolic illnesses such as diabetes and obesity, according to an *in vivo* mouse model study (Haghani et al. 2022; Misawa et al. 2012).

The leaves of *A. vera* contain triterpenes, which are essential structural elements of the membranes of plants, and their free forms are used to support phospholipid bilayers in plant cell membranes. A triterpenoid with pharmacological activity called lupeol derived from *A. vera* has potential bioactive properties such as anti-inflammatory and anti-carcinogenic activities (Saleem 2009).

Saponins are an additional integral component of *A. vera* leaves. Saponins' benefits in biological functions like antiviral and antidiabetic qualities are supported by a substantial body of research

(Choudhury et al. 2018). Saponins may improve the permeability of cell membranes, control nutrient absorption in the stomach and inhibit protein digestion. Saponins reduce blood cholesterol by limiting cholesterol reabsorption in the digestive system (Vinarova et al. 2015). These chemicals have antimutagenic and antitumour properties and may lower the risk of human cancer by inhibiting the proliferation of cancer cells (Sampedro et al. 2004).

Aloe vera contains a variety of polysaccharide compounds. Mannan, arabinan, galactan, cellulose, pectic acid, glucuronic acid, and xylan are the main polysaccharides isolated from *A. vera*. Moreover, acemannan, an acetylated-polysaccharide extracted from *A. vera*, exhibits anti-inflammatory, antiviral, antibacterial, antitumour, and wound healing effects, mediated mostly via mechanisms that activate the immune system (Sierra-García et al. 2014).

Bioactive compounds include fatty acids that make up a significant portion of the chemical composition of *A. vera*. According to studies, oleic acid (omega-9 fatty acid) and α-linolenic acid (omega-3 fatty acid) are the most important fatty acids in various *A. vera* subspecies (Andrea et al. 2020). Essential fatty acids include omega-3 fatty acids which cannot be synthesized by the human body. It seems that omega-3 fatty acids are beneficial for the heart, immunological system and neurological system (Stark et al. 2008). Among omega-9 fatty acids, there is just one double bond; hence they are classified as monounsaturated fatty acids (MUFA). Omega-9 fatty acids are not exactly "necessary", since our body can synthesize them. An investigation revealed that large dosages of MUFA improved mice's insulin sensitivity and reduced inflammation (Finucane et al. 2015).

Melatonin is another bioactive chemical found in *A. vera* (Chen et al. 2003). N-acetyl methoxytryptamine is the chemical name for melatonin, an antioxidant that occurs naturally. Melatonin is known to lower blood sugar levels via boosting insulin and hepatic glycogen production (Li et al. 2018).

5.5 PHARMACOLOGICAL STUDIES

A. vera is often regarded as a miraculous gift from nature due to its vast medical potential. Phytoconstituents of *A. vera* offer significant promise for preventing and treating several disorders, some of which are mentioned later (Kumar et al. 2019).

5.5.1 Anticancer Properties

According to a study by Wang et al. (2020), aloin reduces NOX2-ROS-mediated activation of the Akt-mTOR, Stat3 and NF-B signalling pathways, which prevents HGC-27 and BGC-823 gastric cancer cells from proliferating and migrating. By stopping the cell cycle in the S phase and G2/M phase through the upregulation of p53 and p21 expression, aloe-emodin derived from *A. vera* also reduced the proliferation of U87 malignant glioma, H460 and HL60 cells (Shalabi et al. 2015).

5.5.2 Anti-Microbial Effects

Anthraquinones and lectin, which have antiviral characteristics and stop cytomegalovirus reproduction, are present in *A. vera* (Haghani et al. 2022). These substances are also efficient against a variety of gram-positive and gram-negative pathogenic bacteria (Hamman 2008). According to a study, *A. vera* has antifungal properties since it inhibited the growth of *Malassezia furfur* and *Candida albicans* over time (Parihar et al. 2019).

5.5.3 Wound-Healing Properties

Coordination between dermal and epidermal cells is necessary for the wound-healing process. Mucopolysaccharides in *A. vera* facilitate the absorption of moisture by the skin and activate fibroblasts, which leads to the formation of collagen and elastin fibres, hence minimizing the appearance of wrinkles (West and Zhu 2003). Through its cohesive impact on superficial epidermal cells, it also softens the skin. By releasing growth factors, increasing cell division and encouraging regeneration in the deepest skin layers, *A. vera* aids in wound healing and lessens scarring. It has been demonstrated that the isolated glycoprotein fraction of aloe vera gel can treat skin wounds and stomach ulcers in diabetic mice (Choi et al. 2001; Tarameshloo et al. 2012). Due to the presence of various inorganic electrolytes, including iron, potassium and magnesium, as well as important amino acids, *Aloe vera* helps the healing process of wounds (Alven et al. 2021).

5.5.4 Anti-Inflammatory Properties

Bradykinin is one of the key mediators of inflammation causing pain in the body, and *Aloe vera* is proved to stimulate an enzyme bradykinase which breaks down bradykinin (Sahu et al. 2013). Furthermore, it is suggested that it reduces inflammation by downregulation of the C-reactive protein, interleukin 1β, interleukin-6, polymorphonuclear leukocyte infiltration and TNF-α (Babu and Noor 2020).

5.5.5 Hepatoprotective Activity

Aqueous extract of *A. vera* greatly reduced liver damage generated by carbon tetrachloride (CCl4) in mice and reversed biochemical parameters associated with this damage. According to histological studies (Chandan et al. 2007), the aqueous extract of *A. vera* repaired CCl4-induced liver damage by reversing macrovascular fatty alterations and dispersed lymphomononuclear cell infiltration in hepatic parenchyma. The extract also increased bile flow and bile solids, which promoted the secretory activity of liver cells. Its hepatoprotective effects were also aided by its antioxidant qualities (Kumar et al. 2019).

5.5.6 Antilipaemia Effects

Hyperlipidaemia has been reported to be closely connected to the advancement of diabetic nephropathy (Chen and Tseng 2013). *A. vera* gel may be an effective treatment for hyperlipidaemic people who do not respond to dietary interventions because it is claimed to have antihyperlipidaemic characteristics (Mulay et al. 2013). According to Kumar et al. (2013), the combination of *Lactobacillus rhamnosus* and *Aloe vera* gel was an effective treatment for hypercholesterolaemic rats and may lower the prevalence of cardiovascular diseases. Recent studies showed that *A. vera*'s effects on lipid modification and renal oxidative stress may decrease the progression of nephropathy (Arora et al. 2019).

5.6 ANTIDIABETIC RESPONSE

Diabetes is a chronic metabolic illness characterized by decreased or discontinuation of insulin secretion in response to physiological stimuli, or decreased insulin sensitivity in peripheral tissues (Gerstein et al. 2007). The adverse medical side effects of modern synthetic antidiabetic pharmacological therapies compel us to return to the past and pay more attention to herbal remedies and other alternative therapies. Herbal remedies have been shown to aid in illness prevention and treatment (Choudhury et al. 2018). Recently, there has been a resurgence of interest in plant-based medications, particularly in the hunt for natural anti-hyperglycaemic substances with little to negligible side effects. There are around 1200 plant species known to have been utilized as traditional remedies to treat a variety of ailments, including diabetes mellitus (Babu and Noor 2020).

Aloe vera has been used in traditional therapies since ancient times. Due to its diverse medicinal activities, it has acquired quite significant popularity among researchers. Several *in vitro* and *in vivo* studies have been performed with *Aloe vera* in order to decipher its therapeutic prospects, including hypoglycaemic activity.

5.6.1 *In Vivo* Studies

Mohamed (2011) explored the antidiabetic effect of *Aloe vera* gel extract in 40 alloxan-induced diabetic mice. The blood sugar level nearly restored to normal after five weeks of oral administration of *Aloe vera* gel extract (0.5 ml per day). The considerable reduction in blood glucose in the diabetic group was accompanied by reductions in serum cholesterol and triacylglycerols levels of 31% and 20.61%, respectively, compared to the control group. Sethi et al. (2012) carried out a related study as well. They examined the antioxidant and antidiabetic effects of *Aloe vera* gel extract on 24 alloxan-induced diabetic rabbits for 21 days, which effectively reduced the blood glucose level close to normal state. Several similar studies had been conducted by numerous schools of researchers. Some of the notable examples are presented in the Table 5.2.

TABLE 5.2
In-Vivo* Antidiabetic Activity of *Aloe vera

Serial No.	Sample	Intervention	Result (blood glucose level)	Studied by
1	108 male C57BL/6J high-fat-diet-induced	Oral administration of processed *Aloe vera* gel for 8 weeks	• **Group I** (Normal condition rats) 7.6 ± 0.7 mM • **Group II** (Negative control diabetic rats with phosphate buffered saline) 14.8 ± 0.1 mM • **Group III** (Diabetic rats with processed *Aloe vera* gel 25 mg/kg) 7.1 ± 0.7 mM • **Group IV** (Diabetic rats with processed *Aloe vera* gel 50 mg/kg) 7.2 ± 0.7 mM • **Group V** (Diabetic rats with processed *Aloe vera* gel 100 mg/kg) 7.1 ± 0.6 mM • **Group VI** (Positive control diabetic rats with pioglitazone 2.5 mg/kg) 6.8 ± 0.7 mM	Kim et al. (2009)
2	24 male albino streptozotocin-induced diabetic rats	Diabetic rats were treated with *Aloe vera* gel diluted with phosphate-buffered saline (10 ml/kg) daily using an intragastric tube for 14 days	• **Group I** (Healthy rats "normal control") 82.24 ± 5.17 mg/dL • **Group II** (Diabetics rats) 331.88 ± 29.72 mg/dL • **Group III** (Diabetic rats with glimepiride treatment "positive control") 117.43 ± 21.96 mg/dL • **Group IV** (Diabetics rats with *Aloe vera* treatment) 93.66 ± 26.92 mg/dL	Abo-Youssef and Hussein (2013)
3	25 male Wistar 12 weeks old streptozotocin-induced diabetic rats	Rats are given oral administration of diluted *Aloe* extract (400mg/kg) for 8 weeks	• Group I (Healthy rats, control): 109.7 ± 3 mg/dl • Group II (Healthy rats + *Aloe* gel treated): 107.2 ±3.7 mg/dl • Group III (Diabetic): 541.5 ± 20.5 mg/dl • Group IV (Diabetic+ *Aloe* extract): 255.4 ± 32.5 mg/dl • Group V (Diabetic + insulin): 204.7 ± 44.2 mg/dl	Mahabady et al. (2021b)

(Continued)

TABLE 5.2
(Continued)

Serial No.	Sample	Intervention	Result (blood glucose level)	Studied by
4	34 adult Long-Evans female streptozotocin-induced diabetic rats	Diabetic rats received concentrated gel extract (Gel-C), ethanol (80%) gel extract (Gel-Et), ethanol (80%) skin extract of *Aloe vera* (Skin-Et) 1.25 g/kg for 28 days	• Group I (Water-treated control rats) >9 mmol/L • Group II (glibenclamid control rats) 8 mmol/L • Group III (Gel-Et treated rats) >8 mmol/L • Group IV (Gel-C treated rats) ≥ 8 mmol • Group V (Skin-Et treated rats) >8 mmol	Moniruzzaman et al. (2012)
5	40 male albino alloxan-induced diabetic rats	Diabetic rats treated with 300 mg/kg *Aloe vera* gel orally for 21 days	• Group I (Normal untreated control rats) 130 mg/dl • Group II (Diabetic untreated rats) >500 mg/dl • Group III (Diabetic rats treated with 300 mg/kg polysaccharide equivalent *Aloe vera* gel orally) 145 mg/dl • Group IV (Diabetic rats treated with 2 mg/kg metformin orally) 120 mg/dl • Group V (Diabetic rats co-administered with 300 mg/kg PE *Aloe vera* gel and 2 mg/kg metformin orally 155 mg/dl	Atanu et al. (2018)
6	Alloxan-induced 24 Swiss albino diabetic mice of either sex	Diabetic mice received *Aloe vera* gel extract 300 mg/kg and 500 mg/kg for 21 days	• Group I (Normal control "saline") 77.24 ± 6.78 mg/dl • Group II (Alloxan-treated control) 226.4 ± 8.88 mg/dl • Group III (*Aloe vera* extract 300 mg/kg) 117.2 ± 12.17 mg/dl • Group IV (*Aloe vera* extract 500 mg/kg) 98.06 ± 5.06 mg/dl	Sharma et al. (2013)
7	Alloxan-induced 105 female CD-1 mice	Diabetic mice received polysaccharide-rich fractions UP780 2000 mg/kg and UP 394 80 mg/kg for 4 weeks	• Group I (Healthy mice without alloxan injection) 98.3 ± 5.4 mg/dL • Group II (Healthy mice without alloxan injection with UP780 administered) 85.3 ± 3.3 mg/dL • Group III (Alloxan-induced diabetic rats) 431.9 ± 21.2 mg/dL • Group IV (Alloxan-induced diabetic rats with glyburide administered) 398.7 ± 19.1 mg/dL • Group V (Alloxan-induced diabetic rats with UP780 administered) 315.1 ± 34.9 • Group VI (Alloxan-induced diabetic rats with UP394 administered) 442.1 ± 48.0 mg/dL • Group VII (Alloxan-induced diabetic rats administered with Qmatrix) 443.2 ± 42.0 mg/dL	Yimam et al. (2014)
8	24 male Charles Foster alloxan-induced diabetic rats	Diabetic rats received *Aloe vera* extract (500 mg/kg) for 30 days	• Group I (Control rats with normal saline) 91.22 ± 7.31 mg/dL • Group II (Alloxan-induced diabetic rats with normal saline) 340.48 ± 22.60 mg/dL • Group III (Alloxan-induced diabetic rats with *Aloe vera* extract) 256.27 ± 21.93 mg/dL • Group IV (Alloxan-induced diabetic rats with glibenclamide) 238.77 ± 21.14 mg/dL	Verma et al. (2016)

(Continued)

TABLE 5.2
(Continued)

Serial No.	Sample	Intervention	Result (blood glucose level)	Studied by
9	30 albino alloxan-induced diabetic rats of either sex	Diabetic rats received *Aloe vera* leaf extract in three graded doses of 100, 200 and 400 mg/kg, respectively, for five weeks	• Group I (Control rats with 1 ml distilled water) 251.5±7.69 mg/dL • Group II (Control positive with metformin 50 mg/kg) 94±6.8 mg/dL • Group III (Alloxan-induced diabetics rats with 100 mg/kg extract treated) 222.5±9.17 mg/dL • Group IV (Alloxan-induced diabetics rats with 200 mg/kg extract treated) 106±3.8 mg/dL • Group V (Alloxan-induced diabetics rats with 300 mg/kg extract treated) 76.6±1.8 mg/dL	Manjunath et al. (2016)
10	24 Wistar male albino streptozotocin-induced diabetic rats	Diabetic rats treated with *Aloe vera* extract (300 mg/kg) for 21 days	• Group I (Control rats received 0.1 M citrate buffer) 85.3 ± 5.8 mg/dL • Group II (Diabetic controls, streptozotocin-induced diabetic rats) 270 ± 13 mg/dL • Group III (Streptozotocin-induced diabetic rats treated with *Aloe vera* extract 300 mg/kg) 102.5 ± 10.5 mg/dL • Group IV (Streptozotocin-induced diabetic rats treated with glibenclamide 600mg/kg) 128.5 ± 12 mg/dL	Rajasekaran et al. (2010)
11	18 Wistar male alloxan-induced diabetic rats	Diabetic rats received 400 IU/day/adult instant *Aloe vera* for five weeks	• Group I (Normal control) around >50 mg/dl • Group II (Diabetic rats) around >200 mg/dl • Group III (Diabetic rats with vitamin E) 100mg/dl • Group IV (Diabetic rats with *Aloe vera* treatment) >100 mg/dl	Riyanto and Wariyah (2018)
12	24 Albino alloxan-induced diabetic rabbits of either sex	Diabetic rabbits received *Aloe vera* leaf gel extract (300 mg/kg) in an aqueous solution for 21 days	• Group I (Healthy rabbits "normal control") 150 ± 16.08 mg/dL • Group II (Diabetic rabbits) 266.17 ± 14.08 mg/dL • Group III (Diabetic rabbits with *Aloe vera*) 182 ± 12.26 mg/dL • Group IV (Diabetic rabbits with glibenclamide "positive control") 160 ± 13.84 mg/dL	Sethi et al. (2012)
13	40 male albino alloxan-induced	Diabetic rats fed on a pellet diet plus *Aloe vera* gel extract (0.5 ml/day) for five weeks	• Group I (Healthy rats "normal control") 82.58 ± 9.8 mg/dL • Group II (Diabetic rats) 202.43 ± 7.31 mg/dL • Group III (Healthy rats with *Aloe vera* gel "positive control") 83.45 ± 4.59 mg/dL • Group IV (Diabetic rats with *Aloe vera* gel) 96.23 ± 9.05 mg/dL	Mohamed (2011)

5.6.2 Mechanism behind the Hypoglycaemic Activity

Oxidative stress may directly increase the establishment of diabetes mellitus by reducing insulin sensitivity and damaging pancreatic cells that produce insulin. Reactive oxygen species (ROS) contribute to mitochondrial DNA damage in the pancreatic cells leading to reduced pancreatic β-cell activity (Nieto-Vazquez et al. 2007). Molecular pathways that mediate oxidative stress are intricately connected to diabetic complications (Yang et al. 2011). Therefore, drug research should not just concentrate on insulin-centric targets; rather, it should also include oxidant-scavenging tactics, thus protecting β-cells against free radical damage. This may also promote the repair of β-cells damaged by oxidative stress. Several articles describe that *Aloe vera* is rich in antioxidant agents such as α-tocopherol, ascorbic acid, carotenoids and many other phenolic substances (Radha and Laxmipriya 2015). A study by Kang et al. (2014) shows that purified polysaccharide (APS) from *Aloe vera* scavenged free radicals including DPPH, hydroxyl, and alkyl radicals, with high efficiency in *in vitro* experiments. In addition, APS demonstrated a protective impact against oxidative stress induced by 2,2-azobis(2-amidinopropane) dihydrochloride in both Vero cells and the *in vivo* zebrafish model. The effectiveness of ethanolic extract of *A. vera* is also proved in neutralizing oxidative stress through its reducing power (Miladi and Damak 2008). Based on decades of research, it is strongly believed by many researchers that the hypoglycaemic effect of *A. vera* is mediated by its powerful antioxidant properties (Rajasekaran et al. 2004). *A. vera* also showed an anti-lipid peroxidation effect, which in turn prevents membrane lipid damage in β-cells in the pancreas (Parihar et al. 2004). Diabetic condition is often correlated with high cellular levels of 8-oxo-2'-deoxyguanosine and DNA fragmentation. *A. vera* extract brings down the 8-oxo-dG level and DNA fragmentation considerably in streptozotocin-induced diabetes (Christijanti et al. 2019). Unfortunately, the molecular mechanism behind the antidiabetic activities of *Aloe vera* is still understudied. Kumar et al. (2011) showed that aqueous *Aloe* extract (1 mg/ml) upregulated the GLUT-4 mRNA synthesis in mouse embryonic NIH/3T3 cells (Mohammed et al. 2020). Another interesting compound found in this plant in moderate quantities is melatonin (A64). By triggering melatonin receptors (MT1 and MT2), melatonin may influence insulin secretion. Melatonin is also demonstrated to trigger the phospholipase-C/IP3 pathway, which mobilizes calcium from organelles and promotes insulin production (Sharma et al. 2013).

5.7 TRADITIONAL AND OTHER POTENTIAL USES

For thousands of years, traditional and folk medicines have employed *A. vera*, also known as Barbados/Curaçao aloe, to treat and cure a wide range of illnesses. Traditional remedies have been found in India, Mexico, China, West Indies and Middle America as well as the Caribbean (Grundmann 2012). *A. vera* has played a significant traditional role in endemic medical systems including Unani, Siddha, homeopathy and Ayurveda (Pathak and Sharma 2017). Uses of *A. vera* in folk medicine, for embalming and beauty regimes, are found in Greek, Egyptian and Roman literatures as well as in Assyrians and Mediterranean civilizations. Many pharmacopoeias still include aloe as a laxative since it has a very long history of use as a potent laxative for treating chronic constipation (Manvitha and Bidya 2014). Some African tribes use the transparent, mucilaginous gel to treat chronic conjunctivitis, while hunters in the Congo purportedly put it on their bodies to reduce sweat. The gel is also used to treat asthma in India (Morton 1961). Aloe has traditionally been used to treat burns, particularly to speed up healing, lessen inflammation and prevent tissue scarring. Dioscorides describes the gel as being used to heal ulcers, alleviate itching and treat mouth infections. Trinidad & Tobago uses *Aloe vera* gel as an alternative treatment for hypertension (Grindlay and Reynolds 1986). *Aloe vera* gel has been used as a common household medication in the United States since reports of its effectiveness in treating radiation dermatitis and is still widely grown and used for abrasions and burns (Lans 2006). Aloe is thought to restore youthful vitality and femininity. In traditional Indian medicine, it is used as a tonic for the female reproductive system and as a cure for constipation, colic, skin conditions, worm infestations and infections. It also has revitalizing, purgative and vulnerary effects. In addition, it has other uses as a laxative, anthelminthic, anti-haemorrhoidal and uterine stimulant. Along with liquorice root, aloe extract is also applied topically to treat psoriasis or eczema. Aloe can be consumed as food. In Tamil Nadu, India, people frequently make a curry with *A. vera*, which is eaten with rice or nan bread (Indian bread) (Ghazanfar 1994; Lanka 2018)

5.8 SAFETY ISSUES

Before use of the *A. vera* products in cosmetics or in medicinal trial, investigations must be conducted, confirmed and recognized by government regulatory agencies to ensure safe use (Ahlawat and Khatkar 2011). The safety and toxicological characteristics of *A. vera* products used in food applications need to be investigated by regulatory organizations as well (Eshun and He 2004). One group of scientists advises utilizing it carefully and cautiously to prevent aloin contamination from the yellow exudates because aloin has been linked to cancer and DNA damage (Lachenmeier et al. 2005). *A. vera* gel is reportedly safe to apply externally, allergies are uncommon and harmful drug interactions have not been documented. Aloe should not be used internally by anybody experiencing stomach discomfort, appendicitis or intestinal blockage or when pregnant or nursing (Ahlawat and Khatkar 2011). In an experiment using up to a maximum dose of 3330 mg/kg bodyweight, *A. vera* soft capsules did not exhibit any overt subacute hazardous effects. Additionally, the outcomes of the tests for genotoxicity and acute toxicity were both negative (Wu et al. 2021). At dosages of 200, 400, and 600 mg/kg, aqueous extracts of the leaves and roots of *A. vera* did not result in any harmful side effects or animal deaths. Similarly, chronic and subchronic treatment showed that administering a methanolic gel extract of *A. vera* to animals at dosages of 1000, 2000, 4000, 8000, and 16,000 mg/kg did not result in any mortalities or modifications to any of the parameters under investigation. Similarly, over the course of therapy, an aqueous leaf extract and a supercritical carbon dioxide gel extract of *A. vera* did not result in any deaths or modifications to the parameters under study (Saritha and Anilakumar 2010; Tanaka et al. 2012; Sehgal et al. 2013).

5.9 CULTIVATION PRACTICES

Aloe-based products have grown in popularity around the world, particularly in the realms of medicine, food and cosmetics. Aloes thrive in a wide range of environments, including deserts, grasslands, coastal locations and even mountains. Suckers near the roots form new plants. Seeds are not typically utilized to propagate it (Manvitha and Bidya 2014). The plants take little attention and should not be overwatered. For the growth of this plant, the soil should ideally be porous; it appears that a coarse, sandy loam to which some manure has been applied is suitable (Rathore and Mathur 2019). Saks and Ish-shalom-Gordon proposed that *A. vera* may be grown in environments other than sandy soils and subtropical climates (Saks and Ish-shalom-Gordon 1995). A varied range of climatic adaptability (like dry, humid and extreme cold) can be observed for the cultivation process of *A. vera* (Das and Chattopadhay 2004; Datta et al. 2012). Generally, a warm and dry climate is best known for its cultivation (Banik and Sharangi 2019). The suitable time of cultivation is March–June, with nominal annual rainfall of 50 to 300 mm and proper protective irrigation (Datta et al. 2012). There are varied types of soil for the growth of this plant, such as sandy coastal soils, loamy soils, light soils, black cotton soils with tolerable high pH (7.0 to 8.5) and excessive Na, K salts (Maiti and Chandra 2002; Datta et al. 2012). A typical land preparation for cultivation should be maintained. Depending on the soil type and agro-climatic condition, 1–2 ploughing for levelling of soil and ridges and furrows with 45 cm distance will make up the best result. As *A. vera* propagates through axillary shoots, the cut suckers or rhizomes are to be planted in the soil keeping a two-thirds portion under the ground. For the proper growth of the plants, organic manures such as cow-dung compost, vermicompost and wood ash are best (Das and Chattopadhay 2004; Maiti and Chandra 2002; Datta et al. 2012). During the cultivation process an attack of mealy bug, termite or fungi causing anthracnose, leaf spot, might be observed, which may affect the yield and quality of gel in leaves. To get rid of this problem, a regular spraying of fungicide (to control fungal infections) and light irrigation of water (to remove termites) is to be maintained. Commercial yield of plantation is generally done from the second to the fifth year of transplantation and thereby large healthy outer leaves from the base of the plant are to be harvested. Aggarwal and Barna (2004) reported that the manual production of leaves does not meet the industry demand. The vegetative production of *A. vera* is rather a very slow multiplication along with presence of male sterility among some species of *Aloe*. An alternate process to overcome this problem is becoming a practice, where *in vitro* techniques are offering opportunities. There have been several attempts to conduct *in vitro* cultivation with stem nodes, shoot tips and microshoots analyses (Bawankar et al. 2014).

5.10 FUTURE REMARKS

The ancient Indian herb *A. vera* has a long history of therapeutic use in diverse cultures, and its pharmacological characteristics have been proved via several clinical trials. Furthermore, *A. vera* is becoming increasingly popular as a health and cosmetic product. The efficacy and mechanism of *A. vera* products in these areas must be elucidated, and their usage in new areas must be broadened. However, there have been conflicting findings in the safety studies, so it would be beneficial to undertake additional research on the toxicological profile of *A. vera* to help avoid any potential negative effects that could arise from its use and to promote the growth of this natural product. More clinical research is also needed to validate treatment efficacy. Despite its benefits, *A. vera* consumption is not a risk-free way to decrease blood sugar in diabetics. It is crucial to remember that *A. vera*'s benefits in decreasing blood sugar levels can have other adverse effects. The bioactivity of the chemicals from *A. vera* must be further investigated, and as quickly as feasible, their pure forms must be isolated. There is a good possibility to develop *A. vera* economically, but it requires succinct information about cultivation and management.

REFERENCES

Abo-Youssef, A. M. and Hussein, B. A. S. M. 2013. Beneficial effects of Aloe vera in treatment of diabetes: Comparative in vivo and in vitro studies. *Bull. Fac. Pharm. Cairo Univ.* 51: 7–11. http://doi.org/10.1016/j.bfopcu.03.002

Aggarwal, D. and Barna, K. S. 2004. Tissue culture propagation of elite plant of *Aloe vera* Linn. *J. Plant Biochem. Biotechnol.* 13: 77–79.

Ahlawat, K. S. and Khatkar, B. S. 2011. Processing, food applications and safety of aloe vera products: A review. *J. Food Sci. Technol.* 48(5): 525–533. http://doi.org/10.1007/s13197-011-0229-z.

Ajabnoor, M. A. 1990. Effect of aloes on blood glucose levels in normaland alloxan diabetic mice. *J. Ethno. Pharmacol.* 28: 215–220.

Almeida, H. G., Campos, J. J., Kfouri, C., Tanita, M. T., Dias, A. E. and Souza, M. M. 2002. Profile of patients with diabetes type 1: Insulin therapy and self-monitoring. *Rev. Assoc. Med. Bras.* 48(2): 151–1555. http://doi.org/10.1590/s0104-42302002000200035.

Alven, S., Khwaza, V., Oyedeji, O. O. and Aderibigbe, B. A. 2021. Polymer-based scaffolds loaded with *Aloe vera* extract for the treatment of wounds. *Pharmaceutics.* 13(7). https://doi.org/10.3390/pharmaceutics13070961

Andrea, B., Dumitriţa, R., Florina, C., Francisc, D., Anastasia, V., Socaci, S. and Adela, P. 2020. Comparative analysis of some bioactive compounds in leaves of different *Aloe* species. *BMC Chem.* 14(1). https://doi.org/10.1186/s13065-020-00720-3

Arora, M. K., Sarup, Y., Tomar, R., Singh, M. and Kumar, P. 2019. Amelioration of diabetes-induced diabetic nephropathy by *Aloe vera*: Implication of oxidative stress and hyperlipidemia. *J. Dietary Suppl.* 16(2): 227–244. https://doi.org/10.1080/19390211.2018.1449159

Atanu, F. O., Avwioroko, O. J. and Momoh, S. 2018. Anti-diabetic effect of combined treatment with *Aloe vera* gel and metformin on alloxan-induced diabetic rats. *J. Ayurvedic Herb. Med.* 4(1): 1–5.

Babu, S. N. and Noor, A. 2020. Bioactive constituents of the genus *Aloe* and their potential therapeutic and pharmacological applications: A review. *J. App. Pharm. Sci.* 10(11): 133–145. https://doi.org/10.7324/JAPS.2020.101118

Bahrami, T. M., Mazaheri, Y., Khaksary, M. M., Fatemi, T. S. R. and Tabandeh, M. R. 2017. The Aloe vera gel improves structural disorders in the brain of streptozotocin-induced diabetic male rats. *J. Isfahan Med. Sci.* 35(422): 235–242.

Banik, S., Baran, S. A. and Sharangi, A. B. 2019. Phytochemistry, health benefits and toxicological profile of aloe. *J. Pharmacogn. Phytochem.* 8(3): 4499–4506.

Banik, S. and Sharangi, A. B. 2019. Phytochemistry, health benefits and toxicological profile of Aloe. *J. Pharma. Phytochem.* 8(3): 4499–4506.

Baruah, A., Bordoloi, M. and Deka Baruah, H. P. 2016. *Aloe vera*: A multipurpose industrial crop. *Ind. Crops Prod.* 94: 951–963. https://doi.org/10.1016/j.indcrop.2016.08.034

Bawankar, R., Deepti, V. C., Singh, P., Subashkumar, R., Vivekanandhan, G. and Babu, S. 2013. Evaluation of bioactive potential of an *Aloe vera* sterol extract. *Phytother. Res.* 27(6): 864–868. https://doi.org/10.1002/ptr.4827

Bawankar, R., Singh, P. and Subramanian, B. 2014. Bioactive compounds and medicinal properties of *Aloe vera* L. An update. *J. Plant Sci.* 2(3): 102–107. http://doi.org/10.11648/j.jps.20140203.11

Boudreau, M. D. and Beland, F. A. 2006. An evaluation of the biological and toxicological properties of Aloe barbadensis (Miller), Aloe vera. *J. Environ. Sci. Health.* 24: 103–154.

Bozzi, A., Perrin, C., Austin, S. and Arce, V. F. 2007. Quality and authenticity of commercial Aloe vera gel powders. *Food Chem.* 103: 22–30.

Budiastutik, I., Subagio, W. H., Kartasurya, I. M., Widjanarko, B., Kartini, A., Soegiyanto, B. and Suhartono, S. 2022. The effect of *Aloe vera* on fasting blood glucose levels in pre-diabetes and type 2 diabetes mellitus: A systematic review and meta-analysis. *J. Pharm. Pharmacogn. Res.* 10(4): 737–747. https://jppres.com

Can, A., Akev, N., Ozsoy, N., Bolken, S., Arda, B. P., Yanardag, R. and Okyar, A. 2004. Effect of Aloe vera leaf gel and pulp extracts on the liver in type-II diabetic rat models. *Biol. Pharm. Bull.* 27(5): 694–698.

Chacón, O., Forno, N., Lapierre, L., Muñoz, R., Fresno, M. and San Martín, B. 2019. Effect of Aloe barbadensis Miller (Aloe vera) associated with beta-lactam antibiotics on the occurrence of resistance in strains of Staphylococcusaureus and Streptococcus uberis. *Eur. J. Integr. Med.* 32: 100996.

Chandan, B. K., Saxena, A. K., Shukla, S., Sharma, N., Gupta, D. K., Suri, K. A., Suri, J., Bhadauria, M. and Singh, B. 2007. Hepatoprotective potential of *Aloe* barbadensis Mill. against carbon tetrachloride induced hepatotoxicity. *J. Ethnopharmacol.* 111(3): 560–566. https://doi.org/10.1016/j.jep.2007.01.008

Chen, G., Huo, Y., Tan, D. X., Liang, Z., Zhang, W. and Zhang, Y. 2003. Melatonin in Chinese medicinal herbs. *Life Sci.* 73(1): 19–26. https://doi.org/10.1016/S0024-3205(03)00252-2

Chen, S. C. and Tseng, C. H. 2013. Dyslipidemia, kidney disease, and cardiovascular disease in diabetic patients. *Rev. Diabet. Stud.* 10(2–3): 88–100. https://doi.org/10.1900/RDS.2013.10.88

Choi, S. W., Son, B. W., Son, Y. S., Park, Y. I., Lee, S. K. and Chung, M. H. 2001. The wound-healing effect of a glycoprotein fraction isolated from *Aloe vera*. *Br. J. Dermatol.* 145: 535–545.

Choudhary, M., Anita, K. and Jaswinder, S. 2014. Hypoglycemic and hypolipidemic effect of Aloe vera L. in non-insulin dependent diabetics. *J. Food Sci. Technol.* 51(1): 90–96. http://doi.org/10.1007/s13197-011-0459-0

Choudhury, H., Pandey, M., Hua, C. K., Mun, C. S., Jing, J. K., Kong, L., Ern, L. Y., Ashraf, N. A., Kit, S. W., Yee, T. S., Pichika, M. R., Gorain, B. and Kesharwani, P. 2018. An update on natural compounds in the remedy of diabetes mellitus: A systematic review. *J. Tradit. Complement. Med.* 8(3): 361–376. https://doi.org/10.1016/j.jtcme.2017.08.012

Christijanti, W., Juniarto, A. Z. and Suromo, L. B. 2019. Sperm DNA fragmentation index andmalondialdehyde of diabetic rats treated by *Aloe vera* peel extract. *Pakistan J. Med. Heal. Sci.* 13(4): 1237–1240. https://doi.org/10.4103/2230-8210.104057

Dammak, I., Lasram, S., Hamdi, Z., ben Moussa, O., Mkadmini Hammi, K., Trigui, I., Houissa, H., Mliki, A. and Hassouna, M. (2018). In vitro antifungal and anti-ochratoxigenic activities of *Aloe vera* gel against Aspergillus carbonarius isolated from grapes. *Ind. Crops Prod.* 123: 416–423. https://doi.org/10.1016/j.indcrop.2018.07.023

Das, N. and Chattopadhay, R. N. 2004. Commercial cultivation of aloe. *Nat. Prod. Radiance* 3: 85–87.

Datta, A. K., Mandal, A., Jaime, A., da Silva, T., Saha, A., Paul, R., Sengupta, S., Dubey, P. K. and Sandip Halder, S. 2012. An updated overview on *Aloe vera* (L.) Burm. f. *Med. Aromat. Plant. Sci. Biotechnol.* 6(1): 1–11.

Dey, P., Dutta, S., Chowdhury, A., Das, A. P. and Chaudhuri, T. K. 2017. Variation in phytochemical composition reveals distinct divergence of *Aloe vera* (L.) Burm.f. from other *Aloe* species: Rationale behind selective preference of *Aloe vera* in nutritional and therapeutic use. *J. Evid. Based Complementary Altern. Med.* 22(4): 624–631. https://doi.org/10.1177/2156587217698292

Eshun, K. and He, Q. 2004. *Aloe vera*: A valuable ingredient for the food, pharmaceutical and cosmetic industries – a review. *Crit. Rev. Food Sci. Nutr.* 44(2): 91–96.

Farid, A., Tawfik, A., Elsioufy, B. and Safwat, G. 2021. In vitro and in vivo anti-Cryptosporidium and anti-inflammatory effects of *Aloe vera* gel in dexamethasone immunosuppressed mice. *Int. J. Parasitol. Drugs Drug Res.* 17: 156–167. https://doi.org/10.1016/j.ijpddr.2021.09.002

Finucane, O. M., Lyons, C. L., Murphy, A. M., Reynolds, C. M., Klinger, R., Healy, N. P., Cooke, A. A., Coll, R. C., Mcallan, L., Nilaweera, K. N., O'Reilly, M. E., Tierney, A. C., Morine, M. J., Alcala-Diaz, J. F., Lopez-Miranda, J., O'Connor, D. P., O'Neill, L. A., Mcgillicuddy, F. C. and Roche, H. M. 2015. Monounsaturated fatty acid-enriched high-fat diets impede adipose NLRP3 inflammasome-mediated IL-1β secretion and insulin resistance despite obesity. *Diabetes*. 64(6): 2116–2128. https://doi.org/10.2337/db14-1098

Foadoddini, M. and Samaneh, A. M. 2020. Effect of Aloe vera extract on depression in people with prediabetes. *Mod. Care J.* 7(2): e100927. http://doi.org/10.5812/modernc.100927

Gerstein, H. C., Santaguida, P., Raina, P., Morrison, K. M., Balion, C., Hunt, D., Yazdi, H. and Booker, L. 2007. Annual incidence and relative risk of diabetes in people with various categories of dysglycemia: A systematic overview and meta-analysis of prospective studies. *Diabetes Res. Clin. Pract.* 78(3): 305–312. https://doi.org/10.1016/j.diabres.2007.05.004

Ghazanfar, S. A. 1994. *Handbook of Arabian Medicinal Plants*. CRC Press, Boca Raton.

Grindlay, D. and Reynolds, T. 1986. The *Aloe vera* phenomenon: A review of the properties and modern uses of the leaf parenchyma gel. *J. Ethnopharmacol.* 16(2–3): 117–151.

Grundmann, O. 2012. *Aloe vera* gel research review. *J. Nat. Med.* 4(9).

Gupta, V. and Malhotra, S. 2012. Pharmacological attribute of *Aloe vera*: Revalidation through experimental and clinical studies. *AYU.* 33(2): 193–196. https://doi.org/10.4103/0974-8520.105237

Haghani, F., Arabnezhad, M. R., Mohammadi, S. and Ghaffarian-Bahraman, A. 2022. *Aloe vera* and streptozotocin-induced diabetes mellitus. *Rev. Bras. Farmacogn.* 32(2): 174–187. https://doi.org/10.1007/s43450-022-00231-3

Hamman, J. H. 2008. Composition and applications of *Aloe vera* leaf gel. *Molecules*. 13(8): 1599–1616. https://doi.org/10.3390/molecules13081599

Hardy, K. 2021. Paleomedicine and the evolutionary context of medicinal plant use. *Rev. Bras. Farmacogn.* 31: 1–15. https://doi.org/10.1007/s43450-020-00107-4/Published

Hirata, T. and Suga, T. 1977. Biologically active constituents of leaves and roots of Aloe arborescens var. natalensis. *Z. Nat. C Biosci.* 32: 731–744.

Hutapea, A. M. and Chandra, S. 2021. Hypoglycemic potential of *Aloe vera* in diabetes mellitus induced by diabetogenic substances and high fat diet: A systematic meta-analysis review. *Int. J. App. Dent. Sci.* 7(3): 360–368. http://doi.org/10.22271/oral.2021.v7.i3f.1322.

Jangra, A., Garima, S., Sonia, S. and Vinod, C. 2021. The dark side of miracle plant-*Aloe vera*: A review. *Mol. Biol. Rep.* 49(6): 5029–5040. http://doi.org/10.1007/s11033-022-07176-9.

Kahramanoğlu, I., Chen, C., Chen, J. and Wan, C. 2019. Chemical constituents, antimicrobial activity and food preservative characteristics of *Aloe vera* gel. *Agronomy*. 9(12). https://doi.org/10.3390/agronomy9120831

Kang, M. C., Kim, S. Y., Kim, Y. T., Kim, E. A., Lee, S. H., Ko, S. C., Wijesinghe, W. A. J. P., Samarakoon, K. W., Kim, Y. S., Cho, J. H., Jang, H. S. and Jeon, Y. J. 2014. In vitro and in vivo antioxidant activities of polysaccharide purified from *Aloe vera* (*Aloe* barbadensis) gel. *Carbohydr. Polym.* 99: 365–371. https://doi.org/10.1016/j.carbpol.2013.07.091

Kaur, R., Willis, S., Shackelford, L., Walker, L. T. and Verghese, M. 2022. Determination of antioxidant potential of selected parts of *Aloe vera* plant. *J. Food Nutr. Sci.* 4(1): 50–67.

Khajeeyana, R., Salehia, A., Movahhedi, D. M., Farajee, H. and Amin, K. M. 2019. Physiological and yield responses of Aloe vera plant to biofertilizers under different irrigation regimes. *Agric. Water Manag.* 225: 105768.

Kim, K., Kim, H., Kwon, J., Lee, S., Kong, H., Im, S. A., Lee, Y. H., Lee, Y. R., Oh, S. T., Jo, T. H., Park, Y. I., Lee, C. K. and Kim, K. 2009. Hypoglycemic and hypolipidemic effects of processed *Aloe vera* gel in a mouse model of non-insulin-dependent diabetes mellitus. *Phytomedicine*. 16(9): 856–863. https://doi.org/10.1016/j.phymed.2009.02.014

Kumar, M., Rakesh, S., Nagpal, R., Hemalatha, R., Ramakrishna, A., Sudarshan, V., Ramagoni, R., Shujauddin, M., Verma, V., Kumar, A., Tiwari, A., Singh, B. and Kumar, R. 2013. Probiotic Lactobacillus rhamnosus

GG and *Aloe vera* gel improve lipid profiles in hypercholesterolemic rats. *Nutrition*. 29(3): 574–579. https://doi.org/10.1016/j.nut.2012.09.006

Kumar, R., Sharma, B., Tomar, N. R., Roy, P., Gupta, A. K. and Kumar, A. 2011. In vivo evaluation of hypoglycemic activity of *Aloe* spp. and identification of its mode of Action on GLUT-4 gene expression in vitro. *Appl. Biochem. Biotechnol.* 164(8): 1246–1256. https://doi.org/10.1007/s12010-011-9210-6

Kumar, R., Singh, A. K., Gupta, A., Bishayee, A. and Pandey, A. K. 2019. Therapeutic potential of *Aloe vera* – A miracle gift of nature. *Phytomedicine*. 60. https://doi.org/10.1016/j.phymed.2019.152996

Kumar, S. and Yadav, J. P. 2014. Ethnobotanical and pharmacological properties of Aloe vera: A review. *J. Med. Plant Res.* 8(48): 1387–1398. http://doi.org/10.5897/JMPR2014.5336x

Lachenmeier, K., Kuepper, U., Musshoff, F., Madea, B., Reusch, H. and Lachenmeier, D. W. 2005. Quality control of *Aloe vera* beverages. *Elec. J. Env. Agricult. Food Chem.* 4(4): 1033–1042.

Lanka, S. 2018. A review on Aloe vera-the wonder medicinal plant. *J Drug Deliv Ther.* 8(5-s): 94–99.

Lans, C. A. 2006. Ethnomedicines used in Trinidad and Tobago for urinary problems and diabetes mellitus. *J. Ethnobiol. Ethnomed.* 2(1): 1–11.

Li, T., Ni, L., Zhao, Z., Liu, X., Lai, Z., Di, X., Xie, Z., Song, X., Wang, X., Zhang, R. and Liu, C. 2018. Melatonin attenuates smoking-induced hyperglycemia via preserving insulin secretion and hepatic glycogen synthesis in rats. *J. Pineal Res.* 64(4). https://doi.org/10.1111/jpi.12475.

Mahabady, M. K., Naeem, E. M., Mohammad, B. T. and Yazdan, M. 2021a. Evaluating the effects of *Aloe vera* gel on cerebellum istomorphometrical changes in diabetic male rats. *Zahedan J. Res. Med. Sci.* 23(3): e102483. http://doi.org/10.5812/zjrms.102483.

Mahabady, M. K., Tapebur, M. B., Mazaheri, Y., Tabandeh, R. M. and Tabatabaei, S. R. F. 2021b. Effect of *Aloe vera* on the expression of nerve factors, p75 and TrkA receptors in the hippocampus of diabetic rats efecto del *Aloe vera* sobre la Expresi6n de Factores de Crecimiento Nervioso, Receptores p75 y TrkA en el Hipocampo de Ratas Diabéticas. *Int. J. Morphol.* 39(2): 577–586.

Mahor, G. and Sharique, A. A. 2016. Recent update on the medicinal properties and use of Aloe vera in the treatment of various ailments. *Bios. Biotech. Res. Comm.* 9(2): 273–288.

Maiti, S. and Chandra, R. 2002. *Cultivation of Aloe vera. National Research Centre for Medicinal and Aromatic Plants, Extension Bulletin.* Anand Press, Gujarat.

Majumder, R., Das, C. K. and Mandal, M. 2019. Lead bioactive compounds of *Aloe vera* as potential anticancer agent. *Pharmacol. Res.* 148. https://doi.org/10.1016/j.phrs.2019.104416

Manjunath, K., Bhanu Prakash, G., Subash, K. R., Tadvi, N. A., Manikanta, M. and Umamaheswara Rao, K. 2016. Effect of *Aloe vera* leaf extract on blood glucose levels in alloxan induced diabetic rats. *Natl. J. Physiol. Pharm. Pharmacol.* 6(5): 471–474. https://doi.org/10.5455/njppp.2016.6.0718613072016

Manvitha, K. and Bidya, B. 2014. *Aloe vera*: A wonder plant its history, cultivation and medicinal uses. *J. Pharmacogn. Phytochem.* 2(5): 85–88.

Martínez-Sánchez, A., María, E., López-Cañavate, J. G. M., María, J. R. and Encarna, A. 2020. Aloe vera flowers, a byproduct with great potential and wide application, depending on maturity stage. *Foods.* 9(11): 1542. http://doi.org/10.3390/foods9111542.

Miladi, S. and Damak, M. 2008. In vitro antioxidant activities of *Aloe vera* leaf skin extracts. *J. de La Soc. Chimique de Tunisie.* 10: 101–109.

Misawa, E., Tanaka, M., Nabeshima, K., Nomaguchi, K., Yamada, M., Toida, T. and Iwatsuki, K. 2012. Administration of dried *Aloe vera* gel powder reduced body fat mass in diet-induced obesity (DIO) rats. *J. Nutr. Sci. Vitaminol.* 58: 195–201.

Misir, J., Brishti, F. H. and Hoque, M. M. 2014. *Aloe vera* gel as a novel edible coating for fresh fruits: A review. *Am. J. Food Sci. Technol.* 2(3): 93–97. https://doi.org/10.12691/ajfst-2-3-3

Mohamed, A. K. M. 2011. Antidiabetic, antihypercholestermic and antioxidative effect of *Aloe vera* gel extract in alloxan induced diabetic rats. *Aust. J. Basic Appl. Sci.* 5(11): 1321–1327.

Mohammed, A., Ibrahim, M. A., Tajuddeen, N., Aliyu, A. B. and Isah, M. B. 2020. Antidiabetic potential of anthraquinones: A review. *Phytotherapy Res.* 34(3): 486–504. https://doi.org/10.1002/ptr.6544

Moniruzzaman, M., Rokeya, B., Ahmed, S., Bhowmik, A., Khalil, M. I. and Gan, S. H. 2012. In vitro antioxidant effects of *Aloe* barbadensis miller extracts and the potential role of these extracts as antidiabetic and antilipidemic agents on streptozotocin-induced type 2 diabetic model rats. *Molecules.* 17(11): 12851–12867. https://doi.org/10.3390/molecules171112851

Morton, J. F. 1961. Folk uses and commercial exploitation of *Aloe* leaf pulp. *Econ. Bot.* 15(4): 311–319.

Mukherjee, P. K., Nema, N. K., Maity, N., Mukherjee, K. and Harwansh, R. K. 2014. Phytochemical and therapeutic profile of *Aloe vera*. *J. Nat. Rem.* 14(1). www.jnronline.com

Mulay, S., Borade, A., Bele, A. and Khale, A. 2013. *Aloe vera* – A review. *Int. J. Pharm. Phytopharmacol. Res.* 3(3): 203–211. www.eijppr.com

Nieto-Vazquez, I., Fernández-Veledo, S., de Alvaro, C., Rondinone, C. M., Valverde, A. M. and Lorenzo, M. 2007. Protein-tyrosine phosphatase 1B-deficient myocytes show increased insulin sensitivity and protection against tumor necrosis factor-α-induced insulin resistance. *Diabetes*. 56(2): 404–413. https://doi.org/10.2337/db06-0989

Parihar, A. S., Kumar, V., Sinha, A. and Sharma, S. 2019. Characterization of malassezia furfur and its control by using plant extracts. *Int. J. Chem. Stud.* 550(5): 550–553.

Parihar, M. S., Chaudhary, M., Shetty, R. and Hemnani, T. 2004. Susceptibility of hippocampus and cerebral cortex to oxidative damage in streptozotocin treated mice: Prevention by extracts of Withania somnifera and *Aloe vera*. *J. Clin. Neurosci.* 11(4): 397–402. https://doi.org/10.1016/j.jocn.2003.09.008

Pathak, D. and Sharma, R. 2017. Review on "*Aloe vera*-medicinal plant". *Int. J. Adv. Res. Innov. Ideas Educ.* 3(1): 661.

Quispe, C., Villalobos, M., Bórquez, J. and Simirgiotis, M. 2018. Chemical composition and antioxidant activity of *Aloe vera* from the pica oasis (Tarapacá, Chile) by UHPLC-Q/Orbitrap/MS/MS. *J. Chem.* 2018: 6123850. https://doi.org/10.1155/2018/6123850

Radha, M. H. and Laxmipriya, N. P. 2015. Evaluation of biological properties and clinical effectiveness of *Aloe vera*: A systematic review. *J. Tradit. Complement. Med.* 5(1): 21–26. https://doi.org/10.1016/j.jtcme.2014.10.006

Rajasekaran, S., Sivagnanam, K., Ravi, K. and Subramanian, S. 2004. Hypoglycemic effect of *Aloe vera* gel on streptozotocin-induced diabetes in experimental rats. *J. Med. Food.* 7(1): 61–66.

Rajasekaran, S., Sivagnaman, K. and Subramanian, S. 2005. Modulatoryeffects of aloe vera leaf gel extract on oxidative stress in ratstreated with streptozotocin. *J. Pharm. Pharmacol.* 57: 241–246.

Rajasekaran, S., Sivagnanam, K. and Subramanian, S. 2010. Modulatory effects of *Aloe vera* leaf gel extract on oxidative stress in rats treated with streptozotocin. *J. Pharm. Pharmacol.* 57(2): 241–246. https://doi.org/10.1211/0022357055416

Rathore, R. and Mathur, A. 2019. Scope of cultivation and value chain perspectives of medicinal herbs in India: A case study on aloe Vera and Isabgol. *J. Pharmacogn. Phytochem.* 8: 243–246.

Reynolds, T. 1985. The compounds in *Aloe* leaf exudates: A review. *Bot. J. Linn. Soc.* 90: 157–177.

Reynolds, T. and Dweck, A. C. 1999. Aloe vera gel leaf, a review update. *J. Ethnopharm.* 68: 3–37.

Riyanto, A. and Wariyah, C. 2018. Hypoglycemic effect of instant *Aloe vera* on the diabetic rats. *Food Res.* 2(1): 46–50. https://doi.org/10.26656/fr.2017.2(1).119

Sahu, P. K., Singh, R., Dayal Giri, D., Pandey, P., Gupta, S., Kumar Shrivastava, A., Kumar, A. and Dev Pandey, K. 2013. Therapeutic and medicinal uses of *Aloe vera*: A review. *Pharm. Pharmacol.* 4: 599–610. www.scirp.org/journal/pp

Saks, Y. and Ish-shalom-Gordon, N. 1995. Aloe vera L., a potential crop for cultivation under conditions of low-temperature winter and basalt soils. *Ind. Crops Prod.* 4(2): 85–90.

Saleem, M. 2009. Lupeol, a novel anti-inflammatory and anti-cancer dietary triterpene. *Cancer Lett.* 285(2): 109–115. https://doi.org/10.1016/j.canlet.2009.04.033

Salehi, B., Albayrak, S., Antolak, H., Kręgiel, D., Pawlikowska, E., Sharifi-Rad, M., Uprety, Y., Fokou, P. V. T., Yousef, Z., Zakaria, Z. A., Varoni, E. M., Sharopov, F., Martins, N., Iriti, M. and Sharifi-Rad, J. 2018. *Aloe* genus plants: From farm to food applications and phytopharmacotherapy. *Int. J. Mol. Sci.* 19(9). https://doi.org/10.3390/ijms19092843

Sampedro, M. C., Artola, R. L., Murature, M., Murature, D., Ditamo, Y., Roth, G. A. and Kivatinitz, S. 2004. Mannan from *Aloe* saponaria inhibits tumoral cell activation and proliferation. *Int. Immunopharmacol.* 4(3): 411–418. https://doi.org/10.1016/j.intimp.2003.12.016

Saritha, V. and Anilakumar, K. R. 2010. Toxicological evaluation of methanol extract of *Aloe vera* in rats. *Int. J. Pharm. Biol. Res. (IJPBR).* 1(5): 142–149.

Sehgal, I., Winters, W. D., Scott, M., David, A., Gillis, G., Stoufflet, T., Nair, A. and Kousoulas, K. 2013. Toxicologic assessment of a commercial decolorized whole leaf *Aloe vera* juice, lily of the desert filtered whole leaf juice with aloesorb. *J. Toxicol.*: 802453. https://doi.org/10.1155/2013/802453.

Sethi, J., Gupta, A., Sood, S., Dahiya, K., Singh, G. and Gupta, R. 2012. Antioxidant effect of *Aloe vera* in experimentally induced diabetes mellitus. *IJPSR*. 3(8): 8. www.ijpsr.com

Shalabi, M., Khilo, K., Zakaria, M. M., Elsebaei, M. G., Abdo, W. and Awadin, W. 2015. Anticancer activity of *Aloe vera* and Calligonum comosum extracts separetely on hepatocellular carcinoma cells. *Asian Pac. J. Trop. Biomed.* 5(5): 375–381. https://doi.org/10.1016/S2221-1691(15)30372-5

Sharma, B., Siddiqui, S., Ram, G., Chaudhary, M. and Sharma, G. 2013. Hypoglycemic and hepatoprotective effects of processed *Aloe vera* gel in a mice model of alloxan induced diabetes mellitus. *J. Diabetes Metab.* 4(9). https://doi.org/10.4172/2155-6156.1000303

Sierra-García, G. D., Castro-Ríos, R., González-Horta, A., Lara-Arias, J., Chávez-Montes, A. and González, J. E. 2014. Acemannan, an extracted polysaccharide from *Aloe vera*: A literature review. *Nat. Prod. Comm.* 9(8): 1217–1221.

Stark, A. H., Crawford, M. A. and Reifen, R. 2008. Update on alpha-linolenic acid. *Nutri. Rev.* 66(6): 326–332. https://doi.org/10.1111/j.1753-4887.2008.00040.x

Surjushe, A., Vasani, R. and Saple, D. G. 2008. Aloe vera: A short review. *Indian J. Dermatol.* 53(4): 163–166. https://doi.org/10.4103/0019-5154.44785

Tanaka, M., Yamada, M., Toida, T. and Iwatsuki, K. 2012. Safety evaluation of supercritical carbon dioxide extract of *Aloe vera* gel. *J. Food Sci.* 77(1): T2–T9.

Tarameshloo, M., Norouzian, M., Zarein-Dolab, S., Dadpay, M., Mohsenifar, J. and Gazor, R. 2012. *Aloe vera* gel and thyroid hormone cream may improve wound healing in Wistar rats. *Anat. Cell Biol.* 45(3): 170. https://doi.org/10.5115/acb.2012.45.3.170

Tong, X., Lao, C., Li, D., Du, J., Chen, J., Xu, W., Li, L., Ye, H., Guo, X. and Li, J. 2022. An acetylated mannan isolated from *Aloe vera* induce colorectal cancer cells apoptosis via mitochondrial pathway. *Carbohydr. Polym.* 291. https://doi.org/10.1016/j.carbpol.2022.119464

Unuofin, J. O. and Sogolo, L. L. 2020. Antioxidant effects and mechanisms of medicinal plants and their bioactive compounds for the prevention and treatment of type 2 diabetes: An updated review. *Oxid. Med. Cell. Long.*: Article ID 1356893. http://doi.org/10.1155/2020/1356893

Verma, P., Kumar, V., Rathore, B., Singh, R. K. and Mahdi, A. A. 2016. Anti-diabetic and anti-oxidant properties of A*loe vera* in alloxan induced diabetic rats. *Int. J. Pharmacogn.* 3(7): 319–324. https://doi.org/10.13040/IJPSR.0975-8232.IJP.3(7).319–24

Vinarova, L., Vinarov, Z., Atanasov, V., Pantcheva, I., Tcholakova, S., Denkov, N. and Stoyanov, S. 2015. Lowering of cholesterol bioaccessibility and serum concentrations by saponins: In vitro and in vivo studies. *Food Funct.* 6(2): 501–512. https://doi.org/10.1039/c4fo00785a

Wang, Z., Tang, T., Wang, S., Cai, T., Tao, H., Zhang, Q., Qi, S. and Qi, Z. 2020. Aloin inhibits the proliferation and migration of gastric cancer cells by regulating NOX2–ROS-mediated pro-survival signal pathways. *Drug Des. Dev. Ther.* 14, 145–155. https://doi.org/10.2147/DDDT.S219247

West, D. P. and Zhu, Y. F. 2003. Evaluation of *Aloe vera* gel gloves in the treatment of dry skin associated with occupational exposure. *Am. J. Infect. Control.* 31(1): 40–42. https://doi.org/10.1067/mic.2003.12

Wu, J., Zhang, Y., Lv, Z., Yu, P. and Shi, W. 2021. Safety evaluation of *Aloe vera* soft capsule in acute, subacute toxicity and genotoxicity study. *PLoS ONE.* 16(3): 0249356.

Xu, X. L., Li, S., Zhang, R. and Le, W. D. 2022. Neuroprotective effects of naturally sourced bioactive polysaccharides: An update. *Neural Regen. Res.* 17(9): 1907–1912. https://doi.org/10.4103/1673-5374.335142

Yagi, A., Hegazy, S., Kabbash, A. and Wahab, E. A. 2009. Possible hypoglycemic effect of Aloe vera L. high molecular weight fractions on type 2 diabetic patients. *Saudi Pharm. J.* 17: 209–215.

Yang, H., Jin, X., Lam, C. W. K. and Yan, S. K. 2011. Oxidative stress and diabetes mellitus. *Clin. Chem. Lab. Med.* 49(11): 1773–1782. https://doi.org/10.1515/CCLM.2011.250

Yeh, G. Y., Eisenberg, D. M., Kaptchuk, T. J. and Phillips, R. S. 2003. Systematic review of herbs and dietary supplements for glycemic control in diabetes. *Diabetes Care.* 26(4): 1277–1294. http://doi.org/10.2337/diacare.26.4.1277

Yimam, M., Zhao, J., Corneliusen, B., Pantier, M., Brownell, L. and Jia, Q. 2014. Blood glucose lowering activity of *Aloe* based composition, UP780, in alloxan induced insulin dependent mouse diabetes model. *Diabetes Metab. Syndr.* 6(61). www.dmsjournal.com/content/6/1/61

Zhang, Y., Wen, L. D. L., Tieyun, Z. and Haoming, T. 2016. Efficacy of aloe vera supplementation on prediabetes and early non-treated diabetic patients: A systematic review and meta-analysis of randomized controlled trials. *Nutrients.* 8(7): 388. http://doi.org/10.3390/nu8070388

6 Phytochemical Constituents and Antidiabetic Features of Black Cumin (*Nigella sativa* L.)

Rajib Hossain, Rasel Ahmed Khan, Nabarun Mukhopadhyay, Divya Jain, and Muhammad Torequl Islam

CONTENTS

ABBREVIATIONS

ACC: Acetyl-CoA carboxylase
ADAM-17: A disintegrin and metalloprotease 17
AMPK: Adenosine monophosphate kinase
eNOS: Endothelial nitric oxide synthase
GPx: Glutathione peroxidase
HbA_{1C}: Haemoglobin A1c
IGF-1: Insulin-like growth factor-1
LOX-1: Lectin-like oxidized low-density lipoprotein (LDL) receptor-1
MDA: Malondialdehyde
NF-κB: Nuclear factor kappa-light-chain-enhancer of activated B cells
NSO: *Nigella sativa* oil

DOI: 10.1201/b23347-6

P3K:	Phosphoinositide-3 kinase
PPAR-γ:	Peroxisome proliferator-activated receptor-γ
T3DM:	Type-3 diabetes mellitus
TGF-β1:	Transforming growth factor-β1
TNF-α:	Tumour necrosis factor-α
S100B:	S100 calcium binding protein B
SOD:	Superoxide dismutase
VCAM-1:	Vascular cell adhesion protein 1
VEGF-A:	Vascular endothelial growth factor A

6.1 INTRODUCTION

Nigella sativa (*N. sativa*, NS) is an herbaceous annual flowering plant belonging to the family Ranunculaceae. This plant is widely cultivated in many countries, such as Bangladesh, India, Pakistan, Syria, Turkey, and Saudi Arabia (Ahmad et al., 2013). NS is also called black cumin, black caraway seeds, kalonji, krishnajirika, and kalajira, among other names (Khan, 1999), and has an important herbal medicinal value in different traditional systems of medicine like Ayurveda, Unani, Chinese, and Arabic (Goreja, 2003; Sharma et al., 2005). According to one of the Prophetic hadiths, this plant is considered one of the most important healing medicines available and can be used for the treatment of all diseases except death. *N. sativa* seeds consist of fixed oil, volatile oil, quinones, alkaloids, saponins, and other compounds (Liu et al., 2011; Botnick et al., 2012). The important active constituents that are present in this plant include thymoquinone (TQ), thymohydroquinone, dithymoquinone, p-cymene, carvacrol, 4-terpineol, t-anethol, thymol, melantin, nijillin, thymoquinone, dithymyquinone, and damascening. Two different types of alkaloids are also present in *N. sativa* seeds; i.e. isoquinoline alkaloids named as nigellicimine and nigellicimine N-oxide and pyrazol alkaloids named as nigellidine and nigellicine. The seeds also consist of a water soluble pentacyclic triterpene and saponin named as alpha-hederin, which can be used as an anticancer agent (Al-Jassir et al., 1992), antioxidant, antidiabetic, anti-inflammatory, antidote, and some others. Large quantities of vitamins and minerals are also present in the seeds of NS (Nickavar et al., 2003).

Diabetes is a serious metabolic condition that is growing more prevalent across the globe. Multiple variables contribute to the disease's genesis. Medicine and medical nutrition therapy are the most important aspects of diabetic management. The use of antioxidants and anti-inflammatory spices, also known as spices, has been more common since the advent of medical nutrition therapy in the area, and their application in diabetes control has gotten a lot of attention in recent years (Sanlier and Gencer, 2020). *N. sativa* seed and its oil are especially essential for their antidiabetic effects, in addition to their many other health benefits. In addition to being ingested as a seed, NS is now accessible in capsule form and other commercial forms as a nutritional supplement (Ahmad and Alkreathy, 2018; Babaeibonab, 2019; Rahmani et al., 2022). Linoleic acid, melantin, nijillin, thymoquinone, dithiocinine, damascenin, and tannins are among the black cumin bioactive chemicals described. By inducing mitochondrial apoptosis, thymoquinone has an antidiabetic and antioxidant impact (Houcher et al., 2007; Kanter, 2008; Al-Logmani, 2009). Fasting and postprandial blood glucose and HbA1c levels were significantly higher in those with type 2 diabetes and metabolic syndrome, according to clinical research (Kanter et al., 2003; Al-Trad et al., 2016).

6.2 BOTANICAL DESCRIPTION

Originally, NS was the common name for *Bunium persicum*, later renamed *Carum bulbocastanum*, which is now near extinction, and *Carum carvi* gradually graduated to the name NS, and due to the species' inability to spread throughout India, *N. sativa* was adopted from Portuguese or Turkish merchants (Botnick et al., 2012). Punjab, Himachal Pradesh, the Gangetic plains, Bihar, Bengal, Assam, and Maharashtra are among the states where the plant is grown and distributed. Outside of India, the plant is cultivated in Syria, Lebanon, Israel, and southern Europe (Al-Jassir, 1992), as well as Bangladesh,

FIGURE 6.1 Parts of black cumin (A) leaves, (B) flower, (C) shoot, (D) unripe fruit, (E) ripe fruit, and (F) seed.

Turkey, the Middle East, and the Mediterranean basin. In several countries *N. sativa* has been introduced as different names like kalo jeeray (Bangladesh), fennel flower, nutmeg flower, Roman coriander, blackseed or black caraway, black sesame (England/USA/Australia), Assamese – kaljeera or kolajeera, Bengali – kalo jeeray, Kannada – Krishna jeerige, Tamil – karum jeerakam, Hindi – kalaunji (India), mangrail (Pakistan), chernushka (Russia), ketzakh (Hebrew), çörek out (Turkey), Habbat Al-Barakah (Saudi Arabia), siyâh dâne (Iran), jintan hitam (Indonesia), þurekot (Bosnia) (Botnick et al., 2012), nigelle de Crète or toute épice (France), schwarzkümmel (Germany), cominho-negro (Portugal), ajenuz, arañuel (Spain), and svartkummin (Sweden) (Liu et al., 2011; Naz, 2011).

The species is an erect annual herb (Figure 6.1) attaining 30.0 cm to 67.6 cm at maturity. The number of primary branches per plant ranges from 4 to 10; leaf arrangement alternate, leaf phyllotaxy 1–2, pinnae of leaves broad, number of pinna per rachis 5–6; total branches per plant (6–48); flower hermaphrodite with determinate flowing patterns, main axis terminate with a solitary flower, delicate; flower size 2.74 cm × 2.78 cm; colour – French blue (43/3 – Horticultural Colour Chart); flowers without any involucre of bracts, pedunculate; peduncle long, erect; petaloid sepals broad, ovate in a single whorl, 4–6 mostly 5 and characterized by the presence of nectarines; flower fertility 89.89%; stamens in 3 to 4 whorls, numerous (32 to 66) and shed their pollen as the filament bent outward during male phase; gynoecium 5, completely united follicles, each with a long indehiscent style and composed of variable number of multi-ovule carpel, developing into a follicle after pollination; fruit single partially connected to form a capsule-like structure (capsule 5 to 45; capsule fertility 94.5%) dehiscence through suture; fruits (length – 0.4 to 1.7 cm, seta per capsule 4 to 8, with numerous seeds; average seed production/plant; seed yield – 1.91 gm; seed viability 80% to 90%); seeds ovate, tetrangular, angles sharp, acute, more tapering at the end, colour black (000021 – British Atlas of Colour, 2007); seed size 2.33 mm ± 0.1 × 1.14 mm ± 0.02 (Mandal et al., 2011).

6.3 DISTRIBUTION

The species is cultivated and distributed all over India, especially in Punjab, Himachal Pradesh, Gangetic plains, Bihar, Bengal, Assam, and Maharashtra. Apart from India, the species is also grown in Syria, Lebanon, Israel, and southern Europe (Javidi et al., 2016), as well as in Bangladesh, Turkey, the Middle East, and the Mediterranean basin. NS is widely distributed throughout Australia, Bosnia, England, USA, China, Pakistan, Russia, Turkey, Saudi Arabia, Iran, Indonesia (Botnick et al., 2012), France, Germany, Portugal, Spain, and Sweden (Liu et al., 2011; Naz, 2011).

6.4 PHYTOCHEMICAL CONSTITUENTS

Biochemical constituents of NS seeds are fixed oil – 32% to 40% (saturated fatty acids – 30%; palmitic acid, stearic, and myristic acid; unsaturated fatty acids: arachidonic, eicosadienoic – 3%, linoleic – 50% to 60%; oleic acid – 20%; dihomolinoleic fatty acids – 10%), volatile oil – 0.4% to 0.45% (nigellone, thymoquinone, thymohydroquinone, dithymoquinone, thymol, carvacrol, Į and ǜ-pinene, d-limonene, d-citronellol, p-cymene), proteins 16%–19.9% (arginine, glutamic acid, leucine, lysine, methionine, tyrosine, proline, threonine), minerals 1.79%–3.74% (calcium, phosphorus, potassium, sodium, iron), carbohydrate 33.9%, fibre 5.50%, and water 6.0% (Bharat and Ajaikumar, 2009). Categorically different phytochemical compounds are found in NS like alkaloids carbohydrates, terpenes and terpenoids, phytosterols, tocols, polyphenols, glycerolipids phospholipids, vitamins, minerals, and alkane hydrocarbons. All the classes are provided in Table 6.1 and Figure 6.2.

TABLE 6.1
Phytochemical Constituents of Black Cumin

Class	Compound	References
Alkaloids	Nigellicimine, nigellicimine-N-oxide, pyrazole, nigellidine, nigellicine, nigelamines A1–A5	Veeramani et al. (2022)
Amino acids	Valine, phenylalanine, threonine, methionine, histidine, tryptophan, leucine, isoleucine, lysine	Sadiq et al. (2021)
Carbohydrates	Rhamnose, xylose, arabinose	Enomoto et al. (2001)
Coumarins	6-methoxy-coumarin, 7-hydroxy-coumarin, 7-oxy-coumarin	Tembhurne et al. (2014)
Fixed oil	Oleic acid, linoleic acid, dihomolinoleic acid, eicodadienoic acid; palmitic acid, stearic acid	Naz (2011), Sadiq et al. (2021)
Glycerolipids	Monoacylglycerols, diacylglycerols, triacylglycerols	Nickavar et al. (2003)
Minerals and alkane hydrocarbons	Calcium, iron, and potassium, phosphorus, zinc, n-nonane, 2-undecanone, n-octyl isobutyrate, 8-heptadecene	Sharma et al. (2005), Sadiq et al. (2021)
Phospholipids	Phosphatidylinositol, phosphatidylcholine, phosphatidylglycerol	Bourgou et al. (2008)
Phytosterols	Stigmasterol, β-sitosterol, δ7-stigmasterol, δ 5-avenasterol, campesterol, and cholesterol	Cheikh-Rouhou et al. (2008)
Polyphenols	Caftaric acid, gentisic acid, caffeic acid, chlorogenic acid, p-coumaric acid, ferulic acid, sinapic acid, cichoric acid, hyperoside, isoquercitrin, rutin, myricetin, fisetin, quercitrin, quercetin, patuletin, luteolin, kaempferol, apigenin, kaempferol 3-glucosyl galactosyl glucoside, quercetin 3-galactosyl glucoside, trigillin quercetin-3-glucoside, vanillic acid, hydroxybenzoic acid, syringic acid	Rajkapoor et al. (2002), Dabeek et al. (2019)
Saponins	α-hedrin, steryl glucosides, acetyl-steryl-glucoside	Sadiq et al. (2021)
Terpenes and terpenoids	Thymoquinone, carvacrol, 4-terpineol, α-pinene, thymol, t-anethol, thymohydroquinone, dithymoquinone, p-cymene, longifolene, thymoquinone, p-cymene, α-pinene, dithymoquinone, thymohydroquinone, carvacrol, carvone, limonene, 4-terpineol, citronellol, anethol	Enomoto et al. (2001), Cheikh-Rouhou et al. (2008), Naz (2011), Sadiq et al. (2021), Veeramani et al. (2022)
Tocols	α, β, γ, δ- tocopherol	Cheikh-Rouhou et al. (2007), Kiralan et al. (2014)
Vitamins	Vitamin A, E, and C, folic acid, thiamin, riboflavin, pyridoxine, niacin	Sharma et al. (2005)

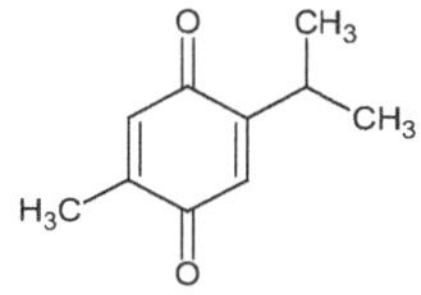
Thymoquinone

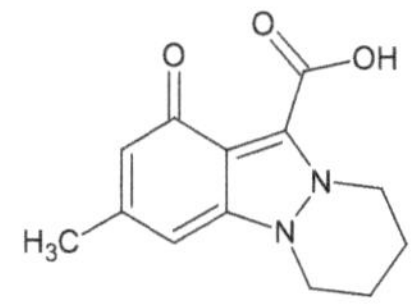
Nigellicine

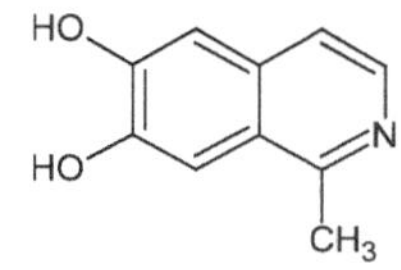
Nigellimine

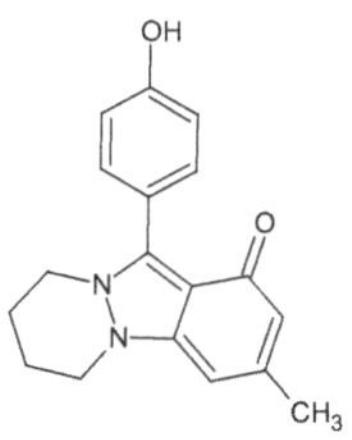
Nigellidine

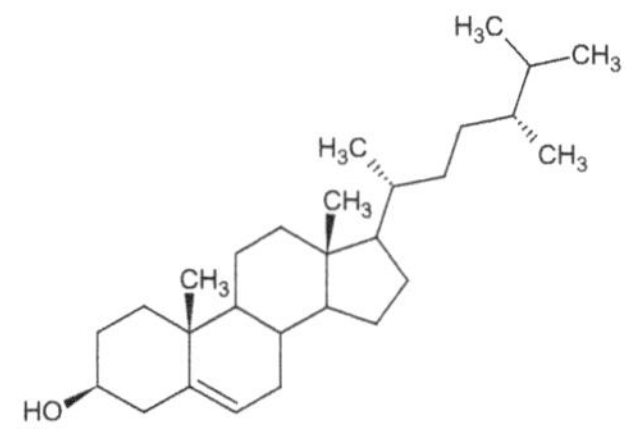
Campesterol

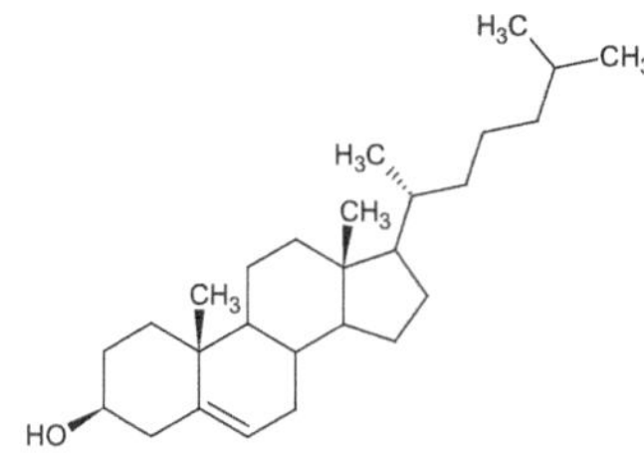
Cholesterol

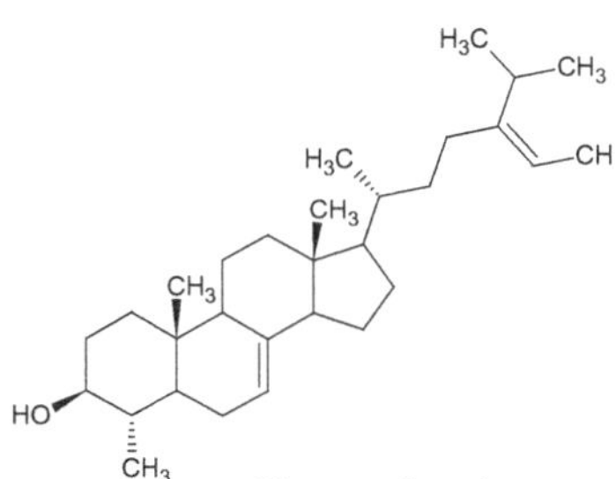
Citrostadienol

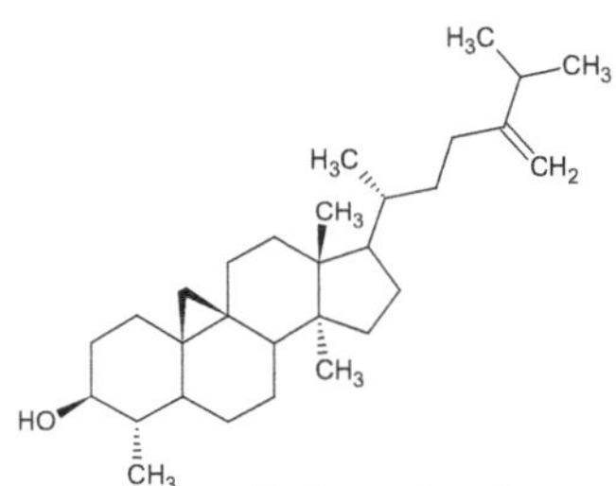
Cycloeucalenol

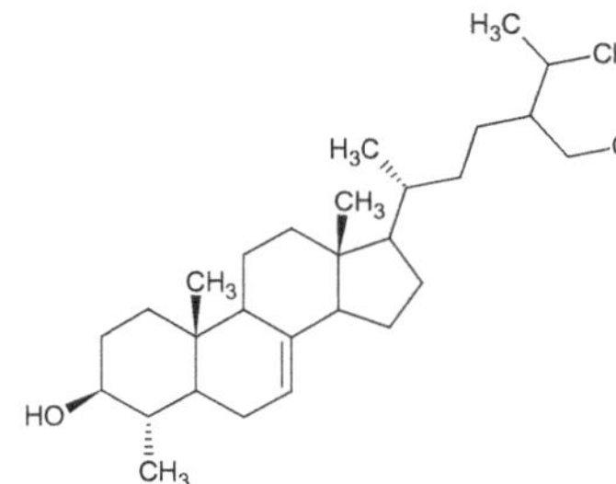
24-Ethyllophenol

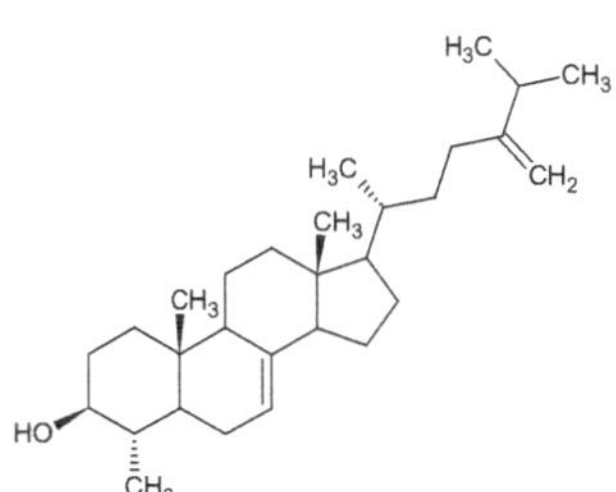
Gramisterol

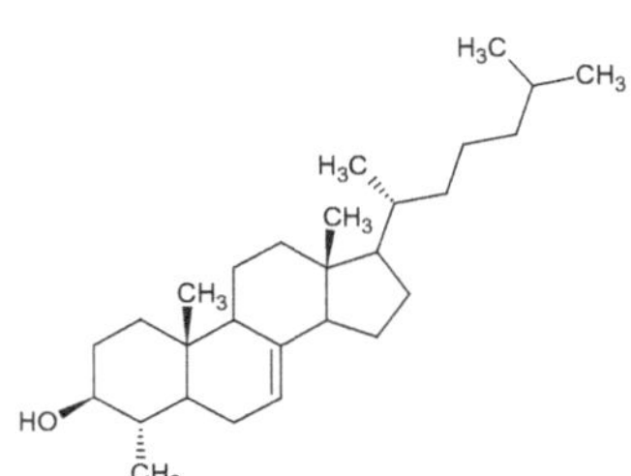
Lophenol

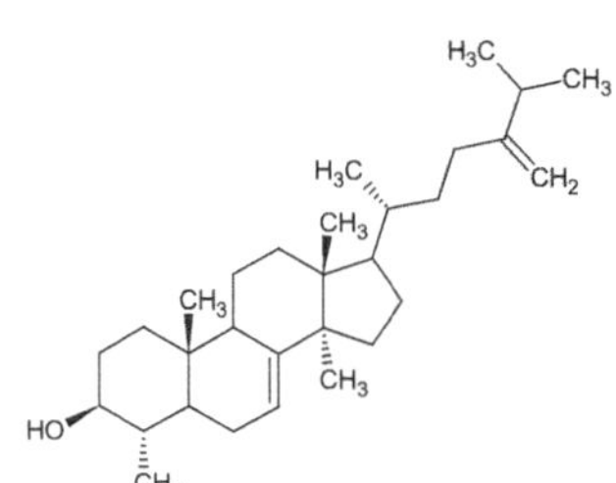
Obtusifoliol

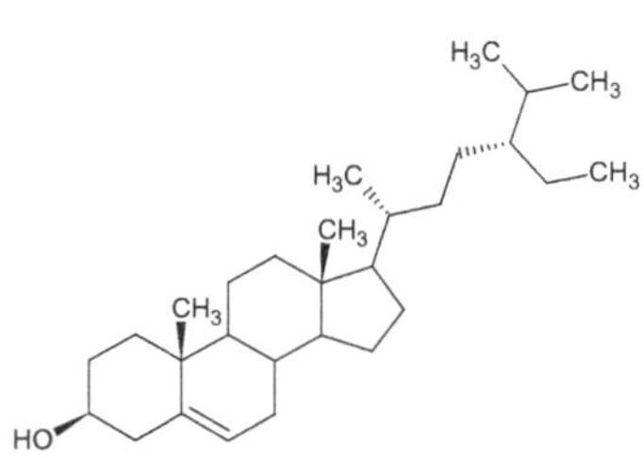
Sitosterol

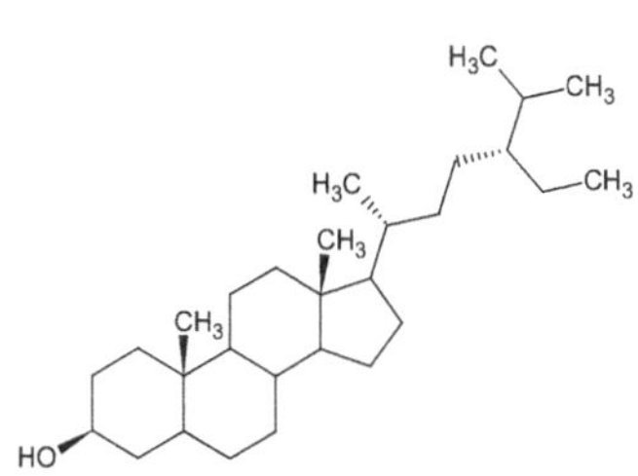
Sitgmastanol

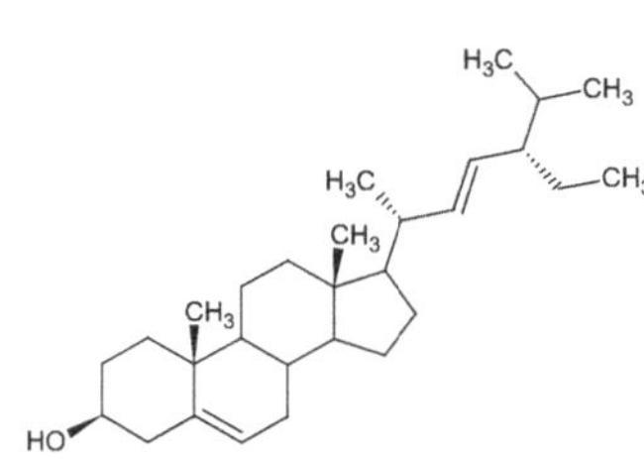
Sitgmasterol

FIGURE 6.2 Chemical compounds of black cumin.

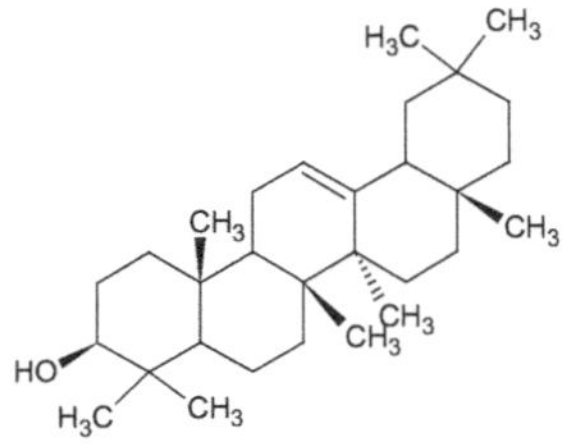

β-Amyrin

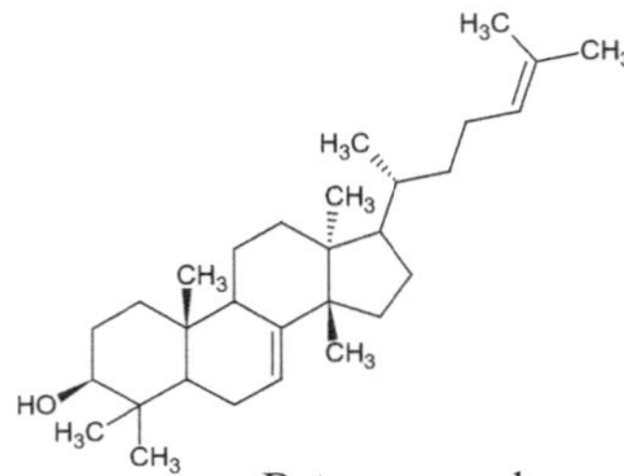

Butyrospermol

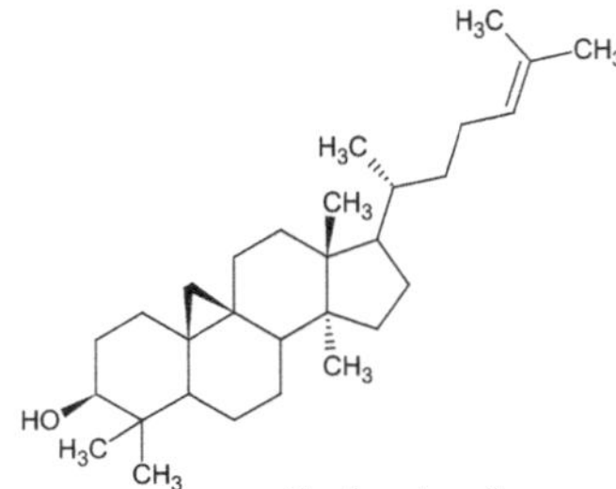

Cycloartenol

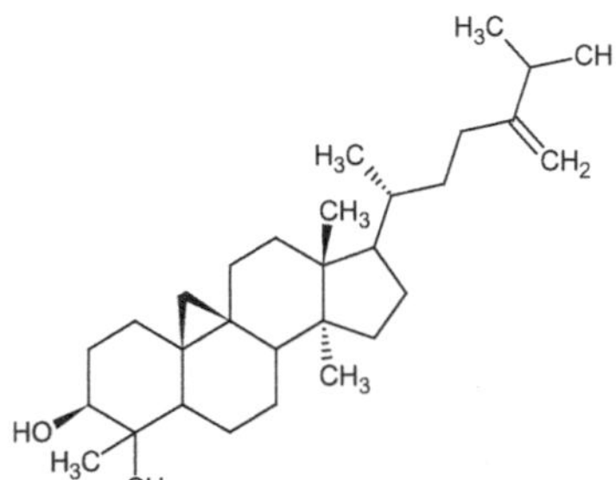

24-Methylenecycloartenol

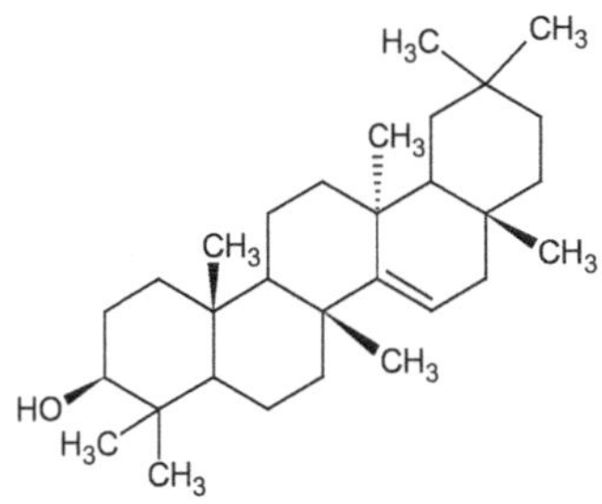

Taraxerol

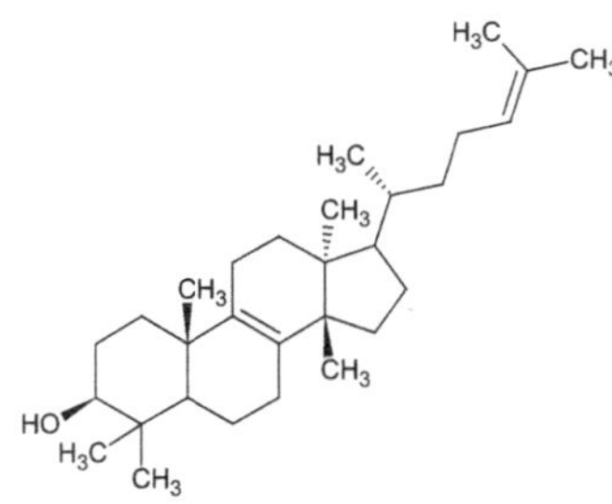

Tirucallol

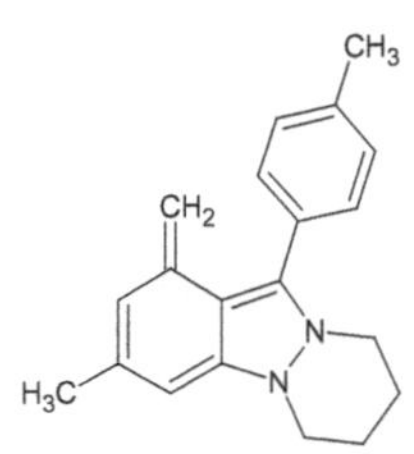

Nigellidine

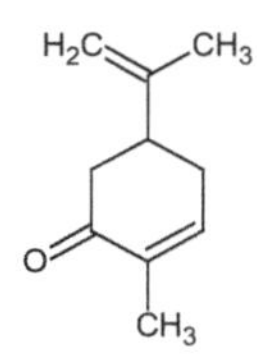

Carvone

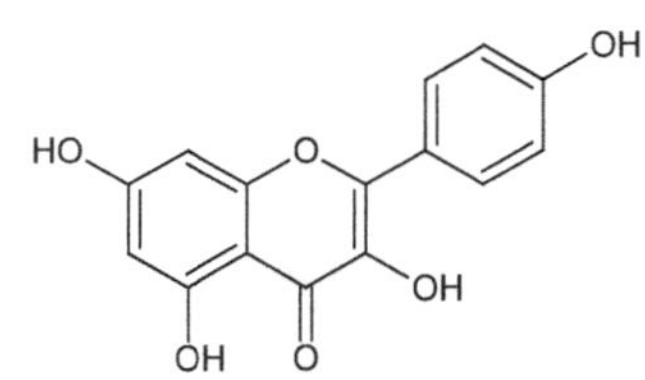

Kaempferol

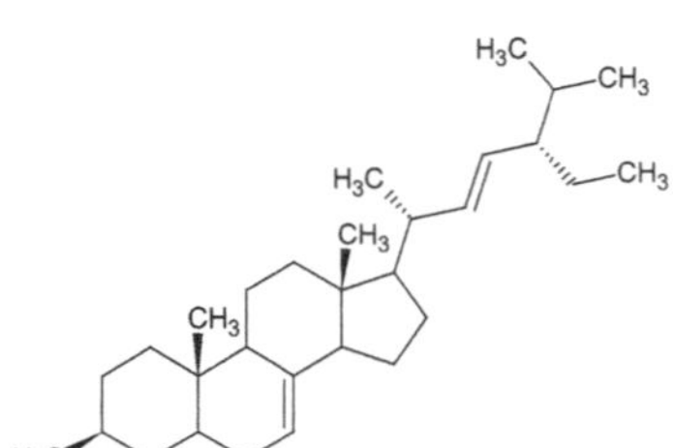

α-Spinasterol

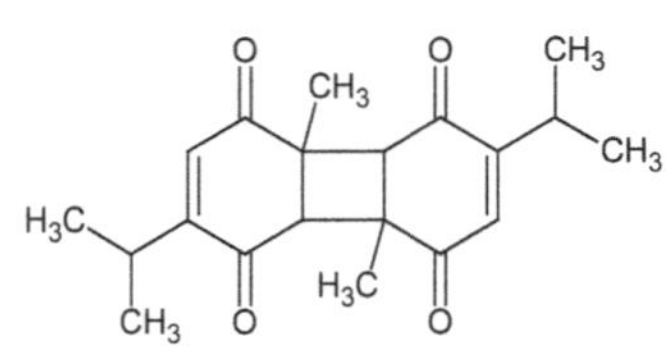

Nigellone

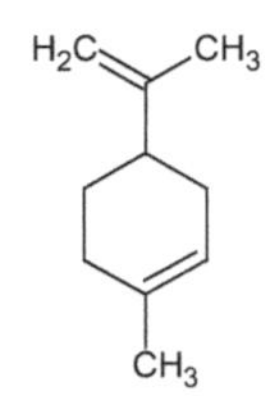

Limonene

FIGURE 6.2 (Continued)

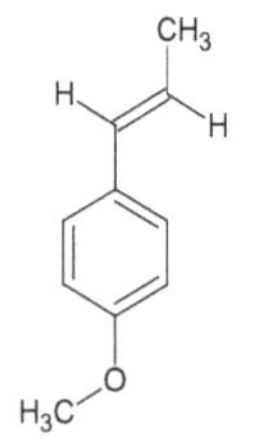
Trans-anethole

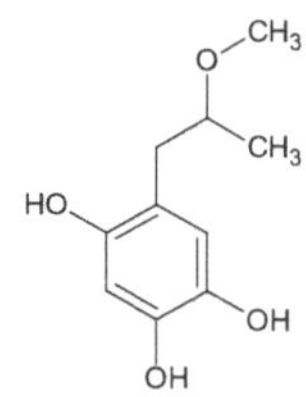
2-(2-Methoxypropyl)-5-methyl-1,4-benzenediol

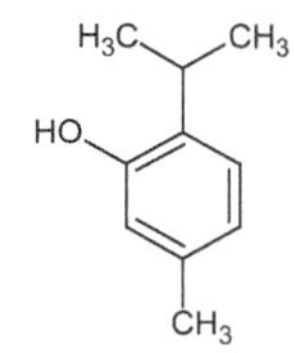
Thymol

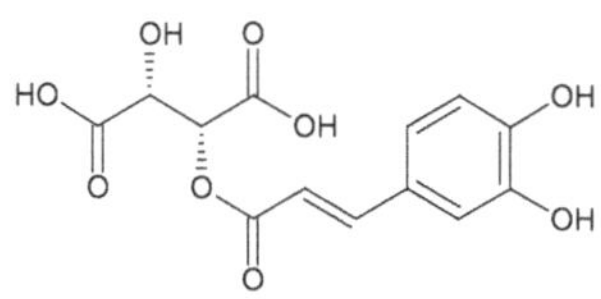
Caftaric acid

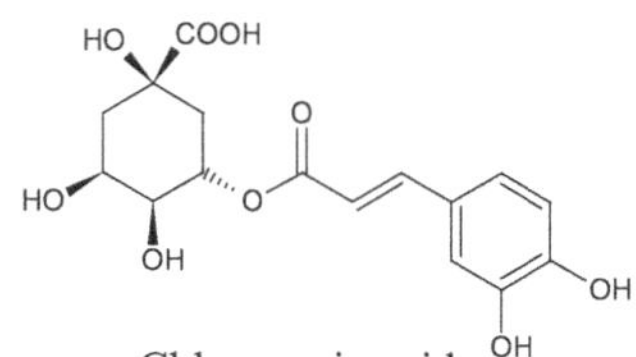
Chlorogenic acid

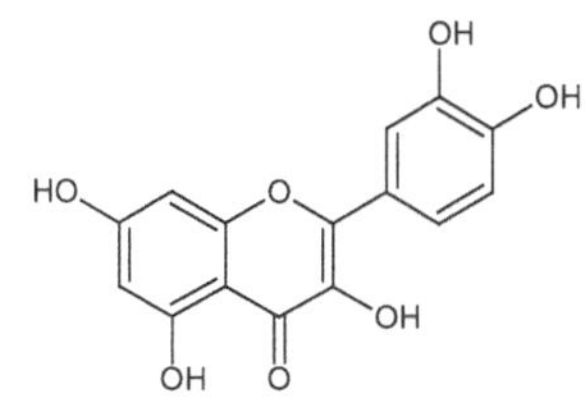
Quercetin

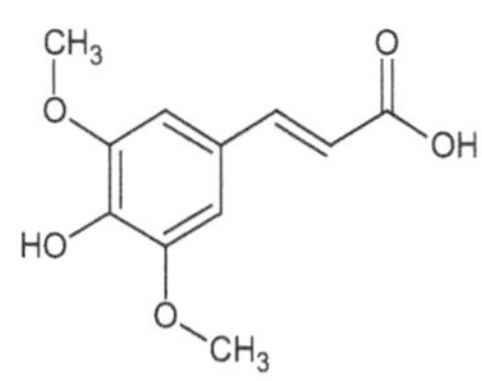
Sinapinic acid

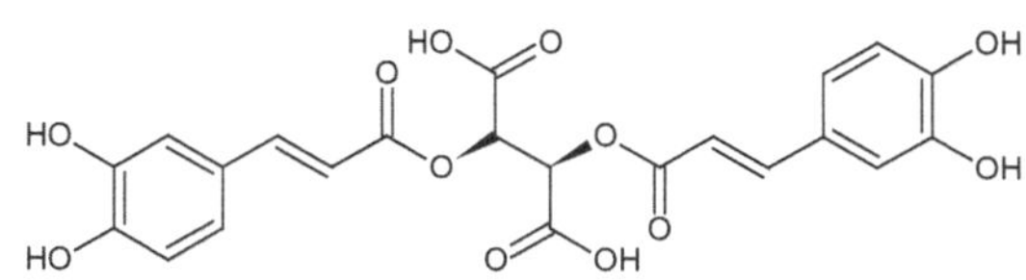
Chicoric acid

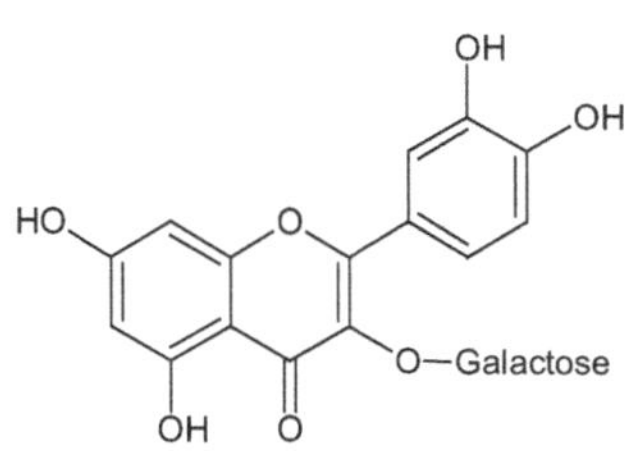
Hyperoside

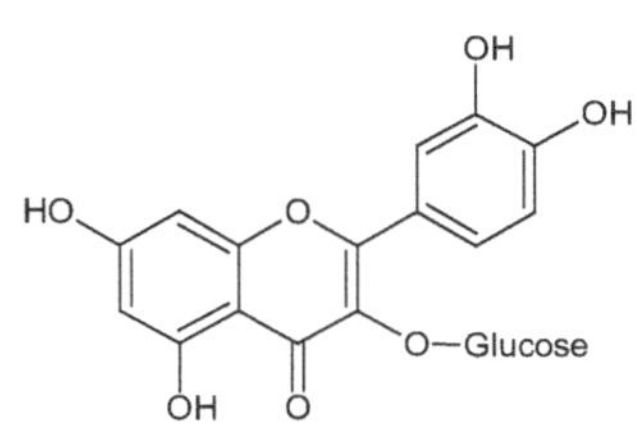
Isoquercitrin

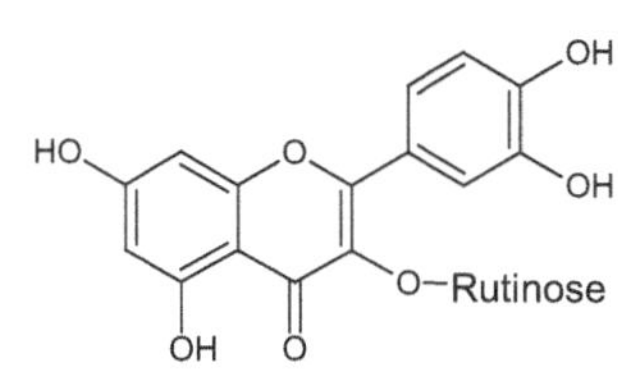
Rutin

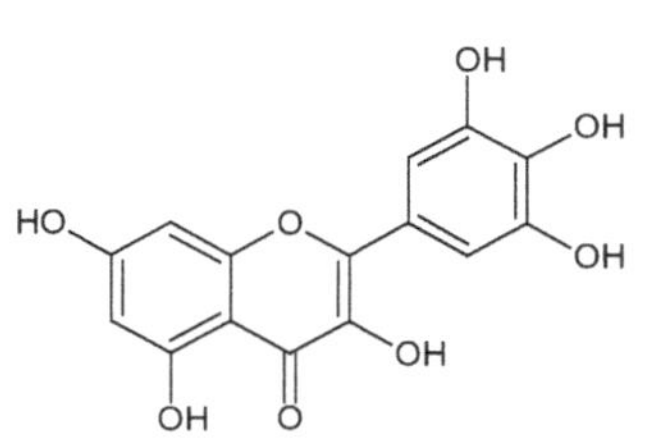
Myricetin

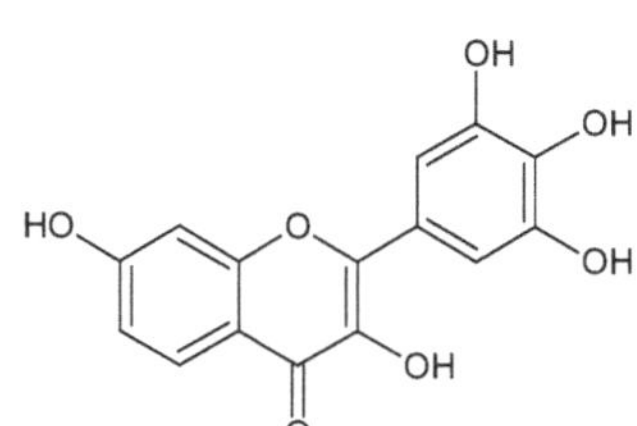
Fisetin

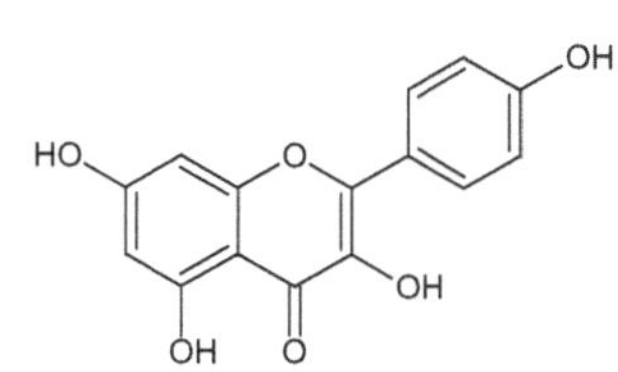
Apigenin

FIGURE 6.2 (Continued)

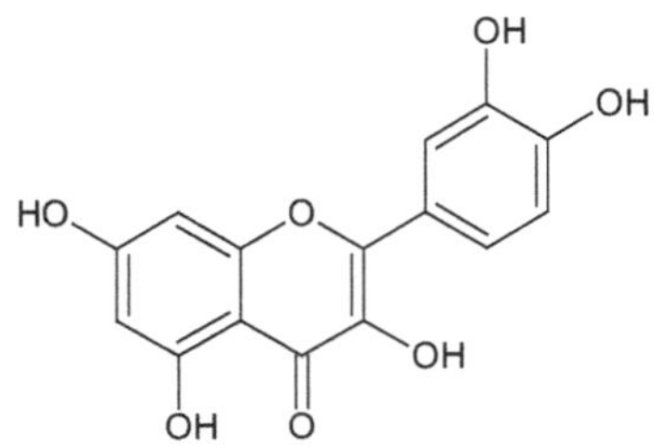

Luteolin

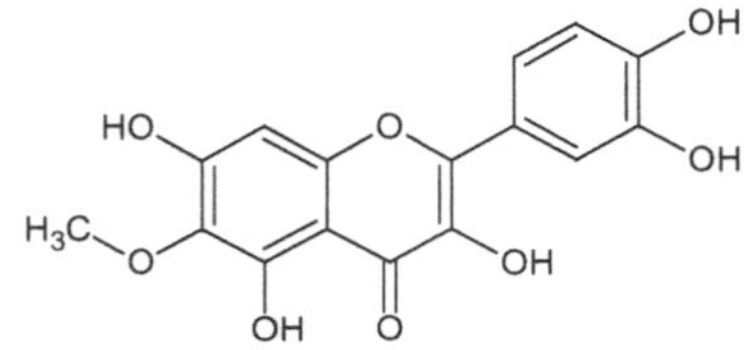

Patuletin

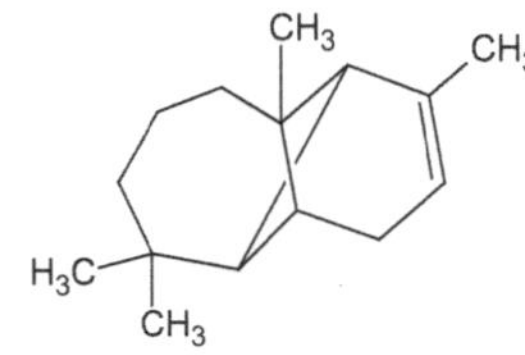

α–Longipinene

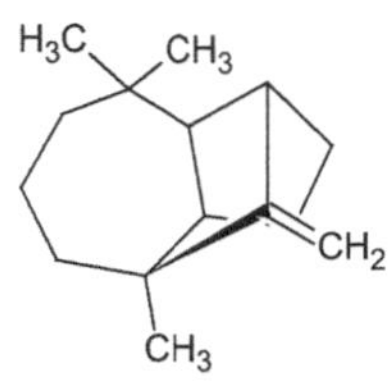

Longifolene

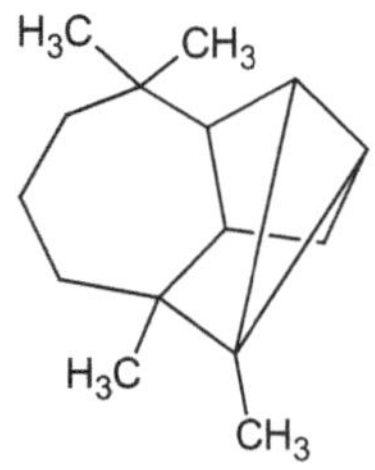

α–Longicyclene

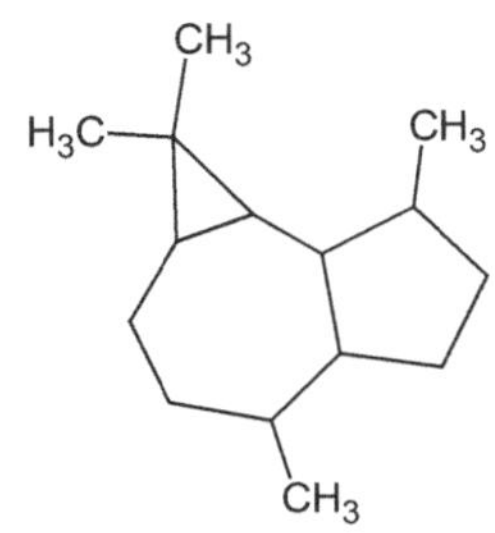

Aromadendrene

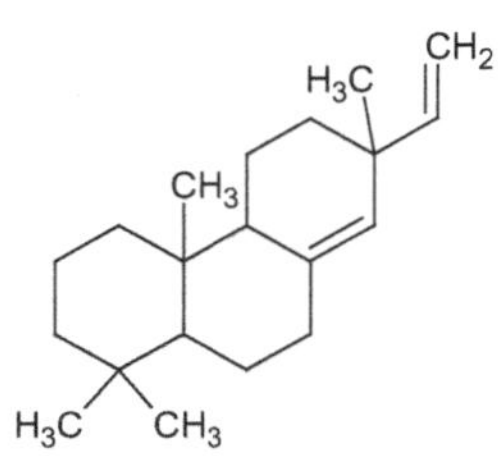

Pimara-8(14),15-diene

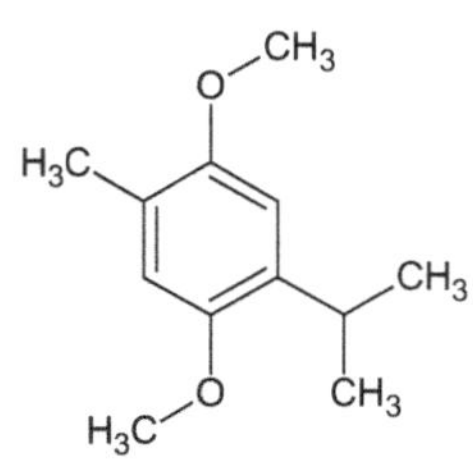

Thymohydroquinone dimethyl ether

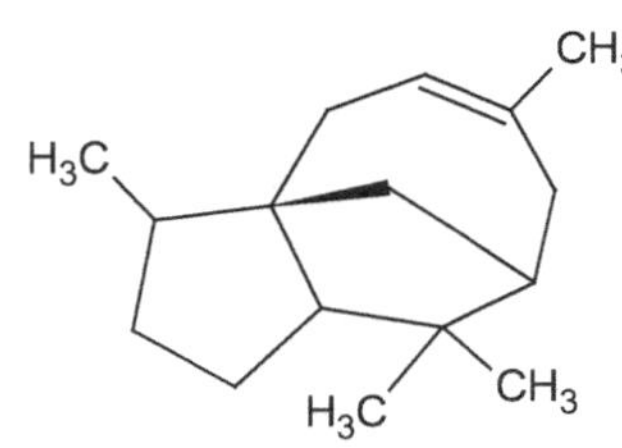

α-Longipinene

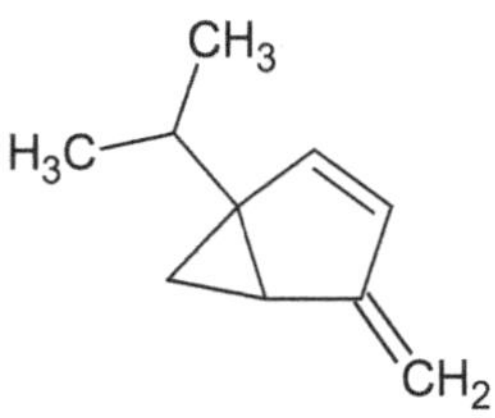

2,4,(10)-Thujadiene

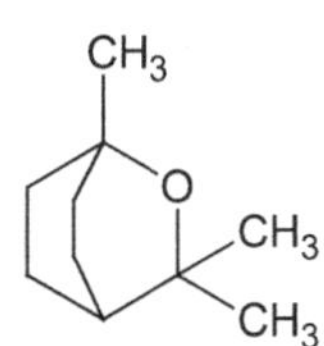

1,8-Cineole

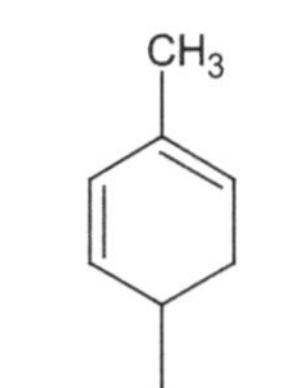

p-Mentha-1,5,8-triene

FIGURE 6.2 (Continued)

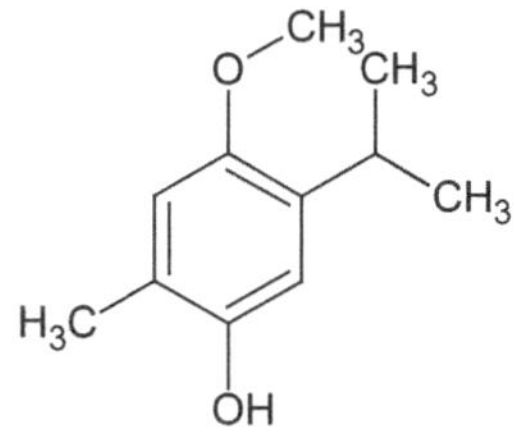

Thymohydroquinone methyl ether

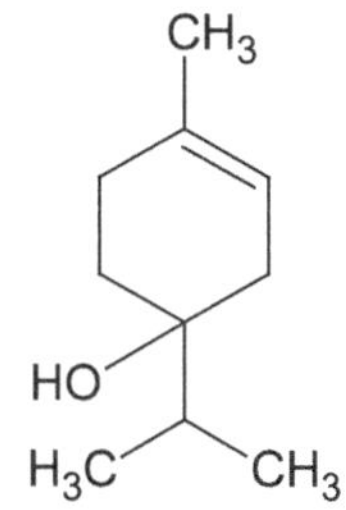

Terpinene 4-ol

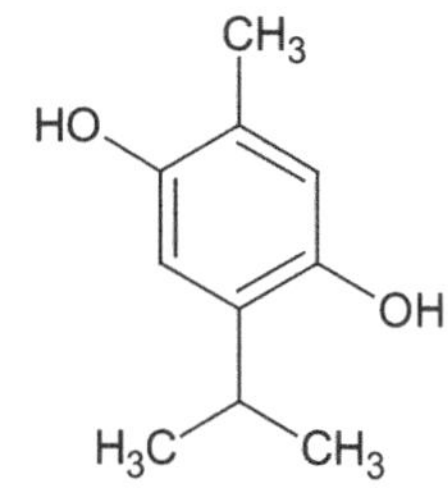

Thymohydroquinone

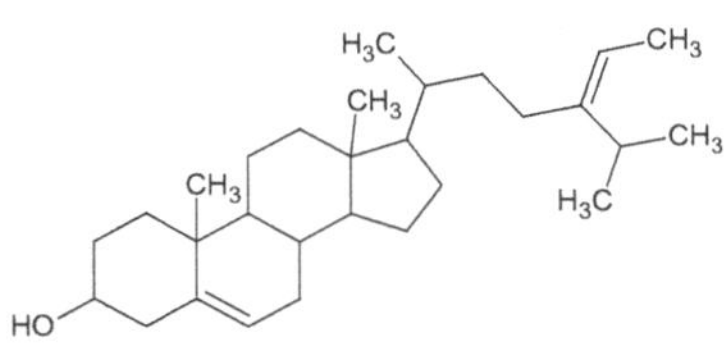

δ5-Avenasterol

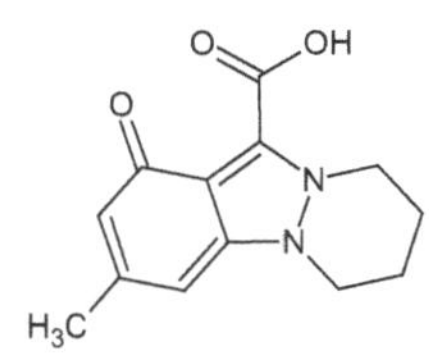

Nigellicine

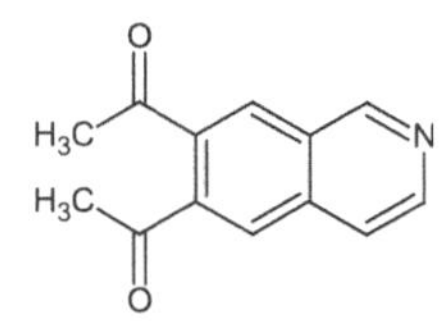

Nigellicimine

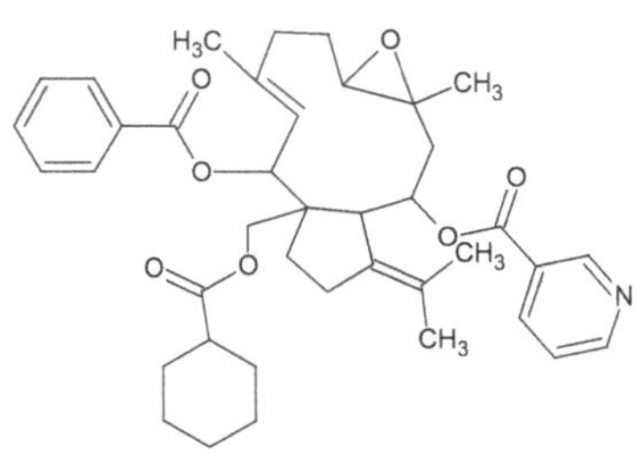

Nigellamines A1-A5

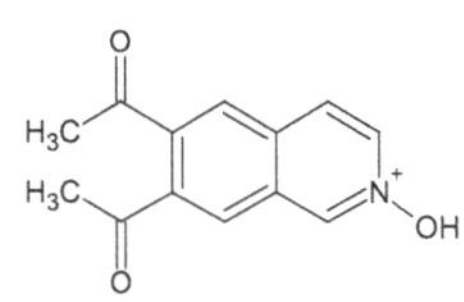

Nigellicimine-N-oxide

FIGURE 6.2 (Continued)

6.5 PHARMACOLOGICAL STUDIES

It has been reported that NS seeds contain a number of bioactive compounds rich in unsaturated fatty acids like linoleic acid, oleic acid, eicodadienoic acid, and dihomolinoleic acid (Bourgou et al., 2008; Cheikh-Rouhou et al., 2008; Mehta et al., 2008). NS seeds have been used for thousands of years for the treatment of variety of diseases related to the respiratory system, digestive tract, kidney and liver function, cardiovascular system, and immune system (Goreja, 2003). Islamic scholar Avicenna mentioned the importance of black cumin seeds in his book *The Canon of Medicine*. He explained that the seeds stimulate the body's energy and help recovery from fatigue and dispiritedness (Warrier et al., 2004; Yarnell and Abascal, 2011). The seeds are used traditionally to treat asthma, bronchitis, rheumatism, and related inflammatory diseases (Padhye et al., 2008; Yarnell and Abascal, 2011). The seeds are used as antioxidant, antidiuretic, antihypertensive, antidiabetic, anticancer, analgesic, immunomodulatory, antimicrobial, anthelmintic, anti-inflammatory, spasmolytic, bronchodilator, renal protective, gastro protective, and hepatoprotective, and so on (Goreja, 2003; Assayed, 2010; Boskabady et al., 2010; Abdel-Zaher et al., 2011; Abel-Salam et al., 2012). Major therapeutic activities of this plant are due to the presence of TQ, which is a chief active constituent of the essential oil. NS is also used in food as a flavouring additive in breads and pickles because of its low toxicity (Al-Ali et al., 2008).

6.6 ANTIDIABETIC RESPONSE

Much research has proved that this plant is very helpful for the reduction of blood sugar. According to one study, NS reduced blood sugar level probably due to the presence of essential oil (Al Yahya, 1986) (Table 6.2). Other studies reported that the antidiabetic properties of NS are induced by activation of adenosine monophosphate kinase (AMPK), affecting cellular uptake of proteins with

TABLE 6.2
Antidiabetic Effects of Black Cumin

Extract/Compound	Concentration/ Dose	Experimental model	Study design	Mode of action	Reference
Silver nanoparticles prepared from NSO	Unknown	biochemical assay	*In vitro*	↓α-amylase activity	Preety et al. (2020)
Seed extract	125 and 250 mg/kg	Alloxan-induced diabetic rats	*In vivo*	↓Glucose and MDA levels; ↑SOD and GPx; ↑diameter and amount of islets of Langerhans cells	Widodo et al. (2016)
	2 gm/kg	Alloxan-induced diabetic mice	*In vivo*	↓Blood glucose, TG, T-cholesterol, LDL-C, and TBARS; ↑ HDL-C	Bensiameur-Touati et al. (2017)
	100, 200, and 400 mg/kg	STZ-induced diabetic rat	*In vivo*	↓Serum glucose and lipids; ↑AIP; ↑eNOS expression; ↓VCAM-1 and LOX-1 expressions	Abbasnezhad et al. (2019)
Green synthesis of silver nanoparticles using seed extract	200 mg/kg	STZ-induced model of diabetic neuropathy in rats	*In vivo*	↓Glucose and AGE and aldose reductase level; ↓TNF-α, NF-κB and S100B; ↓MDA and NO; ↑SOD and GSH; ↑TKr A; ↑nitrotyrosine	Alkhalaf et al. (2020)
NSO	400 mg/kg	STZ-induced diabetic rats	*In vivo*	↓Myositis, hyaline degeneration and Zenker's necrosis; ↑Bcl-2 expression	Altun et al. (2019)
	(0.2 mg/kg)	STZ-induced diabetic rats	*In vivo*	↓Blood glucose; ↓Bax and caspase-3 expression in aortic medial layer	Cüce et al. (2016)
	(100 mg/Kg)	STZ-induced diabetic rats	*In vivo*	↑Insulin-like growth factor-1 and phosphoinositide-3 kinase; ↓ADAM-17; ↓blood glucose level, lipid profile, TBARS, NO, serum insulin/ insulin receptor ratio, and tumour necrosis factor	Balbaa et al. (2016)
	(2 mL/kg)	STZ-induced diabetic rats	*In vivo*	↓FBG; ↑insulin levels; ↑pancreatic and hepatic CAT and GSH; ↑insulin immunoreactive parts	Abdelrazek et al. (2018)

(Continued)

TABLE 6.2
(Continued)

Extract/Compound	Concentration/Dose	Experimental model	Study design	Mode of action	Reference
NSO and TQ	(2 mL/kg) + (50 mg/kg)	STZ-induced model of diabetic nephropathy	*In vivo*	↓collagen IV, TGF-β1 and VEGFA	Al-Trad et al. (2016)
NSO	(2.5 mL/kg)	Alloxan-induced diabetic rabbits	*In vivo*	↓CAT activity, TC, TGs, LDL-cholesterol and VLDL-cholesterol levels, serum blood glucose levels and lipid contents; ↑ HDL-cholesterol, vitamin C levels	Akhtar et al. (2020)
	2.5 mL/kg	Randomized clinical trial on patients with diabetic nephropathy	Clinical trail	↓Blood glucose, serum creatinine, blood urea	Ansari et al. (2017)
	(1 g as two capsules)	Double-blind randomized clinical trial on T2DM patients	Clinical trail	↓Lipid profile and glucose level, C-reactive protein level, and lipid peroxidation	Kooshki et al. (2020)
Seed capsules	(2 g daily)	Single-blind, nonrandomized controlled clinical trials on T2DM patients	Clinical trail	↓TC, LDL-C, TC/HDL-C and LDL-C/HDL-C ratios; ↑serum HDL-C; ↓DBP, MAP and HR	Badar et al. (2017)
Nigella sativa flavonoids surface coated gold NPs	100, 200, 300, 400, and 500 μg/mL	α-amylase inhibition assay	*In vitro*	↓α-amylase	Veeramani et al. (2022)
TQ	65 mg/kg	STZ-induced diabetic rat	*In vivo*	↓HbA1c, lipid peroxidase, and NO	Faisal Lutfi et al. (2021)

hypolipidaemic and antidiabetic properties (Haddad et al., 2003; Sanz, 2008). It was observed that the blood glucose level in Balb/c mice was significantly reduced when the volatile oil of NS was administered orally (Bakathir and Abbas, 2011). The volatile oil present in the seeds of NS showed protective effect on insulin immune activity and ultra-structural changes of pancreatic β-cells in streptozotocin (STZ) induced diabetic rats. For the induction of diabetes, STZ was injected intraperitoneally at a single dose of 50 mg/kg. Increased intensity of staining for insulin and preservation of β-cell numbers were apparent in the NS treated diabetic rats. NS exhibited protective effect on STZ-diabetic rats and moderate increase in the lowered secretory vesicles with granules, and slight destruction with loss of cristae within the mitochondria of β-cell was observed when compared to control rats. It was proved that treatment of NS exhibited therapeutic protective effect in diabetes by decreasing morphological changes and preserving pancreatic β-cell integrity (Kanter et al., 2009). The hypoglycaemic potential of thymoquinone was evaluated in STZ-nicotinamide (NA) induced diabetic rats. TQ was administered orally in diabetic rats at the dose of 20, 40, 80 mg/kg body weight for 45 days, and it was observed that glycaemic status in STZ-NA induced rats was improved. The results showed that TQ exerts significant hypoglycaemic effect (Pari and Sankaranarayanan, 2009). NS extract showed an insulin-sensitizing action by enhancing acetyl-CoA carboxylase (ACC) phosphorylation, a major component of the insulin-independent AMPK signalling pathway, and by enhancing muscle Glut4 content (Al-Hader et al., 1993). The extract of NS also causes regeneration and relative proliferation in beta cells and a decrement in free radicals' production in streptozotocin-diabetic rats (Ramadan et al., 2003; Benhaddou-Andaloussi et al., 2011). The levels of insulin in Hb increased with significant reduction in glucose and HbA_{1C} levels. The altered activities of carbohydrate metabolic enzymes were restored to near normal. These results proved that TQ at 80 mg/kg body weight is associated with beneficial changes in hepatic enzyme activities and thereby exerts potential hypoglycaemic effects (Umar et al., 2012). NS at 1350 mg/day possesses antidiabetic effect on type-3 diabetes mellitus (T3DM) patients (Moustafa et al., 2019). Furthermore, silver nanoparticles with *Nigella sativa* oil (NSO) exert hypoglycaemic effect by inhibiting α-amylase enzyme activity (Preety et al., 2020) and glucose and AGE and aldose reductase, tumour necrosis factor (TNF)-α, nuclear factor kappa-light-chain-enhancer of activated B cells (NF-κB), and S100 calcium binding protein B (S100B) (Alkhalaf et al., 2020), whereas in alloxan-induced diabetic rats (*in vivo*) study, seed extract of NS (125 and 250 mg/kg) attenuated glucose, malondialdehyde (MDA), and nitric oxide (NO) levels (Alkhalaf et al., 2020), along with increasing antioxidant enzymes such as superoxide dismutase (SOD) and glutathione peroxidase (GPx) and the proliferation of islets of Langerhans cells (Widodo et al., 2016). In the same study model, seed extract (2 gm/kg) of NS has shown reduced blood glucose, triglycerides (TGs), T-cholesterol (TC), low-density lipoprotein cholesterol (LDL-C), and thiobarbituric acid reactive substances (TBARS) and increased high-density lipoprotein cholesterol (HDL-C) (Bensiameur-Touati et al., 2017). Abbasnezhad and co-workers (2019) demonstrated that NS seed extract at doses 100, 200, and 400 mg/kg reduce serum glucose and lipids, vascular cell adhesion protein 1 (VCAM-1), and lectin-like oxidized low-density lipoprotein (LDL) receptor-1 (LOX-1) expressions while inducing ATP; endothelial nitric oxide synthase (eNOS) expression (Abbasnezhad et al., 2019). Additional study has revealed that NSO (0.2 mg/kg) reduces the blood glucose level via increasing the apoptotic-related proteins such as Bax and caspase-3 expression. Moreover, Balbaa et al. (2016) found that NSO has the ability to upregulate insulin-like growth factor-1 (IGF-1) and phosphoinositide-3 kinase (P3K) with downregulating A disintegrin and metalloprotease 17 (ADAM-17), blood glucose level, lipid profile, TBARS, NO, serum insulin/insulin receptor ratio, and TNF-α (Balbaa et al., 2016; Faisal Lutfi et al., 2021). NSO at 2 mL/kg concentration blocks FBG activity, increases pancreatic and hepatic antioxidant enzymes such as catalase (CAT) and glutathione (GSH) which protect pancreatic and liver cell lines from oxidative damage, and accelerates insulin levels in blood by increasing the insulin immunoreactive parts (Abdelrazek et al., 2018). Combinational treatment of NSO and TQ (2 mL/kg + 50 mg/kg) in diabetic nephropathy rat model inhibited collagen IV, transforming growth factor (TGF)-β1 and vascular endothelial growth factor A (VEGF-A) (Al-Trad et al., 2016). NSO (2.5 mL/kg) in a diabetic rabbit model decreased CAT activity, TC, TGs, LDL-C, and VLDL-C levels, serum blood glucose levels, and lipid contents; whereas it increases HDL-C and vitamin C (Akhtar et al., 2020). The antidiabetic effect of NSO (2.5 mL/kg) in phase I and II clinical trials has been shown by a significant reduction of blood glucose, serum creatinine, blood urea (Ansari et al., 2017), lipid profile, and glucose level, C-reactive protein level, and lipid peroxidation (Kooshki et al., 2020), TC, LDL-C, TC/HDL-C, and LDL-C/HDL-C ratios and increased essential lipid HDL-C (Badar et al., 2017).

Benhaddou-Andaloussi et al. (2010) demonstrated that NS seed increases Akt, peroxisome proliferator-activated receptor (PPAR)-γ, and AMPK pathway activation which meanwhile increases insulin secretion from pancreas cells in a dose-dependent manner (Benhaddou-Andaloussi et al., 2010). Further, the active biochemical compound of NS, TQ reduced serum glucose level in diabetic rat model (Sangi et al., 2015). TQ also increases insulin and haemoglobin (Hb) and reduces glucose and haemoglobin A1c (HbA_{1C}), which may be responsible for antihyperglycaemic effects (Pari and Sankaranarayanan, 2009). Fararh et al. (2004) revealed that TQ (50mg/kg) attenuated hepatic gluconeogenesis *through reducing* blood glucose and glycated Hb in serum (Fararh et al., 2004; Fararh et al., 2010). Moreover, TQ blocks the activity of glycogen phosphorylase enzyme, which plays a vital role in glycogenesis. Glycogen phosphorylase releases glucose-1-phosphate from the terminal alpha-1,4-glycosidic bond to catalyse the rate-limiting step in glycogenolysis in mammals

(El-Ameen et al., 2015; Abdelmeguid et al., 2010). At other times, TQ in combination with hypoglycaemic drugs (metformin) can reduce glucose, cholesterol, and TGs, HbA1c, and MDA and induce antioxidant capacity (GPx, SOD, and CAT) and insulin (Rchid et al., 2004; Fouad and Alwadani, 2015; El-Aarag et al., 2017). Other study has exhibited that TQ nanoparticles have a potential antihyperglycaemic effect. They reduce blood glucose and glycated haemoglobin and induce the lipid profile in rats (Fararh et al., 2004; Rani et al., 2018). TQ has shown antihyperglycaemic effect by its antioxidant properties, for example it increases antioxidant enzymes such as CAT, GPx, GST, and GSH (Kaleem et al., 2006), which help to block β-cell lipid peroxidation (Meral et al., 2001; Kanter et al., 2003). Thus, it reduces β-cell damage with increasing serum insulin levels and reduces blood glucose (El-Dakhakhny et al., 2002; Houcher et al., 2007; Kanter, 2008). Glycation of SOD is also responsible for diabetic complications and TQ blocks the SOD glycation (Anwar et al., 2014). NS seed extract as gold nanoparticles *reduces the glycogenesis enzymes' effect* like α-amylase and α-glucosidase (Tiji et al., 2021; Veeramani et al., 2022), which reduce blood glucose and increase the number of β islet cells in the pancreas (Alimohammadi et al., 2013; Sadiq et al., 2021; Sutrisna et al., 2022). Babaeibonab (2019) asserted that NS improves serum lipid profile (LDL, HDL, TC, and TG), FBG (Al-Hader et al., 1993), HbA1c, and insulin resistance (Ahmad and Alkreathy, 2018; Babaeibonab, 2019). NS reduces glycated HbA1c level due to the presence of TQ (active compounds) combinational treatment with hypoglycaemic agent metformin (Fararh et al., 2005; Ali et al., 2021). TQ significantly reduces blood sugar levels in animals by affecting their SG, serum insulin levels, and BW, which may block gluconeogenesis in liver cells (Fararh et al., 2005; Abduallah et al., 2017; Bule et al., 2020). A dose-dependent oral TQ administration tweaked glycaemic status in STZ-NA persuaded diabetic rats. It provokes insulin level, and Hb escalate with drastically lowered glucose and HbA_{1C} levels (Pari and Sankaranarayanan, 2009). Several studies have shown that extract of NS and NSO has the ability to reduce serum glucose and induce TAC (Houcher et al., 2007; Kanter, 2008; Al-Logmani, 2009) by inhibiting glucose absorption (Kanter et al., 2003; Al-Trad et al., 2016). Additionally, NSO diminished glucose from serum and increased insulin, c-peptide, and TAC (Ayed and Talal, 2011; Salama, 2011). The ethanolic and methanolic extract of NS can block FBS, TG, TC, LDL, and FBG, while increasing HDL (Alimohammadi et al., 2013; Asaduzzaman et al., 2015; Mohamed et al., 2015). In some clinical trials, studies have revealed that NS exerts an antioxidant effect by increasing the antioxidant enzymes such as SOD, MDA, TAC, and hs-CRP and reduces FBS (Rahmani et al., 2022). NS reduced FBG, 2hPG, and HbA1c that's increase β-cell function in pancreas and showed hypoglycaemic effect (Bamosa et al., 2010). From the overall discussion it has been found that NS has shown antidiabetic effect through antioxidant, antihyperglycaemic, and protective effect of β-cell from damage.

6.7 TRADITIONAL AND OTHER POTENTIAL USES

In traditional system of medicine NS seeds are effective against cough, bronchitis, asthma, chronic headache, migraine, dizziness, chest congestion, dysmenorrhoeal, obesity, diabetes, paralysis, hemiplegia, back pain, infection, inflammation, rheumatism, hypertension, and gastrointestinal problems such as dyspepsia, flatulence, dysentery, and diarrhoea (Tariq, 2008). It has also been used as a stimulant, diuretic, emmenagogue, lactagogue, anthelmintic, and carminative, and it is also applied to abscesses, nasal ulcers, orchitis, eczema, and swollen joints (Tariq, 2008). Seed oil is considered to be a local anaesthetic (Warrier et al., 1996). Indian and Middle Eastern cuisines use it as a spice. It was employed as a preservative in mummification in ancient Egypt. It has long been used to treat asthma, diabetes, hypertension, fever, inflammation, bronchitis, dizziness, eczema, and gastrointestinal problems (Srinivasan, 2018). Traditionally, NS has been used in cardiovascular disease, scabies, emmenagogue, abortifacient, tonic, galactagogue, induce menses, condiment, diuretic, stimulant, stomach ache, stimulated the kidneys, antispasmodic, diaphoretic, vermifuge, cough (Egypt, Iran, Nepal, Saudi Arabia, Thailand), carminative (Jordan), depurative, pectoral (Jordan), antitussive (Morocco), antiasthmatic (Morocco), sinusitis, a poison antidote, anti-influenza

(Morocco), paralysis (Oman), conjunctivitis, anti-congestive (Oman), abdominal pain, earache (Turkey), antiemetic (Yemen), tongue inflammation (Ethiopia), eruptions of skin (India), pulmonary catarrh (Iran), diabetes mellitus (Jordan), appetizer, toothache, intestinal parasites (Libya), digestive, antipyretic, antidiarrheal (Saudi Arabia), anti-malaria/pass kidney stones (Sudan), aphrodiastic (United Arab Emirates), eczema, and chronic coughs (Egypt) (Javidi et al., 2016).

6.8 SAFETY ISSUES

The toxicity of the seed extract and its components seems to be minimal. The activities of various enzymes and metabolites indicative of hepatic and renal function were not substantially affected by giving rats 50 mg/kg of NS seed extract intraperitoneally for 5 days (El Daly, 1998). During a 48-hour observation period, oral administration of the seed oil at dosages up to 10 mL/kg in rats and mice caused neither death nor overt toxicity (Khanna et al., 1993). This was recently verified when oral treatment of the fixed oil of NS at a dosage of 10 mL/kg to rats for up to 12 weeks caused no mortality or substantial changes in major liver enzymes (Zaoui et al., 2002). TQ's LD50 was determined to be 2.4 g/kg (range 1.52–3.77). (Badary et al., 1998). Acute high-dose injection (2 g/kg or greater) resulted in hypoactivity and difficulties breathing. GSH concentrations in the liver, kidney, and heart were reduced, and the liver and kidney were damaged, as indicated by substantial increases in plasma metabolites and enzymes (Badary et al., 1998). There was no indication of harm when mice were given thymoquinone in their drinking water at concentrations up to 0.03% for 90 days, save for a substantial reduction in fasting plasma glucose content (Badary et al., 1998). Two cases of allergic contact dermatitis were documented in two people who developed maculopapular eczema after applying pure NS oil topically. Surprisingly, the oil was promoted as a treatment for 'skin dysfunction, inflammation, acne, and eczema'. Contact dermatitis has previously been linked to the usage of essential oils used in cosmetics and fragrances. Topical corticosteroids were used to treat these instances (Steinmann et al., 1997; Zedlitz et al., 2002).

6.9 CULTIVATION PRACTICE

In the entire world, NS is mostly grown once in a year as rabi (yearly) crop during the months of October (late)–November to March–April in the plains, while rarely in the hills in May–June (ŞEN et al., 2010).

6.9.1 Area of Cultivation and Production

Area of cultivation and annual production were reported to be in India, Turkey, USA, and UK.

6.9.2 Climate

Grows well in a cool-dry climate with light snowfall areas to warm, humid areas. Cool and humid weather favours flowering and seed setting (Datta et al., 2012).

6.9.3 Soil

Sandy loam rich in microbial activity is the most suitable soil for cultivation. The sloppy soils of heavy rainfall areas and levelled and well drained soils of moderate rainfall areas are quite suitable for cultivation. Soil pH 7.0 to 7.5 is favourable for cultivation (Datta et al., 2012).

6.9.4 Preparation of Land

One ploughing followed by 2–3 harrowing and levelling is suitable (Datta et al., 2012).

6.9.5 Method of Sowing

Seed sowing or replanting previous year root stocks. Seed sowing is done during October–November by broadcasting (1.5 kg/hectare) or seed drill method or by line sowing, keeping space between lines (30, 40, or 50 cm) and at a depth of 2 cm. After 20 days of sowing, thinning of the plant to a distance of 20 cm is done. Sowing by bulbs (previous year root stock) is possible when soil moisture content of the field is favourable for deep ploughing, i.e. neither too wet nor too dry (ŞEN et al., 2010).

6.9.6 Manure and Fertilizer

Nitrogen, phosphorus, and potassium (NPK = 5:3:2) is generally applied every year along the side of the planted bulbs (Seidavi et al., 2020).

6.9.7 Weed Control

Frequent weeding reduces weed competition and produces good environmental condition for growth and development. About 3–5 weedings at an interval of 20 to 25 days is recommended by hand hoe or khurpi (Al-Jassir, 1992; Kabir et al., 2020).

6.9.8 Irrigation

One or two irrigations at flowering and seed formation stage are helpful to increase grain size and oil content (Datta et al., 2012).

6.9.9 Harvesting

NS grown as rabi crop in the West Bengal plains are generally harvested late March to the first week of April. The crop harvested before shedding at the little green stage gives high aromatic oil content, providing good marketability (Seidavi et al., 2020). Black cumin retains seed viability longer when they are fully ripe. It is essential that harvesting is done before shedding (shattering of fruits is a major problem) and therefore two to three or more pickings can be done to avoid loss of seeds due to shattering of the capsules. The harvested crop is dried under the sun and threshed by beating with a stick (Datta et al., 2012).

6.9.10 Post-Harvest Management

NS requires extensive labour in collection and harvest as the capsules (fruit) tend to shatter at maturity. Post-harvest management of the fruits usually involves their harvest, one by one, by hand, and dry storage until natural dehiscence. The mature fruits do not require much attention as they are self-preserving and their essential oil is a great deterrent to fungal attack, insect attack, and rodent infestation (ŞEN et al., 2010).

6.10 FUTURE REMARKS

Plenty of herbs are used in different forms in traditional systems of medicine. NS has been considered an important medicinal plant all over the world. From different research articles it was revealed that NS and its active constituents, particularly TQ, have remarkable pharmacological properties. The studies also confirmed the pharmacological properties of the seeds, oil, and extracts of NS. Overall, it was observed that NS possesses remarkable *in vitro* and *in vivo* pharmacological properties against a large variety of diseases and is found to be relatively safe. To date, many research studies have been done on the chemical composition and pharmacological properties of NS. Further studies need to be done for the determination of the mechanism of actions of NS seeds and their constituents by which they exert their therapeutic effects. More effective and safer drugs for the treatment of a wide variety of diseases in the future can be developed by chemically altering the molecular structure of TQ, α-hederin, and other constituents of NS seeds. For the development of effective therapies of many infectious diseases, the seeds of NS, NSO, and other constituents like TQ and α-hederin could be used in suitable combinations with existing chemotherapeutic agents. Further, the specific cellular and molecular targets of various constituents of NS (specifically TQ) can be explored in order to get more information regarding their pharmacological properties. This review article may be useful for the scientific community as it covers all the aspects of this important medicinal plant. This article also helps those researchers who are working on this miracle herb, and it would help them for performing preclinical and clinical studies on the use of NS for the treatment of various diseases.

REFERENCES

Abbasnezhad, A., et al. (2019) "Nigella sativa L. Seed regulated eNOS, VCAM-1 and LOX-1 genes expression and improved vasoreactivity in aorta of diabetic rat," *Journal of Ethnopharmacology*, 228, pp. 142–147. http://doi.org/10.1016/j.jep.2018.09.021.

Abdelmeguid, N. E., et al. (2010) "Effects of Nigella sativa and thymoquinone on biochemical and subcellular changes in pancreatic β-cells of streptozotocin-induced diabetic rats," *Journal of Diabetes*, 2(4), pp. 256–266. http://doi.org/10.1111/j.1753-0407.2010.00091.x.

Abdelrazek, H. M. A., et al. (2018) "Black seed thymoquinone improved insulin secretion, hepatic glycogen storage, and oxidative stress in streptozotocin-induced diabetic male Wistar rats," *Oxidative Medicine and Cellular Longevity*, 2018, pp. 1–10. http://doi.org/10.1155/2018/8104165.

Abdel-Zaher, A. O., Abdel-Rahman, M. S. and Elwasei, F. M. (2011) "Protective effect of Nigella sativa oil against tramadol-induced tolerance and dependence in mice: Role of nitric oxide and oxidative stress," *Neurotoxicology*, 32(6), pp. 725–733. http://doi.org/10.1016/j.neuro.2011.08.001.

Abduallah, A. M., et al. (2017) "The effect of Nigella sativa extract (thymoquinone) on glucose insulin levels and body weight of induced diabetic female rats," *American Journal of Life Sciences*, 5(2).

Abel-Salam, B. K. A. (2012) "Immunomodulatory effects of black seeds and garlic on alloxan-induced Diabetes in albino rat," *Allergologia et immunopathologia*, 40(6), pp. 336–340. http://doi.org/10.1016/j.aller.2011.07.002.

Ahmad, A. and Alkreathy, H. M. (2018) "Comparative biochemical and histopathological studies on the efficacy of metformin and Nigella sativa oil against thioacetamide-induced acute hepatorenal damage in rats," *Biomedical Research*, 29(15), pp. 3106–3116.

Ahmad, A., et al. (2013) "A review on therapeutic potential of Nigella sativa: A miracle herb," *Asian Pacific Journal of Tropical Biomedicine*, 3(5), pp. 337–352. http://doi.org/10.1016/S2221-1691(13)60075-1.

Akhtar, M. T., et al. (no date) "Antidiabetic potential of Nigella sativa L seed oil in alloxan-induced diabetic rabbits," *Tropical Journal of Pharmaceutical Research*, 2020, pp. 283–289.

Al-Ali, A., et al. (2008) "Oral and intraperitoneal LD50 of thymoquinone, an active principle of Nigella sativa, in mice and rats," *Journal of Ayub Medical College, Abbottabad: JAMC*, 20(2), pp. 25–27.

Al-Hader, A., Aqel, M. and Hasan, Z. (1993) "Hypoglycemic effects of the volatile oil of nigella sativa seeds," *International Journal of Pharmacognosy*, 31(2), pp. 96–100. http://doi.org/10.3109/13880209309082925.

Ali, S. M., et al. (2021) "Thymoquinone with metformin decreases fasting, post prandial glucose, and HbA1c in type 2 diabetic patients," *Drug Research*, 71(6), pp. 302–306. http://doi.org/10.1055/a-1388-5415.

Alimohammadi, S., et al. (2013) "Retracted article: Protective and antidiabetic effects of extract from Nigella sativa on blood glucose concentrations against streptozotocin (STZ)-induced diabetic in rats: An experimental study with histopathological evaluation," *Diagnostic pathology*, 8(1), pp. 1–7.

Al-Jassir, M. S. (1992) "Chemical composition and microflora of black cumin (Nigella sativa L.) seeds growing in Saudi Arabia," *Food Chemistry*, 45(4), pp. 239–242. http://doi.org/10.1016/0308-8146(92)90153-s.

Alkhalaf, M. I., Hussein, R. H. and Hamza, A. (2020) "Green synthesis of silver nanoparticles by Nigella sativa extract alleviates diabetic neuropathy through anti-inflammatory and antioxidant effects," *Saudi Journal of Biological Sciences*, 27(9), pp. 2410–2419. http://doi.org/10.1016/j.sjbs.2020.05.005.

Al-Logmani, A. S. (2009) "Effects of Nigella sativa L. and *Cinnamomum zeylanicum* Blume oils on some physiological parameters in streptozotocin-induced diabetic rats," *Boletín latinoamericano y del caribe de plantas medicinales y aromáticas*, 8, pp. 86–96.

Al-Trad, B., et al. (2016) "Nigella sativa oil and thymoquinone ameliorate albuminuria and renal extracellular matrix accumulation in the experimental diabetic rats," *European Review for Medical and Pharmacological Sciences*, 20(12), pp. 2680–2688.

Altun, E., Avci, E., Yildirim, T. and Yildirim, S. (2019) "Protective effect of Nigella sativa oil on myocardium in streptozotocin-induced diabetic rats," *Acta Endocrinologica (Bucharest)*, 15(3), p. 289.

Ansari, Z. M., et al. (2017) "Protective role of Nigella sativa in diabetic nephropathy: A randomized clinical trial," *Saudi Journal of Kidney Diseases and Transplantation: An Official Publication of the Saudi Center for Organ Transplantation, Saudi Arabia*, 28(1), pp. 9–14. http://doi.org/10.4103/1319-2442.198093.

Anwar, S., et al. (2014) "A structural study on the protection of glycation of superoxide dismutase by thymoquinone," *International Journal of Biological Macromolecules*, 69, pp. 476–481. http://doi.org/10.1016/j.ijbiomac.2014.06.003.

Asaduzzaman, M., et al. (2015) "Hypoglycemic and hypolipidemic potential of Nigella sativa L. seed extract in streptozotocin (STZ)-induced diabetic rats," *Journal of Plant Biochemistry & Physiology*, 3(158), p. 2.

Assayed, M. E. (2010) "Radioprotective effects of black seed (Nigella sativa) oil against hemopoietic damage and immunosuppression in gamma-irradiated rats," *Immunopharmacology and Immunotoxicology*, 32(2), pp. 284–296. http://doi.org/10.3109/08923970903307552.

Ayed, A. L. and Talal, Z. (2011) "Long-term effects of Nigella sativa L. oil on some physiological parameters in normal and streptozotocin-induced diabetic rats," *Journal of Diabetes Mellitus*, 1(3), pp. 46–53.

Babaeibonab, S. (2019) "Effect of 8 weeks aerobic training and nigella supplement on insulin resistance, lipid profile and plasma level of hba1c in type 2 diabetic rats," *Journal of Advances in Medical and Biomedical Research*, 27(120), pp. 20–29.

Badar, A., et al. (2017) "Effect of Nigella sativa supplementation over a one-year period on lipid levels, blood pressure and heart rate in type-2 diabetic patients receiving oral hypoglycemic agents: Nonrandomized clinical trial," *Annals of Saudi Medicine*, 37(1), pp. 56–63. http://doi.org/10.5144/0256-4947.2017.56.

Badary, O. A., et al. (1998) "Acute and subchronic toxicity of thymoquinone in mice," *Drug Development Research*, 44(2–3), pp. 56–61. http://doi.org/10.1002/(sici)1098-2299(199806/07)44:2/3<56::aid-ddr2>3.0.co;2-9.

Bakathir, H. A. and Abbas, N. A. (2011) "Detection of the antibacterial effect of Nigella sativa ground seeds with water," *African Journal of Traditional, Complementary, and Alternative Medicines*, 8(2), pp. 159–164. http://doi.org/10.4314/ajtcam.v8i2.63203.

Balbaa, M., et al. (2016) "Nigella sativa relieves the altered insulin receptor signaling in streptozotocin-induced diabetic rats fed with a high-fat diet," *Oxidative Medicine and Cellular Longevity*, 2016, p. 2492107. http://doi.org/10.1155/2016/2492107.

Bamosa, A. O., et al. (2010) "Effect of Nigella sativa seeds on the glycemic control of patients with type 2 diabetes mellitus," *Indian Journal of Physiology and Pharmacology*, 54(4), pp. 344–354.

Benhaddou-Andaloussi, A., et al. (2010) "Multiple molecular targets underlie the antidiabetic effect of Nigella sativa seed extract in skeletal muscle, adipocyte and liver cells," *Diabetes, Obesity & Metabolism*, 12(2), pp. 148–157. http://doi.org/10.1111/j.1463-1326.2009.01131.x.

Benhaddou-Andaloussi, A., et al. (2011) "The in vivo antidiabetic activity of Nigella sativa is mediated through activation of the AMPK pathway and increased muscle Glut4 content," *Evidence-based Complementary and Alternative Medicine: eCAM*, 2011, p. 538671. http://doi.org/10.1155/2011/538671.

Bensiameur-Touati, K., et al. (2017) "In vivo subacute toxicity and antidiabetic effect of aqueous extract of nigella sativa," *Evidence-Based Complementary and Alternative Medicine*. https://doi.org/10.1155/2017/8427034.

Bharat, B. A. and Ajaikumar, B. K. (2009) *Molecular Targets and Therapeutic Uses of Spices: Modern Uses for Ancient Medicine*. World Scientific Publishing Company.

Boskabady, M. H., Mohsenpoor, N. and Takaloo, L. (2010) "Antiasthmatic effect of Nigella sativa in airways of asthmatic patients," *Phytomedicine: International Journal of Phytotherapy and Phytopharmacology*, 17(10), pp. 707–713. http://doi.org/10.1016/j.phymed.2010.01.002.

Botnick, I., et al. (2012) "Distribution of primary and specialized metabolites in Nigella sativa seeds, a spice with vast traditional and historical uses," *Molecules (Basel, Switzerland)*, 17(9), pp. 10159–10177. http://doi.org/10.3390/molecules170910159.

Bourgou, S., et al. (2008) "Phenolic composition and biological activities of Tunisian Nigella sativa L. shoots and roots," *Comptes Rendus Biologies*, 331(1), pp. 48–55. http://doi.org/10.1016/j.crvi.2007.11.001.

Bule, M., et al. (2020) "The antidiabetic effect of thymoquinone: A systematic review and meta-analysis of animal studies," *Food Research International (Ottawa, Ont.)*, 127(108736), p. 108736. http://doi.org/10.1016/j.foodres.2019.108736.

Cheikh-Rouhou, S., et al. (2007) "Nigella sativa L.: Chemical composition and physicochemical characteristics of lipid fraction," *Food Chemistry*, 101(2), pp. 673–681. http://doi.org/10.1016/j.foodchem.2006.02.022.

Cheikh-Rouhou, S., et al. (2008) "Sterol composition of black cumin (Nigella sativa L.) and Aleppo pine (Pinus halpensis Mill.) seed oils," *Journal of Food Composition and Analysis*, 21(2), pp. 162–168.

Cüce, G., Sözen, M. E., Çetinkaya, S., Canbaz, H. T., Seflek, H. and Kalkan, S. (2016) "Effects of Nigella sativa L. seed oil on intima-media thickness and Bax and Caspase 3 expression in diabetic rat aorta," *Anatolian Journal of Cardiology*, 16(7), pp. 460–466. http://doi.org/10.5152/AnatolJCardiol.2015.6326.

Dabeek, W. M. and Marra, M. V. (2019) "Dietary quercetin and kaempferol: Bioavailability and potential cardiovascular-related bioactivity in humans," *Nutrients*, 11(10), p. 2288. http://doi.org/10.3390/nu11102288.

Datta, A. K., et al. (2012) "Black cumin (Nigella sativa L.)-a review," *Journal of Plant Development Sciences*, 4(1), pp. 1–43.

El-Aarag, B., et al. (2017) "Thymoquinone improves anti-diabetic activity of metformin in streptozotocin-induced diabetic male rats," *Journal of Diabetes & Metabolic*, 8(12).

El-Ameen, N. M. H., et al. (2015) "Anti-diabetic properties of thymoquinone is unassociated with glycogen phosphorylase inhibition," *Pharmacognosy Journal*, 6.

El-Dakhakhny, M., et al. (2002) "The hypoglycemic effect of nigella sativa oil is mediated by extrapancreatic actions," *Planta Medica*, 68(5), pp. 465–466. http://doi.org/10.1055/s-2002-32084.

El Daly, E. S. (1998) "Protective effect of cysteine and vitamin E, Crocus sativus and Nigella sativa extracts on cisplatin-induced toxicity in rats," *Journal de pharmacie de Belgique*, 53(2), pp. 87–93; discussion 93–95.

Enomoto, S., et al. (2001) "Hematological studies on black cumin oil from the seeds of Nigella sativa L," *Biological & Pharmaceutical Bulletin*, 24(3), pp. 307–310. http://doi.org/10.1248/bpb.24.307.

Faisal Lutfi, M., et al. (2021) "Thymoquinone lowers blood glucose and reduces oxidative stress in a rat model of diabetes," *Molecules (Basel, Switzerland)*, 26(8), p. 2348. http://doi.org/10.3390/molecules26082348.

Fararh, K. M., Ibrahim, A. K. and Elsonosy, Y. A. (2010) "Thymoquinone enhances the activities of enzymes related to energy metabolism in peripheral leukocytes of diabetic rats," *Research in Veterinary Science*, 88(3), pp. 400–404. http://doi.org/10.1016/j.rvsc.2009.10.008.

Fararh, K. M., et al. (2004) "Mechanisms of the hypoglycaemic and immunopotentiating effects of Nigella sativa L. oil in streptozotocin-induced diabetic hamsters," *Research in Veterinary Science*, 77(2), pp. 123–129. http://doi.org/10.1016/j.rvsc.2004.03.002.

Fararh, K. M., et al. (2005) "Thymoquinone reduces hepatic glucose production in diabetic hamsters," *Research in Veterinary Science*, 79(3), pp. 219–223. http://doi.org/10.1016/j.rvsc.2005.01.001.

Fouad, A. A. and Alwadani, F. (2015) "Ameliorative effects of thymoquinone against eye lens changes in streptozotocin diabetic rats," *Environmental Toxicology and Pharmacology*, 40(3), pp. 960–965. http://doi.org/10.1016/j.etap.2015.09.010.

Goreja, W. (2003) "Black seed," *Nature's Miracle*, pp. 1–64.

Haddad, P. S., et al. (2003) "Comparative study on the medicinal plants most recommended by traditional practitioners in morocco and Canada," *Journal of Herbs, Spices & Medicinal Plants*, 10(3), pp. 25–45. http://doi.org/10.1300/j044v10n03_04.

Houcher, Z., et al. (2007) "Effects of methanolic extract and commercial oil of Nigella sativa L. on blood glucose and antioxidant capacity in alloxan-induced diabetic rats," *Pteridines*, 18(1), pp. 8–18. http://doi.org/10.1515/pteridines.2007.18.1.8.

Javidi, S., Razavi, B. M. and Hosseinzadeh, H. (2016) "A review ofneuropharmacology effects of Nigella sativa and its main component, thymoquinone," *Phytotherapy Research*, 30(8), pp. 1219–1229.

Kabir, Y., et al. (2020) "Volatile compounds of black cumin (Nigella sativa L.) seeds cultivated in Bangladesh and India," *Heliyon*, 6(10), p. e05343. http://doi.org/10.1016/j.heliyon.2020.e05343.

Kaleem, M., et al. (2006) "Biochemical effects of Nigella sativa L seeds in diabetic rats," *Indian Journal of Experimental Biology*, 44(9), pp. 745–748.

Kanter, M. (2008) "Effects of Nigella sativa and its major constituent, thymoquinone on sciatic nerves in experimental diabetic neuropathy," *Neurochemical Research*, 33(1), pp. 87–96. http://doi.org/10.1007/s11064-007-9419-5.

Kanter, M., Akpolat, M. and Aktas, C. (2009) "Protective effects of the volatile oil of Nigella sativa seeds on beta-cell damage in streptozotocin induced diabetic rats: A light and electron microscopic study," *Journal of Molecular Histology*, 40(5–6), pp. 379–385.

Kanter, M., et al. (2003) "Partial regeneration/proliferation of the β-cells in the Islets of Langerhans by Nigella sativa L. in streptozotocin-induced diabetic rats," *The Tohoku Journal of Experimental Medicine*, 201(4), pp. 213–219.

Khan, M. A. (1999) "Chemical composition and medicinal properties of Nigella sativa Linn," *Inflammopharmacology*, 7(1), pp. 15–35. http://doi.org/10.1007/s10787-999-0023-y.

Khanna, T., Zaidi, F. A. and Dandiya, P. C. (1993) "CNS and analgesic studies on Nigella sativa," *Fitoterapia-Milano*, 64, pp. 407–407.

Kiralan, M., et al. (2014) "Physicochemical properties and stability of black cumin (Nigella sativa) seed oil as affected by different extraction methods," *Industrial Crops and Products*, 57, pp. 52–58. http://doi.org/10.1016/j.indcrop.2014.03.026.

Kooshki, A., et al. (2020) "Effect of Nigella sativa oil supplement on risk factors for cardiovascular diseases in patients with type 2 diabetes mellitus," *Phytotherapy Research: PTR*, 34(10), pp. 2706–2711. http://doi.org/10.1002/ptr.6707.

Liu, X., El-Aty, A. and Shim, A. M. (2011) "Various extraction and analytical techniques for isolation and identification of secondary metabolites from Nigella sativa seeds." *Mini-Reviews in Medicinal Chemistry*, 11, pp. 947–955.

Mandal, A., Datta, A. K. and Bhattacharya, A. (2011) "Evaluation of pollen and productive parameters, their interrelationship and clustering of eight Corchorus spp. (Tiliaceae)," *The Nucleus; an International Journal of Cytology and Allied Topics*, 54(3), pp. 147–152. http://doi.org/10.1007/s13237-011-0044-y.

Mansi, K. M. S. (2005a) "Effects of oral administration of water extract of Nigella sativa on serum concentrations of insulin and testosterone in alloxan-induced diabetic rats," *Pakistan Journal of Biological Sciences*, 8(8), pp. 1152–1156.

Mansi, K. M. S. (2005b) "Effects of oral administration of water extract of Nigella sativa on the hypothalamus pituitary adrenal axis in experimental diabetes," *International Journal of Pharmacology*, 2(1), pp. 104–109. http://doi.org/10.3923/ijp.2006.104.109.

Mehta, B. K., Verma, M. and Gupta, M. J. (2008) "Novel lipid constituents identified in seeds of Nigella sativa Linn," *Journal of the Brazilian Chemical Society*, 19(3), pp. 458–462.

Meral, I., et al. (2001) "Effect of Nigella sativa on glucose concentration, lipid peroxidation, anti-oxidant defence system and liver damage in experimentally-induced diabetic rabbits," *Journal of Veterinary Medicine. A, Physiology, Pathology, Clinical Medicine*, 48(10), pp. 593–599. http://doi.org/10.1046/j.1439-0442.2001.00393.x.

Mohamed, W. S., et al. (2015) "Effects of fenugreek, Nigella, and termis seeds in nonalcoholic fatty liver in obese diabetic albino rats," *Arab Journal of Gastroenterology: The Official Publication of the Pan-Arab Association of Gastroenterology*, 16(1), pp. 1–9. http://doi.org/10.1016/j.ajg.2014.12.003.

Moustafa, H. A. M., et al. (2019) "Effect of Nigella Sativa oil versus metformin on glycemic control and biochemical parameters of newly diagnosed type 2 diabetes mellitus patients," *Endocrine*, 65(2), pp. 286–294. http://doi.org/10.1007/s12020-019-01963-4.

Naz, H. (2011) "Nigella sativa: The miraculous herb," *Pakistan Journal of Biochemistry and Molecular Biology*, 44, pp. 44–48.

Nickavar, B., et al. (2003) "Chemical composition of the fixed and volatile oils of Nigella sativa L. from Iran," *Zeitschrift für Naturforschung. C, Journal of Biosciences*, 58(9–10), pp. 629–631. http://doi.org/10.1515/znc-2003-9-1004.

Padhye, S., et al. (2008) "From here to eternity-the secret of Pharaohs: Therapeutic potential of black cumin seeds and beyond," *Cancer Therapy*, 6.

Pari, L. and Sankaranarayanan, C. (2009) "Beneficial effects of thymoquinone on hepatic key enzymes in streptozotocin-nicotinamide induced diabetic rats," *Life Sciences*, 85, pp. 830–834.

Preety, R., et al. (2020) "Anti-diabetic activity of silver nanoparticles prepared from cumin oil using alpha amylase inhibitory assay," *International Journal of Research in Pharmaceutical Sciences*, 11(2), pp. 1267–1269. http://doi.org/10.26452/ijrps.v11i2.1978.

Rahmani, A., et al. (2022) "Effect of Nigella Sativa oil on oxidative stress, inflammatory, and glycemic control indices in diabetic hemodialysis patients: A randomized double-blind, controlled trial," *Evidence-Based Complementary and Alternative Medicine: eCAM*, 2022, p. 2753294. http://doi.org/10.1155/2022/2753294.

Rajkapoor, B., Anandan, R. and Jayakar, B. (2002) "Anti-ulcer effect of Nigella sativa Linn. against gastric ulcers in rats," *Current Science*, pp. 177–179.

Ramadan, M. F., Kroh, L. W. and Mörsel, J.-T. (2003) "Radical scavenging activity of black cumin (Nigella sativa L.), coriander (Coriandrum sativum L.), and niger (Guizotia abyssinica Cass.) crude seed oils and oil fractions," *Journal of Agricultural and Food Chemistry*, 51(24), pp. 6961–6969. http://doi.org/10.1021/jf0346713.

Rani, R., et al. (2018) "Improvement of antihyperglycemic activity of nano-thymoquinone in rat model of type-2 diabetes," *Chemico-Biological Interactions*, 295, pp. 119–132. http://doi.org/10.1016/j.cbi.2018.02.006.

Rchid, H., et al. (2004) "Nigella sativa seed extracts enhance glucose-induced insulin release from rat-isolated Langerhans islets," *Fundamental & clinical pharmacology*, 18(5), pp. 525–529. http://doi.org/10.1111/j.1472-8206.2004.00275.x.

Sadiq, N., et al. (2021) "Antidiabetic effect of Nigella sativa compared with metformin on blood glucose levels in streptozotocin induced diabetic albino Wistar rats," *International Journal of Basic and Clinical Pharmacology*, 10(4), p. 361. http://doi.org/10.18203/2319-2003.ijbcp20211016.

Salama, R. H. M. (2011) "Hypoglycemic effect of lipoic Acid, carnitine and nigella sativa in diabetic rat model," *International Journal of Health Sciences*, 5(2), pp. 126–134.

Sangi, S. M. A., et al. (2015) "Antihyperglycemic effect of thymoquinone and oleuropein, on streptozotocin-induced diabetes mellitus in experimental animals," *Pharmacognosy Magazine*, 11(Suppl 2), pp. S251–S257. http://doi.org/10.4103/0973-1296.166017.

Sanlier, N. and Gencer, F. (2020) "Role of spices in the treatment of diabetes mellitus: A minireview," *Trends in Food Science & Technology*, 99, pp. 441–449. http://doi.org/10.1016/j.tifs.2020.03.018.

Sanz, P. (2008) "AMP-activated protein kinase: Structure and regulation," *Current Protein & Peptide Science*, 9(5), pp. 478–492. http://doi.org/10.2174/138920308785915254.

Seidavi, A. R., et al. (2020) "Feeding of black cumin (Nigella sativa L.) and its effects on poultry production and health. World's," *Poultry Science Journal*, 76(2), pp. 346–357.

Şen, N., Kar, Y. and Tekelï, Y. (2010) "Antioxidant activities of black cumin (Nigella Sativa l.) seeds cultivating in different regions of turkey: Antioxidant potency of black cumin from turkey," *Journal of Food Biochemistry*, 34, pp. 105–119. http://doi.org/10.1111/j.1745-4514.2009.00309.x.

Sharma, P. C., Yelne, M. B. and Dennis, T. J. (2005) *Database on Medicinal Plants Used in Ayurveda*. New Delhi: CCRAS, Dept. of AYUSH, Ministry of Health and Family Welfare, Govt. of India, pp. 265–266.

Sharma, Y., et al. (2005) "Effects of acute dimethoate administration on antioxidant status of liver and brain of experimental rats," *Toxicology*, 206(1), pp. 49–57. http://doi.org/10.1016/j.tox.2004.06.062.

Srinivasan, K. (2018) "Cumin (Cuminum cyminum) and black cumin (Nigella sativa) seeds: Traditional uses, chemical constituents, and nutraceutical effects," *Food Quality and Safety*, 2(1), pp. 1–16. http://doi.org/10.1093/fqsafe/fyx031.

Steinmann, A., et al. (1997) "Allergic contact dermatitis from black cumin (Nigella sativa) oil after topical use," *Contact Dermatitis*, 36(5), pp. 268–269. http://doi.org/10.1111/j.1600-0536.1997.tb00219.x.

Sutrisna, E., Azizah, T. and Wahyuni, S. (2022) "Potency of Nigella sativa linn. Seed as antidiabetic (preclinical study)," *Research Journal of Pharmacy and Technology*, pp. 381–384. http://doi.org/10.52711/0974-360x.2022.00062.

Tariq, M. (2008) "Nigella sativa seeds: Folklore treatment in modern day medicine," *Saudi Journal of Gastroenterology: Official Journal of the Saudi Gastroenterology Association*, 14(3), pp. 105–106. http://doi.org/10.4103/1319-3767.41725.

Tembhurne, V., et al. (2014) "A review on therapeutic potential of Nigella sativa (kalonji) seeds," *Journal of Medicinal Plant Research*, 8(3), pp. 167–177. http://doi.org/10.5897/jmpr10.737.

Tiji, S., et al. (2021) "Linking the phytochemicals and the α-glucosidase and α-amylase enzyme inhibitory effects of Nigella sativa seed extracts," *Foods (Basel, Switzerland)*, 10(8), p. 1818. http://doi.org/10.3390/foods10081818.

Umar, S., et al. (2012) "Modulation of the oxidative stress and inflammatory cytokine response by thymoquinone in the collagen induced arthritis in Wistar rats," *Chemico-biological Interactions*, 197(1), pp. 40–46. http://doi.org/10.1016/j.cbi.2012.03.003.

Veeramani, S., et al. (2022) "Nigella sativa flavonoids surface coated gold NPs (Au-NPs) enhancing antioxidant and anti-diabetic activity," *Process Biochemistry (Barking, London, England)*, 114, pp. 193–202. http://doi.org/10.1016/j.procbio.2021.01.004.

Warrier, P. K. and Nambiar, V. P. K. (2004) *Indian Medicinal Plants-a Compendium of 500 Species*. Chennai: Orient Longman Pvt Ltd, pp. 139–142.

Warrier, P. K., Nambiar, V. P. K. and Ramankutty, C. (1996) "Indian medicinal plants: A compendium of 500 species," *Orient Longman Pvt. Ltd*, 4, pp. 139–142.

Widodo, G. P., et al. (2016) "Antihyperglycemic, antioxidant, and pancreas regeneration activities of black cumin (Nigella sativa L.) seeds ethanol extract in alloxan-induced diabetic rats," *International Journal of Pharmacy and Pharmaceutical Sciences*, 8, pp. 37–40.

Yahya, A. (1986) "Phytochemical studies of the plants used in traditional medicine of Saudi Arabia," *Fitoterapia*, 57, pp. 179–182.

Yarnell, E. and Abascal, K. (2011) "Nigella sativa: Holy herb of the middle east," *Alternative & Complementary Therapies: A New Bimonthly Publication for Health Care Practitioners*, 17(2), pp. 99–105. http://doi.org/10.1089/act.2011.17203.

Zaoui, A., et al. (2002) "Acute and chronic toxicity of Nigella sativa fixed oil," *Phytomedicine: International Journal of Phytotherapy and Phytopharmacology*, 9(1), pp. 69–74. http://doi.org/10.1078/0944-7113-00084.

Zedlitz, S., Kaufmann, R. and Boehncke, W.-H. (2002) "Allergic contact dermatitis from black cumin (Nigella sativa) oil-containing ointment: Contact point," *Contact Dermatitis*, 46(3), p. 188. http://doi.org/10.1034/j.1600-0536.2002.460318.x.

7 Bioactive Components of Bitter Melon (*Momordica charantia* L.) and Their Antidiabetic Response

Kishoree Krishna Kumaree and Anchalee Prasansuklab

CONTENTS

7.1 INTRODUCTION

Diabetes mellitus is a metabolic disorder that affects insulin production, insulin action, or both and is one of the five major causes of death in the world. The occurrence of diabetic patients is increasing at an alarming statistical rate and is a major public health concern. It is projected that in the year 2000, nearly 171 million people had diabetes, and this is anticipated to double by the end of the year 2030 (Walker & Colledge, 2013).

Type II diabetes is the main form of diabetes, holding for approximately 90%–95% of all diabetic cases. It is a condition of malfunctioned metabolism, usually due to a mixture of hereditary and/or environmental causes, which leads to abnormally high blood glucose levels (Shaw et al., 2010). Etiologically, it begins with insulin insensitivity, a condition in which muscle, liver, and fat cells do not respond to insulin properly. The pancreas eventually loses the ability to produce and secrete enough insulin followed by food intake. Diabetes or hyperglycaemia is linked to several other complications like damage to nerves and blood vessels, leading to complications such as heart disease, stroke, kidney dysfunction, blindness, and nerve problems (Kitabchi et al., 2006).

DOI: 10.1201/b23347-7

Currently, accessible treatments have some shortcomings, including drug resistance side effects and even toxicity (Lada & Idrees, 2005). Diabetes is creating a substantial economic burden. However, due to undesirable side effects and their economic challenges, the effectiveness of many currently existing treatments is controversial and there is a requirement for more accessible and affordable options for the treatment of diabetes.

Apart from the huge market revolving around mainstream diabetic treatment using Western medicine, several alternative therapies including natural herbs are gaining wide acceptance among people. Natural products have been an important therapeutic aid for alleviating several ailments in humankind (Bnouham et al., 2006; Joseph & Raj, 2010).

Plant-based treatment has been practiced for a very long time and is a cost-effective way to treat diabetes. In several underdeveloped countries, treating with herbal plants could be a solo way of treating diabetes. Ayurveda and other traditional medicine systems have mentioned the use of herbal drugs for holistically treating diabetes (Ayodhya et al., 2010; Patel et al., 2012). The most common pathway elucidated from a few studies is that the active compounds in these herbal drugs tend to regenerate beta cells, induce insulin production, and overcome insulin-related resistances (Kavishankar et al., 2011). These herbs, also referred to as hypoglycaemic herbs, are responsible for inducing insulin, increasing glucose uptake, and inhibiting glucose absorption from the gut and glucose synthesis from the liver (Hui et al., 2009).

Bitter gourd (bitter melon) or *Momordica charantia* (Cucurbitaceae) is used in the Ayurvedic system of medicine for treating several health-related conditions. Diabetes is a complex ailment that distresses millions of people. Bitter melon may have a valuable role to play in addressing its treatment. There are only a few clinical studies on bitter melon use for treating diabetes, and the studies are small and non-comprehensive. In this chapter, the aim is to summarize the widespread use of bitter melon in addressing diabetes. Thus, this chapter will mainly focus on the antidiabetic properties of *M. charantia* and its future potential.

7.2 BOTANICAL DESCRIPTION OF *M. CHARANTIA*

Various names exist for the plant and its fruit, including bitter melon, bitter gourd, goya from the Japanese, or karela from Hindi. *M. charantia*, which belongs to the family Cucurbitaceae, is a tropical and subtropical flowering climber and is popularly known for its bitter fruit. The plant is an annual to perennial monoecious crawling herb, which can grow up to 2–3 m in height. The leaves are carried singly along the stem, and each leaf is 4–12 cm across, rounded in outline, with 3–7 deeply separated lobes (Figure 7.1).

Flowers occur singly at the top leaf axis and have unisexual properties. Male flowers have a slender basal swelling as compared to the female flower, where the basal swelling appears warty.

The fruits are ovoid, spindle-shaped, or egg-shaped in appearance and have a distinct warty-like outer cover. The fruit has a hollow centre during the cross-sectioning with soft flesh and contains large, flattened seeds (Figure 7.1).

7.3 GEOGRAPHICAL DISTRIBUTION OF *M. CHARANTIA*

M. charantia is widely distributed in tropical and subtropical countries of all continents, and the wild form of this specific vine grows in Asian countries, East Africa, the Caribbean, the Amazon, Australia, and several parts of China. The cultivation of this plant is widespread throughout the world and is used for vegetables as well as for medicinal purposes. *M. charantia* has been traditionally practiced as a medicinal plant in several countries, mostly used for diabetes, colic, and intestinal worms and parasites. It is also used topically as an antibiotic paste for healing external wounds, as an anti-inflammatory, and for several other ailments (Hussain et al., 2016; Joseph & Jini, 2013; Satyavati et al., 1987).

FIGURE 7.1 *M. charantia* (bitter melon) flower, leaves, and fruits on the plant.

7.4 *M. CHARANTIA* FEATURES, AND BIOACTIVE COMPONENTS

Bitter melon (*M. charantia*) is an exceptional bitter herbaceous medicinal plant that grows in tropical and subtropical regions of many countries; all parts of the plant taste bitter, including the fruits. It is an herbal plant, which is rich in bioactive components such as flavonoids, phytosterols, and saponins. The most active chemical compositions are cucurbitane-type triterpenoids; the bitterness of *M. charantia* is the consequence of cucurbitane-type triterpenoids: cucurbitacins, momordicines I and II, and triterpene glycosides: momordicosides, displaying a wide range of biological actions, mainly acting against diabetes and inflammation (Ríos et al., 2005).

7.4.1 Phytochemistry of *M. charantia*

The major components of bitter melon are classified as (i) heteropolysaccharides, which are composed of galactose, glucose, mannose, rhamnose, and arabinose; (ii) proteins and peptides like momordins, momorcharins, and *M. charantia* lectin; (iii) terpenoids and saponins like cucurbitanes and cucurbitacins; and (iv) flavonoids and phenolic compounds (Akihisa et al., 2007; Ayeni et al., 2015; L. Ma et al., 2014; Zhang et al., 2016). Major bioactive components present in *M. charantia* which are widely studied for their benefits in human physiology are presented in Table 7.1.

7.4.2 Bioactive Components of *M. charantia* for Antidiabetic Response

Based on the fact that *M. charantia* has the potential to be a natural therapeutic for several health conditions, studies are conducted to identify its bioactive compounds and their mechanism of action on the body. However, there has been a particular emphasis on its antidiabetic attributes and its hypoglycaemic property. Several clinical reports have shown that extracts from bitter melon fruits, seeds, and leaves have bioactive compounds that have hypoglycaemic ability in both humans and animals (McCarty et al., 2022; Saeed et al., 2021).

TABLE 7.1
Major Bioactive Components of *M. charantia* and Their Related Physiological Roles

Major bioactive compounds	Physiological roles	Distribution	Key references
Polypeptides	RNA N-glycosidase, polynucleotide adenosine glycosidase (PAG), DNase-like phospholipase, antitumour, antimicrobial	Seeds	(Fang et al., 2012; Jabeen & Khanum, 2017; Leung et al., 1987; Meng et al., 2012; Puri et al., 2009)
Polysaccharides	Antidiabetic, immune enhancement, neuroprotection, antitumour	Various parts of the plant	(Cai et al., 2010; Y.-Y. Deng et al., 2014; Duan et al., 2015; Xu et al., 2015; Zhang et al., 2016)
Lipids	Antioxidant, antitumour	Seeds and fruit	(Dhar et al., 2007; Tsuzuki et al., 2004)
Phytosterols	Antioxidant, neuroprotection	Fruit	(Pattarachotanant et al., 2021)
Terpenoids	Anticancer, hypoglycaemic, antioxidant, antidiabetic	Stem, leaves, fruit	(Agrawal & Beohar, 2010; Akihisa et al., 2007)
Phenolics	Antioxidant, anti-inflammation, immunity enhancer	Fruit, seeds	(Bajpai et al., 2005; Qader et al., 2011)

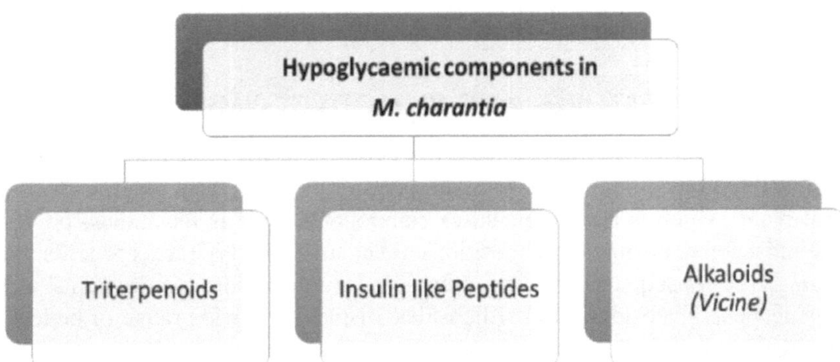

FIGURE 7.2 Most common reported hypoglycaemic agents of *M. charantia* plant.

Several animal studies have confirmed the hypoglycaemic properties of bitter melon and treatment due to the existence of abundant hypoglycaemic agents such as alkaloids, flavonoids, saponin, charantin, catechins, vicine, and polypeptide fractions (Figure 7.2) (Tan et al., 2016). In a recent study, mice fed with a high-fat diet were provided with a pure saponin-rich fraction from *M. charantia*; it was reported that mice developed a better glucose tolerance, whereas it did not affect the β-cell growth in the pancreas (Keller et al., 2021).

7.4.2.1 Triterpenoids

The triterpenoids like phytochemicals also contribute to the hypoglycaemic activity of *M. charantia*. Until now around 200 triterpenoids have been reported (Sun et al., 2021), including charantinin and saponins; these two have been widely reported for their effective antidiabetic property (Sun et al., 2021).

Charantinin is a terpenoid found in Cucurbitaceae types of plants. Terpenoids are a type of biological component that is responsible for several biological functions in plants. Charantin is a steroidal glycoside and exists in nature with an equal combination of β-stigmasterol glucoside and sitosterol glucoside. Charantin was first isolated from *M. charantia* leaf extract by Lotlikar et al. with a yield of 0.01% (Lolitkar & Rao, 1962). In an *in vivo* application of charintinin to nondiabetic rabbits, a reduction in blood glucose levels was observed. However, the downside of that experiment was a smaller sample size for each treatment group. Whereas in an elaborative different experiment, charintinin administrated to nondiabetic rabbits intravenously significantly reduced blood glucose levels. It has been identified as a potential blood sugar lowering agent that is equivalent to insulin (Pitipanapong et al., 2007).

Charantin extracted from bitter melon fruits has shown to possess strong hypoglycaemic properties. In a comparative research, this compound showed more effective hypoglycaemic properties than the practiced medicine tolbutamide (Cousens, 2007).

Other Cucurbitaceae terpenoids have also proven to have hypoglycaemic effects. In a recent study on *M. charantia* fruits, novel triterpenoids were extracted (momordicoside Y), along with three earlier reported triterpenoids (charantoside-C, momordicoside-G, and momordicoside-F1), that could inhibit hepatic gluconeogenesis (Y. Deng et al., 2022).

7.4.2.2 Polypeptides

Polypeptide-p was first identified as an insulin-like protein from *M. charantia* fruit, seed, and tissue culture (Khanna et al., 1981). Bitter melon is one of the vegetables with an abundant amount of polypeptide-p which is useful in controlling diabetes naturally. It is a naturally occurring protein like human insulin and has hypoglycaemic properties in humans as well as in animal models when treated subcutaneously (Khanna et al., 1981; Virdi et al., 2003). Structurally, polypeptide-p mimics human insulin, so it could be a possible way of plant-based replacement in patients with type 1 diabetes. A group of researchers have cloned the polypeptide-p protein from *M. charantia* and have also proved it to have a hypoglycaemic effect in alloxan-induced diabetic mice (B. L. Wang et al., 2011).

Apart from polypeptide-p, newly identified polypeptide-K (PPK) was extracted from the seeds of *M. charantia* and it also has shown the ability to reduce the blood glucose level (Kanna, 2004). It was also reported to be more effective compared to polypeptide-p in its hypoglycaemic potential. In an *in vitro* experiment, PPK showed potent inhibition of the α-glucosidase enzyme (35.5%) and α-amylase enzyme (79.1%) (Sirn et al., 2021).

Another newly found polypeptide named inhibitor against trypsin (TI) was also found to have a strong affinity with insulin receptors (IR). TI was extracted from the aqueous extract of *M. charantia* seeds. TI was identified to regulate the insulin signalling pathway (Lo et al., 2013).

In a recent study, a novel protein was also identified from the seeds of *M. charantia*, consisting of double heterodimeric proteins; further characterization revealed their similarities with napin-like proteins. Using such purified extracts on diabetic rat models showed a significant reduction in blood glucose levels with no toxicity (Murugesan et al., 2022).

7.4.2.3 Vicine

Vicine, a glycol alkaloid, was also first isolated from the seeds of karela with a yield of 0.6% (Handa et al., 1990). When it was administered intraperitoneally to nondiabetic albino rats, it had a significant impact on the reduction of blood sugar levels (Kedar & Chakrabarti, 1982). It was also reported that 95% ethanolic extract of *M. charantia* seeds showed the most potent inhibitory effect against the α-glucosidase enzyme, under further investigation via HPLC, it was confirmed to be vicine. Vicine reported from other plant extracts demonstrated toxicity and has shown to induce favism, whereas no such study related to *M. charantia* is reported yet, but further studies are needed to clarify the nontoxicity use of vicine as a hypoglycaemic agent.

7.5 ANTIDIABETIC MECHANISM OF BITTER MELON AND PHARMACOLOGICAL STUDIES

Drugs obtained from herbal plants are used all over the world to treat diabetes. Bitter melon is used to treat many diseases with good medicinal values, but more emphasis is given to its antidiabetic properties. As an antidiabetic drug, bitter melon has been widely used in different countries for thousands of years and is suggested as a remedy for the treatment of diabetes (Sharma et al., 1960).

7.5.1 Enhancing Glucose Uptake, Consumption, and Utilization

To identify a treatment method for being a potent hypoglycaemic agent, several cell-based experiments are based on the insulin signalling pathways of insulin-targeted cells (skeletal cells, hepatic cells, and adipocytes). The mechanism of *M. charantia* extract is through improving insulin resistance by enhancing glucose uptake and its utilization.

Insulin receptor binding protein from *M. charantia* (mcIRBP-9)	• The protein interacts with insulin receptors (IR), which are different from insulin binding sites. The protein increases the glucose consumption of cells. The protein operates on IR/phosphoinositide-3-kinase (PI3K)/protein kinase B (Akt) pathways. Promotes the translocation of glucose transporter 4 (GLUT4) into the plasma (Lo et al., 2014). • In another study with type 1 diabetic mice, administration of mcIRB for a month lead to the reduction of excess blood sugar levels and a significant reduction in the level of haemoglobin A1c (HbA1c) (Lo et al., 2017).
Cucurbitane glycosides	• Newly found glycosides during this research (kuguaosides A–D) managed to reduce glucose uptake by cells (Hsiao et al., 2013). • In another study, glycosides activated the AMP-activated protein kinase (AMPK) pathway when compared to troglitazone (Chang et al., 2011). • One study demonstrated that triterpenoids stimulated glucose uptake in myotubes through the insulin receptor substrate pathway (IRS-1) (Han et al., 2018).
Insulin-like compounds	A study reported the hypoglycaemic effect of *M. charantia* extract was due to three insulin-like active components, which could regulate the glycogen synthase kinase 3(GSK-3) by blocking its activity; henceforth restoring the glycogen content (Hazarika et al., 2012).
Saponins	In an animal trial, using ALX-induced diabetic mice, saponins from *M. charantia* has a significant effect on the upregulation of AMPK, which helps in the upregulation of glycogen synthesis using glycogen synthase (Q. Wang et al., 2019).

Others:

- In an animal trial with induced diabetes using streptozotocin (STZ), SOCS-3 and JNK pathways were downregulated, whereas the upregulation of IRS-1/PI3K, AKT-2, and GLUT4 in various tissues was improved (C. Ma et al., 2017).
- Aqueous extract of *M. charantia* on type 2 diabetic-induced mice model showed the upregulation of AMPK eventually affecting glucose uptake in muscle tissues (Miura et al., 2009).
- In addition, studies have shown that *M. charantia* extract can influence (PPAR) PPARα/PPARc and further regulate acyl-CoA oxidase (ACO) (Chao & Huang, 2003), leptin, and resistin, which enhanced the glucose utilization of adipocytes in high-fat diet-fed mice consequently (Shih et al., 2008).

7.5.2 Protection of Pancreatic β-Cells

Physiological studies have proven the efficacy of *M. charantia*, which has the potential to induce insulin secretion from the pancreas. Studies have also shown the improvement of insulin-producing cells followed by insulin secretion could be one of the pathways adapted by *M. charantia* in being a

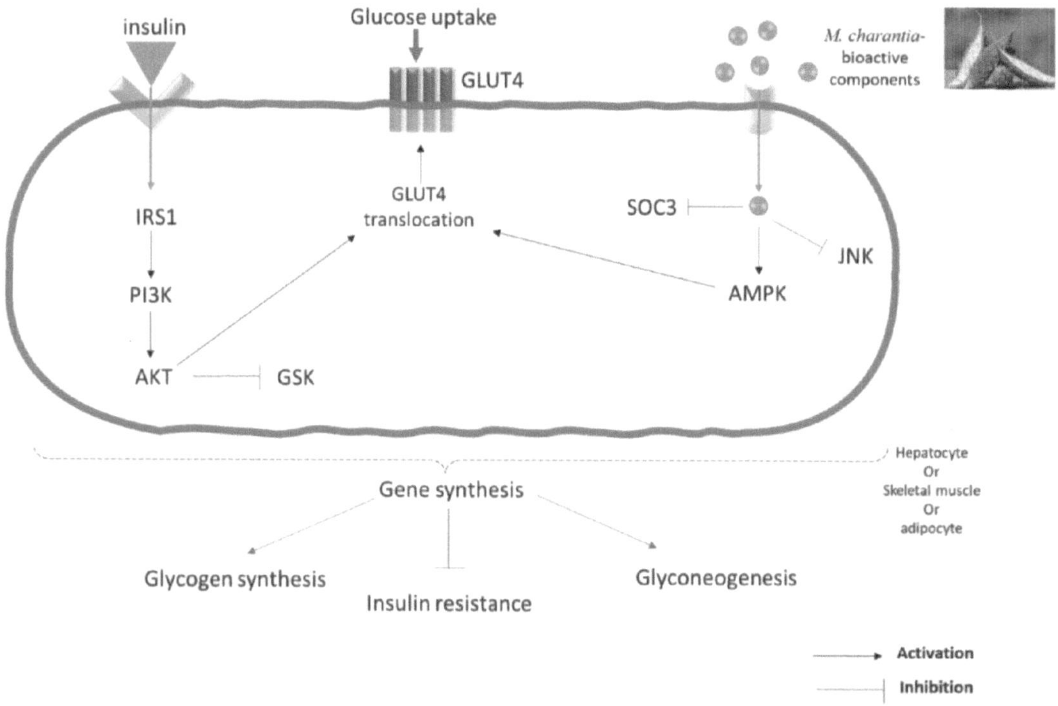

FIGURE 7.3 Overall summary of *M. charantia* antidiabetic mechanism at the cellular level.

hypoglycaemic agent (Ahmed et al., 2004). In an *in vitro* study, it showed that four new triterpenoids had a protective effect on damaged pancreatic cells. In another study, it was reported that with an administration of bitter melon juice to diabetic-induced rats, there was an increase in the number of g cells in the pancreas (Ahmed et al., 1998), whereas the exact mechanism of *M. charantia* in protecting β-cells is still unknown. One study showed the involvement of GLP-1 secretion could be a possible mechanism behind beta-cell proliferation (Huang et al., 2013). Another article stated that *M. charantia* extracts protect β-cells by regulating the gene for pancreatic duodenal homeobox-1(Pdx-1) (Malekshahi et al., 2019), as it was proven earlier that Pdx-1 is a known factor to control pancreatic development. Another study also supported the involvement of *M. charantia* extracts in the upregulation of GLP-1 and insulin production (Y. Zhu et al., 2017), Another study reported a decrease in pancreatic malondialdehyde (MDA) but increasing pancreatic glutathione levels (GSH) (Mahmoud et al., 2017).

Overall, pancreatic protection by *M. charantia* extract is through the regulation of several pathways (Figure 7.3), such as upregulation of GLP-1, Pdx-1 genes, and genes related to pancreatic MDA, GSH, and serum TAOC.

7.5.3 Delaying Glucose Uptake and Inhibiting Gluconeogenesis and Glycogenolysis

Earlier studies have shown that different extracts of *M. charantia* can work against the enzymatic activity, those involved in the glycolysis together with glucose 6-phosphatase, fructose 1, 6-diphosphatase; these plant extracts upregulated glucose 6-phosphatase dehydrogenase (Jia et al., 2017). The protein extract of bitter melon may lead to competitive inhibition of α-amylase and α-glucosidase activities.

Momordicine, a hypoglycaemic agent of *M. charantia* extract, inhibited the action of α-amylase. Similar results were also observed in cucurbitane-type bioactive compounds, they could inhibit

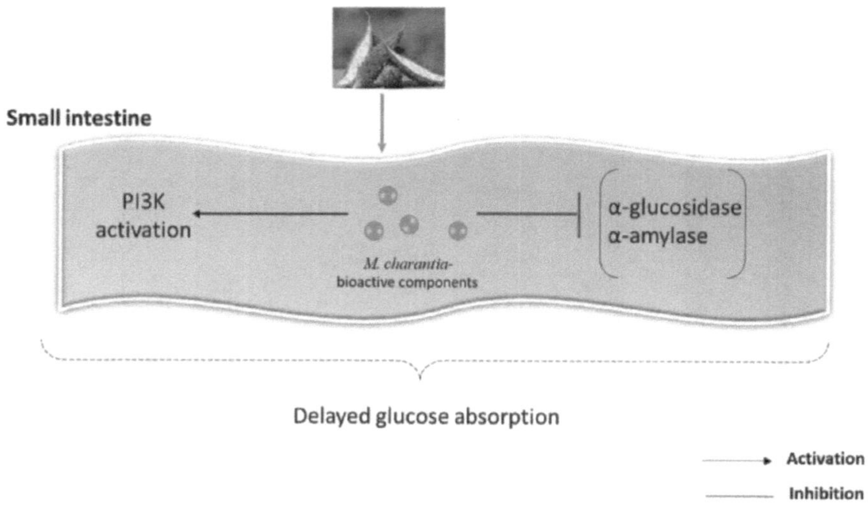

FIGURE 7.4 Overall summary of *M. charantia* antidiabetic mechanism in the small intestine.

and work against the enzymatic activity of α-glucosidase and α-amylase (Nhiem et al., 2010; Shivanagoudra et al., 2019). Moreover, it was reported that the fermented juice of *M. charantia* with lactic acid bacteria improved the inhibition of α-glucosidase enzyme action. In another animal trial, STZ-induced diabetic rats when administered with *M. charantia* juice had reduced ions (Na^+ and K^+) which were responsible for glucose absorptions by affecting the PI3K pathway (Ahmed et al., 2004). It was also shown that the direct fibre content of the *M. charantia* powder or fruit can also affect glucose absorption (Shetty et al., 2005). In general, the antidiabetic mechanism of *M. charantia* in the small intestine is presented at Figure 7.4.

When *M. charantia* extract was given to ALX-induced diabetic rats, it was shown that glycogenolysis was inhibited in liver tissues (Mahmoud et al., 2017). Saponin from *M. charantia* extract also can upregulate AMPK phosphorylation, eventually avoiding gluconeogenesis in diabetic-induced mice models (Q. Wang et al., 2019).

7.5.4 Anti-Inflammatory Effect of *M. charantia*

Patients having type 2 diabetes are also a victim of chronic inflammation conditions, and such a state can affect insulin signal transduction which leads to the formation of insulin resistance (Esser et al., 2014). *In vitro* studies have shown that cucurbit-type triterpenes have an inhibitory effect on the level of various cytokines like tumour necrosis factor-α (TNF-α), IL-1, IL-6, IL-1β (interferons), and CCL-5 (Kobori, Nakayama, et al., 2008). Similar results were confirmed in HepG2 cells. It was identified that the presence of 17 triterpenoids had an inhibitory effect on the activity of NF-κB, which is mediated by TNF-α and stimulated PPARγ (peroxisome proliferator-activated receptor-gamma) significantly. Similarly in an animal trial on diabetic-induced mice, when fed with *M. charantia* extract, there was a significant reduction in the mRNA levels on IL-1β, TNF-α, and inducible nitric oxide synthase (iNOS). It has been reported that *M. charantia* juice linked with the high fibre content (pectin) could help improve the diabetic status in STZ-induced diabetic mice (Shetty et al., 2005). Supposedly, high fibre content can regulate good intestinal microflora and upregulate the short-chain fatty acid content (SCFA) (Rohajatien et al., 2018). Apart from several antidiabetic mechanisms reported in animal and *in vitro* models, *M. charantia* has also been used for human clinical trials and has shown promising results in the treatment of diabetic patients (Table 7.2).

TABLE 7.2
Recent Clinical Trials Conducted Using *M. charantia* as a Hypoglycaemic Agent (Last Five Years)

Study design	Subjects	Form of *M. charantia* administered	Treatment duration	Outcome measure	Key references
Randomized clinical trial in two groups	24 patients (male and female) with type 2 diabetes both Mean avg years – 48.3	• Group 1 – 2 g of *M. charantia* capsule per day • Group 2 – Placebo	12 weeks	• Significant reduction in HbA1C in the group of *M. charantia*. Increase in insulin secretion. • Significant increase in insulin secretion in the *M. charantia* group.	(Cortez-Navarrete et al., 2018)
Cross-over randomized clinical trial	52 diabetic adults Mean avg years – 47.5	*M. charantia* juice (2.5 g/day)	8 weeks	• Significant reduction in fasting plasma glucose. • But non-significant changes in insulin levels and lipid profile.	(Krawinkel et al., 2018)
Double-blinded, placebo-controlled randomized clinical trial	114 presumptive prediabetics Median age – 48 years	Group 1 – Polyherbal constituents containing *M. charantia* extract (2 g/day) + lifestyle management (LSM) Group 2 – Placebo + lifestyle management	6 months	• Participants in group 1 showed a statistically significant reduction in their blood glucose level (fasting and PP), HbA1c, and fasting serum insulin. • A combination of herbal therapy, along with LSM, is associated with chances of conversion to diabetic patients.	(Nakanekar et al., 2019)
Open-label, four parallel-groups, prospective interventional clinical trial	48 type 2 diabetic patients 4 groups	Group 1 – Allopathic drug alone Group 2 – Allopathic drug + supplements from *M. charantia* (200 ml/day) Group 3 – Allopathic drug + fenugreek seeds (6–7 g/day) Group 4 – Allopathic drug+ *M. charantia* (200 ml/day) + fenugreek seeds (6–7 g/day)	90 days	• Group 4 had a significant reduction in fasting and postprandial blood sugar levels. • Groups 3 and 4 had also a significant reduction in serum cholesterol, triglycerides, and LDL-cholesterol.	(Sinha et al., 2020)

(*Continued*)

TABLE 7.2
(Continued)

Study design	Subjects	Form of *M. charantia* administered	Treatment duration	Outcome measure	Key references
Randomized, double-blind, placebo-controlled trial	100 elderly presumptive prediabetic individuals with impaired fasting glucose levels	Group 1 – Received polyherbal formulation containing *M. charantia* Group 2 – Placebo	12 weeks	• Group 1 had a significant reduction in post-prandial glucose levels, also HbA1C levels were lower in group 1. • Elderly patients with normal post-prandial glucose but presumptive fasting glucose levels had significant improvement in both cases.	(J. Zhu et al., 2019)
Blind randomized clinical trial in two groups	96 subjects with type 2 diabetes Group 1 – *M. charantia* extract Group 2 – placebo	Group 1 – Capsules of *M. charantia* extract (2.38 g/day) Group 2 – Placebo	12 weeks	• Non-significant changes in HbA1c. • Significant reduction in the fasting glucose level in the group given *M. charantia* extract.	(Kim et al., 2020)
A comparative *M. charantia* extract and allopathic drug	22 type 2 diabetic patients	Group A – Patients with allopathic drugs Group B – 200 ml of bitter melon juice along with allopathic drugs	90 days	• Significant reduction in fasting glucose as well in post-prandial blood glucose level in group B compared to group A patients.	(Rauniyar et al., 2021)
A randomized, double-blinded, and placebo-controlled trial	30 diabetic patients with a foot ulcer	Treatment group 1 – *M. charantia* leaves' extract orally (6 g/day) Treatment group 2 – Placebo	4 weeks	• Insignificant changes in VEFG serum levels within the two groups.	(Rosyid et al., 2021)
Randomized control trial to determine the effect of *M. charantia* on cardiovascular disease of diabetic patients	36 men with type 2 diabetes with 4 groups	• Group 1 – Placebo • Group 2 – *M. charantia* extract (4 g/day) • Group 3 – Only physical activity • Group 4 – Physical activity + *M. charantia* extract (4 g/day)	8 weeks	• LDL-C, TG, and TC decrease significantly in the experimental group compared to the control group. • Aerobic training and *M. charantia* consumption can reduce cardiovascular lipid profile in T2DM patients.	(Amini et al., 2020)

7.6 OTHER USES OF *M. CHARANTIA*

The plant has been known for ages for its several other health benefits apart from its hypoglycaemic properties. Inflammation could lead to the pathogenesis of several diseases: a clear link has been established with obesity, neurodegenerative diseases, cancer, cardiovascular diseases, metabolic syndrome, and diabetes (Minihane et al., 2015). *M. charantia* extracts have the beneficial properties of controlling anti-inflammatory and antioxidative stress. Various studies have shown that *M. charantia* extracts regulate inflammation through the NF-κB pathway (Kobori, Nakayama, et al., 2008); bitter melon also reduces TNF-α, IL-1β, and IL-6 levels. Dietary supplementation of *M. charantia* to high-fat obese mice showed lowered TNF-α and IL-6 in serum (Bai et al., 2016). Several other studies have also shown their anti-inflammatory properties have benefited patients with arthritis.

Evidence also supports that chronic inflammation can trigger tumour initiation. Several phytochemicals present in *M. charantia* are reported to possess potent anticancer therapies. *M. charantia* extracts have been investigated for their anticancer properties, through their anti-proliferative and immunomodulatory effects. Several *in vitro* as well as *in vivo* studies have reported the anti-proliferative effects of *M. charantia* extracts, whereas human clinical trial is still lacking (Pitchakarn et al., 2010; Weng et al., 2013).

Bitter melon intake has also been linked to the treatment of high-fat diet-induced obesity in *in vivo* models (Bai et al., 2016; Bao et al., 2013). Reports have suggested that *M. charantia* phytochemicals can modulate fat-metabolizing kinases such as AMPKs genes, nuclear factors like PPARs, and liver X receptor (LXR) in skeletal and liver tissues and can affect adipocyte differentiation (Alam et al., 2015).

7.7 TOXICITY AND SIDE EFFECTS

There has been no significant study found to date reporting side effects related to *M. charantia* consumption. There have been few reports on hypoglycaemic coma induced due to bitter melon juice. However, seeds of bitter melon have not been recognized as safe for consumption, because of their possible side effects (headache, infertility, abortion, etc.) (Basch et al., 2003; Chan et al., 1984). Moreover, *M. charantia* lectin had a cytotoxicity impact which eventually affected protein and DNA synthesis in human normal or cancerous cells (Kobori et al., 2008; Licastro et al., 1980). On the contrary, to evaluate the subchronic toxicity of *M. charantia* seeds, a recent study was conducted on Wistar rats, and no observable adverse effects were found during the study (Chung et al., 2022).

7.8 CONCLUSION AND FUTURE PERSPECTIVE

Due to advanced techniques, several novel compounds have been discovered to be present in *M. charantia*. The characterization and isolation of such bioactive components encourage more attention and still maintain an upward trend. The exact mechanisms and their involvement in the physiological functions, however, need to be explored further. Bitter melon has been used as a dietary supplement for centuries for relieving the symptoms of diabetes. Despite the abundant data on its biochemical characterization, clinical data present to date are often flawed by sample size or experimental designs. Henceforth, elaborative clinical studies are required for considering *M. charantia* as an alternative therapy to modern Western medicine. Bitter melon is nature's gift and has gained widespread popularity for its hypoglycaemic agents and several other bioactive components. The current advancements in modern science have made it possible to make these bioactive compounds available to common people in the form of nutraceutical supplements. Considering its antidiabetic, antioxidant, and antitumour properties and several other beneficial impacts, it holds as an excellent health supplement for such cases. Though its consumption does not have any severe adverse effects, further human clinical trials are very much needed. Moreover, the upcoming generation is looking more for alternative, cost effective, and safer natural therapy for diabetes. As such, genetic engineering to create a higher content of hypoglycaemic bioactive compounds from bitter melon could be the next goal for such treatment.

REFERENCES

Agrawal, R., & Beohar, T. (2010). Chemopreventive and anticarcinogenic effects of Momordica charantia extract. *Asian Pacific Journal of Cancer Prevention*, *11*(2), 371–375.

Ahmed, I., Adeghate, E., Cummings, E., Sharma, A., & Singh, J. (2004). Beneficial effects and mechanism of action of Momordica charantia juice in the treatment of streptozotocin-induced diabetes mellitus in rat. *Molecular and Cellular Biochemistry*, *261*(1), 63–70.

Ahmed, I., Adeghate, E., Sharma, A., Pallot, D., & Singh, J. (1998). Effects of Momordica charantia fruit juice on islet morphology in the pancreas of the streptozotocin-diabetic rat. *Diabetes Research and Clinical Practice*, *40*(3), 145–151.

Akihisa, T., Higo, N., Tokuda, H., Ukiya, M., Akazawa, H., Tochigi, Y., . . . Nishino, H. (2007). Cucurbitane-type triterpenoids from the fruits of Momordica charantia and their cancer chemopreventive effects. *Journal of Natural Products*, *70*(8), 1233–1239.

Alam, M. A., Uddin, R., Subhan, N., Rahman, M. M., Jain, P., & Reza, H. M. (2015). Beneficial role of bitter melon supplementation in obesity and related complications in metabolic syndrome. *Journal of Lipids*, *2015*: 496169.

Amini, M., Abdi, A., & Abbassi-Daloii, A. (2020). Effects of moderate-intensity exercise with Momordica charantia L. Consumption on serum reverse cholesterol transport elements and lipid profile in men with type 2 diabetes. *KAUMS Journal (FEYZ)*, *24*(4), 374–386.

Ayeni, M., Oyeyemi, S., Kayode, J., & Peter, G. (2015). Phytochemical, proximate and mineral analyses of the leaves of Gossypium hirsutum L. and Momordica charantia L. *Journal of Natural Sciences Research*, *5*(6), 99–107.

Ayodhya, S., Kusum, S., & Anjali, S. (2010). Hypoglycaemic activity of different extracts of various herbal plants. *International Journal of Research in Ayurveda and Pharmacy (IJRAP)*, *1*(1), 212–224.

Bai, J., Zhu, Y., & Dong, Y. (2016). Response of gut microbiota and inflammatory status to bitter melon (Momordica charantia L.) in high fat diet induced obese rats. *Journal of Ethnopharmacology*, *194*, 717–726.

Bajpai, M., Pande, A., Tewari, S., & Prakash, D. (2005). Phenolic contents and antioxidant activity of some food and medicinal plants. *International Journal of Food Sciences and Nutrition*, *56*(4), 287–291.

Bao, B., Chen, Y.-G., Zhang, L., Na Xu, Y. L., Wang, X., Liu, J., & Qu, W. (2013). Momordica charantia (Bitter Melon) reduces obesity-associated macrophage and mast cell infiltration as well as inflammatory cytokine expression in adipose tissues. *PLoS ONE*, *8*(12), e84075.

Basch, E., Gabardi, S., & Ulbricht, C. (2003). Bitter melon (Momordica charantia): A review of efficacy and safety. *American Journal of Health-System Pharmacy*, *60*(4), 356–359.

Bnouham, M., Ziyyat, A., Mekhfi, H., Tahri, A., & Legssyer, A. (2006). Medicinal plants with potential antidiabetic activity-A review of ten years of herbal medicine research (1990–2000). *International Journal of Diabetes and Metabolism*, *14*(1), 1.

Cai, Y., Liu, M., Wu, X., Wang, Z., Liang, C., & Yang, Y. (2010). Study on the antitumor and immune-stimulating activity of polysaccharide from Momordica charantia. *Pharmaceutical and Clinical Research*, *18*(2), 131–134.

Chan, W., Tam, P., & Yeung, H. (1984). The termination of early pregnancy in the mouse by β-momorcharin. *Contraception*, *29*(1), 91–100.

Chang, C.-I., Tseng, H.-I., Liao, Y.-W., Yen, C.-H., Chen, T.-M., Lin, C.-C., & Cheng, H.-L. (2011). In vivo and in vitro studies to identify the hypoglycaemic constituents of Momordica charantia wild variant WB24. *Food Chemistry*, *125*(2), 521–528.

Chao, C.-Y., & Huang, C.-J. (2003). Bitter gourd (Momordica charantia) extract activates peroxisome proliferator-activated receptors and upregulates the expression of the acyl CoA oxidase gene in H4IIEC3 hepatoma cells. *Journal of Biomedical Science*, *10*(6), 782–791.

Chung, W.-Y., Jadhav, S., Hsu, P.-K., & Kuan, C.-M. (2022). Evaluation of acute and sub-chronic toxicity of bitter melon seed extract in Wistar rats. *Toxicology Reports*, *9*, 1024–1034.

Cortez-Navarrete, M., Martinez-Abundis, E., Perez-Rubio, K. G., Gonzalez-Ortiz, M., & Méndez-del Villar, M. (2018). Momordica charantia administration improves insulin secretion in type 2 diabetes mellitus. *Journal of Medicinal Food*, *21*(7), 672–677.

Cousens, G. (2007). *There is a cure for diabetes: The tree of life 21-day+ program*. CA: North Atlantic Books, pp. 191–192.

Deng, Y.-Y., Ma, Y., Liu, H., Zhang, Y., Wei, Z., Liu, G., . . . Jia, X. (2022). Structure determination, bitterness evaluation and hepatic gluconeogenesis inhibitory activity of triterpenoids from the Momordica charantia fruit. *Food Chemistry*, *372*, 131224.

Deng, Y.-Y., Yi, Y., Zhang, L.-F., Zhang, R.-F., Zhang, Y., Wei, Z.-C., . . . Zhang, M.-W. (2014). Immunomodulatory activity and partial characterisation of polysaccharides from Momordica charantia. *Molecules*, *19*(9), 13432–13447.

Dhar, P., Chattopadhyay, K., Bhattacharyya, D., Roychoudhury, A., Biswas, A., & Ghosh, S. (2007). Antioxidative effect of conjugated linolenic acid in diabetic and non-diabetic blood: An in vitro study. *Journal of Oleo Science*, *56*(1), 19–24.

Duan, Z.-Z., Zhou, X.-L., Li, Y.-H., Zhang, F., Li, F.-Y., & Su-Hua, Q. (2015). Protection of Momordica charantia polysaccharide against intracerebral hemorrhage-induced brain injury through JNK3 signaling pathway. *Journal of Receptors and Signal Transduction*, *35*(6), 523–529.

Esser, N., Legrand-Poels, S., Piette, J., Scheen, A. J., & Paquot, N. (2014). Inflammation as a link between obesity, metabolic syndrome and type 2 diabetes. *Diabetes Research and Clinical Practice*, *105*(2), 141–150.

Fang, E. F., Zhang, C. Z. Y., Wong, J. H., Shen, J. Y., Li, C. H., & Ng, T. B. (2012). The MAP30 protein from bitter gourd (Momordica charantia) seeds promotes apoptosis in liver cancer cells in vitro and in vivo. *Cancer Letters*, *324*(1), 66–74.

Han, J. H., Tuan, N. Q., Park, M. H., Quan, K. T., Oh, J., Heo, K. S., . . . Myung, C. S. (2018). Cucurbitane triterpenoids from the fruits of Momordica charantia improve insulin sensitivity and glucose homeostasis in streptozotocin-induced diabetic mice. *Molecular Nutrition & Food Research*, *62*(7), 1700769.

Handa, G., Singh, J., Sharma, M., Kaul, A., & Zafar, R. (1990). Hypoglycemic principle of Momordica charantia seeds. *Indian Journal of Natural Products*, *6*(1), 16–19.

Hazarika, R., Parida, P., Neog, B., & Yadav, R. N. S. (2012). Binding energy calculation of GSK-3 protein of Human against some anti-diabetic compounds of Momordica charantia linn (Bitter melon). *Bioinformation*, *8*(6), 251.

Hsiao, P.-C., Liaw, C.-C., Hwang, S.-Y., Cheng, H.-L., Zhang, L.-J., Shen, C.-C., . . . Kuo, Y.-H. (2013). Antiproliferative and hypoglycemic cucurbitane-type glycosides from the fruits of Momordica charantia. *Journal of Agricultural and Food Chemistry*, *61*(12), 2979–2986.

Huang, T.-N., Lu, K.-N., Pai, Y.-P., Hsu, C., & Huang, C.-J. (2013). Role of GLP-1 in the hypoglycemic effects of wild bitter gourd. *Evidence-Based Complementary and Alternative Medicine*, *2013*.

Hui, H., Tang, G., & Go, V. L. W. (2009). Hypoglycemic herbs and their action mechanisms. *Chinese Medicine*, *4*(1), 1–11.

Hussain, T., Tan, B., Yin, Y., Blachier, F., Tossou, M. C., & Rahu, N. (2016). Oxidative stress and inflammation: What polyphenols can do for us? *Oxidative Medicine and Cellular Longevity*, *2016*.

Jabeen, U., & Khanum, A. (2017). Isolation and characterization of potential food preservative peptide from Momordica charantia L. *Arabian Journal of Chemistry*, *10*, S3982–S3989.

Jia, S., Shen, M., Zhang, F., & Xie, J. (2017). Recent advances in Momordica charantia: Functional components and biological activities. *International Journal of Molecular Sciences*, *18*(12), 2555.

Joseph, B., & Jini, D. (2013). Antidiabetic effects of Momordica charantia (bitter melon) and its medicinal potency. *Asian Pacific Journal of Tropical Disease*, *3*(2), 93–102.

Joseph, B., & Raj, S. J. (2010). Phytopharmacological properties of Ficus racemosa Linn-An overview. *International Journal of Pharmaceutical Sciences Review and Research*, *3*(2), 134–138.

Kanna, P. (2004). Protein/polypeptide-k obtained from Momordica charantia. *US Patent*, *6*(831), 162.

Kavishankar, G., Lakshmidevi, N., Murthy, S. M., Prakash, H., & Niranjana, S. (2011). Diabetes and medicinal plants-A review. *International Journal of Pharmaceutical and Bio-Medical Science*, *2*(3), 65–80.

Kedar, P., & Chakrabarti, C. (1982). Effects of bittergourd (Momordica charantia) seed and glibenclamide in streptozotocin induced diabetes mellitus. *Indian Journal of Experimental Biology*, *20*(3), 232–235.

Keller, A. C., He, K., Brillantes, A.-M., & Kennelly, E. J. (2021). A characterized saponin-rich fraction of Momordica charantia shows antidiabetic activity in C57BLK/6 mice fed a high fat diet. *Phytomedicine Plus*, *1*(4), 100134.

Khanna, P., Jain, S., Panagariya, A., & Dixit, V. (1981). Hypoglycemic activity of polypeptide-p from a plant source. *Journal of Natural Products*, *44*(6), 648–655.

Kim, S. K., Jung, J., Jung, J. H., Yoon, N., Kang, S. S., Roh, G. S., & Hahm, J. R. (2020). Hypoglycemic efficacy and safety of Momordica charantia (bitter melon) in patients with type 2 diabetes mellitus. *Complementary Therapies in Medicine*, *52*, 102524.

Kitabchi, A. E., Umpierrez, G. E., Murphy, M. B., & Kreisberg, R. A. (2006). Hyperglycemic crises in adult patients with diabetes: A consensus statement from the American Diabetes Association. *Diabetes Care*, *29*(12), 2739–2748.

Kobori, M., Nakayama, H., Fukushima, K., Ohnishi-Kameyama, M., Ono, H., Fukushima, T., . . . Hoshi, Y. (2008). Bitter gourd suppresses lipopolysaccharide-induced inflammatory responses. *Journal of Agricultural and Food Chemistry*, *56*(11), 4004–4011.

Kobori, M., Ohnishi-Kameyama, M., Akimoto, Y., Yukizaki, C., & Yoshida, M. (2008). α-Eleostearic acid and its dihydroxy derivative are major apoptosis-inducing components of bitter gourd. *Journal of Agricultural and Food Chemistry*, *56*(22), 10515–10520.

Krawinkel, M. B., Ludwig, C., Swai, M. E., Yang, R.-y., Chun, K. P., & Habicht, S. D. (2018). Bitter gourd reduces elevated fasting plasma glucose levels in an intervention study among prediabetics in Tanzania. *Journal of Ethnopharmacology*, *216*, 1–7.

Lada, P., & Idrees, U. (2005). Toxicity of oral agents used to treat diabetes. *Journal of Pharmacy Practice*, *18*(3), 145–156.

Leung, S., Yeung, H., & Leung, K. (1987). The immunosuppressive activities of two abortifacient proteins isolated from the seeds of bitter melon (Momordica charantia). *Immunopharmacology*, *13*(3), 159–171.

Licastro, F., Franceschi, C., Barbieri, L., & Stirpe, F. (1980). Toxicity of Momordica charantia lectin and inhibitor for human normal and leukaemic lymphocytes. *Virchows Archiv B*, *33*(1), 257–265.

Lo, H.-Y., Ho, T.-Y., Li, C.-C., Chen, J.-C., Liu, J.-J., & Hsiang, C.-Y. (2014). A novel insulin receptor-binding protein from Momordica charantia enhances glucose uptake and glucose clearance in vitro and in vivo through triggering insulin receptor signaling pathway. *Journal of Agricultural and Food Chemistry*, *62*(36), 8952–8961.

Lo, H.-Y., Ho, T.-Y., Lin, C., Li, C.-C., & Hsiang, C.-Y. (2013). Momordica charantia and its novel polypeptide regulate glucose homeostasis in mice via binding to insulin receptor. *Journal of Agricultural and Food Chemistry*, *61*(10), 2461–2468.

Lo, H.-Y., Li, C.-C., Chen, F.-Y., Chen, J.-C., Hsiang, C.-Y., & Ho, T.-Y. (2017). Gastro-resistant insulin receptor-binding peptide from momordica charantia improved the glucose tolerance in streptozotocin-induced diabetic mice via insulin receptor signaling pathway. *Journal of Agricultural and Food Chemistry*, *65*(42), 9266–9274.

Lolitkar, M., & Rao, M. R. (1962). Note on a hypoglycaemic principle isolated from the fruits of Momordica charantia. *Journal of the University of Bombay*, *29*, 223–224.

Ma, C., Yu, H., Xiao, Y., & Wang, H. (2017). Momordica charantia extracts ameliorate insulin resistance by regulating the expression of SOCS-3 and JNK in type 2 diabetes mellitus rats. *Pharmaceutical Biology*, *55*(1), 2170–2177.

Ma, L., Yu, A.-H., Sun, L.-L., Gao, W., Zhang, M.-M., Su, Y.-L., . . . Ji, T. (2014). Two new bidesmoside triterpenoid saponins from the seeds of Momordica charantia L. *Molecules*, *19*(2), 2238–2246.

Mahmoud, M. F., El Ashry, F. E. Z. Z., El Maraghy, N. N., & Fahmy, A. (2017). Studies on the antidiabetic activities of Momordica charantia fruit juice in streptozotocin-induced diabetic rats. *Pharmaceutical Biology*, *55*(1), 758–765.

Malekshahi, H., Bahrami, G., Miraghaee, S., Ahmadi, S. A., Sajadimajd, S., Hatami, R., . . . Keshavarzi, S. (2019). Momordica charantia reverses type II diabetes in rat. *Journal of Food Biochemistry*, *43*(11), e13021.

McCarty, M. F., Lewis Lujan, L., & Iloki Assanga, S. (2022). Targeting Sirt1, AMPK, Nrf2, CK2 and soluble guanylate cyclase with nutraceuticals: A practical strategy for preserving bone mass. *International Journal of Molecular Sciences*, *23*(9), 4776.

Meng, Y., Liu, S., Li, J., Meng, Y., & Zhao, X. (2012). Preparation of an antitumor and antivirus agent: Chemical modification of α-MMC and MAP30 from Momordica Charantia L. with covalent conjugation of polyethyelene glycol. *International Journal of Nanomedicine*, *7*, 3133.

Minihane, A. M., Vinoy, S., Russell, W. R., Baka, A., Roche, H. M., Tuohy, K. M., . . . Vauzour, D. (2015). Low-grade inflammation, diet composition and health: Current research evidence and its translation. *British Journal of Nutrition*, *114*(7), 999–1012.

Miura, T., Kawata, T., Takagi, S., Nanpei, M., Nakao, H., Ishihara, E., & Ishida, T. (2009). Effect of Momordica charantia on adenosine monophosphate-activated protein kinase in genetically type 2 diabetic mice muscle. *Journal of Health Science*, *55*(5), 805–808.

Murugesan, A., Yadav, S. K. R., & Dixit, A. (2022). Anti-hyperglycemic activity of HPLC-fractionated Momordica charantia seed extract enriched in a novel napin-like protein in experimental diabetic rats and its validation with recombinant napin-like protein. *Current Research in Biotechnology*, *4*, 179–189.

Nakanekar, A., Kohli, K., & Tatke, P. (2019). Ayurvedic polyherbal combination (PDBT) for prediabetes: A randomized double blind placebo controlled study. *Journal of Ayurveda and Integrative Medicine*, *10*(4), 284–289.

Nhiem, N. X., Van Kiem, P., Van Minh, C., Ban, N. K., Cuong, N. X., Tung, N. H., . . . Quang, T. H. (2010). α-Glucosidase inhibition properties of cucurbitane-type triterpene glycosides from the fruits of Momordica charantia. *Chemical and Pharmaceutical Bulletin*, *58*(5), 720–724.

Patel, P., Harde, P., Pillai, J., Darji, N., & Patel, B. (2012). Antidiabetic herbal drugs a review. *Pharmacophore*, *3*(1), 18–29.

Pattarachotanant, N., Prasansuklab, A., & Tencomnao, T. (2021). Momordica charantia L. Extract protects hippocampal neuronal cells against PAHs-induced neurotoxicity: Possible active constituents include stigmasterol and vitamin E. *Nutrients*, *13*(7), 2368.

Pitchakarn, P., Ogawa, K., Suzuki, S., Takahashi, S., Asamoto, M., Chewonarin, T., . . . Shirai, T. (2010). Momordica charantia leaf extract suppresses rat prostate cancer progression in vitro and in vivo. *Cancer Science*, *101*(10), 2234–2240.

Pitipanapong, J., Chitprasert, S., Goto, M., Jiratchariyakul, W., Sasaki, M., & Shotipruk, A. (2007). New approach for extraction of charantin from Momordica charantia with pressurized liquid extraction. *Separation and Purification Technology*, *52*(3), 416–422.

Puri, M., Kaur, I., Kanwar, R., Gupta, R., Chauhan, A., & Kanwar, J. (2009). Ribosome inactivating proteins (RIPs) from Momordica charantia for anti viral therapy. *Current Molecular Medicine*, *9*(9), 1080–1094.

Qader, S. W., Abdulla, M. A., Chua, L. S., Najim, N., Zain, M. M., & Hamdan, S. (2011). Antioxidant, total phenolic content and cytotoxicity evaluation of selected Malaysian plants. *Molecules*, *16*(4), 3433–3443.

Rauniyar, G., Sinha, R., Chapagain, K., Maskey, R., & Pandey, D. (2021). Effects of Momordica charantia (karela/bittergourd) in type 2 diabetic patients taking allopathic drugs: A pilot study. *Kathmandu University Medical Journal*, *74*(2), 243–247.

Ríos, J. L., Escandell, J. M., & Recio, M. C. (2005). New insights into the bioactivity of cucurbitacins. *Studies in Natural Products Chemistry*, *32*, 429–469.

Rohajatien, U., Estiasih, T., & Sriwahyuni, E. (2018). Bitter melon (Momordica charantia L) fruit decreased blood glucose level and improved lipid profile of streptozotocin induced hyperglycemia rats. *Current Research in Nutrition and Food Science Journal*, *6*(2), 359–370.

Rosyid, F. N., Prasetyo, T. A., Muhtadi, M., & HS, K. (2021). The effects of oral supplementation of bitter melon (Momordica charantia L.) leaves extract on the VEGF serum level in diabetic foot ulcers: Randomized controlled trial. In *Proceedings of the 2nd Borobudur International Symposium on Humanities and Social Sciences, BIS-HSS 2020*, 18 November 2020, Magelang, Central Java, Indonesia.

Saeed, F., Sultan, M. T., Riaz, A., Ahmed, S., Bigiu, N., Amarowicz, R., & Manea, R. (2021). Bitter melon (Momordica charantia L.) fruit bioactives charantin and vicine potential for diabetes prophylaxis and treatment. *Plants*, *10*(4), 730.

Satyavati, G., Raina, M., & Sharma, M. (1987). *Medicinal plants of India* (Vol. 2). New Delhi: Indian Council of Medical Research.

Sharma, V., Sogani, R., Arora, R., & Bhargava, K. (1960). Some observations on hypoglycaemic activity of Momordica charantia. *Indian Journal of Medical Research*, *48*, 471–477.

Shaw, J. E., Sicree, R. A., & Zimmet, P. Z. (2010). Global estimates of the prevalence of diabetes for 2010 and 2030. *Diabetes Research and Clinical Practice*, *87*(1), 4–14.

Shetty, A., Kumar, G. S., Sambaiah, K., & Salimath, P. (2005). Effect of bitter gourd (Momordica charantia) on glycaemic status in streptozotocin induced diabetic rats. *Plant Foods for Human Nutrition*, *60*(3), 109–112.

Shih, C.-C., Lin, C.-H., & Lin, W.-L. (2008). Effects of Momordica charantia on insulin resistance and visceral obesity in mice on high-fat diet. *Diabetes Research and Clinical Practice*, *81*(2), 134–143.

Shivanagoudra, S. R., Perera, W. H., Perez, J. L., Athrey, G., Sun, Y., Jayaprakasha, G., & Patil, B. S. (2019). Cucurbitane-type compounds from Momordica charantia: Isolation, in vitro antidiabetic, anti-inflammatory activities and in silico modeling approaches. *Bioorganic Chemistry*, *87*, 31–42.

Sinha, R., Maskey, R., Adhikari, S., & Sarraf, D. (2020). Effects of Momordica charantia (Bitter gourd) and *Trigonella foenumgraecum* (Fenugreek) supplements in type-2 diabetics taking allopathic drugs. *Journal of Drug Delivery and Therapeutics*, *10*(6), 110–119.

Sirn, Y. Y., Lok, L. C., Weng, P. H., Ahmad, Z., & Hakim, M. N. (2021). Anti-diabetic activity of polypeptide-K isolated from Momordica charantia: A retrospective study of 142 cases. *Biomedical and Pharmacology Journal*, *14*(4), 2007–2016.

Sun, L., Zhang, X., Dong, L., Zhang, C., Guo, P., & Wu, C. (2021). The triterpenoids of the bitter gourd (Momordica Charantia) and their pharmacological activities: A review. *Journal of Food Composition and Analysis*, *96*, 103726.

Tan, S. P., Kha, T. C., Parks, S. E., & Roach, P. D. (2016). Bitter melon (Momordica charantia L.) bioactive composition and health benefits: A review. *Food Reviews International*, *32*(2), 181–202.

Tsuzuki, T., Tokuyama, Y., Igarashi, M., & Miyazawa, T. (2004). Tumor growth suppression by α-eleostearic acid, a linolenic acid isomer with a conjugated triene system, via lipid peroxidation. *Carcinogenesis*, *25*(8), 1417–1425.

Virdi, J., Sivakami, S., Shahani, S., Suthar, A., Banavalikar, M., & Biyani, M. (2003). Antihyperglycemic effects of three extracts from Momordica charantia. *Journal of Ethnopharmacology*, *88*(1), 107–111.

Walker, B. R., & Colledge, N. R. (2013). *Davidson's principles and practice of medicine e-book*. 22nd edition. Elsevier Health Sciences.

Wang, B. L., Zhang, W. J., Zhao, J., Wang, F. J., Fan, L. Q., Wu, Y. X., & Hu, Z. B. (2011). Gene cloning and expression of a novel hypoglycaemic peptide from Momordica charantia. *Journal of the Science of Food and Agriculture*, *91*(13), 2443–2448.

Wang, Q., Wu, X., Shi, F., & Liu, Y. (2019). Comparison of antidiabetic effects of saponins and polysaccharides from Momordica charantia L. in STZ-induced type 2 diabetic mice. *Biomedicine & Pharmacotherapy*, *109*, 744–750.

Weng, J.-R., Bai, L.-Y., Chiu, C.-F., Hu, J.-L., Chiu, S.-J., & Wu, C.-Y. (2013). Cucurbitane triterpenoid from Momordica charantia induces apoptosis and autophagy in breast cancer cells, in part, through peroxisome proliferator-activated receptor γ activation. *Evidence-Based Complementary and Alternative Medicine*, *2013*.

Xu, X., Shan, B., Liao, C.-H., Xie, J.-H., Wen, P.-W., & Shi, J.-Y. (2015). Anti-diabetic properties of Momordica charantia L. polysaccharide in alloxan-induced diabetic mice. *International Journal of Biological Macromolecules*, *81*, 538–543.

Zhang, F., Lin, L., & Xie, J. (2016). A mini-review of chemical and biological properties of polysaccharides from Momordica charantia. *International Journal of Biological Macromolecules*, *92*, 246–253.

Zhu, J., Xing, G., Shen, T., Xu, G., Peng, Y., Rao, J., & Shi, R. (2019). Postprandial glucose levels are better associated with the risk factors for diabetes compared to fasting glucose and glycosylated hemoglobin (HbA1c) levels in elderly prediabetics: Beneficial effects of polyherbal supplements – A randomized, double-blind, placebo controlled trial. *Evidence-Based Complementary and Alternative Medicine*, *2019*.

Zhu, Y., Liu, Q., Zhou, Z., & Ikeda, Y. (2017). PDX1, Neurogenin-3, and MAFA: Critical transcription regulators for beta cell development and regeneration. *Stem Cell Research & Therapy*, *8*(1), 1–7.

8 Antidiabetic and Other Responses of Bitter Apple [*Citrullus colocynthis* (L.) Schrad.]

Sana Nafees, Ranjeet Kaur, Jamal Akhtar, Nighat Anjum, and Jasbir Kaur

CONTENTS

8.1 INTRODUCTION

Bitter apple (*Citrullus colocynthis*) belongs to the family Cucurbitaceae and is believed to be the wild gourd of the Old Testament in ancient medicine. Ancient Greeks and Roman physicians were well aware of its vicious purgative property and its bitter taste (Lloyd, 1898). In English, it is popularly known as bitter cucumber, bitter apple, bitter gourd and colocynth (Bedevian, 1936). It is a well-known medicinal plant in various traditional system of medicines like traditional Unani medicine, Ayurveda and traditional Iranian medicine (TIM), and it has been used alone and in many compound formulations for medicinal purposes.

The fruit pulp is the main medicinal part of the plant that is used, and when it is administered it is better to remove the pulp. This removal decreases the potency over time and after two years it is completely lost. However, the pulp remains potent for about four years when it remains in the fruit (Hussain, 1855; Kabiruddin, 2007). Other parts of the plant that are used as medicines are

DOI: 10.1201/b23347-8

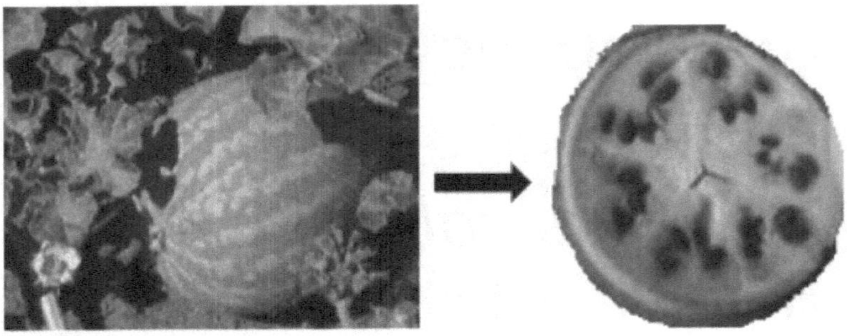

FIGURE 8.1 (a) Fruit of *Citrullus colocynthis*; (b) transverse section of fruit.

leaf, seed and root (Hussain, 1855). The plant belongs to the subkingdom Tracheabionta, division Magnoliophyta, class Magnoliopsida, order Cucurbitales, family Cucurbitaceae, genus *Citrullus* and species *colocynthis*. Common names of this plant include bitter apple, bitter gourd, bitter cucumber (English), coloquinte (French), Koloquinthe (German) (De Smet, 1997), Indravaruni, Indrayan (Hindi), Chedupuchcha (Telugu), Cinnapapara (Kannada), Indravana (Punjabi) and Tumtikayi (Tamil). This traditional plant medicine is well-known for the treatment of diabetes, asthma and jaundice (Baquar and Tasnif, 1984; Kirtikar et al., 1984; Qureshi et al., 2010).

8.2 BOTANICAL DESCRIPTION

Bitter apple (*C. colocynthis*) resembles the watermelon and is an annual herbaceous plant. The stems are covered with rough hairs with alternately arranged leaves on long petioles. They are triangular in shape with a dark green color on the upper surface and underneath they are rough and pale in color. The leaf has many clefts and sinuses, obtuse and hairy. Flowers are yellow in color and appear singly at the axis of the leaf. The fruit is globular, similar to the size of an orange, yellow in color and smooth. On ripening, the fruit becomes hard with a coriaceous rind and white spongy pulp and contains many ovate flattened seeds that are white or brown in color (Figure 8.1 a and b). The fruit is very bitter in taste (Shishkin and Bobrov, 1957).

8.3 DISTRIBUTION

Citrullus colocynthis grows in Asia and Africa (Rahimi et al., 2012). The plant is indigenous to arid sandy areas of West Asia, Arabia, tropical Africa and the Mediterranean region (Pravin et al., 2013). Its origins are in Asia and the Mediterranean basin, mainly in Turkey and Nubia, Egypt and Sahara in the east, and Africa's western coastal regions. It is also widely distributed in the desert areas of Pakistan, India, southeast, east, southwest and central area of Iran and the northern coastal regions of the Mediterranean and Caspian seas (Kamran et al., 2018).

8.4 PHYTOCHEMICAL CONSTITUENTS

The chief chemical constituents of this plant are citrulonol, citrulluin, citrollic acid, methyl lengenol and inositol (Gurudeeban et al., 2010), whereas the pulp contains "colocynthin," a bitter compound, and the resins colocynthetin, colocynthein, pectin and gum. In seeds, fixed oil and albuminoids are found (Panda, 1999). The chemical constituents present in this plant are mostly glycosides and, on enzymatic hydrolysis, it produces "elaterin" (cucurbitacin E), "elatericin B" (cucurbitacin I) and "dihydroelatericin B" (cucurbitacin L) (Lavie et al., 1964). In addition to these compounds, it also encompasses quercetin, caffeic acid derivatives, cucurbitacin E-, J-, L-glucosides, chlorogenic acid, phenolic acid, flavonoids, tocopherols, alkaloids and fatty acids (Fleming, 2000; Meena and Patni, 2008; Hussain et al., 2013).

8.5 PHARMACOLOGICAL STUDIES

Numerous pharmacological studies have been inspired by the traditional medicinal uses of *C. colocynthis*. The biological activities of a number of extracts and isolated compounds have been studied, particularly their effects on diabetes and cancer. Some of the important pharmacological studies have described in the following sections.

8.5.1 Antilipidemic Activity

A study by Dardaka et al. (2007) examined the effect of ethanolic extract of *Citrullus colocynthis* in rabbit that consumed 1.2 g/kg of body weight daily. Other *Citrullus colocynthis* antilipidemic properties were demonstrated in Rahbar and Nabipour's (2010) study, which involved nondiabetic individuals ingesting *Citrullus colocynthis* seed powder (300 mg) daily; the results showed a lower level of cholesterol and triglyceride concentration. *Citrullus colocynthis* has been shown to have antilipidemic properties on both human and animal hyperlipidemia.

8.5.2 Antimicrobial Activity

Extracts of *Citrullus colocynthis* were found to have antibacterial and anticandidal activities according to various research, and antimicrobial efficacy varied depending on the population (Marzouk et al., 2009; Marzouk et al., 2011). *Citrullus colocynthis* extract affects both gram-positive and gram-negative bacteria and has inhibitory action against *Bacillus subtilis*; however, it has poor action against two-gram negative bacteria (*Pseudomonas aeruoginosa* and *E. coli*). Bioactive components including flavonoids, glycoside and tannins may also be responsible for the potent antibacterial effects of fruit extracts on the bacterial stains tested (Memon et al., 2003). *Staphylococcus aureus*, *S. epidermidis*, *B. subtilis*, *E. coli*, *S. typhimurium*, *P. aeruginosa*, *Proteus vulgaris* and *Mycobacterium smegmatis* were some of the pathogenic bacteria that were studied for the antibacterial activity of *Citrullus colocynthis* (Sharma et al., 2010) against all bacterial pathogens. Most of the extracts exhibit moderate MIC in the 20–100 g/ml range. Another study assessed the antibacterial activity of *Citrullus colocynthis* fruit extracts (water and ethanol) against hospital-isolated strains of *S. aureus* and a standard strain (ATCC 25923) from patients receiving novobiocin treatment (Najafi et al., 2010). The results showed that inhibitory activities of *Citrullus colocynthis* fruit were comparable to those of the widely used antibiotic (novobiocin).

8.5.3 Anti-Inflammatory Activity

The use of nonsteroidal anti-inflammatory drugs (NSAIDs) in "modern therapeutics" results in gastrointestinal tracts ulcers (Graham et al., 1988). Selective COX-2 inhibitor NSAIDs are less ulcerogenic, but some of them have been taken off the market due to their fatal cardiac toxicity. As a results, the demand for safer anti-inflammatory drugs is increasing. *Citrullus colocynthis* is one of most common plants used in traditional medicine because of its anti-inflammatory properties (Wasfi et al., 1995; Marzouk et al., 2011). The aqueous extract of the fruit is investigated for anti-inflammatory activity and to validate this as a remedy for rheumatoid arthritis as well as an analgesic and anti-inflammatory (Marzouk et al., 2011). Marzouk et al. (2011) also validates the anti-inflammatory potential of this plant. Zaidi et al. (2012) stated that this plant may be free of ulcerogenic activity as plants are rich in anti-ulcerogenic components; however, the anti-ulcerogenic activity has not been yet investigated.

8.5.4 Antioxidant Activities

Citrullus colocynthis is a good source of antioxidants, as this plant has polyphenols and plant sterol (Sebbagh et al., 2009). The excellent antioxidant activity of fruit extracts was confirmed by some *in vitro* investigations (Kumar et al., 2008). The fruit's ability to scavenge free radicals is improved when the extract concertation is increased (Kumar et al., 2008). The 2,2-diphenyl-1-picrylhydrazyl (DPPH) assay was used to determine the antioxidant activity of three flavonoids (isoponarine, isovitexin and isorientin, 3-O-methyl ether) isolated from *Citrullus colocynthis.* IC50 (concentration that scavenges 50% of the DPPH radical) values were in the range of 5.62104 to 7.13102 mg/mL. Whereas in that study, quercetin, the positive control, had an IC50 value of 2.78105 mg/mL (Kumar et al., 2008). According to Marzouk et al. (2011), aqueous extracts of *Citrullus colocynthis* seed obtained from Tunisia had a very high DPPH radicals scavenging activity with an IC50 of 0.021 mg/mL. *Citrullus colocynthis* cucurbitacin glycoside displayed ABTS radicals scavenging properties (IC50, 145M), which are likely due to direct scavenging effects on many free radicals (Tannin-Spitz et al., 2007). These results provided substantial evidence in favor of *C. colocynthis* as a source of natural antioxidant.

8.5.5 Anticancer Activity

There are several publications on the anti-proliferative activity of *C. colocynthis* extract and its isolates (Chen et al., 2005; Tannin-Spitz et al., 2007; Liu et al.,2008; Yin et al., 2008). Cucurbitacin glucoside inhibits the proliferation of human breast cancer cell line at a dose of 20 μM (Tannin-Spitz et al., 2007). Cucurbitacin B administration inhibits the cellular proliferation in laryngeal cancer cell line (Hep-2) in a dose- and time-dependent manner (Liu et al., 2008). Flow cytometry results reveal the arrest of cells at the G2/M phase of cell cycle and cell apoptosis. Hoechst 33258 staining made it possible to see the distinct morphological changes that occur during cell death, such as chromatin fragmentation, fragmentation of nucleus and the condensation of apoptotic bodies. The expression of Bcl-2, cyclin B1 and p-STAT3 was dramatically decreased, as seen in western blotting expression. According to *in vivo* investigations, it also inhibits the laryngeal squamous carcinoma with the low, medium and high dose groups (32.43%, 43.24% and 70.27%, respectively). The most common malignant brain tumor in adults is glioblastoma multiforme (GBM) and in human GBM cell lines, cucurbitacin B produced a 50% growth reduction at 0.1 M (ED50). Cucurbitacin B appeared to inhibit nearly all GBM clonogenic cells in soft-gel studies, indicating that clinical trials for this drug may be a promising alternative (Yin et al., 2008). Cucurbitacin E also has attracted much attention due to its anticancer and cytotoxic properties. Cucurbitacin L was effective against HeLa and KB cell lines, despite being less potent than cucurbitacin I, isolated from *Citrullus colocynthis* (Lavie et al., 1964; Chen et al., 2005).

8.5.6 Anti-Allergic Activity

Cucurbitacin E is the anti-allergic compound that is isolated from the methanolic fruit extract of *Citrullus colocynthis.* It prevents the development of passive cutaneous anaphylaxis in type 1 allergic rats at the doses of 100 to 125 mg/kg (Yoshikawa et al., 2007).

8.5.7 Larvicidal and Pesticidal

A natural insecticide, "4-methylquinoline" derived from *C. colocynthis* may be used to control spider mites and stored grain mites, along with its structural counterpart (Jeon and Lee, 2014). Additionally, isolated component of *C. colocynthis* exhibited strong larvicidal properties. Rahuman et al. (2008) investigated the larvicidal activity against mosquito larvae by bioassay-guided fractionation of petroleum ether extracts and found "oleic acid" and "linoleic acid" to be larvicidal chemicals, as they were effective against the larvae of four insects (*Aedes aegypti*, *Anopheles stephensi*, *Liston* and *Culex quinquefasciatus*). The finding of this investigation clearly demonstrated that *Citrullus colocynthis* extract and fraction containing "oleic and linoleic acids" displayed a high larval mortality.

8.5.8 Other Pharmacological Activities

In light of its Ayurvedic use as tonic for hair, albino rats were used as a model to study the effects of *Citrullus colocynthis* on hair growth. Results showed that the treatment with petroleum ether extract shortened the duration for hair growth. Additionally, this was effective in achieving a higher percentage of hair follicles (47%) in the anagenic phase, compared with the standard drug minoxidil (67%) (Roy et al., 2007).

8.6 ANTIDIABETIC RESPONSE

One of the metabolic disorders with the quickest rate of growth is diabetes mellitus. The hunt for more efficient and safe antidiabetic medications continues because the current treatment is symptomatic and necessitates lifelong use of pharmacological treatments, which have several adverse effects in addition to being expensive. In several nations, *Citrullus colocynthis* is used extensively as an antidiabetic (Al-Ghaithi et al., 2004) and intriguingly attracted several investigations on both humans and animals. Blood sugar levels and some of the harmful effects of streptozotocin may be lessened by taking the aqueous extract orally (Abdel-Hassan et al., 2000; Al-Ghaithi et al., 2004). The insulinotropic action of fruit extracts was also documented in literature (Nmila et al., 2000; Bnouham et al., 2006). Different seed extracts administered to isolated pancreas and islets of rats with glucose (8.3 Mm) dramatically increased insulin production *in vitro*. In rabbits with normal glycemia level, the treatment of aqueous extract in dose of 300 mg/kg orally had significantly decreased the glycemia. The phytoconstituents, namely glycosides, saponins, and tertiary and quaternary alkaloids, extracted from the fruits of *C. colocynthis* were investigated in normoglycemic rabbits at a dose of 50 mg/kg orally. The glycosidic component considerably reduced the glycemia, however none of the alkaloid possesses hypoglycemic effect. In experimentally induced diabetic rats by alloxan, the saponin component decreased glycemia at significantly lower dosages (10–20 mg/kg), suggesting that the saponin and glycosides component present in the rind of *C. colocynthis* may be responsible for the lowering of blood sugar levels (Bnouham et al., 2006). Some of the pharmacological studies on antidiabetic activity of different parts of the *C. colocynthis* have been listed in Table 8.1.

TABLE 8.1
Pharmacological Studies on Antidiabetic Activity

S. No.	Part used	Type of extract	Reference
1	Fruit pulp	Ethanolic	Dallak and Bin-Jaliah (2010)
2	Fruit pulp	Petroleum ether	Jayaraman et al. (2009)
3	Fruit pulp	Ethanolic	Babiker et al. (2012)
4	Seeds	Ethanolic	Lakshmi et al. (2013)
5	Seeds	Ethanolic	Sebbagh et al. (2009)
6	Leaf	Suspension	Gurudeeban et al. (2010)
7	Rind	Aqueous	Jeyanthi and Mary (2009)
8	Rind	Ethanolic	Aghanouri et al. (2009)
9	Root	Aqueous	Agarwal et al. (2012)
10	Root	Aqueous	Atole et al. (2009)

The effectiveness of the *Citrullus colocynthis* fruit was examined in a clinical trial of 50 type II diabetic patients over the course of two months. Capsules of *Citrullus colocynthis* fruit (100 mg) or placebos were given three times in a day, to two groups of 25 people each receiving "standard anti-diabetic" therapy. Fasting blood glucose level, glycosylated hemoglobin (HbA1c), total cholesterol, high density lipoprotein (HDL), low density lipoprotein (LDL), triglyceride, alkaline phosphatase, alanine transaminase, aspartate transaminase, creatinine and urea were assessed before and after two months of treatment. The outcomes demonstrated a substantial reduction in fasting blood glucose levels and in HbA1c levels in patients receiving *C. colocynthis*. Other serological parameter levels did not significantly differ between the two groups, and there were no significant gastrointestinal side effects in either group (Huseini et al., 2009).

The fruit of bitter apple possesses insulin-enhancing activity and this elucidates the antidiabetic effect in traditional systems of medicines. It recognizes *C. colocynthis* as a potential source of novel insulin-enhancing agent that may be useful in treating hyperglycemia in type 2 diabetes mellitus patients. The pulp and seed extracts improved the insulin-induced translocation of the glucose transporter (GLUT4) from intracellular storage sites to the plasma membrane, which in turn boosted the insulin-induced glucose uptake. The enhanced glucose absorption is more pronounced in pulp extract than in seeds, and it is caused by the same intracellular signaling pathway that insulin uses to promote GLUT4 translocation and glucose uptake (Drissi et al., 2021). The molecular mechanisms (Figure 8.2) of hypoglycemic and hypolipidemic activities of *Citrullus colocynthis* showed that it affects both sugar and fat metabolic pathways and there is G6Pase and PEPCK's differential expressions as gluconeogenic enzymes. The anti-fatty liver effect of *C. colocynthis* in early-stage diabetes is explained by PPAR activation by *C. colocynthis* without changing CPT1 expression. The preventive effect of *C. colocynthis* against hepatic lipid accumulation in late-stage diabetes is explained by the activation of PPAR expression and the suppression of CPT1 expression in the liver. In early-stage diabetes, it prevents glycogenolysis by inhibiting G6Pase expression. However, its ability to lower blood sugar in late-stage diabetes was not brought about by downregulation of PEPCK gene expression; rather, it did so primarily by diverting PEPCK activity in the direction of glyceroneogenesis (Afshari et al., 2021).

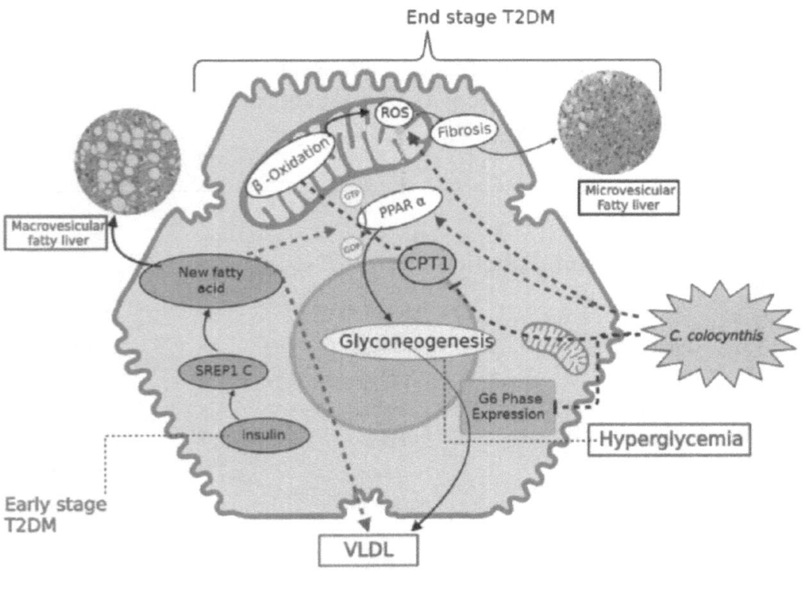

FIGURE 8.2 *Citrullus colocynthis* antidiabetic mechanism at the molecular level.

8.7 TRADITIONAL AND OTHER POTENTIAL USES

Bitter apple (*Citrullus colocynthis*) is used extensively for the treatment of various diseases that includes diabetes, jaundice, cancer, mastitis and joint pain in different parts of the world (Chopra, 1958; Perveen et al., 2007; Abo et al., 2008; Asyaz et al., 2010). This plant is used as medicine in traditional systems of medicine in Africa and Asian countries such as India, Pakistan and China, which reported its use in gastrointestinal disorders such as dysentery, indigestion, colic pain and gastroenteritis, as well as diabetes, common cold, cough, toothache and wounds (Qureshi et al., 2010; Amamou et al., 2011; Gurudeeban et al., 2010). The Indian Ayurvedic pharmacopoeia recommended using the fruit for jaundice, the root for diseases of liver and spleen, and the leaf for alopecia and other skin conditions (Lakhan et al., 2017).

8.8 SAFETY ISSUES

Evidence on systemic toxicity and safety evaluation of *Citrullus colocynthis* is rare, as it has a long history of use as medicine. Concerning acute toxicity, Diwan et al. (2000) reported the median lethal dose (LD50) of saponin extracted from *C. colocynthis* was 200 mg/kg, and on comparing with the LD50 values of other bioactive pharmaceuticals used in therapeutics, it was found to be nontoxic. It is detrimental for patients with hot temperament (Shahabi et al., 2008). In winters, *C. colocynthis* causes abdominal cramps, headache and nausea while it causes inflammation and dysentery in hot weather (Hussain, 1855). Therefore, the intake of this should be avoided in summer and winter (Kabeeruddin, 2007). It has been reported that the toxic dose 600–1000 mg causes diarrhea, colic, vomiting, hematochezia and nephrosis. Lethal dose of 2 g causes paralysis, convulsions and possibly death by circulatory collapse (Duke, 2007). Herbal medicine sellers also reported adverse events of bitter apple such as inaction of intestine, diarrhea and liver impairment, and they do not prescribe this medicine for more than two months. Barth et al. (2002) found that *C. colocynthis* extract showed no hepatotoxicity in concentrations up to 100 μg/mL. In another study on rat liver, however, it was disclosed that the alcoholic extract of bitter apple has a toxic effect on liver cells in a concentration above 100 mg/kg and causes necrosis of hepatocyte and fibrosis of liver (Dehghani and Panjehshahin, 2006). In traditional medicine, especially in Iranian medicine, for prevention of adverse effects it is recommended that equal weights of Arabic gum, starch, myrrh gum or tragacanth gum should be administered (Harvi, 1894; Kabeeruddin, 2007). Administration of opium tincture to be followed by stimulation and intake of mucilaginous drinks is recommended in case of poisoning (Duke, 2007). It is harmful for slender people, so it is better that it should be given with fruit juices. It is contraindicated during pregnancy as it has abortifacient action (Hussain, 1855). In addition to this, care should be taken during powdering of this drug – it should be completely powdered, as the ungrounded particle may cause inflammation of the gastrointestinal tract (Ibn Sina, 2014). Soufane et al. (2013) investigated that rat's bone marrow, liver and kidneys have some negative effects after ingesting the extract of ripe bitter apple fruit. Intestinal injury can result because of membranolytic action of *C. colocynthis* (Javadzadeh et al., 2013).

8.9 CULTIVATION PRACTICES

In most of the native regions, the crop of this plant is mainly harvested from wild plants. This is easily cultivated from seed, is rapidly grown, requires no attention, and has a yield of 6,700 kg/ha. Fully developed but unripe fruits are collected by hand picking. The outer ring (pericarp), which is thin and hard, is removed by peeling, while the inner white spongy pulp filled with seeds is to be dried in the oven or in sun. Of the dried product weight, 75% is constituted by the seeds. Both wild and cultivated varieties are commercially supplied. Commercially, colocynth is available in two forms – pulp and bitter apple. Pulp is available without seeds, whereas the bitter apples or masses are made into balls, filled with seeds. Both forms are usually shipped in boxes. In the United States,

Sudan is the main supplier, whereas it is also imported from Spain and Turkey, which supplies the finest quality of colocynth. However, it is not cultivated in Egypt, but wild plants supply the small quantity of pulp (Duke, 1978).

8.10 FUTURE REMARKS

Published material and research works elaborate *C. colocynthis* as a potential antiglycemic medicinal plant; however, additional studies are needed to support its promise as insulin enhancer for the treatment of diabetes mellitus, and more studies are required to assess the mechanism of action. Attempts should be made to encourage people to protect this valuable medicinal plant, which grows freely in the arid regions of the northwestern Punjab, Sind, and central and southern India. It is evident that additional research should be done on phytopharmaceutical and nutraceutical applications, as this plant has a huge potential for this; however, in-depth toxicity studies are still needed.

REFERENCES

Abdel-Hassan, I.A., Abdel-Barry, J.A., Tariq Mohammeda, S. 2000. The hypoglycaemic and antihyperglycaemic effect of *Citrullus colocynthis* fruit aqueous extract in normal and alloxan diabetic rabbits. *Journal of Ethnopharmacology* 71, 325–330.

Abo, K.A., Fred-Jaiyesimi, A.A., Jaiyesimi, A.E.A. 2008. Ethnobotanical studies of medicinal plants used in the management of diabetes mellitus in South Western Nigeria. *Journal of Ethnopharmacology* 115, 67–71.

Afshari, A., Salimi, F., Nowrouzi, A., Khalili, M.B., Bakhtiyari, S., Hassanzadeh, G., Shabani, M., Ahadi, A., Farhadiannezhad, M. 2021. Differential expression of gluconeogenic enzymes in early- and late-stage diabetes: the effect of *Citrullus colocynthis* (L.) Schrad. Seed extract on hyperglycemia and hyperlipidemia in Wistar-Albino rats model. *Clinical Phytoscience* 7, 88.

Agarwal, V., Sharma, A.K., Upadhyay, A., Singh, G., Gupta, R. 2012. Hypoglycemic effects of *Citrullus colocynthis* roots. *Acta Poloniae Pharmaceutica – Drug Research* 69, 75–84.

Aghanouri, Z., Noureddini, M., Salami, M., Feyz, J. 2009. Effect of *Citrullus colocynthis* on diabetic rats plasma glucose. *Kashan University of Medical Sciences* 12, 1.

Al-Ghaithi, F., El-Ridi, M.R., Adeghate, E., Amiri, M.H. 2004. Biochemical effects of *Citrullus colocynthis* in normal and diabetic rats. *Molecular and Cellular Biochemistry* 261, 143–149.

Amamou, F., Bouafia, M., Chabane-Sari, D., Meziane, R.K., Nani, A. 2011. *Citrullus colocynthis*: A desert plant native in Algeria, effects of fixed oil on blood homeostasis in Wistar rat. *Journal of Natural Product and Plant Resources* 1, 1–7.

Asyaz, A., Salimi, F., Nowrouzi, A., Khalili, M.B., Bakhtiyari, S., Hassanzadeh, G., Shabani, M., Ahadi, A., Farhadiannezhad, M. 2010. Differential expression of gluconeogenic enzymes in early- and late-stage diabetes: The effect of *Citrullus colocynthis* (L.) Schrad. Seed extract on hyperglycemia and hyperlipidemia in Wistar-Albino rats model. *Clinical Phytoscience* 7, 88–88.

Atole, S.K., Jangde, C.R., Philip, P., Rekhe, D.S., Aghav, D.V., Waghode, H.J. 2009. Safety evaluation studies of *Citrullus colocynthis* for diabetes in rats. *Veterinary World* 2, 423–425.

Babiker, H.A., Eldin, I.M.T., Abd-Elwahab, H.M. 2012. The effects of ethanolic extract of Handal (*Citrullus colocynthis* L.) fruit pulp on blood glucose level in normal fasting rat. *Gezira Journal of Health Sciences* 8, 1.

Baquar, S.R., Tasnif, M. 1984. *Medicinal Plants of Southern West Pakistan, D-42*. Delhi: Periodical Expert Book Agency, 4, 108.

Barth, A., Müller, D., Dürrling, K. 2002. In vitro investigation of a standardized dried extract of *Citrullus colocynthis* on liver toxicity in adult rats. *Experimental Toxicologic Pathology* 54, 223–230.

Bedevian, A.K. 1936. *Illustrated Polyglottic Dictionary of Plant Names*. Cairo, Egypt: Argus D Papazian Presses.

Bnouham, M., Ziyyat, A., Mekhfi, H., Tahri, A., Legssyer, A. 2006. Medicinal plants with potential antidiabetic activity: A review of ten years of herbal medicine research (1990–2000). *International Journal of Diabetes and Metabolism* 14, 1–25.

Chen, J.C., Chiu, M.H., Nie, R.L., Cordell, G.A., Qiu, S.X. 2005. Cucurbitacins and cucurbitane glycosides: Structures and biological activities. *Natural Product Report* 22, 386–399.

Chopra, R.N. 1958. *Indigenous Drugs of India*. Calcutta, India: U.N Dhur and Sons, Ltd., 167.

Dallak, M., Bin-Jaliah, I. 2010. Antioxidant activity of *Citrullus colocynthis* pulp extract in the Rbc's of alloxan-induced diabetic rats. *Pakistan Journal of Physiology* 6, 1.

Daradka, H., Almasad, M.M., Qazan, W.S., El-Banna, N.M., Samara, O.H. 2007. Hypolipidamic effects of *Citrullus colocynthis* L. in rabits. *Pakistan Journal of Biological Sciences* 10, 2768–2771.

Dehghani, F., Panjehshahin, M.R. 2006. The toxic effect of alcoholic extract of *Citrullus colocynthis* on rat liver. *Iranian Journal of Pharmacology and Therapeutics* 2, 117–119.

De Smet, P. 1997. Citrullus colocynthi. In *Adverse Effects of Herbal Drugs*. Berlin, Heidelberg: Springer, 29–36.

Diwan, F.H., Abdel-Hassan, I.A., Mohammad, S.T. 2000. Effect of saponin on mortality and histopathological changes in mice. *Eastern Mediterannean Health Journal* 6, 345–351.

Drissi, F., Lahfa, F., Gonzalez, T., Peiretti, F., Tanti, J.F., Haddad, M., Fabre, N., Govers, R. 2021. A *Citrullus colocynthis* fruit extract acutely enhances insulin-induced GLUT4 translocation and glucose uptake in adipocytes by increasing PKB phosphorylation. *Journal of Ethnopharmacology* 270, 113772.

Duke, J.A. 1978. The quest for tolerant germplasm. *American Society of Agronomy*, 1–61.

Duke, J.A. 2007. *Duke's Handbook of Medicinal Plants of the Bible*. Fulton, MD: CRC Press, 115–118

Fleming, T. 2000. *PDR for Herbal Medicines*. NJ: Medical Economics.

Graham, D.Y., Agrawal, N.M., Roth, S.H. 1988. Prevention of NSAID induced gastric ulcer with misoprostol: Multicentre, double-blind placebo- controlled trial. *Lancet* 2, 1277–1280.

Gurudeeban, S., Satvavani, K., Ramanathan, T. 2010. Bitter apple (*Citrullus colocynthis*) an overview of chemical composition and biomedical potential. *Asian Journal of Plant Science* 9, 394–401.

Harvi, M.Y. 1894. *Behrul Jawahar*. Delhi: Mujtaba Press, 145, 67.

Huseini, H.F., Darvishzadeh, F., Heshmat, R., Jafariazar, Z., Raza, M., Larijani, B. 2009. The clinical investigation of *Citrullus colocynthis* (L.) Schrad fruit in treatment of type II diabetic patients: A randomized, double blind, placebo controlled clinical trial. *Phytotherapy Research* 23, 1186–1189.

Hussain, A.I., Rathore, H.A., Sattar, M.Z., Chatha, S.A., Ahmad, F., Ahmad, A., Johns, E.J. 2013. Phenolic profile and antioxidant activity of various extracts from *Citrullus colocynthis* (L.) from the Pakistani flora. *Industrial Crops and Products* 45, 416–422.

Hussain, M.M. 1855. *Tohfatul Momineen (Persian)*. Delhi: Matbah Husaini Publicaion, 164.

Ibn Sina. 2014. *Al-Qanoon Fil Tibb* (Urdu translation by Ghulam Hussain Kantoori). Delhi: Idara Kitab-us-Shifa, 745–746.

Javadzadeh, H.R., Davoudi, A., Davoudi, F., Valizadegan, G., Goodarzi, H., Mahmoodi, S., Ghane, M.R., Faraji, M. 2013. *Citrullus colocynthis* as the cause of acute rectorrhagia. Case report. *Emergency Medicine* 1–5.

Jayaraman, R., Shivakumar, A., Anitha, T., Joshi, V.D., Palei, N.N. 2009. Antidiabetic effect of petroleum ether extract of *Citrullus colocynthis* fruits against streptozotocin-induced hyperglycemic rats. *Romanian Journal of Biology – Plant Biology* 4, 127–134.

Jeon, J.H., Lee, H.S. 2014. Biofunctional constituent isolated from *Citrullus colocynthis* fruits and structure-activity relationships of its analogues show acaricidal and insecticidal efficacy. *Journal of Agricultural and Food Chemistry* 62, 8663–8667.

Jeyanthi, K.A., Mary, C.A. 2009. Antihyperglycemic effect of *Citrullus colocynthis* seeds powder in alloxan induced diabetic rats. *The IUP Journal of Biotechnology* 3, 30.

Kabeeruddin, M. 2007. *Makhzanul Mufridat yaani Kitabul Advia*. New Delhi, India: Idara Kitabul Shifa, 256.

Kamran, Z., Ruby, T., Hussain, M., Ali, S., Ahmad, S., Koutoulis, K.C., et al. 2018. Comparative efficacy of *Citrullus colocynthis* fruit powder and popular antibiotic growth promoters in broilers diet. *Animal Nutrition and Feed Technology* 21, 421–428.

Kirtikar, K.R., Basu, B.D., Blatter, E., Caius, J.F., Mhaskar, K.S. 1984. *Indian Medicinal Plants*. Allahabad, India: Lalit Mohan Basu, 127.

Kumar, S., Kumar, D., Manjusha, K., Saroha, K., Singh, N., Vashishta, B. 2008. Antioxidant and free radical scavenging potential of *Citrullus colocynthis* (L.) Schrad. methanolic fruit extract. *Acta Pharmaceutica* 58, 215–220.

Lakhan, D., Singh, C., Yadav, C. 2017. Review on Indravaruni (*Citrullus colocynthis* Schrad): An ethno-medicinally rich plant w.s.r. to its therapeutic uses. *International Journal of Ayurveda and Pharma Research* 5, 32–36.

Lakshmi, B., Sendrayaperumal, V., Subramanian, S. 2013. Beneficial effects of *Citrullus colocynthis* seeds were studied in alloxan-induced diabetic rats. *International Journal of Pharmaceutical Sciences Review and Research* 19, 47–55.

Lavie, D., Willner, D., Merenlender, Z. 1964. Constituents of *Citrullus colocynthis* (L.) Schrad. *Phytochemistry* 3, 51–56.

Liu, T., Zhang, M., Zhang, H., Sun, C., Yang, X., Deng, Y., Ji, W. 2008. Effects of cucurbitacin B on cell proliferation and apoptosis in Hep-2 cells. *Lin Chung Er Bi Yan Hou Tou Jing Wai Ke Za Zhi* 22, 403–407.

Lloyd, J.U., Cincinnati, O. 1898. *Citrullus colocynthis*. Chicago: The Western Druggist.

Marzouk, B., Marzouk, Z., Décor, R., Edziri, H., Haloui, E., Fenina, N., Aouni, M. 2009. Antibacterial and anticandidal screening of Tunisian *Citrullus colocynthis* Schrad. From medenine. *Journal Ethnopharmacol* 125, 344–349.

Marzouk, B., Marzouk, Z., Fenina, N., Bouraoui, A., Aouni, M. 2011. Anti-inflammatory and analgesic activities of Tunisian *Citrullus colocynthis* Schrad. Immature fruit and seed organic extracts. *European Review for Medical and Pharmacological Sciences* 15, 665–672.

Meena, M.C., Patni, V. 2008. Isolation and identification of flavonoid "quercetin" from *Citrullus colocynthis* (Linn.) Schrad. *Asian Journal of Experimental Sciences* 22, 137–142.

Memon, U., Brohi, A.H., Ahmed, S.W., Azhar, I., Bano, H. 2003. Antibacterial screening of *Citrullus colocynthis*. *Pakistan Journal of Pharmaceutical Sciences* 16, 1–6.

Najafi, S., Sanadgol, N., Nejad, B.S., Beiragi, M.A., Sanadgol, E. 2010. Phytochemical screening and antibacterial activity of *Citrullus colocynthis* (Linn.) Schrad against Staphylococcus aureus. *Journal of Medicinal Plants Research* 4, 2321–2325.

Nmila, R., Gross, R., Rchid, H., Roye, M., Manteghetti, M., Petit, P., Tijane, M., Ribes, G., Sauvaire, Y. 2000. Insulinotropic effect of *Citrullus colocynthis* fruit extracts. *Planta Medica* 66, 418–423.

Panda, H. 1999. *Herbs Cultivation and Medicinal Uses*. Delhi, India: National Institute of Industrial Research, 240–241.

Perveen Upadhyay, B., Roy, S., Kumar, A. 2007. Traditional uses of medicinal plants among the rural communities of Churu district in the Thar desert, India. *Journal of Ethnopharmacology* 113, 387–399.

Pravin, B., Tushar, D., Vijay, P., Kishanchnad, K. 2013. Review on *Citrullus colocynthis*. *International Journal of Research in Pharmacy and Chemistry* 3, 46–53.

Qureshi, R., Bhatti, G.R., Memon, R.A. 2010. Ethnomedicinal uses of herbs from northern part of Nara desert, Pakistan. *Pakistan Journal of Botany* 42, 839–851.

Rahbar, A.R., Nabipour, I. 2010. The hypolipidemic effect of *Citrullus colocynthis* on patients with hyperlipidemia. *Pakistan Journal of Biological Sciences* 13, 1202–1207.

Rahimi, R., Amin, G., Ardekani, M.R.S. 2012. A review on *Citrullus colocynthis* Schrad.: From traditional Iranian medicine to modern phytotherapy. *The Journal of Alternative and Complementary Medicine* 8, 551–554.

Rahuman, A.A., Venkatesan, P., Gopalakrishnan, G. 2008. Mosquito larvicidal activity of oleic and linoleic acids isolated from *Citrullus colocynthis* (Linn.) Schrad. *Parasitology Research* 103, 1383–1390.

Roy, R.K., Thakur, M., Dixit, V.K. 2007. Effect of *Citrullus colocynthis* on hair growth in albino rats. *Pharmaceutical Biology* 45, 739–744.

Sebbagh, N., Cruciani-Guglielmacci, C., Ouali, F., Berthault, M.F., Rouch, C., Sari, D.C., et al. 2009. Comparative effects of *Citrullus colocynthis*, sunflower and olive oil-enriched diet in streptozotocin-induced diabetes in rats. *Diabetes & Metabolism* 35, 178–184.

Shahabi, S., Hassan, Z.M., Mahdavi, M., Dezfouli, M., Rahvar, M.T., Naseri, M., Jazani, N.H., Khalkhali, H.R. 2008. Hot and cold natures and some parameters of neuroendocrine and immune systems in traditional Iranian medicine: A preliminary study. *Journal of Alternative and Complementary Medicine* 14, 147–156.

Sharma, A., Singh, S., Nag, T.N. 2010. Antibacterial activity of *Citrullus colocynthis* and *Tribulus terrestris* against some pathogenic bacteria. *Asian Journal of Microbiology and Biotechnology and Environmental Sciences* 12, 633–637.

Shishkin, B.K., Bobrov, E.G. 1957. Flora of the U.S.S.R. *Moskow-Leningrad: Publishing House of the USSR Academy of Sciences*, 82–83.

Soufane, S., Bedda, A., Mahdeb, N., Bouzidi, A. 2013. Acute toxicity study on *Citrullus colocynthis* fruit methanol extract in albino rats. *Journal of Applied Pharmaceutical Science* 3, 88–88.

Tannin-Spitz, T., Grossman, S., Dovrat, S., Gottlieb, H.E., Bergman, M. 2007. Growth inhibitory activity of cucurbitacin glucosides isolated from *Citrullus colocynthis* on human breast cancer cells. *Biochemical Pharmacology* 73, 56–67.

Wasfi, I.A., Bashir, A.K., Abdalla, A.A., Banna, N.R., Tanira, M.O.M. 1995. Antiinflammatory activity of some medicinal plants of the United Arab Emirates. *International Journal of Pharmaceutics* 33, 124–128.

Yin, D., Wakimoto, N., Xing, H., Lu, D., Huynh, T., Wang, X., Black, K.L., Koeffler, H.P. 2008. Cucurbitacin B markedly inhibits growth and rapidly affects the cytoskeleton in glioblastoma multiforme. *International Journal of Cancer* 123, 1364–1375.

Yoshikawa, M., Morikawa, T., Kobayashi, H., Nakamura, A., Matsuhira, K., Nakamura, S., Matsuda, H. 2007. Bioactive saponins and glycosides. XXVII. Structures of new cucurbitane-type triterpene glycosides and antiallergic constituents from *Citrullus colocynthis*. *Chemical and Pharmaceutical Bulletin* 55, 428–434.

Zaidi, S.F., Muhammad, J.S., Shahryar, S., Usmanghani, K., Gilani, A.H., Jafri, W., Sugiyama, T. 2012. Anti-inflammatory and cytoprotective effects of selected Pakistani medicinal plants in Helicobacter pylori-infected gastric epithelial cells. *Journal of Ethnopharmacology* 14, 403–410.

9 *Trigonella foenum-graecum* L. *A Useful Medicinal Plant for Diabetes and Other Diseases*

Shakeelur Rahman and Azamal Husen

CONTENTS

9.1 INTRODUCTION

Trigonella foenum-graecum L., which is commonly known as fenugreek, is an annual herbaceous plant belonging to the family Fabaceae. The herb is grown in the Indian subcontinent, North Africa, and Mediterranean Europe. The important phytochemical constituents found in fenugreek seeds are proteins, carbohydrates, vitamins, minerals, lipids, alkaloids, flavonoids, fibres, nitrogen compounds, saponins and steroidal saponins that are categorized under volatile and non-volatile constituents (Sharma et al. 1990). The sun-dried leaves, fresh leaves and seeds are used as medicinal herbs, spice, nutritious vegetables, bakery products, dairy products, pickles and beverages (Acharya et al. 2008). Fenugreek seeds are a popular spice due to their specific aroma and medicinal properties (Max 1992). The dietary fibres of fenugreek seeds control the production of cholesterol in the liver as they contain 45.4% dietary fibre, of which 32% is insoluble and 13.3% is soluble (Roberts 2011). Fenugreek plant is taken to improve the strength and stamina of an individual (Ikeuchi et al. 2006). The herb is traditionally used to enhance the milk production of breastfeeding women (Ghedira et al. 2010). It also stimulates appetite and physical attractiveness of women (Rguibi and Belahsen 2006). It was reported that fenugreek seeds are effective in gastric ulcer. Similarly, the seed oil is used as an emollient and improves skin glaze, purifies blood and detoxifies the lymphatic system. The seeds of fenugreek are used to treat hay fever, sinusitis, heart disease, constipation, diabetes and anaemia and as an aphrodisiac (Al-Habori and Raman 1998; Bellakhdar 1997; Campbell-Tofte et al. 2012). It has also been reported in some scientific research that fenugreek has various therapeutic uses such as anti-inflammatory, hepatoprotective, antiulcer, antilithigenic, anticarcinogenic, neuroprotective and antibacterial effects (Nathiya et al. 2014).

The seeds have an estrogenic action that disturbs the endometrial lining system and interferes with foetal development (Sreeja et al. 2010). Here, we discuss a review of traditional and scientific findings of *T. foenum-graecum* in diabetes and other diseases.

DOI: 10.1201/b23347-9

9.2 BOTANICAL DESCRIPTION

Fenugreek plant is an erect, smooth, annual herbaceous legume reaching a height up to 50 cm. The tap root is a mass of finger-like structures. The leaves are alternate, compound, trifoliate, obovate to oblanceolate, tripartite, toothed, grey-green obovate, 7–12 cm long and light green in colour. The leaflets are oval, up to 5 cm long and hairy on their lower face. The plants bear white or yellow sessile flowers in leaf axils that are white, lemon-yellow or purplish blue in colour. The fruits are long, slender and yellow to brown and occur as straight or sickle-like pods of 2–10 cm, long, thin sword-shaped and pointed, and contain 10–20 seeds. At maturity the pods contain hard brown seeds of fenugreek, which is known and utilized for its medicinal use (Figure 9.1). The seeds are 6–8 mm long, oblong or square, green-olive or brownish in colour, with a very strong and spicy odour (Alaoui 2005).

9.3 DISTRIBUTION

Trigonella foenum-graecum is native to Asia and Eastern Europe but is now broadly cultivated almost all over the world. Fenugreek is suitable for areas with dry to moderate or low rainfall. (Rastogi and Mehrotra 1990). Fenugreek optimally grows in places where annual temperatures are in the range of 8–27 °C and where annual rainfall is between 400–1500 mm. It is a full-sunlight species (Alaoui 2005).

It is said that the plant originated from south-eastern Europe and western Asia (Ecocrop 2017; Alaoui 2005). In North Africa, it has been grown for fodder in Saharan oases from very early times (Alaoui 2005). The Greek named the plant "telis", which means "green", and the Romans learned from the Greeks that this plant was a valuable fodder (Petropoulos 2002). Fenugreek is now extensive in India and neighbouring countries, in northern Africa, the Near East, western

FIGURE 9.1 *Trigonella foenum-graecum* L.

Asia, Ethiopia, Chile, Argentina, China and the USA. In the semi-arid regions of North America, it is considered a high yielding niche crop (Acharya et al. 2008). In Europe, it is grown in Austria, Belgium, France, Hungary and Spain (Alaoui 2005). Fenugreek is naturally found in field verges, uncultivated ground, dry grasslands and hillsides in semi-highland and highland regions (Alaoui 2005). Fenugreek is grown as a cool season crop in India and the Mediterranean region, both irrigated and as a rain-fed crop. It grows on a wide range of preferably well drained soils with a pH ranging from 5.3 to 8.2. Wet soils are not suitable. In cooler areas, growth is slow and weak during cold periods, and it is better grown as a summer crop. The seeds require warm dry weather for ripening and harvest (Alaoui 2005).

9.4 PHYTOCHEMICAL CONSTITUENTS

A large number of valuable phytochemical constituents are found in the seeds of fenugreek, such as saponins including disogenin, homorientin saponaretin, gitogenin, neogitogenin, trigogenin and neogigogenin, as well as amino acids, fatty acids, fibres, vitamins, flavonoids, polysaccharides and fixed oils, along with alkaloids such as choline and trigonelline (Jayaweera 1981; Yoshikawa et al. 1997). These phytochemicals have different properties, for example steroid saponins have anti-inflammatory and lactation-stimulating properties (Petit et al. 1995). Polysaccharides such as galactomannans contain antidiabetic effects (Madar and Shomer 1990). Amino acid 4-hydroxyisoleucine has been shown to possess insulin-mimetic properties (Broca et al. 2004). Polyphenolic flavonoids are antiperoxidative, hypoglycaemic, hypocholesterolaemic and hypotriglyceridaemic (Gupta and Nair 1999). The potential medicinal properties of the phytochemical constituents were examined by researchers, who found that isolation of furostanol saponins called trigoneoside Ia, Ib, IIa, IIb, IIIa and IIIb held to increase food consumption and induce hypocholesterolaemia in experimental diabetic rats (Petit et al. 1995; Yoshikawa et al. 1997; Petit et al. 1995). Extracts of fenugreek seeds modulate the expression of glucose transporter 4 (GLUT-4) in skeletal muscle (Mohammad et al. 2006), and alkaloids like trigonelline inhibit glucose uptake *in vitro* (Al-Habori et al. 2001). The five flavonoids (apigenin-7-O-rutinoside, naringenin, quercetin, vitexin and kaempferol-3-O-glucoside) of seed extract have been identified using LC-MS/MS. The antiperoxidative action of fenugreek seeds in the brain during diabetes could also be credited to its hypoglycaemic property (Raju and Bird 2006; Siddiqui et al. 2005). It is likely that inhibition of lipid peroxides (LPO) and subsequently of lactate dehydrogenase (LDH) could be due to anti-free radical and antioxidant potential of polyphenolic flavonoids of *Trigonella* seeds emphasized through *in vitro* and *in vivo* experiments (Genet et al. 2002; Kaviarasan et al. 2007). Muraki et al. (2011) determined the effective, safe and tolerable dose of fenugreek extract to be around 2.50% (w/w) in experimental rats. Hypoglycaemic and insulinotropic properties are displayed by the amino acid (4-hydroxyisoleucine) extracted and purified from fenugreek seed (Broca et al. 1999).

9.5 PHARMACOLOGICAL STUDIES

Antioxidant activity: It is important to suppress the free radicals that cause various diseases by some natural process which is safe and effective. The antioxidant potential of fenugreek seeds has been determined. It was reported in an experiment that the methanolic seed extract of fenugreek reduces free radicals. Rat liver was damaged using ethanol, but when treated with fenugreek seed polyphenol extract (200 mg kg^{-1} day^{-1}) there was a significant reduction in the levels of lipid peroxidation products and protein carbonyl content. Also, there was an increase in the activities of antioxidant enzymes along with restoration of the levels of thiol groups (Kaviarasan et al. 2008). Similar work in rats was carried by applying aqueous extract of fenugreek seeds. Ethanol was fed for 60 days to induce toxicity in rats, which resulted in augmentation in the activities of alkaline phosphatase, serum aspartate transaminase and alanine transaminase. However, simultaneously administrating aqueous extract of fenugreek seeds resulted in a boosting of antioxidant level and barred further rise in lipid peroxidation. The administration of aqueous seed extract could result in prevention of the enzymatic leakage and the rise in lipid peroxidation and enrichment of the antioxidant potential. Histopathological studies related to the rat liver and brain revealed the protective role of the seed extract against ethanol-induced toxicity (Thirunavukkarasu et al. 2003). The phytochemical constituents that are understood to be responsible were flavonoids and phenolic compounds, which usually mark their presence in the polar solvent system due to their self-polar nature. Thus, because of the ability of fenugreek extracts to reduce the radicals, it can be a helpful factor in eradicating the harmful effects of various diseases.

Antilipidemic effect: It was reported that the gum contained in fenugreek seed as a functional food and nutraceutical has potential to affect glycaemia and lipidemia and attributed hypocholesterolaemia (Roberts 2011). The seeds have been shown to regulate the lipid levels in experimental models (Bordia et al. 1997). Fenugreek showed lower serum thyroglobulin and total cholesterol and hepatic lipid concentrations (Hannan et al. 2003; Raju and Bird 2006). It was found in a study that *Trigonella*-derived soluble dietary fibre (SDF) showed a useful impact on dyslipidaemia and a property to inhibit platelet aggregation in type 2 diabetic rats (Hannan et al. 2003). Fenugreek seeds given at a dose of 2.5 g twice daily for three months to healthy individuals showed no effect on the blood lipids and fasting or postprandial blood sugar. However, fenugreek given in similar trend to coronary artery disease patients with or without type 2 diabetes significantly decreased total cholesterol, blood lipids and triglycerides, without affecting the high-density lipoprotein (HDL) cholesterol (Bordia et al. 1997).

Atherogenic lipids, that is, triglycerides, cholesterol and low-density lipoprotein (LDL) cholesterol have been found to decrease significantly in SDF-fed rats, while HDL cholesterol increased. The hepatic lipid-lowering effect of fenugreek seeds can be attributed to its role in modulating the activity of several glucose and lipid metabolism enzymes (Raju et al. 2001). It was reported in a study that fenugreek has been verified to decrease the hepatic triglyceride and total cholesterol levels dose-dependently and enhance the secretion of cholesterol and total bile acids into the faeces (Muraki et al. 2011). Saponins affect lipid metabolism, including liver and plasma triglyceride and plasma cholesterol concentrations (Li et al. 2008). Feeding ethanol extract of *Trigonella* seed to hypercholesteraemic rats showed 18% to 26% reduced plasma cholesterol levels and lowered concentrations of liver cholesterol, suggesting its hypocholesterolaemic effect. Saponin-like active components in the ethanol extract of fenugreek seeds may have interacted with bile salts in the digestive tract and modified lipid metabolism (Stark and Madar 1993). Fenugreek seed-derived steroid saponins also changed feeding behaviour by increasing food consumption and motivation to eat and reduced plasma cholesterol levels (Roberts 2011).

Anticarcinogenic activity: Researchers have studies on anticancer properties of phytochemical constituents of fenugreek and found positive results. One constituent of alkaloids, called "trigonelline," has revealed potential for use in the treatment of cancer (Bhalke et al. 2009). Saponins and phytoestrogens are two important phytochemicals found in fenugreek which have anticarcinogenic properties (Raju et al. 2004). The saponins inhibit the tumour cell division and assist in the activation of apoptotic programs that lead to cell death (Francis et al. 2002). The anticancer activity of diosgenin was reported in bone cancer (Shishodia and Aggarwal 2006). It suppressed cell propagation and growth of bone cells through inhibition of tumour necrosis factor. Protodioscin, a furostanol saponin isolated from fenugreek, also induces apoptotic changes leading to death in a leukemic cell line (HL-60) (Hibasami et al. 2003). In an *in vivo* study that was conducted on rats, azoxymethane was applied to induce colon cancer. The impact of fenugreek seed powder along with its phytoconstituents diosgenin was tested, and it was reported that both the crude extract and diosgenin were able to inhibit the formation of aberrant crypt foci (ACF) which can be observed as preneoplastic lesion (Raju et al. 2004). Breast cancer in the female Wistar rats was induced by a polycyclic aromatic hydrocarbon. Inhibition of the mammary hyperplasia and a reduction in its incidence were seen after aqueous seed extract of fenugreek was given daily to the rats at a dose of 200 mg/kg b.wt for 120 days. Diosgenin was shown to suppress osteoclastogenesis induced by a cytokine, through activation of NF-$\kappa\beta$. An increasing dose of diosgenin inhibited TNF-α activated transcription factors such as NF-$\kappa\beta$ and Akt. Thus, decrease in cell proliferation by diosgenin was due to its inhibition of NF-$\kappa\beta$ regulated gene products. The cell line used in the experiment was human chronic myelogenous leukaemia cells (Amin et al. 2005). Intraperitoneal administration of the extract resulted in change in number and growth pattern of ascites cells, and tumour growth also seemed to be significantly inhibited (Ardelean et al. 2010). It was found that diosgenin shows an inhibitory effect on the human osteosarcoma cell line (Moalic et al. 2001). *In vitro* studies of the ethanolic fenugreek seed extract expressed its cytotoxic effect on a number of cancer cell lines such as pancreatic, breast and prostate cancer cell lines (Shabbeer et al. 2009).

Anticholesterol activity: It was observed that the addition of fenugreek seeds as a diet component for the mice helped in reducing cholesterol level up to 42% and 58% both in the control group and in the hypocholesterolaemic group, respectively (Singhal et al. 1982). Similarly, the dried leaves of fenugreek help in reducing blood cholesterol level (Hannan et al. 2003). Fenugreek seed administration and its extracts significantly decreased plasma cholesterol, triglyceride and low density lipoprotein (LDL) cholesterol. However, high density lipoprotein (HDL) cholesterol level was observed to be constant (Chatterjee et al. 2013), but there was a reduction in phospholipids, triacylglycerols and free fatty acids (Moorthy et al. 2010). The phytochemical constituents responsible for the activity are diosgenin, saponins, galactomannan and fibre.

Antigenotoxic activity: The seeds of fenugreek are well known for their antigenotoxic effect. The root tip meristem cells of *Allium cepa* were treated with toxic chromium trioxide. Methanolic extract of the leaves of fenugreek showed dose-dependent decrease in chromosomal aberration in *Allium cepa* roots. The aqueous extract of fenugreek seeds checked the mutagenic activity of the direct acting mutagens against *Salmonella typhimurium* (Sharma et al. 2012).

Anti-inflammatory activity: The alkaloids, saponins and flavonoids are important phytochemical constituents that are effective in anti-inflammatory activity. Inflammation in terms of oedema was induced in Wistar rats by applying carrageenan, and an anti-inflammatory effect was observed both by intraperitoneal administration and by topical application in cream form (Sharififara et al. 2009). An anti-inflammatory and antimelanogenic effect was observed in an *in vitro* system by using a human monocytic cell line (THP-1) (Kawabata et al. 2011). The fenugreek seed extract was further subjected to the extraction of bioactive compounds like saponin and two other compounds which were also found to inhibit other cytokines like IL-1 and IL-6 along with TNF-α. (Raju and Bird 2006). In another study, immersion stress and indomethacin were used to induce ulcer in rats. The seeds and leaves of fenugreek have antipyretic and anti-inflammatory effect (Sumanth et al. 2011). The chloroform fraction of seeds and aqueous extract from fenugreek seed and leaves are effective as an anti-inflammatory agent (Ravichandiran and Jayakumari 2013).

Antimicrobial activity: The plant parts of fenugreek are very effective against different microbial diseases such as bacterial, viral and fungal diseases (Dash et al. 2011). The disc diffusion method was studied to identifythe antibacterial effect on three bacteria, namely *E. coli*, *P. putida*, and *M. furfur*. The extract was found to be helpful against *E. coli* and *M. furfur* but showed no response against *P. putida* (Chandra et al. 2011). The extract of fenugreek seeds using methanol and acetone was tested against four gram-negative bacteria, where it was observed that the methanolic extract showed that *S. typhi* is resistant towards the acetonic extract. *E. coli* was found to show the highest sensitivity towards acetonic extract, while methanolic extract reported a high response against *Pseudomonas* spp. It was also observed that sprouted or germinated seeds had enhanced antimicrobial activity specifically against *H. pylori* (Moradi and Moradi 2013). Fungus has also reported its sensitivity towards one of the proteins called defensin isolated from fenugreek leaves. Defensin not only inhibited the mycelial spread of *Rhizoctonia solani* but also inhibited spore germination and consequential hyphal growth of *Phaeoisariopsis* (Olli and Kirti 2006).

Gastroprotective effect: The oil extracts from fenugreek seed have good pharmacological properties. One of the important properties is gastroprotective activity. Gastric ulceration, mean ulcer score and ulcer index were reported to be significantly decreased in a group of mice subjected to

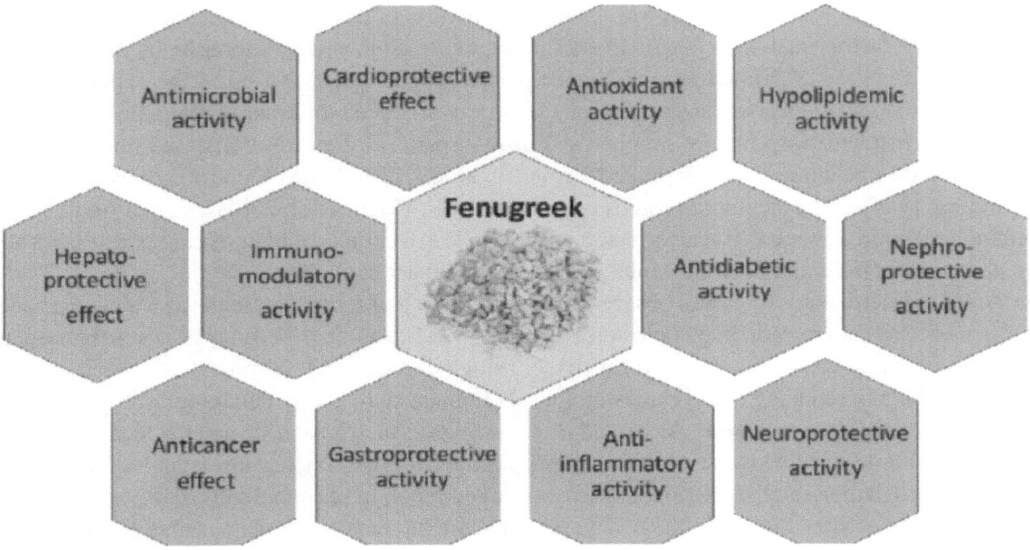

FIGURE 9.2 *Trigonella foenum-graecum* L. (fenugreek) and its various responses.

indomethacin to induce ulcer (Figure 9.2). The decrease in the gastric ulcer can be attributed to saponins, phytic acid and trigonelline present in the essential oil of fenugreek (Kamel et al. 2014).

9.6 ANTIDIABETIC RESPONSE

The seeds, leaves and extracts of the fenugreek plant are a good means to fight against diabetes (Raju et al. 2001). It has been investigated that *Trigonella foenum-graecum* is effective in decreasing the blood glucose level of diabetic patients because of its hypoglycaemic or antihyperglycaemic properties (Bordia et al. 1997). As experimental models, different types of extractive methods of fenugreek seeds have demonstrated their hypoglycaemic effect. It was observed in a study that when extracts of fenugreek seeds were fed to normal mice orally, it created a hypoglycaemic effect and reduced blood glucose levels (Zia et al. 2001). The aqueous extract of fenugreek leaves given either intraperitoneally or orally has a hypoglycaemic effect in normoglycaemic rats (Abdel-Barry et al. 1997). The induced diabetes by streptozotocin in rats and the result of fenugreek aqueous seed extract was determined via three levels of doses by intragastric intubation. It was found that there was a weight gain in fenugreek-treated mice compared to the group that received only streptozotocin. In addition, blood glucose level reduced to a greater extent compared to the group that received streptozotocin only (Xue et al. 2007). A similar result was observed by Abdelatif et al. (2012) by an experiment conducted on rabbits where it was observed that the plasma glucose level reduced by the oral intake of fenugreek seed powder in nondiabetic rabbits. In a study, the effect of fenugreek seeds on alloxan-induced diabetic rats was observed where the histopathological analysis of pancreas of placebo controls was done in which normal acini and cytosol in the islets of Langerhans was reported. But there was extensive damage to the islets of Langerhans and reduced dimensions of islets in alloxan-induced diabetes. Islets of Langerhans in diabetic rats that were treated with fenugreek extract were found to be restored (Ramesh et al. 2010). In an *in vitro* study it was observed that the extract of fenugreek in a dose-dependent manner was able to inhibit α-amylase activity. A further *in vivo* study concluded and reported *in vitro* inhibition as it showed inhibition of starch digestion and absorption in normal rats. It indicates that the hypoglycaemic effect of the plant extract used was mediated through an insulin mimetic effect (Gad et al. 2006). Moorthy et al. (2010) isolated antihyperglycaemic compound from the aqueous extract of fenugreek seeds. This compound was able to decrease blood glucose in a glucose tolerance test in subdiabetic and moderately diabetic rabbits. The seeds and leaves of the fenugreek plant reduced blood glucose concentration of a diabetic patient. The important constituents that are found to be responsible for generating the antidiabetic effects are galactomannan-rich soluble fibre fraction, saponin and an amino acid known as 4-hydroxyleucine that helped in raising insulin in hyperglycaemic rats and humans (Sharma 1986; Sharma et al. 1990; Al-Habori and Raman 1998). In a recent study on type 2 diabetic rats, the daily oral administration of *Trigonella* seed-extracted SDF for 28 days reduced serum glucose and raised liver glycogen components as well as increased the total antioxidant status, but the serum insulin and insulin discharge remained unaffected (Hannan et al. 2007). Enhanced glucose level has a causal association with atherosclerotic changes in diabetic patients. A useful vascular result of aqueous leaf extract of fenugreek has been reported in diabetic rats. The diabetic rats got fenugreek leaf extract (200 mg/kg; i.p.) every alternate day for one month, and contractile reactivity of the thoracic aorta to potassium chloride and noradrenalin and relaxation reaction to acetylcholine was determined. An important enhancement in the maximum contractile response to potassium chloride and sodium and a significant reduction in maximum relaxation due to acetylcholine in diabetic rats have been reported as compared to controls. The treatment with fenugreek extract significantly increased these changes, suggesting its ameliorative effects on the vascular system in diabetic rats (Mahdavi et al. 2008). Ameliorative effects of *Trigonella* seed extract was demonstrated on painful peripheral neuropathy in rats. *Trigonella* seed-derived fraction has been purified and standardized by high-performance liquid chromatography (HPLC) to a marker compound trigonelline. Daily oral administration was observed that restored motor nerve conduction velocity in rats (Morani et al. 2012). Some of the researchers have observed insulin-sensitizing measures of fenugreek seed-derived polyphenols and reported that it is similar to a well-known diabetic drug metformin in a rat model. In experimental rats, fructose feeding caused increased levels of glucose, insulin, triglycerides (TG) and free fatty acid (FFA), altered insulin sensitivity indices and enzyme activities and reduced glycogen content. Further, fructose-rich diet leads to higher protein tyrosine phosphatase (PTP) and lower protein tyrosine kinase (PTK) activities, suggesting decreased tyrosine phosphorylation status in these rats. Administration of fenugreek seed polyphenolic extract (FPEt) improved insulin sensitivity and tyrosine phosphorylation status in fructose-fed animals compared to metformin treated rats. This indicates that FPEt improved insulin signalling and sensitivity and thereby promoted the cellular actions of insulin (Kannappan and Anuradha 2009). The aforementioned information points out that *Trigonella foenum-graecum* not only controls glucose levels but also blocks the development of secondary diabetic complications. These results are associated with improved control of blood glucose that checks the development of secondary complications. Therefore, it looks like the antidiabetic impacts of *Trigonella* may derive from its hypoglycaemic properties.

9.7 TRADITIONAL AND OTHER POTENTIAL USES

The green leaves of *Trigonella foenum-graecum* are used as a vegetable; the dried leaves are a good additive in special food preparations in the Indian subcontinent. In Ayurveda, fenugreek has been used as an important medicinal herb to treat digestive and mucosal problems and has properties of demulcent, carminative, expectorant, stomachic agent and laxative (Escot 1994; Passano 1995). Fenugreek is known as one of the oldest medicinal herbs with many traditional uses. A poultice made of the herb can be applied to relieve swelling, and the leaves are also made into paste and applied over burns or to the scalp to prevent premature greying of the hair (Elizabeth 2002). Fenugreek is generally used as a condiment and as a lactation stimulant in India (Patil et al. 1997). In traditional system of Chinese medicine, fenugreek seeds have been used as a tonic, as well as a treatment for weakness and oedema of the legs (Yoshikawa et al. 1997). The herb was used in ancient Egypt in incense and to embalm mummies. In modern Egypt, fenugreek seed is still used as a supplement in wheat and maize flour for bread-making (Morcos et al. 1981). Fenugreek is traditionally used by Chinese people for problems in kidney and male reproductive tracts. The nutritious seeds are given during convalescence, anorexia and to increase weight gain. The herb helps in decreasing fever and treating gastritis and gastric ulcers. The seeds restore a poor sense of taste and remove bad breath. The aerial parts of the plant are used traditionally for abdominal cramps caused by gastroenteritis and menstrual pain. Fenugreek has been used to cure peptic ulcers and inflamed conditions of the stomach and bowel. They are also used to ease labour pains. Fenugreek is effective in cleansing the blood to detoxify the body. The healing and soothing action creates a defensive coating, like a lubricant, over swollen areas. The somewhat bitter characteristics of the seed are useful for digestion. Because of its richness in mucilage, it soothes irritated or inflamed tissue as a powerful demulcent. The seeds are soaked overnight and rinsed, then the water is drunk and the sprouts blended to a liquid, which gives relief from the agonizing symptoms of diverticulitis, irritable bowel syndrome, colitis. The oil in the seeds is used as a skin softener and emollient. In China, fenugreek seeds are used as a pessary to treat cervical cancer. Fenugreek is also used as a lymphatic cleansing herb. Fenugreek is a natural herb for all mucus conditions of the body, especially to the lungs, by assisting to clean congestion. It is a dominant antioxidant that acts as a mucus solvent and throat cleanser. Fenugreek is used for influenza, catarrh, constipation, bronchial complaints, asthma, emphysema, pneumonia, pleurisy, tuberculosis, sore throat, laryngitis, hay fever and sinusitis (Home Remedies Guide). Fenugreek reduces fever when it is given with lemon juice and honey. Herbal tea is made by using the seeds of fenugreek, as it help in the balancing of women's hormones (Global Herbal Supplies). Fenugreek seeds contain hormone precursors that increase milk supply in lactating women. Fenugreek seeds are used in the removal of kidney stones. The seeds are also used as a preservative and are added to pickles, chutneys and other similar products (Vortex Health, *Fenugreek*). In modern food practice, the seed extract is used in bakery products, frozen dairy products, meat products, relish, condiments, candy, gravy sauces, gelatin puddings and in alcoholic and non-alcoholic beverages (Indian Food, *Fenugreek*).

9.8 SAFETY ISSUES

It has been scientifically approved that every medicine has potential side effects, and even herbal medicines are not totally safe in this regard. There must be scientific evaluation of the drugs concerning health and potential side effects. It is important to study the pharmacological and toxicological impacts of herbal medicines to examine their clinical efficacy and safety. The study of side effects of fenugreek and its various formulations was carried out, and it was found that the herb is absolutely safe. Although no key clinical trials have been undertaken to study the use of *Trigonella* as normal or alternative herbal medicine, most of the side effects known today are the result of either user-reported symptoms such as stomach upset, diarrhoea and bloating in animal studies (Muraki et al. 2011). Therefore, some of the experts suggest the possibility of side effects such as signs of low blood sugar, nervousness, fast heartbeat and sweating. Although a very serious allergic reaction to this herb is rare, minor rash itching/swelling, especially of the face/tongue/throat, severe dizziness and trouble breathing may occur in some cases. Patients taking oral drugs for diabetes or using insulin should be monitored closely by a health care professional while using fenugreek. There is some confirmation that fenugreek may reduce potassium levels in the blood. Although it has not been extensively studied in humans, fenugreek may alter the levels of thyroid hormones. Consumption of fenugreek seeds during pregnancy has been associated with a range of congenital malformations, including hydrocephalus, anencephaly and spina bifida. It was evaluated that fenugreek has potential toxic effects on pregnant mice and foetal development (Khalki et al. 2010). The herbal isolation was given to the mated female mice during the entire period of pregnancy, at doses of 500 and 1000 mg/kg/day. The mothers revealed no obvious symptoms of toxicity, but an increase in the foetal death rate, a decrease in the litter size and a reduction in the foetal body weight and increase in the incidence of morphological abnormalities were reported in offspring (Khalki et al. 2010). This study suggested that fenugreek may have harmful toxic effects on reproductive performance and potential teratogenic effects in foetuses. However, more detailed study is required to determine these findings (Khalki et al. 2012).

9.9 CULTIVATION PRACTICES

Trigonella foenum-graecum L. has a broad adaptability and is effectively cultivated both in the tropics as well as in temperate regions. The plant is tolerant to cold, frost and freezing weather. It flourishes well in places receiving moderate or low rainfall, but the plant does not survive in heavy rainfall. The plant can be grown on different types of soils but clayey loam is relatively suitable. The optimum soil pH should be 6.0 to 7.0 for its better growth and development. Land is prepared by ploughing thrice for the cultivation of the plant, and beds of standardized size are prepared. Broadcasting the seed in the bed and raking the surface to cover the seeds is normally followed. But line sowing is suggested in rows at 20 to 25 cm apart which facilitates the intercultural operations. Sowing of fenugreek seeds in the plains are usually taken up in September to November, while in the hills it is grown starting in March. Approximately 20 to 25 kg of seed is required for one hectare, and the seed takes about 6–8 days to complete its germination. Besides 15 tonnes of farm yard manure, a fertilizer dose of 25 kg nitrogen, 25 kg phosphorus and 50 kg potash per hectare is recommended. Half of the nitrogen dose and the entire quantity of phosphorus and potash are applied basally, and the remaining nitrogen is applied 30 days after sowing. To attain more successful leafy growth, nitrogen should be applied after each cutting. First irrigation is given immediately after sowing, and subsequent irrigation is applied at 7 to 10 days interval. Root rot (*Rhizoctonia solani*) is a serious disease and can be controlled by drenching carbendazim 0.05% first at the onset of the disease and another after one month. In about 25 to 30 days, young shoots are nipped off 4 to 5 cm above ground level and following cuttings of leaves may be taken after 15 days. It is advisable to take one to two cuttings before the crop is allowed to flower and fruit. After drying the pods, the plants are pulled out and dried in the sun and seeds are threshed by beating with stick or by rubbing with hands. Seeds are winnowed, cleaned and dried in the sun. The seeds can be stored in gunny bags lined with paper. A yield of 1200 to 1500 kg of seeds and about 800 to 100 kg of leaves may be obtained per hectare (indiaagronet).

9.10 FUTURE REMARKS

The fenugreek herb has enormous international demand in the related pharmaceutical, nutraceutical and functional food industries. A wide range of agro-ecosystems are suitable for the cultivation of fenugreek, including dry, arid and semiarid climatic regimes. In addition, being a good forage legume and a natural nitrogen fixer, it could be easily incorporated in the local crop cycles of different geological regions for replenishing the soil naturally. Therefore, the aforementioned information provides clues to the researchers to identify compounds and their targets, which, sooner or later, can be translated into drug development. Because fenugreek is cultivated in different climates around the world, it may have a large range of genotypes. An extensive genomic characterization could further help in identifying the genes responsible for its medicinal effect. As a result, more high-quality research is needed to fully demonstrate the fenugreek plant's clinical usefulness. Once this barrier is overcome, it can be properly acknowledged as a good plant candidate with a high prospect of being used as a credible medicinal plant to derive new drugs.

REFERENCES

Abdel-Barry, J. A., I. A. Abdel-Hassan & M. H. Al-Hakiem. 1997. Hypoglycaemic and antihyperglycaemic effects of *Trigonella foenum-graecum* leaf in normal and alloxan induced diabetic rats. *Journal of Ethnopharmacology* 58:149–155.

Abdelatif, A. M., M. Y. Ibrahim & A. S. Mahmoud. 2012. Antidiabetic effects of fenugreek (*Trigonella foenum-graecum*) seeds in the domestic rabbit (*Oryctolagus cuniculus*). *Research Journal of Medicinal Plant* 6:6:449–455.

Acharya, S. N., J. E. Tomas & S. K. Basu. 2008. Fenugreek, an alternative crop for semiarid regions of North America. *Crop Science* 48:3:841–853.

Alaoui, S. B. 2005. *Trigonella foenum-graecum*. Ecoport Database.

Al-Habori, M. & A. Raman. 1998. Antidiabetic and hypocholesterolemic effects of fenugreek. *Phytotherapy Research* 12:4:233–242.

Al-Habori, M., A. Raman, M. J. Lawrence, P. Skett. 2001. In vitro effect of fenugreek extracts on intestinal sodium-dependent glucose uptake and hepatic glycogen phosphorylase A. *International Journal of Experimental Diabetes Research* 2:91–99.

Amin, A., A. Alkaabi, S. Al-Falasi & S. A. Daoud. 2005. Chemopreventive activities of *Trigonella foenum graecum* (Fenugreek) against breast cancer. *Cell Biology International* 29:8:687–694.

Ardelean, A., G. Pribac & A. Hermenean. 2010. Cytostatic and cytotoxic effects of *Trigonella foenum graecum* (fenugreek) seed extract. *Studia Universitatis Vasile Goldis Seria Stiintele Vietii* 20:1:25–29.

Bellakhdar, J. 1997. *La Pharmacopee Marocaine Traditionnelle Medecine Arabe Ancienne et Savoirs Populaires*. Paris, Italy: IBIS Press.

Bhalke, R. D., S. J. Anarthe, K. D. Sasane, S. N. Satpute, S. N. Shinde & V. S. Sangle. 2009. Antinociceptive activity of *Trigonella foenum graecum* leaves and seeds (Fabaceae). *International Journal of Pharmacy and Technology* 8:2:57–59.

Bordia, A., S. K. Verma & K. C. Srivastava. 1997. Effect of ginger (*Zingiber officinale Rosc.*) and fenugreek (*Trigonella foenum-graecum* L.) on blood lipids, blood sugar and platelet aggregation in patients with coronary artery disease. *Prostaglandins Leukot Essent Fatty Acids* 56:379–384.

Broca, C., V. Breil & C. Cruciani-Guglielmacci. 2004. Insulinotropic agent ID-1101 (4-hydroxyisoleucine) activates insulin signaling in rat. *American journal of Physiology-Endocrinology and Metabolism* 287:E463–E471.

Broca, C., R. Gross & P. Petit. 1999. 4-hydroxyisoleucine: Experimental evidence of its insulinotropic and antidiabetic properties. *American Journal of Physiology* 277:617–623.

Campbell-Tofte, J. I., P. Molgaard & K. Winther. 2012. Harnessing the potential clinical use of medicinal plants as anti-diabetic agents. *Botanics: Targets and Therapy* 2:7–19.

Chandra, R., V. Dwivedi, K. Shivam & A. K. Jha. 2011. Detection of antimicrobial activity of *Oscimum sanctum* (Tulsi) &*Trigonella foenum graecum* (Methi) against some selected bacterial & fungal strains. *Research Journal of Pharmaceutical, Biological and Chemical Sciences* 2:4:809–813.

Chatterjee, S., N. Goswami, P. Bhatnagar, M. Kumar & A. Kumar. 2013. Antimutagenic and chemopreventive potentialities of fenugreek (*Trigonella foenum graecum*) seed extract. *Oxidants and Antioxidants in Medical Science* 2:1:45–53.

Dash, B. K., S. Sultana & N. Sultana. 2011. Antibacterial activities of methanol and acetone extracts of Fenugreek (*Trigonella foenumgraecum*) and Coriander (*Coriandrum sativum*). *Life Sciences and Medicine Research*: 1–8.

Ecocrop. 2017. *Ecocrop database*. Rome, Italy: FAO.

Escot, N. 1994. Fenugreek. *Atoms Summar*: 7–12.

Francis, G., Z. Kerem, P. S. Makkar & K. Becker. 2002. The biological action of saponins in animals systems: A review. *British Journal of Nutrition* 88:587–605.

Gad, M. Z., M. M. El-Sawalhi, M. F. Ismail & N. D. El-Tanbouly. 2006. Biochemical study of the antidiabetic action of the egyptian plants fenugreek and balanites. *Molecular and Cellular Biochemistry* 281:1–2:173–183.

Genet, S., R. K. Kale & N. Z. Baquer. 2002. Alterations in antioxidant enzymes and oxidative damage in experimental diabetic rat tissues: Effect of vanadate and fenugreek (*Trigonella foenum-graecum*). *Molecular and Cellular Biochemistry* 236:7–12.

Ghedira, K., P. Goetz & R. Le Jeune. 2010. Fenugrec: *Trigonella foenum-græcum* L. (Fabaceae ex. Leguminosae). *Phytothérapie* 8:3:180–184.

Global Herbal Supplies. *Information about the herb fenugreek*. Available from: www.globalherbalsupplies.com/herb_information/fenugreek.htm.

Gupta, R. & S. Nair. 1999. Antioxidant flavonoids in common Indian diet. *South Asian Journal of Preventive Cardiology* 3:83–94.

Hannan, J. M., L. Ali & B. Rokeya. 2007. Soluble dietary fibre fraction of *Trigonella foenum-graecum* (fenugreek) seed improves glucose homeostasis in animal models of type 1 and type 2 diabetes by delaying carbohydrate digestion and absorption, and enhancing insulin action. *British Journal of Nutrition* 97:514–521.

Hannan, J. M. A., B. Rokeya & O. Faruque. 2003. Effect of soluble dietary fibre fraction of *Trigonella foenum graecum* on glycemic, insulinemic, lipidemic and platelet aggregation status of type 2 diabetic model rats. *Journal of Ethnopharmacology* 88:1:73–77.

Hibasami, H., H. Moteki & K. Ishikawa. 2003. Protodioscin isolated from fenugreek (*Trigonella foenum graecum* L.) induces cell death and morphological change indicative of apoptosis in leukemic cell line H-60, but not in gastric cancer cell line KATO III. *International Journal of Molecular Medicinex* 11:1:23–26.

Home Remedies Guide. *Fenugreek medicinal properties and benefits.* Available from: www.homeremedies-guide.com/herbs/fenugreek.htm.

Ikeuchi, M., K. Yamaguchi & T. Koyama. 2006. Effects of fenugreek seeds (*Trigonella foenum-greaecum*) extract on endurance capacity in mice. *Journal of Nutraceutical Science and Vitaminology (Tokyo)* 52:287–292.

Indian Food. *Fenugreek.* Available from: www.indianetzone.com/1/fenugreek.htm; https://indiaagronet.com/Horticulture/CONTENTS/fenugreek.htm.

Jayaweera, D. M. A. 1981. *Medicinal Plant: Part III.* Peradeniya: Sri Lanka Royal Botanic Garden, 225.

Kamel, M. A., R. Z. Hamza, N. E. Abdel-Hamid & F. A. Mahmoud. 2014. Anti-ulcer and gastro protective effects of fenugreek, ginger and peppermint oils in experimentally induced gastric ulcer in rats. *Journal of Chemical and Pharmaceutical Research* 6:2:451–468.

Kannappan, S. & C. V. Anuradha. 2009. Insulin sensitizing actions of fenugreek seed polyphenols, quercetin and metformin in a rat model. *Indian Journal of Medical Research* 129:401–408.

Kaviarasan, S., G. H. Naik, R. Gangabhagirathi, C. V. Anuradha & K. I. Priyadarsini. 2007. In vitro studies on antiradical and antioxidant activity of (*Trigonella foenum graecum*) fenugreek seeds. *Food Chemistry* 103:31–37.

Kaviarasan, S., R. Sundarapandiyan & C. V. Anuradha. 2008. Protective action of fenugreek (*Trigonella foenum graecum*) seed polyphenols against alcohol-induced protein and lipid damage in rat liver. *Cell Biology and Toxicology* 24:5:391–400.

Kawabata, T., M.-Y. Cui, T. Hasegawa, F. Takano & T. Ohta. 2011. Anti-inflammatory and anti-melanogenic steroidal saponin glycosides from fenugreek (*Trigonella foenum graecum* L.) seeds. *Planta Medica* 77:7:705–710.

Khalki, L., M. Bennis, Z. Sokar & S. B. M'hamed. 2012. The developmental neurobehavioral effects of fenugreek seeds on prenatally exposed mice. *Journal of Ethnopharmacology* 139:672–677.

Khalki, L., S. B. M'hamed & M. Bennis. 2010. Evaluation of the developmental toxicity of the aqueous extract from *Trigonella foenum-graecum* (L.) in mice. *Journal of Ethnopharmacology* 131:321–325.

Li, H., Q. J. Wang, D. N. Zhu & Y. Yang. 2008. A triterpene saponin of *Polygala aureocauda* Dunn, exerts hypolipidemic effect on hyperlipidemic mice. *Phytotherapy Research* 22:159–164.

Madar, Z. & I. Shomer. 1990. Polysaccharide composition of a gel fraction derived from fenugreek and its effect on starch digestion and bile acid absorption in rats. *Journal of Agricultural and Food Chemistry* 38:1535–1539.

Mahdavi, M. R., M. Roghani & T. Baluchnejadmojarad. 2008. Mechanisms responsible for the vascular effect of aqueous *Trigonella foenum-graecum* leaf extract in diabetic rats. *Indian Journal of Pharmacology* 40:59–63.

Max, B. 1992. This and that: The essential pharmacology of herbs and spices. *Trends in Pharmacological Sciences* 13:15–20.

Moalic, S., B. Liagre & C. Corbiere. 2001. A plant steroid, diosgenin, induces apoptosis, cell cycle arrest and COX activity in osteosarcoma cells. *FEBS Letters* 506:3:225–230.

Mohammad, S., A. Taha, R. N. Bamezai & N. Z. Baquer. 2006. Modulation of glucose transporter (GLUT4) by vanadate and *Trigonella* in alloxan-diabetic rats. *Life Science* 78:820–824.

Moorthy, R., K. M. Prabhu & P. S. Murthy. 2010. Anti-hyperglycemic compound (GII) from fenugreek (*Trigonella foenumgraecum* linn.) seeds, its purification and effect in diabetes mellitus. *Indian Journal of Experimental Biology* 48:11:1111–1118.

Moradi, N. & K. Moradi. 2013. Physiological and pharmaceutical effects of fenugreek (*Trigonella foenum graecum* L.) as a multipurpose and valuable medicinal plant. *Global Journal of Medicinal Plant Research* 1:2:199–206.

Morani, A. S., S. L. Bodhankar, V. Mohan & P. A. Thakurdesai. 2012. Ameliorative effects of standardized extract from *Trigonella foenum-graecum* L. seeds on painful peripheral neuropathy in rats. *Asian Pacific Journal of Tropical Medicine* 5:385–390.

Morcos, S. R., Z. Elhawary & G. N. Gabrial. 1981. Protein rich food mixtures for feeding the young in Egypt. Formulation. *Z Ernahrungswiss.* 20:275–282.

Muraki, E., Y. Hayashi & H. Chiba. 2011. Dose-dependent effects, safety and tolerability of fenugreek in diet-induced metabolic disorders in rat. *Lipids, Health and Diseases* 10:240–246.

Nathiya, S., M. Durga & T. Devasena. 2014. Therapeutic role of *Trigonella foenum-graecum* (fenugreek) a review. *International Journal of Pharmaceutical Sciences Review and Research* 24:7:74–80.

Olli, S. & P. B. Kirti. 2006. Cloning, characterization and antifungal activity of defensin Tfgd1 from *Trigonella foenum graecum* L. *Journal of Biochemistry and Molecular Biology* 39:3:278–283.

Passano, P. 1995. The many uses of methi. *Manushi*, November, 31–34.

Patil, S. P., P. V. Niphadkar & M. M. Bapat. 1997. Allergy to fenugreek (*Trigonella foenum graecum*). *Annals of Allergy, Asthma & Immunology* 78:297–300.

Petit, P. R., Y. D. Sauvaire & D. M. Hillaire-Buys. 1995. Steroid saponins from fenugreek seeds: Extraction, purification, and pharmacological investigation on feeding behavior and plasma cholesterol. *Steroids* 60:674–680.

Petropoulos, G. A. 2002. *Fenugreek: The Genus Trigonella.* Medicinal and Aromatic Plants – Industrial Profiles. London and New York: Taylor & Francis, 255.

Raju, J. & R. P. Bird. 2006. Alleviation of hepatic steatosis accompanied by modulation of plasma and liver TNF-α levels by *Trigonella foenum graecum* (fenugreek) seeds in Zucker obese (fa/fa) rats. *International Journal of Obesity* 30:8:1298–1307.

Raju, J., D. Gupta, A. R. Rao, P. K. Yadava & N. Z. Baquer. 2001. *Trigonella foenum graecum* (fenugreek) seed powder improves glucose homeostasis in alloxan diabetic rat tissues by reversing the altered glycolytic, gluconeogenic and lipogenic enzymes. *Molecular and Cellular Biochemistry* 224:1–2:45–51.

Raju, J., J. M. R. Patlolla, M. V. Swamy & C. V. Rao. 2004. Diosgenin, a steroid saponin of *Trigonella foenum graecum* (Fenugreek), inhibits azoxymethane-induced aberrant crypt foci formation in F344 rats and induces apoptosis in HT-29 human colon cancer cells. *Cancer Epidemiology Biomarkers and Prevention* 13:8:1392–1398.

Ramesh, B. K., Yogesh, H. L. Raghavendra, S. M. Kantikar & K. B. Prakash. 2010. Antidiabetic and histopathological analysis of fenugreek extract on alloxan induced diabetic rats. *International Journal of Drug Development and Research* 2:2:356–364.

Rastogi, R. P. & B. N. Mehrotra. 1990. *Compendium of Indian Medicinal Plants.* New Delhi: PID, vol. II, 422.

Ravichandiran, V. & S. Jayakumari. 2013. Comparative study of bioactive fraction of *Trigonella foenum graecum* L. leaf and seed extracts for inflammation. *International Journal of Frontiers in Science and Technology* 1:2:128–148.

Rguibi, M. & R. Belahsen. 2006. Fattening practices among Moroccan Saharawi women. *Eastern Mediterranean Health Journal* 12:5:619–624.

Roberts, K. T. 2011. The potential of fenugreek (*Trigonella foenumgraecum*) as a functional food and nutraceutical and its effects on glycemia and lipidemia. *Journal of Medicinal Food* 14:1485–1489.

Shabbeer, S., M. Sobolewski & R. K. Anchoori. 2009. Fenugreek: A naturally occurring edible spice as an anticancer agent. *Cancer Biology and Therapy* 8:3:272–278.

Sharififara, F., P. Khazaelia & N. Alli. 2009. Evaluation of antiinflammatory activity of topical preparations from Fenugreek (*Trigonella foenum graecum* L.) seeds in a cream base. *Iranian Journal of Pharmaceutical Sciences: Summer* 5:3:157–162.

Sharma, A., M. Kumar, M. Chandel & S. J. Kaur. 2012. Modulation of chromium trioxide induced genotoxicity by methanol extract of leaves of *Trigonella foenum-graecum* L. *Journal of Experimental and Integrative Medicine* 29:1:77–83.

Sharma, R. D. 1986. Effect of fenugreek seeds and leaves on blood glucose and serum insulin responses in human subjects. *Nutrition Research* 6:12:1353–1364.

Sharma, R. D., T. C. Raghuram & N. S. Rao. 1990. Effect of fenugreek seeds on blood glucose and serum lipids in type I diabetes. *European Journal of Clinical Nutrition* 44:4:301–306.

Shishodia, S. & B. B. Aggarwal. 2006. Diosgenin inhibits osteoclastogenesis, invasion, and proliferation through the downregulation of Akt, IκB kinase activation and NF-κB-regulated gene expression. *Oncogene* 25:10:1463–1473.

Siddiqui, M. R., A. Taha & K. Moorthy. 2005. Amelioration of altered antioxidant status and membrane linked functions by vanadium and Trigonella in alloxan diabetic rat brains. *Journal of Bioscience* 30:483–490.

Singhal, P. C., R. K. Gupta & L. D. Joshi. 1982. Hypocholesterolmic effect of seeds. *Current Science* 51:136–137.

Sreeja, S., V. S. Anju & S. Sreeja. 2010. In vitro estrogenic activities of fenugreek *Trigonella foenum-graecum* seeds. *Indian Journal of Medical Research* 131:814–819.

Stark, A. & Z. Madar. 1993. The effect of an ethanol extract derived from fenugreek (*Trigonella foenum-graecum*) on bile acid absorption and cholesterol levels in rats. *British Journal of Nutrition* 69:277–287.

Sumanth, M., P. Kapil & P. Mihir. 2011. Screening of aqueous extract of *Trigonella foenum graecum* seeds for its antiulcer activity. *International Journal of Research in Pharmaceutical and Biomedical Sciences* 2:3:1085–1089.

Thirunavukkarasu, V., C. V. Anuradha & P. Viswanathan. 2003. Protective effect of fenugreek (*Trigonella foenum graecum*) seeds in experimental ethanol toxicity. *Phytotherapy Research* 17:7:737–743.

Vortex Health. *Fenugreek*. Available from: www.vortexhealth.net/fenugreek.html

Williamson, E. M. 2002. *Major herbs of Ayurveda*. Churchill Livingstone, China: Elsevier Science, 315–320.

Xue, W. L., X. S. Li, J. Zhang, Y. H. Liu, Z. L. Wang & R. J. Zhang. 2007. Effect of *Trigonella foenum-graecum* (Fenugreek) extracts on blood glucose, blood lipid and hemorheological properties in streptozotocin-induced diabetic rats. *Asia Pacific Journal of Clinical Nutrition* 16:1:422–426.

Yoshikawa, M., T. Murakami & H. Komatsu. 1997. Medicinal Foodstuffs: IV. Fenugreek seeds (1): Structures of trigoneosides Ia, Ib, IIb, IIIa and IIIb new furostanol saponins from the seeds of Indian *Trigonella foenum-graecum* L. *Chemical Pharmacology Bulletin* 45:81–87.

Zia, T., S. N. Hasnain & S. K. Hasan. 2001. Evaluation of the oral hypoglycaemic effect of *Trigonella foenum-graecum* L. (methi) in normal mice. *Journal of Ethnopharmacology* 75:191–195.

10 Phytochemical Constituents, Antidiabetic and Other Activities of Gurmar (*Gymnema sylvestre* R. Br.)

Shakeelur Rahman and Azamal Husen

CONTENTS

10.1 INTRODUCTION

Gymnema sylvestre R. Br. is a valuable herb belonging to the family Apocynaceae (previously Asclepiadaceae). It is a well-known medicinal herb in Ayurvedic, Siddha, Unani and homeopathic systems of medicine advised to the diabetic patients (Mitra et al. 1996). The plant is called Periploca of the woods, cow plant and miracle fruit in English. In Hindi it is called *Gurmar*, meaning "demolisher of sugar" (Anonymous 1956), as it is supposed to neutralize the excess sugar contents from the diabetic patient (Keshavamurthy and Yoganarasimhan 1990). The herb is also helpful in the treatment of eye complaints, asthma, birth control, urinary problems, snakebite, digestive problems, constipation, colic pain, hepatosplenomegally, chronic cough, breathing troubles, cardiopathy, piles, dyspepsia and hemorrhoids. The plant also has antihypercholesterolemic, antimicrobial and anti-inflammatory properties (Potawale et al. 2008; Kumar et al. 2008). The present chapter focuses on the antidiabetic and other characteristics features of *G. sylvestre*.

10.2 BOTANICAL DESCRIPTION

G. sylvestre is a slow growing perennial, moderately pubescent, light brown, cylindrical, woody climber up to 8 m length and usually needs support for upward movement and growth. The leaves are green, opposite, elliptical to ovate, 2–6 cm in length and 1–4 cm in width. The leaves are simple, petiolate, rounded to cordate base, margin entire, opposite with acute apex, reticulate venation and pubescent on both the surfaces. Taste of the leaf is astringent and slightly bitter. The leaf has the amazing characteristic of paralyzing the sense of taste for sweet substances (Madhurima et al. 2009).

DOI: 10.1201/b23347-10

FIGURE 10.1 *Gymnema sylvestre.*

Flowers of the herb are yellow, small in size, auxiliary and lateral umbel in cymes. The follicles are lanceolate and terete maximum length is 3 inches. The calyx lobes are ovate, pubescent, long and obtuse. The corolla is valvate, pale yellow, campanulate, corona single, with five fleshy scales. Scales are adnate to throat of corolla tube between lobes; the anther connective is produced into a membranous tip, two pollinia, erect, two carpels, unilocular having many ovuled in locules (Gurav et al. 2007). Seeds are cotyledons, ovate, margined, ending in a silky coma, elliptic and the radicle is cylindric (Mathew and Rani 1983). Flowering occurs in August and March and the fruiting season is from October onwards (Figure 10.1). The *Gymnema* species are diploid with a chromosome number of 2n = 22 (Sredeevi and Namboodiri 1977).

10.3 DISTRIBUTION

An altitude from 100 to 1000 m is suitable for *G. sylvestre* to grow well (Sabitha et al. 2012). The plant species is widely distributed in tropical and subtropical and humid climatic zones and is commonly found in hills of evergreen forests in India, Malaysia, Sri Lanka, Australia, Indonesia, Japan, Vietnam, tropical Africa and the southwestern region of China. The plant is also found in western, northern, central and southern parts of India (Yadav et al. 2001).

10.4 PHYTOCHEMICAL CONSTITUENTS

The phytochemical constituents found in *G. sylvestre* are the mixture of different bioactive molecules including gymnemic acids, gymnemasaponins, gymnemosides, gurmarin, gymnemanol, stigmasterol, d-quercitol, β-amyrin related glycosides, anthraquinones, lupeol, hydroxycinnamic acids and coumarols group. The leaves of the plantcontain triterpene saponins, which are part of oleanane and dammarene classes. Oleanane saponins are gymnemic acids and gymnemasaponins, while dammarene saponins are gymnemasides (Khramov et al. 2008). A new flavonol glycoside namely kaempferol 3-O-beta-D-glucopyranosyl-(1–>4)- alpha-L-rhamnopyranosyl-(1–>6)-beta-D-galactopyranoside has also been found in aerial parts of *G. sylvestre* (Masayuki et al. 1997). The leaves also contain albumin, resins, chlorophyll, carbohydrates, organic acid, tartaric acid, formic acid, butyric acid, anthraquinone derivatives, inositole alkaloids, parabin, calcium oxalate, lignin and cellulose (Sinsheimer et al. 1970). The gymnemic acids contain several acylates, such as tiglolyl and methylbutyroyl, which are derivatives of deacylgymnemic acid. The gymnemic acids consist of gymnemic acids I to VII, gymnemasaponins and gymnemosides A–F. The presence of

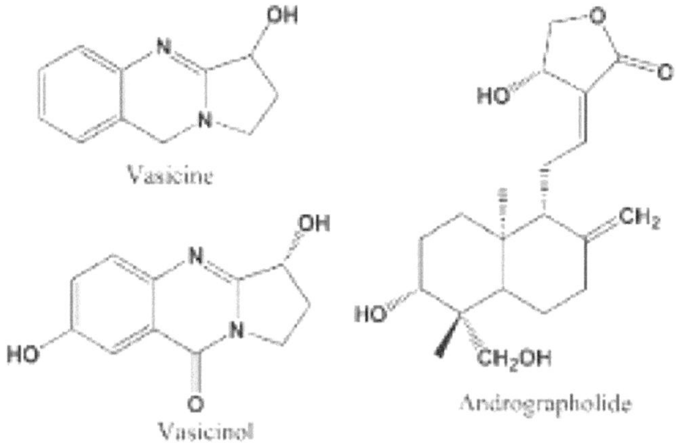

FIGURE 10.2 Gurmarin.

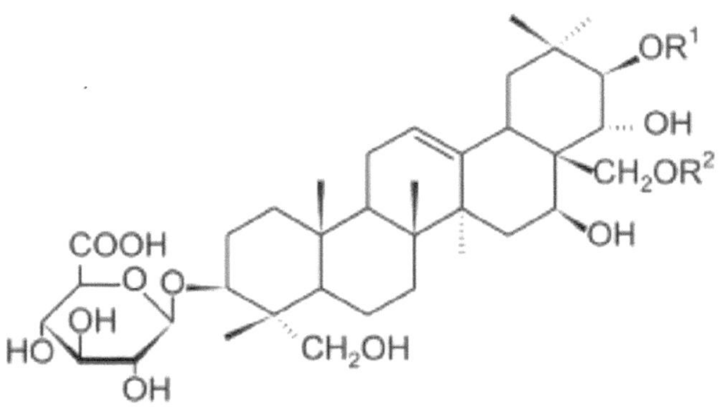

FIGURE 10.3 Gymnamic acid.

gymnemic acids, (+) quercitol, lupeol, (–) amyrin, stigma sterols and so on have been observed from *G. sylvestre.* It was found in the ethanol extract of the leaves that there are three new oleanane type triterpene glycosides, i.e. beta-O-benzoylsitakisogenin 3-O-beta-D-glucopyranosyl (1–>3)-beta-D-glucuronopyranoside, the potassium salt of longiospinogenin 3-O-beta-D-glucopyranosyl (1–>3)-beta-D-glucopyranoside and the potassium salt of 29-hydroxylongispinogenin 3-O-beta-D-glucopyranosyl (1–>3)-beta-D-glucopyranoside along with sodium salt of alternoside II (Ye et al. 2001). Similarly, four new triterpenoid saponins, gymnemasins A, B, C and D, were isolated from the leaves of *G. sylvestre* (Rao and Sinsheimer 1971), these are 3-O-[beta-D-glucopyranosyl(1–>3)-beta-D-glucopyranosyl]-22-O-tiglyol-gymnemanol,3-O-[beta-D-glucopyranosyl(1–>3)-beta-D-glucuro-nopyranosyl]-gymnemanol, 3-O-beta-D-glucuronopyranosyl-22-O-tigloyl-gymnemanol and 3-O-beta-D-glucopyranosyl-gymnemanol respectively (Figure 10.2 and 10.3). The new compounds gymnemanol and aglycone have been characterized as 3 beta-16 beta-22 alpha-23-28-pentahydroxyolean-12-ene. A new pentahydroxytriterpene called gymnestrogenin has been obtained from the leaves of *G. sylvestre* (Sahu et al. 1996). Similarly, gurmarin is an important 35 aminoacid peptide or polypeptide isolated from *G. sylvestre* (Imoto et al. 1991). It has an antisweet property specifically to the sweet taste on the tongue which is affected by the pH change. Gurmarin impacts maximum anti-sweetener effect near its isoelectric point (Chattopadhyay 1999). The hydrophobic communication plays an important role in accurate binding of gurmarin to the target molecules (Arai et al. 1995). Gymnemasins A, B, C and D and alkaloids are also valuable phytochemicals extracted from the leaves of *G. sylvestre* (Suttisri et al. 1995).

10.5 PHARMACOLOGICAL STUDIES

It was observed in several studies that the leaves of *G. sylvestre* cause hypoglycemia and, therefore, it is used in the treatment of diabetes mellitus. The leaf extract of the herb causes pancreatic stimulation to increase insulin release as well as fecal secretion of cholesterol (Persaud et al. 1999). The mechanisms of hypoglycemic acid effects of gymnemic acid are regeneration of islet cells, increase in the secretion of insulin, inhibition of glucose absorption from intestine and increase in the utilization of glucose as it enhances the function of enzymes responsible for utilization of glucose by insulin-dependent pathways, an increase in phosphorylase activity, decrease in sorbitol dehydrogenase and gluconeogenic enzymes (Kanetkar et al. 2007).

Anticancer: It was found in several studies that phytoconstituents such as saikosaponins, soyasaponins and ginsenosides demonstrate significant anticancer activity. The same anticancer potential was reported by applying gymnemagenol of *G. sylvestre* on HeLa cancer cell lines in *in vitro* conditions (Jain et al. 2007). But it was not toxic to the growth of normal cells under *in vitro* conditions (Khanna and Kannabiran 2009).

Antiobesity: *G. sylvestre* is effective in weight loss probably because of its characteristics of decreasing feelings of sweetness, which control blood sugar levels. It happens due to the property of gurmarin peptide to block the sweet taste ability (Pierce 1999). The extracts of *G. sylvestre* along with hydroxycitric acid and niacin-bound chromium were found effective in antiobesity activity. Similarly, the hexane extract of the herb helps to reduce increased body weight (Preuss et al. 2004).

Antimicrobial: The antimicrobial study on the leaves of *G. sylvestre* found that their ethanolic extract is effective against *Staphylococcus aureus*, *Bacillus pumilis*, *B. subtilis* and *Pseudomonas aeruginosa* (Yogisha and Raveesha 2009; Satdive et al. 2003). The aqueous extract of the leaf is more effective against the *Salmonella* species (Pasha et al. 2009), while the methanolic extract showed moderate effect against *Salmonella paratyphi*, *S. typhi* and *S. typhimurium*. The ethanolic, ethyl acetate and chloroform extract of the aerial part of the herb have good antibacterial effects against *P. aeroginosa*, *S. aureus*, *Klebsiella pneumoniae*, *P. vulgaris* and *E. coli* (Paul and Jayapriya 2009).

Anti-inflammatory: Tannins and saponins derived from *G. sylvestre* are use as anti-inflammatory agents (Diwan et al. 1995). The aqueous extract of leaves of the herb was found effective as anti-inflammatory activity in rats (Malik et al. 2007).

Antiarthritic: The aqueous and petroleum base leaf extracts of *G. sylvestre* was found significantly effective as an antiarthritis because of the nature of steroids, saponin glycosides and triterpenoids (Malik et al. 2010). The extract from the petroleum ether was found to significantly reduce paw swelling probably because of inhibiting the response of inflammatory cells or blocking the release of mediators like cytokines (IL-Ib and TNF-a), GM-CSF, interferons and PGDF, which are responsible for causing pain due to damage of bone and cartilage (Eric and Lawrence 1996). It is also effective against joint cartilage and bone damage in chronic arthritis (Malik et al. 2010).

Antihyperlipidemic: Coronary artery disease has a high incidence of mortality. The chances of coronary heart disease can be lowered by reducing the level of serum cholesterol (Hardman et al. 2001). Hyperlipidemia is the key factor causing coronary artery diseases and atherosclerosis (Kaushik et al. 2011). It was reported that gymnemic acid is effective against obesity (Yoshikawa et al. 1997). The hydroalcoholic extract of *Gymnema* leaves significantly reduces the levels of lipids and increases HDL-C as compared to a high cholesterol diet control given to rats. Similarly, the hexane leaf extract of the herb has antiobesity activity as the hexane extract improves cholesterol, triglyceride, LDL, and HDL levels. The hexane extract has the potential to treat obesity comparable with that of standard drug atorvastatin (Rachh et al. 2010; Bishayee and Chatterjee 1994).

Anticaries: Dental caries is the infection of the tooth caused by different types of gram-positive cariogenic bacteria (Marsh and Martin 1992). It was found that the petroleum ether, chloroform and methanolic leaf extracts of *G. sylvestre* at various concentrations are effective against microbial dental infections. Gurmar-based tooth powder has also been reported effective in dental caries (Devi and Ramasubramaniaraja 2010).

Immunostimulatory: Immunomodulation is referred to as the regulation or control of the immunity which involves the enhancement or reduction in the immune responses. The body response to a particular condition might be regulated by an agent that enhances or suppresses its action (Trease and Evans 1983). *G. sylvestre* is reported to be an immunostimulatory plant, and the leaves possess immunostimulatory effect (Gupta et al. 2010). Aqueous leaf extract of *G. sylvestre* showed remarkable immunostimulatory activity at 10, 25, 50, 100 and 1000 μg/mL on human neutrophils under *in vitro* conditions (Malik et al. 2009).

Hepatoprotective: The hydroalcoholic extract of fresh leaf of *G. sylvestre* was found effective in rat as a hepatoprotective and reduces the D-galactosamine-induced hepatotoxicity. The cells showed revival of the altered biochemical parameters towards the normal when compared to D-galactosamine treated groups in a dose-dependent manner (Srividya et al. 2010).

Wound healing: The alcoholic extract of leaves of *G. sylvestre* was found to demonstrate good wound healing property in rats as compared with control group (Malik et al. 2009). It was reported in a TLC analysis that the leaf extract is effective in wound healing. Especially the flavonoids help in the healing of wounds (Alam et al. 2011; Kiranmai et al. 2011).

10.6 ANTIDIABETIC RESPONSE

The incidence of diabetes has grown worldwide at an unprecedented rate. The pharmacological study related to role of phytoconstituents in diabetes is encouraging. Bioactive molecules from natural sources are effective and have fewer side effects. About 150 plant species of 50 families have been enlisted to treat diabetes (Handa et al.1989). Most of the plant species are effective in the balancing of blood glucose level; however, *G. sylvestre* stimulates to increase insulin level by regenerating the pancreas to maintain the homeostasis of the blood glucose (Shanmugasundaram et al. 1990). It is a unique property of the herb over other medicinal plants and chemicals used for the treatment of diabetes. The scientific study of *G. sylvestre* confirmed that the herb is useful in diabetes as it reduces urine glucose (Charpurey 1926). In case of animals, it was observed that the plant has good effect to reduce the blood glucose level (Paliwal et al. 2009). The gurmarine, crude saponin and five triterpene have good antihyperglycemic effects as they help to reduce the fasting and postprandial blood glucose levels without any adverse effect (Sugihara et al. 2000). Some of the experimental trials have confirmed that *G. sylvestre* has hypoglycemic effect on those rats treated with streptozotocin and beryllium nitrate. The ethanolic extract leaf of the herb has antioxidant properties in diabetic rats (Kang et al. 2012). The action of transaminases in gluconeogenesis and ketogenesis in diabetes like glutamate pyruvate transaminase in serum and glutathione peroxidase in cytosolic liver returned to normal levels after the administration of ethanolic leaf extract in diabetic rats (Patil et al. 2012). The impact of *G. sylvestre* leaf extract was administered to nondiabetic and alloxan-diabetic rats. It was reported that the leaf extracts of *G. sylvestre* have no effect on the alleviated glycemia caused by balanced meal. However, in nondiabetic and alloxan diabetic rats, the subacute and chronic treatment with *Gymnema* extract had no effect on the ingestion of food and water, gain of bodyweight, and the level of glucose and lipid in blood. But the herbal formulation requires clinical approval and scientific validation before being used for the treatment of diabetes and hyperlipidemia (Galletto et al. 2004). It was studied that gymnemic acid IV enhanced plasma insulin levels in STZ-diabetic mice at a concentration of 13.4mg/kg but did not create an inhibitory effect on α-glucosidase activity in the brush border membrane vesicles of the small intestine in normal rats. The extracts of the herb work to stimulate insulin secretion from the pancreas (Patel et al. 2009). The phytochemicals delay the absorption of glucose in the blood. The atomic arrangements of gymnemic acids to the taste buds are similar to sugar molecules, which fill the receptors in the taste buds preventing its activation by the sugar molecule in the food. Gurmarin distinguishes sweet and bitter tastes by interfering with the ability of taste buds on the tongue. Similarly, in the intestine it attaches to the receptor present in the external layer of the intestine, thereby preventing the absorption of sugar molecules by the intestine, leading to reduction in blood sugar levels (Sahu et al. 1996). Gymnemic acid has the properties of hypoglycemic effect which adds a flow of actions starting from modulation of incretion activity that stimulates insulin secretion. The pancreatic islet cells regenerated through the stimulation of insulin to increase enzyme mediated uptake of glucose. This procedure reduced glucose and fatty acid assimilation in the small intestine and intervenes in the capacity of receptors in the mouth and intestine to sense sweetness (Bone 2002). Gymnemic acid fuses with glyceraldehyde-3-phosphate dehydrogenase, which is a key enzyme in the glycolysis pathway (Bone 2002; Sugihara et al. 2000).

10.7 TRADITIONAL AND OTHER POTENTIAL USES

It has been mentioned in Ayurveda and Susruta that *G. sylvestre* is used in glycosuria and urinary disorder. The herb is bitter, astringent, acrid, thermogenic and diuretic (Kokate 1999), generally is used to treat dyspepsia, cardiac problems, cough, inflammation, digestive problems, liver disease, jaundice, stomachache, hemorrhoids, renal and vesicle calculi, asthma, bronchitis, amenorrhea, conjunctivitis and leucoderma, and is used as a stimulant, laxative, antipyretic, anthelminthic and uterine tonic (Chopra et al. 1992; Nadkarni 1993). The herb is used in Ayurveda to treat malaria and snakebites (Singh et al. 2008), while the bark and flowers are used to treat colds and phlegm (Kirtikar and Basu 1975). A few green and fresh leaves of the herb, when chewed in the morning, is effective to reduce glycosuria and clean urine. The juice extracted from the herb root is given to treat vomiting and dysentery, and traditionally the leaf paste is applied with breast milk to treat mouth ulcer (Ekka and Dixit 2007; Agnihotri et al. 2004). The leaf of *G. sylvestre*, along with *Trigonella foenum-graecum*, *Galega officinalis* and *A. indica*, is given for the treatment of diabetes, and with *Cynara cardunculus* for weight loss. The herb is recommended in hypercholesterolemia mixing with turmeric, hawthorn, *Silybum*, garlic and globe artichoke (Kerry 2007). It is also given to patients suffering from Parkinson's disease (Anis et al. 2000).

10.8 SAFETY ISSUES

Availability of authentic and pure biological materials is the important factor of efficacy of the medicinal or herbal drugs. The toxic effect of the extracts of *G. sylvestre* was not found when it was taken in adequate dose. In high doses it may cause side effects such as muscular dystrophy, hypoglycemia, shakiness, excessive sweating and weakness. Sometimes the herb was reported in drug-induced liver injury or toxic hepatitis (Shiyovich et al. 2010). It was reported in experiments that 1% basal powder mix in the food of Wistar rat for 52 weeks has no any toxic or side effects (Ogawa et al. 2004). The pharmacological use of gymnemic acid is a key for the *in vivo* parameter. Gymnemic acid has complex structure and poor lipid solubility; therefore it is hard to pass through the biomembranes for absorption in the circulatory system. An herbal formulation of gymnemic acid was prepared to improve its pharmacokinetics and bioabsorption. A phytosome shows good absorption because of its increased ability to cross lipid biomembranes and reach the systemic circulation. The complex expresses antiapoptotic potential in doxorubicin-induced cardiotoxicity in rats and shows cardioprotective effect (Pathan et al. 2012).

10.9 CULTIVATION PRACTICES

The seeds of *G. sylvestre* are sown in November–December and harvested from September to February. However, its propagation through seeds is not easy because of low seed viability, therefore cut root is mostly used in the rainy season (June–July) as an alternative method of propagation (Reddy et al. 2004). Terminal cutting of the stem with three or four nodes is also a good vegetative propagation method, which is generally planted in February–March (Anonymous 2008). The herb can be grown in low fertile upland.

10.10 FUTURE REMARKS

The herbal therapeutics of *G. sylvestre* is gaining popularity in pharmacological use and as molecular targets in the development of diabetic drugs. People are moving towards herbal systems because of complications with commercial medications. *G. sylvestre* has fewer side effects because of the presence of safe phytoconstituents. However, it is important to screen the herb with its pharmacological significance to develop an effective and safe drug. The scientific validation of the herb must be carried out to manufacture drugs for specific diseases. The foremost problem is the lack of scientific standards for herbal formulations of the herb. Like other medicinal plants, gurmar is also facing threat because of its high commercial demand, destruction of its natural habitats and climate change. Therefore, it is the need of times to focus on the conservation and large-scale cultivation of the herb as well as sustainable use of the herb for pharmacological purpose. As an alternative, the *in vitro* propagation of the herb in plant tissue culture can be a promising alternative for the production of important secondary metabolites. Several studies have been conducted to understand the complex nature of plant cell *in vitro* cultures (Anturlikar et al. 1995). The bioelicitor-based strategies for increased production of gymnemic acids have been in use (Veerashree et al. 2012) and the technique finds relevance for large-scale production of these bioactive compounds in bioreactor-based industrial applications. This new technology will be a smart way to further produce and utilize the herb as an antidiabetic herbal cure (Anturlikar et al. 1995; Patel et al. 2009). The whole genome sequencing project and functional elucidation of pathway genes will also play a good role in the future to develop effective biomolecules. The development of genetic transformation systems must provide a revolution in the propagation and maintenance of pharmacologically important plants for applications in drug discovery and development. The aim of this review is to emphasize the prospects of this important herb as a potential medication for treatment of diseases such as diabetes, obesity and cardiovascular disorders as well as a very important dietary supplement in the food industry.

REFERENCES

Agnihotri, A.K., S. Khatoon, M. Agarwal, A.S. Rawat, S. Mehrota & P. Pushpangadan. 2004. *Natural Product Sciences* 10:4:168–172.

Alam, G., M.P. Singh & A. Singh. 2011. Wound healing potential of some medicinal plants. *International Journal of Pharmaceutical Sciences Review and Research* 9:136–145.

Anis, M., M.P. Sharma & M. Iqbal. 2000. Herbal ethnomedicine of the Gwalior forest division in Madhya Pradesh, India. *Pharmaceutical Biology* 38:4:241–253.

Anonymous. 1956. *The Wealth of India: A Dictionary of Indian Raw Materials and Industrial Products.* Council of Scientific and Industrial Research, New Delhi, 276–277.

Anonymous. 2008. *National Medicinal Plants Board, Department of Ayush, Ministry of Health and Family Welfare.* Government of India.

Anturlikar, S.D., S. Gopumadhavan, B.L. Chauhan & S.K. Mitra. 1995. Effect of D-400, a herbal formulation, on blood sugar of normal and alloxan-induced diabetic rats. *Indian Journal of Physiology and Pharmacology* 39:2:95–100.

Arai, K., R. Ishima & S. Morikawa. 1995. Three-dimensional structure of gurmarin, a sweet taste-suppressing polypeptide. *Journal of Biomolecular NMR* 5:3:297–305.

Bishayee, A. & M. Chatterjee. 1994. *Journal of Phytological Research* 8:2:118–120.

Bone, K. 2002. *Gymnema*: A key herb in the management of diabetes. *Phytotherapy Review and Commentary*. National Institute of Herbalists: National Herbalists Association of Australia.

Charpurey, K.G. 1926. *Indian Medical Gazette*. New Delhi, 155.

Chattopadhyay, R.R. 1999. A comparative evaluation of some blood sugar lowering agents of plant origin. *Journal of Ethnopharmacology* 67:3:367–372.

Chopra, R.N., S.L. Nayar & I.C. Chopra. 1992. *Glossary of Indian Medicinal Plants*. Council of Scientific and Industrial Research, New Delhi, 319–322.

Devi, B.P. & R. Ramasubramaniaraja. 2010. Pharmacognostical and antimicrobial screening of *Gymnema sylvestre* R.Br, and evaluation of Gurmar herbal tooth paste and powder, composed of *Gymnema sylvestre* R.Br, extracts in dental caries. *International Journal of Pharmacological and Biological Sciences* 1:3:1–16.

Diwan, P.V., I. Margaret & S. Ramakrishna. 1995. Influence of *Gymnema sylvestre* on inflammation. *Inflammopharmacology* 3:271–277.

Ekka, N.R. & V.K. Dixit. 2007. *International Journal of Green Pharmacy* 1:1–4.

Eric, G.B. & J.L. Lawrence. 1996. Rheumatoid arthritis and its therapy. In *The Text Book of Therapeutics: Drug and Disease Management*, vol. 16. Williams and Wilkins Company, Baltimore, MD, 16th edition.

Galletto, R., V.L.D. Siqueira, E.B. Ferreira, A. Oliveira & R. Bazotte. 2004. Absence of antidiabetic and hypolipidemic effect of *Gymnema sylvestre* in non-diabetic and alloxan-diabetic rats. *Brazilian Archives of Biology and Technology* 47:4:545–551.

Gupta, S.N., S. Pramanik, O.P. Tiwari, N. Thacker, M. Pande & N. Upmanyu. 2010. Immunomodulatory activity of *Gymnema sylvestre* leaves. *Internetional Journal of Pharmacology* 8:2:1531–2976.

Gurav, S., V. Gulkari, N. Durgkar & A. Patil. 2007. Systematic review: Pharmacognosy, Phytochemistry and clinical application of *Gymnema sylvestre* R.Br. *Pharmacognocy Review* 1:2:338–343.

Handa, S.S., A.S. Chawla & Maninder. 1989. Hypoglycaemic plants – A review. *Fitoterapia* 60:195–224.

Hardman, J.G., L.E. Limbird, L.S. Goodman & A.G. Gilman. 2001. *Goodman and Gilman's the Pharmacological Basis of Therapeutics*. Mc Graw Hill, New York, 10th edition.

Imoto, T., A. Miyasaka, R. Ishima & K. Akasaka. 1991. Anovel peptide isolated from the leaves of *Gymnema sylvestre* – I. Characterization and its suppressive effect on the neural responses to sweet taste stimuli in the rat. *Comparative Biochemistry and Physiology* 100:2:309–314.

Jain, K.S., M.K. Kathiravan, R.S. Somani & C.J. Shishoo. 2007. The biology and chemistry of hyperlipidemia. *Bioorganic and Medicinal Chemistry* 15:14:4674–4699.

Kanetkar, P., R. Singhal & M. Kamat. 2007. *Journal of Clinical Biochemistry and Nutrition* 41:77–81.

Kang, M.H., M.S. Lee, M.K. Choi, K.S. Min & T. Shibamoto. 2012. Hypoglycemic activity of *Gymnema sylvestre* extracts on oxidative stress and antioxidant status in diabetic rats. *Journal of Agricultural and Food Chemistry* 60:10:2517–2524.

Kaushik, M., A. Kaushik, R. Arya, G. Singh & P. Malik. 2011. Antiobesity property of hexane extract from the leaves of *Gymnema sylvestre* in high fed cafeteria diet induced obesity rats. *International Research Journal of Pharmacy* 2:112–116.

Kerry, B. 2007. Gymnema: A key herb in the management of diabetes. *Phytotherapy Review & Commentary*. National Institute of Herbalists, National Herbalists Association of Australia.

Keshavamurthy, K.R. & S.N. Yoganarasimhan. 1990. *Flora of Coorg–Karnataka*. Vimsat Publishers, Banglore, 282.

Khanna, V. & K. Kannabiran. 2009. Anticancer-cytotoxic activity of saponins isolated from the leaves of *Gymnema sylvestre* and *Eclipta prostrata* on *HeLa* cells. *International Journal of Green Pharmacy* 3:3:227–229.

Khramov, V.A., A.A. Spasov & M.P. Samokhina. 2008. Chemical composition of dry extracts of *Gymnema sylvestre* leaves. *Pharmaceutical Chemistry Journal* 42:1:30–32.

Kiranmai, M., S.M. Kazim & M. Ibrahim. 2011. Combined wound healing activity of *Gymnema sylvestre* and *Tagetes erecta linn. International Journal of Pharmaceutical Applications* 2:135–140.

Kirtikar, K.R. & B.D. Basu. 1975. *Indian Medicinal Plants*, vol. 3. Periodicals Experts, Delhi, India.

Kokate, C.K. 1999. *Pharmacognosy*, vol. 12. Nirali Prakashan.

Kumar, C.S., D.V.L. Sarada & R. Rengasamy. 2008. *Indian Journal of Science and Technology* 1:3:1–5.

Madhurima, S.H., P. Ansari, S. Alam, M. Ahmad & S. Akhtar. 2009. Pharmacognistic and phytochemical analysis of *Gymnema sylvestre* R. (BR.) leaves. *Journal of Herbal Medcie and Toxicology* 3:1:73–80.

Malik, J.K., F.V. Manvi & K.R. Alagawadi. 2007. Evaluation of anti-inflammatory activity of *Gymnema sylvestre* leaves extract in rats. *International Journal of Green Pharmacy* 2:114–115.

Malik, J.K., F.V. Manvi & B.R. Nanjware. 2009. Wound healing properties of alcoholic extract of *Gymnema sylvestre* R.Br. leaves in rats. *Journal of Pharmacy Research* 2:1029–1030.

Malik, J.K., F.V. Manvi, B.R. Nanjware, D.K. Dwivedi, P. Purohit & S. Chouhan. 2010. Anti-arthritic activity of leaves of *Gymnema sylvestre* R.Br. leaves in rats. *Der Pharmacia Lettre* 2:336–341.

Marsh, P. & M. Martin. 1992. *Oral Microbiology*, vol. 3. Chapman and Hall, London.

Masayuki, Y., M. Toshiyuki, K. Masashi, L. Yuhao, M. Nubotoshi & Y. Johji. 1997. *Chemical and Pharmaceutical Bulletin* 45:10:1671–1676.

Mathew, K.M. & N. Rani. 1983. Asclepiadaceae. In *The Flora of the Tamil Nadu Carnatic Part Two*. Rapinat herbarium. Diocesan Press, Madras, India, 945–947.

Mitra, S.K., S.G. Madhavan, T.S. Muralidhar & S.J. Seshadri. 1996. Effect of D-400, a herbomineral formulation a liver glycogen content and microscopic structure of pancreas and liver in streptozotocin induced diabetes in rats. *Indian Journal of Experimental Biology* 34:964–967.

Nadkarni, K.M. 1993. *Indian Materia Medica*. Popular Prakashan, Bombay, 596–599.

Ogawa, Y., K. Sekita & T. Umemura. 2004. *Gymnema sylvestre* leaf extract: A 52-week dietary toxicity study in Wistar rats. *Shokuhin Eiseigaku Zasshi* 45:1:8–18.

Paliwal, R., S. Kathori & B. Upadhyay. 2009. Effect of Gurmar (*Gymnema sylvestre*) power intervention on the blood glucose levels among diabetics. *Ethno-Medicine* 3:2:133–135.

Pasha, C., S. Sayeed, S. Ali & Z. Khan. 2009. *Turkish Journal of Biology* 33:59–64.

Patel, S.S., R.S. Shah & R.K. Goyal. 2009. Antihyperglycemic, antihyperlipidemic and antioxidant effects of Dihar, a polyherbal ayurvedic formulation in streptozotocin induced diabetic rats. *Indian Journal of Experimental Biology* 47:7:564–570.

Pathan, R.A., U. Bhandari, S. Javed & T.C. Nag. 2012. Anti-apoptotic potential of gymnemic acid phospholipid complex pretreatment in Wistar rats with experimental cardiomyopathy. *Indian Journal of Experimental Biology* 50:2:117–127.

Patil, P.M., P.D. Chaudhari, N.J. Duragkar & P.P. Katolkar. 2012. Formulation of anti-diabetic liquid preparation of *Gymnema sylvestre* and qualitative estimated by TLC. *Asian Journal of Pharmaceutical and Clinical Research* 5:1:16–19.

Paul, J.P. & K. Jayapriya. 2009. *Pharmacologyonline* 3:832–836.

Persaud, S.J., H.A. Majed, A. Raman & P.M. Jones. 1999. *Journal of Endocrinology* 163:207–212.

Pierce, A. 1999. *Gymnema Monograph: Practical Guide to Natural Medicine*. Stone Song Press Book, New York, 324–326.

Potawale, S.E., V.M. Shinde, L. Anandi, S. Borade, H. Dhalawat & R.S. Deshmukh. 2008. *Pharmacology Online* 2:144–157.

Preuss, H.G., D. Bagchi, M. Bagchi, C.V. Rao, D.K. Dey & S. Satyanarayana. 2004. *Diabetes Obesity and Metabolim* 6:3:171–180.

Rachh, P.R., M.R. Rachh & N.R. Ghadiya. 2010. Antihyperlipidemic activity of *Gymenma sylvestre* R.Br. leaf extract on rats fed with high cholesterol diet. *International Journal of Pharmacology* 6:2:138–141.

Rao, G.S. & J.E. Sinsheimer. 1971. Constituents from *Gymnema sylvestre* leaves. Isolation, chemistry, and derivatives of gymnemagenin and gymnestrogenin. *Journal of Pharmaceutical Sciences* 60:2:190–193.

Reddy, P.S., G.R. Gopal & G.L. Sita. 2004. *In vitro* multiplication of *Gymnema sylvestre* R.Br., an important medicinal plant. *Current Science* 10:1–4.

Sabitha, R.A, V. Nagamani, S. Patnaik & B. Saidulu. 2012. *Gymnema sylvestre* R. Br. An important antidiabetic plant of India: A review. *Plant Sciences Feed* 2:12:174–179.

Sahu, N.P., S.B. Mahato, S.K. Sarkar & G. Poddar. 1996. Triterpenoid saponins from *Gymnema sylvestre*. *Phytochemistry* 41:4:1181–1185.

Satdive, R.K., P. Abhilash & P.F. Devanand. 2003. *Fitoterapia* 74:699–701.

Shanmugasundaram, E.R.B., K.G. Leela, K.S. Radha & V.M. Rajendran. 1990. Possible regeneration of the islets of Langerhans in streptozotocin-diabetic rats given *Gymnema sylvestre* leaf extracts. *Journal of Ethnopharmacology* 30:265–279.

Shiyovich, A., I. Sztarkier & L. Nesher. 2010. Toxic hepatitis induced by *Gymnema sylvestre*, a natural remedy for type 2 diabetes mellitus. *American Journal of the Medical Sciences* 340:6:514–517.

Singh, V.K., S. Umar, S.A. Ansari & M. Iqbal. 2008. *Gymnema sylvestre* for diabetics. *Journal of Herbs, Spices and Medicinal Plants* 14:1–2:88–106.

Sinsheimer, J.E. & G.S. Rao. 1970. *Journal of Pharmaceutical Sciences* 59:5:622–628.

Sinsheimer, J.E., G.S. Rao & H.M. McIlhenny. 1970. Constituents from *G. sylvestre* leaves V: Isolation and preliminary characterization of Gymnemic acid. *Journal of Pharmaceutical Science* 59:5:622–628.

Sredeevi, P. & A.N. Namboodiri. 1977. IOPB chromosome numbers reports. LVI, *Taxon*, 26: 257–274.

Srividya, A.R., S.K. Varma, S.P. Dhanapal, R. Vadivelan & P. Vijayan.2010. *In vitro* and *in vivo* evaluation of hepatoprotective activity of *Gymnema sylvestre. International Journal of Pharmaceutical Sciences and Nanotechnology* 2:768–773.

Sugihara, Y., H. Nojima, H. Matsuda, T. Murakami, M. Yoshikawa & I. Kimura. 2000. Antihyperglycemic effects of gymnemic acid IV, a compound derived from *Gymnema sylvestre* leaves in streptozotocin-diabetic mice. *Journal of Asian Natural Products Research* 2:4:321–327.

Suttisri, R., I.S. Lee & A.D. Kinghorn. 1995. Plant-derived triterpenoid sweetness inhibitors. *Journal of Ethnopharmacology* 47:1:9–26.

Trease, G.E. & W.C. Evans. 1983. *Pharmacognosy*, vol. 12. Balliere Tindale, London.

Veerashree, V., C.M. Anuradha & V. Kumar. 2012. Elicitor-enhanced production of gymnemic acid in cell suspension cultures of *Gymnema sylvestre* R.Br. *Plant Cell, Tissue and Organ Culture* 108:1:27–35.

Yadav, S., J.K. Grover & V. Vats. 2001. Medicinal plants of India with anti-diabetic potential. *Journal of Ethanopharmacology* 81:1: 81–100.

Ye, W., X. Liu, Q. Zhang, C.T. Che & S. Zhao. 2001. *Journal of Natural Products* 64:2:232–35.

Yogisha, S. & K.A. Raveesha. 2009. *In vitro* antibacterial effect of selected medicinal plant extracts. *Journal of Natural Products* 2:64–69.

Yoshikawa, K., T. Murakami & H. Matsuda. 1997. Medicinal food stuffs. IX. The inhibitors of glucose absorption fromthe leaves of *Gymnema sylvestre* R.Br. (Asclepiadaceae): Structures of gymnemosides A and B. *Chemical and Pharmaceutical Bulletin* 45:10:1671–1676.

11 Role of Green Chiretta [*Andaphis paniculata* (Burm. F.)] Nees in Diabetes and Other Ailments

Shakeelur Rahman and Azamal Husen

CONTENTS

11.1 INTRODUCTION

The phytochemicals of traditionally useful medicinal plants are used effectively in various ailments. *Andrographis paniculata* is one of the valuable herbal plants of the Acanthaceae family. Also called Kalmegh or king of bitters, it is used for diabetes and other diseases in traditional as well as modern pharmaceutical systems (Kavishankar et al. 2011). At the primary health care level, *A. paniculata* is very popular because of its availability, affordability, compatibility and acceptability in India, Bangladesh, China, Hong Kong, Pakistan, Philippines, Malaysia, Indonesia and Thailand (Akbar 2011). In the ethnobotanical system, this herb is given to treat diabetes, dysentery, fever, snake bite, bug bite and malaria (Burkill et al. 1966). Commercial production of the extracts of *A. paniculata* is effective forvarious ailments. The aerial part of the herbextracts constitutes lactones, diterpenoids, diterpene glycosides, flavonoids and flavonoid glycosides. This herb has a wide range of pharmacological uses such as antihepatitis (Sharma et al. 1991), cytotoxic, anti-HIV (Nanduri et al. 2004), anticancer (Ajaya et al. 2004), antidiarrhoeal (Gupta et al. 1990), antihyperglycaemic (Yu et al. 2003), anti-inflammatory (Sheeja et al. 2006), antimicrobial, antimalarial (Misra et al. 1992), antioxidant (Akowuah et al. 2008), cardiovascular (Tan and Zhang 2004), hepatoprotective (Trivedi and Rawal 2000), sexual dysfunctions (Akbarsha and Murugaian 2000) and immunostimulatory (Iruretagoyena et al. 2005). Since the herb is given in the treatment of various kinds of diseases in the traditional system, its anticipated benefits need to be validated carefully. Hence, this chapter reviews the botanical description, distribution, phytochemical constituents, pharmacological properties, traditional use, safety issues and cultivation practices of *A. paniculata*.

DOI: 10.1201/b23347-11

11.2 BOTANICAL DESCRIPTION

A. paniculata is an annual, branched, erect herbaceous plant; the stem is solid, woody, tetrangular, slightly winged, glabrous andgreen; leaves are opposite decussate, simple, 6.0–10.0 cm long and 3.5–5.0 cm wide, acute to acuminate at apex, entire at margin, curved at maturity, cuneate at base, herbaceous, unicostate reticulate with 5 to 6 pairs of usually alternate pinnate secondary, glabrous on both surface, green to bottle green and often with copper shade coloration above at edge, especially during flowering, petiolate; petioles are slender 0.4–0.5 mm long, glabrous, greenish, ex-stipulate, inflorescence paniculate cyme, both axillary and terminal, upper 5 to 6 axils with peduncles; peduncles are 3.0 to 5.0 cm long, tetrangular, angles are hairy along the four ridges; hairs dentate to scabrid, glabrescent at maturity, green, bractiate, two bracts, opposite decussate, linear oblong to lanceolate, lower larger, upper smaller, 2.0 to 3.0 mm long and 0.5 to 1.0 mm wide, often hairy, green, persistent; secondary peduncles 2.0 to 3.0 cm long alike to primary peduncle; bracts of secondary peduncles alike to primary peduncles; flowers are complete, bisexual, hypogynous, irregular (zygomorphic), pentamerous, pedicillate; pedicels tetrangular, 2.0 to 3.0 mm long with glandular hairs, dull green often enlarged up to fruit, 2 mm long; sepals are five, linear dentate with glandular hairs; calyx tube turbinate with glandular hairs on the outer surface, corolla two-lipped about 9.0 mm to 10.0 mm long, colour pinkish violet, hairy on the outer surface more towards posterior lobe and apex; posterior lip oblong, hooded over the anterior lip; anterior lips are three-lobed, median lobe larger broad, lateral smaller, outer surface with glandular hairs, glabrous within, having purple oval spot on the median lobe within and two purple lines to each lateral lobe within, corolla tube narrowed 5.0 mm long; androecium's stamens are two; filaments declinate, epipetalous, alternipetalous, pubescent throughout and more bristly at the base of anther attachment, whitish with purple lines, anthers are two-celled oblong, dark purple, gynoeciums are two, syncarpous, ovary superior,

FIGURE 11.1 Green chiretta.

oblong, slightly two-lobed, many ovules per chamber in axile placentation, glandular hairy throughout, style is one, terminal, purple, glabrous, stigma shortly 2-fid; fruits simple, dry dehiscent, capsules laterally flat with twodistinct lobes, marked by central depression along the septation, 12.0 mm to 17.0 mm long and 2.0 mm to 3.5 mm wide, spinous at the apex, narrowed at base, dull brown colour at maturity; dehiscent loculicidal, many seeded (12–16); seeds attached on retinocula, seeds flat and often incurved, compressed, translucent, size 1.4 mm ± 0.34 × 0.9 mm ± 0.25 (length × breadth), golden brown in colour with tumid testal surface. *A. paniculata* is hermaphrodite self-compatible and a habitual inbreeder. The somatic chromosome number in *A. paniculata* was noted to be $2n = 50$. Both stigma and anthers are in intimate proximity, showing synchronization of anther dehiscence and stigma receptivity respectively, thus providing autonomous selfing in the species (Figure 11.1). The flowering and fruiting period of the species in India is December to April (Williamson 2002; Anju et al. 2012; Ghosh et al. 2012; Lattoo et al. 2006; Roy and Datta 1988).

11.3 DISTRIBUTION

A. paniculata flourishes well in hedgerows throughout the waste ground, plains, agricultural fields, moist habitat, seashores, roadsides and hill slopes. The species inhabits India, China and Taiwan. It is also commonly found in a wide range of tropical and subtropical Asia, Southeast Asia, Brunei, Cambodia, Caribbean islands, Indonesia, Laos, Pakistan, Malaysia, Myanmar, Java, Sri Lanka, Thailand and Vietnam. This species is also found in different phytogeographical and edaphic zones of America, Christmas Island and the West Indies (Lattoo et al. 2006). *A. paniculata* performs well in moist shady places, forests and wastelands area (Mishra et al. 2007).

11.4 PHYTOCHEMICAL CONSTITUENTS

Whole plant parts of *Andrographis paniculata* have been reported for its large number of uses in the extraction of important phytoconstituents. The phytochemicals extracted from stems, leaves, flowers, seeds and roots are applied in different pharmacological activities. There are impacts of geographical area, season and time of harvesting on the compositions and efficacy of the phytochemicals (Sharma and Sharma 2013). For example, a key bioactive compound andrographolide of the plant species is extracted in high amount when the herbs are harvested just after the flowering stage. The bioactive compounds can be extracted with different kinds of solvents including ethanol, methanol, acetone, acetone-water, hexane, chloroform and dichloromethane from the whole plant parts of *A. paniculata.* A total of 32 bioactive compounds with 12flavonoids, seven *ent*-labdane diterpenoids and two quinic acid derivatives have been extracted and characterized by the methanol procedure. Through the methanol extraction procedure, whole plant material isshade dried, ground and extracted with methanol (10 L × 6) under reflux for 8 h and filtered to give three different types of residues. These are chloroformresidue, water-insoluble residue and water-soluble residue. Other studies havereported more than 55 *ent*-labdane diterpenoids, 8 quinic acids, 30 flavonoids and 4 xanthones (Subramanian et al. 2012; Xu et al. 2010) as well as 5 noriridoids including andrographidoids A, B, C, D and E (Sareer et al. 2012). Similarly, Zhang et al. (2006) has reported 3 new *ent*-labdane diterpenoids, 19-norandrographolides A, B, and C from the ethanol extracts of the aerial parts of the herb. Andrographolide and echiodinin extracted from acetone and methanol extracts of *in vitro* leaf callus have shown moderately potent antimicrobial and antioxidant activity (Arifullah et al. 2013). A novel flavonoid, 7, 8-dimethoxy-2–hydroxy-5-O-β-d-glucopyranosyloxy flavones, together with 15 known flavonoids, was isolated from the aerial parts of the herb. Four xanthones (1,2-dihydroxy-6,8-dimethoxyxanthone; 1,8-dihydroxy-3,7-dimethoxyxanthone; 3,7,8-trimethoxy-1-hydroxyxanthone; 4,8-dihydroxy-2,7-dimethoxyxanthone) were isolated from the roots of the herb using chloroformfraction and purity confirmed by HPLC (Dua et al. 2004). Five rare types of noriridoids with a known iridoid curvifloruside F were found from the ethanol extraction of roots (Xu et al. 2012).

11.5 PHARMACOLOGICAL STUDIES

Andrographis paniculata is a popular herb used in traditional systemsof medicines that proves its efficacy in various diseases. It was confirmed by clinical (human), *in vitro* and *in vivo* studies that the herb extracts and products are effective in various pharmacological activities. Some of the authentic research works on pharmacology are mentioned as follows.

Anti-inflammatory: The andrographolide and neoandrographolide of *A. paniculata* are the important bioactive compounds found effective in anti-inflammatory activity (Sheeja et al. 2006). The andrographolide showed several anti-inflammatory activities like inhibition of intercellular adhesion molecule-1 expression in monocytes activated by tumour necrosis factor-α(Habtemariam 1998). Oral administration of andrographolide was observed significant analgesic activity on acetic-induced writhing in mice (Madav et al. 1995).

Hepatoprotective: *A. paniculata* is very popular in traditional systems as a hepatoprotective, which means that it stimulatesmultiple enzymes of the liver. In Unani and Ayurvedic systems, the herb is used as an ingredient in the herbal formulations for curing hepatic disorders (Akbar 2011). The bioactive compounds such as andrographolide, neoandrographolide, 14-dexoyandrographolide and 14-deoxy-11,12-didehydroandrographolide used with different herbal extracts have good hepatoprotective results (Chander et al. 1995). It was reported by Handa and Sharma (1990) that andrographolide, methanol extract of whole plant and andrographolide-free methanol extract improved liver histology in rats by 48.6%, 32%, and 15%, respectively, after CCl4-induced liver injury.

Common cold: *A. paniculata* is generally used in the treatment of the common cold in rural communities. It showed significant reduction in clinical symptoms like sore throat, shivering, tiredness, muscular ache, rhinitis, headache and sinus pains (Hancke et al. 1995).

Antibacterial: *A. paniculata* was evaluated by some researchers where they found that the herb is effective in bacterial diseases. The dry powder of the herb dissolved in water to be devoid of *in vitro* antibacterial activity against *Shigella*, *Salmonella*, *Escherichia coli*, *Staphylococcus aureus*, group A *Streptococcus* (Leelarasamee et al. 1990). The aqueous methanol (50%v/v) crude extracts of the whole plant parts have antibacterial activity against *Proteus vulgaris* and *Bacillus subtilis* (Nakanishi et al. 1965). Mishra et al. (2009) investigated that the ethanol extracts of aerial parts of *A. paniculata* were more effective in inhibiting the growth of *E. coli* and ten more gram-positive and gram-negative species of bacteria. The collective effects of the extracted andrographolides and arabinogalactan proteins through the aqueous extraction have good antibacterial activity. The menthol extract showed high antibacterial activity against *Enterococcus faecalis* (Singha et al. 2003).

Antiviral: In addition to other pharmacological activities, *A. paniculata* has significant antiviral effect. Some of the reported antiviral activities of the herb are protection from dengue virus serotype 1 (Tang et al. 2012), herpes simplex virus type 1 (Aromdee et al. 2011), human papilloma virus type 16 (Fangkham et al. 2012), HIV (Reddy et al. 2005) and influenza A virus (Chen et al. 2009). The hot aqueous extraction of the aerial parts was found more significant to reduce the percentage of HIV antigen-positive H9 cells (Chang et al. 1991). The methanol extract of the herb is also effective in the inhibition of dengue virus serotype 1 (Tang et al. 2012). It was found in a study that andrographolide suppressed HPV16 transcription activity (Fangkham et al. 2012).

Antiparasitic: Bioactive compound xanthones derived from the roots of *A. paniculata* have good antimalarial properties (Dua et al. 2004). The aqueous dried leaf derivative is effective against adult worms *Brugia malayi* (Zaridah et al. 2001). The compounds extracted through menthol and aqueous methods are effective against *Pheretima posthuma* (Padma et al. 2011).

Anticancer: Andrographolide is one of the important bioactive constituents of *A. paniculata* which was observed effective on cancer cells by developing the immune system against cancer cells as well as inducting apoptosis and necrosis of cancer cells. It also reduces the proliferation of cancer cell cycle (Vojdani and Erde 2006). Andrographolide is an important bioactive compound of the herb isolated from dichloromethane which reducesthe growth of various kinds of human cancers (Ajaya et al. 2004), whereas the methanol-extracted compounds are less effective against breast and colon cancers because of the low diffusion power of the bioactive constituents. The ethanol extract of the herb has antiproliferative effect against human leukaemia (Chen et al. 2008). It was reported in a study that the flavonoid isolated from the aerial part of the plant has potent antiproliferative activity against human leukaemia HL-60 cells (Chen et al. 2014). Andrographolide and its analogues apply a direct inhibitory effect on cancer cells by inducing expression of cell cycle inhibitory proteins and depressing cyclin-dependent kinase, which blocksthe cell cycle progression (Jada et al. 2007).

Immunomodulatory: *A. paniculata* has good immunomodulatory properties as andrographolide has a key effect on stimulating the common defence activities of the immune system by activating the production of antibodies and nonspecific immune responses such asincreased macrophage phagocytosis (Amroyan et al. 1999). Rajagopal et al. 2003 has reported that increased proliferation of lymphocytes and production of interleukin-2 verify the immune-stimulatory role of the herb. The aqueous extract of the herb also enhancesthe immune functions as well as curesanaemia and multiple myeloma (Bukoye and Musbau 2011). Similarly, andrographolide stopstumour growth in animals by stimulating the production of cytotoxic T lymphocytes (Sheeja and Kuttan 2007).

Cardiovascular: It has been documented in traditional system that *A. paniculata* is effective in improvingheart health. Zhang and Tan (1997) have reported that the bioactive components of the herb are useful in heart diseases. It was investigated that *A. paniculata* has the potential to increase the cyclic guanosine monophosphate, nitric oxide and superoxide dismutase functions by declining lipid peroxide and endothelin in an atherosclerotic rabbit model (Wang et al. 1997). Wang and Zhao (1994) found that the extracts of *A. paniculata* prevent constriction of blood vessels and increase blood clotting time significantly in pre- and post-angioplasty procedures. The aqueous extract of the herb has an antihypertensive role in both spontaneously hypertensive and normotensive rats (Zhang and Tan 1996).

Antihyperlipidaemic: Hyperlipidaemia causes atherosclerosis that leads to obstruction in the coronary arteries and in the arteries of the brain (Al-Attar 2010). The andrographolide worksas a therapeutic factor for atherosclerosis (Chen et al. 2004). It was found in a study that andrographolide and neoandrographolide have antihyperlipidaemic properties. These compounds demonstrate their lipid and lipoprotein reducing effects (Yang et al. 2013).

Sexual functions and contraceptive effect: Akbarsha et al. (1990) investigated that *A. paniculata* has contraceptive effects by terminating spermatogenesis in male albino rats. There were no pregnant female mice that consumed *A. paniculata* mixed food daily after mating with the untreated male of potential fertility, which reveals that the herb has contraceptive effect on female mice (Zoha et al. 1989). Andrographolide was found effective to prevent cytokinesis of dividing spermatogenic cell lines that checksspermatogenesis (Kamal et al. 2003). The antifertility role of andrographolide is due to a decrease of protein content along with significant increases of acid phosphatase, alkaline phosphatase and cholesterol levels with manifestation of fructose in the reproductive systems of rats (Janarthanan 1990). The negative effect of *A. paniculata* in blood progesterone content in rats has also been investigated (Panossian et al. 1999).

11.6 ANTIDIABETIC RESPONSE

The andrographolide functions as an antihyperglycaemic by reducing blood glucose level through inhibition of α-glycosidase and α-amylase (Chao and Lin 2010). Inhibitions of these enzymes and stimulation of insulin sensitivity are considered as effective factors to lower the level of postprandial blood glucose. These enzymes also increase insulin sensitivity and stimulateglucose uptake and oxidation by peripheral tissues (Subramanian et al. 2008). It helps in digestion and absorption of carbohydrates (Kajaria et al. 2013). Andrographolide has properties to control abnormal lipid metabolism and reducefree radicals that interrupt the plasma membrane integrity, resulting in a decreased number of efficient plasma membrane receptors or transporter proteins necessary to uptake glucose from the bloodstream (Augustine et al. 2014). Andrographolide at a dose of 50 mg/kg effectively decreased blood glucose level, stimulated glucose transporter type 4 (GLUT4) translocation (Zhang et al. 2009) and improved diabetic rat's islet and beta cell functions (Nugroho et al. 2014). The aqueous extract of the herb checksthe glucose-induced hyperglycaemia in nondiabetic rats without affecting epinephrine-induced hyperglycaemia (Borhanuddin et al. 1994). Similarly, ethanol extracts of the herb significantly reduced the fasting blood glucose in humans (Subramanian et al. 2008). The bioactive compounds such as 14-deoxy-11, 12-didehydroandrographolide also demonstrated the antihyperglycaemic activity (Lee et al. 2010). The andrographolide effectively prevented the beginning of insulitis and suppressed the development of diabetes in 30-week-old NOD mice. Andrographolide also regulates the Th1/Th2/Th17 homeostasis through which it may prevent β-cell death and inhibit T-cell penetration into pancreatic islets and thus prevent development of type 1 diabetes (Zhang et al. 2013). It was reported that *A. paniculata* decreases blood glucose by increasing glucose utilization and oxidation, restoringinsulin signalling molecules in liver and decreasing the serum lipid levels in high fat and sucrose-induced type 2 diabetic rats without showing hypoglycaemic effect (Augustine et al. 2014). A mixture of n-hexane unsolvable part of *A. paniculata* with curcuminoids part of *Curcuma xanthorrhiza* rhizome also reported the antihyperglycaemic effect on high-fructose fat-fed rats (Nugroho et al. 2014). The combination of *A. paniculata* and *S. cumini* has been used in different *Syzigium* products in Indonesia (Elfahmi et al. 2014). The mixture of extracts of *A. paniculata* and *C. sappan* had reasonable antihyperglycaemic effects (Wediasari et al. 2020). The combination of aerial parts extract of *A. paniculata*, the leaves extract of *S. cumini* and the heartwoods extract of *C. sappan* with the ratio 1:1:1 have showed antidiabetic effect. Therefore, the recognition of more antihyperglycaemic compositions of *A. paniculata* and other herbal plants would be a focusing point of researchers for better treatment alternativesfordiabetic patients.

11.7 TRADITIONAL AND OTHER POTENTIAL USES

Traditionally, *A. paniculata* is generally used for medicinal purposes by tribal practitioners and ethnic communitiesin Asia, Africa and Europe (Kabir et al. 2014). The whole plant parts treateddifferent health ailments such as liver diseases, torpid liver, constipation, oedema, headache, helminthiasis, indigestion, dysentery, intestinal worms, leucorrhoea, low sperm count, sexual disorder, lower urinary tract infections, lung infections, malaria, mucus, chicken pox, pharyngotonsillitis, vertigo, diabetes, diarrhoea, anorexia, hypertension, leprosy, gonorrhoea, scabies, eczema, boils, skin eruptions, herpes zoster, bloating with burning sensations in the chest, snake bites, common cold and chronic seasonal fever and acted as blood purifiers (Akbar 2011). *A. paniculata* lowered body temperature in the treatment of infectious diseases associated with cold symptoms (Melchior et al. 1997). The traditional uses of the plant species are also mentioned in pharmacopoeia of diverse traditional systems of medicine (WHO 2003).

11.8 SAFETY ISSUES

Andrographis paniculata has long been known to be a safe and effective medicinal plant. It has been traditionally used in China, India, Thailand and many other Asiatic countries. Generally, the medicinal uses of the herb have been tested to be safe in various research studies on rats, mice and rabbits as well as in *in vitro* clinical examination. A small number of studies reported the toxic effect of the herb on reproductive system by damaging the Sertoli cellsin male gonads in albino rats (Akbarsha and Murugaian 2000). The safety of *A. paniculata* extracts in relation to oral acute toxicity (Burgos et al. 1997), testicular toxicity (Allan et al. 2009) and genotoxicity (Chandrasekaran et al. 2009) has been investigated. Due to excessive bitterness of the herb, it may create emesis. Many adverse consequences, including gastric instability, allergic reaction, fatigue, headache, loss of appetite, lymphadenopathy, diarrhoea, nausea and metallic taste, are also observed in overdosing of the extracts of the herb (Anju et al. 2012). Therefore, it is advised to avoid this herb during pregnancy due to ovulation preventive effects (Zoha et al. 1989).

11.9 CULTIVATION PRACTICES

A. paniculata is distributed in a large range because it can be cultivated well in all types of soil. The herb performs well in poor and unfertile metallic soil called serpentine soil (Samantaray et al. 2001). The plant was also reported to grow extravagantly in low humid land, tropical regions and in areas of high rainfall. Exceptionally, it cannot grow well in flooded or wetland areas (Kasetklangklung 1996). Seed is the conventional source of cultivation of the plant species. Generally the recommended time for seed sowing is May to July with a spacing of 30 × 15 cm plants/ha that produce average biomass of 3 t/ha (Singh et al. 2011). It was observed that the seed dormancy is a major hindrance for the commercial cultivation of the species (Talei et al. 2012). Seed dormancy can be brokenthrough hot water and hormonal media treatment (Kumar et al. 2011), but the required commercial quantities cannot be metby this technique because of variability among the seed-derived progenies as well as insufficient and delayed rooting of seedlings (Kataky and Handique 2010). Hence, nonconventional methods of propagation like plant tissue culture techniques are good alternative methods to produce an abundance of saplings in a short time and improve phytochemical content (Dandin and Murthy 2012). Planting and harvesting time are the two important factors thatinfluence plant yield. The maximum harvest of total diterpene lactones was found from the aerial part of the plant during flowering. The finest harvesting period of *A. paniculata* leaves is at 3–5 months old or at 50% flowering, whereupon the highest amount of active lactone was reported followed by final harvesting after the next 2–3 months, with a fresh weight yield of 2–3 ton per hectare or 0.5–1 tons per hectare of dried leaves (Bhan et al. 2006; Arora et al. 2000). The first and second stage of processing of the harvested plant is done by cleaning and drying in a hot air oven at 46°C to 50°C for 8 hours or until properly dried. The dried plant parts are stored in airtight bags and kept in a cool and clean place for a maximum period of one year to avoid decline in the quantity of diterpene lactone. The stability of andrographolide was determined using a heat-accelerated experiment to reveal a second-order kinetics of degradation (Plubrukarn et al. 2006).

11.10 CONCLUSION AND FUTURE REMARKS

A. paniculata has effective pharmacological activities using either isolated bioactive compounds or crude extracts. Conventional extraction processes are mostly used by the researchers. However, selection of proper extraction methods is important for quantitative and qualitative findings of bioactive compounds (Smith 2003). The investigated pharmacology of bioactive compounds of *A. paniculata* is based on conventional extraction methods with some exceptions such as using different solvents, like methanol, ethanol, water, acetone, acetone-water, chloroform and dichloromethane. It was observed that the conventional extraction methods have some limitations, including longer extraction time, the costly requirement of high purity solvent, a huge amount of solvent evaporation, low selectivity of extraction and thermal decomposition of thermolabile compounds. Time is needed to overcome these problems –modern promising extraction techniques such as enzyme-assisted extraction, ultrasound-assisted extraction, microwave-assisted extraction, pressurized liquid extraction, pulsed electric field-assisted extraction and supercritical fluid extraction have been introduced andare called nonconventional extraction methods (Azmir et al. 2013). The advantage of the techniques are that they are considered "green techniques" and are relevant for high yield and more purified compounds within a short time compared to conventional methods (Li et al. 2006). The tissue culture techniques have also been applied effectively to form new flavones by differentiating callus culture (Jalal et al. 1979). *In vitro* root culture system can be exploited as a renewable source of minerals essential for designing effective drugs (Behera et al. 2010). Adventitious roots from vegetative propagation of *A. paniculata* can also be an alternative source of plant materials for extraction of bioactive compounds from roots. These adventitious roots can be further investigated to see the quality and content of bioactive compounds and pharmacological effects. Therefore, adventitious rooting and plant tissue culture techniques can be good alternative methods to fulfil the commercial demand of the herb. Enhancement of the yield of andrographolide and neoandrographolide content and investigation of other important bioactive constituents of *A. paniculata* under abiotic stresses could be important to meet the commercial demand of bioactive compounds.

REFERENCES

Ajaya, K. R., K. Sridevi, N. Vijaya Kumar, S. Nanduri & S. Rajagopal. 2004. Anticancer and immunostimulatory compounds from *Andrographis paniculata. Journal of Ethnopharmacology* 92:2–3:291–295.

Akbar, S. 2011. *Andrographis paniculata*: A review of pharmacological activities and clinical effects. *Alternative Medicine Review* 16:1:66–77.

Akbarsha, M. A. & P. Murugaian. 2000. Aspects of the male reproductive toxicity/male antifertility property of andrographolide in albino rats: Effect on the testis and the cauda epididymidal spermatozoa. *Phytotherapy Research* 14:6:432–435.

Akbarsha, M. A., B. Manivannan, K. S. Hamid & B. Vijayan. 1990. Antifertility effect of *Andrographis paniculata* (Nees) in male albino rat. *Indian Journal of Experimental Biology* 28:5:421–426.

Akowuah, G. A., I. Zhari & A. Mariam. 2008. Analysis of urinary andrographolides and antioxidant status after oral administration of *Andrographis paniculata* leaf extract in rats. *Food and Chemical Toxicology* 46:12:3616–3620.

Al-Attar, A. M. 2010. Hypolipidemic effects of coenzyme Q10 in experimentally induced hypercholesterolemic model in female rats. *The American Journal of Pharmacology and Toxicology* 5:1:14–23.

Allan, J. J., M. P. Pore, M. Deepak, B. Murali, A. S. Mayachari & A. Agarwal. 2009. Reproductive and fertility effects of an extract of andrographis paniculata in male Wistar rats. *International Journal of Toxicology* 28:4:308–317.

Amroyan E., E. Gabrielian, A. Panossian, G. Wikman & H. Wagner. 1999. Inhibitory effect of andrographolide from *Andrographis paniculata* on PAF-induced platelet aggregation. *Phytomedicine* 6:1:27–31.

Anju, D., G. Jugnu, S. Kavitha, N. Arjun & D. Sandeep. 2012. A review on medicinal prospective of *Andrographis paniculata* Nees. *Journal of Pharmaceutical and Scientific Innovation* 1:1:1–4.

Arifullah, M., N. D. Namsa, M. Mandal, K. K. Chiruvella, P. Vikrama & G. R. Gopal. 2013. Evaluation of anti-bacterial and anti-oxidant potential of andrographolide and echiodinin isolated from callus culture of *Andrographis paniculata* Nees. *Asian Pacific Journal of Tropical Biomedicine* 3:8:604–610.

Aromdee, C., S. Suebsasana, T. Ekalaksananan, C. Pientong & S. Thongchai. 2011. Stage of action of naturally occurring andrographolides and their semisynthetic analogues against herpes simplex virus type 1 in vitro. *Planta Medica* 77:9:915–921.

Arora, J. R., R. Swarup & S. V. Gupta. 2000. Dept of biotechnology, ministry of science and technology, Govt of India. *The BIOME News* 1:2.

Augustine, A. W., A. Narasimhan, M. Vishwanathan & B. Karundevi. 2014. Evaluation of antidiabetic property of *Andrographis paniculata* powder in high fat and sucrose-induced type-2 diabetic adult male rat. *Asian Pacific Journal of Tropical Disease* 4:1:S140–S147.

Azmir, J., I. S. M. Zaidul & M. M. Rahman. 2013. Techniques for extraction of bioactive compounds from plant materials: A review. *Journal of Food Engineering* 117:4:426–436.

Behera, P. R., P. Nayak, D. P. Barik, T. R. Rautray, M. Thirunavoukkarasu & P. K. Chand. 2010. ED-XRF spectrometric analysis of comparative elemental composition of in vivo and in vitro roots of *Andrographis paniculata* (Burm.f.) Wall. ex Nees-a multi-medicinal herb. *Applied Radiation and Isotopes* 68:12:2229–2236.

Bhan, M. K., A. K. Dhar, S. Khan, S. K. Lattoo, K. K. Gupta & D. K. Choudhary. 2006. Screening and optimization of *Andrographis paniculata* (Burm.f.) Nees for total andrographolide content, yield and its components. *Scientia Horticulturae* 107:4:386–391.

Borhanuddin, M., M. Shamsuzzoha & A. H. Hussain. 1994. Hypoglycaemic effects of *Andrographis paniculata* Nees on non-diabetic rabbits. *BangladeshMedical Research Council Bulletin* 20:1:24–26.

Bukoye, O. & A. Musbau. 2011. Immune modulation potentials of aqueous extract of *Andrographis paniculata* leaves in male rat. *Researcher* 3:1:48–57.

Burgos, R. A., E. E. Caballero, N. S. Sánchez, R. A. Schroeder, G. K. Wikman & J. L. Hancke. 1997. Testicular toxicity assessment of *Andrographis paniculata* dried extract in rats. *Journal of Ethnopharmacology* 58:3:219–224.

Burkill, I. H., W. Birtwistle, F. Foxworthy, J. Scrivenor & J. Watson. 1966. *A Dictionary of the Economic Products of the Malay Peninsula*, Ministry of Agriculture and Co-operatives, Kuala Lumpur, Malaysia.

Chander, R., V. Srivastava, J. S. Tandon & N. K. Kapoor. 1995. Antihepatotoxic activity of diterpenes of *Andrographis paniculata* (Kal-Megh) against *Plasmodium berghei*-induced hepatic damage in Mastomys natalensis. *International Journal of Pharmacognosy* 33:2:135–138.

Chandrasekaran, C. V., P. Thiyagarajan & K. Sundarajan. 2009. Evaluation of the genotoxic potential and acute oral toxicity of standardized extract of *Andrographis paniculata* (KalmCold). *Food and Chemical Toxicology* 47:8:1892–1902.

Chang, R. S., L. Ding, C. Gai-Qing, P. Qi-Choa, Z. Ze-Lin & K. M. Smith. 1991. Dehydroandrographolide succinic acid monoester as an inhibitor against the human immunodeficiency virus. *Proceedings of the Society for Experimental Biology and Medicine* 197:1:59–66.

Chao, W. W. & B. F. Lin. 2010. Isolation and identification of bioactive compounds in *Andrographis paniculata* (Chuanxinlian). *Chinese Medicine* 5:17:1–15.

Chen, J. H., G. Hsiao, A. R. Lee, C. C. Wu & M. H. Yen. 2004. Andrographolide suppresses endothelial cell apoptosis via activation of phosphatidyl inositol-3-kinase/Akt pathway. *Biochemical Pharmacology* 67:7:1337–1345.

Chen, J. X., H. J. Xue & W. C. Ye. 2009. Activity of andrographolide and its derivatives against influenza virus *in vivo* and *in vitro*. *Biological and Pharmaceutical Bulletin* 32:8:1385–1391.

Chen, L. X., H. He, G. Y. Xia, K. L. Zhou & F. Qiu. 2014. A new flavonoid from the aerial parts of *Andrographis paniculata*. *Natural Product Research* 28:3:138–143.

Chen, L. X., H. Zhu, R. Wang, K. Zhou, Y. Jing & F. Qiu. 2008. Entlabdane diterpenoid lactone stereoisomers from *Andrographis paniculata*. *Journal of Natural Products* 71:5:852–855.

Dandin, V. S. & H. N. Murthy. 2012. Regeneration of *Andrographis paniculata* Nees: Analysis of genetic fidelity and andrographolide content in micropropagated plants. *African Journal of Biotechnology* 11:61:12464–12471.

Dua, V. K., V. P. Ojha & R. Roy. 2004. Anti-malarial activity of some xanthones isolated from the roots of *Andrographis paniculata*. *Journal of Ethnopharmacology* 95:2–3:247–251.

Elfahmi, H., J. Woerdenbag & O. Kayser. 2014. Jamu: Indonesian traditional herbal medicine towards rational phytopharmacological use. *Journal of Herbal Medicines*: 1–23.

Fangkham, S., T. Ekalaksananan & C. Aromdee. 2012. The effect of andrographolide on Human papillomavirus type 16 (HPV16) positive cervical cancer cells (SiHa). *International Journal of Infectious Diseases* 16:1:80.

Ghosh, K. B., D. K. Animesh, M. Aninda, D. K. Priyanka & H. Sandip. 2012. An overview on *Andrographis paniculata* (Burm. F.) Nees. *International Journal of Research in Ayurveda and Pharmacy* 3:6:752–760.

Gupta, S., M. A. Choudhry, J. N. S. Yadava, V. Srivastava & J. S. Tandon. 1990. Antidiarrhoeal activity of diterpenes of *Andrographis paniculata* (Kal-Megh) against *Escherichia coli* enterotoxin in *in vivo* models. *International Journal of Crude Drug Research* 28:4:273–283.

Habtemariam, S. 1998. Andrographolide inhibits the tumour necrosis factor-α-induced upregulation of ICAM-1 expression and endothelial-monocyte adhesion. *Phytotherapy Research* 12:1:37–40.

Hancke, J., R. Burgos, D. Caceres & G. Wikman. 1995. A double blind study with a new mono drug Kan Jang: Decrease of symptoms and improvement in the recovery from common colds. *Phytotherapy Research* 9:8:559–562.

Handa, S. S. & A. Sharma. 1990. Hepatoprotective activity of andrographolide from *Andrographis paniculata* against carbon tetrachloride. *The Indian Journal of Medical Research* 92:276–283.

Iruretagoyena, M. I., J. A. Tobar & P. A. González. 2005. Andrographolide interferes with T cell activation and reduces experimental autoimmune encephalomyelitis in the mouse. *Journal of Pharmacology and Experimental Therapeutics* 312:1:366–372.

Jada, S. R., G. S. Subur & C. Matthews. 2007. Semisynthesis and in vitro anticancer activities of andrographolide analogues. *Phytochemistry* 68:6:904–912.

Jalal, M. A. F., K. H. Overton & D. S. Rycroft. 1979. Formation of three new flavones by differentiating callus cultures of *Andrographis paniculata*. *Phytochemistry* 18:1:149–151.

Janarthanan, S. 1990. Antifertility effects of andrographolide in rats. *Journal of Ecobiology* 2:4:325–329.

Kabir, M. H., N. Hasan & M. M. Rahman. 2014. A survey of medicinal plants used by the Deb barma clan of theTripura tribe of Moulvibazar district, Bangladesh. *Journal of Ethnobiology and Ethnomedicine* 10:1:19.

Kajaria, D., J. Ranjana, Y. B. Tripathi & S. Tiwari. 2013. In vitro a amylase and glucosidase inhibitory effect of ethanolic extract of antiasthmatic drug-Shirishadi. *Journal of Advanced Pharmaceutical Technology and Research* 4:4:206–209.

Kamal, R., R. S. Gupta & N. K. Lohiya. 2003. Plants for male fertility regulation. *Phytotherapy Research* 17:6:579–590.

Kasetklangklung. 1996. *Andrographis paniculata*, Thai medicinal plant. *Journal of Transfer Technology Agricultural* 2:14:6–17.

Kataky, A. & P. Handique. 2010. Micropropagation and screening of antioxidant potential of *Andrographis paniculata* (Burm. f) Nees. *Journal of Hill Agriculture* 1:1:13–18.

Kavishankar, G., N. Lakshmidevi, S. M. Murthy, H. Prakash & S. Niranjana. 2011. Diabetes and medicinal plants – A review. *Journal of Pharmaceutical and Biomedical Sciences* 2:3:65–80.

Kumar, B., S. K. Verma & H. P. Singh. 2011. Effect of temperature on seed germination parameters in Kalmegh (*Andrographis paniculata* Wall. ex Nees.). *Industrial Crops and Products* 34:1:1241–1244.

Lattoo, S. K., S. Khan, A. K. Dhar, D. K. Choudhary & K. K. Gupta. 2006. Genetic and mechanism of induced male sterility in *Andrographis paniculata* (Burm. f.) Nees and its significance. *Current Science* 9:4:515–519.

Lee, M. J., Y. K. Rao, K. Chen, Y. C. Lee, Y. S. Chung & Y. M. Tzeng. 2010. Andrographolide and 14-deoxy-11,12-didehydroandrographolide from *Andrographis paniculata* attenuate high glucose-induced fibrosis and apoptosis in murine renal mesangeal cell lines. *Journal of Ethnopharmacology* 132:2:497–505.

Leelarasamee, A., S. Trakulsomboon & N. Sittisomwong. 1990. Undetectable anti-bacterial activity of *Andrographis paniculata* (Burma) wall. ex ness. *Journal of the Medical Association of Thailand* 73:6:299–304.

Li, B. B., B. Smith & M. M. Hossain. 2006. Extraction of phenolics from citrus peels: II. Enzyme-assisted extraction method. *Separation and Purification Technology* 48:2:189–196.

Madav, S., H. C. Tripathi & S. K. Mishra. 1995. Analgesic, antipyretic and antiulcerogenic effects of andrographolide. *Indian Journal of Pharmaceutical Sciences* 57:3:121–125.

Melchior, J., S. Palm & G. Wikman. 1997. Controlled clinical study of standardized *Andrographis paniculata* extract in common cold – A pilot trial. *Phytomedicine* 3: 315–318.

Mishra, S. K., N. S. Sangwan & R. S. Sangwan. 2007. *Andrographis paniculata* (Kalmegh): A review. *Pharmacognosy Reviews* 1:2:283–298.

Mishra, U. S., A. Mishra, R. Kumari, P. N. Murthy & B. S. Naik. 2009. Antibacterial activity of ethanol extract of *Andrographis paniculata*. *Indian Journal of Pharmaceutical Sciences* 71:4:436–438.

Misra, P., N. L. Pal, P. Y. Guru, J. C. Katiyar, V. Srivastava & J. S. Tandon. 1992. Antimalarial activity of *Andrographis paniculata* (Kalmegh) against *Plasmodium berghei* NK 65 in Mastomys natalensis. *International Journal of Pharmacognosy* 30:4:263–274.

Nakanishi, K., S. Sasaki & A. K. Kiang. 1965. Phytochemical survey of Malaysian plants preliminary chemical and pharmacological screening. *Chemical and Pharmaceutical Bulletin* 13:7:882–890.

Nanduri, S., V. K. Nyavanandi & S. Sanjeeva Rao Thunuguntla. 2004. Synthesis and structure-activity relationships of andrographolide analogues as novel cytotoxic agents. *Bioorganic and Medicinal Chemistry Letters* 14:18:4711–4717.

Nugroho, A. E., I. R. Rais & I. Setiawan. 2014. Pancreatic effect of andrographolide isolated from *Andrographis paniculata* (Burm. f.) Nees. *Pakistan Journal of Biological Sciences* 17:1:22–31.

Padma, Y., C. L. Narasimhudu, S. Devi, N. M. B. Natha, R. B. Naga & G. H. Philip. 2011. In vitro anthelmintic activity of *Andrographis paniculata* (burm.f.) nees. *International Journal of Pharmaceutical Research and Development* 3:3:202–205.

Panossian, A., A. Kochikian & E. Gabrielian. 1999. Effect of *Andrographis paniculata* extract on progesterone in blood plasma of pregnant rats. *Phytomedicine* 6:3:157–162.

Plubrukarn, A., S. Pinsuwan, S. Ingkatawornwong & T. Supavita. 2006. Stability of andrographolide in powdered *Andrographis* herb under accelerated conditions. *Planta Medica* 72:10:954–956.

Rajagopal, S., R. A. Kumar, D. S. Deevi, C. Satyanarayana & R. Rajagopalan. 2003. Andrographolide, a potential cancer therapeutic agent isolated from *Andrographis paniculata*. *Journal of Experimental Therapeutics and Oncology* 3:3:147–158.

Reddy, V. L. N., S. M. Reddy & V. Ravikanth. 2005. A new bisandrographolide ether from *Andrographis paniculata* nees and evaluation of anti-HIV activity. *Natural Product Research* 19:3:223–230.

Roy, S. K. & P. C. Datta. 1988. Chromosomal biotypes of *Andrographis paniculata* in India and Bangladesh. *Cytologia* 53:369–378.

Samantaray, S., G. R. Rout & P. Das. 2001. Heavy metal and nutrient concentration in soil and plants growing on a metalliferous chromite minespoil. *Enviro Technologies* 22:1147–1154.

Sareer, O., A. Ahad & S. Umar. 2012. Prophylactic and lenitive effects of *Andrographis paniculata* against common human ailments: An exhaustive and comprehensive reappraisal. *Journal of Pharmaceutical Research and Opinion* 2:10:138–162.

Sharma, A., R. T. Singh, V. Sehgal & S. S. Handa. 1991. Antihepatotoxic activity of some plants used in herbal formulations. *Fitoterapia* 62:2:131–138.

Sharma, M. & R. Sharma. 2013. Identification, purification and quantification of andrographolide from *Andrographis paniculata* (burm. F.) Nees by HPTLC at different stages of life cycle of crop. *Journal of Current Chememical and Pharmaceutical Sciences* 3:1:23–32.

Sheeja, K. & G. Kuttan. 2007. Activation of cytotoxic T lymphocyte responses and attenuation of tumor growth in vivo by *Andrographis paniculata* extract and andrographolide. *Immunopharmacology and Immunotoxicology* 29:1:81–93.

Sheeja, K., P. K. Shihab & G. Kuttan. 2006. Antioxidant and anti-inflammatory activities of the plant *Andrographis paniculata* nees. *Immuno Pharmacology and Immuno Toxicology* 28:1129–1140.

Singh, M., A. Singh & R. Tripathi. 2011. Growth behavior, biomass and diterpenoid lactones production in Kalmegh (*Andrographis paniculata* Nees.) strains at different population densities. *Agricultural Journal* 6:3:115–118.

Singha, P. K., S. Roy & S. Dey. 2003. Antimicrobial activity of *Andrographis paniculata*. *Fitoterapia* 74:7–8:692–694.

Smith, R. M. 2003. Before the injection – Modernmethods of sample preparation for separation techniques. *Journal of Chromatography A* 1000:1–2:3–27.

Subramanian, R., M. Z. Asmawi & A. Sadikun. 2008. Effect of ethanolic extract of *Andrographis paniculata* (Burm. F.) nees on a combination of fat-fed diet and low dose streptozotocin induced chronic insulin resistance in rats. *Diabetologia Croatica* 37:1:13–22.

Subramanian, R., M. Z. Asmawi & A. Sadikun. 2012. A bitter plant with a sweet future? A comprehensive review of an orientalmedicinal plant: *Andrographis paniculata*. *Phytochemistry Reviews* 11:1:39–75.

Talei, D., A. Valdiani, M. P. Abdullah & S. A. Hassan. 2012. A rapid and effective method for dormancy breakage and germination of king of bitters (*Andrographis paniculata* Nees) seeds. *Maydica* 57:2:98–105.

Tan, B. H. & A. Zhang. 2004. *Andrographis paniculata* and the cardiovascular system. *Oxidative Stress and Disease* 14:441–456.

Tang, L. I. C., A. P. K. Ling, R. Y. Koh, S. M. Chye & K. G. L. Voon. 2012. Screening of anti-dengue activity in methanolic extracts of medicinal plants. *BMC Complementary and Alternative Medicine* 12:3.

Trivedi, N. & U. M. Rawal. 2000. Hepatoprotective and toxicological evaluation of *Andrographis paniculata* on severe liver damage. *Indian Journal of Pharmacology*, 32:5:288–293.

Vojdani, A. & J. Erde. 2006. Regulatory T cells, a potent immunoregulatory target for CAM researchers: Modulating tumor immunity, autoimmunity and alloreactive immunity (III). *Evidence-Based Complementary and Alternative Medicine* 3:3:309–316.

Wang, D. W. & H. Y. Zhao. 1994. Prevention of atherosclerotic arterial stenosis and restenosis after angioplasty with *Andrographis paniculata* nees and fish oil. Experimental studies of effects and mechanisms. *Chinese Medical Journal* 107:6:464–470.

Wang, H. W., H. Y. Zhao & S. Q. Xiang. 1997. Effects of *Andrographis paniculata* component on nitric oxide, endothelin and lipid peroxidation in experimental atherosclerotic rabbits. *Chinese Journal of Integrated Traditional and Western Medicine* 17:9:547–549.

Wediasari, F., G. A. Nugroho, Z. Fadhilah, B. Elya, H. Setiawan & T. Mozef. 2020. Hypoglycemic effect of a combined *Andrographis paniculata* and *Caesalpinia sappan* extract in streptozocin-induced diabetic rats. *Advanced Pharmacological and Pharmaceutical Science*: 1–9.

WHO. 2003. Herba andrographidis. In *WHO Monographs on Selected Medicinal Plants, WHO.* Geneva, Switzerland: World Health Organization, 12–25.

Williamson, E., ed. 2002. *Major Herbs of Ayurveda.* Edinburgh: Churchill Livingstone.

Xu, C., G. X. Chou, C. H. Wang & Z. T. Wang. 2012. Rare noriridoids from the roots of *Andrographis paniculata. Phytochemistry* 77:275–279.

Xu, C., G. X. Chou & Z. T. Wang. 2010. A new diterpene from the leaves of *Andrographis paniculata* nees. *Fitoterapia* 81:6:610–613.

Yang, T., H. X. Shi, Z. T. Wang & C. H. Wang. 2013. Hypolipidemic effects of andrographolide and neoandrographolide in mice and rats. *Phytotherapy Research* 27:4:618–623.

Yu, B. C., C. R. Hung, W. C. Chen & J. T. Cheng. 2003. Antihyperglycemic effect of andrographolide in streptozotocin-induced diabetic rats. *Planta Medica* 69:12:1075–1079.

Zaridah, M. Z., S. Z. Idid, A. Wan Omar & S. Khozirah. 2001. In vitro antifilarial effects of three plant species against adult worms of sub periodic *Brugia malayi. Journal of Ethnopharmacology* 78:1:79–84.

Zhang, C., L. Gui, Y. Xu, T. Wu & D. Liu. 2013. Preventive effects of andrographolide on the development of diabetes in autoimmune diabetic NOD mice by inducing immune tolerance. *International Immunopharmacology* 16:4:451–456.

Zhang, C. Y. & B. K. H. Tan. 1996. Hypotensive activityof aqueous extract of *Andrographis paniculata* in rats. *Clinical and Experimental Pharmacology and Physiology* 23:8:675–678.

Zhang, C. Y. & B. K. H. Tan. 1997. Mechanisms of cardiovascular activity of *Andrographis paniculata* in the anaesthetized rat. *Journal of Ethnopharmacology* 56:2:97–101.

Zhang, X. Q., G. C. Wang, W. C. Ye, Q. Li, G. X. Zhou & X.-S. Yao. 2006. New diterpenoids from *Andrographis paniculata* (Burm. f.) nees. *Journal of Integrative Plant Biology* 489:1122–1125.

Zhang, Z., J. Jiang, P. Yu, X. Zeng, J. W. Larrick & Y. Wang. 2009. Hypoglycemic and beta cell protective effects of andrographolide analogue for diabetes treatment. *Journal of Translational Medicine* 7:62.

Zoha, M. S., A. H. Hussain & S. A. Choudhury. 1989. Antifertility effect of andrographis paniculata in mice. *Bangladesh Medical Research Council Bulletin* 15:1:34–37.

12 Phytochemical Constituents and Antidiabetic Properties Aspects of *Gongronema latifolium* Benth

Alessandra Durazzo, Vita Di Stefano, Andrea Copetta, Paolo Gabrielli, Antonello Santini, and Massimo Lucarini

CONTENTS

12.1 INTRODUCTION

Natural products represent a sustainable and valuable resource with the potential to prevent and manage several diseases (Durazzo et al., 2018; Durazzo and Lucarini, 2019; Durazzo et al., 2020, 2021; Santini and Novellino, 2014, 2017, 2018; Santini et al., 2017; Singla et al., 2023; Yeung et al., 2018, 2021). The evaluation of bioactive compounds and the assessments of their interactions could be defined as the first step for the determination of the beneficial potential of a plant (Durazzo et al., 2018; Durazzo and Lucarini, 2018, 2019). *Gongronema latifolium* Benth (Asclepiadaceae) is a tropical rainforest plant primarily used in traditional folk medicine in the treatment of malaria, diabetes, and hypertension and as laxative. The main features of *G. latifolium* are explored as follows: botanical description, distribution and cultivation practice, quantitative research literature analysis, phytochemical constituents, pharmacological studies with a focus on antidiabetic properties, and toxicological aspects.

12.2 BOTANICAL DESCRIPTION

G. latifolium is a perennial crop commonly called "madumaro or arokeke" and "utazi" in the western part and southeastern part of Nigeria respectively (Etim et al., 2008) (Figure 12.1). This plant is a climbing shrub or liana up to 5 m long with woody base and fleshy roots, containing latex. Its stems are hollow, and all epidermic parts are hairy or glabrous. Leaves are opposite, simple, and entire; the stipules are absent; the petiole is up to 2.5–3 cm long; the blade is broadly ovate to almost circular, long 6–12 cm and large 3–10 cm. The leaf base is deeply cordate, and the apex is acuminate. The inflorescence is a terminal and axillary cymose panicle up to 13 cm long. The flowers are bisexual, small, pentamerous, regular, yellow-green, fragrant; pedicel 2–4 mm long; calyx

DOI: 10.1201/b23347-12

FIGURE 12.1 *Gongronema latifolium.*

lobes (2 mm long) are elliptical to rounded, hairy at apex; the corolla is tubular up to 5 mm long, campanulate at apex, with or without trichomes inside, the lobes 2 mm long are triangular-ovate, spreading; the corona lobes are fleshy, color cream/brown at base, shorter than stamina column; stamens with five short appendages, resting on the short, conical style apex. The anthers are erect with membranous apical appendages. The ovary is superior. The fruit consists of an oblong-lanceolate dehiscent seed pod called a follicle whose color changes from green in small fruits to dark brown, then to black at maturity. During maturity stage, the fruit splits open lengthwise, along the seam, releasing the seeds. The seeds are flat, comma shaped, and attached to a white silky tuft (pappus) that aids dispersal. They are strongly compressed and measure about 0.5 cm in length (Hutchinson and Dalziel, 1931; Mosango, 2011).

12.3 DISTRIBUTION AND CULTIVATION PRACTICES

G. latifolium has a very widespread distribution in the tropical and subtropical regions, especially in West African countries, in which it originates, such as Cote d'Ivoire, Ghana, Nigeria, Senegal, and Sierra Leone; in America; and in northern and southeastern Asia with an average abundance (Balogun et al., 2016).

The availability of *G. latifolia* in Nigeria is on the decline and in some places threatened to extinction despite the nutritional and medicinal values of the plant. It can be propagated by seed or stem (softwood, semi-hardwood, and hardwood) cuttings. Seed germination is difficult because the plant exhibits short-lived seed viability (less than one year) and dormancy sets in. Cutting is the quickest way to propagate *G. latifolia* clones. Hardwood and semi-hardwood nodal cuts thus have higher potential of producing new plants with more shoots (Agbo and Omaliko, 2006; Agbo and Obi, 2006, 2007).

12.4 QUANTITATIVE RESEARCH LITERATURE ANALYSIS

On 24 October 2022, the Scopus online database was used to search to retrieve *G. latifolium* publications. The search string "*Gongronema latifolium*" OR "adumaro" OR "arokeke" OR "utazi" was used to extract bibliometric data from the database (www.scopus.com/home.uri, accessed on 24 October 2022) and bibliographic data, i.e., publication year, publication count, document type, countries/territories of origin, and institutions, were recorded. The functions of the Scopus web online platform named "Analyze" and "Create Citation Report" were utilized for basic analyses. A single database was selected to extract the data. Therefore, possible publications not indexed in this database are missing from this analysis. The "full records and cited references" were exported to VOSviewer software (version 1.6.16, www.vosviewer.com, accessed on 24 October 2022) (van Eck and Waltman, 2010; Waltman et al., 2010) for further bibliometric analyses and additional processing. One hundred thirty-five publications ranged from 1980 to 2022, found by literature search, were collectively cited in 1816 documents. The main subject areas are medicine, biochemistry, genetics and molecular biology, pharmacology, toxicology and pharmaceutics, agricultural and biological sciences, and so on. Publications trends are reported in Figure 12.2. The oldest work is published by Ekundayo, O. in *Pharmaceutical Biology* in 1980 and it addressed constituents of *G. latifolium*.

The distribution of documents by type was mainly as follows: "Article" for 91.9%, followed by "Review" (4.5%), and "Book chapter" (1.5%), etc., as reported in Figure 12.3.

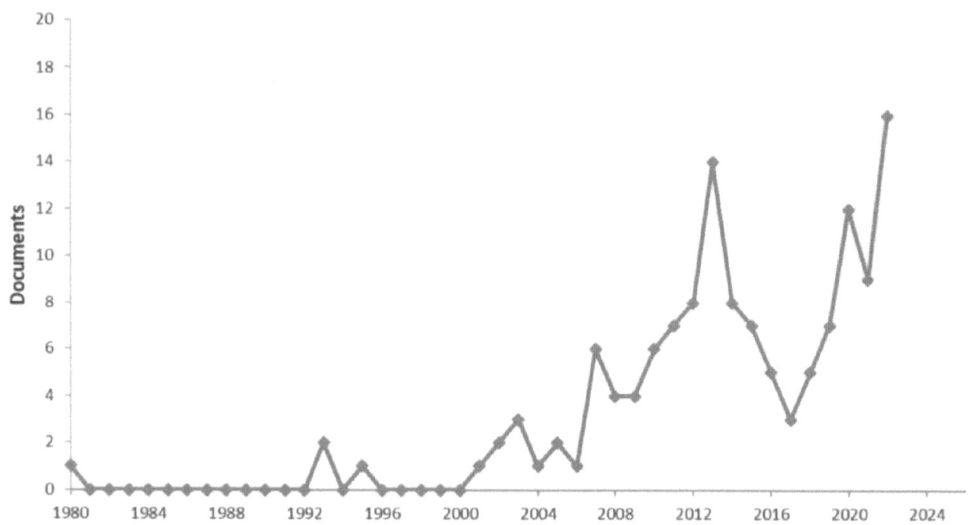

FIGURE 12.2 Publication trends of *G. latifolium* search (based on data from Scopus).

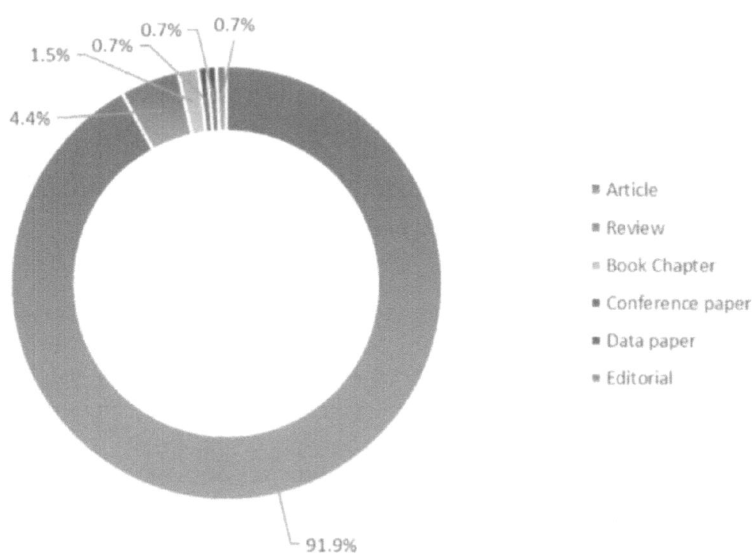

FIGURE 12.3 Distribution of documents by type concerning *G. latifolium* publications (based on data from Scopus).

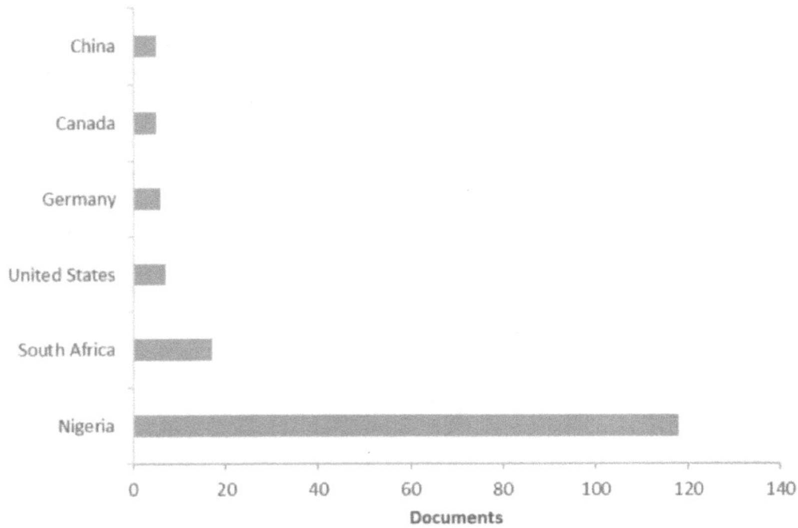

FIGURE 12.4 Most productive countries/territories (based on data from Scopus).

The most cited "Article" is an ethnobotanical survey of Akwa Ibom State of Nigeria (Ajibesin et al., 2007), whereas the most recent is on the hepatoprotective potential of flavonoid-rich extracts from *G. latifolium* leaf in type 2 diabetic rats via fetuin-A and tumor necrosis factor-alpha (Ajiboye et al., 2022). The most cited "Review" is addressed on antioxidant potential of African medicinal plants (Atawodi, 2005). As "Data paper," Omonhinmin et al. (2022) published a dataset on rbcL-based intraspecific diversity of *G. latifolium* in South-East Nigeria, in line with the need to understand its genetic diversity for breeding and conservation: the data consist of 14 partial rbcL gene sequences, nucleotide compositions, and amino acid profiles of *G. latifolium* and the dataset provides insight on the species genetic diversity and evolution.

The most productive authors in this sector reported are Ajiboye, B.O. and Oyinloye, B.E. with nine documents (data from Scopus). Figure 12.4 reports the most productive countries/territories. Regarding countries/territories, Nigeria (n = 117) was the most productive country, followed by South Africa (n = 17) and the United States (n = 7). The most productive institution is the University of Nigeria with 36 documents (data from Scopus).

12.5 NUTRIENTS, PHYTOCHEMICAL CONSTITUENTS, PHARMACOLOGICAL STUDIES, WITH FOCUS ON ANTIDIABETIC PROPERTIES, AND TOXICOLOGICAL ASPECTS

G. latifolium leaves have been used in folklore medicine to manage diabetes mellitus and alleviate dyspepsia.

G. latifolium is widely believed to have strong nutritional and medicinal values. The leaves can be consumed fresh to enrich salads, cooked in soup, or dried and applied like a powdery spice. Both root and stem are used as chewing stick or liquor (Balogun et al., 2016).

The phytochemical profile of *G. latifolium* showed the presence of different types of alkaloids, flavonoids, lignan, terpenes, sterol, hydroxycinnamic acids, saponin, and carotenoid, among others (Imo and Uhegbu, 2015).

Some studies on the aqueous and methanolic leaf extracts of *G. latifolium* have reported the anti-inflammatory activity by significantly inhibiting the carrageenan-induced edema of the rat paw consolidated protocol to test acute inflammation (Morebise et al., 2002, 2005a, 2006). Its antioxidant, antihypercholestrolemic, antilipidemic, and antihyperglycemic effects have been reported too. *G. latifolium* is often used to treat and/or prevent diabetes (Akah et al., 2011). The same extracts inhibited acetic acid-induced vascular permeability and leukocyte migration. Other tests used to confirm chronic anti-inflammatory activity, such as induced arthritis in experimental rats, showed that *G. latifolium* extracts lowered the serum levels of gamma glutamyl transferase, alanine aminotransferase, and aspartate aminotransferase (Morebise et al., 2005b), while the fraction albumin and glucose were reduced in serum (Ribeiro et al., 1991). Balogun et al. (2016) hypothesize that *G. latifolium* can be used to produce nutritional supplements with concentration of nutritionally important minerals and vitamins because the vegetable form is perishable and cannot be stored for a long time. Furthermore, the application of *G. latifolium* extract can be used in order to make drinks.

The methanolic extract of the leaves of *G. latifolium* showed antimicrobial action and showed inhibitory activity against *Salmonella enteritidis*, *Salmonella typhimurium*, *Pseudomonas aeruginosa*, and *Listeria monocytogenes* (Morebise and Fafunso, 1998). The aqueous extract of the leaves, on the other hand, showed inhibitory activity against *E. coli* and *P. aeruginosa* (Chinyere et al., 2015). The essential oil of leaves of *G. latifolium* also showed moderate inhibitory activity against *Staphylococcus* sp., *Escherichia coli*, *Shigella* sp., *Salmonella* sp., *Klebsiella pneumonia*, *Pseudomonas aeruginosa*, and *Candida albicans*, comparable to that of ampicillin but inferior to ciprofloxacin and chloramphenicol (Adeleye et al., 2011).

The main activity of the *G. latifolium* leaf extracts is the antidiabetic activity. Sylvester et al. (2015) evaluated the activity of leaf extracts on rats subjected to induced diabetes. The data reported show the reduction in blood glucose in 66.34% of diabetic rats. The induction of diabetes also induced an increase in total cholesterol and LDL cholesterol in the same rats (54.42% and 55.0%, respectively). Simultaneously with the reduction in glucose, a significant reduction of 58.70% was also observed for total cholesterol and 71.70% for LDL (Sylvester et al., 2015). Other studies report the activity of *G. latifolium* leaf extracts on rats with alloxan-induced diabetes. The more significant effects were on glycogen content, glucose transporter (GLUT-2 and GLUT-4) levels, and relative gene expression of hexokinase, compared to control, as well as a reduction of weight of male Wistar rats (Edet et al., 2011; Ajiboye et al., 2019, 2021). In animals treated with streptozotocin or in human diabetic states, mobilization of glucose transporters and hexokinase activity causes persistent hyperglycemia which in turn causes low hepatic glycogen content (Ajiboye et al., 2022). The administration of alcoholic extracts of *G. latifolium* showed in the animals treated a significant increase in the levels of GLUT-2/-4 and in the activity of the hexokinase with a consequent increase in hepatic glycogen (Ajiboye et al., 2022). Iweala et al. (2015) have shown a strong inhibitory activity against human lung cancer and human breast adenocarcinoma in *in vitro* studies, correlated with the antiradical activity of the leaf extracts. TNF-α (tumor necrosis factor) is a proinflammatory cytokine linked to the development of autoimmune diseases, rheumatoid arthritis, septic shock, and other inflammatory diseases. Much research has recently focused on the potential physiological or pathological involvement of TNF-α in adipocyte differentiation, lipid metabolism, and insulin sensitivity *in vivo*. In rats treated with STZ, the level of TNF-α is always increased compared to control animals. This result gives credence to the idea that TNF-α plays a role in the development of insulin resistance, type 2 diabetes, and lipid abnormalities (Ajiboye et al., 2022).

Administration of the ethanolic extracts of *G. latifolium* to rats with streptozotocin-induced diabetes significantly reduced the level of TNF-α. Moreover, histoarchitecture of the liver of diabetic rats administered ethanolic extract (especially at the low dose) was also ameliorated (Ajiboye et al., 2022). In rats with streptozotocin-induced diabetes, aqueous and ethanolic extracts of the leaves significantly increased the activity of superoxide dismutase, glutathione reductase, glutathione peroxidase, and glucose-6-phosphate dehydrogenase; at the same time the same study showed increased reduced glutathione levels while lipid peroxidation was reduced (Ugochukwu et al., 2002). Prolonged use of glucocorticoids in the treatment of inflammatory disorders has been linked to cardiovascular diseases such as hypertension and myocardial damage linked to increased oxidative stress. In a recent study, the administration of ethanolic extract of *G. latifolium* throats resulted in an increase in serum NO, antioxidant activity, and reduction of oxidative stress as well as a decrease in cardioinflammatory molecules. Furthermore, *G. latifolium* protected and improved the cytoarchitecture of cardiac and liver muscles in glucocorticoid-induced myocardial damage (Sulaiman et al., 2022a).

Regarding toxicological study, interestingly, Sylvester et al. (2015) reported that acute toxicity of the ethanolic extract of the plant on mice showed 0% lethality when administered intraperitoneally at 1000 mg/kg body weight, but 100% mortality when administered at 2000 mg/kg. This clearly suggests that the plant material is not toxic at the safe doses being consumed. Sulaiman et al. (2022b), by studying the biochemical effects of *G. latifolium* ethanolic extracts in male Wistar rats, showed how rat serum bilirubin and albumin levels were comparable to those of the control, suggesting the absence of hepatic assault. The ethanolic extracts increased rat serum urea levels compared with control. The increase in serum urea levels might indicate the ensuing impairment of kidney function (Sulaiman et al., 2022b).

12.6 CONCLUSION

This chapter presents a picture of the updated state of the art of *G. latifolium*. *G. latifolium* Benth. represents a rich source of bioactive compounds that make it a great potential to potentially prevent diseases. *In vivo* and clinical studies needs to be addressed. Due to its contents of nutrients and bioactive molecules, *G. latifolium* raises interest for use in different fields.

REFERENCES

Adeleye IA, Omadime ME, Daniels FV. Antimicrobial activity of essential oil and extracts of Gongronema latifolium on bacterial isolates from bloodstream of HIV infected patients. *J Pharmacol Toxicol.* 2011;6(3):312–320.

Agbo CU, Obi IU. Macro-propagation technique for different physiological ages of Gongronema latifolia Benth cuttings. *Afr J Biotechnol.* 2006;5(13):1254–1258.

Agbo CU, Obi IU. Studies on seed germination of Gongronema latifolia Benth. *Bio-Research.* 2007;5(1):175–178.

Agbo CU, Omaliko CM. Initiation and growth of shoots of Gongronema latifolia Benth stem cuttings in different rooting media. *Afr J Biotechnol.* 2006;5(5):425–428.

Ajibesin KK, Ekpo BA, Bala DN, Essien EE, Adesanya SA. Ethnobotanical survey of Akwa Ibom State of Nigeria. *J Ethnopharmacol.* 2007;115(3):387–408.

Ajiboye BO, Oyinloye BE, Agboinghale PE, Onikanni SA, Asogwa E, Kappo AP. Antihyperglycaemia and related gene expressions of aqueous extract of Gongronema latifolium leaf in alloxan-induced diabetic rats. *Pharm Biol.* 2019;57(1):604–611. https://doi.org/10.1080/13880209.2019.1657907.

Ajiboye BO, Oyinloye BE, Udebor EA, et al. Hepatoprotective potential of flavonoid-rich extracts from Gongronema latifolium benth leaf in type 2 diabetic rats via fetuin-A and tumor necrosis factor-alpha. *Mol Biol Rep.* 2022;49(9):8391–8400. https://doi.org/10.1007/s11033-022-07657-x.

Ajiboye, BO, et al. Hepatoprotective and haematoprotective roles of gongronema latifolium benth aqueous extract in alloxan-induced diabetic rats. *Biointerface Res Appl Chem.* 2021;12(1):537–546.

Akah PA, Uzodinma SU, Okolo CE. Antidiabetic activity of aqueous and methanol extract and fractions of Gongronema latifolium (Asclepidaceae) leaves in alloxan diabetic rats. *J Appl Pharm Sci.* 2011;1:99–102.

Atawodi SE. Antioxidant potential of African medicinal plants. *Afr J Biotechnol.* 2005;4(2):128–133.

Balogun ME, Besong EE, Obimma JN, Mbamalu OS, Djobissie SFA. Gongronema latifolium: A phytochemical, nutritional and pharmacological review. *J Phys Pharm Adv.* 2016;6(1):811–824. https://doi.org/10.5455/jppa.1969123104000.

Durazzo A, D'Addezio L, Camilli E, Piccinelli R, Turrini A, Marletta L, Marconi S, Lucarini M, Lisciani S, Gabrielli P, Gambelli L, Aguzzi A, Sette S. From plant compounds to botanicals and back: A current snapshot. *Molecules.* 2018;23.

Durazzo A, Lucarini M. Current shot and re-thinking of antioxidant research strategy. *Braz J Anal Chem.* 2018;5:9–11.

Durazzo A, Lucarini M. Editorial: The state of science and innovation of bioactive research and applications, health, and diseases. *Front Nutr.* 2019;6:178. https://doi.org/10.3389/fnut.2019.00178

Durazzo A, Lucarini M. Environmental, ecological and food resources in the biodiversity overview: Health benefits. *Life*. 2021;11.

Durazzo A, Lucarini M, Santini A. Nutraceuticals in human health. *Foods*. 2020;9(3):370. https://doi.org/10.3390/foods9030370

Durazzo A, Souto EB, Lombardi-Boccia G, Santini A, Lucarini M. Metrology, agriculture and food: Literature quantitative analysis. *Agriculture*. 2021;11.

Edet EE, Akpanabiatu MI, Uboh F, Edet TE, Eno AE, Itam EH et al. Gongronema latifolium crude leaf extract reverses alterations in haematological indices and weight-loss in diabetic rats. *J Pharmacol Toxicol*. 2011;6(2):174–181.

Ekundayo O. Constituents of gongronema latifolium benth. In hook (asclepiadaceae). *Pharm Biol*. 1980;18(3):127–129.

Etim OE, Akpan EJ, Usoh IF. Hepatoxicity of carbon tetrachloride: Protective effect of gongronema latifolium. *Pak J Pharm Sci*. 2008;21:268–274.

Hutchinson J, Dalziel JM. *Flora of West Tropical Africa*. The Crown Agents for the Colonies 4, Millbank, Westminster, 1931, 60.

Ilodibia CV, Ezeja IJ, Akachukwu EE, Chukwuma MU, Egboka TP, Emeka AN. Phytochemical screening and antimicrobial effects of aqueous and ethanol leaf and stem extracts of *Gongronema latifolium* benth. *Res J Bot*. 2015;10(2):50–60.

Imo C, Uhegbu FO. Phytochemical analysis of *Gongronema latifolium* benth leaf using gas chromatographic flame ionization detector. *Int J Chem Biomol Sci*. 2015;1(2):60–68. www.aiscience.org/journal/ijcbs

Iweala EEJ, Liu F, Cheng R, Li Y. Anticancer and free radical scavenging activity of some Nigerian food plants in vitro. *Int J Cancer Res*. 2015;11(1):41–51.

Morebise O, Fafunso MA. Antimicrobial and phytotoxic activities of saponin extracts from two Nigerian edible medicinal plants. *Biokemistri*. 1998;8(2):69–77.

Morebise O, Fafunso MA, Makinde JM. Membrane stabilizing activity: A possible mechanism of action for the antiinflammatory property of Gongronema latifolium leaves. *Int J Biomed Hlth Sci*. 2005a;1(1):15–19.

Morebise O, Fafunso MA, Makinde JM, Ajani RA. Biochemical studies on the activity of Gongronema latifolium leaves in experimental arthritis. *Int J Biomed Health Sci*. 2005b;1(2):71–76.

Morebise O, Fafunso MA, Makinde JM, Olajide OA. Evaluation of the bioactivity of Gongronema latifolium leaf extract in rodents. *Science Focus*. 2006;11(1):27–30.

Morebise O, Fafunso MA, Makinde JM, Olajide OA, Awe EO. Antiinflammatory property of the leaves of Gongronemalatifoslium. *Phytother Res*. 2002;16:75–77.

Mosango DM. *Gongronema latifolium benth*. Record from PROTA4u. Schmelzer GH, Gurib-Fakim A (Eds.). PROTA (Plant Resources of Tropical Africa), 2011. http://www.prota4u.org/search.asp, accessed 28 November 2015.

Omonhinmin CA, Onuselogu CC, Olomukoro E. Dataset on rbcL-based intra-specific diversity of Gongronema latifolium Benth: (Apocynaceae) in South-East Nigeria. *Data in Brief*. 2022;41:107870.

Ribeiro RA, Frores CA, Cunha FQ, Ferreira SH. IL-8 causes in vivo neutrophil migration by a cell-dependent mechanism. *Immunology*. 1991;73:473–477.

Santini A, Novellino E. Nutraceuticals: Beyond the diet before the drugs. *Curr Bioact Compd*. 2014;10:1–12.

Santini A, Novellino E. Nutraceuticals – shedding light on the grey area between pharmaceuticals and food. *Expert Rev Clin Pharmacol*. 2018;11:545–547.

Santini A, Novellino E. To nutraceuticals and back: Rethinking a concept. *Foods*. 2017;6.

Santini A, Tenore GC, Novellino E. Nutraceuticals: A paradigm of proactivemedicine. *Eur J Pharm Sci*. 2017;96:53–61.

Singla RK, De R, Efferth T, Mezzetti B, Uddin MS, Sanusi S, Ntie-Kang F, Wang D, Schultz F, Kharat KR, Devkota HP, Battino M, Sur D, Lordan R, Patnaik SS, Tsagkaris C, Sai CS, Tripathi SK, Găman MA, Ahmed MEO, . . . Shen B. The International Natural Product Sciences Taskforce (INPST) and the power of Twitter networking exemplified through #INPST hashtag analysis. *Phytomed Int J Phytother Phytopharmacol*. 2023;108:154520. https://doi.org/10.1016/j.phymed.2022.154520

Sulaiman FA, Yusuf BO, Omar SA, Muritala HF, Adisa JM, Olopade AA, Babajamu FI, Jimba AT, Babatunde AL, Adeniyi BA, Opaleye BR, Maimako RF, Otohinoy DA. *Gongronema latifolium* leaf extract protects against dexamethasone-induced myocardial cell injury via cardiac oxido-inflammatory molecules modulation. *J Food Biochem*. 2022a;46(12):1–11.

Sulaiman FA, Yusuf BO, Omar SA, Muritala HF, Adisa JM, Olopade AA, Babajamu FI, Jimba AT, Babatunde AL, Adeniyi BA, Opaleye BR, Maimako RF, Otohinoyi DA, Bello KO, Rotimi D, Osemwegie OO, Adeyemi OS. Ethanolic extracts of the *Gongronema latifolium* stem and leaves caused mild renal injury and modulated serum triglycerides in rats. *Biointerface Res Appl Chem.* 2022b;12(4):5045–5053. https://doi.org/10.33263/BRIAC124.50455053

Sylvester EG, Israel EU, Olajumoke AD. The effect of Gongronema latifolium leaf extract on blood biochemical assay in diabetic rats. *J Sci Res Rep.* 2015;6(7):514–522.

Ugochukwu NH, Babady NE. Antioxidant effects of Gongronema latifolium in hepatocytes or rat models of non-insulin dependent diabetes mellitus. *Fitoterapia.* 2002;73(7–8):612–618.

van Eck NJ, Waltman L. Software survey: VOSviewer, a computer program for bibliometric mapping. *Scientometrics.* 2010;84(2):523–538.

Waltman L, van Eck NJ, Noyons ECM. A unified approach to mapping and clustering of bibliometric networks. *J Informetr.* 2010;4(4):629–635.

Yeung AWK, Aggarwal BB, Barreca D, Battino M, Belwal T, Horbańczuk OK, Berindan Neagoe I, Bishayee A, Daglia M, Devkota HP, Echeverría J, El-Demerdash A, Orhan IE, Godfrey KM, Gupta VK, Horbańczuk JO, Modliński JA, Huber LA, Huminiecki L, Jóźwik A, Marchewka J, Miller MJS, Mocan A, Mozos I, Nabavi SF, Nabavi SM, Pieczynska MD, Pittalà V, Rengasamy KRR, Sanches A, Silva AS, Sheridan H, Stankiewicz AM, Strzałkowska N, Sureda A, Tewari D, Weissig V, Zengin G, Atanasov AG. Dietary natural products and their potential to influence health and disease including animal model studies. *Anim Sci Pap Rep.* 2018;36:345–358.

Yeung AWK, Atanasov AG, Sheridan H, Klager E, Eibensteiner F, Völkl-Kernsock S, Kletecka-Pulker M, Willschke H, Schaden E. Open innovation in medical and pharmaceutical research: A literature landscape analysis. *Front Pharmacol.* 2021;11.

13 Pharmacological Studies, with Focus on Antidiabetic Response of *Perilla frutescens* (L.) Britt

Massimo Lucarini, Vita Di Stefano, Andrea Copetta, Paolo Gabrielli, Antonello Santini, and Alessandra Durazzo

CONTENTS

13.1 INTRODUCTION

The biological functions of each plant matrix derive from the concerted action of biologically active compounds (Durazzo et al., 2018; Durazzo and Lucarini, 2018, 2019). *Perilla frutescens* (L.) Britt. is an Asian multifunctional plant: it is used in edible dishes, for dyeing material, and as an ornamental plant, an oil seed crop, and a medicinal herb (Dhyani et al., 2019). This chapter presents an update of *P. frutescens* from botanical description and distribution to phytochemical profile and related potential beneficial properties.

13.2 BOTANICAL DESCRIPTION

P. frutescens is an annual herb, belonging to the family Lamiaceae (Figure 13.1). It is an aromatic medicinal plant with roots grown in gardens or in pots for ornamental purposes. The robust square stem is covered by short hairs and grows up to 60–150 cm tall. The downy-hair ovate to round leaves are opposite, green to purple lavender or reddish, with margins pointed and serrated; they emit a pleasant scent ranging from mint and anise to lemon balm. Between July and October, this plant produces not particularly showy apical inflorescences like spikes. The tubular purplish to white flowers are hermaphrodite, small (3–4 mm in length), two lipped and bell shaped. *Perilla* seeds are gray-brown, small, and weigh approximately 4 g/1000 seeds (Nitta et al., 2003; Pandey et al., 2008).

Different varieties are cultivated throughout the world, but their classification is controversial. Based on uses and morphological characters, the genus of *Perilla* L. consists of only two different cultivars: *P. frutescens* (L.) Britton var. *frutescens* used for oil seed production and *P. frutescens* (L.) Britton var. *crispa* (Thunb.) used for fresh vegetables and medicinal purposes. In both varieties, green and purple leaf phenotypes can be found. Based on palynologic characters (decorations and

DOI: 10.1201/b23347-13

FIGURE 13.1 *Perilla frutescens.*

size of pollen grains), the genus of *Perilla* L. consists of five varieties: var. *frutescens* and var. *acuta* used for leaf fresh consumption, var. *arguta*, used for seed oil extraction, var. *crispa* used for its medicinal properties, and var. *auriculato-dentata* (Ahmed, 2019).

13.3 DISTRIBUTION AND CULTIVATION PRACTICES

The origin of *Perilla* traces back to East Asian countries (China, Japan, Korea, Taiwan, Vietnam, and India), where it has been used as a valuable source of culinary and traditional medicinal uses. The leaves, seeds, and stems of *P. frutescens* are used for various therapeutic applications in folk medicine. *P. frutescens* is mainly cultivated in China, Japan, Korea, and India.

As one of the important economic crops, the cultivation of *P. frutescens* has more than 2000 years of history in China as well as other countries in Asia (Lee and Kim, 2007).

Classically, *Perilla* is reproduced via seeds at the end of winter/early spring; the different varieties are generally grown using direct sowing or produced in a nursery bed and transplanted in early summer into their permanent position. Seed germination is usually quick. The farmers use fresh seeds because the seed viability is lost relatively quickly over time: in three years the seed germination can decrease from 93% to 7%. The optimal seed germination temperature is 20 °C, while the plants can be grown at slightly lower temperatures. The plant prefers a full sun zone (not the shade), moist sandy/loamy soils, rich and well-drained with pH between 5.5 and 6.0. It is relatively drought tolerant but is sensitive to water deficiency. Edible young leaves can be collected during the entire growing season; in July–October the plant blooms, while the seeds ripen from October to November (Nitta et al., 2003; Choi and Yang, 2005).

13.4 TRADITIONAL AND OTHER POTENTIAL USES

P. frutescens leaves are widely used in traditional Chinese medicine to treat common cold, diarrhea, fever, cough, vomiting, and more. *P. frutescens* has been used as traditional herbal medicine for treating diseases from ancient times in China.

Perilla oil is considered a high-quality oil in nutritional sciences, which are concerned with lifestyle-related diseases.

Except for the edible applications, the plant of *P. frutescens* has also traditionally been used as a medicinal herb in China for thousands of years. The leaves, seeds, and stems of *P. frutescens* are recommended by the Chinese pharmacopeia as three medicinal materials for various therapeutic applications. In the past decades, a number of investigations have been done about different aspects of *P. frutescens*.

In 2002, *P. frutescens* and *P. frutescens* seed have been listed in the 87 medicine-food plants by the Ministry of Health of the People's Republic of China (MOH) (2002) which suggests the potential edible and medicinal values of *P. frutescens* plant.

The traditional uses of *P. frutescens* include two aspects: edible and medicinal. Both edible young leaves and seedlings (microgreens) can be used raw or cooked in salads or hot dishes. Two chemo-varietal phenotypes are used for the production of microgreens or edible leaves: "Zi-So" is widely grown in China and characterized by anthocyanin red leaves and "Shizo" or "Shisoyo" is mainly used in Japan and characterized by green leaves (Rouphael et al., 2019; Dimita et al., 2022).

13.5 PHYTOCHEMICAL CONSTITUENTS

Modern pharmacological studies have demonstrated that *P. frutescens* leaves contain rich bioactive components, like phenolics, flavonoids, anthocyanins, tannins, and essential oil, (Ghimire et al., 2019; Wang et al., 2021) and exhibit a variety of effects, including antioxidant, antiallergy, anti-inflammation, antitumor, and antibacterial (Hashimoto et al., 2020).

Many researchers have extensively investigated the phytochemistry and pharmacology of *P. frutescens*. Various compounds from this plant have been isolated and identified, including flavonoids, volatile oils, fatty acids, triterpenes, phenolic compounds, and others. It is worth mentioning the current review of Hou et al. (2022) describing *P. frutescens* as a rich source of pharmacological active compounds.

Among active constituents of *Perilla*, perillaldehyde and rosmarinic acid have received the most attention, suggesting that these compounds are important constituents. Perillaldehyde contributes to aroma and accounts for approximately 50% of the oil refined from *Perilla*.

Perilla contains several essential oils, including (–) perillaldehyde, (–) perillylalcohol, (+) limonene, alphapinene, and trans-shisool. Moreover, *Perilla* contains the purple pigments shisonin and cyanin. Other constituents of *Perilla* include rosmarinic acid, adenine, and arginine.

13.6 PHARMACOLOGICAL STUDIES, WITH FOCUS ON ANTIDIABETIC RESPONSE

The contemporary studies have demonstrated that *P. frutescens* has a number of bioactive properties, such as anti-inflammatory, antiallergic, and more.

P. frutescens showed strong anti-inflammatory, antidepressant, antispasmodic, anticancer, antioxidant, antimicrobial, insecticidal, neuroprotective, and hepatoprotective effects.

Wang et al. (2020) reported how *Perilla* oil regulates intestinal microbiota and alleviates insulin resistance through the PI3K/AKT signaling pathway in type 2 diabetic KKAy mice and may be a potential functional food for diabetic treatment.

Some crude extracts and essential oil of *Perilla* leaf are used for prevention and treatment of various ailments, and their efficacy has been evaluated by a number of researchers. Documented activities of extracts from *P. frutescens* are as follows.

The antioxidant properties of *P. frutescens* seed, leaf, and extract were investigated. *Perilla* seed and leaf extracts have been reported to exhibit concentration-dependent antioxidant activity based on the radical dosage of 2,2-diphenyl-1-picril-hydrazil-hydrate (DPPH) and 2,2'-azino-bis-(3-ethyl-benzothiazoline-6-sulfonic acid) (ABTS) radical cation assay (Zhou et al., 2014).

A recent study compared the antioxidant activity of the various extracts of *P. frutescens* leaves. The ethyl acetate extracts showed the best antioxidant activity and was significantly greater than the aqueous extract, with values of 1006.33, 1682.80, and 1957.73 μg of trolox/mg of dry extract DPPH, ABTS +, and of the superoxide anion scavenging, respectively. For FRAP, it was 4181.13 μg of FeSO4/mg dry extract (Wang et al., 2021). The ethyl acetate extracts showed high inhibitory effect on MCF-7 and HepG2 tumor cell proliferation (Wang et al., 2021).

In vivo, protective leaf activity of *P. frutescens* has been demonstrated on lipopolysaccharide (LPS)-induced liver injury of mice sensitized to d-GalN due to superoxide or peroxynitrite scavenging or reducing activities and inhibition of factor-producing tumor necrosis (Osakabe et al., 2002).

A study demonstrated that the essential oil of *P. frutescens* produces an antidepressant-like effect, through restoring brain-derived neurotrophic factor (BDNF) hippocampal expression in mild stress-induced mouse model (Yi et al., 2013), suggesting that *Perilla* might be useful for the prevention of depression. These results showed that *Perilla* essential oil is an effective treatment for depression because it can reduce pro-inflammatory cytokines, as well as modulate 5-HT metabolism, and thus ameliorate depression-like behavior (Ji et al., 2014).

Different bioactive compounds of *P. frutescens* showed significant anti-inflammatory activity. *Perilla* oil is rich in omega-3 fatty acids. Omega-3 fatty acids act as inhibitors of conversion of arachidonic acid by cyclooxygenase (COX) and pro-inflammatory oxygenated inflammatory mediators known as eicosanoids. *Perilla* oils have been shown to reduce the production of the inflammatory cytokines IL-1β and TNF-α by monocytes stimulated *in vitro* (Liu et al., 2022).

These cytokines are important effector molecules in inflammatory responses, and TNF-α blocking agents are now used widely to treat rheumatoid disease.

The anti-inflammatory activities of glycoglycerolipids isolated from the leaves of *P. frutescens* were investigated in LPS-stimulated murine macrophage cells, and they did exhibit the anti-inflammatory activity through the inhibition of NO production and pro-inflammatory cytokines secretion (Zi et al., 2021).

Several studies have examined the effects of different *P. frutescens* extracts on tumor necrosis factor (TNF-α) levels in mice with induced inflammation and found that the aqueous extracts of *Perilla* had higher inhibitory activity than organic extracts (Ueda et al., 1997).

Different bioactive components of *P. frutescens*, including rosmarinic acid, caffeic acid, luteolin, apigenin, methoxyflavanone, and α-linolenic acid, were found to possess antiallergic activity.

Kamei et al. (2017) reported the presence of 8-hydroxy-5,7- dimethoxyflavanone (PDMF) in *P. frutescens* leaves. PDMF may be used to prevent IgE-driven type I hypersensitivity reactions. Akt (serine/threonine kinase that plays a critical role in cell growth, differentiation, and survival) phosphorylation and intracellular Ca2+ influx, two critical molecular events involved in mast cell degranulation in allergic reactions, are suppressed by PDMF, which acts as a potent antiallergic compound.

The aqueous *P. frutescens* extract inhibited histamine released by rat peritoneal mast cells *in vitro* (Shin et al., 2000). The glycoprotein derived from the hot aqueous extract of *P. frutescens* has demonstrated the effective component of *P. frutescens* in inhibiting mast cell degranulation (Asada et al., 1999). These studies have suggested the potential antiallergic effect of *P. frutescens.*

Perilla seed oil increases lung function and it may be beneficial to asthma sufferers.

A clinical study in patients with asthma has shown that *Perilla* seed oil improves lung function and also suppresses the release of leukotriene LB4 and LC4 from leukocytes responsible for asthma, possibly due to the presence of α-linolenic acid (omega-3) (Chang et al., 2008).

In another study, the asthmatic effect of *Perilla* seed oil was studied both *in vitro* and *in vivo* in guinea pigs. The study pointed out that *Perilla* seed oil inhibits airway constriction and also improves lung function by decreasing lung resistance and increasing dynamic lung compliance (Deng et al., 2007).

Perilla leaves and its seed oil showed anticancerous properties. In an experiment it was reported that *Perilla* leaf extract (PLE) has strong growth inhibitory activities in both HCT116 and H1299 cancer cell lines (Kwak and Ju, 2015). This is in line with previous reports on the potent antiproliferative effects of PLE at a range of concentrations (50–500 μg/ml) in human hepatoma (Lin et al., 2007) and leukemia cells (Kwak et al., 2009). Kwak and Ju (2015) demonstrated that the ethanol extract of *Perilla* leaf inhibited colorectal cancer (HCT116) and non-small cell lung cancer (H1299), in a dose-dependent manner, with IC50 values of 50 μg/mL and 67 μg/mL, respectively. The ethanol extract treated cells exhibited nuclear fragmentation and chromatin condensation. Ethanol extract treatment significantly enhanced the sub-G1 cell population and inhibited the migration of non-small cell lung cancer cells and their adhesion (Kwak and Ju, 2015).

A methoxyflavone derived from *P. frutescens* synergizes with tyrosine kinase inhibitors to increase tumor suppressive potency on human adenocarcinoma A549 *in vitro* and *in vivo*. Mechanically, the antitumor synergistic effect is satisfied by the induction of a two-stage cell cycle arrest in the G1 and G2/M phases (Abd El-Hafeez et al., 2018).

α-Glucosidase is a hydrolytic enzyme that breaks down starch, oligosaccharides, and disaccharides into simple sugars, such as glucose molecules, by acting on α (1–4) bonds to facilitate intestinal absorption of carbohydrates. Their action drastically increases post-meal blood sugar levels in diabetes patients. α-Glucosidase inhibitors prevent or delay the breakdown of carbohydrates into simple sugars (glucose molecules), which delays the intestinal absorption of glucose and consequently the increase in blood sugar levels after meals.

P. frutescens leaf extracts showed promising hypoglycemic activity *in vitro* and *in vivo*, which could be exploited against diabetes (Hou et al., 2022). *P. frutescens* extracts revealed the inhibitory activity of α-glucosidase *in vitro*, twice higher than acarbose, suggesting a hypoglycemic activity (Wang et al., 2021). Rat model treated with streptozocin and high fat diet, simulating the common manifestation of the metabolic abnormalities and resembling the natural history of type 2 diabetes in human population, was used to evaluate the hypoglycemic activity of *P. frutescens* extracts. Histochemical changes in the pancreatic islets were examined. The results showed that extracts significantly reduced serum glucose levels and improved their total cholesterol, triglycerides, high-density lipoprotein cholesterol, and low-density lipoprotein cholesterol levels. Histopathological analysis of the liver and intestinal tissue of rats revealed that treatments with the extracts could actually improve diabetes-related pathophysiological conditions in the liver and intestinal tissue, such as tissue lesions and ruptured villi. PAS staining analysis showed that taking the extract could increase glycogen accumulation in the liver and intestines, and thus prevent glucose from entering the bloodstream, which in turn causes hyperglycemia and diabetes.

As safety issues, it was reported that pregnant women should avoid eating *Perilla* leaves.

13.7 CONCLUSION

This chapter provides updates concerning *P. frutescens*. Different studies have reported bioactive components and related potential beneficial properties of *P. frutescens*. Clinical studies need to be addressed as well as applications in different types of fields in line with needs for natural products as healthy preventive tools.

REFERENCES

Abd El-Hafeez AA, Fujimura T, Kamei R, Hirakawa N, Baba K, Ono K, Kawamoto S. Synergistic tumor suppression by a Perilla frutescens-derived methoxyflavanone and anti-cancer tyrosine kinase inhibitors in A549 human lung adenocarcinoma. *Cytotechnology*. 2018;70(3):913–919. https://doi.org/10.1007/s10616-017-0124-1

Ahmed HM. Ethnomedicinal, phytochemical and pharmacological investigations of *Perilla frutescens* (L.) britt. *Molecules*. 2019;24(1):102. https://doi.org/10.3390/molecules24010102

Asada M, Fukumori Y, Inoue M, et al. Glycoprotein derived from the hot water extract of mint plant, Perilla frutescens Britton. *J Agric Food Chem*. 1999;47(2):468–472.

Chang HH, Chen CS, Lin JY. Dietary Perilla oil inhibits proinflammatory cytokine production in the bronchoalveolar lavage fluid of ovalbumin-challenged mice. *Lipids*. 2008;43:499–506. https://doi.org/10.1007/s11745-008-3171-8.

Choi SK, Yang DC. Leaf and steam production of Perilla frutescens by fertilizing of Sta-Green. *Korean J Plant Res*. 2005;8(2):167–170.

Deng YM, Xie QM, Zhang SJ, Yao HY, Zhang H. Anti-asthmatic effects of Perilla seed oil in the guinea pig in vitro and in vivo. *Planta Med*. 2007;73:53–58. https://doi.org/10.1055/s-2006-957062

Dhyani A, Chopra R, Garg M. A review on nutritional value, functional properties and pharmacological application of perilla (*Perilla frutescens* L.). *Biomed Pharmacol J*. 2019;12(2).

Dimita R, Min Allah S, Luvisi A, Greco D, De Bellis L, Accogli R, Mininni C, Negro C. Volatile compounds and total phenolic content of *Perilla frutescens* at microgreens and mature stages. *Horticulturae*. 2022;8(1):71. https://doi.org/10.3390/horticulturae8010071

Durazzo A, D'Addezio L, Camilli E, Piccinelli R, Turrini A, Marletta L, Marconi S, Lucarini M, Lisciani S, Gabrielli P, Gambelli L, Aguzzi A, Sette S. From plant compounds to botanicals and back: A current snapshot. *Molecules*. 2018;23.

Durazzo A, Lucarini M. Current shot and re-thinking of antioxidant research strategy. *Braz J Anal Chem*. 2018;5:9–11.

Durazzo A, Lucarini M. Editorial: The state of science and innovation of bioactive research and applications, health, and diseases. *Front Nutr*. 2019;6:178. https://doi.org/10.3389/fnut.2019.00178

Ghimire BK, Yoo JH, Yu CY, Kim SH, Chung IM. Profiling volatile and phenolic compound composition and characterization of the morphological and biological activities of *Perilla frutescence* Britton var. Japonica accessions. *Acta Physiol Plant*. 2019;41:108.

Hashimoto M, Tanabe Y, Hossain S, Matsuzaki K, Ohno M, Kato S, Katakura M, Shido O. Intake of alpha-linolenic acid-rich *Perilla frutescens* leaf powder dec reases home blood pressure and serum oxidized low-density lipoprotein in Japanese adults. *Molecules*. 2020;25:2099.

Hou T, Netala VR, Zhang H, Xing Y, Li H, Zhang Z. Perilla frutescens: A rich source of pharmacological active compounds. *Molecules*. 2022;27(11):3578. https://doi.org/10.3390/molecules27113578

Ji WW, Li RP, Li M, Wang SY, Zhang X, Niu XX, Li W, Yan L, Wang Y, Fu Q, Ma SP. Antidepressant-like effect of essential oil of Perilla frutescens in a chronic, unpredictable, mild stress-induced depression model mice. *Chin J Nat Med*. 2014;12(10):753–759. https://doi.org/10.1016/S1875-5364(14)60115-1

Kamei R, Fujimura T, Matsuda M, Kakihara K, Hirakawa N, Baba K, Ono K, Arakawa K, Kawamoto S. A flavanone derivative from the Asian medicinal herb (Perilla frutescens) potently suppresses IgE-mediated immediate hypersensitivity reactions. *Biochem Biophys Res Commun*. 2017;483:674–679. https://doi.org/10.1016/j.bbrc.2016.12.083

Kwak CS, Yeo EJ, Moon SC, Kim YW, Ahn HJ, Park SC. Perilla leaf, Perilla frutescens, induces apoptosis and G1 phase arrest in human leukemia HL-60 cells through the combinations of death receptor mediated, mitochondrial, and endoplasmic reticulum stress-induced pathways. *J Med Food*. 2009;12:508–517.

Kwak Y, Ju J. Inhibitory activities of Perilla frutescens britton leaf extract against the growth, migration, and adhesion of human cancer cells. *Nutr Res Pract*. 2015;9(1):11–16. https://doi.org/10.4162/nrp.2015.9.1.11

Lee JK, Kim, NS. Genetic diversity and relationships of cultivated and weedy types of *Perilla frutescens* collected from East Asia revealed by SSR markers. *Korean J Breed Sci*. 2007;39:491–499.

Lin CS, Kuo CL, Wang JP, Cheng JS, Huang ZW, Chen CF. Growth inhibitory and apoptosis inducing effect of Perilla frutescens extract on human hepatoma HepG2 cells. *J Ethnopharmacol*. 2007;112: 557–567.

Liu D, Wang Y, Lu Z, Lv F, Bie X, Zhao H. Separation, characterization and anti-inflammatory activities of galactoglycerolipids from *Perilla frutescens* (L.) Britton. *Nat Prod Res*. 2022:1–6. https://doi.org/10.1080/14786419.2022.2095637

The Ministry of Health of the People's Republic of China (MOH). Notice of the Ministry of Health to further standardize the management of health food ingredients; 2002.

Nitta M, Lee JK, Ohnishi O. Asian *Perilla* crops and their weedy forms: Their cultivation, utilization and genetic relationships. *Econ Bot*. 2003;57:245–253.

Osakabe N, Yasuda A, Natsume M, Sanbongi C, Kato Y, Osawa T, Yoshikawa T. Rosmarinic acid, a major polyphenolic component of *Perilla frutescens*, reduces lipopolysaccharide (LPS)-induced liver injury in D-galactosamine (D-GalN)-sensitized mice. *Free Radic Biol Med*. 2002;33:798–806.

Pandey A, Bhatt KC. Diversity distribution and collection of genetic resources of cultivated and weedy type in *Perilla frutescens* (L.) Britton var. frutescens and their uses in Indian Himalaya. *Genet Resour Crop Evol.* 2008;55:883–892.

Rouphael Y, Kyriacou MC, Carillo P, Pizzolongo F, Romano R, Sifola MI. Chemical eustress elicits tailored responses and enhances the functional quality of novel food Perilla frutescens. *Molecules.* 2019;24(1):185. https://doi.org/10.3390/molecules24010185

Shin T-Y, Kim S-H, Kim S-H, et al. Inhibitory effect of mast cell-mediated immediate-type allergic reactions in rats by Perilla frutescens. *Immunopharmacol Immunotoxicol.* 2000;22(3):489–500.

Ueda H, Yamazaki M. Inhibition of tumor necrosis factor-α production by orally administering a Perilla leaf extract. *Biosci Biotechnol Biochem.* 1997;61(8):1292–1295. https://doi.org/10.1271/bbb.61.1292

Wang J, He Y, Yu D, Jin L, Gong X, Zhang B. Perilla oil regulates intestinal microbiota and alleviates insulin resistance through the PI3K/AKT signaling pathway in type-2 diabetic KKAy mice. *Food Chem Toxicol.* 2020;135:110965. https://doi.org/10.1016/j.fct.2019.110965

Wang Z, Tu Z, Xie X, Cui H, Kong KW, Zhang L. Perilla frutescens leaf extract and fractions: Polyphenol composition, antioxidant, enzymes (α-glucosidase, acetylcholinesterase, and tyrosinase) inhibitory, anticancer, and antidiabetic activities. *Foods.* 2021;10:315. https://doi.org/10.3390/foods10020315

Yi LT, Li J, Geng D, et al. Essential oil of *Perilla frutescens*induced change in hippocampal expression of brain-derived neurotrophic factor in chronic unpredictable mild stress in mice. *J Ethnopharmacol.* 2013;147 (1):245–253.

Zhou XJ, Yan LL, Yin PP, Shi LL, Zhang JH, Liu YJ, Ma C. Structural characterisation and antioxidant activity evaluation of phenolic compounds from cold-pressed Perilla frutescens var. arguta seed flour. *Food Chem.* 2014;164:150–157.

Zi Y, Yao M, Lu Z, Lu F, Bie X, Zhang C, Zhao H. Glycoglycerolipids from the leaves of Perilla frutescens (L.) Britton (Labiatae) and their anti-inflammatory activities in lipopolysaccharide-stimulated RAW264.7 cells. *Phytochemistry.* 2021;184:112679.

14 Beneficial Role of Safflower (*Carthamus tinctorius* L.) against Diabetes

Raya Algonaiman and Hassan Barakat

CONTENTS

14.1 INTRODUCTION

Carthamus tinctorius, known as safflower in English, is an old crop grown over 4,000 years ago in many countries. It is characterized as a bushy or thistle-like herbaceous plant with several branches ending with yellowish-reddish flowers in a globular structure (Lee et al., 2021; Singh & Nimbkar, 2006; Tanwar & Goyal, 2021). In some regions, it is known as false saffron (*Crocus sativus*), as the dried florets look similar to saffron but differ in price and aroma. Saffron is a highly expensive spice due to the limited crop yield; one kilogram of saffron can be obtained from 150,000 flowers. In other words, only 8% of the total flower yield is used to produce saffron spice. In the case of *C. tinctorius*, the raw material is the whole flower (Adamska & Biernacka, 2021).

The chemical structure of *C. tinctorius* consists of different phytochemical constituents such as flavonoids, alkaloids, steroids, and organic acids. Among flavonoids, serotonin derivatives were first identified in *C. tinctorius* and are known for playing several health-promoting activities (Adamska & Biernacka, 2021; Asgarpanah & Kazemivash, 2013; Kang et al., 2009; L.-L. Zhang et al., 2016).

Since ancient times, *C. tinctorius* has been used in traditional medicine; for instance, in Chinese traditional medicine, *C. tinctorius* oil has been used for treating amenorrhea and gastric tumors and healing wounds. In other cultures, it has been used to treat various health issues such as arthritis and rheumatism, relieve and sedate menstrual pain, and improve fertility (Dajue & Mündel, 1996;

DOI: 10.1201/b23347-14

Delshad et al., 2018). Such traditional uses have raised the researchers' interest in looking toward *C. tinctorius*'s potential health benefits. Many researchers have reported different pharmacological effects of *C. tinctorius* both *in vivo* and *in vitro. C. tinctorius* seed extracts or its oil can significantly exert several health-promoting effects against many diseases such as cardiovascular diseases, osteoporosis, cancer, brain diseases, and diabetes (Delshad et al., 2018; L.-L. Zhang et al., 2016).

Antidiabetic activities of *C. tinctorius* have been shown in human and animal studies. The administration of *C. tinctorius* seed extracts or its oil could promote beneficial effects in the homeostasis of glucose and insulin levels. *C. tinctorius* interventions could prevent or ameliorate several diabetic complications, such as promoting hepatoprotective and nephroprotective effects, enhancing dyslipidemia, and accelerating diabetic wound healing. Furthermore, *C. tinctorius* have been shown *in vitro* to play roles in different pathophysiological mechanisms involved in treating and managing diabetes. In this chapter, the role of *C. tinctorius* in the prevention and/or management of diabetes will be discussed in depth.

14.2 BOTANICAL DESCRIPTION

Safflower (*C. tinctorius)* belongs to the family Asteraceae, an annual herbaceous plant mainly grown in dry climates. It can tolerate challenging conditions such as drought and salinity. *C. tinctorius* is characterized as bushy or thistle-like, with several branches classified into three sections: primary, secondary, and tertiary (Lee et al., 2021; Singh & Nimbkar, 2006). Each branch ends with a flower in a globular structure called a "capitulum" (Figure 14.1). The main stem and branches are surrounded by leaves with numerous little spines. Branches can reach 100 to 130 cm (Adamska & Biernacka, 2021). Under the ground, the roots can elongate to 200 to 300 cm. This deep root system allows the plant to extract water and nutrients from the deeper soil layers compared to many other crops. Therefore, it is considered an ideal rainfed crop (Singh & Nimbkar, 2006).

C. tinctorius seeds (Figure 14.2) are characterized by a shiny and smooth structure, four-sided, with a thick pericarp colored white, may or may not have pappus (tufts of hair presented on the top of the seed), and each seed weighs 0.01 to 0.1 g (Singh & Nimbkar, 2006).

The flowering period of *C. tinctorius* lasts for a month. The capitulum of the primary branches flowers first, followed by the secondaries and the tertiaries. Each capitulum produces many flowers

FIGURE 14.1 The safflower (*C. tinctorius*) plant shows the different branches ending with the flower capitulum.

FIGURE 14.2 Safflower (*C. tinctorius*) seeds (A) and safflower dried flower petals (B).

ranging from 25 to 250; each flower consists of petals attached to the flower tube called the "corolla tube" (Figure 14.1). In the bloom period, flowers are colored either white, yellow, or orange. The most common ones are those colored yellow in bloom and red in drying. After drying (Figure 14.2), the yellow flowers turn red or stay yellow, the orange turns dark red, and the white stays white. At the end of the flowering period, *C. tinctorius* attains maturity in 30 to 35 days (Singh & Nimbkar, 2006).

14.3 DISTRIBUTION

C. tinctorius can be found in many different regions worldwide. It was grown over 4,000 years ago in the areas from China to the Mediterranean countries and all over the Nile border countries, including Egypt and up to Ethiopia (Singh & Nimbkar, 2006). The most reported countries for producing *C. tinctorius* are located in six main regions: (*i*) countries of the Far East, including Russia and China, (*ii*) South, Central, and Western Asian countries, including India, Pakistan, Kazakhstan, Uzbekistan, Tajikistan, Iran, and Turkey, (*iii*) African countries including Ethiopia, (*iv*) European countries including Spain, (*v*) North and South American countries including the United States, Mexico, Canada, and Argentina, and lastly (*vi*) Australia (Singh & Nimbkar, 2006; Tanwar & Goyal, 2021).

Some of the reported countries may not be a center of origin for the *C. tinctorius* but commercially produce *C. tinctorius* in huge amounts. For example, it was introduced to the United States in 1925, and commercial production started in the 1950s (Adamska & Biernacka, 2021; Nazir et al., 2021). The largest production of *C. tinctorius* seeds reported in 2019, accounting for 200,000, 88,000, and 81,000 tons, were commercially produced in Kazakhstan, the United States, and Russia, respectively. Many other countries, such as China, India, and Argentina, produce *C. tinctorius* seeds in amounts ranging from 50,000 to 3,000 tons (ÖZÇINAR, 2021).

14.4 PHYTOCHEMICAL CONSTITUENTS

The phytochemical constituents of *C. tinctorius* have been well investigated. The most reported isolated components from *C. tinctorius* are flavonoids, alkaloids, steroids, lignanoids, polyacetylenes, phenylethanoid glycosides, coumarins, organic acids, and polysaccharides (Adamska & Biernacka, 2021; Asgarpanah & Kazemivash, 2013; L.-L. Zhang et al., 2016). The most active constituents in the *C. tinctorius* plant are considered to be flavonoids (L.-L. Zhang et al., 2016). Flavonoids of the C-glucosylquinochalcone group represent most of the colorful pigments in the *C. tinctorius* petals (Adamska & Biernacka, 2021). Carthamine (also known as safflower yellow or carthamus red) and carthamidin (known as carthamic acid) are most of the known flavonoids in *C. tinctorius*. Carthamine generates the red color and compromises about 3%–6% of petal composition, whereas carthamidin generates a yellowish color and compromises 24%–30% of the total *C. tinctorius* chemical constituents. Other common flavonoids isolated from *C. tinctorius* are hydroxysafflor yellow A (HYA or HSYA), kaempferol, and quercetin. HYA has been widely investigated both *in vivo* and *in vitro* (Adamska & Biernacka, 2021; Asgarpanah & Kazemivash, 2013). Serotonin derivatives such as feruloylserotonin and 4-coumaroylserotonin are also present in *C. tinctorius*. The first serotonin derivatives identification was found in *C. tinctorius* (Kang et al., 2009).

14.5 PHARMACOLOGICAL STUDIES

Over the last two decades, *C. tinctorius* has exerted several pharmacological effects. *C. tinctorius* florets, oil, and seed extracts are rich in bioactive components that have been confirmed in several *in vivo* and *in vitro* studies to promote various pharmacological effects (Figure 14.3).

14.5.1 Antioxidant Effects

High phenolic compounds are present in *C. tinctorius*, increasing its antioxidant capacity. Serotonin derivatives represent the major phenolic and bioactive components of *C. tinctorius* and are known for their strong antioxidant capacity. Indeed, several *in vivo* and *in vitro* studies have demonstrated strong antioxidant activities of *C. tinctorius* (Lee et al., 2020; Nimrouzi et al., 2020; Yu et al., 2013). *C. tinctorius* aqueous extract showed scavenging radical activity on 2,2-diphenyl-1-picrylhydrazyl

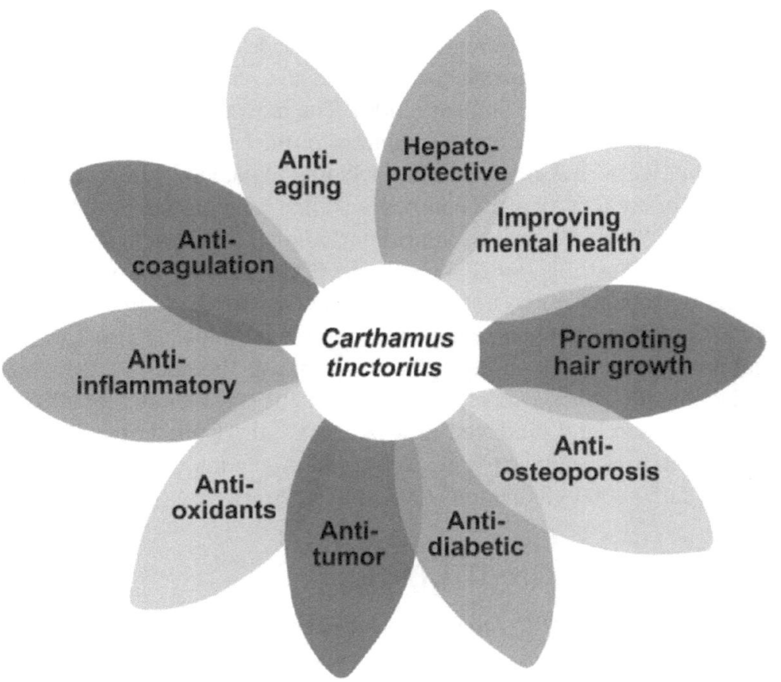

FIGURE 14.3 Pharmacological effects promoted by safflower (*C. tinctorius*) in both *in vivo* and *in vitro* studies.

(DPPH) by almost 97% (Gautam et al., 2014). Furthermore, consuming *C. tinctorius* tea significantly improved oxidative stress markers in post-menopausal women (Cho et al., 2011).

14.5.2 Anti-Inflammatory Effects

In lipopolysaccharide (LPS)-induced acute respiratory distress syndrome in mice, HYA treatment suppressed the expression of some pro-inflammatory cytokines such as tumor necrosis factor-alpha (TNF-α) and interleukin 1 beta (IL)-1β, as well as promoted the blocking of TLR-4/NF-κB signaling pathway (Y. Zhang et al., 2017). Consistent results were shown in LPS-induced inflammatory response in murine macrophage cells (RAW 264.7) treated with *C. tinctorius* extract; results showed inhibitory effects on the production of nitric oxide (NO), prostaglandin E2 (PGE2), and IL-1β (C. C. Wang et al., 2011). In human fetal lung fibroblast cells (MRC-5) induced with TNF-α, the HYA treatment promoted anti-inflammatory effects through binding with the TNF-α receptor (TNFR1), thus, suppressing its activity (S. Liu et al., 2019). Similarly, in LPS-induced macrophage cells (RAW 264.7), *C. tinctorius* showed inhibitory effects on the expression of the pro-inflammatory cytokines. Such results were attributed to *C. tinctorius* serotonin derivatives (D.-H. Kim et al., 2015).

14.5.3 Anti-Tumor Effects

Several *in vivo* and *in vitro* studies demonstrated the efficiency of *C. tinctorius* in promoting anti-tumor activities (Gautam et al., 2014; Mani et al., 2020; Tu et al., 2015). HYA was shown to inhibit the proliferation and migration of hepatocellular carcinoma and stimulate its apoptosis (J. Zhang et al., 2019). Similarly, HYB (Qu et al., 2019), as well as *C. tinctorius* polysaccharide (Luo et al., 2015), showed antiproliferative effects on human breast cancer cells (MCF-7). Consistently, in an *in vitro* experiment, safflower yellow inhibited breast cancer migration into the lungs (Fu et al., 2016). Furthermore, HYA could suppress tumor growth by inhibiting the secretion of angiogenesis factors (Yang et al., 2015). The possible mechanisms of such anti-tumor effects of *C. tinctorius* were related to regulating gene expressions involved in the intrinsic pathway of apoptosis (Tu et al., 2015).

14.5.4 Anti-Osteoporosis Effects

C. tinctorius seed extract was shown to prevent bone loss in mouse pre-osteoblast cells (MC3T3-E1) (Jang et al., 2007). Preventive effects on bone loss were also observed in ovariectomy-induced osteoporosis in rats treated with *C. tinctorius* seed oil (Alam et al., 2006). In addition, in glucocorticoid-induced osteoporosis in zebrafish, treatment with HYA promoted bone mineralization, osteoblast viability, and bone collagen expression and inhibited bone resorption (L. Liu et al., 2018). *C. tinctorius* was shown to exert anti-osteoporosis effects by inhibiting the activity of matrix metalloproteinases (MMPs), key enzymes involved in the progression of osteoarthritis through promoting cartilage degradation. The inhibition of MMPs was mediated by *C. tinctorius* lowering effects on the expression of some pro-inflammatory cytokines involved in activating signaling pathways that plays a role in increasing the expression of MMPs, and therefore, promoting osteoarthritis development (Han et al., 2021).

14.5.5 Cognitive- and Memory-Enhancing Effects

In scopolamine-induced memory impairment in mice, *C. tinctorius* seed extract showed inhibitory effects on cholinergic dysfunction, indicating promising results in protecting against Alzheimer's disease (J. H. Kim et al., 2019). Indeed, in Alzheimer's disease rodent models, safflower yellow showed improvements in cognitive functions (Shi et al., 2018; L. Zhang et al., 2019). In addition, HYA attenuated brain injury induced by lymphostatic encephalopathy in rat models (Pan et al., 2012). The possible mechanisms beyond these protective effects are related to antioxidative stress activities and effects on upregulating the expression of some key enzymes involved in various cellular processes, as well as the impact on decreasing cell apoptosis and structural damage of nervous tissues (Tu et al., 2015).

14.5.6 Other Effects

Hepatoprotective effects were observed in carbon tetrachloride (CCl_4)-induced liver damage in rats treated with HYA or carthamus red. Such results could be mediated by enhancing the antioxidant defense system and targeting the fibrogenic pathways (Tu et al., 2015). Anti-coagulation effects of HYA were also reported; in rats induced with focal cerebral ischemia, the injection of HYA showed enhancements by promoting inhibitory effects on thrombosis formation and platelet aggregation (Zhu et al., 2005). Consistently, in rats induced with cerebral infarction, the intranasal administration of *C. tinctorius* extracts showed therapeutic activities via improving blood circulation in the central nervous system (Y. Wang et al., 2020).

Moreover, *C. tinctorius* was shown to prevent skin damage in mice exposed to ultraviolet (UV) irradiation; the topical application of HYA promoted significant skin damage recovery mediated by HYA effects in promoting endogenous collagen synthesis (Kong et al., 2013). In addition, in D-galactose (D-gal)-induced aging in mice, the injection of HYA promoted anti-aging effects by suppressing the p16/Rb pathway, which is involved in stimulating cellular senescence by slowing down cell cycling (Min et al., 2020).

On the other hand, *C. tinctorius* was shown to exert hair growth-promoting effects by inhibiting 5α-reductase activity, an enzyme involved in the progression of scalp hair loss (Kumar et al., 2012). The topical application of *C. tinctorius* extract on hair-shaved skin of female C57BL/6 mice promoted hair growth. The possible mechanisms were mediated by significant stimulating activities of *C. tinctorius* on hair growth-promoting genes and suppressing the expression of hair loss-related genes (Junlatat & Sripanidkulchai, 2014).

14.6 ANTIDIABETIC RESPONSE

Besides the different pharmacological activities demonstrated for *C. tinctorius*, several studies *in vivo* and *in vitro* have investigated different *C. tinctorius* applications and strongly showed antidiabetic effects.

The consumption of *C. tinctorius* oil in obese diabetic post-menopausal women at a dose of 8 g/day for 16 weeks resulted in a significant reduction in fasting blood glucose (FBG) (Norris et al., 2009) and glycosylated hemoglobin (HbA_{1c}) and an increase in high-density lipoprotein (HDL) cholesterol levels (Asp et al., 2011). *C. tinctorius* oil at an equivalent dose (8 g/day) given to patients diagnosed with metabolic syndrome for 12 weeks showed a significant reduction in FBG levels and improved insulin resistance compared to a placebo group. In addition, abdominal obesity was commonly observed among those metabolic syndrome patients at the baseline. After the intervention, it was significantly reduced (Ruyvaran et al., 2022).

Furthermore, in a systematic review and meta-analysis, safflower yellow has shown significant efficiency in improving or preventing diabetic nephropathy, which is one of the most severe diabetic complications and a leading cause of death. Multiple factors related to diabetic nephropathy, including FBG, urinary albumin excretion rate, blood urea nitrogen, and high-sensitivity C-reactive protein, have been significantly improved due to interventions of safflower yellow (X. Wang et al., 2019).

In animal models, *C. tinctorius* oil was administered orally to alloxan-induced diabetic rats at a 200 mg/kg BW dose for 28–30 days. The results showed that *C. tinctorius* oil could reverse multiple metabolic abnormalities that occurred due to alloxan injection, including FBG and serum lipids (Rahimi et al., 2014; Rahimi et al., 2015). Rahimi et al. (2014) also reported hepatoprotective effects based on the improvements in liver function, including serum enzyme levels of alanine aminotransferase (ALT), aspartate aminotransferase (AST), and alkaline phosphatase (ALP). Consistently, an extract of *C. tinctorius* was orally administered to alloxan-induced diabetic rats at 200 mg/kg BW for six weeks. The results showed significant improvements in FBG, total cholesterol, and low-density lipoprotein (LDL) cholesterol levels, whereas insulin levels increased compared to diabetic rats receiving no treatment. In addition, considerable hepatoprotective effects were also observed, which indicates that *C. tinctorius* therapy could play a strong beneficial role in treating diabetic complications (Asgary et al., 2012).

Similarly, in streptozotocin (STZ)-induced diabetic rabbits, orally administered with *C. tinctorius* extract at an equivalent dose (200 mg/kg BW) for 30 days, a significant reduction in FBG levels and an increase in insulin levels have been shown (Qazi et al., 2014). In another study, fructose-induced metabolic syndrome in rats was treated by oral administration of either *C. tinctorius* seed extract or its oil at different doses for eight weeks; both treatments, in a dose-dependent manner, significantly reversed the abnormalities that occurred as a result of fructose administration, including levels of FBG, serum lipids, and insulin (Nimrouzi et al., 2020). The leaves extract of *C. tinctorius* was also investigated in STZ-induced diabetic rats. The oral administration at a dose of 30 mg/kg BW showed similar positive results. Serum glucose levels decreased gradually from the onset of administration up to the seventh hour. The results showed a significant difference after the fifth and seventh hours compared to the non-treated diabetic rats (Mahadevappa & Kuppast, 2009).

Furthermore, in STZ-induced diabetic rats injected with *C. tinctorius* extract in combination with aceglutamide for 21 days, the results showed that this treatment significantly reversed the decrease in the corneal nerve fibers that occurred as a result of diabetic peripheral neuropathy, indicating that it could strongly prevent nerve damage associated with diabetes (D. Li et al., 2017). Another study investigated the efficiency of HYA in accelerating wound healing in STZ-induced diabetic rats; it was found that the local application of HYA solution on wound sites significantly improved the healing process after 14 days, and the wound had completely recovered after 17 days. The standard treatment showed complete recovery after 30 days (Gao et al., 2018). This highly indicates the potential efficiency of HYA as a therapeutic intervention in chronic non-healing diabetic foot ulcers.

14.6.1 Mechanisms of Action

From the perspective of pathophysiology, several mechanisms mediated by *C. tinctorius* can promote significant antidiabetic effects through inhibiting some key digestive enzymes and activating major signaling pathways, as well as anti-inflammatory and antioxidative stress effects (Figure 14.4).

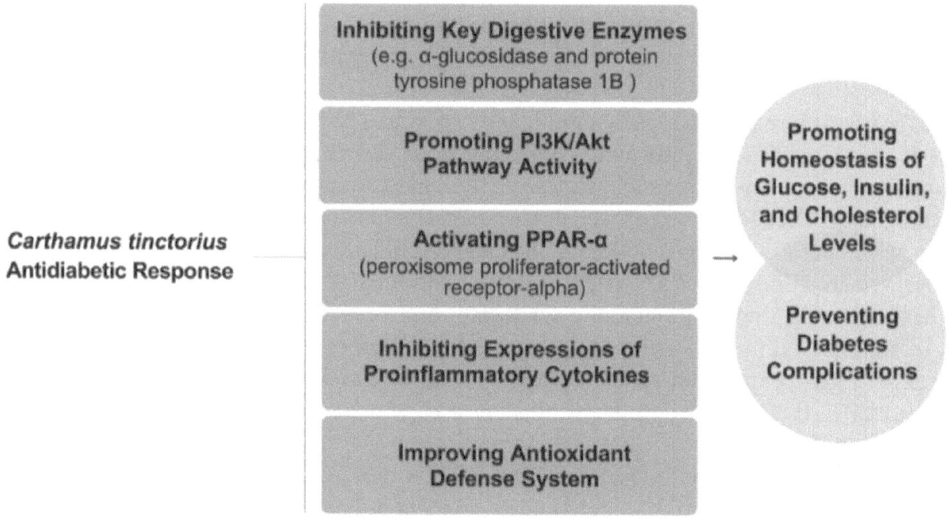

FIGURE 14.4 Safflower (*C. tinctorius*) mechanism of action against diabetes.

The beneficial effects of *C. tinctorius* in lowering glucose concentrations have been attributed to its inhibitory effects on α-glucosidase (Salehi et al., 2019), which is an exoenzyme located in the brush border of the intestine enterocytes lining and responsible for regulating the availability of glucose for intestinal absorption as well as the extent of postprandial hyperglycemia (Bedekar et al., 2010). Inhibitory effects of α-glucosidase are related to the high content of serotonin derivatives in *C. tinctorius*, which are reported to play roles in inhibiting α-glucosidase (Salehi et al., 2019; Takahashi & Miyazawa, 2012). In addition, significant inhibition in the activity of protein tyrosine phosphatase 1B (PTP1B) has been reported *in vitro* after injecting an *n*-hexane *C. tinctorius* essential oil extract. PTP1B is a group of enzymes negatively regulating the insulin signaling pathway (L. Li et al., 2012).

Furthermore, enhancements in FBG and serum lipids, as well as in insulin resistance in HFD- and STZ-induced type 2 diabetic rats, have been related to direct or indirect effects of HYA in promoting the activity of PI3K/Akt (phosphatidylinositol-3-kinase and protein kinase B), which is a well-known intracellular signaling pathway that plays a major role in regulating the homeostasis of glucose and lipid metabolism. The insulin secretion from β-cells is promoted by activating the PI3K/Akt pathway (Huang et al., 2018). Other studies indicated that *C. tinctorius* treatment could activate peroxisome proliferator-activated receptor (PPAR)-alpha (Guo et al., 2013; Hsu & Huang, 2006). It could be considered a transcription factor that controls gene expression of several pathways, including glucose and lipid metabolic pathways (Bougarne et al., 2018).

From another point of view, inflammatory responses can play key roles in the progression of diabetes. Inflammatory responses are upregulated by pro-inflammatory cytokines, mainly produced by activated macrophages. Overexpression or overproduction of these cytokines can result in significant adverse effects. TNF-α and IL-1β are two of the most well-known pro-inflammatory cytokines (Khan et al., 2018). The overexpression of both TNF-α and IL-1β was reduced under hyperglycemic stress in macrophage cells (RAW 264.7) after the injection of two of the most key components in *C. tinctorius*, kaempferol and HYA, individually. In addition, the treatment prevented podocyte apoptosis, which occurs due to hyperglycemic stress (Y. Li et al., 2020). The overexpression of IL-1β is well evidenced to play a major role in the loss of beta-cell mass in type 1 and 2 diabetes. Elevated IL-1β expression was observed in the pancreas, insulin, and blood circulation of type 2 diabetic patients (Aharon-Hananel et al., 2015; Dinarello et al., 2010). In addition, IL-1β was reported to play a significant role in delayed wound healing in diabetic patients (Dai et al., 2021). As for TNF-α, its overexpression can promote dyslipidemia and insulin resistance (J. Kim & Bajaj, 2014). Therefore, the beneficial effects of *C. tinctorius* in enhancing insulin resistance and glucose concentrations are highly attributed to the lowering effects of both TNF-α and IL-1β expressions. Decreased expression of TNF-α was also observed in type 2 diabetic animal models treated with HYA (Lee et al., 2020). The model was established by a high-fat diet (HFD) for four weeks, followed by streptozotocin injection, then treated with HYA (120 mg/kg BW) by oral administration for a further eight weeks. The results significantly showed a reduction in TNF-α and some inflammatory products such as free fatty acids (FFA) and lactic dehydrogenase (LDH). These effects were positively related to enhancements in levels of FBG and serum lipids, including triglyceride, total cholesterol, and LDL cholesterol, as well as to enhancements in kidney functions based on improvements in serum creatinine and blood urea nitrogen. Significant reduction in TNF-α expression was also observed by Nimrouzi et al. (2020) after the administration of *C. tinctorius* seed extract or its oil to the fructose-induced metabolic syndrome in rats correlated with enhancements in serum lipids. In addition, Nimrouzi et al. (2020) reported that such enhancements were also mediated by an increase in gene expression of CD36 (cluster of differentiation 36), fatty acyl-CoA synthetase (ACS), and CPT-1β (carnitine palmitoyl transferase I), which are involved in controlling beta-oxidation capacity of the liver.

Furthermore, the beneficial effects of *C. tinctorius* can also relate to its antioxidative stress mechanisms; oxidative stress is a widely known concept defined by the excess production of reactive oxygen species, known as free radicals (Shankar & Mehendale, 2014). Studies have shown that

C. tinctorius can significantly improve antioxidative stress markers responsible for eliminating the production of free radicals, such as superoxide dismutase (SOD), reduced glutathione peroxidase (GSH), and malondialdehyde (MDA) (Lee et al., 2020; Nimrouzi et al., 2020). *C. tinctorius* oil is rich in tocopherol, mainly alpha-tocopherol, which is known for its antioxidant activity through scavenging free radicals and increasing the body's antioxidant defense (Tucker & Townsend, 2005). In addition, serotonin derivatives in *C. tinctorius* are also known to play significant antioxidant activities by preventing lipid peroxidation (Cho et al., 2011). Lipid peroxidation is a chain of reactions that dysregulates the body's lipids and is mediated by free radicals, leading to oxidative deterioration (Fuchs et al., 2014). Moreover, it was reported that *C. tinctorius* could enhance the homeostasis of trace elements, such as zinc, manganese, iron, and copper (Nimrouzi et al., 2020). These elements play roles in improving the antioxidant defense system and glucose and lipid metabolism (Pizent et al., 2010).

14.7 TRADITIONAL AND OTHER POTENTIAL USES

Different parts of *C. tinctorius* were applied in traditional medicine over 4,000 years ago. In various cultures, the decoction of the seeds was known as a laxative remedy, relieving and sedating menstrual pain, improving fertility, and being an anti-inflammatory agent. In other cultures, such as Central Asian countries, it has been used to flush urinary tracts and improve liver ailments. The liver and heart ailments were also treated using *C. tinctorius* oil in traditional Iranian medicine. In India, the oil has been used to treat various health issues such as sores, rheumatism, scabies, arthritis, and mastalgia, whereas the decoction was used to prevent abortion. In China, *C. tinctorius* oil has been traditionally used for treating amenorrhea, gastric tumors, and healing wounds. *C. tinctorius* materials were also used as an antidote for poisons and curing fever (Dajue & Mündel, 1996; Delshad et al., 2018).

Besides using *C. tinctorius* in traditional medicine, *C. tinctorius* was mainly grown since ancient times for its colorful florets for extracting its dye. In Egyptian culture, *C. tinctorius* dye was used to color cotton, silk, and ceremonial ointments. *C. tinctorius* seeds and wreaths of the florets were found with 4,000-year-old mummies. *C. tinctorius* dyes were introduced into European countries in the 18th century and were used for coloring cheese. They had also used the dye in carpet-weaving industries. The application of *C. tinctorius* dye in such sectors has vanished since the 19th century due to the development of much cheaper synthetic dyes (Dajue & Mündel, 1996; Delshad et al., 2018). The *C. tinctorius* dye has also been applied in food preparations. Recently, it has been used as a substitute for saffron, one of the most expensive spices worldwide, to color rice, soups, sauces, and many other recipes yellow to bright orange. In different cultures, the *C. tinctorius* dye was used for coloring hair and manufacturing cosmetics such as eyeliners and lipsticks (Dajue & Mündel, 1996; Delshad et al., 2018; Nazir et al., 2021).

Besides using the florets in food preparations as a coloring agent, the Chinese folks used the florets as herbal tea. Although *C. tinctorius* has a bitter herbal taste, the Chinese have developed a non-bitter, sweet-smelling *C. tinctorius* tea (Dajue & Mündel, 1996).

14.8 SAFETY ISSUES

No adverse effects have been reported with the proper administration of biologically active substances of *C. tinctorius*; doses up to 2,000 mg/kg BW showed nontoxic effects in an *in vivo* acute toxicity study. However, in long-term interventions of a 200 mg/kg BW dose for 35 days, *C. tinctorius* showed some adverse effects such as allergic reactions, spermatogenetic failure, fatty liver, pharyngitis, nosebleeds, and nephrotoxicity (Adamska & Biernacka, 2021; Tu et al., 2015). In addition, hepatotoxicity was reported in a case report of three patients with a history of consuming *C. tinctorius* oil for weight loss; they have been diagnosed with acute liver failure requiring liver transplantation for treatment. Other possible reasons for such diagnosis (e.g., viral hepatitis, autoimmune diseases, or any other drug usage) were excluded. Thus, *C. tinctorius* oil was the only possible triggering factor (de Ataide et al., 2018). Moreover, *C. tinctorius* has stimulatory effects on uterine contraction, thus should not be consumed during pregnancy. Therefore, *C. tinctorius* should be used cautiously; the average daily dose of *C. tinctorius* florets infusion or decoction ranges from 3 to 9 g (Al-Snafi, 2015).

14.9 CULTIVATION PRACTICES

C. tinctorius is a multiple-purpose crop; since ancient times, it has been cultivated for its seeds, meals, and florets. The seeds were mainly used to extract a high-quality oil in different non-food industries, such as manufacturing soaps, waterproof leather buckets, and the paint and varnish industry. However, with developing technologies in recent times, the presence of cheaper petroleum products, and the shift to water-based paints, the application of *C. tinctorius* oil in such industries have vanished (Singh & Nimbkar, 2006). Recently, *C. tinctorius* oil has been chosen as an excellent edible oil; it is considered highly nutritious similar to olive oil (Singh & Nimbkar, 2006). Linoleic acid compromises about 70% of the total produced oil. It also contains 10% oleic acid and small amounts of stearic acid (Asgarpanah & Kazemivash, 2013; Delshad et al., 2018). *C. tinctorius* oil has also been applied to many pharmaceutical and cosmetic products due to its high quality (ÖZÇINAR, 2021).

C. tinctorius seeds have also been cultivated as birdseed due to their high-grade vegetable protein (Asgarpanah & Kazemivash, 2013). The meals, which are the residue of the oil extraction process, are rich in protein and fiber and can be used for poultry and cattle feed (Abrham et al., 2018). Furthermore, colorful flowers were cultivated in ancient times for dye extraction in the food and fabric industries. However, since the 20th century, the application of *C. tinctorius* dyes as an edible dye has vanished due to cheaper synthetic dyes such as aniline (Y. Liu et al., 2022; Nazir et al., 2021; Singh & Nimbkar, 2006). Nowadays, flowers are cultivated for pharmacological and traditional applications (Nazir et al., 2021).

14.10 FUTURE REMARKS

Applying C. *tinctorius* florets as a decoction or using C. *tinctorius* oil in food preparations can play a role in preventing or managing diabetes. Studies have proven the beneficial effects of *C. tinctorius* both *in vivo* and *in vitro*. In future approaches, systemic information on the pharmacokinetics of *C. tinctorius* edible parts and estimates of its target-organ toxicity are not well understood enough. As a result, an additional cutting-edge study must be focused on in the future to evaluate its potential therapeutic effects. Further research is needed to understand its bioactive components and their mechanism of action using gene-expression studies in depth. Research suggests that *C. tinctorius* is not harmful when consumed orally. However, due to a lack of research, there are no studies about the interventions with the gut consortium. Conversely, *C. tinctorius* has not yet been extensively studied *in vivo* with any drug delivery nanocarriers. Therefore, investigations examining its encapsulation and therapeutical efficacy in the encapsulated form should be prioritized to enhance the therapeutic benefits.

REFERENCES

Abrham, Z., Awuk, A., Teshager, N., & Wondifraw, Z. (2018). Effects of substituting safflower (*Carthamus tinctorius*) meal with soya bean meal on the performance of SASSO X RIR crossbred chicken. *Poultry, Fisheries & Wildlife Sciences*, *6*(193), 2.

Adamska, I., & Biernacka, P. (2021). Bioactive substances in safflower flowers and their applicability in medicine and health-promoting foods. *International Journal of Food Science*, *2021*.

Aharon-Hananel, G., Jörns, A., Lenzen, S., Raz, I., & Weksler-Zangen, S. (2015). Antidiabetic effect of interleukin-1β antibody therapy through β-Cell protection in the cohen diabetes-sensitive rat. *Diabetes*, *64*(5), 1780–1785.

Alam, M. R., Kim, S. M., Lee, J. I., Chon, S. K., Choi, S. J., Choi, I. H., & Kim, N. S. (2006). Effects of safflower seed oil in osteoporosis induced-ovariectomized rats. *The American Journal of Chinese Medicine*, *34*(4), 601–612.

Al-Snafi, A. E. (2015). The chemical constituents and pharmacological importance of Carthamus tinctorius-An overview. *Journal of Pharmaceutical Biology*, *5*(3), 143–166.

Asgarpanah, J., & Kazemivash, N. (2013). Phytochemistry, pharmacology and medicinal properties of *Carthamus tinctorius* L. *Chinese Journal of Integrative Medicine*, *19*(2), 153–159.

Asgary, S., Rahimi, P., Mahzouni, P., & Madani, H. (2012). Antidiabetic effect of hydroalcoholic extract of *Carthamus tinctorius* L. in alloxan-induced diabetic rats. *Journal of Research in Medical Sciences: The Official Journal of Isfahan University of Medical Sciences*, *17*(4), 386.

Asp, M. L., Collene, A. L., Norris, L. E., Cole, R. M., Stout, M. B., Tang, S.-Y., . . . Belury, M. A. (2011). Time-dependent effects of safflower oil to improve glycemia, inflammation and blood lipids in obese, post-menopausal women with type 2 diabetes: A randomized, double-masked, crossover study. *Clinical Nutrition*, *30*(4), 443–449.

Bedekar, A., Shah, K., & Koffas, M. (2010). Chapter 2 – Natural products for type II diabetes treatment. In *Advances in Applied Microbiology* (Vol. 71, pp. 21–73). Academic Press.

Bougarne, N., Weyers, B., Desmet, S. J., Deckers, J., Ray, D. W., Staels, B., & De Bosscher, K. (2018). Molecular actions of PPAR α in lipid metabolism and inflammation. *Endocrine Reviews*, *39*(5), 760–802.

Cho, S.-H., Jang, J.-H., Yoon, J. Y., Han, C.-D., Choi, Y., & Choi, S.-W. (2011). Effects of a safflower tea supplement on antioxidative status and bone markers in postmenopausal women. *Nutrition Research and Practice*, *5*(1), 20–27.

Dai, J., Shen, J., Chai, Y., & Chen, H. (2021). IL-1β impaired diabetic wound healing by regulating MMP-2 and MMP-9 through the p38 pathway. *Mediators of Inflammation*, *2021*.

Dajue, L., & Mündel, H.-H. (1996). *Safflower, Carthamus tinctorius L* (Vol. 7). Bioversity International.

de Ataide, E. C., Reges Perales, S., de Oliveira Peres, M. A., Bastos Eloy da Costa, L., Quarella, F., Valerini, F. G., Chueiri Neto, F., Silveira Bello Stucchi, R., & de Fátima Santana Ferreira Boin, I. (2018). Acute liver failure induced by Carthamus tinctorius oil: Case reports and literature review. *Transplantation Proceedings*, *50*(2), 476–477.

Delshad, E., Yousefi, M., Sasannezhad, P., Rakhshandeh, H., & Ayati, Z. (2018). Medical uses of *Carthamus tinctorius* L. (safflower): A comprehensive review from traditional medicine to modern medicine. *Electronic Physician*, *10*(4), 6672.

Dinarello, C. A., Donath, M. Y., & Mandrup-Poulsen, T. (2010). Role of IL-1β in type 2 diabetes. *Current Opinion in Endocrinology, Diabetes and Obesity*, *17*(4), 314–321. https://doi.org/10.1097/MED.0b013e32833bf6dc

Fu, H., Wu, R., Li, Y., Zhang, L., Tang, X., Tu, J., . . . Shou, Q. (2016). Safflower yellow prevents pulmonary metastasis of breast cancer by inhibiting tumor cell invadopodia. *The American Journal of Chinese Medicine*, *44*(7), 1491–1506.

Fuchs, P., Perez-Pinzon, M. A., & Dave, K. R. (2014). Chapter 2 – Cerebral ischemia in diabetics and oxidative stress. In V. R. Preedy (Ed.), *Diabetes: Oxidative Stress and Dietary Antioxidants* (pp. 15–23). San Diego: Academic Press.

Gao, S.-Q., Chang, C., Niu, X.-Q., Li, L.-J., Zhang, Y., & Gao, J.-Q. (2018). Topical application of hydroxysafflor yellow A accelerates the wound healing in streptozotocin induced T1DM rats. *European Journal of Pharmacology*, *823*, 72–78.

Gautam, S., Bhagyawant, S. S., & Srivastava, N. (2014). Detailed study on therapeutic properties, uses and pharmacological applications of safflower (*Carthamus tinctorius* L.). *International Journal of Ayurveda and Pharma Research*, *2*(3), 1–12.

Guo, J. X., Wang, C. L., & Wu, Z. J. (2013). Study on hypolipidemic activities of safflower seed oil. In *Applied Mechanics and Materials* (Vol. 275, pp. 1640–1645). Trans Tech Publications Ltd.

Han, S. J., Lim, M. J., Lee, K. M., Oh, E., Shin, Y. S., Kim, S., . . . Kang, L.-J. (2021). Safflower seed extract attenuates the development of osteoarthritis by blocking NF-κB signaling. *Pharmaceuticals*, *14*(3), 258.

Hsu, S.-C., & Huang, C.-J. (2006). Reduced fat mass in rats fed a high oleic acid–rich safflower oil diet is associated with changes in expression of hepatic PPARα and adipose SREBP-1c–regulated genes. *The Journal of Nutrition*, *136*(7), 1779–1785. https://doi.org/10.1093/jn/136.7.1779

Huang, X., Liu, G., Guo, J., & Su, Z. (2018). The PI3K/AKT pathway in obesity and type 2 diabetes. *International Journal of Biological Sciences*, *14*(11), 1483.

Jang, H.-O., Park, Y.-S., Lee, J.-H., Seo, J.-B., Koo, K.-I., Jeong, S.-C., . . . Yun, I. (2007). Effect of extracts from safflower seeds on osteoblast differentiation and intracellular calcium ion concentration in MC3T3-E1 cells. *Natural Product Research*, *21*(9), 787–797.

Junlatat, J., & Sripanidkulchai, B. (2014). Hair growth-promoting effect of *Carthamus tinctorius* floret extract. *Phytotherapy Research*, *28*(7), 1030–1036.

Kang, K., Park, S., Kim, Y. S., Lee, S., & Back, K. (2009). Biosynthesis and biotechnological production of serotonin derivatives. *Applied Microbiology and Biotechnology*, *83*(1), 27–34.

Khan, H. A., Khan, I., & Lee, Y. K. (2018). Role of immune factors in bioavailability and toxicity of carbon nanomaterials. In *Fullerens, Graphenes and Nanotubes* (pp. 601–630). William Andrew Publishing.

Kim, D.-H., Moon, Y.-S., Park, T.-S., & Son, J.-H. (2015). Serotonins of safflower seeds play a key role in anti-inflammatory effect in lipopolysaccharide-stimulated RAW 264.7 macrophages. *Journal of Plant Biotechnology*, *42*(4), 364–369.

Kim, J. H., & Bajaj, M. (2014). Normal adipose tissue biology: Adipocytokines and inflammation. In L. M. McManus & R. N. Mitchell (Eds.), *Pathobiology of Human Disease* (pp. 488–497). Academic Press. https://doi.org/10.1016/B978-0-12-386456-7.02006-2

Kim, J. H., He, M. T., Kim, M. J., Yang, C. Y., Shin, Y. S., Yokozawa, T., . . . Cho, E. J. (2019). Safflower (*Carthamus tinctorius* L.) seed attenuates memory impairment induced by scopolamine in mice via regulation of cholinergic dysfunction and oxidative stress. *Food & Function*, *10*(6), 3650–3659.

Kong, S.-Z., Shi, X.-G., Feng, X.-X., Li, W.-J., Liu, W.-H., Chen, Z.-W., . . . Zhang, X.-J. (2013). Inhibitory effect of hydroxysafflor yellow a on mouse skin photoaging induced by ultraviolet irradiation. *Rejuvenation research*, *16*(5), 404–413.

Kumar, N., Rungseevijitprapa, W., Narkkhong, N.-A., Suttajit, M., & Chaiyasut, C. (2012). 5α-reductase inhibition and hair growth promotion of some Thai plants traditionally used for hair treatment. *Journal of Ethnopharmacology*, *139*(3), 765–771.

Lee, M., Zhao, H., Liu, X., Liu, D., Chen, J., Li, Z., . . . Deng, Y. (2020). Protective effect of hydroxysafflor yellow A on nephropathy by attenuating oxidative stress and inhibiting apoptosis in induced type 2 diabetes in rat. *Oxidative Medicine and Cellular Longevity*, *2020*.

Lee, Y. J., Lee, Y. P., Seo, C. S., Choi, E. S., Han, B. H., Yoon, J. J., Jang, S. H., Jeong, C. G., Mun, Y. J., & Kang, D. G. (2021). The modulation of Nrf-2/HO-1 signaling axis by *Carthamus tinctorius* L. alleviates vascular inflammation in human umbilical vein endothelial cells. *Plants*, *10*(12), 2795.

Li, D., Chen, Y.-G., Zhang, C.-J., Tian, J., & Li, X. (2017). Safflower extract and aceglutamide injection promoting recovery of peripheral innervations via vascular endothelial growth factor-B signaling in diabetic mice. *Chinese Medical Journal*, *130*(23), 2829–2835.

Li, L., Wang, Q., Yang, Y., Wu, G., Xin, X., & Aisa, H. (2012). Chemical components and antidiabetic activity of essential oils obtained by hydrodistillation and three solvent extraction methods from *Carthamus tinctorius* L. *Acta Chromatographica*, *24*(4), 653–665.

Li, Y., Zheng, D., Shen, D., Zhang, X., Zhao, X., & Liao, H. (2020). Protective effects of two safflower derived compounds, kaempferol and hydroxysafflor yellow A, on hyperglycaemic stress-induced podocyte apoptosis via modulating of macrophage M1/M2 polarization. *Journal of Immunology Research*, *2020*.

Liu, L., Tao, W., Pan, W., Li, L., Yu, Q., Zhang, D., & Jiang, J. (2018). Hydroxysafflor yellow A promoted bone mineralization and inhibited bone resorption which reversed glucocorticoids-induced osteoporosis. *BioMed Research International*, *2018*.

Liu, S., Wang, Y., Wen, H., Sun, X., & Wang, Y. (2019). Hydroxysafflor yellow a inhibits TNF-α-induced inflammation of human fetal lung fibroblasts via NF-κB signaling pathway. *Evidence-based Complementary and Alternative Medicine*, *2019*.

Liu, Y., Wang, M., Cao, Y., Zeng, M., Zhang, Q., Ren, Y., . . . Zheng, X. (2022). Chemical constituents from the flowers of *Carthamus tinctorius* L. and their lung protective activity. *Molecules*, *27*(11), 3573.

Luo, Z., Zeng, H., Ye, Y., Liu, L., Li, S., Zhang, J., & Luo, R. (2015). Safflower polysaccharide inhibits the proliferation and metastasis of MCF-7 breast cancer cell. *Molecular Medicine Reports*, *11*(6), 4611–4616.

Mahadevappa, P., & Kuppast, I. (2009). Antihyperglycemic activity of methanolic extract of *Carthamus tinctorius* L., Annigere-2. *Asian Journal of Experimental Sciences*, *23*(3), 1–6.

Mani, V., Lee, S.-K., Yeo, Y., & Hahn, B.-S. (2020). A metabolic perspective and opportunities in pharmacologically important safflower. *Metabolites*, *10*(6), 253.

Min, F., Sun, H., Wang, B., Ahmad, N., Guo, H., Gao, H., . . . Li, H. (2020). Hepatoprotective effects of hydroxysafflor yellow A in D-galactose-treated aging mice. *European Journal of Pharmacology*, *881*, 173214.

Nazir, M., Arif, S., Ahmed, I., & Khalid, N. (2021). Safflower (*Carthamus tinctorius*) seed. In *Oilseeds: Health Attributes and Food Applications* (pp. 427–453). Singapore: Springer.

Nimrouzi, M., Ruyvaran, M., Zamani, A., Nasiri, K., & Akbari, A. (2020). Oil and extract of safflower seed improve fructose induced metabolic syndrome through modulating the homeostasis of trace elements, TNF-α and fatty acids metabolism. *Journal of Ethnopharmacology*, *254*, 112721.

Norris, L. E., Collene, A. L., Asp, M. L., Hsu, J. C., Liu, L.-F., Richardson, J. R., . . . Jackson, R. D. (2009). Comparison of dietary conjugated linoleic acid with safflower oil on body composition in obese post-menopausal women with type 2 diabetes mellitus. *The American Journal of Clinical Nutrition*, *90*(3), 468–476.

Özçinar, A. B. (2021). Safflower (*Carthamus tinctorius* L.): Utilisation, genetics and agronomy. *MAS Journal of Applied Sciences*, *6*(5), 1130–1136.

Pan, Y., Zheng, D. Y., Liu, S. M., Meng, Y., Xu, H. Y., Zhang, Q., . . . Li, H. Y. (2012). Hydroxysafflor yellow A attenuates lymphostatic encephalopathy-induced brain injury in rats. *Phytotherapy Research*, *26*(10), 1500–1506.

Pizent, A., Pavlovic, M., Jurasovic, J., Dodig, S., Pasalic, D., & Mujagic, R. (2010). Antioxidants, trace elements and metabolic syndrome in elderly subjects. *The Journal of Nutrition, Health & Aging*, *14*(10), 866–871.

Qazi, N., Khan, R. A., Rizwani, G. H., & Feroz, Z. (2014). Effect of *Carthamus tinctorius* (Safflower) on fasting blood glucose and insulin levels in alloxan induced diabetic rabbits. *Pakistan Journal of Pharmaceutical Sciences*, *27*(2), 377–380.

Qu, C., Zhu, W., Dong, K., Pan, Z., Chen, Y., Chen, X., . . . Zheng, Q. (2019). Inhibitory effect of hydroxysafflor yellow B on the proliferation of human breast cancer MCF-7 cells. *Recent Patents on Anti-Cancer Drug Discovery*, *14*(2), 187–197.

Rahimi, P., Asgary, S., & Kabiri, N. (2014). Hepatoprotective and hypolipidemic effects of *Carthamus tinctorius* oil in alloxan-induced type 1 diabetic rats. *Journal of HerbMed Pharmacology*, *3*(2), 107–111.

Rahimi, P., Setorki, M., & Doudi, M. (2015). Effects of safflower oil on FBS and lipid profile in alloxan induced diabetic rats and its mechanism. *Journal of Medicinal Herbs*, *6*(1), 7–12.

Ruyvaran, M., Zamani, A., Mohamadian, A., Zarshenas, M. M., Eftekhari, M. H., Pourahmad, S., . . . Nimrouzi, M. (2022). Safflower (*Carthamus tinctorius* L.) oil could improve abdominal obesity, blood pressure, and insulin resistance in patients with metabolic syndrome: A randomized, double-blind, placebo-controlled clinical trial. *Journal of Ethnopharmacology*, *282*, 114590.

Salehi, B., Ata, A., V. Anil Kumar, N., Sharopov, F., Ramirez-Alarcon, K., Ruiz-Ortega, A., . . . Amiruddin Zakaria, Z. (2019). Antidiabetic potential of medicinal plants and their active components. *Biomolecules*, *9*(10), 551.

Shankar, K., & Mehendale, H. M. (2014). Oxidative stress. In P. Wexler (Ed.), *Encyclopedia of Toxicology* (3rd ed., pp. 735–737). Oxford: Academic Press.

Shi, X.-M., Zhang, H., Zhou, Z.-J., Ruan, Y.-Y., Pang, J., Zhang, L., . . . Hu, Y.-L. (2018). Effects of safflower yellow on beta-amyloid deposition and activation of astrocytes in the brain of APP/PS1 transgenic mice. *Biomedicine & Pharmacotherapy*, *98*, 553–565.

Singh, V., & Nimbkar, N. (2006). Safflower (*Carthamus tinctorius* L.). *Chapter*, *6*, 167–194.

Takahashi, T., & Miyazawa, M. (2012). Potent α-glucosidase inhibitors from safflower (*Carthamus tinctorius* L.) seed. *Phytotherapy Research*, *26*(5), 722–726.

Tanwar, B., & Goyal, A. (2021). *Oilseeds: Health Attributes and Food Applications*. Berlin/Heidelberg, Germany: Springer.

Tu, Y., Xue, Y., Guo, D., Sun, L., & Guo, M. (2015). Carthami flos: A review of its ethnopharmacology, pharmacology and clinical applications. *Revista Brasileira de Farmacognosia*, *25*, 553–566.

Tucker, J., & Townsend, D. (2005). Alpha-tocopherol: Roles in prevention and therapy of human disease. *Biomedicine & Pharmacotherapy*, *59*(7), 380–387.

Wang, C. C., Choy, C. S., Liu, Y. H., Cheah, K. P., Li, J. S., Wang, J. T. J., . . . Hu, C. M. (2011). Protective effect of dried safflower petal aqueous extract and its main constituent, carthamus yellow, against lipopolysaccharide-induced inflammation in RAW264. 7 macrophages. *Journal of the Science of Food and Agriculture*, *91*(2), 218–225.

Wang, X., Xu, Y., Chu, C., Li, H., Mi, J., Wen, Z., . . . Quan, S. (2019). Effect of safflower yellow on early type II diabetic nephropathy: A systematic review and meta-analysis of randomized controlled trials. *Journal of Pediatric Endocrinology and Metabolism*, *32*(7), 653–665.

Wang, Y., Shi, Y., Zou, J., Zhang, X., Wang, M., Guo, D., . . . Wang, T. (2020). The intranasal administration of *Carthamus tinctorius* L. extract/phospholipid complex in the treatment of cerebral infarction via the TNF-α/MAPK pathway. *Biomedicine & Pharmacotherapy*, *130*, 110563.

Yang, F., Li, J., Zhu, J., Wang, D., Chen, S., & Bai, X. (2015). Hydroxysafflor yellow A inhibits angiogenesis of hepatocellular carcinoma via blocking ERK/MAPK and NF-κB signaling pathway in H22 tumor-bearing mice. *European Journal of Pharmacology*, *754*, 105–114.

Yu, S.-Y., Lee, Y.-J., Kim, J.-D., Kang, S.-N., Lee, S.-K., Jang, J.-Y., . . . Lee, O.-H. (2013). Phenolic composition, antioxidant activity and anti-adipogenic effect of hot water extract from safflower (*Carthamus tinctorius* L.) seed. *Nutrients*, *5*(12), 4894–4907.

Zhang, J., Li, J., Song, H., Xiong, Y., Liu, D., & Bai, X. (2019). Hydroxysafflor yellow a suppresses angiogenesis of hepatocellular carcinoma through inhibition of p38 MAPK phosphorylation. *Biomedicine & Pharmacotherapy*, *109*, 806–814.

Zhang, L., Zhou, Z., Zhai, W., Pang, J., Mo, Y., Yang, G., . . . Hu, Y. (2019). Safflower yellow attenuates learning and memory deficits in amyloid β-induced Alzheimer's disease rats by inhibiting neuroglia cell activation and inflammatory signaling pathways. *Metabolic Brain Disease*, *34*(3), 927–939.

Zhang, L.-L., Tian, K., Tang, Z.-H., Chen, X.-J., Bian, Z.-X., Wang, Y.-T., & Lu, J.-J. (2016). Phytochemistry and pharmacology of *Carthamus tinctorius* L. *The American Journal of Chinese Medicine*, *44*(2), 197–226.

Zhang, Y., Song, L., Pan, R., Gao, J., Zang, B.-X., & Jin, M. (2017). Hydroxysafflor yellow a alleviates lipopolysaccharide-induced acute respiratory distress syndrome in mice. *Biological and Pharmaceutical Bulletin*, *40*(2), 135–144.

Zhu, H.-B., Zhang, L., Wang, Z.-H., Tian, J.-W., Fu, F.-H., Liu, K., & Li, C.-L. (2005). Therapeutic effects of hydroxysafflor yellow a on focal cerebral ischemic injury in rats and its primary mechanisms. *Journal of Asian Natural Products Research*, *7*(4), 607–613.

15 Role of Sage (*Salvia officinalis* L.) in Promoting Antidiabetic Activities

Raya Algonaiman and Hassan Barakat

CONTENTS

15.1 INTRODUCTION

Dalmatian sage (*Salvia officinalis*), known as common sage or just "sage," is a medical and aromatic plant of the mint family, Lamiaceae. A bushy, evergreen subshrub characterized by woody stems, green-grayish leaves, and violet-blueish flowers, it is native to the Middle Eastern and Mediterranean regions and has been naturalized in many countries worldwide (Gali-Muhtasib et al., 2000; Grdiša et al., 2015). It is commonly cultivated as a culinary herb and a source of essential oils, mostly used in the food, perfumery, and cosmetic industries and for medical purposes (Jedidi et al., 2021; Sharma et al., 2019).

In traditional medicine, the dried leaves of *S. officinalis* were mainly used as a tea for treating several health issues such as kidney and gall bladder stones, heart disorders, and nerve-related disorders, as well as for relieving headaches, stomachaches, abdominal pain, and many other health issues. The fresh leaves were also used in some cultures for treating hypotensive and respiratory disorders (Gali-Muhtasib et al., 2000; Sharma et al., 2019). The genus *Salvia* is originally derived from the Latin word "salvere" which means "to save," indicating the given appreciation for *S. officinalis* for its power in promoting numerous health effects (Grdiša et al., 2015). Such privileges are

DOI: 10.1201/b23347-15

attributed to the intensity of the phytochemical composition of *S. officinalis*; it contains different types of bioactive components such as alkaloids, glycosidic derivatives, phenolic compounds, steroids, and terpenes or terpenoids. Phenolic acid concentrations are presented at high levels; ferulic acid, benzoic acid, rosmarinic acid, and vanillic acid are most reported in *S. officinalis* leaf extracts.

S. officinalis has growing evidence of its potential in preventing or treating various health disorders. It could exert different health-promoting activities such as anti-inflammatory, antimicrobial, anti-tumor, and antidiabetic activities. It has also been shown to promote enhancements in cognitive and memory functions; some studies have reported its efficiency in preventing or improving Alzheimer's disease (Ghorbani & Esmaeilizadeh, 2017; Miroddi et al., 2014). Eliminating diarrhea and menopausal symptoms have also been attributed to *S. officinalis* as promising effects (Khan et al., 2011) (Grdiša et al., 2015). Furthermore, the antidiabetic activities of *S. officinalis* have been reported in both *in vivo* and *in vitro* studies. This chapter aims to summarize and discuss the available scientific data on the efficiency of *S. officinalis* in promoting antidiabetic effects and the potential mechanism of action, as well as provide a brief comprehensive profile of *S. officinalis*.

15.2 BOTANICAL DESCRIPTION

S. officinalis, or sage (Figure 15.1), Dalmatian sage, or common sage, is an annual/perennial bushy evergreen subshrub with woody stems, green-grayish leaves, and violet-blueish flowers in the mint family, Lamiaceae. The *Salvia* genus is the largest in the family, compromising more than 900 species. It is characterized by a strong root system, and its height can reach 100 cm long but mostly ranges from 40 to 60 cm. The stems can be erect or procumbent with several fine hairy dark green branches, surrounded by numerous leaves with tiny spines. The leaves (Figure 15.2) are 1–4 cm in diameter, 1–4 cm long, and may or may not contain basal lobes. White hairs cover the lower leaf surface, whereas greenish or greenish-grey hairs cover the upper surface, as shown in Figure 15.2 (A). The flowers are 2–4 mm and are found in clusters with 2–10 flowers in each cluster and may reach 40 flowers in rare times (Altindal & Altindal, 2016; Grdiša et al., 2015; Jakovljević et al., 2019; Mafakheri & Kordrostami, 2021).

FIGURE 15.1 Full sage plant (*Salvia officinalis*).

FIGURE 15.2 Sage (*Salvia officinalis*) leaves. (A) A close-up of the leaves, the upper surface on top and lower on the bottom, (B) fresh leaves, and (C) dried leaves.

S. officinalis is preferably grown in cool temperatures to subtropical conditions, generally warm to dry climate, with plentiful sunshine, and in fertile loamy soil with slightly acidic pH and a proper drainage facility. It could tolerate drought and low soil fertility but produces many bushy leaves in preferable conditions. The beginning of the spring season is the ideal time for planting *S. officinalis* (Sharma et al., 2019), and the flowers start to bloom from March to July, depending on the habitat's climatic conditions (Grdiša et al., 2015; Jakovljević et al., 2019).

15.3 DISTRIBUTION

S. officinalis is distributed widely worldwide; natively, it is grown in the Middle East and Mediterranean regions. Middle Eastern countries mostly include Lebanon, Syria, and Palestine (Gali-Muhtasib et al., 2000). Mediterranean Europe countries mostly include the Western Balkans areas in southern and southeastern Europe, such as Croatia, Bosnia and Herzegovina, Montenegro, and Albania. It is also grown in northern Greece, farther from the seaside. In recent times, *S. officinalis* has been naturalized in many other countries with continental climates; including other European countries, such as Ukraine, Moldavia, Germany, Slovakia, Bulgaria, Romania, Italy, and Britain, and Asian and some African countries such as India, Japan, Indonesia, and Tanzania, as well as in North America and Australia (Ghorbani & Esmaeilizadeh, 2017; Grdiša et al., 2015; Jug-Dujaković et al., 2020).

15.4 PHYTOCHEMICAL CONSTITUENTS

Numerous phytochemicals are present in *S. officinalis* flowers, leaves, and stems. The most reported components are alkaloids, glycosidic derivatives, phenolic compounds, polyacetylenes, steroids, terpenes/terpenoids, and waxes. Most of the reported components have been isolated from *S. officinalis* essential oil, alcoholic extract, or aqueous extract, as well as from butanol fraction and infusion preparation (Ghorbani & Esmaeilizadeh, 2017). Higher concentrations of phenolic compounds have been identified in *S. officinalis* leaf extracts; the major phenolic acids were ferulic acid, benzoic acid, rosmarinic acid, and vanillic acid. Flavonoids are also present in high concentrations; the major ones are kaempferol, resveratrol, and chrysin (Alharbi et al., 2022). In the essential oil, more than 200 components were found. The most reported were borneol, camphor, caryophyllene, cineole, elemene, humulene, ledene, pinene, and thujone (Ghorbani & Esmaeilizadeh, 2017). Khedher et al. (2017) isolated more than 40 components from *S. officinalis* essential oil, and the major ones were camphor, α-thujone, 1,8-cineole, viridiflorol, β-thujone, and β-caryophyllene. The chemical composition of *S. officinalis* can vary depending on climatic conditions, including temperature, sunlight, drought, and salinity (Ghorbani & Esmaeilizadeh, 2017).

15.5 PHARMACOLOGICAL STUDIES

Several pharmacological effects have been demonstrated for *S. officinalis*. The current section highlights the most reported effects (Figure 15.3).

15.5.1 ANTI-INFLAMMATORY EFFECTS

Anti-inflammatory effects of *S. officinalis* have been demonstrated in several studies. Suppressing the expression of pro-inflammatory cytokines such as tumor necrosis factor-alpha (TNF-α) and interleukin 6 (IL-6) were observed after *S. officinalis* treatment (Imanshahidi & Hosseinzadeh, 2006). Moreover, topical application of *S. officinalis* extracts, *n*-hexane or chloroform, was shown to promote significant anti-inflammatory effects in croton oil-induced ear edema in mice. These effects were attributed to the high ursolic acid content in *S. officinalis* (Baricevic et al., 2001). *S. officinalis* is also rich in rosmarinic acid, which promotes anti-inflammatory effects. Enhancements in epidermal inflammation were observed after topical application of rosmarinic acid (Ghorbani & Esmaeilizadeh, 2017). In addition, flavonoids and terpenes are also present in high concentrations in *S. officinalis*,

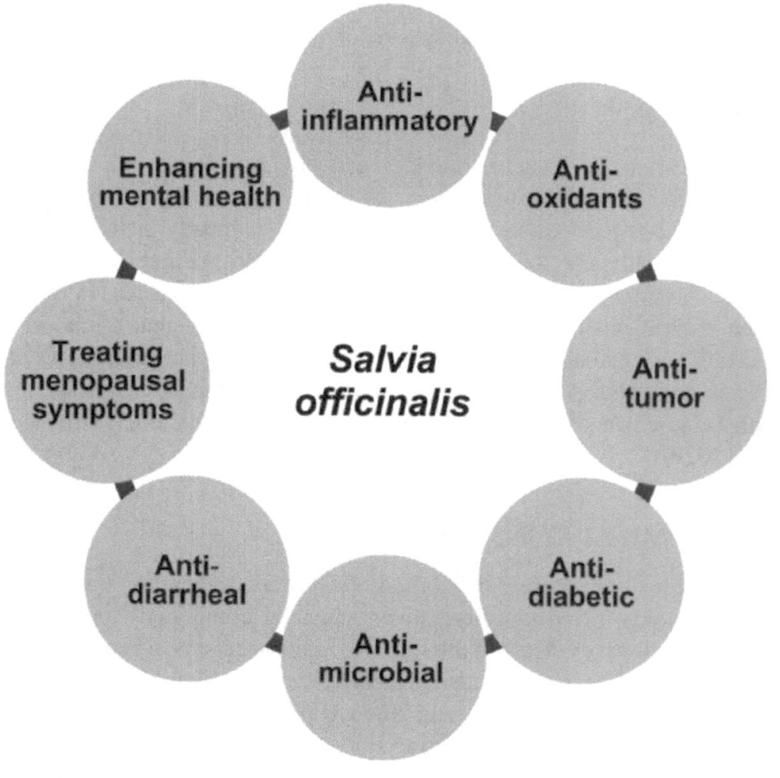

FIGURE 15.3 Pharmacological effects of sage (*Salvia officinalis*).

which are considered the most active components in anti-inflammatory effects. Furthermore, *S. officinalis* essential oil contains borneol, which was reported to play a significant role in suppressing the expression of pro-inflammatory cytokines involved in colonic inflammation (Jakovljević et al., 2019).

15.5.2 Antioxidant Effects

The endogenous antioxidant defense system plays a key role in preventing various diseases. *S. officinalis* is rich in several bioactive components that exert antioxidant activities via playing as free radical scavengers or improving the antioxidant defense systems. Carnosol, rosmarinic acid, and carnosic acid were reported as the most effective antioxidants in *S. officinalis*, followed by caffeic acid, rosmanol, rosmadial, genkwanin, and cirsimaritin. Carnosol exerts radical scavenging effects similar to α-tocopherol, whereas rosmarinic acid was shown to scavenge radicals 15–20 times more than trolox. Other flavonoids, such as quercetin and rutin, are also present in *S. officinalis* and are known for their strong antioxidant activities (Ghorbani & Esmaeilizadeh, 2017; Hamidpour et al., 2014). The antioxidant effects of *S. officinalis* have been shown in animal models; the administration of *S. officinalis* extract in drinking water showed an increase in the antioxidant defense against oxidative damage in rats' hepatocytes (Horváthová et al., 2016).

15.5.3 Anti-Tumor Effects

In several cancerous cell lines, *S. officinalis* has been demonstrated to exert inhibitory effects on the proliferation of tumors (Ghorbani & Esmaeilizadeh, 2017). In animal models, inhibitory effects on tumor growth were also reported. In rats administrated with *S. officinalis* extract before azoxymethane injection, a colon carcinogen, the results showed preventative effects against colon carcinogenesis (Pedro et al., 2016). The protective effects of *S. officinalis* against tumor growth could be attributed to the presence of different bioactive components, such as terpenes and terpenoids, which were reported to promote anti-tumor effects (El Hadri et al., 2010; Lage et al., 2010). In addition, rosmarinic acid was also shown to promote anti-tumor effects; in rats induced with a skin carcinogen, the application of rosmarinic acid showed inhibitory effects on the formation of skin tumors (Sharmila & Manoharan, 2012). Moreover, inhibition of angiogenesis was reported *in vivo* after the application of ursolic acid derived from *S. officinalis* (Hamidpour et al., 2014).

15.5.4 Cognitive- and Memory-Enhancing Effects

Emerging evidence on the effectiveness of *S. officinalis* in improving cognitive and memory functions has increased. In a systematic review, the administration of *S. officinalis* to healthy individuals and patients with dementia or cognitive impairment showed enhancements in cognitive performance, and the use of *S. officinalis* regarding this issue was safe (Miroddi et al., 2014). Ethanolic extract of *S. officinalis* showed similar effects on improving memory and attention in older individuals (Scholey et al., 2008). The aroma of *S. officinalis* essential oil was also found to exert some beneficial effects on improving memory's quality (L. Moss et al., 2010; M. Moss et al., 2014). Furthermore, *S. officinalis* has improved cognitive functions in patients with mild to moderate Alzheimer's disease (AD) (Akhondzadeh et al., 2003). Such effects are attributed to *S. officinalis*'s inhibitory effects on the activity of acetylcholinesterase, an enzyme involved in the progression of AD and considered a therapeutic target in most clinical medications used for treating AD (Ghorbani & Esmaeilizadeh, 2017).

15.5.5 Anti-Microbial Effects

S. officinalis essential oil and methanolic extracts exert inhibitory effects on the growth of several bacteria, such as *Enterococcus faecalis*, *E. coli*, and *Salmonella*. *S. officinalis* could also promote antifungal, antiviral, and anti-malarial effects (Ghorbani & Esmaeilizadeh, 2017). In addition,

S. officinalis was reported to inhibit the growth of some dental caries-causing bacteria such as *Streptococcus mutans* and *Actinomyces viscosus*. Moreover, it has been suggested that *S. officinalis* aqueous extract could be used as a natural preservative in the food industry (Hamidpour et al., 2014). Such anti-microbial effects were attributed to the presence of terpenes and terpenoid components in *S. officinalis* (Ghorbani & Esmaeilizadeh, 2017).

15.5.6 Other Effects

Hyperactive gut disorders such as abdominal colic and diarrhea have been enhanced using *S. officinalis* crude extracts. *S. officinalis* could promote inhibitory effects on gut motility, thus providing protective effects against diarrhea (Khan et al., 2011). Moreover, *S. officinalis* extract effectively suppresses menopausal symptoms (Grdiša et al., 2015).

15.6 ANTIDIABETIC RESPONSE

Apart from the different pharmacological effects of *S. officinalis*, several studies have demonstrated strong antidiabetic activities of *S. officinalis* in humans and animals, as well as in *in vitro* studies.

15.6.1 Human Interventions

The intake of *S. officinalis* capsules (500 mg) three times a day for two months among 50 diabetic patients who were under a therapeutic strategy of glyburide, metformin, and atorvastatin resulted in significant improvements in their fasting blood glucose (FBG), 2-h postprandial glucose, and glycosylated hemoglobin (HbA_{1c}), as well as in serum lipids as compared to a placebo group. These results strongly indicate the beneficial effects of *S. officinalis* consumption combined with drug therapy (Kianbakht et al., 2016). In another controlled trial, in 80 diabetic patients who were not following a strict therapy or were irregularly using medications, the intake of *S. officinalis* tablet at a dose of 150 mg three times a day for three months resulted in a reduction in the 2-h postprandial glucose and serum lipids levels as compared to a placebo group. However, FBG and HbA_{1c} did not change significantly (Behradmanesh et al., 2013). Such results may be attributed to the low dose of *S. officinalis* used in this study (150 mg kg^{-1}) as compared to the dose used by Kianbakht et al. (2016) (500 mg kg^{-1}). In addition, regular medications combined with *S. officinalis* intake could show much more effectiveness (Kianbakht et al., 2016). Still, *S. officinalis* tablets in this study showed significant enhancement in 2-h postprandial glucose and serum lipids levels, again indicating the beneficial effects of *S. officinalis* in managing diabetes complications. However, medication usage should not be excluded. On the other hand, it should be considered that consuming *S. officinalis* tablets at these doses did not exert any adverse side effects (Behradmanesh et al., 2013; Kianbakht et al., 2016; Sá et al., 2009).

Enhancements in serum lipids were also observed among women aged 40–50 who consumed 300 mL of *S. officinalis* tea twice a day for four weeks. From the second week, results showed enhancements in the ratio of low-density lipoprotein (LDL) to high-density lipoprotein (HDL) cholesterol levels (Sá et al., 2009). Consistently, serum lipids showed enhancements in hyperlipidemic patients taking *S. officinalis* tablets (500 mg kg^{-1}) every eight hours daily for two months compared to placebo tablets (Kianbakht et al., 2011). Kianbakht and Dabaghian (2013) conducted a similar trial in hyperlipidemic type 2 diabetic patients; taking *S. officinalis* tablets (500 mg) for three months resulted in improvements in FBG, HbA_{1c}, and serum lipids based on a decrease in total cholesterol, triglycerides, and LDL and an increase in HDL as compared to a placebo group.

Furthermore, *S. officinalis* could also enhance insulin resistance; among euglycemic women diagnosed with polycystic ovary syndrome, the consumption of *S. officinalis* extract at a dose of 330 mg/day for eight weeks significantly improved their insulin resistance as compared to the placebo group (Amini et al., 2020).

15.6.2 Animal Models

Hypoglycemic effects of *S. officinalis* were reported in several animal studies (Al-Fartosi et al., 2015; Belhadj et al., 2018; Eidi et al., 2005; Kanana et al., 2020; Lamia & Al-Mashhady, 2016; Moradabadi et al., 2013). The oral administration of *S. officinalis* decoction at 300 mg kg^{-1} BW for five weeks was reported to promote significant hypoglycemic and hypolipidemic effects in alloxan-induced diabetic mice (Salah et al., 2016). Significant improvements in FBG and serum lipids were also observed in streptozotocin (STZ)-induced diabetic rabbits administrated orally with *S. officinalis* extract at 250 and 500 mg kg^{-1} BW daily for 21 days (Mokogwu et al., 2022). Similar results were reported in alloxan-induced diabetic rats treated with *S. officinalis* leaf extract at a dose of 1 mL/kg BW for 30 days (Al-Fartosi et al., 2015). Consistently, the administration of *S. officinalis* extract at a dose of 100 mg kg^{-1} BW for 14 days to alloxan-induced diabetic rats showed a significant reduction in FBG and improved serum lipid profile. The reduction in FBG was more effective than glibenclamide administration (Khashan & Al-Khefaji, 2015). Others reported that *S. officinalis* treatment was efficient in a similar manner to acarbose (Mahdi et al., 2020). Furthermore, the fortification of *S. officinalis* leaf powder to fermented camel milk at different concentrations showed significant enhancements in lowering serum glucose and lipid profile in STZ-induced diabetic rats (Alharbi et al., 2022) and in alloxan-induced diabetic rats (Shahein et al., 2022).

S. officinalis was shown to prevent or delay β-cell destruction; in rats administrated with *S. officinalis* extract for seven days before STZ injection, their β-cell mass was significantly higher by almost 188% compared to rats that did not receive the *S. officinalis* extracts (Ashkani-Esfahani et al., 2022). Moreover, *S. officinalis* was reported to promote hepatoprotective and nephroprotective effects in diabetic animals. In mice fed with a high-fat diet (HFD) and *S. officinalis* essential oil (4 mg kg^{-1} BW), results showed enhancements in their liver and kidney functions based on improvements in activities of aspartate transaminase (AST), alanine transaminase (ALT), gamma-glutamyltranspeptidase (GGTP), and lactate dehydrogenase activities (LDH) (Koubaa-Ghorbel et al., 2020). In another study, alloxan-induced diabetic rats were administrated orally with *S. officinalis* essential oil (2.5 μL per rat) for 30 days. The results showed improvements in liver and kidney functions based on a reduction in serum AST, ALT, and LDH by 35%, 79%, and 43%, respectively, for the liver, and a reduction in serum creatinine and uric acid by 47% and 62.5%, respectively, for kidneys (Belhadj et al., 2018).

Furthermore, *S. officinalis* was shown to prevent cognitive deficits in STZ-induced diabetic rats; the administration of *S. officinalis* extract at 800 mg kg^{-1} BW reversed the cognitive deficits resulting from STZ injection. However, the lower doses did not show such effects (Hasanein et al., 2016). Although in a pre-type 2 diabetes rat model established by feeding with HFD for 15–16 weeks, the oral administration of *S. officinalis* extract at 100 mg/kg BW for five weeks promoted beneficial effects in enhancing glucose tolerance and insulin sensitivity much more effectively than did a higher dose (400 mg/kg BW). Such results might be related to a high intake of palmitic acid in the higher dose of *S. officinalis* extract, as high concentrations of palmitic acid were reported to promote insulin resistance. Therefore, in long-term interventions, lower doses of *S. officinalis* are recommended (Khedher et al., 2018).

15.6.3 Mechanisms of Action

Hypoglycemic efficiency of *S. officinalis* could be attributed to several pathophysiological mechanisms; the most reported mechanisms are inhibiting key digestive enzymes, activating and upregulating some signaling pathways, and promoting anti-inflammatory and antioxidant activities (Figure 15.4).

The inhibition of key digestive enzymes such as the pancreatic α-amylase and intestinal α-glucosidase was reported after applying different *S. officinalis* treatments (Chehade et al., 2022; Javid et al., 2021; Mahdi et al., 2020; Moradabadi et al., 2013). These enzymes, α-amylase and α-glucosidase, are

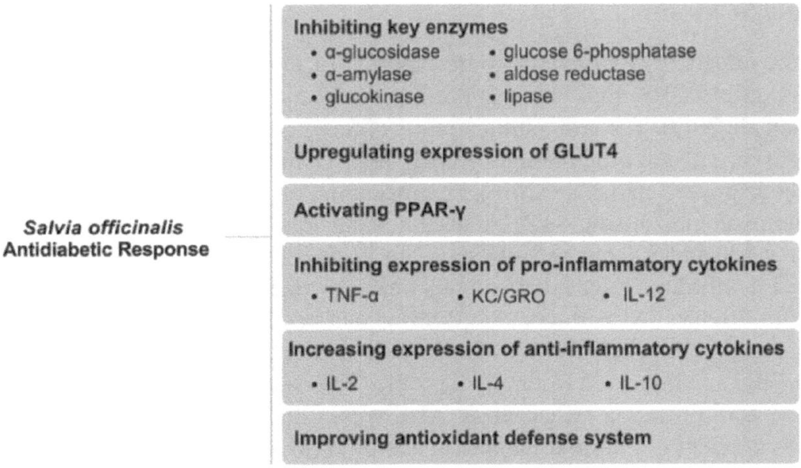

FIGURE 15.4 Sage (*S. officinalis*) mechanism of action against diabetes.

responsible for hydrolyzing carbohydrate fractions into monosaccharides. Therefore, inhibiting these enzymes limits glucose uptake in the bloodstream, a therapeutic target used in managing type 2 diabetes (Eskandani et al., 2016). Even at low concentrations, *S. officinalis* essential oil was reported to inhibit the activities of α-amylase and α-glucosidase compared to acarbose (Al-Mijalli et al., 2022). However, acarbose, a standard drug used to eliminate the rate of intestinal glucose absorption by delaying digestion, can result in some side effects, such as diarrhea, flatulence, soft stools, and bowel bloating. Therefore, *S. officinalis* as a natural substance could be more beneficial (Holmbäck et al., 2020). The inhibitory effects of such key digestive enzymes have been strongly correlated with plants' phytochemicals content; several biochemical components such as flavonoids, steroids, and terpenoids are present in *S. officinalis* and are known for their inhibitory effects on the activity of both α-amylase and α-glucosidase (Bahadori et al., 2016; Javid et al., 2022; Mahdi et al., 2020).

Moreover, *S. officinalis* derivatives such as quercetin, lutein, and kaempferol have been demonstrated to inhibit the activities of glucokinase and glucose 6-phosphatase enzymes, which are known for playing key roles in glucose homeostasis through regulating glucose production via the gluconeogenesis pathway in hepatocytes. Therefore, inhibiting these enzymes leads to gluconeogenesis and increases hepatocyte sensitivity to insulin (Gasparin et al., 2003; Wang et al., 2000). Inhibition of gluconeogenesis activity was reported in hepatocytes isolated from normal rats administrated with *S. officinalis* tea for 14 days. However, such results were not observed in hepatocytes isolated from STZ-induced diabetic rats. *S. officinalis* tea did not alter or modify the effectiveness of the standard metformin in reducing the glucose production in hepatocytes of the STZ-induced diabetic rats, which indicates that *S. officinalis* would not interfere negatively with metformin therapy (Lima et al., 2006).

Furthermore, inhibitory effects on lipase activity were also reported (Koubaa-Ghorbel et al., 2020). Belhadj et al. (2018) reported a reduction in lipase activity in blood serum by 32.1% after *S. officinalis* essential oil intervention. Lipases are the group of enzymes responsible for hydrolyzing fat fractions. Thus, inhibiting its activity leads to limiting dietary fat absorption, which could explain the hypolipidemic efficiency of *S. officinalis* (Koubaa-Ghorbel et al., 2020). In addition, *S. officinalis* could also promote hypolipidemic effects by regulating total cholesterol fecal excretion (Koubaa-Ghorbel et al., 2020) and lowering cholesterol production by inhibiting lipogenesis activity in adipocytes (Khedher et al., 2018).

The inhibitory effects on lipase might be attributed to the high level of carnosic acid and carnosol present in *S. officinalis*, which are reported to play a role as lipid absorption inhibitors (Koubaa-Ghorbel et al., 2020). In addition, lectin, saponin, and sterols are present in *S. officinalis* and are reported to promote hypolipidemic effects. Lectin can lower both serum and hepatic cholesterol; saponin can inhibit intestinal fat absorption and stimulate the conversion of cholesterol into bile acid in hepatocytes (Khashan & Al-Khefaji, 2015).

On the other hand, hypolipidemic and hypoglycemic effects of *S. officinalis* have also been attributed to the activation of peroxisome proliferator-activated receptor gamma (PPAR-γ) (Christensen et al., 2010; Rau et al., 2006), which is a nuclear receptor protein responsible for controlling glucose and lipid metabolism and is considered a potential therapeutic target in the prevention and treatment of metabolic disorders (Tobin & Freedman, 2006). The activation of PPAR-γ has been correlated with the presence of 12-O-methyl carnosic acid and α-linolenic acid in *S. officinalis* (Christensen et al., 2010). Moreover, *S. officinalis* was shown to play a role in upregulating the expression of the insulin-responding glucose transporter type 4 (GLUT4), which is a key transporter in glucose homeostasis (Moradabadi et al., 2013).

From another point of view, oxidative stress and inflammatory state play major roles in the development and progression of diseases, including diabetes (Ghorbani & Esmaeilizadeh, 2017). Oxidative stress is a widely known concept defined by the excess production of reactive oxygen species, known as free radicals (Shankar & Mehendale, 2014). Plants' phytochemical contents have also been demonstrated for their antioxidant activities via stimulating the endogenous antioxidant defense systems or acting as reactive species scavengers (Zhang et al., 2015). *S. officinalis* is rich in antioxidant components such as flavanones, flavonols, tannins, triterpenes, alkaloids, saponins, sterols, and phenolic acids (Kanana et al., 2020).

The administration of *S. officinalis* in different studies showed significant enhancements in the oxidative stress markers; Sá et al. (2009) reported significant enhancements in serum levels of catalase (CAT) and superoxide dismutase (SOD) among women who were consuming *S. officinalis* tea for four weeks. In alloxan-induced diabetic rabbits, serum malondialdehyde (MDA) levels were significantly improved after the oral administration of *S. officinalis* extracts for four weeks (Lamia & Al-Mashhady, 2016). In addition, Alharbi et al. (2022) reported enhancements in CAT, SOD, MDA, and glutathione (GSH) by 89.93%, 63.06%, 58.69%, and 53.75%, respectively, in STZ-diabetic rats after the oral administration of fermented camel milk fortified with *S. officinalis* leaf powder. Moreover, *S. officinalis* was shown to exert inhibitory effects on the activity of aldose reductase, a key enzyme that plays a role in the progression of oxidative stress. Aldose reductase can eliminate major non-enzymatic antioxidants, such as vitamin E and carotenoids, which lead to oxidative stress in several organs (Amraee & Bahramikia, 2019).

Furthermore, inflammatory responses can play major roles in glucose and insulin homeostasis. In an *in vitro* experiment, *S. officinalis* extract showed significant effects in reducing the pro-inflammatory cytokines TNF-α, KC/GRO (keratinocyte-derived chemoattractant/human growth-regulated oncogene), and IL-12 (interleukin-12) and in increasing the anti-inflammatory cytokines IL-2, IL-4, and IL-10 (Khedher et al., 2018).

15.7 TRADITIONAL AND OTHER POTENTIAL USES

Since ancient times, many cultures have used *S. officinalis* in traditional medicine to cure several health disorders. In the Middle Eastern culture, including the Lebanese, Syrian, Jordanian, and Palestinian, the dried leaves of *S. officinalis* are boiled as a tea to relieve headaches, stomachaches, abdominal pain or indigestion, and ulcer pain; it was also used for treating heart disorders. In the Mediterranean to Eastern Europe culture, the Turkish folks use *S. officinalis* tea for treating kidney and gall bladder stones and for relieving colds, coughs, and influenza (Gali-Muhtasib et al., 2000). In other European countries, *S. officinalis* tea is used to reduce excessive sweating, treat skin, mouth, and throat inflammations, and improve memory (Gali-Muhtasib et al., 2000; Ghorbani & Esmaeilizadeh, 2017; Grdiša et al., 2015). In Asian and Latin American cultures, *S. officinalis* tea and the essential oil have been used for nerve-related disorders and high blood pressure (Sharma et al., 2019).

Fresh leaves are used for treating hypotensive and respiratory disorders (Sharma et al., 2019). Many other health disorders, such as seizures, ulcers, rheumatism, dizziness, tremor, paralysis, and diarrhea, were treated traditionally in the Asian and Latin American cultures using *S. officinalis* (Ghorbani & Esmaeilizadeh, 2017). In other cultures, the essential oil has been used as a local disinfectant, anti-inflammatory, antifungal, and antispasmodic and also for treating endocrine diseases. *S. officinalis* ground leaf paste was also applied locally to relieve allergic swelling, insect bites, and skin inflammation. A leaf decoction was also used traditionally as an antidepressant (Sharma et al., 2019).

Besides using *S. officinalis* in traditional medicine, Chinese folks prefer *S. officinalis* tea over ordinary tea in winter. In the Brazilian culture, the fresh leaves were used for refreshing mouth breath. Other cultures used the *S. officinalis* decoction in hair oils and dark hair dyes (Sharma et al., 2019). On the other hand, *S. officinalis* essential oil is used in other cultures in steam baths due to its strong aromatic property (Gali-Muhtasib et al., 2000). In addition, the essential oil has been reported for its potential effects in protecting against pathogenic microorganisms and environmental problems related to exposure to synthetic chemicals (Grdiša et al., 2015).

15.8 SAFETY ISSUES

S. officinalis has been consumed since ancient times, and there were no reports of severe adverse health effects. However, the long-term consumption or overdosing of *S. officinalis* oil extract or the irrational usage of *S. officinalis* essential oil may cause some adverse health effects. The most noticed side effects in such cases are vomiting, salivation, tachycardia, allergic reactions, tongue swallowing, cyanosis, spasmolytic, loss of equilibrium, dementia, seizures, and other effects related to the nervous system. The most reported toxic compounds related to these side effects found in *S. officinalis* essential oil are camphor, thujone, and terpene ketones. These compounds may also negatively affect fetuses and newborns. Therefore, *S. officinalis* must be avoided during pregnancy and lactating (Gali-Muhtasib et al., 2000; Ghorbani & Esmaeilizadeh, 2017; Grdiša et al., 2015).

S. officinalis should be consumed carefully without exceeding the recommended doses, and essential oil must be considered a drug handled with high precaution. The recommended dose for dry *S. officinalis* leaves is 4–6 g daily (Gali-Muhtasib et al., 2000; Martins et al., 2015).

15.9 CULTIVATION PRACTICES

The main final material of *S. officinalis* is the dried leaves, cultivated worldwide for application in medicine, perfumery, and cosmetic industries, as well as a culinary herb. The application of *S. officinalis* in food preparations is mostly known in Western cooking for seasoning meats, poultry, and fish (Sharma et al., 2019). Due to its aromatic properties, *S. officinalis* is also cultivated for its essential oil. It contains a significant amount of essential oil ranging from 0.7% to 5.2% and consists mainly of volatile organic compounds known as terpenes, recognized for their biological activities (Đurović et al., 2022). In the past two decades, the essential oil was mostly applied in perfumes, deodorants, insecticidal treatments, and commercial pharmaceuticals (Jedidi et al., 2021; Sharma et al., 2019).

15.10 FUTURE REMARKS

The administration of *S. officinalis* has shown promising effects in managing diabetes and preventing or attenuating diabetes complications. So far, it has been demonstrated that the intervention of *S. officinalis* capsules or tablets in combination with a therapeutic medication plan could significantly promote the homeostasis of glycemia, insulinemia, and lipidemia and manage other diabetes complications. However, future investigations are still needed to estimate its encapsulation safety and establish a well-designed dosage limit, particularly for diabetic individuals, to state a safe and guaranteed recommendation. In addition, *S. officinalis* was shown to promote different pathophysiological mechanisms involved in preventing and managing diabetes. Additional in-depth research is still needed to understand its bioactive components and their mechanism of action.

REFERENCES

Akhondzadeh, S., Noroozian, M., Mohammadi, M., Ohadinia, S., Jamshidi, A., & Khani, M. (2003). *Salvia officinalis* extract in the treatment of patients with mild to moderate Alzheimer's disease: A double blind, randomized and placebo-controlled trial. *Journal of Clinical Pharmacy and Therapeutics*, *28*(1), 53–59.

Al-Fartosi, K. G., Merdas, S. M., & Abd Alameer, A. S. (2015). Lipid profile of diabetic male rats treated with leaves extracts of Salvia officinalis. *Journal of Chemical, Biological and Physical Sciences*, *14*, 59–70.

Alharbi, Y. M., Sakr, S. S., Albarrak, S. M., Almundarij, T. I., Barakat, H., & Hassan, M. F. (2022). Antioxidative, antidiabetic, and hypolipidemic properties of probiotic-enriched fermented camel milk combined with *Salvia officinalis* leaves hydroalcoholic extract in streptozotocin-induced diabetes in rats. *Antioxidants*, *11*(4), 668.

Al-Mijalli, S. H., Assaggaf, H., Qasem, A., El-Shemi, A. G., Abdallah, E. M., Mrabti, H. N., & Bouyahya, A. (2022). Antioxidant, antidiabetic, and antibacterial potentials and chemical composition of *Salvia officinalis* and mentha suaveolens grown wild in Morocco. *Advances in Pharmacological and Pharmaceutical Sciences*, *2022*.

Altindal, D., & Altindal, N. (2016). Chapter 81 – Sage (*Salvia officinalis*) oils. In V. R. Preedy (Ed.), *Essential Oils in Food Preservation, Flavor and Safety* (pp. 715–721). San Diego: Academic Press.

Amini, L., Mojab, F., Jahanfar, S., Sepidarkish, M., Raoofi, Z., & Maleki-Hajiagha, A. (2020). Efficacy of *Salvia officinalis* extract on the prevention of insulin resistance in euglycemic patients with polycystic ovary syndrome: A double-blinded placebo-controlled clinical trial. *Complementary Therapies in Medicine*, *48*, 102245.

Amraee, S., & Bahramikia, S. (2019). Inhibitory effect of effective fraction of *Salvia officinalis* on aldose reductase activity: Strategy to reduce complications of type 2 diabetes. *Oriental Pharmacy and Experimental Medicine*, *19*(2), 211–216. https://doi.org/10.1007/s13596-018-0354-6

Ashkani-Esfahani, S., Noorafshan, A., Ebrahimi, A., Bahmani-Jahromi, M., Imanieh, M.-H., Ebrahimi, S., . . . Tanideh, N. (2022). *Salvia officinalis* protects pancreatic beta-cells against streptozotocin-induced damage; A stereological study. *Jundishapur Journal of Natural Pharmaceutical Products*, *17*(1).

Bahadori, M. B., Valizadeh, H., Asghari, B., Dinparast, L., Bahadori, S., & Moridi Farimani, M. (2016). Biological activities of *Salvia santolinifolia* Boiss. A multifunctional medicinal plant. *Current Bioactive Compounds*, *12*(4), 297–305.

Baricevic, D., Sosa, S., Della Loggia, R., Tubaro, A., Simonovska, B., Krasna, A., & Zupancic, A. (2001). Topical anti-inflammatory activity of *Salvia officinalis* L. leaves: The relevance of ursolic acid. *Journal of Ethnopharmacology*, *75*(2–3), 125–132.

Behradmanesh, S., Derees, F., & Rafieian-Kopaei, M. (2013). Effect of *Salvia officinalis* on diabetic patients. *Journal of Renal Injury Prevention*, *2*(2), 51.

Belhadj, S., Hentati, O., Hammami, M., Hadj, A. B., Boudawara, T., Dammak, M., . . . El Feki, A. (2018). Metabolic impairments and tissue disorders in alloxan-induced diabetic rats are alleviated by *Salvia officinalis* L. essential oil. *Biomedicine & Pharmacotherapy*, *108*, 985–995.

Chehade, S., Kobeissy, M., Kanaan, H., & Haddad, M. (2022). Comparison between the chemical compositions and the in-vitro antidiabetic and anti-inflammatory activities of *Salvia libanotica*'and *Salvia officinalis*'leaves essential oils. *European Journal of Pharmaceutical and Medical Research*, *9*(3), 34–43.

Christensen, K. B., Jørgensen, M., Kotowska, D., Petersen, R. K., Kristiansen, K., & Christensen, L. P. (2010). Activation of the nuclear receptor PPARγ by metabolites isolated from sage (*Salvia officinalis* L.). *Journal of Ethnopharmacology*, *132*(1), 127–133.

Đurović, S., Micić, D., Pezo, L., Radić, D., Bazarnova, J. G., Smyatskaya, Y. A., & Blagojević, S. (2022). The effect of various extraction techniques on the quality of sage (*Salvia officinalis* L.) essential oil, expressed by chemical composition, thermal properties and biological activity. *Food Chemistry: X, 13*, 100213.

Eidi, M., Eidi, A., & Zamanizadeh, H. (2005). Effect of *Salvia officinalis* L. leaves on serum glucose and insulin in healthy and streptozotocin-induced diabetic rats. *Journal of Ethnopharmacology, 100*(3), 310–313.

El Hadri, A., del Rio, M. G., Sanz, J., Coloma, A. G., Idaomar, M., Ozonas, B. R., . . . Reus, M. I. S. (2010). Cytotoxic activity of α-humulene and transcaryophyllene from *Salvia officinalis* in animal and human tumor cells. *An R Acad Nac Farm, 76*(3), 343–356.

Eskandani, M., Babak Bahadori, M., Zengin, G., Dinparast, L., & Bahadori, S. (2016). Novel natural agents from Lamiaceae family: An evaluation on toxicity and enzyme inhibitory potential linked to diabetes mellitus. *Current Bioactive Compounds, 12*(1), 34–38.

Gali-Muhtasib, H., Hilan, C., & Khater, C. (2000). Traditional uses of Salvia libanotica (East Mediterranean sage) and the effects of its essential oils. *Journal of Ethnopharmacology, 71*(3), 513–520.

Gasparin, F., Spitzner, F., Ishii-Iwamoto, E., Bracht, A., & Constantin, J. (2003). Actions of quercetin on gluconeogenesis and glycolysis in rat liver. *Xenobiotica, 33*(9), 903–911.

Ghorbani, A., & Esmaeilizadeh, M. (2017). Pharmacological properties of *Salvia officinalis* and its components. *Journal of Traditional and Complementary Medicine, 7*(4), 433–440.

Grdiša, M., Jug-Dujaković, M., Lončarić, M., Carović-Stanko, K., Ninčević, T., Liber, Z., . . . Šatović, Z. (2015). Dalmatian sage (*Salvia officinalis* L.): A review of biochemical contents, medical properties and genetic diversity. *Agriculturae Conspectus Scientificus, 80*(2), 69–78.

Hamidpour, M., Hamidpour, R., Hamidpour, S., & Shahlari, M. (2014). Chemistry, pharmacology, and medicinal property of sage (Salvia) to prevent and cure illnesses such as obesity, diabetes, depression, dementia, lupus, autism, heart disease, and cancer. *Journal of Traditional and Complementary Medicine, 4*(2), 82–88.

Hasanein, P., Felehgari, Z., & Emamjomeh, A. (2016). Preventive effects of *Salvia officinalis* L. against learning and memory deficit induced by diabetes in rats: Possible hypoglycaemic and antioxidant mechanisms. *Neuroscience Letters, 622*, 72–77.

Holmbäck, U., Forslund, A., Grudén, S., Alderborn, G., Söderhäll, A., Hellström, P. M., & Lennernäs, H. (2020). Effects of a novel combination of orlistat and acarbose on tolerability, appetite, and glucose metabolism in persons with obesity. *Obesity Science & Practice, 6*(3), 313–323.

Horváthová, E., Srančíková, A., Regendová-Sedláčková, E., Melušová, M., Meluš, V., Netriová, J., . . . Kozics, K. (2016). Enriching the drinking water of rats with extracts of *Salvia officinalis* and Thymus vulgaris increases their resistance to oxidative stress. *Mutagenesis, 31*(1), 51–59.

Imanshahidi, M., & Hosseinzadeh, H. (2006). The pharmacological effects of Salvia species on the central nervous system. *Phytotherapy Research: An International Journal Devoted to Pharmacological and Toxicological Evaluation of Natural Product Derivatives, 20*(6), 427–437.

Jakovljević, M., Jokić, S., Molnar, M., Jašić, M., Babić, J., Jukić, H., & Banjari, I. (2019). Bioactive profile of various *Salvia officinalis* L. preparations. *Plants, 8*(3), 55.

Javid, H., Moein, S., & Moein, M. R. (2021). Evaluating the inhibitory effects of dichloromethane and methanol extracts of salvia macilenta and salvia officinalis on the diabetes marker enzyme, alpha-glucosidase: An approach for the treatment of diabetes. *Jundishapur Journal of Natural Pharmaceutical Products, 16*(3).

Javid, H., Moein, S., & Moein, M. R. (2022). An investigationof the inhibitory effects of dichloromethane and methanol extracts of *Salvia macilenta*, *Salvia officinalis*, *Salvia santolinifola* and *Salvia mirzayanii* on diabetes marker enzymes, an approach for the treatment diabetes. *Clinical Phytoscience, 8*(1), 1–8.

Jedidi, S., Sammari, H., Selmi, H., Hosni, K., Rtibi, K., Aloui, F., . . . Sebai, H. (2021). Strong protective effects of *Salvia officinalis* L. leaves decoction extract against acetic acid-induced ulcerative colitis and metabolic disorders in rat. *Journal of Functional Foods, 79*, 104406.

Jug-Dujaković, M., Ninčević, T., Liber, Z., Grdiša, M., & Šatović, Z. (2020). *Salvia officinalis* survived in situ Pleistocene glaciation in 'refugia within refugia'as inferred from AFLP markers. *Plant Systematics and Evolution, 306*(2), 1–12.

Kanana, F. M., Maina, C. M., Kibet, J. M., & Clement, J. M. (2020). Hypoglycaemic effects of *Salvia officinalis* extracts on alloxan-induced diabetic Swiss albino mice. *Journal of Medicinal Plants Research, 14*(10), 518–525.

Khan, A., ur Rehman, N., AlKharfy, K. M., & Gilani, A.-H. (2011). Antidiarrheal and antispasmodic activities of *Salvia officinalis* are mediated through activation of K+ channels. *Bangladesh Journal of Pharmacology, 6*(2), 111–116.

Khashan, K. T., & Al-Khefaji, K. A. (2015). Effects of Salvia officinalis L.(sag) leaves extracts in normal and alloxan-induced diabetes in white rats. *International Journal of Scientific and Engineering Research, 6*, 20–28.

Khedher, M. R. B., Hammami, M., Arch, J. R., Hislop, D. C., Eze, D., Wargent, E. T., . . . Zaibi, M. S. (2018). Preventive effects of *Salvia officinalis* leaf extract on insulin resistance and inflammation in a model of high fat diet-induced obesity in mice that responds to rosiglitazone. *PeerJ, 6*, e4166.

Khedher, M. R. B., Khedher, S. B., Chaieb, I., Tounsi, S., & Hammami, M. (2017). Chemical composition and biological activities of *Salvia officinalis* essential oil from Tunisia. *EXCLI Journal, 16*, 160.

Kianbakht, S., Abasi, B., Perham, M., & Hashem Dabaghian, F. (2011). Antihyperlipidemic effects of *Salvia officinalis* L. leaf extract in patients with hyperlipidemia: A randomized double-blind placebo-controlled clinical trial. *Phytotherapy Research, 25*(12), 1849–1853.

Kianbakht, S., & Dabaghian, F. H. (2013). Improved glycemic control and lipid profile in hyperlipidemic type 2 diabetic patients consuming *Salvia officinalis* L. leaf extract: A randomized placebo. Controlled clinical trial. *Complementary Therapies in Medicine, 21*(5), 441–446.

Kianbakht, S., Nabati, F., & Abasi, B. (2016). *Salvia officinalis* (Sage) leaf extract as add-on to statin therapy in hypercholesterolemic type 2 diabetic patients: A randomized clinical trial. *International Journal of Molecular and Cellular Medicine, 5*(3), 141.

Koubaa-Ghorbel, F., Chaâbane, M., Turki, M., Makni-Ayadi, F., & El Feki, A. (2020). The protective effects of *Salvia officinalis* essential oil compared to simvastatin against hyperlipidemia, liver, and kidney injuries in mice submitted to a high-fat diet. *Journal of Food Biochemistry, 44*(4), e13160.

Lage, H., Duarte, N., Coburger, C., Hilgeroth, A., & Ferreira, M. (2010). Antitumor activity of terpenoids against classical and atypical multidrug resistant cancer cells. *Phytomedicine, 17*(6), 441–448.

Lamia, A. K. R., & Al-Mashhady, A. (2016). Hypoglycemic effect of *Salvia officinalis* L. Extracts on induced diabetic rabbits. *International Journal of PharmTech Research, 9*(9), 250.

Lima, C. F., Azevedo, M. F., Araujo, R., Fernandes-Ferreira, M., & Pereira-Wilson, C. (2006). Metformin-like effect of *Salvia officinalis* (common sage): Is it useful in diabetes prevention? *British Journal of Nutrition, 96*(2), 326–333.

Mafakheri, M., & Kordrostami, M. (2021). Chapter 23 – Recent advances toward exploiting medicinal plants as phytoremediators. In M. Hasanuzzaman & M. N. V. Prasad (Eds.), *Handbook of Bioremediation* (pp. 371–383). Cambridge, MA: Academic Press.

Mahdi, S., Azzi, R., & Lahfa, F. B. (2020). Evaluation of in vitro α-amylase and α-glucosidase inhibitory potential and hemolytic effect of phenolic enriched fractions of the aerial part of *Salvia officinalis* L. *Diabetes & Metabolic Syndrome: Clinical Research & Reviews, 14*(4), 689–694. https://doi.org/10.1016/j.dsx.2020.05.002

Martins, N., Barros, L., Santos-Buelga, C., Henriques, M., Silva, S., & Ferreira, I. C. (2015). Evaluation of bioactive properties and phenolic compounds in different extracts prepared from *Salvia officinalis* L. *Food Chemistry, 170*, 378–385.

Miroddi, M., Navarra, M., Quattropani, M. C., Calapai, F., Gangemi, S., & Calapai, G. (2014). Systematic review of clinical trials assessing pharmacological properties of salvia species on memory, cognitive impairment and A lzheimer's disease. *CNS Neuroscience & Therapeutics, 20*(6), 485–495.

Mokogwu, A., Ikpefan, J., Eyenubo, O., & Avwioro, G. (2022). Hypoglycaemic and hypolipidemic effect of alcoholic extract of common sage (*Salvia officinalis*) in streptozotocin–induced diabetic rabbits. *African Journal of Biomedical Research, 25*(2), 243–247.

Moradabadi, L., Kouhsari, S. M., & Sani, M. F. (2013). Hypoglycemic effects of three medicinal plants in experimental diabetes: Inhibition of rat intestinal α-glucosidase and enhanced pancreatic Insulin and cardiac Glut-4 mRNAs expression. *Iranian Journal of Pharmaceutical Research, 12*(3), 387.

Moss, L., Rouse, M., Wesnes, K. A., & Moss, M. (2010). Differential effects of the aromas of Salvia species on memory and mood. *Human Psychopharmacology: Clinical and Experimental, 25*(5), 388–396.

Moss, M., Rouse, M., & Moss, L. (2014). Aromas of salvia species enhance everyday prospective memory performance in healthy young adults. *Advances in Chemical Engineering and Science*, *4*, 339–346.

Pedro, D. F., Ramos, A. A., Lima, C. F., Baltazar, F., & Pereira-Wilson, C. (2016). Colon cancer chemoprevention by sage tea drinking: Decreased DNA damage and cell proliferation. *Phytotherapy Research*, *30*(2), 298–305.

Rau, O., Wurglics, M., Paulke, A., Zitzkowski, J., Meindl, N., Bock, A., . . . Schubert-Zsilavecz, M. (2006). Carnosic acid and carnosol, phenolic diterpene compounds of the labiate herbs rosemary and sage, are activators of the human peroxisome proliferator-activated receptor gamma. *Planta Medica*, *72*(10), 881–887.

Sá, C. M., Ramos, A. A., Azevedo, M. F., Lima, C. F., Fernandes-Ferreira, M., & Pereira-Wilson, C. (2009). Sage tea drinking improves lipid profile and antioxidant defences in humans. *International Journal of Molecular Sciences*, *10*(9), 3937–3950.

Salah, M., Hussein, M., Rana, I., & Khalid, L. B. (2016). Effect of *Salvia officinalis* L.(Sage) aqueous extract on liver and testicular function of diabetic albino male rats. *JUBPAS*, *242*, 390–399.

Scholey, A. B., Tildesley, N. T., Ballard, C. G., Wesnes, K. A., Tasker, A., Perry, E. K., & Kennedy, D. O. (2008). An extract of Salvia (sage) with anticholinesterase properties improves memory and attention in healthy older volunteers. *Psychopharmacology*, *198*(1), 127–139.

Shahein, M. R., Elmeligy, A. A., & El-Sayed, M. I. (2022). The biological effects of fermented camel milk fortified with sage (*Salvia officinalis* L.) and mint (*Mentha piperita*) leaves powder on alloxan-induced diabetic rats. *Preprints*, 1–38. https://doi.org/10.21203/rs.3.rs-1362024/v1

Shankar, K., & Mehendale, H. M. (2014). Oxidative stress. In P. Wexler (Ed.), *Encyclopedia of Toxicology* (3rd ed., pp. 735–737). Cambridge, MA: Academic Press.

Sharma, Y., Fagan, J., Schaefer, J., & Yashaswini Sharma, C. (2019). Ethnobotany, phytochemistry, cultivation and medicinal properties of Garden sage (*Salvia officinalis* L.). *Journal of Pharmacognosy and Phytochemistry*, *8*, 3139–3148.

Sharmila, R., & Manoharan, S. (2012). Anti-tumor activity of rosmarinic acid in 7, 12-dimethylbenz (a) anthracene (DMBA) induced skin carcinogenesis in Swiss albino mice. *Indian Journal of Experimental Biology*, *50*, 187–194.

Tobin, J. F., & Freedman, L. P. (2006). Nuclear receptors as drug targets in metabolic diseases: New approaches to therapy. *Trends in Endocrinology & Metabolism*, *17*(7), 284–290.

Wang, M., Kikuzaki, H., Zhu, N., Sang, S., Nakatani, N., & Ho, C.-T. (2000). Isolation and structural elucidation of two new glycosides from sage (*Salvia officinalis* L.). *Journal of Agricultural and Food Chemistry*, *48*(2), 235–238.

Zhang, Y.-J., Gan, R.-Y., Li, S., Zhou, Y., Li, A.-N., Xu, D.-P., & Li, H.-B. (2015). Antioxidant phytochemicals for the prevention and treatment of chronic diseases. *Molecules*, *20*(12), 21138–21156.

16 *Swertia chirata* Buch.-Ham. Ex Wall., a Traditional Medicinal Plant with Antidiabetic Potential

Protha Biswas, Ikbal Hasan, Shatakshi Mitra, Tuyelee Das, Sujata Mandal, Abdel Rahman Al-Tawaha, and Abhijit Dey

CONTENTS

ABBREVIATIONS

ADP: adenosine diphosphate
AYUSH: Ayurveda, yoga and naturopathy, Unani, Siddha and homeopathy
CCl_4: carbon tetrachloride
$CHCl_3$: Chloroform
DM: Diabetes mellitus
GC/MS: Gas chromatography/mass spectrometry
IKK-β: Inhibitor of nuclear factor kappa-B kinase subunit beta
IL: Interleukin
IUCN: International Union for Conservation of Nature
LPS: Lipopolysaccharides
MAPK: Mitogen-activated protein kinase
MPTP: 1-methyl-4-phenyl-1,2,3,6-tetrahydropyridine
NDEA: N-nitrosodiethylamine
NF-κB: Nuclear factor kappa-light-chain-enhancer of activated B cells
PGE2: Prostaglandin E2
RAW: Murine macrophage cell line
STZ: NAD – Streptozotocin-nicotinamide
TGF: Transforming growth factor
TNF: Tumor necrosis factor

DOI: 10.1201/b23347-16

16.1 INTRODUCTION

The global community views diabetes mellitus as a serious health risk. By the year 2025, there will be 300 million adult cases of diabetes mellitus worldwide. China and India in particular will see a surge in the number of diabetic people (57 million in India) (Liu et al., 2013). Persistent hyperglycemia and abnormalities in glucose, protein, and lipid metabolism are hallmarks of diabetes, a chronic endocrine condition brought on by impaired insulin secretion and/or elevated cellular resistance (Kuzuya et al., 2002). Hyperglycemia, a metabolic disturbance brought on by abnormalities in insulin secretion, action, or both, characterizes diabetes mellitus. Type I and type II make up its two types. Only 5% of people with diabetes have type I diabetes, often known as juvenile diabetes, which is insulin dependent. Adults over the age of 40 typically develop the non-insulin-dependent type II. Chronic diabetes-related hyperglycemia has already been linked to long-term harm, dysfunction, and finally organ failure, particularly in the eyes, kidneys, nerves, heart, and blood vessels (Huang et al., 2005). First-class diabetes is most often identified by insufficient production or a diminished response to the vital regulating hormone insulin during the body's metabolic process. In order to convert carbs like sugar and starches into energy, insulin is thought to be the most important hormone. When this hormone is not produced, blood glucose levels rise (Patel et al., 2012). Insulin and other oral antibiotics including sulfonylureas, biguanides, and glinides are common therapies and medications that are now used to treat the condition, however most of them have significant drawbacks and negative side effects (Arumugam et al., 2013). Another important factor that intensifies research on traditional herbal medicines with increased hypoglycemic activity is the rising expense of conventional treatment in poor nations.

Plants have long been an important source for creating traditional medicines, and they continue to serve this role today (Husen and Iqbal, 2021; Husen, 2021, 2022). About 800 plants were specifically classified by WHO as having potential to treat diabetes (Alarcon-Aguilara et al., 1998). *Swertia chirata*, a member of the Gentianaceae family and one of these herbal plants, is particularly well known and well recognized for its antihyperglycemic properties (Kumar et al., 2016). The *Swertia* genus contains high amounts of terpenoids, xanthones, alkaloids, iridoids, and flavonoids. Many researchers conducted experiments to test the hypoglycemic effects of extracts made from this plant's bark, leaves, shoots, roots, and even the entire plant. Different researchers conducted experiments to determine the anticarcinogenic, antibacterial, anthelmintic, antimalarial, antihepatitis, analgesic, and anti-inflammatory characteristics of plant extract made from various plant sections (Dey et al., 2020). The purpose of this chapter is to describe the current state of diabetes, *S. chirata*'s active constituents and efficacy against diabetes and related conditions, and the plant's habitat, distribution, cultivation practices, traditional uses, other pharmacological properties, toxicity profile, and polyherbal formulations.

16.2 BOTANICAL DESCRIPTION

S. chirata is an erect, annual or biannual herb that grows to a height of 0.6–1.5 m (Figure 16.1). Stem is approximately 6 mm long, width 1 mm, cylindrical, glabrous, quadrangular at upper portion having a prominent decurrent line at each angle, and the middle portion is cylindrical, orangish brown or purplish in color with large continuous yellow pith (Joshi and Dhawan, 2005). Leaves are opposite, 4 cm long, without stalks, lanceolate, sessile, apex acuminate, base cordate, broad, five to seven nerved, green, glabrous, and membranous (Scartezzini and Speroni, 2000). Flowers are numerous, small, 3–4 mm, tetramerous, with large greenish-yellow leafy panicles, tinged with purple and green or white hairs, ovoid. Corolla is lobed, four lobes twisted, superimposed, and united at the base, nectarines present; calyx gamophyllous attached with four lobes; stamens four, opposite to corolla lobe, attached at the base of the corolla; stigma two; ovary superior, bicarpellary, unilocular; ovules are laminal; placentation parietal. Fruits are egg shaped capsules, with two valves separated by a transparent yellow pericarp (Scartezzini and Speroni, 2000; Joshi and Dhawan, 2005). The

FIGURE 16.1 Chiretta (*Swertia chirata*).

plant has a simple, geniculate root, yellowish, short, and tapering. It is around 7–8 cm long, width 0.5 cm (Bentley and Trimen, 1880; Scartezzini and Speroni, 2000).

16.3 DISTRIBUTION

Swertia chirata, belonging to the Gentianaceae family, is listed as a critically endangered medicinal herb in the IUCN list. It grows on the slopes of moist shady places and is indigenous to the sub-temperate regions of the Himalayas. Its distribution ranges from an altitude of 1200 m to 2100 m, encompassing the Indian states of Kashmir, Himachal Pradesh, and Uttarakhand and is also found in Nepal and Bhutan. Khasi Hills also boasts of a rich population of the herb ranging from altitudes of 1200–1500 m which includes the northeastern state of Meghalaya (Kumar and Van Staden, 2016). It is also found in higher altitudes of regions in tropical Asia, Africa, and Europe (Negi et al., 2011).

16.4 PHYTOCHEMICAL CONSTITUENTS

Herbal plants may contain a variety of chemical compounds with therapeutic values. Plants containing phytochemical constituents with high therapeutic values are used in traditional and folk medicines throughout the world. The *S. chirata* plant was discovered to contain xanthones, flavonoids, lignans, iridoids, terpenoids, palmitic acid, secoiridoid glycosides, saponins, tannins, and phenols in many studies (Pant et al., 2000; Roy et al., 2015; Bhandari et al., 2019). Chiratanin was the first dimeric xanthone to be isolated from *S. chirata*. *S. chirata*'s pharmacological efficacy has been connected to many significant phytoconstituents, including amarogentin, amaroswerin, and gentiopicrin, including mangiferin, swerchirin, sweroside, and amaroswerin (Kumar and Staden, 2016; Mahendran et al., 2022). These components have been discovered to have various medicinal attributes; for example, swerchirin is hypoglycemic, hepatoprotective, pro-hematopoietic, and slightly chemoprotective, as well as a potent antimalarial agent. Swerchirin also significantly boosted insulin release in response to glucose (Saxena et al., 1991; Dey et al., 2020). Using gas chromatography/mass spectrometry (GC/MS), Gyawali et al. (2006) extracted 77 volatiles from *S. chirata*, including 4 acids, 21 alcohols, 15 aldehydes, 3 esters, 3 furans, 7 hydrocarbons, and 17 aromatic hydrocarbons ketones and seven other kinds of substances. The ketone was the predominant component among these volatiles (28.58%), followed next by alcohol (27.44%) (Gyawali et al., 2006). Some significant chemical components of *S. chirata*, including the bitter glucosides ophelic acid, chiratin, and amarogentin, were reported by researchers. The herb also contains swerta-7, 9 (11)-dien-3-ol and pichierenol, as well as swertanone, episwertinol, swertenol, chiratenol, pichierenol, β-amyrin, y-taraxasterol, 21-a-H-hop22(29)-en-3-ol, taraxerol, gammacer-16-en-3-ol, oleanolic acid, ursolic acid, and lupeol. A novel xanthone, 1,5-dihydroxy-3,8-dimethoxyxanthone (chiratol), was isolated from the herb in addition to swerchirin, 7-O-Me swertiarin, monohydroxy terephthalic acid, and 2,5-dihydroxy terephthalic acid. Additionally, the plant produces the alkaloids gentianine, gentiocrucine, and enicoflavine, as well as xanthones such as chiratanin, a novel dimeric xanthone, 1,5,8-trihydroxy-3-methoxyxanthone, 1,8-dihydroxy-3,5-dimethoxy-xanthone, 1,8-dihydroxy-3,7-dimethoxy-xanthone, 1-hydroxy-3,5,8- trimethoxyxanthone,1,3,8-trihydroxy-5-methoxy xanthone, 1,3,7,8-tetrahydroxy xanthone, and 1,3,5,8-tetrahydroxy xanthone (Joshi and Dhawan, 2005; Kaloo et al., 2020).

16.5 PHARMACOLOGICAL STUDIES

The *S. chirata* plant contains many medicinally important phytochemical constituents that have broad spectrum of pharmacological activities. Amarogentin, obtained from the methanolic extracts of *S. chirata*, was found to exhibit an antileishmanial property by inhibiting the topoisomerase I enzyme's catalytic activity, which prevents the formation of binary complex (Ray et al., 1996). Amarogentin also showed anticancer property in carbon tetrachloride (CCl_4)/N-nitrosodiethylamine (NDEA)-induced liver carcinogenesis mouse model, where a dose of 0.2 mg/kg was able to prevent progression of liver carcinogenesis at mild dysplastic stage, through modulation of the cell cycle and significant induction of apoptosis through upregulation of the Bax-Bcl2 ratio, activation of caspase-3 and poly ADP ribose polymerase cleavage (Pal et al., 2012). Mangiferin, obtained from *S. chirata* extracts, was used to investigate its neuroprotective effect against MPTP mouse model of Parkinson's disease. Male C57BL/6 mice were treated with mangiferin at dose of 10, 20, and 40 mg/kg for 14 days. The result showed prevention of MPTP induced apoptosis and loss of dopominergic neurons (Kavitha et al., 2013). Two compounds (1-hydroxy-2,3,4,6-tetramethoxy xanthone and 1,5,8-trihydroxy-3-methoxy xanthone) isolated from *S. chirata* were initially considered as primary natural neuroprotective antioxidants. These two compounds (20 mg/kg) significantly decreased infarct size to below 5% and exhibited protective effects against cerebral damages induced by ischemia-reperfusion (Shi et al., 2013). Xanthones (bellidifolin and swerchirin) from *S. chirata* were observed to potently inhibit pro-inflammatory cytokines, IL-6 and TNF-α and prostaglandin E2 (PGE2) in LPS stimulated RAW 264.7 macrophages by blocking the expression of COX-2 and phosphorylation of Akt, IKK-β, MAPK and NF-κB, activation (TY Hu et al., 2019). The methanolic extracts of the leaves of *S. chirata* was observed to have hepatoprotective activity against hepatotoxicity induced by paracetamol (500 mg/kg) in Swiss Wistar albino rats (Kumar et al., 2019). $CHCl_3$ soluble, a crude extract of the whole *S. chirata* plant, inhibited the expression of viral protein R (Vpr) in HeLa cells harboring the TREx plasmid encoding full-length Vpr (TREx-HeLa-Vpr cells). Investigations revealed that a xanthone called bellidifolin and a triterpenoid called oleanolic acid, at an effective dose of 10 μM, are potent Vpr inhibitors (Woo et al., 2019). *S. chirata* also showed antiviral activity against herpes simplex virus type-1 and anti-hepatitis B virus (Verma et al., 2008; Zhou et al., 2015). Broad spectrum antibacterial activity of *S. chirata* was evaluated against *Staphylococcus aureus* and *Escherichia coli* in albino mice. The result showed that that the *S. chirata* plant extract was significantly better ($P < 0.05$) against *S. aureus* than was *E. coli* (Sati and Bhatt, 2022). Figure 16.2 describes the pharmacological activities and its phytochemical constituents.

16.6 ANTIDIABETIC ACTIVITY

Numerous studies have already been conducted to determine the antidiabetic potential of plant extracts made from different sections of the herb (Figure 16.3). The presence of flavonoids and secoiridoids in the plant extract makes it extremely effective at preventing problems from hyperglycemia (Rastogi et al., 1990). Both methanolic and aqueous extract of *S. chirata* leaf, stem, and root showed antidiabetic activity *in vitro* in terms of α-amylase inhibition with highest inhibition of 75%

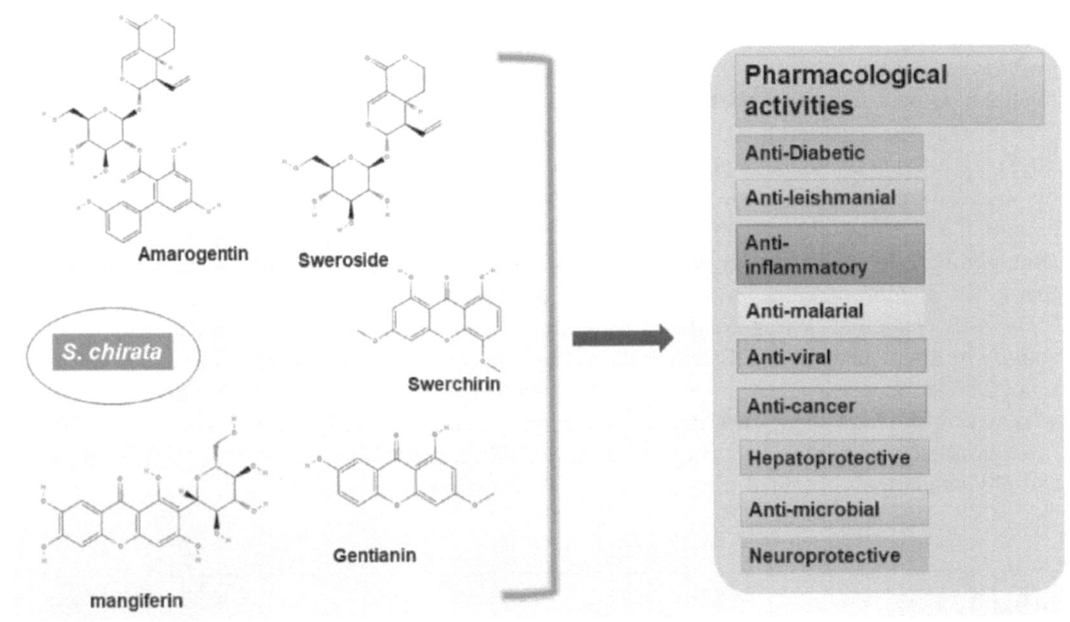

FIGURE 16.2 *Swertia chirata*: phytochemicals and biological activities.

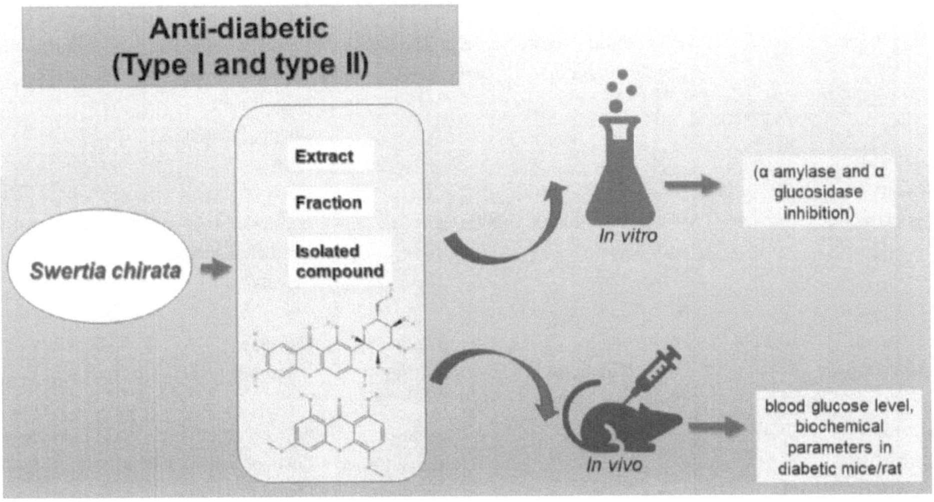

FIGURE 16.3 Antidiabetic activity of *Swertia chirata*: *in vitro* and *in vivo*.

by leaf aqueous extract (Roy et al., 2015). Crude aqueous and ethanolic extract of the plant along with standard mangiferin were analyzed for the antidiabetic and antidiabetic linked antihyperglycemic potential. The result showed α-glucosidase inhibitory activity (highest inhibition of 43% by ethanol extract) and angiotensin I-converting enzyme inhibition assay (highest inhibition of 76.3% by ethanol extract) (Phoboo et al., 2013). Aqueous *S. chirata* bark extract was investigated to see the antidiabetic properties by the effects on insulin production, cellular glucose absorption, and protein glycation. The findings show that *S. chirata* aqueous bark extracts have antidiabetic effects via enhancing insulin action (28%–59%), increasing insulin production ($P < 0.001$), as well as inhibiting protein glycation ($P < 0.05$–$P < 0.001$) (Thomson et al., 2014).

The antidiabetic activity of aqueous extract of *S. chirata* was studied on the blood glucose level in streptozotocin-induced diabetes in albino rats. Treatment of 200 mg/kg body weight dose of the *S. chirata* aqueous extract for 21 days showed significant antidiabetic activity (Kavitha and Dattatri, 2013). Significant antidiabetic activity was also noticed in Swiss albino mice with diabetes induced by streptozotocin-nicotinamide (STZ-NAD), administered 200 mg/kg and 500 mg/kg ethanolic extract of *S. chirata*. The extract also indicated positive effects on triglyceride and cholesterol levels (Renu et al., 2011). The hypoglycemic effect of ethanolic extract of leaf and its different fractions (pet-ether, dichloromethane, and methanol) of *S. chirata* at a dose of 250 mg/kg body weight on Swiss albino mice at fasting condition were studied. The result showed significant hypoglycemic activity by about 32% and 47.2% (ethanolic extract and pet-ether fraction) respectively for the reduction of blood glucose level (Alam et al., 2011). A 95% ethanol extract and four fractions of *S. chirata* were tested for blood sugar lowering activity in rats when given orally at 250 mg/kg and the hexane fraction caused maximum lowering (Sekar et al., 1987). Antidiabetic property of different extract and phytochemicals obtained from *S. chirata* is mentioned in Table 16.1.

TABLE 16.1
Antidiabetic Potential of *Swertia chirata*

Extract/isolated compound	*In vitro /in vivo*	Model	Method	Result	Key references
95% ethanol extract and four fractions	*In vivo*	rats	250mg/kg oral dose	lowered blood glucose level	Sekar et al. (1987)
ethanolic extract of leaf and its different fractions (pet-ether, dichloromethane, and methanol)	*in vivo*	Swiss albino mice at fasting condition	dose of 250 mg/kg body weight	reduction of blood glucose level, significant hypoglycemic activity	Alam et al. (2011)
ethanolic extract	*in vivo*	Swiss albino mice with diabetes induced by STZ-NAD	200 mg/kg and 500 mg/kg dose	controlled triglyceride and cholesterol levels	Renu et al. (2011)
aqueous extract	*in vivo*	streptozotocin-induced diabetes in albino rats	200 mg/kg body weight dose for 21 days	significant antidiabetic activity	Kavitha and Dattatri (2013)
crude aqueous, ethanolic extract, and mangiferin	*in vitro*	_	α-glucosidase inhibition and α-glucosidase inhibitory	43% by ethanol extract and 76.3% by ethanol extract	Phoboo et al. (2013)
aqueous bark extract	*in vitro*	_	insulin production, cellular glucose absorption, and protein glycation	enhanced insulin action (28%–59%), increased insulin production ($P < 0.001$), inhibited protein glycation ($P < 0.05$–$P < 0.001$)	Thomson et al. (2014)
methanolic and aqueous extract of leaf	*in vitro*	_	α-amylase inhibition	75% inhibition by leaf aqueous extract	Roy et al. (2015)

16.7 POLYHERBAL FORMULATIONS

Polyherbal is a term used in traditional medicine to describe the use of more than one herb to treat a disease or condition (Rezghi et al., 2021). Polyherbal formulations that comprise *S. chirata*, *Ficus dalhousie*, *Ichnocarpus frutescens*, *Creteva magna*, and *Alpinia galanga* have been shown to alleviate flatulence and gastrointestinal illnesses induced by metformin and other glycosidase inhibitors (Patwekar et al., 2022). *Alstoniascholaris*, *Picrorhiza kurroa* Royle, *S. chirata*, and *Caesalpinia crista* are the primary constituents of AYUSH 64, which is effective and may be used for various viral illnesses that cause a fever (Gundeti et al., 2020). The polyherbal ointment made from the combination of *S. chirata*, *Ficus dalhousie*, *Ichnocarpus frutescens*, *Crateva magna*, and *Alpinia galanga* may be used therapeutically to treat diabetic and nondiabetic wounds (Quazi et al., 2022). Ingredients of Jawarish jalinoos in Unani are made up of herbs such as *Pistacia lentiscus*, *Cinnamomum caeruleum*, *Nardostachys jatamansi*, *S. chirata*, and others with sugar or honey which has antioxidant, anti-inflammatory, and gastroprotective properties (Aleem, 2022). Several medicinal herbs, including *S. chirata*, are used in Sudarshan Churna, which has antipyretic and antibacterial action (Sharma et al., 2020).

16.8 TRADITIONAL AND OTHER POTENTIAL USES

Swertia chirata is a very well-known medicinal herb in many ancient medication systems such as Ayurvedic, Unani, Siddha, and alternative mainstream medical systems, and are still used today. Its beneficial uses are well documented in the British, American, and Indian pharmacopoeias as well as the Indian Pharmaceutical Codex. This plant is endemic to the temperate Himalayas and is used in traditional medicines to treat a variety of illnesses such liver diseases, malaria, and polygenic disease (Amit and Singh, 2021). *S. chirata* is used as a bitter tonic owing to its numerous bioactive components. There is a widespread practice among locals to utilize the whole plant as a remedy for hepatitis, inflammation, and digestive disorders (Pandey and Sharma, 2021). Mahasudarshanachurna, a remedy including more than 50 herbs, is perhaps the most well-known use of *S. chirata* in Indian medicine (Kumar and Van Staden, 2016). Both male and female responders widely use *S. chirata* in Sikkim to cure fever, cold, and cough. Sherpa communities utilize *S. chirata* paste to cure skin ailments (Pradhan and Badola, 2015). In Unani, they use this plant extract and other parts to treat vomiting, ulcer, gastrointestinal tract complaints, intestinal worms, and constipation (Hoq, 2018). *S. chirata* as tinctures and infusions are used in the British and American pharmacopoeias (Amit and Singh, 2021). The indigenous people of Assam, India, employ *S. chirata* and lime juice, according to prevent dysentery. Stomach worms may be treated using the twigs of *S. chirata*, soaked in hot water. The twigs are also effective in warding off scabies (Das et al., 2008). The paste of the whole plant is applied to treat skin diseases such as eczema and pimples, liver and stomach disorders like dyspepsia and diarrhea, intestinal worms, hiccups and vomiting, ulcers, gastrointestinal infections, and kidney diseases, excessive vaginal discharge, and in cases of scorpion bite in combination with other drugs. The plant is used as an effective tonic for general weakness, fever, cough, joint pain, asthma, and the common cold (Joshi and Dhawan, 2005; Malla et al., 2015; Kirtikar and Basu, 1984; Jadhav and Bhutani, 2005). *S. chirata* has been used as a blood purifier in the Himalayas to treat malaria, as well as for mental illnesses such as epilepsy, respiratory tract issues such as asthma, and flu symptoms (Kunwar et al., 2021). The leaves and cut stems are steeped in water overnight for headaches and blood pressure. One glass of water is used to make a paste, which is then filtered. For two to three days, the preparation is ingested once daily (de Rus Jacquet et al., 2014). For the treatment of malaria, one cup of the plant decoction is boiled in water (Shah et al., 2014).

16.9 SAFETY USES

One of the vital issues faced by pharmaceutical industries is concern regarding the safety of conventional drugs (Verschaeve and Van Staden, 2008). *S. chirata*, despite being used in traditional medicine for a very long time, still lacks scientific information concerning its safety. It can be traced through the medicinal history as a nontoxic and safe ethnomedicinal herb that has been used since long to expel various ailments. No obvious toxic effects were caused by *S. chirata* extracts in mice because there were no significant differences in body weight and body temperature between the treated and control groups (Alam et al., 2011; Das et al., 2012). Since traditional usage is insufficient to guarantee the safety of herbal treatments, a history of herbal use demonstrates that not all herbs are benign and can occasionally be lethal. Furthermore, because so few herbal remedies have undergone extensive testing for toxicity or carcinogenicity, we cannot be certain that they are all risk-free (Grollman and Marcus, 2016). Studies have found that *S. chirata* revealed no evidence of toxicity in both liposomal and niosomal forms (Medda et al., 1999). The male KM mice were injected with 2000 mg/kg b.w. of *S. chirata* ethanolic extract over a 72-hour period, however no deaths or toxic symptoms were noted (Chen et al., 2011). Amarogentin, a secoiridoid glycoside from *S. chirata*, was administered orally to female Swiss albino mice for 15 days (twice weekly) and did not show any adverse effects in another study on subacute toxicity (Pal et al., 2012). Stringent efforts are required to further investigate the safety levels of the plant, to confirm its safe and effective usage in traditional medicine as well as usage in pharmaceuticals for modern medicines.

16.10 CULTIVATION PRACTICES

The plant may grow in a wide range of soils, including sandy loam to soils rich in humus and carbon. Additionally, it can be found in bare areas and slash-and-burn woodlands (Edwards, 1993). Natural plant regeneration occurs through the production of seeds, which are biologically mature and have a high chance of germination in November. Seeds have poor viability before November, and after next October. According to reports, up to 90% of seeds are collected after November and properly cleaned for germination. Sowing for seeds is typically done in the spring, when the temperature is below 10°C and the soil has a healthy amount of humus. When the seedlings are big enough to handle, they are removed one at a time and put into various pots or containers. Early in the summer, the young plants are replanted outside (Patil et al., 2013).

16.11 CONCLUSION AND FUTURE REMARKS

Swertia chirata, the medicinal herb, is becoming more and more important pharmacologically as a result of the valuable phytochemicals that make up its composition. The treatment of diabetes and several other related conditions has benefited greatly from its beneficial effects. Despite having a wide range of therapeutic benefits, particularly as a strong antidiabetic herb, this plant has not attained the prominence it deserves in the commercial world. It might be because there aren't enough clearly defined dosages or specifications for the medications that are generated from them, and people believe in commercially available ones. Additionally, not enough research has been done to demonstrate the plant's use in mechanism of action and functions of its active ingredients specifically against diseases that affect humans. On the other hand, expanding commercial uses and demand for plant species result in unfettered overexploitation and unethical overharvesting of wild plant life. However, the biggest obstacle to its widespread commercial applications on people is the absence of sufficient evidence concerning safety. Toxicological and histopathological research is still needed to standardize the plant product's response to human illnesses and to better understand its mechanism. Incorporating various cutting-edge plant biotechnological techniques, such as micro-propagation, hairy root culture technologies, and synthetic seed manufacturing, may increase the supply of these plant species and satisfy future demand.

REFERENCES

Alam, K.D., Ali, M.S., Mahjabeen, S., Hassan, M.R., Rahman, M.F. and Chowdhury, R.M.A.A., 2011. Potential hypoglycemic effect of *Swertia chirata* – An Indian subcontinent herb with important medicinal value. *Pharmacologyonline*, *2*, 642–647.

Alarcon-Aguilara, F.J., Roman-Ramos, R., Perez-Gutierrez, S., Aguilar-Contreras, A., Contreras-Weber, C.C. and Flores-Saenz, J.L., 1998. Study of the anti-hyperglycemic effect of plants used as antidiabetics. *Journal of Ethnopharmacology*, *61*(2), 101–110.

Aleem, M., 2022. Jawarish jalinoos: A polyherbal unani formulation for gastrointestinal disorders. *International Journal of Green Pharmacy (IJGP)*, *16*(1), 1–13.

Amit, A. and Singh, A.K., 2021. Ethanobotanical uses of *Swertia chirata*. *Advances in Pharmaceutical Sciences*, *1*(1), 46–57.

Arumugam, G., Manjula, P. and Paari, N., 2013. A review: Anti diabetic medicinal plants used for diabetes mellitus. *Journal of Acute Disease*, *2*(3), 196–200.

Bentley, R. and Trimen, H., 1880. *Medicinal plants: Being descriptions with original figures of the principal plants employed in medicine and an account of the characters, properties, and uses of their parts and products of medicinal value*. Vol. 4. London: J. & A. Churchill.

Bhandari, L., Bista, B.B., Bhatta, M.R., Khanal, C., Khanal, S., Ranjitkar, R. and Bhandari, D.P., 2019. Phytoconstituents, antioxidant and bitterness value of *Swertia chirata* from four different geographical region of Nepal. *Journal of Plant Resources*, *17*(1), 104–111.

Chen, Y., Huang, B., He, J., Han, L., Zhan, Y. and Wang, Y., 2011. In vitro and in vivo antioxidant effects of the ethanolic extract of *Swertia chirata*. *Journal of Ethnopharmacology*, *136*(2), 309–315.

Das, F.A., Barua, I. and Das, D.D., 2008. Ethno-medicinal practices: A case study among the Sonowal Kacharis of Dibrugarh, Assam. *Studies on Ethno-Medicine*, *2*(1), 33–37.

Das, S.C., Bhadra, S., Roy, S., Saha, S.K., Islam, M.S. and Bachar, S.C., 2012. Analgesic and anti-inflammatory activities of ethanolic root extract of *Swertia chirata* (Gentianaceae). *Jordan Journal of Biological Science*, *5*, 31–36.

de Rus Jacquet, A., Subedi, R., Ghimire, S.K. and Rochet, J.C., 2014. Nepalese traditional medicine and symptoms related to Parkinson's disease and other disorders: Patterns of the usage of plant resources along the Himalayan altitudinal range. *Journal of Ethnopharmacology*, *153*(1), 178–189.

Dey, P., Singh, J., Suluvoy, J.K., Dilip, K.J. and Nayak, J., 2020. Utilization of *Swertia chirata* plant extracts for management of diabetes and associated disorders: Present status, future prospects and limitations. *Natural Products and Bioprospecting*, *10*(6), 431–443.

Edwards, D.M., 1993. The marketing of non-timber forest products from the Himalayas: The trade between East Nepal and India. *Network Paper-Rural Development Forestry Network (United Kingdom)*, *15b*, 1–21.

Grollman, A.P. and Marcus, D.M., 2016. Global hazards of herbal remedies: Lessons from *Aristolochia*: The lesson from the health hazards of *Aristolochia* should lead to more research into the safety and efficacy of medicinal plants. *EMBO Reports*, *17*(5), 619–625.

Gundeti, M.S., Bhurke, L.W., Mundada, P.S., Murudkar, S., Surve, A., Sharma, R., Mata, S., Rana, R., Singhal, R., Vyas, N. and Khanduri, S., 2020. AYUSH 64, a polyherbal Ayurvedic formulation in Influenza-like illness-Results of a pilot study. *Journal of Ayurveda and Integrative Medicine*, *13*(1), 100325.

Gyawali, R., Ryu, K.Y., Shim, S.L., Kim, J.H., Seo, H.Y., Han, K.J. and Kim, K.S., 2006. Essential oil constituents of *Swertia chirata* buch.-ham. *Preventive Nutrition and Food Science*, *11*(3), 232–236.

Hoq, M.O., 2018. Medicinal importance of *Swertia chirayita* in Unani system of medicine. *International Journal of Unani and Integrative Medicine*, *2*(2), 47–51.

Hu, T.Y., Ju, J.M., Mo, L.H., Ma, L., Hu, W.H., You, R.R., Chen, X.Q., Chen, Y.Y., Liu, Z.Q., Qiu, S.Q. and Fan, J.T., 2019. Anti-inflammation action of xanthones from *Swertia chirayita* by regulating COX-2/NF-κB/MAPKs/Akt signalling pathways in RAW 264.7 macrophage cells. *Phytomedicine*, *55*, 214–221.

Huang, T.H., Peng, G., Kota, B.P., Li, G.Q., Yamahara, J., Roufogalis, B.D. and Li, Y., 2005. Anti-diabetic action of Punica granatum flower extract: Activation of PPAR-γ and identification of an active component. *Toxicology and Applied Pharmacology*, *207*(2), 160–169.

Husen, A., 2021. *Traditional herbal therapy for the human immune system*. Boca Raton, FL: CRC Press. https://doi.org/10.1201/9781003137955

Husen, A., 2022. *Herbs, shrubs and trees of potential medicinal benefits*. Boca Raton, FL: CRC Press. https://doi.org/10.1201/9781003205067

Husen, A. and Iqbal, M., 2021. Plant-based potential nutraceuticals for improving the human immune system. In *Traditional herbal therapy for the human immune system* (pp. 1–12). Boca Raton, FL: CRC Press.

Jadhav, A.N. and Bhutani, K.K., 2005. Ayurveda and gynecological disorders. *Journal of Ethnopharmacology*, *97*(1), 151–159.

Joshi, P. and Dhawan, V., 2005. *Swertia chirata* – an overview. *Current Science*, 635–640.

Kaloo, M.A. and Bhat, B.A., 2020. Chemical composition and pharmacology of a medicinal herb: *Swertia chirata*. *Internationale Pharmaceutica Sciencia 2020*, *11*(1), 105.

Kavitha, K.N. and Dattatri, A.N., 2013. Experimental evaluation of antidiabetic activity of *Swertia chirata*–aqueous extract. *Journal of Public Health and Medical Research*, *1*(2), 71–75.

Kavitha, M., Nataraj, J., Essa, M.M., Memon, M.A. and Manivasagam, T., 2013. Mangiferin attenuates MPTP induced dopaminergic neurodegeneration and improves motor impairment, redox balance and Bcl-2/Bax expression in experimental Parkinson's disease mice. *Chemico-Biological Interactions, 206*(2), 239–247.

Kirtikar, K.R. and Basu, B.D., eds., 1984. *Indian medicinal plants.* Vol. 3. Allahabad: LM Basu Publishers.

Kumar, A., Pai, M.O. and Rai, N., 2019. Evaluation of hepatoprotective activity of *Albizia lebbeck Linn., Cassia occidentalis Linn. and Swertia chirata Roxb.* against paracetamol induced liver injury in Wistar rats. *Medicinal Plants-International Journal of Phytomedicines and Related Industries, 11*(4), 455–463.

Kumar, V. and Van Staden, J., 2016. A review of *Swertia chirata* (Gentianaceae) as a traditional medicinal plant. *Frontiers in Pharmacology, 6,* 308.

Kunwar, R.M., Sher, H. and Bussmann, R.W., eds., 2021. *Ethnobotany of the Himalayas.* Switzerland: Springer Nature.

Kuzuya, T., Nakagawa, S., Satoh, J., Kanazawa, Y., Iwamoto, Y., Kobayashi, M., Nanjo, K., Sasaki, A., Seino, Y., Ito, C. and Shima, K., 2002. Report of the committee on the classification and diagnostic criteria of diabetes mellitus. *Diabetes Research and Clinical Practice, 55*(1), 65–85.

Liu, Y., Sun, J., Rao, S., Su, Y. and Yang, Y., 2013. Antihyperglycemic, antihyperlipidemic and antioxidant activities of polysaccharides from *Catathelasma ventricosum* in streptozotocin-induced diabetic mice. *Food and Chemical Toxicology, 57,* 39–45.

Mahendran, G., Verma, N., Singh, S., Parveen, S., Singh, M., Luqman, S., Shanker, K. and ur Rahman, L., 2022. Isolation and characterization of a novel xanthone from the hairy root cultures of *Swertia chirata* (Roxb.) H. Karst. and its biological activity. *Industrial Crops and Products, 176,* 114369.

Malla, B., Gauchan, D.P. and Chhetri, R.B., 2015. An ethnobotanical study of medicinal plants used by ethnic people in Parbat district of western Nepal. *Journal of Ethnopharmacology, 165,* 103–117.

Medda, S., Mukhopadhyay, S. and Basu, M.K., 1999. Evaluation of the in-vivo activity and toxicity of amarogentin, an antileishmanial agent, in both liposomal and niosomal forms. *Journal of Antimicrobial Chemotherapy, 44*(6), 791–794.

Negi, J.S., Singh, P. and Rawat, B., 2011. Chemical constituents and biological importance of Swertia: A review. *Current Research in Chemistry, 3*(1), 1–15.

Pal, D., Sur, S., Mandal, S., Das, A., Roy, A., Das, S. and Panda, C.K., 2012. Prevention of liver carcinogenesis by amarogentin through modulation of G 1/S cell cycle check point and induction of apoptosis. *Carcinogenesis, 33*(12), 2424–2431.

Pandey, S. and Sharma, S., 2021. Review on *Swertia chirata* as traditional medicine. *Asian Journal of Current Research,* 1–5.

Pant, N., Jain, D.C. and Bhakuni, R.S., 2000. Phytochemicals from genus *Swertia* and their biological activities. *Indian Journal of Chemistry, 39B,* 565–586.

Patel, D.K., Prasad, S.K., Kumar, R. and Hemalatha, S., 2012. An overview on antidiabetic medicinal plants having insulin mimetic property. *Asian Pacific Journal of Tropical Biomedicine, 2*(4), 320–330.

Patil, K., Dhande, S. and Kadam, V., 2013. Therapeutic *Swertia chirata* – an overview. *Research Journal of Pharmacognosy and Phytochemistry, 5,* 199–207.

Patwekar, M., Quazi, A., Faheem, I.P., Kamal, M.A., Mukim, M., Rather, G.A. and Patil, A.P., 2022. *In vitro* inhibitory effect on alpha amylase enzyme by polyherbal dip tea in diabetes. *Indo Global Journal of Pharmaceutical Sciences (IGJPS), 12,* 156–165.

Phoboo, S., Pinto, M.D.S., Barbosa, A.C.L., Sarkar, D., Bhowmik, P.C., Jha, P.K. and Shetty, K., 2013. Phenolic-linked biochemical rationale for the anti-diabetic properties of *Swertia chirayita* (Roxb. ex Flem.) Karst. *Phytotherapy Research, 27*(2), 227–235.

Pradhan, B.K. and Badola, H.K., 2015. Local knowledge on the use of Swertia chirayita as traditional medicine: Conservation challenges in Sikkim Himalaya, India. *Ethnobotany Research and Applications, 14,* 345–355.

Quazi, A., Patwekar, M., Patwekar, F., Mezni, A., Ahmad, I. and Islam, F., 2022. Evaluation of wound healing activity (excision wound model) of ointment prepared from infusion extract of polyherbal tea bag formulation in diabetes-induced rats. *Evidence-Based Complementary and Alternative Medicine, 2022.*

Rastogi, R.P., Mehrotra, B.N., Sinha, S., Pant, P. and Seth, R., 1990. *Compendium of Indian medicinal plants: 1985–1989.* Vol. 4. New Delhi: Central Drug Research Institute and Publications & Information Directorate.

Ray, S., Majumder, H.K., Chakravarty, A.K., Mukhopadhyay, S., Gil, R.R. and Cordell, G.A., 1996. Amarogentin, a naturally occurring secoiridoid glycoside and a newly recognized inhibitor of topoisomerase I from *Leishmania donovani*. *Journal of Natural Products*, *59*(1), 27–29.

Renu, A., Sunil, K., Dinesh, K., Ajay, M. and Tarun, K., 2011. Antidiabetic activity of ethanolic extract of Swertia chirayita Buch-Ham. *International Research Journal of Pharmacy*, *2*(1), 230–232.

Rezghi, M., Mortazavi, S.A., Fahimi Sh, C.R., Sheikholeslami, M.A. and Hamzeloo-Moghadam, M., 2021. Polyherbal tablet based on Iranian traditional medicine. *Journal of Medicinal Plants*, *20*(77), 15–25.

Roy, P., Abdulsalam, F.I., Pandey, D.K., Bhattacharjee, A., Eruvaram, N.R. and Malik, T., 2015. Evaluation of antioxidant, antibacterial, and antidiabetic potential of two traditional medicinal plants of India: *Swertia cordata* and *Swertia chirata*. *Pharmacognosy Research*, *7*(1), S57–S62.

Sati, B. and Bhatt, S., 2022. Antibacterial effect of *Swertia chirata* against multi-drug resistant strains S. aureus and E. coli: In vivo and in vitro study. *European Journal of Molecular & Clinical Medicine*, *9*(3), 1106–1115.

Saxena, A.M., Bajpai, M.B. and Mukherjee, S.K., 1991. Swerchirin induced blood sugar lowering of streptozotocin treated hyperglycemic rats. *Indian Journal of Experimental Biology*, *29*(7), 674–675.

Scartezzini, P. and Speroni, E., 2000. Review on some plants of Indian traditional medicine with antioxidant activity. *Journal of Ethnopharmacology*, *71*(1–2), 23–43.

Sekar, B.C., Mukherjee, B., Chakravarti, R.B. and Mukherjee, S.K., 1987. Effect of different fractions of *Swertia chirata* on the blood sugar level of albino rats. *Journal of Ethnopharmacology*, *21*(2), 175–181.

Shah, G.M., Abbasi, A.M., Khan, N., Guo, X., Khan, M.A., Hussain, M., Bibi, S., Nazir, A. and Tahir, A.A., 2014. Traditional uses of medicinal plants against malarial disease by the tribal communities of Lesser Himalayas–Pakistan. *Journal of Ethnopharmacology*, *155*(1), 450–462.

Sharma, N., Sharma, A., Ravat, D. and Sharma, R., 2020. Comparative assessment of anti-microbial property of different brand of polyherbal preparation. *For Healthcare Professionals: Scope and Opportunities*, *138*.

Shi, G.F., Wang, G.Y. and Chen, X.F., 2013. Screening of radical-scavenging natural neuroprotective antioxidants from *Swertia chirata*. *Acta Biologica Hungarica*, *64*(3), 267–278.

Thomson, H., Ojo, O., Flatt, P. and Abdel Wahab, Y., 2014. Antidiabetic actions of aqueous bark extract of *Swertia chirata* on insulin secretion, cellular glucose uptake and protein glycation. *Journal of Experimental and Integrative Medicine*, *4*(4), 268–272.

Verma, H., Patil, P.R., Kolhapure, R.M. and Gopalkrishna, V., 2008. Antiviral activity of the Indian medicinal plant extract, *Swertia chirata* against herpes simplex viruses: A study by in-vitro and molecular approach. *Indian Journal of Medical Microbiology*, *26*(4), 322–326.

Verschaeve, L. and Van Staden, J., 2008. Mutagenic and antimutagenic properties of extracts from South African traditional medicinal plants. *Journal of Ethnopharmacology*, *119*(3), 575–587.

Woo, S.Y., Win, N.N., Oo, W.M.N., Ngwe, H., Ito, T., Abe, I. and Morita, H., 2019. Viral protein R inhibitors from *Swertia chirata* of Myanmar. *Journal of Bioscience and Bioengineering*, *128*(4), 445–449.

Zhou, N.J., Geng, C.A., Huang, X.Y., Ma, Y.B., Zhang, X.M., Wang, J.L. and Chen, J.J., 2015. Anti-hepatitis B virus active constituents from *Swertia chirata*. *Fitoterapia*, *100*, 27–34.

17 Ethnopharmacological Efficacy of *Stevia rebaudiana* (Bert.) *An Antidiabetic Herb*

Umakanta Dash, Ashwath M N, Ankita Bhardwaj, Susmita Shil, Sidhartha T N, Vijay Kumar, Rohan M S, and Kajal Samantara

CONTENTS

DOI: 10.1201/b23347-17

17.1 INTRODUCTION

Diabetes mellitus (DM) has become one of the most widespread chronic diseases throughout the world (Leon and Maddox, 2015). This disease transmission has surged from 30 million in 1985 to 382 million in 2014, and the current projections indicate that these rates will only continue to rise (Wild et al., 2004). One in ten people, or 592 million people globally, are expected to develop diabetes by 2035, according to the International Diabetes Federation forecast (Aguiree et al., 2013). A significant financial burden is placed on the healthcare system and the patient due to rising DM prevalence. The average yearly cost of DM in the US is $2,108 per patient, over twice as much as the cost for people without diabetes (Bahia et al., 2011). There is a significant link between DM and cardiovascular disease (CVD), which remains the major reason behind morbidity and mortality in diabetic humans. Obesity, hypertension and dyslipidaemia are common cardiovascular causative factors in individuals with DM that increases their risk for cardiac events such as strokes, heart failure, coronary heart disease (CHD), myocardial infarction, and hypertension due to blood coagulation (Artham et al., 2009). Based on these factors, plants with antidiabetic and anti-coagulant properties will aid in the treatment of these types of chronic disorders.

Since antiquity, these plants have been chief source of therapeutic agents. Increasing recognition of conventional herbal systems of medicine, like *Ayurveda*, within and outside India has resulted in the resumption of ancient traditions of medicine (Bala, 1982). Declining efficacy of synthetic medications and proven therapeutic values of plants in curing many health maladies with minimal side effects has led to a rise in the use of medicinal plants worldwide. Even as per the assessment of World Health Organization (WHO), 80% of the world's population prefers conventional medications. Also, the majority of treatments make use of plant-based active ingredients and extracts (Das et al., 2015). The category of medicinal plant comprises nearly 8,000 species and accounts for about 50% of total higher flowering plant species of India. Numerous studies suggest that these medicinal species have considerable antioxidant activity and may assist in cell protection from oxidative damage brought on by free radicals since they include flavonoids, terpenoids, alkaloids, diterpenes, tannins, glycosides, and other phenolic chemicals (Krishnaiah et al., 2011). The production of these free radicals and reactive oxygen species (ROS), which include the superoxide anion, hydroxyl radical, and hydrogen peroxide in the human body as well as from exogenous sources, is crucial for the emergence of a number of diseases, including arthritis, asthma, dementia, mongolism, carcinoma, and Parkinson's disease (Halliwell and Gutteridge, 1990). There is an imbalance between oxidants and antioxidants as a result of the interaction of these free radicals with diverse biological components (lipids, proteins, and deoxyribonucleic acids). The medicinal plants scavenge these free radicals because of the antioxidants and secondary metabolites, thereby protecting human health from a number of ailments (Figure 17.1).

Likewise, *Stevia rebaudiana* Bertoni. is such a medicinally potent plant species belonging to the family Asteraceae/Compositae (among the largest plant families with approx. 24,000 accepted plant species with potency to cure various health ailments). In addition, it is known as the "cosmopolitan family" since 1,600 to 1,700 of its genera are dispersed on all continents of the planet (except Antarctica). Asteroideae, Barnadesioideae, and Cichorioideae are the three subfamilies that make up the genus Asteraceae (Gasmalla et al., 2014). English names for *S. rebaudiana* include stevia, sugar herb, candy leaf, sweet honey leaf, non-caloric bio sweetener, and sweet herb of Paraguay; Sanskrit and Hindi names include Madhu patra and Meethi Patti, respectively (Khan et al., 2018). It gained economic significance as a result of its significant contribution to the sugar industries, which liberally employs its very sweet leaves to sweeten a wide range of drinks (Gantait et al., 2015). There are 150–200 species in the genus *Stevia* (Compositae family), 20 of which are widely cultivated (*S. rebaudiana*, *S. plummerae*, *S. rhombifolia*, *S. salicifolia*, *S. serrata*, *S. tevia*, *S. commixta*, *S. satureiaefilia*, *S. leptophylla*, *S. myriadenia*, *S. ophryphylla*, *S. selloi*, *S. nepetifolia*, *S. oligophylla*, *S. origanoides*, *S. eupatoria*, *S. lemmonii*, *S. micrantha*, *S. ovate*, and *S. triflora*) (Yadav et al., 2011). It was reported that *S. rebaudiana* produced the sweetest fragrance out of these 20 species (Soejarto et al., 1983). The Paraguayan people have been employing this remarkable plant as an artificial sweetener and in several herbal medicines since the dawn of time. According to early sources,

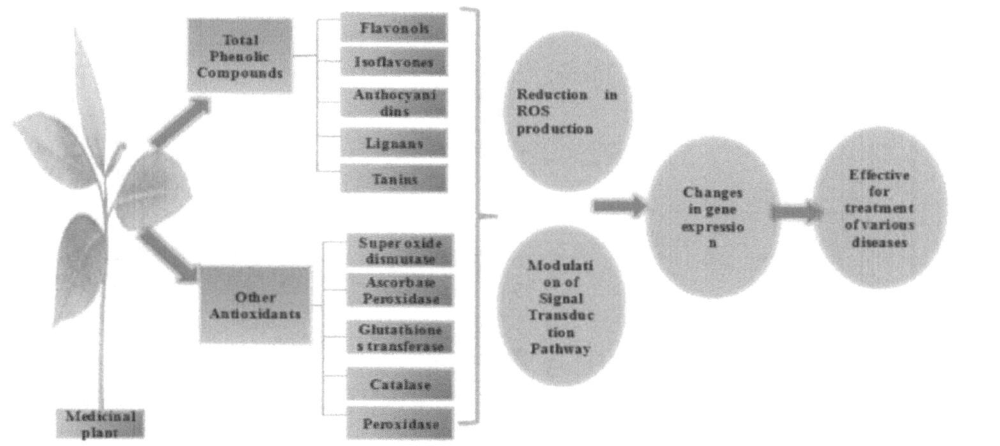

FIGURE 17.1 Mechanism behind therapeutic effects of medicinal plants (adopted from Singh et al., 2016).

stevia was also acknowledged by the Spanish people throughout the fifteenth century; however, its significance was not made clear to the Europeans until 1888, when it was once more brought to their notice. Finally, the eminent Swiss botanist M.S. Bertoni, who obtained it from the Paraguayan Indians and Mestizos, rediscovered the pleasant characteristic of this plant's leaves. Steviol glycosides, a class of sweet-tasting compounds found in the leaves, may be processed commercially into powdered non-caloric sweeteners and used fresh or dried to sweeten drinks or desserts. In addition to its industrial applications, numerous studies have reported various health benefits of stevia including antidiabetic, anti-obesity, anti-tumour, anti-hypertensive, antimicrobial, anti-caries, and antioxidant properties (Kim et al., 2011). Therefore, here an attempt has been to provide thorough and up to date information on the phytography, health benefits, phytochemical composition, industrial, and pharmacological aspects of *S. rebaudiana*.

17.2 OCCURRENCE

The genus *Stevia* can be found throughout the world starting from the southern United States to Argentina and the Brazilian highlands, via Mexico, the Central American states, and the Andes in South America (Figure 17.2). Stevia is mostly found between 500 and 3,500 metres above sea level (Midmore and Rank, 2002). It may be found in a variety of environments, including grasslands, scrub woods, conifer forests, mountain slopes, and subalpine vegetation. As purported by Ramesh et al. (2006), Sunk specified the range in 1975 to be between 22° and 24°S latitudes and 55° to 56°W longitudes, as well as within 200–700 m altitudinal regions, whereas Bertoni estimated the plant's distribution range in 1905 to be between 22°30'–25°30'S latitudes and 55°–57°W longitudes. The natural habitat of stevia is located in north-eastern Paraguay (Rio Monday Valley of Paraguayan highlands) at a latitude of 25°S in a subtropical area between 500 and 1,500 m above mean sea level, with a constant annual temperature of 25 °C and an average precipitation of 1,375 mm (Gantait et al., 2018). This area, which appears to be located in an imaginary triangle formed by Peru, Bolivia, Southern Brazil, and Northern Argentina, is home to at least 120 species of stevia (Soejarto, 2002).

Considering its immense therapeutic values, stevia is recognized as a significant shrub crop in different nations, including Brazil, Canada, Indonesia, India, Japan, Korea, Mexico, the United States, and Tanzania (Figure 17.2). Stevia production is mostly concentrated in China, and the main market has historically been Japan (Kinghorn and Soejarto, 2002). In recent years, stevia has been grown in India's Karnataka, Rajasthan, Maharashtra, and Odisha (Amarakoon, 2021). The farmers in India have been compelled to engage in extensive stevia farming due to the rising demand for natural sweeteners (Megeji et al., 2005).

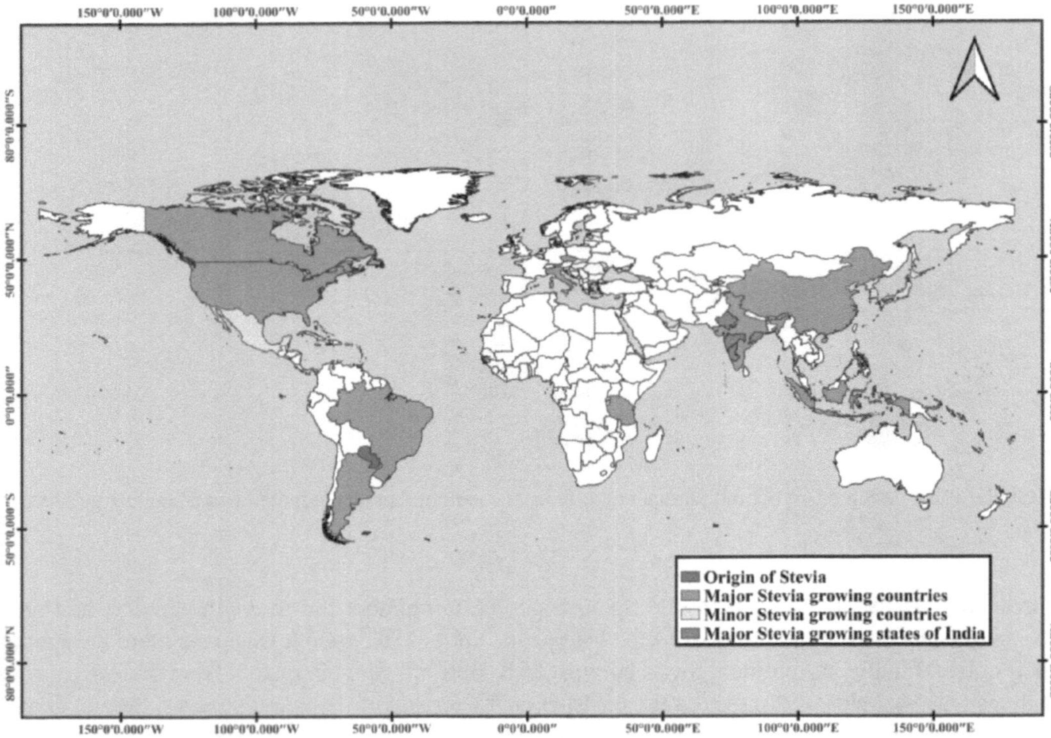

FIGURE 17.2 Global distribution of *Stevia rebaudiana*.

17.3 PHYTOGRAPHY

Stevia a small perennial shrub reaching upto 65–80 cm in height (Figure 17.3). It has stringy, thread-like perennial roots, generating copious stock that is rarely branched and is the sole portion that lacks stevioside (Skaria et al., 2004). The thinner roots spread out roughly on the earth's top, while the bigger roots stay in the deeper zones of soil. The stem lacks distinct edges and is annual, sublinguous, and slightly pubescent. The plant generates several secondary basal branches, which perish and reappear each year. The plant is thinner and less branching in its natural habitat than it is in a cultivation setting. The leaves, branches, and stem are all green and coated with tiny, white hairs. The cultivated plant's stem often generates a large number of lateral branches, which together form a crown that is roughly rounded and thick (Soejatro et al., 1983). Its leaves are tiny, lanceolate, oblong, serrate, and subsessile with an alternating leaf arrangement and herbaceous growth habit (Dwivedi, 1999).

The blades are subcoriaceous and have whole edges that are commonly serrated on the top half and entire on the bottom half. From the leaf base, three principal veins appear, and the secondary venation is reticulate. The blades have an olive green to brownish green colour in their dried form, with the upper side usually being darker. Both the surfaces are subscabrous along with black glandular spots on the bottom side (Soejarto, 2002). Also, trichomes of two different sizes, *viz.*, one large (4–5 lm), and one small (2.5 lm), can be found on the leaf surface. This plant is insect pollinated and is self-incompatible (Gantait et al., 2018). Its seed (achene) is made up of a fluffy/hairy pappus with little endosperm. Viable seeds (very rare, due to sterility) are black in colour (Skaria et al., 2004).

FIGURE 17.3 *Stevia rebaudiana.*

17.4 BIOACTIVE COMPOUNDS

17.4.1 Chemical Description

Stevia leaves are incredibly chemically multifarious. A total of nine essential amino acids, including glutamate, aspartate, methionine, tyrosine, proline, alanine, isoleucine, lysine, and serine, are found in the compounds (Boonkaewwan et al., 2006). All essential amino acids, with the exception of tryptophan, exist in the leaves of stevia. Since stevia contains a shedload of sweet diterpene glycosides, it is utilized as a natural sweetener (Sultana et al., 2020). Steviol glycoside is the collective word used to describe the common backbone structure shared by all the glycosides found in stevia. These substances are four-ringed diterpenes. The C-13 hydroxyl group and C-19 carboxyl group produces the sweet flavour. The main glycosides in the herb are stevioside, steviolbioside, isosteviol, and rebaudiosides A, B, C, D, E, and F. Every plant organ contains glycosides at a different level (Figure 17.4). Their richest suppliers are leaves, which are followed by flowers, stems, and seeds (Marcinek and Krejpcio, 2015).

Rebaudioside A is more abundant in the leaves than the stem (Kumari and Chandra, 2015). Rebaudioside C and A are 250–300, 50–120, and 250–450 times sweeter than sugar, respectively, as are stevioside and rebaudioside. Under alkaline hydrolysis, rebaudioside A and D are converted to rebaudioside B. The pair can be safely digested in the human body due to higher water solubility, pH stability, thermostability (up to 200 °C), and lack of fermentation (Mlambo et al., 2022). Stevioside molecules are extremely stable in aqueous solutions over a wide pH (1–10) and temperature range (up to 198°C) (Marcinek and Krejpcio, 2015). Concurrently, Kroyer (2010) reported that steviosides are stable across a variety of processing and storage settings, as well as in interactions with coffee, organic acids, water-soluble vitamins, and sweeteners. They do not take part in the Maillard reactions when heated and also do not ferment. As per Kumari and Chandra (2015), *S. rebaudiana*'s leaves have 1.2 times greater stevioside levels than those of the stem. Depending on how thoroughly the stevia leaves have been processed, their chemical makeup can differ. The dried stevia leaf extract includes 10% amino acids, 18% proteins, 33% carbs, and 39% reducing sugars, but the quantities are 25%, 19%, 31%, and 25%, respectively, in the extract from fresh leaves (Snehal and Madhukar, 2012). Stevioside, which makes up 4%–13% of the dry weight of stevia leaves, is the most abundant glycoside, followed by rebaudioside (2%–5%), the most stable one, and dulcoside (0%–0.7%) (Ahsan et al., 2020). Rebaudioside is metabolized to stevioside by intestinal bacteria, which then produces glucose and a steviol molecule. Diterpenes and triterpenes are also present in sweet leaves in addition to diterpene glycosides (Marcinek and Krejpcio, 2015).

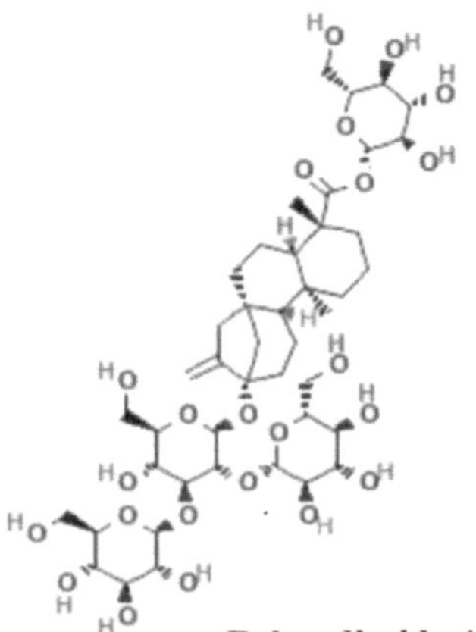

Rebaudioside A

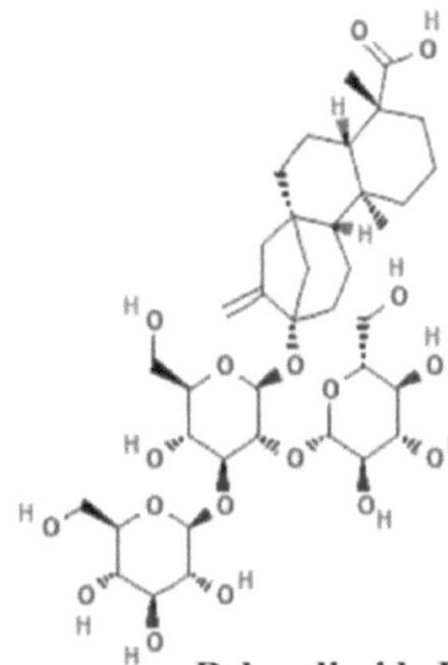

Rebaudioside B

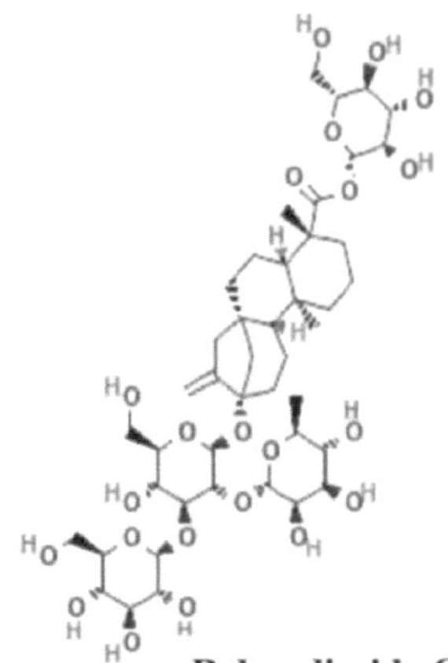

Rebaudioside C

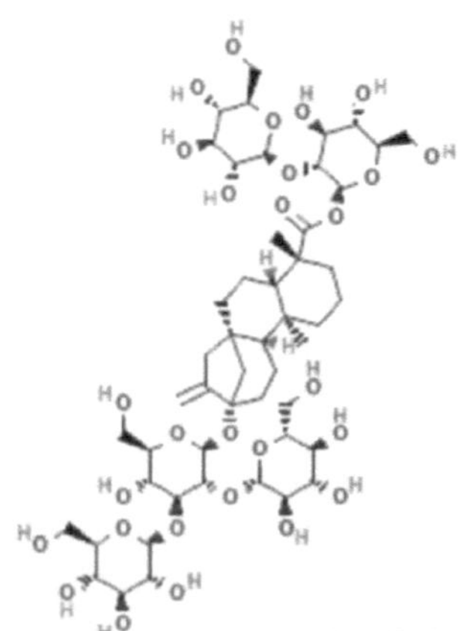

Rebaudioside D

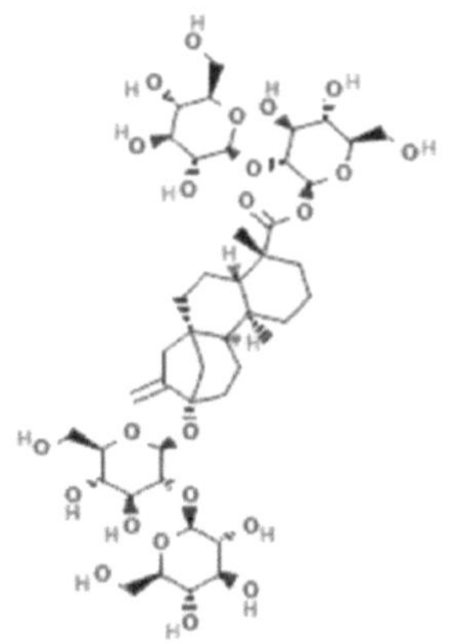

Rebaudioside E

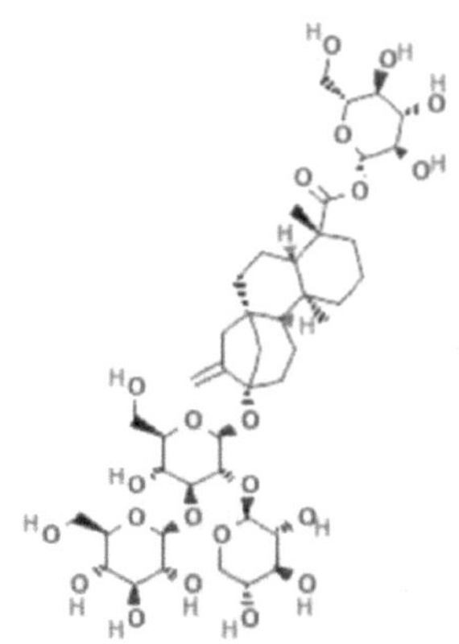

Rebaudioside F

FIGURE 17.4 Types of rebaudioside and their chemical structures.

The stevia extract contains 37 bioactive components with their corresponding ratios, the most significant of which was hydroxydehydrostevic acid (steviol), i.e. 53.96% followed by active phenolic compound 1,2- benzenediol (9.84%), 4-ethylcatechol (2.46%), 9-octade (Khan et al., 2018).

17.4.2 Phytochemical Constituents

Plants build up phytochemicals, which are secondary metabolites, to defend themselves from microbial diseases or pest attacks. Phytochemicals are active compounds that are thought of as drugs or medicines because they have therapeutic characteristics (Shakya, 2016). The plant leaves contained the phytochemicals – carotene, thiamine, steviol, stevioside, riboflavin, rebaudi oxides, nilacin, dulcoside, and austroinullin (Mlambo et al., 2022). The majority of papers and studies claim that stevia leaves include anthraquinones, reducing chemicals, triterpenes, sterols, saponins, cardiac glycosides, alkaloids, and tannins. The chemical structure of stevia leaves is dominated by terpenes and flavonoids. According to Mlambo et al. (2022), phenols and coumarins were found in the stevia leaves. With the use of their appropriate solvent systems, Srivastava et al. (2016) demonstrated the presence of several phytochemicals in the stevia leaf extract. Table 17.1 provides a summary of the phytochemical characteristics of the bioactive compounds found in stevia leaves.

TABLE 17.1
Biological Activities of Bioactive Compounds of *Stevia rebaudiana*

Compound		Biological role
Campesterol	:	inhibits the absorption of cholesterol in the intestine, antioxidant
Phenols	:	anti-apoptotic, anti-inflammatory, anti-aging
Eugenol	:	anti-inflammatory agents and antioxidants
Phytol	:	antimicrobial
Tetradecanoic acid	:	antioxidant, cancer preventive, nematocide, cholesterolaemic, lubricant
Hexadecanoic acid	:	haemolytic, anthelmintic, anti-bacterial, anti-allergic, pesticide, nematicide, lubricant, flavour, 5 alpha-reductase inhibitor
Heptadecanoic acid	:	antioxidant
Octadecanoic acid	:	anti-inflammatory, anti-androgenic, anti-cancer, cholesterolaemic, 5-alpha reductase inhibitor, flavour, insectifuge
Oleic acid	:	prevents metabolic syndrome and cardiovascular disease
Eicosanoic acid	:	anti-inflammatory, anti-therogenic
Isosteviol	:	anti-hyperglycaemic, anti-hypertensive, anti-inflammatory, anti- tumour, anti-diarrhoeal, diuretic
1-Heptatriacotanol	:	antioxidant, anti-cancer, anti-inflammatory, and sex hormone activity
Trans-geranylgeraniol	:	antidiabetic, antimicrobial, anti-viral, anti-fungal, anti-tumour, anti-hypertensive, anti-inflammatory
Coumarins	:	prevent hyper-proliferative skin diseases
Tannins	:	treating diarrhoea and dysentery, and wound healing properties

TABLE 17.2
Proximate Analysis, Mineral and Water Soluble Vitamins Content of Dried Stevia Leaves

Proximate parameters	Contents ($g/100\ g^{-1}$)	Minerals	Contents ($mg/100\ g^{-1}$)	Water soluble vitamins	Contents ($mg/100\ g^{-1}$)
Moisture	6.7	Iron	34.2	C	14.98
Ash	11.5	Sodium	184.3	B2	0.43
Fat	4.2	Potassium	2500	B6	0
Protein	18	Calcium	534.43	Folic acid	52.18
Moisture	14.89	Magnesium	465.35	Niacin	0
Crude fibre	30.4	Phosphorus	305	Thiamine	0
Carbohydrates	6.7	Iron	34.2	-	-

17.4.3 Nutritional Composition

Stevia leaves have 2.7 kcal of energy per gram of dry weight, making them a low-calorie sweetener (Ahsan et al., 2020) (Table 17.2). The nutritional makeup of stevia leaves, which is a considerable source of carbohydrates, protein, and crude fibre that preserves well-being and lowers the risk of numerous diseases, is mostly responsible for their health benefits. Fatty acids, including linoleic acid, linolenic acid, oleic acid, stearic acid, palmitoleic acid, and palmitic acid, were also present (Mlambo et al., 2022). Dry stevia powder has 1.9 to 4.34 g of fat per 100 grams, compared to 52 to 64.06 g of carbs and 10.0 to 18.0 g of protein (Srivastava et al., 2016; Gasmalla et al., 2014).

Due to the presence of poly- and fructo-oligosaccharides, which regulate lipid metabolism and lower blood sugar levels, the primary source of energy in stevia roots and leaves is carbohydrates (Braz-De-Oliveira et al., 2011). Additionally, it contains vitamins like vitamin B12, C, and folic acid (Mlambo et al., 2022). From sweetleaf leaves, Kim et al. (2011) extracted water-soluble vitamins.

17.5 PHARMACOLOGICAL OVERVIEW

17.5.1 Antidiabetic Properties and Effect on Blood Glucose Concentration

In the search for effective antidiabetic drugs, medicinal plants/herbs have been utilized for a long time to manage diabetic problems in many traditional medical systems. To date, over 1,200 different varieties of medicinal plants have been found to have antidiabetic characteristics. One of these plants that is particularly useful in treating diabetes is stevia (Michel et al., 2020). Despite their flavour, stevia-based sweets will not be absorbed by the body and will not increase blood sugar levels. To support this, clinical trial studies, supplementing stevioside, a leaf extract from *Stevia rebaudiana*, in meals significantly reduced blood glucose levels in type 2 diabetic patients (Gregersen et al., 2004).

Stevioside is recommended at various dosages and lowers the blood glucose levels of type 1 and type 2 diabetics in animal models (Marles et al., 1995; Jeppesen et al., 2000; Starratt et al., 2002; Gregersen et al., 2004; Abudula et al., 2004, Chen et al., 2006). Rats pre-fed with stevia leaf powder before receiving an injection of diabetogenicstreptozotocin (STZ) displayed less severe diabetic symptoms, *viz.* polyphagia and weight loss, and their hyperglycaemia was lower than the untreated diabetic rats (Shivanna et al., 2013). The aforementioned study showed that stevia leaf powder and its polyphenol extract boost insulin production from pancreatic islet cells in type 1 diabetic rats and improve glucose tolerance and cellular insulin sensitivity in type 2 diabetic rats. According to recent research, under *in vitro* conditions, the stevia leaf extract inhibits the activity of amylase and glucosidase, significant enzymes used in the digestion of dietary carbohydrates, which is another potential mechanism through which stevia can lower blood glucose levels (Carrera-Lanestosa et al., 2020; Zaidan et al., 2019).

In the work ofHe et al. (2019), it was interesting to note that four phenylethanolyl glycosides and a new phenylethanoid glycoside, called steviophethanoside (4-hydroxyphenyl ethyl-8-O-[-l-arabinopyranosyl-(1 6)] – d-glucopyranoside) isolated from leaves, had low toxicity and significantly stimulated the rat INS-1 islet cells, suggesting that it might be a safe hypoglycaemic agent. *S. rebaudiana* showed lower postprandial glucose and insulin levels (received either an aspartame or sucrose preload) and significant improvements in insulin and liver glycogen levels and reduced their fasting and postprandial blood glucose levels (Ahmad and Ahmad, 2018; Gu et al., 2019).

17.5.2 Anti-Cancer Property

Despite the fact that several anti-tumour agents and chemicals have been invented, natural herbal remedies have gained importance in the medical field. In 1997, Toyoda et al. conducted the first study demonstrating stevia's anti-tumour properties, in which it was proven that stevioside treatment reduced the incidence of mammary adenomas in female rats compared to controls in later years (Talevi, 2016). Konoshima and Takasaki (2002) studied the impacts of stevioside on the *in vivo* induction of two stages of mouse skin cancer by 7,12-dimethylbenz[a]anthracene (DMBA) and 12-O-tetradecanoylphorbol-13-acetate (TPA). This discovery is consistent with previous research that stevioside, steviol, and isosteviol, as well as their metabolites, have been shown to slow the growth of tumours by inhibiting the activation of the Epstein-Barr virus early antigen (Akihisa et al., 2004; Takasaki et al., 2009; Rajesh et al., 2010).

In a different investigation, Caco-2, T84, and HT29 colon cancer cell lines had lower viability at high stevioside (2–5 mM) and steviol (0.2–0.8 mM) concentrations (Boonkaewwan et al., 2008). Stevioside inhibits the proliferation of the ovarian cancer cell line OVCAR-3, probably via inducing apoptosis, deactivating the P13K/AKT pathway, and arresting the cell cycle, according to a more recent study (Li et al., 2017). Steviol, a component found in stevia leaves, was discovered to have a strong anti-cancer effect against human gastrointestinal cancer cells in a more recent investigation (Chen et al., 2018).

Gupta et al. (2017) demonstrated that steviol's dose-dependent action on MCF-7 cells caused the induction of apoptosis, suggesting that this substance may have potential as a treatment for breast cancer. Stevioside was previously discovered to have an anti-proliferative effect on the MCF-7 breast cancer cell line by inducing apoptosis and inhibiting DNA synthesis (Paul et al., 2012). Purified stevioside suppressed the growth of human breast cancer cell lines MDA-MB-231 and SKBR3 in a more recent investigation by yielding similar results (Khare and Chandra, 2019). Gas chromatographic analysis of *Stevia rebaudiana* extract demonstrated the existence of 20 phytocompounds. Using the *insilico* approach, the discovered phytocompounds were evaluated for their anti-cancer activity in which tetradecanoic acid and stigmastan3,5-diene had docking energies of 26.52 and 28.24 Kcal/mol, respectively (Mallu et al., 2019).

17.5.3 Effect on Renal Function

According to early physiological and pharmaceutical studies, stevioside from the leaves of *Stevia rebaudiana* may function as a typical systemic vasodilator. Using clearance techniques, the impact of stevioside on renal function in rats with experimental renal hypertension and normal kidneys was assessed. Both the normal and the hypertensive rats had hypotension, diuresis, and natriuresis in response to stevioside (Melis, 1995). Later, stevia extract has marked increase in free water discharge suggesting that it impaired Na+ transport in the proximal tubule when water-diuresis and anti-diuresis rats were used (Melis, 1999). To substantiate this, Shivanna et al. (2013) demonstrated an experiment in which the glomerular filtration rate (GFR) significantly increased when stevia was fed to diabetic-induced rats. Therefore, it is stevia that probably reduced lipid peroxidation and kidney tissue damage through the action of antioxidant enzymes. Steviol directly affects the function of the cystic fibrosis transmembrane conductance regulator (CFTR) chloride channel and decreases CFTR expression by preventing the Madin-Darbey canine kidney (MDCK) cyst from forming by preventing CFTR proteasome degradation (Yuajit et al., 2013).

Nazari et al. (2017) showed that *Stevia rebaudiana* bitter fraction significantly restored the blood glucose level toward normal level faster than glibenclamide. According to a clinical trial conducted by Rizwan et al. in 2018, stevia has the ability to significantly enhance some biochemical indicators in CKD patients (Stage II) as it increases serum creatinine, reduces serum uric acid, decreases diastolic and systolic blood pressure, fasting blood sugar, and post-prandial blood sugar, and improves microalbumin level. Recently, Mehmood et al. (2022) showed that stevia reduces abnormal increases in uric acid level, which leads to hyperuricaemia-associate renal injuries and gout. Thus, the aforementioned studies suggest that stevia and its polyphenol compounds have renal protective characteristics because they have the ability to suppress apoptosis, oxidative stress, and inflammation.

17.5.4 Immunomodulation Effect

Stevioside lessens inflammation by lowering the expression of the cytokines IL-6, TNF, and IL-1, pro-inflammatory cytokines in cancer cells. Stevioside was discovered to decrease the expression of cytokines *via* deactivating the MAPK pathway, TLR2, and NF-B (Wang et al., 2014). These findings are supported by a number of lines of research, since stevioside has been shown to block pro-inflammatory cytokines in the RAW 264.7 (mouse macrophage) cell line, acute lung damage, and epididymal, epithelial, and intestinal cells by comparable pathways (Zhao et al., 2008; Boonkaewwan et al., 2008; Fengyang et al., 2012; Yingkun et al., 2013).

Basically, stevioside indicates anti-inflammatory properties through downregulation of two major pathways, namely nuclear factor kappa B (NF-κB) and mitogen-activated protein kinase (MAPK) signalling pathways (Meng et al., 2018). Stevioside also hindered necrosis of the *S. aureus*-infected MMECs by downregulating the gene expression for type 2 toll-like receptors (TLRs). Ahirwal et al. (2015) demonstrated in a different study that stevia leaf extract effectively modulated immune responses and prevented immunological diseases. By altering B and T cell responses and enhancing phagocytic activity, researchers were able to validate the similar impact of stevioside (Sehar et al., 2008). In summary, stevioside can suppress pro-inflammatory cytokines at the gene expression level to reduce inflammation and mediate immunomodulation.

17.5.5 Antioxidant Effect

Stevia possesses high bioactive compound concentration, including phenolic compounds, tannins, flavonoids, and vitamin C (Abou-Arab and Abu-Salem, 2010; Kim et al., 2011; Zlabur et al., 2019), and these concentrations vary depending on several factors, including growth conditions and the time of harvest (Tavarini and Angelini, 2013).

Feeding stevia to Wister rats lowered the MDA concentration in liver and enhanced its antioxidant status through antioxidant enzymes, thus reducing the risk of oxidative stress and ameliorating liver damage (Shivanna et al., 2013). By boosting the functions of superoxide dismutase, catalase, glutathione peroxidase, and total antioxidant capacity and lowering acetylcholinesterase activity and malondialdehyde level in the blood, brain, or liver, stevia residue extract, a polyphenol-rich extract made from side products during the production of steviol glycosides, revealed protective actions against oxidative stress in D-galactose-induced ageing mice (Zhao et al., 2019).

17.5.6 Treatment for Obesity

The intake of high-sugar foods and beverages has been identified as one of the prior causes of obesity. Therefore, it makes sense that switching from high sugar to alternative sugar sources with non-calorific effects can result in decreasing total calorie intake and, thus, a decrease in body weight (Ashwell, 2015). It could be argued that using a natural and secure sweetener like stevia is the best alternative (Gupta et al., 2013).

17.5.7 Effect on Tooth Decay/Dental Caries

Finding a healthy alternative to sucrose that does not encourage the development of caries is a crucial first step in preventing dental caries (Matsukubo and Takazoe, 2006). The main secondary metabolites of the non-nutritive sweetener stevia, including stevioside, steviol, isosteviol, and rebaudioside A, B, C, and E, are non-cariogenic. The antimicrobial properties of stevia are reported to have direct impact on tooth health as stevioside is highly active against the microbial activities of several tooth decay microbes such as *S. mutans*, *Curvularialunata*, *Staphylococcus aureus*, *Escherichia coli*, and *Bacillus subtilis* (Mehta et al., 2016; Ahmad et al., 2018). In addition to not being cariogenic, the stevioside and rebaudioside A extracts are non-acidogenic (Brambilla et al., 2014). Childhood tooth decay can be combated by the sweetener of stevia instead of artificial sugars with high calories in snacks (Cocco et al., 2019).

17.5.8 Effect on Hypertension

Oral consumption of stevioside was recommended for treating hypertension, as it helps in inhibiting the activity of the angiotensin-converting enzyme causing hypertension (Chan et al., 2000). Recently, it was reported that ethanol extracts from stevia leaves substantially decreased angiotensin-converting enzyme activity, denoting that these substances may be helpful for the treatment of hypertension (Wang and Wu, 2019).

17.5.9 Antimicrobial

Stevia leaf's crude extract was discovered to have antibacterial properties against bacteria, i.e. *Bacillus subtilis*, *Escherichia coli*, *Streptococcus mutans*, and *Staphylococcus aureus* (Debnath, 2008). The various extraction types, such as chloroform and methanol extract of stevia, were found to be active only against *Sclerotonia minor* and *Curvularialunata* against *Aspergillus niger, Sclerotonia minor, Rhizopus, Alternaria alternate*, and *Microsporiumgypsium*, indicating that their anti-fungal properties were more limited. The acetone, chloroform, and ethyl acetate extracts of stevia showed antimicrobial activity against *Cryptococcus neoformans, Aeromonashydrophila, Candida albicans, Salmonella typhii, Vibrio cholera, Trichophytonmentagrophytes, Epidermophyton, E. coli, B. subtilis, and S. aureus* (Jayaraman et al., 2008). Use of stevia extracts as a preservative for salmon paste and other seafood products is one potential use of the herb's antimicrobial and antioxidant qualities (Ortiz-Viedma et al., 2017).

17.5.10 Cardioprotective Properties

Stevioside has synergistic pharmacological action against cardiotherapeutic regimens. According to Bhatt et al. (2020), pre-treatment with stevioside reduced the leakage of cardiac biomarkers into the extracellular compartments, normalizing the levels of CK-MB, ALT, AST, LDH, and CK-NAC enzymes in the perfusate and serum. Total cholesterol and triglyceride levels were subjected to anti-hyperlipidaemic effects, which were dose dependent. SOD and catalase both showed protective effects from stevioside (Mehmed et al., 2021). By considerably normalizing electrocardiographic parameters, myocardial histology, antioxidant levels, and biomarker levels, a combination of stevioside and diltiazem (200 mg/kg and 17.5 mg/kg, respectively) was discovered to be more effective in terms of pharmacodynamic response (Kumar et al., 2021).

Moreover, it is impossible to draw firm conclusions on the hypoglycaemic potential and other medicinal properties of stevia due to the scarcity of these human trials. Ultimately, based on *in vivo* and clinical research, it is likely that stevia can be regarded as a promising new medication for the management of diabetes and other human illness, which could result in a decrease in complications linked with illness.

17.6 CULTIVATION AND PRODUCTIVITY

17.6.1 Climate

Stevia rebaudiana was first commercially cultivated in Paraguay in 1964, and now it has been spread to a number of other nations (Kumar et al., 2012b). Today, the crop is effectively grown in a wide range of geographical regions and cultivation conditions around the world. The plant also has been successfully cultivated in many Indian states like Karnataka, Maharashtra, Rajasthan, Kerala, Orissa, and Punjab (Singh et al., 2014). According to Akatov et al. (2004), cultivating stevia will enhance the output of sugar equivalent by seven-fold. As the plant is native to South America, a semi-humid subtropical climate with an average temperature of 24 °C is the ideal environment for its growth and development (Libik-Konieczny et al., 2018). Conversely, the physiological, biochemical, molecular, growth, and developmental processes of the plants are impacted by cold stress (Yadav, 2010). The crop's growth, flowering, and sweet glycoside quality are also influenced by radiation, day length, temperature, soil moisture, wind, and the position of the leaves (Jarma-Orozco et al.,2020). The photoperiod has a significant impact on *S. rebaudiana*. This is because the plant is a short-day species, where a crucial day length of 12–13 hours causes early blooming (Kobus-Moryson et al., 2014). Long-day conditions (14–16 hours) are necessary to extend vegetative development and boost leaf yield (Yoneda et al., 2017). Stevia is a subtropical semi-humid plant whose growth and yield will drastically reduce under heavy rainfall because it is highly sensitive to waterlogging. Under such conditions, the crop had shown promising results if planted in furrow-irrigated raised beds (Kumar et al., 2012a).

17.6.2 Soil

Generally, stevia prefers fertile sandy loam soil with good drainage or loam soil with high organic materials (Goyal et al., 2010). For best growth, it prefers lighter acidic to neutral (pH 6–7) soil. It usually requires a regular supply of water but avoids waterlogging conditions. Additionally, stevia is also found to grow naturally on infertile, acidic sand, muck soil, or near the boundaries of marshlands (Ramesh et al., 2006). Zaman et al. (2015) evaluated the effect of seven different soil types found in Bangladesh on stevia production. According to their findings, the plant cultivated in non-calcareous soil had the best crop yield and was identical to the plant grown in acid soil, and the lowest value of yield was found in plants cultivated in acid sulphate soil (Jain and Joshi, 2012).

17.6.3 Propagation

17.6.3.1 Propagation by Seeds

Stevia reproduction in the wild occurs mostly through the seed;, however, germination and establishment from seeds are frequently subpar and occasionally unsuccessful (Shaffert and Chebotar, 1994). The timing of flowering, pollination techniques, and seed harvest timing are all crucial to seed production. Generally, stevia produces two types of seeds – black and tan coloured, with the black seeds being heavier than the tan seeds. Black seeds have a substantially greater viability rate (76.7%) than do tan seeds (8.3%) (Goettemoeller and Ching, 1999). In stevia, seed germination rates vary significantly. Germination can start anywhere between the fifth and the twelfth day after seeds are sown. The germination study of seeds obtained from four potting media treatments: germination percentage in soil media (67.5%), germination percentage in the combination of soil and rice husk (57.4%), germination percentage in sand media (48.4%), and germination percentage in vermiculite (41.1%) suggests that potting has a significant influence on the germination of seeds (Kumar, 2013). The optimum temperature requires for germination varies from 20–25°C; light generally increases germination (Yadav et al., 2011). Ucar et al. (2016) evaluated the germination performance of stevia seeds in two lights (light/darkness) and four different temperatures (15 °C, 20 °C, 25 °C, and 30 °C); the highest seed germination rate (71%) was observed in darkness at 25 °C. The findings showed that regardless of the light treatments, germination rates were considerably reduced at lower (15°C) and higher (30°C) temperatures. Melatonin can also improve seed germination when applied in low concentrations (5 and 20 μm) but retard seed germination when applied in high concentrations (100 and 500 μm) (Simlat et al., 2018).

17.6.4 Vegetative Propagation

Stevia is often propagated using stem cuttings, which root readily, but this is a very labour-intensive process. Gvasaliya et al. (1990) reported that when a cutting is taken from current year growth, the rooting percentage of the cutting is 98%–100%. Additionally, the cutting position on the plant might have an impact on its growth and ability to establish roots. Typically, cuttings taken from the top of the main stem produce the finest results (Kassahun et al., 2011). Environmental conditions during the collection of cuttings also influence the survival of the cutting. Cuttings taken in late winter had a higher rooting percentage than those collected during other seasons (Carvalho and Zaidan, 1995). Growth regulators are often used to encourage the rooting of cuttings (Abdullateef et al., 2012). Treatment of stevia cutting with IBA 7.4 mM or ANA 6.4 mM + IBA 0.3 mM was found to be an effective growth regulator for root production and vegetative propagation (Castañeda-Saucedo et al., 2020).

The best way to produce stevia seems to be through micropropagation, or *in vitro* culture, as it has the capacity to quickly generate a significant number of stevia plantlets. *Stevia rebaudiana* shoot apex, leaf, and nodal explant can regenerate shoot when cultured in 6-benzyladenine (BA; 8.87 μm) and indole-3-acetic acid (5.71 μm) enriched Murashige and Skoog (MS) medium.

FIGURE 17.5 Propagation of *Stevia rebaudiana* (Bert.) through leaf explant (a) callus formation, (b and c) induction of shoot from callus culture, (d) shoot multiplication, (e) shoot elongation, (f) root induction, and (g and h) successful acclimatization of *in vitro* regenerated plantlets (adopted from Khalil et al., 2014).

In vitro-derived shoots can induce roots when subcultured on a medium containing auxin. The maximum sweetener concentration was found in a callus cultivated on agar-solidified MS media supplemented with BA (8.87 μm) and indole-3-butyric acid (9.80 μm) (Sivaram and Mukundan, 2003). The MS medium supplemented with 0.5 mg/l BAP+2.0 mg/l Kn had the maximum induction of numerous shoots from nodal segments. IBA was utilized in various concentrations for rooting; the maximum rooting was observed on MS medium with 1.0 mg/l IBA (Mehta et al., 2012). Khalil et al. (2014) examined the efficacy of stevia propagation techniques, with the results indicating that micropropagation (85%) outperformed seed germination (25.5%–40%) and stemcutting (60%) in terms of productivity (Figure 17.5).

17.6.5 Manures and Fertilizer Application

Among the agronomic practices, reliable nutrient supply is the most important factor for higher crop yield. Among the 17 essential plant nutrients, N, P, and K are the most often limiting macronutrients for plant growth and development. Hence a balanced dose of N, P, and K is recommended in order to obtain the maximum yield. A study conducted by Pal et al. (2015) to standardize the nutrition requirement of this crop showed that the crop responds well to the combination of 90 kg N, 40 kg P_2O_5, and 40 kg K_2O ha^{-1} for maximum dry leaf yield. The application of bio-fertilizers has been found to increase the available N, P, and K contents in soils and plant biomass. Vermicompost and bone meal when applied at the 75:25 ratio was proved to be the best for improving the growth, yield, and quality of stevia (Das et al., 2008).

17.6.6 Irrigation

Since the herb is extremely sensitive to water stress and requires frequent light irrigation, sprinkler irrigation is shown to be useful. The best results come from watering plants every three to five days during the summer. According to studies, stevia requires frequent irrigation to keep the soil moist above the wilting point, i.e. up to 80% of the field's capacity, and moisture stress lowers leaf yield

(Lavini et al., 2008). A study conducted by Behera et al. (2013) shows that drip fertigation improved fresh leaf yield by 4.9%, dry leaf yield by 4.0%, and total biomass yield by 2.04% over conventional surface irrigation with soil application of fertilizer. Aladakatti et al. (2012) investigated the response of stevia under various irrigation schedules and discovered that irrigation scheduled at 1.2 IW/CPE produced the maximum dry leaf yield (10.54 t ha^{-1}). Additionally, mulches should be used around the plant to lessen the effects of drought and excessive temperatures.

17.6.7 Weeding

Numerous weeds compete with the plant, which reduces branching and lowers the output of stevia. The sole methods for weed management are hand-picking and herbicide applications (Harrington et al., 2011). Compared to traditional weed management techniques like manual weeding, cover crops, and mulching (including organic or synthetic mulching), chemical approaches are proven to be more successful and cost-efficient (Ghorai, 2008). The impact of herbicides and mulching on weed control in *Stevia rebaudianses* was assessed by Taak et al., 2021. The study's findings demonstrate that the use of mulches successfully increased crop output and growth characteristics, including plant height (98.2 cm on average), branch number (5.1 $plant^{-1}$), and leaf number (415 $plant^{-1}$).

17.6.8 Harvesting and Yield

Usually, after four months of planting, stevia leaves are ready for the first harvest, and additional harvests can be made after every three to four months. Depending on the soil type and climatic conditions, an average of three to four commercial harvests can be expected each year (Singh et al., 2014). Mid-September to late September, when plants are 50 to 70 cm tall, is the ideal time to harvest leaves. Harvesting should be done prior to the onset of flowering because the total glycoside content of the leaves begins to decline with the onset of flowering (Ahmed and Mukta, 2017). The highest yield is usually observed in the first harvest, and yield decreases subsequently with the second and third harvests because multiple cuts to the same group of plants affect the dry biomass weight (Pereira et al., 2016). Different planting dates also have a significant effect on leaf yield. Khan et al. (2012) evaluated the effect of different planting times on the yield of stevia and the results of studies show that plants planted during April showed the highest yield [plant height (68.50 cm), number of leaves $plant^{-1}$ (142.33), fresh weight of leaves $plant^{-1}$ (14.11 g) and dry weight of leaves $plant^{-1}$ (3.38 g)]. Kumar et al. (2014) evaluated the effect of eight treatments of pinching on the yield, growth, and quality of stevia. According to the study, compared to control plants, stevia plants pinched at 40 DAT and 20 cm tall recorded a considerably larger number of leaves $plant^{-1}$ (213), fresh leaf weight (93.8 g $plant^{-1}$), and 27.7% more total leaf dry biomass and this treatment also had greater net returns (Rs. 1,86,998 ha^{-1}), and net return per rupee (Rs. 2.5) invested.

17.7 INDUSTRIAL APPLICATIONS

Currently, studies are mostly concentrated on the therapeutic effect of stevia, and it has raised its significance globally. In around 20 countries, including the USA, Japan, Canada, China, Brazil, Argentina, Korea, Mexico, Indonesia, and Tanzania, stevia is recommended to be used in food and medicinal applications (Giri et al., 2014). A lot of research has been carried out on the food and medicinal perspective of stevia. The functional components found in stevia such as antioxidant and polyphenols, phytochemicals, vitamins, minerals, and steviol compounds are of utmost use for health benefits and associated with the immune boosting source. When Covid-19 prevalence became a major concern around the globe, immunity boosting was the key factor to reduce chances of becoming infected. Due to the potent activities of compounds found in stevia, researchers recommended the use of stevia specifically in the risk groups for Covid-19 and also for boosting the general health of patients with obesity, diabetes, hypertension, lung disease, and cardiovascular diseases.

Besides the plethora of health benefits of stevia, it has potential scope in industrial applications, particularly food and food ingredients such as sucrose replacer, solubilizing agent, and fertilizers and animal feed (Ahmad et al., 2020). Beverages and foods sweetened with stevia assist diabetic people in settling their blood glucose levels. This substitution for sugar may likewise decrease the quantity of calories that an individual consumes, which is likely to help in weight reduction (Koser and Jaffer, 2022). The European Union has approved steviol glycosides with the E 960 symbol for use in 31 food categories. In order to develop novel formulations for cakes and biscuits with stevia as sucrose substitute, food industries have a challenge to produce products with good quality and achieve consumer acceptability (Kobus and Gramza, 2015). Cakes are the products that are well accepted all over the world due to their organoleptic properties. But these are usually formulated with relatively high levels of sucrose. It was found that low calorie cake can be produced by replacing 20% of sugar with stevia leaf powder or replacing 50% sugar with stevia leaf aqueous extract with good organoleptic properties (Sulaiman et al., 2022). Due to higher biotransformation efficiency (37.5%), it is recommended to use stevia as a substitute for sucrose in functional food products like biscuits, which may be beneficial in managing obesity due to its zero glycaemic value and in maintaining good health (Shahu et al., 2022).

Stevia leaf extract is a potential sugar replacer in sweetened dairy products. It has been proven that the replacement of 100% sugar is possible with the commercially available extract for the production of vanilla-flavoured set yogurt without any change in the quality (Narayana et al., 2022). It reduces the use of artificial non-nutritive sweeteners for those with dietary restrictions and therefore can be used to make lite products (Narayanan et al., 2014). Amarkhand, which is variation of traditional shrikhand (part of Maharashtrian and Gujrati recipe), when prepared with different levels of sugar and stevia extract, shows a decreasing trend in carbohydrates and calorific value (Tondare and Hembade, 2021).

The dietary sources of lutein and zeaxanthin can be increased in the human diet by feeding stevia leaves to laying hens, which has found to increase the concentration of macular carotenoids in egg yolks. It has also been reported that stevia feeds do not impact storage quality (Pirgozliev et al., 2022). Microencapsulation of *Stevia rebaudiana* Bertoni leaf extract with an effective dose of 100 mg/kgBW proved to have anti-hypercholesterolaemic effect on male Wistar strain rats (Bagiana et al., 2022). The studies also found that the addition of *Stevia rebaudiana* Bertoni in the compound feed at 20 g per capita per day increases average daily gain of piglets. Another aspect also studied during pregnancy and nursing periods shows that supplementation of *Stevia rebaudiana* Bertoni has a positive effect on the health status of sows (Radena Nenova, 2022). Studies show that stevia has a potential benefit on the microbiome's alpha diversity. The host's gut microbiota depends upon the amount and frequency of stevia intake as well as simultaneous consumption of other dietary components (Kasti et al., 2022). It is generally accepted that one can consume 5 gm of dried stevia leaves without health problems. It has been also proved to be an alternative to antibiotics through the synthesis of silver nanoparticles (Timotina et al., 2022). Stevia leaves contain diterpene steviol glycosides and other valuable biologically active compounds, which are similar in structure and properties to gibberellins, and when extracted in the form of water are capable of accelerating growth and increasing the yield and quality of plants.

17.8 CURRENT STATUS AND FUTURE REMARKS

Drug therapy using natural herbs are currently increasing in popularity and gaining more and more importance. As discussed, phytochemicals in the herb, such as steviol glycosides, phenols, alkaloids, flavonoids, and lipids, hold a potential application at the industrial level. As a medicinal herb, it has been reported to have nontoxic effects on human health (Abbas Momtazi-Borojeni et al.,

2017). The consumption of stevia, with its major components stevioside and rebaudioside A, has been already approved by the US FDA, EFSA, and JECFA with an amount of 4 to 7.9 mg/kg body weight considering age and health status (EFSA, 2010). Due to the importance of the crop and its components, several attempts have been made at improved extraction processes with high-quality bioactivates. Several advanced extraction procedures include high-pressure assisted and pressurized liquid (Kovačević et al., 2018), microwave-assisted (Ameer et al., 2017), and solid-liquid dynamic extraction (Gallo et al., 2017).

Along with the high extraction technologies, enhancing productivity has become a priority in the research line. The increased demand for stevia put light on genetic improvement for higher productivity, which opened a new window for farmers for growing in a large area. Currently, the varieties are available with a prolonged vegetative phase rather than early flowering because the steviol glycoside content is high during the vegetative phase just before reproductive age (Gantait et al., 2018; Ceunen et al., 2013). Several attempts were made to standardize agronomic practices (Pal and Mahajan, 2017; Munz et al., 2018). *In vitro* propagation through tissue culture for the production of disease-free plants has recently been developed (Razak et al., 2014; Nower, 2014; Thiyagarajan and Venkatachalam, 2012; Wang et al., 2013; Zayova et al., 2017).

The sweetener pathways were explained by genes such as SrDXS, SrDXR, SrCPPS, SrKS, SrKO, SrUGT85C2, SrUGT74G1, and SrUGT76G1 (Kumar et al., 2012a). Breeding efforts currently made varieties suitable for different climatic conditions through inter-cross, mutation breeding, and so on. Current attempts are being made to produce the sweetener from biowaste with yeast and dairy products (Arshad et al., 2022). The demand for herbal remedies is increasing with changing climate and population scenarios (Al-Sultan et al., 2021). Meeting the current high demand for industries is really a hurdle even with new technologies. There is an option to develop alternative varieties, harvest technologies, propagation, *in vitro* secondary metabolite production, enhanced bio-activates with improved quality, and more.

17.9 CONCLUSION

The present review provides a comprehensive outlook on the distribution, origin, pharmacology, and commercial production of *S. rebaudiana*. One of the most studied methods for creating non-caloric glycosides is stevia, which is also a natural alternative for lowering blood sugar and so contributes to the prevention of diabetes. In addition to its usage in the food and pharmaceutical industries, research has suggested that stevia has positive health effects against conditions like diabetes, obesity, hypertension, cancer, dental cavities, oxidative stress, and microbes. Stevia has been shown to have antidiabetic and anti-hyperglycaemic properties due to the presence of steviol glycosides in studies using animal models of diabetes. Along with numerous flavonoids and fatty acids, stevia leaves also contain a variety of glycosides, which combined give the plant its wide range of biological characteristics. However, the majority of the studies either used animals, were underpowered, or were conducted *in vitro*, making them unable to draw firm conclusions about stevia's health advantages. To ascertain the possible function of stevia in various health conditions, its physiological mechanisms, and the primary active components, additional long-term human research with substantial sample sizes are necessary. Stevia and isolated components have generally been shown in toxicology studies to be safe for human ingestion.

REFERENCES

Abbas Momtazi-Borojeni, A., Esmaeili, S.A., Abdollahi, E. & Sahebkar, A. 2017. A review on the pharmacology and toxicology of steviol glycosides extracted from *Stevia rebaudiana*. *Current Pharmaceutical Design* 23:1616–1622.

Abdullateef, R.A. & Osman, M. 2012. Effects of stem cutting types, position and hormonal factors on rooting in *Stevia rebaudiana* Bertoni. *Journal of Agricultural Science* 4(1):49p.

Abou-Arab, E.A. & Abu-Salem, F.M. 2010. Evaluation of bioactive compounds of *Stevia rebaudiana* leaves and callus. *Journal of Food and Dairy Sciences* 1(4):209–224.

Abudula, R., Jeppesen, P.B., Rolfsen, S.E.D., Xiao, J. and Hermansen, K. 2004. Rebaudioside A potently stimulates insulin secretion from isolated mouse islets: Studies on the dose-, glucose-, and calcium-dependency. *Metabolism* 53(10):1378–1381.

Aguiree, F., Brown, A., Cho, N.H., Dahlquist, G., Dodd, S., Dunning, T., Hirst, M., Hwang, C., Magliano, D. and Patterson, C. 2013. *IDF Diabetes Atlas.*

Ahirwal, L., Singh, S., Dubey, M.K., Bharti, V., Mehta, A. & Shukla, S. 2015. In vivo immunomodulatory effects of the methanolic leaf extract of *Gymnema sylvestre* in Swiss albino mice. *Archives of Biological Sciences* 67(2):561–570.

Ahmad, J., Khan, I., Blundell, R., Azzopardi, J. & Fawzi, M. 2020. Trends in food science & technology *Stevia rebaudiana* Bertoni: An updated review of its health benefits, industrial applications and safety. *Trends in Food Science & Technology* 100:177–189. https://doi.org/10.1016/j.tifs.2020.04.030.

Ahmad, J., Khan, I., Johnson, S.K., Alam, I. & Din, Z.U. 2018. Effect of incorporating stevia and moringa in cookies on postprandial glycemia, appetite, palatability, and gastrointestinal well-being. *Journal of the American College of Nutrition* 37(2):133–139.

Ahmad, U. & Ahmad, R.S. 2018. Anti diabetic property of aqueous extract of *Stevia rebaudiana* Bertoni. leaves in Streptozotocin-induced diabetes in albino rats. *BMC Complementary and Alternative Medicine* 18(1):1–11.

Ahmed, S.R. & Mukta, S. 2017. Review on *Stevia rebaudiana* as a non-caloric natural sweetener producing plant. *Journal of Syllabus Agriculture University* 4:15–25.

Ahsan, F., Bashir, S. & Shah, F. 2020. Nutritional and medicinal properties of *Stevia rebaudiana. Current Research in Diabetes & Obesity Journal* 13(4):47–53.

Akatov, P.M., Apasov, I.V. & Podporinova, G.K. 2004. Outlooks for introduction of stevia, a sugar crop. *SakharnayaSvekla* 10:31–32.

Akihisa, T., Hamasaki, Y., Tokuda, H., Ukiya, M., Kimura, Y. & Nishino, H. 2004. Microbial transformation of isosteviol and inhibitory effects on Epstein– Barr virus activation of the transformation products. *Journal of Natural Products* 67(3):407–410.

Aladakatti, Y.R., Palled, Y.B., Chetti, M.B., Halikatti, S.I., Alagundagi, S.C., Patil, P.L., Patil, V.C. & Janawade, A.D. 2012. Effect of irrigation schedule and planting geometry on growth and yield of stevia (*Stevia rebaudiana* Bertoni.). *Karnataka Journal of Agricultural Sciences* 25(1):1–10.

Al-Sultan, R.M., Al-Johani, A.M., Al-Harbi, A.F. & Al-Sowayan, N.S. 2021. Comparative study on the effect of two sweeteners stevia and sucrose on blood glucose levels in healthy studies. *Egyptian Journal of Chemistry and Environmental Health* 7(2):1–9. https://doi.org/10.21608/ejceh.2021.233158.

Amarakoon, S. 2021. *Stevia rebaudiana* – A review on agricultural, chemical and industrial applications. *Journal of Nature and Applied Research* 1(1):14–27.

Ameer, K., Bae, S.W., Jo, Y., Lee, H.G., Ameer, A. & Kwon, J.H. 2017. Optimization of microwave-assisted extraction of total extract, stevioside and rebaudioside-A from *Stevia rebaudiana* (Bertoni) leaves, using response surface methodology (RSM) and artificial neural network (ANN) modelling. *Food Chemistry* 229:198–207.

Arshad, S., Rehman, T., Saif, S., Rajoka, M.S.R., Ranjha, M.M.A.N., Hassoun, A., Cropotova, J., Trif, M., Younas, A. & Aadil, R.M. 2022. Replacement of refined sugar by natural sweeteners: Focus on potential health benefits. *Heliyon* 10:711–712.

Artham, S.M., Lavie, C.J., Milani, R.V. & Ventura, H.O. 2009. Obesity and hypertension, heart failure, and coronary heart disease-risk factor, paradox, and recommendations for weight loss. *Ochsner Journal* 9(3):124–132. PMID: 21603427; PMCID: PMC3096264.

Ashwell, M. 2015. Stevia, nature's zero-calorie sustainable sweetener: A new player in the fight against obesity. *Nutrition Today* 50(3):129p.

Bagiana, I.K., Puspitaningrum, I., Kusmita, L., Franyoto, Y.D., Pharmasi, S.Y., Sarwo, L. & Wibowo, E. 2022. Anti-hypercholesterol effects of microencapsulation of leaf extract *Stevia rebaudiana* 2:1–6.

Bahia, L.R., Araujo, D.V., Schaan, B.D., Dib, S.A., Negrato, C.A., Leão, M.P., Ramos, A.J., Forti, A.C., Gomes, M.B. & Foss, M.C. 2011. The costs of type 2 diabetes mellitus outpatient care in the Brazilian public health system. *Value Health* 14:S137–S140.

Bala, P. 1982. The Ayurvedic and unani systems of medicine in medieval India. *Studies in the History of Medicine* 6(4):282–289.

Behera, M.S., Verma, O.P., Mahapatra, P.K., Singandhupe, R.B., Kumar, A., Kannan, K. & Brahmanand, P.S. 2013. Effect of fertigation on stevia (Stevia rebaudiana) under drip irrigation. *Indian Journal of Agronomy* 58(2):243–250.

Bhatt, L., Amrutia, J., Chakraborty, M. & Kamath, J. 2020. Evaluation of cardioprotection and bio-efficacy enhancement of stevioside and diltiazem in rats. *Future Journal of Pharmaceutical Sciences* 6(1):1–13.

Boonkaewwan, C., Ao, M., Toskulkao, C. & Rao, M.C. 2008. Specific immunomodulatory and secretory activities of stevioside and steviol in intestinal cells. *Journal of Agricultural and Food Chemistry* 56(10):3777–3784.

Boonkaewwan, C., Toskulkao, C. & Vongsakul, M. 2006. Anti-inflammatory and immunomodulatory activities of stevioside and its metabolite steviol on THP-1 cells. *Journal of Agricultural and Food Chemistry* 54(3):785–789.

Brambilla, E., Cagetti, M.G., Ionescu, A., Campus, G. & Lingström, P. 2014. An in vitro and in vivo comparison of the effect of *Stevia rebaudiana* extracts on different caries-related variables: A randomized controlled trial pilot study. *Caries Research* 48(1):19–23.

Braz-De-Oliveira, A., Goncalves, R.A.C., Chierrito, T.P.C., Souza, L. & Gorin, P.A.J. 2011. Structure and degree of polymerisation of fructooligosaccharides present in roots and leaves of *Stevia rebaudiana* bertoni. *Food Chemistry* 129(2):305–311.

Carrera-Lanestosa, A., Coral-Martínez, T., Ruíz-Ciau, D., Moguel-Ordoñez, Y. & MairaRubí Segura-Campos. 2020. "Phenolic compounds and major steviol glucosides by HPLC-DAD-RP and invitro evaluation of the biological activity of aqueous and ethanolic extracts of leaves and stems: *S. rebaudiana* Bertoni (creole variety INIFAP C01): Bioactive compounds and Functionality. *International Journal of Food Properties* 23(1):199–212.

Carvalho, M.A.M. & Zaidan, L.B.P. 1995. Propagation of *Stevia rebaudiana* from stem cuttings. *PesquisaAgropecuariaBrasileira (Brazil)* 30(2):201–206.

Castañeda-Saucedo, M.C., Tapia-Campos, E., Ramírez-Anaya, J.D.P. & Beltrán, J. 2020. Growth and development of stevia cuttings during propagation with hormones in different months of the year. *Plants* 9(3):294 p.

Ceunen, S., Compernolle, F., Mai, A.H. & Geuns, J.M. 2013. Diterpene glycosides from *Stevia phlebophylla* A. Gray. *Carbohydrate Research* 379:1–6.

Chan, P., Tomlinson, B., Chen, Y.J., Liu, J.C., Hsieh, M.H. & Cheng, J.T. 2000. A double-blind placebo-controlled study of the effectiveness and tolerability of oral stevioside in human hypertension. *British Journal of Clinical Pharmacology* 50(3):215–220.

Chen, J., Jeppesen, P.B., Nordentoft, I. & Hermansen, K. 2006. Stevioside counteracts the glyburide-induced desensitization of the pancreatic beta-cell function in mice: Studies in vitro. *Metabolism* 55(12):1674–1680.

Chen, J., Xia, Y., Sui, X., Peng, Q., Zhang, T., Li, J. & Zhang, J. 2018. Steviol, a natural product inhibits proliferation of the gastrointestinal cancer cells intensively. *Oncotarget* 9(41):262–299.

Cocco, F., Cagetti, M.G., Livesu, R., Camoni, N., Pinna, R., Lingström, P. & Campus, G. 2019. Effect of a daily dose of snacks containing maltitol or *Stevia rebaudiana* as sweeteners in high caries risk schoolchildren. A double-blind RCT study. *Oral Health and Preventive Dentistry* 17(6):515–522.

Das, A., Saha, B.N., Maji, A., Kumar, S., Kumar, M. & Mandal, N. 2015. Reproductive phenology and factors affecting reproductive success in stevia (*Stevia rebaudiana* Bert.). *New Agriculturist* 26:247–255.

Das, K., Dang, R. & Shivananda, T.N. 2008. Influence of bio-fertilizers on the availability of nutrients (N, P and K) in soil in relation to growth and yield of *Stevia rebaudiana* grown in South India. *International Journal of Applied Research in Natural Products* 1(1):20–24.

Debnath, M. 2008. Clonal propagation and antimicrobial activity of an endemic medicinal plant *Stevia rebaudiana. Journal of Medicinal Plants Research* 2(2):45–51.

Dwivedi, R.S. 1999. Unnurtured and untapped super sweet nonsacchariferous plant species in India. *Current Science* 76:1454–1461.

EFSA (European Food Safety Authority) Panel on Food Additives and Nutrient Sources added to Food (ANS). 2010. Scientific opinion on the safety of steviol glycosides for the proposed uses as a food additive. *EFSA Journal* 8(4):1537p.

Fengyang, L., Yunhe, F., Bo, L., Zhicheng, L., Depeng, L., Dejie, L., Wen, Z., Yongguo, C., Naisheng, Z., Xichen, Z. & Zhengtao, Y. 2012. Stevioside suppressed inflammatory cytokine secretion by downregulation of NF-κB and MAPK signaling pathways in LPS-stimulated RAW264. 7 cells. *Inflammation* 35(5):1669–1675.

Gallo, M., Vitulano, M., Andolfi, A., DellaGreca, M., Conte, E., Ciaravolo, M. & Naviglio, D. 2017. Rapid solid-liquid dynamic extraction (RSLDE): A new rapid and greener method for extracting two steviol glycosides (stevioside and rebaudioside A) from stevia leaves. *Plant Foods for Human Nutrition* 72(2):141–148.

Gantait, S., Das, A. & Banerjee, J. 2018. Geographical distribution, botanical description and self-incompatibility mechanism of genus Stevia. *Sugar Technology* 20(1):1–10.

Gantait, S., Das, A. & Mandal, N. 2015. Stevia: A comprehensive review on ethnopharmacological properties and in vitro regeneration. *Sugar Technology* 17:95–106.

Gasmalla, M.A.A., Yang, R., Issoufou, A. & Hua, X. 2014. Nutritional composition of *Stevia rebaudiana* Bertoni leaf: Effect of drying method. *Tropical Journal of Pharmaceutical Research* 13:61–65.

Ghorai, A.K. 2008. Integrated weed management in jute (*Corchorus olitorius*). *Indian Journal of Agronomy* 53(2):149–151.

Giri, A., Rao, H.G.R. & Ramesh, V. 2014. Effect of partial replacement of sugar with stevia on the quality of kulfi. *Journal of Food Science and Technology* 51(8):1612–1616.

Goettemoeller, J. & Ching, A. 1999. Seed germination in *Stevia rebaudiana*. In *Perspectives on New Crops and New Uses*, 510:511p. Alexandria, VA: ASHS Press.

Goyal, S., Samsher, & Goyal., R.K. 2010. Stevia a bio-sweetener: A review. *International Journal of Food Science* 61(1):1–10.

Gregersen, S., Jeppesen, P.B., Holst, J.J. & Hermansen, K. 2004. Antihyperglycemic effects of stevioside in type 2 diabetic subjects. *Metabolism* 53(1):73–76.

Gu, W., Rebsdorf, A., Anker, C., Gregersen, S., Hermansen, K., Geuns, J.M. & Jeppesen, P.B. 2019. Steviol glucuronide, a metabolite of steviol glycosides, potently stimulates insulin secretion from isolated mouse islets: Studies in vitro. *Endocrinology, Diabetes & Metabolism*, 2(4):93p.

Gupta, E., Kaushik, S., Purwar, S., Sharma, R., Balapure, A.K. & Sundaram, S. 2017. Anticancer potential of steviol in MCF-7 human breast cancer cells. *Pharmacognosy Magazine* 13(51):345p.

Gupta, P., Gupta, N., Pawar, A.P., Birajdar, S.S., Natt, A.S. & Singh, H.P. 2013. Role of sugar and sugar substitutes in dental caries: A review. *ISRN Dentistry*:1–5. https://doi.org/10.1155/2013/519421 2013.

Gvasaliya, V.P., Kovalenko, N.V. & Garguliya, M.C. 1990. Studies on the possibility of growing honey grass in Abkhazia conditions. *Subtropicheskie Kul Tury* 5:149156.

Halliwell, B. & Gutteridge, J.M.C. 1990. Role of free-radicals and catalitic metal ions in humandisease: An overview. *Methods in Enzymology* 186:1–85.

Harrington, K.C., Southward, R.C., Kitchen, K.L. & He, X.Z. 2011. Investigation of herbicides tolerated by *Stevia rebaudiana* crops. *New Zealand Journal of Crop and Horticultural Science* 39(1):21–33.

He, J., Zhu, N.L., Kong, J., Peng, P., Li, L.F., Wei, X.L., Jiang, Y.Y., Zhang, Y.L., Bian, B.L., She, G.M. & Shi, R.B. 2019. A newly discovered phenylethanoid glycoside from *Stevia rebaudiana* Bertoni affects insulin secretion in rat INS-1 islet β cells. *Molecules* 24(22):4178.

Jain, P.K. & Joshi, H. 2012. Coumarin: Chemical and pharmacological profile. *Journal of Applied Pharmaceutical Science* 2(6):236–240.

Jarma-Orozco, A., Combatt-Caballero, E. & Jaraba-Navas, J. 2020. Growth and development of *Stevia rebaudiana* Bert., in high and low levels of radiation. *Current Plant Biology* 22:100–144.

Jayaraman, S., Manoharan, M.S. & Illanchezian, S. 2008. In-vitro antimicrobial and antitumor activities of *Stevia rebaudiana* (Asteraceae) leaf extracts. *Tropical Journal of Pharmaceutical Research* 7(4):1143–1149.

Jeppesen, P.B., Gregersen, S., Poulsen, C.R. & Hermansen, K. 2000. Stevioside acts directly on pancreatic β cells to secrete insulin: Actions independent of cyclic adenosine monophosphate and adenosine triphosphate–sensitivie K+-channel activity. *Metabolism* 49(2):208–214.

Kassahun, B.M. & Mekonnen, S.A. 2011. Effect of cutting position and rooting hormone on propagation ability of stevia (*Stevia rebaudiana* Bertoni). *Journal of Plant Biochemistry and Biotechnology* 6(1):5–8.

Kasti, A., Nikolaki, M., Synodinou, K.D. & Katsas, K. 2022. The effects of Stevia consumption on gut bacteria: friend or foe? *Microorganisms* 10(4):744.

Khalil, S.A., Zamir, R. & Ahmad, N. 2014. Selection of suitable propagation method for consistent plantlets production in *Stevia rebaudiana* (Bertoni). *Saudi Journal of Biological Sciences* 21(6):566–573.

Khan, A.R., Chowdhury, S.H. & Karim, M.M. 2012. Effect of date of planting on the growth and leaf yield of Stevia (*Stevia rebaudiana*). *Journal of the Bangladesh Agricultural University* 10(2):205–210.

Khan, S., Khan, T. & Shah, A.J. 2018. Total phenolic and flavonoid contents and antihypertensive effect of the crude extract and fractions of *Calamintha vulgaris*. *Phytomedicine* 47:174–183. https://doi.org/10.1016/j.phymed.2018.04.046

Khare, N. & Chandra, S. 2019. Stevioside mediated chemosensitization studies and cytotoxicity assay on breast cancer cell lines MDA-MB-231 and SKBR3. *Saudi Journal of Biological Sciences* 26(7):1596–1601.

Kim, I., Yang, M., Lee, O. & Nam-Suk, K. 2011. The antioxidant activity and the bioactive compound content of *Stevia rebaudiana* water extracts. *LWT – Food Science and Technology*, 44:1328–1332.

Kinghorn, A.D. & Soejarto, D.D. 2002. Discovery of terpenoid and phenolic sweeteners from plants. *Pure and Applied Chemistry* 74:1169–1179.

Kobus-Moryson, M. & Gramza-Michałowska, A. 2015. Directions on the use of stevia leaves (*Stevia rebauidana*) as an additive in food products. *Acta Scientiarum PolonorumTechnologia Alimentaria* 14(1):1–10.

Kobus-Moryson, M., Gramza-Michałowska, A., KobusCisowska, J. & Korczak, J. 2014. Zawartość wybranych pierwiastków w ekstraktach stewii (*Stevia rebaudiana* Bertoni) [Contents of selected elements in extracts from sweetleaf (*Stevia rebaudiana* Bertoni)]. *Problems Hig Epidemiology*, 95(2), 445–448.

Konoshima, T. & Takasaki, M. 2002. Cancer-chemopreventive effects of natural sweeteners and related compounds. *Pure and Applied Chemistry* 74(7):1309–1316.

Koser, N. & Jaffer, H.M. 2022. Role of Stevia in diabetes. *Pakistan BioMedical Journal*, March, 02–02. https://doi.org/10.54393/pbmj.v5i3.345.

Kovačević, D.B., Barba, F.J., Granato, D., Galanakis, C.M., Herceg, Z., Dragović-Uzelac, V. & Putnik, P. 2018. Pressurized hot water extraction (PHWE) for the green recovery of bioactive compounds and steviol glycosides from *Stevia rebaudiana* Bertoni. leaves. *Food Chemistry* 254:150–157.

Krishnaiah, D., Sarbatly, R. & Nithyanandam, R. 2011. A review of the antioxidant potential of medicinal plant Species. *Food Bioproducts Products Process* 89:217–233.

Kroyer, G. 2010. Stevioside and stevia-sweetener in food, application, stability and interaction with food ingredients. *Journal für Verbraucherschutz und Lebensmittelsicherheit* 5:225–229.

Kumar, H., Kaul, K., Bajpai-Gupta, S., Kaul, V.K. & Kumar, S. 2012a. A comprehensive analysis of fifteen genes of steviol glycosides biosynthesis pathway in *Stevia rebaudiana* (Bertoni). *Gene* 492(1):276–284.

Kumar, R. 2013. Seed germination of *Stevia rebaudiana* influenced by various potting media. *Octa Journal of Biosciences* 1(2):1–9.

Kumar, R., Ramesh, K., Tehria, S., Singh, B. & Prasad, R. 2010. Crop weather interaction studies in a natural sweetener plant (*Stevia rebaudiana* (Bert.) Bertoni) in Indian Western Himalaya. *Berichte des MeteorologischenInstituts der Albert-Ludwigs-Universität Freiburg*:38p.

Kumar, R., Sharma, S., Ramesh, K., Prasad, R., Lata Pathania, V., Singh, B. & Singh, R. 2012b. Effect of agro-techniques on the performance of natural sweetener plant-stevia (*Stevia rebaudiana*) under western Himalayan conditions. *Indian Journal of Agronomy* 57(1):74–79.

Kumar, R., Sharma, S. & Sharma, M. 2014. Growth and yield of natural-sweetener plant stevia as affected by pinching. *Indian Journal of Plant Physiology* 19(2):119–126.

Kumari, M. & Chandra, S. 2015. Phytochemical studies and estimation of major steviol glycosides in varied parts of *Stevia rebaudiana*. *International Journal of Pharmaceutical Sciences* 7(7):62–65.

Lavini, A., Riccardi, M., Pulvento, C., De Luca, S., Scamosci, M. & d'Andria, R. 2008. Yield, quality and water consumption of *Stevia rebaudiana* Bertoni grown under different irrigation regimes in Southern Italy. *Italian Journal of Agronomy* 3(2):135–143.

Leon, B.M. & Maddox, T.M. 2015. Diabetes and cardiovascular disease: Epidemiology, biological mechanisms, treatment recommendations and future research. *World Journal of Diabetes* 6(13):1246–1258.

Li, X.Y., Lu, W.M., Shen, W.F., Wu, Y., Liu, Y.P. & Tuo, Y. 2017. Growth inhibitory effect of stevioside on ovarian cancer through AKT/ERK pathway. *Biomedical Research* 28(4):1820–1827.

Libik-Konieczny, M., Capecka, E., Kąkol, E., Dziurka, M., Grabowska-Joachimiak, A., Sliwinska, E. & Pistelli, L. 2018. Growth, development and steviol glycosides content in the relation to the photosynthetic activity of several *Stevia rebaudiana* Bertoni strains cultivated under temperate climate conditions. *Scientiahorticulturae* 234:10–18.

Mallu, M.R., Vemula, S. & Kante, R.K. 2019. Molecular docking studies of bioactive compounds from *Stevia rebaudiana* for its anti-cancer activity. *Journal of Pharmaceutical Sciences and Research* 11(5):2016–2018.

Marcinek, K. & Krejpcio, Z. 2015. *Stevia rebaudiana* Bertoni – chemical composition and functional properties. *Acta Scientiarum Polonorum, Technologia Alimentaria* 14(2):145–152.

Marles, R.J. & Farnsworth, N.R. 1995. Antidiabetic plants and their active constituents. *Phytomedicine* 2(2):137–189.

Matsukubo, T. & Takazoe, I. 2006. Sucrose substitutes and their role in caries prevention. *International Dental Journal* 56(3):119–130.

Megeji, N.W., Kumar, J.K., Singh, V. & Ahuja, P.S. 2005. Introducing *Stevia rebaudiana* a natural zero caloric sweetener. *Current Science* 88:801–804.

Mehmed, A., Enchev, S. & Bozhimirov, S. 2021. Assessment of initial material for stevia selection (*Stevia rebaudiana* b.). *Bulgarian Journal of Agricultural Science* 27(3):536–540.

Mehmood, A., Althobaiti, F., Zhao, L., Usman, M., Chen, X., Alharthi, F., Soliman, M.M., Shah, A.A., Murtaza, M.A., Nadeem, M. & Ranjha, M.M.A.N. 2022. Anti-inflammatory potential of stevia residue extract against uric acid-associated renal injury in mice. *Journal of Food Biochemistry*:e14286.

Mehta, J., Khan, S., Bisht, V., Syedy, M., Rathore, R. & Bagari, L. 2012. High frequency multiple shoots regeneration and callus induction an anti diabetic plant-*Stevia rebaudiana* Bertoni – an important medicinal plant. *American Journal of Pharmtech Research* 2:19–27.

Mehta, R., Kumar Pundir, R., Sakhare, D. & K Sharma, A. 2016. Herbal formulation against dental caries causing microorganisms using extracts of *Stevia rebaudiana* leaves (a natural sweetner). *The Natural Products Journal* 6(2):126–133.

Melis, M.S. 1995. Chronic administration of aqueous extract of *Stevia rebaudiana* in rats: Renal effects. *Journal of Ethnopharmacology* 47(3):129–134.

Melis, M.S. 1999. Effect of crude extract of *Stevia rebaudiana* on renal water and electrolytes excretion. *Phytomedicine* 6(4):247–250.

Meng, J., Zhou, C., Hu, B., et al. 2018. Stevioside prevents wear particle-induced osteolysis by inhibiting osteoclastogenesis and inflammatory response via the suppression of TAK1 activation. *Frontiers in Pharmacology* 9:1053–1063.

Michel, J., Abd Rani, N.Z. & Husain, K. 2020. A review on the potential use of medicinal plants from asteraceae and lamiaceae plant family in cardiovascular diseases. *Frontiers in Pharmacology* 11:852. https://doi.org/10.3389/fphar.2020.00852

Midmore, D.J. & Rank A.H. 2002. A new rural industry stevia to replace imported chemical sweeteners. *Rural Industries Research and Development Corporation*, Pub No. W02/22, p. 55.

Mlambo, R., Wang, J. & Chen, C. 2022. *Stevia rebaudiana*, a versatile food ingredient: The chemical composition and medicinal properties. *Journal of Nanomaterials*:12p. https://doi.org/10.1155/2022/3573005.

Mousumi, D. 2008. Clonal propagation and antimicrobial activity of an endemic medicinal plant *Stevia rebaudiana*. *Journal of Medicinal Plants Research* 2(2):45–51.

Munz, S., Präger, A., Merkt, N., Claupein, W. & Graeff-Hönninger, S. 2018. Leaf area index, light interception, growth and steviol glycoside formation of *Stevia rebaudiana* Bertoni under field conditions in southwestern Germany. *Industrial Crops and Products* 111:520–528.

Narayana, N.K., Govinda, G., Kumari, A.V. & Palliyaguru, O.G. 2022. Effect of sugar replacement with stevia on quality of vanilla flavoured cow milk set yoghurt. *Food Research* 6(5):174–182. https://doi.org/10.26656/fr.2017.6(5).548.

Narayanan, P., Chinnasamy, B., Jin, L. & Clark, S. 2014. Use of just-about-right scales and penalty analysis to determine appropriate concentrations of stevia sweeteners for vanilla yogurt. *Journal of Dairy Science* 97(6):3262–3272. https://doi.org/10.3168/jds.2013-7365.

Nazari, L., Goodarzi, N., Zangeneh, M.M., Zangeneh, A., Tahvilian, R. & Moradi, R. 2017. Stereological study of kidney in streptozotocin-induced diabetic mice treated with ethanolic extract of *Stevia rebaudiana* (bitter fraction). *Comparative Clinical Pathology* 26(2):455–463.

Nower, A.A. 2014. In vitro propagation and synthetic seeds production: An efficient methods for *Stevia rebaudiana* Bertoni. *Sugar Technologies* 16(1):100–108.

Ortiz-Viedma, J., Romero, N., Puente, L., Burgos, K., Toro, M., Ramirez, L., Rodriguez, A., Barros-Velazquez, J. & Aubourg, S.P. 2017. Antioxidant and antimicrobial effects of stevia (*Stevia rebaudiana* Bert.) extracts during preservation of refrigerated salmon paste. *European Journal of Lipid Science and Technology* 119(10):160–467.

Pal, P.K., Kumar, R., Guleria, V., Mahajan, M., Prasad, R., Pathania, V., Gill, B.S., Singh, D., Chand, G., Singh, B. & Singh, R.D. 2015. Crop-ecology and nutritional variability influence growth and secondary metabolites of *Stevia rebaudiana* Bertoni. *BMC Plant Biology* 15(1):1–16.

Pal, P.K. & Mahajan, M. 2017. Tillage system and organic mulch influence leaf biomass, steviol glycoside yield and soil health under sub-temperate conditions. *Industrial Crops and Products* 104:33–44.

Paul, S., Sengupta, S., Bandyopadhyay, T.K. & Bhattacharyya, A. 2012. Stevioside induced ROS-mediated apoptosis through mitochondrial pathway in human breast cancer cell line MCF-7. *Nutrition and Cancer* 64(7):1087–1094.

Pereira, C., Storck, L., Lopes, S.J., Martin, T.N. & Bisognin, D.A. 2016. Dry biomass and glycosides yield from *Stevia rebaudiana* leaves under different harvesting times. *Bioscience Journal* 32(6):1462–1471.

Pirgozliev, V.R., Whiting, I.M., Kljak, K., Mansbridge, S.C., Atanasov, A.G., Rose, S.P. & Enchev, S.B. 2022. Stevia (*Stevia rebaudiana*) improves carotenoid content in eggs when fed to laying hens. *Foods* 11(10). https://doi.org/10.3390/foods11101418.

Radena Nenova, S.E. 2022. Influence of the inclusion of Stevia (*Stevia rebaudiana* Bertoni) in the compound feed of lactating sows on their haematological parameters. *Journal of Mountain Agriculture on the Balkans* 25(1):1–14.

Rajesh, P., Kannan, V.R. & Durai, M.T. 2010. Effect of *Stevia rebaudiana* Bertoni ethanolic extract on anticancer activity of Erlisch's Ascites carcinoma induced mice. *Journal of Current Biotica* 3(4):1.10.

Ramesh, K., Virendra, S. & Megeji, N.W. 2006. Cultivation of stevia [*Stevia rebaudiana* (Bert.)]: A comprehensive review. *Advances in Agronomy* 89:137–177.

Razak, U.N.A.A., Ong, C.B., Yu, T.S. & Lau, L.K. 2014. In vitro micropropagation of *Stevia rebaudiana* Bertoni in Malaysia. *Brazilian Archives of Biology and Technology* 57:23–28.

Rizwan, F., Rashid, H.U., Yesmine, S., Monjur, F. & Chatterjee, T.K. 2018. Preliminary analysis of the effect of stevia (*Stevia rebaudiana*) in patients with chronic kidney disease (stage I to stage III). *Contemporary Clinical Trials Communications* 12:17–25.

Sehar, I., Kaul, A., Bani, S., Pal, H.C. & Saxena, A.K. 2008. Immune up regulatory response of a non-caloric natural sweetener, stevioside. *Chemico-Biological Interactions* 173(2):115–121.

Shaffert, E.E. & Chebotar, A.A. 1994. Structure, topography and ontogeny of *Stevia rebaudiana*. *BotZhurnal* 79:38–48.

Shahu, R., Jha, P. & Jobby, R. 2022. Biotransformation of stevioside to rebaudioside – A in suspension cells of Stevia: Applications of leaf extract in functional biscuits. *Research Journal of Biotechnology* 17(11):101–109.

Shakya, A.K. 2016. Medicinal plants: Future source of new drugs. *The International Journal of Herbal Medicine* 4:59–64.

Shivanna, N., Naika, M., Khanum, F. & Kaul, V.K. 2013. Antioxidant, anti-diabetic and renal protective properties of Stevia rebaudiana. *Journal of Diabetes and its Complications* 27(2):103–113.

Simlat, M., Ptak, A., Skrzypek, E., Warchoł, M., Morańska, E. & Piórkowska, E. 2018. Melatonin significantly influences seed germination and seedling growth of *Stevia rebaudiana* Bertoni. *PeerJ* 6:e5009.

Singh, A., Singh, S. & Prasad, S.M. 2016. Role of medicinal plants for health perspective: Special reference to antioxidant potential. *Journal of Chemical Biology & Therapeutics* 1:106. https://doi.org/10.4172/2572-0406.1000106

Singh, B., Singh, J. & Kaur, A. 2014. Agro-production, processing and utilization of *Stevia rebaudiana* as natural sweetener. *Journal of Agricultural Engineering and Food Technology* 1(1):28–31.

Sivaram, L. & Mukundan, U. 2003. In vitro culture studies on *Stevia rebaudiana*. *In vitro Cellular & Developmental Biology-Plant* 39(5):520–523.

Skaria, B.P., Joseph, R., Malhew, S. & Joy, P.P. 2004. Stevia: A sweet herb. *Indian Journal of Arecanut, Spices and Medicinal Plant* 6:24–27.

Snehal, P. & Madhukar, K. 2012. Quantitative estimation of biochemical content of various extracts of *Stevia rebaudiana* leaves. *Asian Journal of Pharmaceutical and Clinical Research* 5(1):115–117.

Soejatro, D.D. 2002. Botany of stevia and *Stevia rebaudiana*. In *Medicinal and Aromatic Plants-Industrial Profiles, Vol. 19: Stevia*, ed. A.D. Kinghorn, 18–40. London: Taylor & Francis.

Soejarto, D.D., Compadre, C.M., Medon, P.J., Kamath S.K. & Kinghorn, A.D. 1983. Potential sweetening agents of plant origin. II. Field search for sweet tasting Stevia species. *Economic Botany* 37:71–79.

Srivastava, S., Chouhan, D.E.S. & Chawla, P. 2016. Proximate and phytochemical analysis of stevia leaves powder. *International Ayurvedic Medical Journal* 4:2476–2480.

Starratt, A.N., Kirby, C.W., Pocs, R. & Brandle, J.E. 2002. Rebaudioside F, a diterpene glycoside from *Stevia rebaudiana*. *Phytochemistry* 59(4):367–370.

Sulaiman, A.M., Hashem, H.A. & Nassar, A.G. 2022. Utilization of stevia leaves powder or stevia leaves aqueous extract as a substitute of sugar for producing low calorie cake. *Al-Azhar Journal of Agricultural Research* 47(1):8–18.

Sultana, C., Azad, A.K., Rahman, M., Muhit, A. & Rahman, S.M. 2020. Phytochemical and biological investigation of *Stevia rebaudiana* (Bert.) leaves grown in Bangladesh. *Dhaka University Journal of Pharmaceutical Sciences* 19(2):191–197.

Taak, P., Koul, B., Chopra, M. & Sharma, K. 2021. Comparative assessment of mulching and herbicide treatments for weed management in *Stevia rebaudiana* (Bertoni) cultivation. *South African Journal of Botany* 140:303–311.

Takasaki, M., Konoshima, T., Kozuka, M., Tokuda, H., Takayasu, J., Nishino, H., Miyakoshi, M., Mizutani, K. & Lee, K.H. 2009. Cancer preventive agents. Part 8: Chemopreventive effects of stevioside and related compounds. *Bioorganic & Medicinal Chemistry* 17(2):600–605.

Talevi, A. 2016. Beneficial effects of *Stevia rebaudiana* Bertoni and steviol-related compounds on health. *Sweeteners*:1–22. https://doi.org/10.1007/978-3-319-26478-3_24-1

Tavarini, S. & Angelini, L.G. 2013. *Stevia rebaudiana* Bertoni as a source of bioactive compounds: The effect of harvest time, experimental site and crop age on steviol glycoside content and antioxidant properties. *Journal of the Science of Food and Agriculture* 93(9):2121–2129.

Thiyagarajan, M. & Venkatachalam, P. 2012. Large scale in vitro propagation of Stevia rebaudiana (bert) for commercial application: Pharmaceutically important and antidiabetic medicinal herb. *Industrial Crops and Products* 37(1):111–117.

Timotina, M., Aghajanyan, A., Schubert, R., Trchounian, K. & Gabrielyan, L. 2022. Biosynthesis of silver nanoparticles using extracts of *Stevia rebaudiana* and evaluation of antibacterial activity. *World Journal of Microbiology and Biotechnology* 38(11):196. https://doi.org/10.1007/s11274-022-03393-3.

Tondare, J.C. & Hembade, A.S. 2021. Nutritional and physico-chemical measurement and characterization of dietetic amrakhand manufactured by using stevia leaf extract powder. *Vidyabharati International Interdisciplinary Research Journal* 2021:2951–2956.

Toyoda, K., Matsui, H., Shoda, T., Uneyama, C., Takada, K. & Takahashi, M. 1997. Assessment of the carcinogenicity of stevioside in F344 rats. *Food and Chemical Toxicology* 35(6):597–603.

Ucar, E., Özyiğit, Y. & Turgut, K. 2016. The effects of light and temperature on germination of Stevia (*Stevia rebaudiana* Bert.) seeds. *Turkish Journal of Agricultural Research* 3:37–40.

Wang, L. & Wu, W. 2019. Angiotensin-converting enzyme inhibiting ability of ethanol extracts, steviol glycosides and protein hydrolysates from stevia leaves. *Food & Function* 10(12):7967–7972.

Wang, T., Guo, M., Song, X., Zhang, Z., Jiang, H., Wang, W., Fu, Y., Cao, Y., Zhu, L. & Zhang, N. 2014. Stevioside plays an anti-inflammatory role by regulating the NF-κB and MAPK pathways in *S. aureus*-infected mouse mammary glands. *Inflammation* 37(5):1837–1846.

Wang, Y.H., ElSohly, M.A., Khan, I.A., Lata, H. & Chandra, S. 2013, August. In vitro germplasm conservation of elite *Stevia rebaudiana* Bertoni. *InII International Symposium on Plant Cryopreservation* 1039:303–308.

Wild, S., Roglic, G., Green, A., Sicree, R. & King, H. 2004. Global prevalence of diabetes: Estimates for the year 2000 and projections for 2030. *Diabetes Care* 27:1047–1053.

Yadav, A.K., Singh, S., Dhyani, D. & Ahuja, P.S. 2011. A review on the improvement of stevia [*Stevia rebaudiana* (Bertoni)]. *Canadian Journal of Plant Science* 91:1–27.

Yadav, S.K. 2010. Cold stress tolerance mechanisms in plants. A review. *Agronomy for Sustainable Development* 30:515–527.

Yingkun, N., Zhenyu, W., Jing, L., Xiuyun, L. & Huimin, Y. 2013. Stevioside protects LPS-induced acute lung injury in mice. *Inflammation* 36(1):242–250.

Yoneda, Y., Shimizu, H., Nakashima, H., Miyasaka, J. & Ohdoi, K. 2017. Effects of light intensity and photoperiod on improving steviol glycosides content in *Stevia rebaudiana* (Bertoni) Bertoni while conserving light energy consumption. *Journal of Applied Research on Medicinal and Aromatic Plants* 7:64–73.

Yuajit, C., Homvisasevongsa, S., Chatsudthipong, L., Soodvilai, S., Muanprasat, C. & Chatsudthipong, V. 2013. Steviol reduces MDCK Cyst formation and growth by inhibiting CFTR channel activity and promoting proteasome-mediated CFTR degradation. *PLoS ONE*, 8(3):e58871.

Zaidan, U.H., Zen, N.I.M., Amran, N.A., Shamsi, S. & AbdGani, S.S. 2019. Biochemical evaluation of phenolic compounds and steviol glycoside from Stevia rebaudiana extracts associated with in vitro antidiabetic potential. *Biocatalysis and Agricultural Biotechnology* 18:101049.

Zaman, M.M., Chowdhury, M.A.H. & Chowdhury, T. 2015. Growth parameters and leaf biomass yield of stevia (*Stevia rebaudiana*, Bertoni) as influenced by different soil types of Bangladesh. *Journal of the Bangladesh Agricultural University* 13(1):31–37.

Zayova, E., Nedev, T. & Dimitrova, L. 2017. In vitro storage of *Stevia rebaudiana* Bertoni under slow growth conditions and mass multiplication after storage. *Biology Bulletin* 3(1):30–38.

Zhao, L., Yang, H., Xu, M., Wang, X., Wang, C., Lian, Y., Mehmood, A. & Dai, H. 2019. Stevia residue extract ameliorates oxidative stress in d-galactose-induced aging mice via Akt/Nrf2/HO-1 pathway. *Journal of Functional Foods* 52:587–595.

Zhao, Y.T., Guo, J.H., Wu, Z.L., Xiong, Y. & Zhou, W.L. 2008. Innate immune responses of epididymal epithelial cells to Staphylococcus aureus infection. *Immunology Letters* 119(1–2):84–90.

Zlabur, S.J., Dobričević, N., Brnčić, M., Barba, F.J., Lorenzo, J.M., Franco, D., Atanasov, A.G., Voća, S. and RimacBrnčić, S. 2019. Evaluation of the behavior of phenolic compounds and steviol glycosides of sonicated strawberry juice sweetened with stevia (Stevia rebaudianaBertoni). *Molecules* 24(7):1202.

18 Antidiabetic Potential of *Ajuga* sp.

Dolly Kain, Atul Arya, and Sahil Mehta

CONTENTS

18.1 INTRODUCTION

Diabetes is a chronic health condition that badly affects the human body, especially the mechanism of conversion of food into energy (Reed et al., 2021). The food we eat is converted into sugar or glucose and released into the bloodstream resulting in an increased level of blood sugar, and after that the pancreas releases insulin. Insulin is a type of hormone that helps to store unused blood sugar in the body's cells for further use as an energy source. In diabetes, the function of insulin is hampered or the body is unable to synthesize the hormone. Increased blood sugar level over time results in various disorders such as heart disease, kidney disease, and vision loss. The three main types of diabetes are type 1, type 2, and gestational diabetes (diabetes during pregnancy) (Reed et al., 2021).

Type 1 diabetes is an autoimmune disease that stops the production of insulin in the human body. Approximately 5%–10% of diabetic people have type 1diabetes. It usually occurs in children, teens, and young adults. In this condition, the patient needs to take insulin every day to survive (Kousar, 2019). Preventive measures for type 1 diabetes are still not very well known. Common causes include having a parent, brother, or sister with type 1 diabetes. In type 2 diabetes, human body cells are not able to use insulin well and hence blood sugar levels are not at normal levels (Figure 18.1). About 90%–95% of diabetic people have type 2 diabetes. It develops over years and is usually diagnosed in adults (Reed et al., 2021). Type 2 diabetes can be prevented with healthy lifestyle changes that include weight loss, healthy food in the diet, and active hours in the day (Olokoba et al., 2012; Kharroubi et al., 2015; Gaonkar and Hullatti, 2020; Reed et al., 2021).

Gestational diabetes develops during pregnancy in women who never had diabetes in the past; it also puts the baby at a higher risk for health problems. Gestational diabetes soon goes away usually after the birth of a baby, but in later life it increases the risk for type 2 diabetes (McIntyre et al.,

DOI: 10.1201/b23347-18

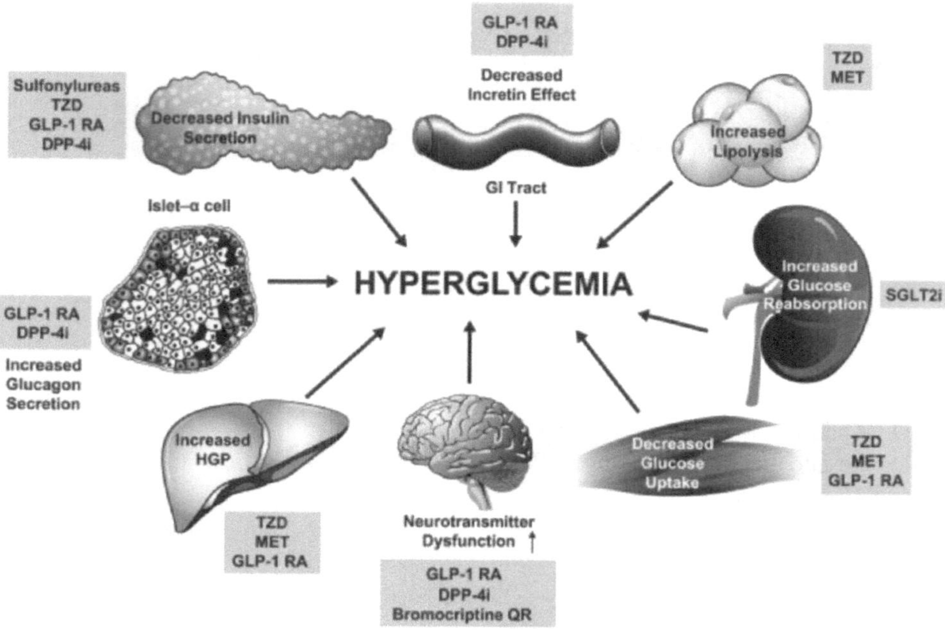

FIGURE 18.1 Overview of type 2 diabetes mellitus (adapted from DeFronzo et al., 2013).

2020). The baby is also at risk of obesity and is more likely to develop type 2 diabetes. Common causes of this type of diabetes include the birth of a baby who weighed over 9 pounds, the mother being overweight, being older than 25 years, and polycystic ovary syndrome (PCOS). Prediabetes is also a serious health condition that increases the chances of having diabetes in the future. The major causes include being overweight, having an age of 45 years or older, maybe having a parent or sibling with type 2 diabetes, having no physical activity, and ever being diagnosed with gestational diabetes. Gestational diabetes and prediabetes can also be reversed with proven lifestyle changes including losing weight, eating a healthy diet, and getting regular physical activity (McIntyre et al., 2020; Kampmann et al., 2015).

The global impact of diabetic mellitus can be understood as it is a very common and serious disease all over the world including in both developed and developing countries. With 25% of the world population, India has the highest number of diabetic patients and is known as the diabetic capital of the world. According to the World Health Organization (WHO), in 1985 about 30 million people suffered from diabetes, and it was estimated that the number increased to more than 171 million in 2000 and up to 366 million by 2030 in developing countries. It has also been estimated that India will spend around USD 2.8 billion on diabetes (Olokoba et al., 2012; Kharroubi et al., 2015; WHO).

Treatment of diabetes is one of the major health concerns in today's world. Many medicinal plants have been reported with antidiabetic potential and can be used as an alternative in the treatment of the disease. *Gymnema sylvestre* (Asclepiadaceae), *Momordica charantia* (Cucurbitaceae), *Trigonella foenum-graceum* (Fabaceae), *Tinospora cordifolia* (Menispermaceae), and *Curcuma longa* (Zingiberaceae) are examples of some famous antidiabetic plants on which *in-vivo* animal studies and human clinical trials are also conducted successfully (Gaonkar and Hullatti, 2020). In addition, one such group of plants falls in the genus *Ajuga* of the Lamiaceae family. Lamiaceae is one of the major plant families that is known for its miraculous medicinal values, including hypoglycemic activity. Different species of *Ajuga* have been reported with antidiabetic properties and other ethnobotanical uses by different tribal communities of India.

18.2 INTO THE MEDICINAL PLANTS

Different medicinal plants were reported in the literature as having significant antidiabetic potential and can be used as an herbal alternative in the treatment of diabetes (Husen 2021, 2022). *In vitro* potential of plants can be determined by using one of the two methods available including α-amylase and α-glucosidase inhibition assay. Some medicinal plants having α-amylase inhibition properties include *Pterocarpus marsupium*, *Catharanthus roseus*, *Carthamus tinctorius*, *Momordica charantia*, *Psidium guajav*, *Gynostemma pentaphyllum*, *Glycyrrhiza glabra*, and *Smilax glabra*. In addition, it has been also found in *Rehmannia glutinosa*, *Santalum spicatum*, *Ocimum tenuiflorum*, *Rhizoma fagopyri*, *Rosa rugosa*, *Rhizoma alpiniae officinarum*, *Caulis polygoni*, *Fructus amomi*, *Folium ginkgo*, and *Cortex cinnamomi*. Methanol extracts of *Marrubium radiatum*, *Aloe vera*, *Salvia acetabulosa*, *Paronychia argentea*, *Terminalia arjuna*, *Aegle marmelos*, *Linum usitatisumum*, *Eugenia cumini*, *Morus alba*, *Moringa stenopetala*, and *Nelumbo nucifera* and aqueous extracts of *Costus pictus* are reported to have α-amylase inhibitory potential and thus all are antidiabetic in nature (Salehi et al., 2019).

α-Glucosidase inhibition activity was reported in *Beyeria leshnaultii*, *Mucuna pruriens*, *Acacia ligulata*, *Pterocarpus marsupium*, *Boerhaavia diffusa*, hydroalcoholic extract of *Juniperus oxycedrus*, *Fructus amomi*, *Rhizoma alpiniae officinarum*, *Fagonia cretica*, *Santalum spicatum*, *Rhizoma fagopyri*, *Rosa rugosa*, *Caulis polygoni*, *Folium ginkgo*, and *Cortex cinnamomi*. Methanol extracts of *Marrubium radiatum Salvia acetabulosa*, methanol-water extracts of *Eugenia polyantha*, hydroalcoholic extracts of *Ludwigia octovalvis*, *Camellia sinensis*, *Aralia elata*, *Iostephane heterophylla*, *Cinnamomum zeylanicum*, and *Nelumbo nucifera*, and aqueous extracts of *Costus pictus* reported in the literature to possess the α-glucosidase activity and carries potential antidiabetic potential (Salehi et al., 2019).

18.3 DISTRIBUTION AND BOTANICAL DESCRIPTION OF *AJUGA* SP.

The genus *Ajuga* which belongs to the familyLamiaceae is a major group of flowering plants. It is also known asbugleweed, ground pine, and carpet bugle. The majority of the species are native to Asia, Africa, and Europe, and two species are native to southeastern Australia. Some species are distributed in subtropical and temperate regions from Kashmir to Bhutan, the western Himalayas, the plains of Punjab, and the upper Gangetic plains of India, Pakistan, Afghanistan, China, and Malaysia (Khare, 2007) at an altitude of 1300 m (Chandel and Bagai, 2010). It is also found along the roadsides, rock crevices, and open slopes up to 1500 m above mean sea level (Chauhan, 1999; Upadhyay et al., 2012; Suryavanshi et al., 2022).

Ajuga species are well distributed throughout the world. *Ajuga bracteosa* is commonly distributed in Afghanistan, Pakistan, Kashmir, the Himalayas to Bhutan, Burma, China, and Chitral, South Japan, and Malaysia and is a synonym of *Ajuga integrifolia*. *Ajuga parviflora* is found in Afghanistan, Pakistan, Kashmir, and Northwest India. *Ajuga brachystemon* is found in the Himalaya region (Kumaun to Nepal). *Ajuga forrestii* occurs in Himalaya (Nepal) and southwest China, mainly Yunnan. *Ajuga lobata* is found in the Himalayas from Nepal to Bhutan, India, Tibet, Burma, and southwest China, mainly in Yunnan. *Ajuga lupulina* is distributed throughout Nepal, Assam, and China. Several varieties of *Ajuga macrosperma* are very well distributed in the Kumaun region to North – East Frontier Agency (NEFA), Burma, Indo-China, and southwest China. *Ajuga parviflora* is well distributed in Afghanistan, Chitral, and the Himalayas (Kashmir to Nepal). *Ajuga forrestii* and *Ajuga lupulina* var. lupulina is available only in China. *Ajuga lobata* is present in Bhutan, India, northern Myanmar, and Nepal.

Ajuga sp. is a perennial herb that is branched at the base, usually 5–50 cm in height, with erect hairy stems, leaves usually in many pairs, the lower leaves are oblong, with a cuneate base and obscure petiole and the upper leaves are obovate-oblong, obtuse, sessile, entire or crenate, slightly hairy. Flowers are borne from the axils of the leaves with short and hairy pedicels. Petals are blue in

TABLE 18.1
Status of *Ajuga* sp. as per "The Plant List"

Status	Total (A)		Total (B)	
Accepted	70	37.2%	94	33.5%
Synonym	110	58.5%	179	63.7%
Unplaced	6	3.2%	6	2.1%
Unassessed	2	1.1%	2	0.7%

color (Journal Storage). As per "The Plant List," 188 scientific plant species in the genus *Ajuga* are reported up to now, of these 70 are accepted, and 93 species are of infraspecific rank for the genus *Ajuga*. The status of the 188 species names for the genus *Ajuga* recorded (A) and the status of the 281 names (including infraspecific names) for the genus *Ajuga* recorded (B) in "The Plant List" are shown in Table 18.1. A few species have been reported as antidiabetic, including *Ajuga bracteosa*, *Ajuga parviflora*, *Ajuga integrifolia*, and *Ajuga remota* (The Plant List; Kew).

18.4 DIVING INTO THE WIDE DIVERSITY OF WELL-KNOWN ANTIDIABETIC *AJUGA* SP. AND THEIR PHYTOCONSTITUENTS

18.4.1 *Ajuga bracteosa*

Ajuga bracteosa Wall. ex Benth. is a perennial hairy herb, commonly known as "bungle weed" in English, "Jan-i-adam" in the Kashmir region, "Nilkanthi" in Ayurveda/Sanskrit, and "Ratapaati" in the Kumaon region of Uttarakhand (Figure 18.1). *Ajuga bracteosa* is reported to have different biological activities and is used in the treatment of various diseases including fever, bronchitis, pneumonia, diarrhea, jaundice, rheumatism, gout, palsy, and amenorrhea (Kartikar and Basu, 1918; Manjunath, 1948; Chopra et al., 1986; Manandhar, 2002; Sharma et al., 2004, Islam et al., 2006), as well as of burns, boils, and syphilis (Johnson, 1999; Sharma et al., 2004). *Ajuga bracteosa* has a high amount of antioxidants and effectively controls diabetes and hypertension (Ziyyat et al., 1997; Eddouks et al., 2007; Pawar et al., 2011). The aqueous extract of leaves also shows diuretic and stimulant potential. Leaf decoction is traditionally used for several diseases including fever, stomach pain, malaria, hypertension, and diabetes (Nadkarni, 1976).

Many compounds like β-sitosterol, γ-sitosterol, tetracosanoic acid, and triacontanyldocosanoate have been isolated from the aerial part (Rastogi and Mehrotra, 1996). Other constituents like phenolic components, cerotic acid, palmitic acid bitter components, arabinose, along with glucosidic constituents, D-glucoside, and anthocyanidin-glucosides, have been found (Chopra et al., 1956). Presence of Lupulin A, 12-bromo-ajugarin I, clerodin, clerodinin A, deacetylajugarin IV, 14,15-dihydroajugapitin, dihydroclerodin-1, bractin A &B, bractic acid, sphingolipids, phthalic acid ester, lignoceric acid, phytoecdysteroids (22-acetylcyasterone, ajugalactone, ajugasterone A, B,&C, cyasterone, 3-epicyasterone, 20-hydroxyecdysone); iridoid glycosides (8-O-acetylharpagide, reptoside), sterols (β-sitosterol, stigmasterol); withanolides (bracteosin A, B,& C) linalyl acetate, sitoglucoside, neo-clerodane diterpenoids (ajugarin I, II, III, IV,&V, ajugapitin, bracteonin A, 14-hydro-15-hydroxyclerodin,14,15-dihydroxyajugapitin, 14-hydroxy-15-hydroxyajugapitin also have been reported) (Kubo et al., 1976; Kubo et al., 1980; Kubo et al., 1982; Kubo et al., 1983; Bhakuni et al., 1990; Bhakuni et al., 1987; Odek-Ogunde et al., 1993; Cantrell et al., 1999; Khan et al., 1999a; Khan et al., 1999c; Nawaz et al., 1999, Kuria et al., 2002; Verma et al., 2002; Fekete et al., 2004; Riaz et al., 2004; Shafi et al., 2004; Coll et al., 2005; Singh et al., 1996; Riaz et al., 2007).

18.4.2 *Ajuga parviflora*

In folk medicine, the leaf powder of *A. parviflora* (Figure 18.2) along with the tuber powder of *Aconitum heterophyllum* (Wall. ex Royle) is used in the treatment of leucorrhoea, high fever, colic, and diabetes. In the Kullu district of Himachal Pradesh (India), the whole plant of *Ajuga parviflora* has been used for reducing blood glucose level. Ethnomedicinal surveys reveal that *A. parviflora* is used by different tribal communities in India, including the Gaddi and Gujjar tribes (Arya et al., 2018), Bhotia tribe (Kumar et al., 2019), and Jaunsari tribe (Yousaf et al., 2018), against diabetes.

The plant contains palmitic acid, oleic acid, linoleic acid, ceryl alcohol, cerotic acid, phenolic acids and neutral bitter components, alkaloids, diterpenoids, and triterpenoids, neo-clerodane diterpenoids (13-β-acetoxy-clerodinin C. dihydroclerodin I, 15-α-ethoxy-14-hydroajugapitin, 15-β-ethoxy-14-hydroajugapitin ajugamarin F4, ajugarin I & II, ajugarin I chlorohydrin, clerodinin C & D, deoxyajugarin I), withanolides (Ajugin A, B, C, D, E, F, G, H,& I and acetylated quinols) (Beauchamp et al., 1996; Khan et al., 1999a; Khan et al., 1999b; Khan et al., 1999c; Khan et al., 1999d; Muhammad et al., 1999; Khan et al., 1999c; Nawaz et al., 2000; Choudhary et al., 2005).

18.4.3 *Ajuga integrifolia*

Ajuga integrifolia is considered a synonym of *A. bracteosa*. *Ajuga integrifolia* occurs in different regions of Ethiopia including Amhara and Southern Nations, Nationalities, and Peoples' Region (SNNPR), Tigray, and Oromia (Bekeri et al., 2018). The plant is locally known as "Tut Astel,"

FIGURE 18.2 Plant morphology of *Ajuga bracteosa* and *Ajuga parviflora* in the field.

"Akorarach," and "Harmegusa" in various parts of Ethiopia (Seifu, 2017; Assefa, 2013; Chekole, 2017). Aqueous and alcoholic infusion of the leaves and roots of the plant is traditionally used in the treatment of diabetes (Bekeri et al., 2018).

A. integrifolia is reported to contain terpenoids, flavonoids, alkaloids, glycosides, tannins, steroids, phenols, and saponins (Bekeri et al., 2018). The roots of *Ajuga integrifolia* contain a high amount of chromium that can be correlated with the treatment of diabetes. Neo-clerodane diterpenoids (ajugacumbin B, ajugamacrin A, B, C, D,&E, ajugamarin C1, ajugapantin A) and pyrrolizidine alkaloids (senecionine, integerrimine) are also reported in the plant (Shimomura et al., 1987; Shen et al., 1993b).

18.4.4 *Ajuga remota*

Ajuga remota Benth. (synonyms: *Ajuga integrifolia* Buch.-Ham and *Ajuga bracteosa* Wall ex Benth.) (Quattrocchi, 2012) grows widely in East Africa, Saudi Arabia, Yemen, and Afghanistan to East Asia at an altitude of 1500–3400 m (Hedberg et al., 2006; Col and Tandrón, 2005). Traditionally the plant is used in Ethiopia to treat various diseases including diabetes, skin disease, high blood pressure, malaria, toothache, stomach pain, pneumonia, liver problem, and swelling of the legs. The plant is bitter in taste, hence taken with honey to make it more palatable, and stored for a long period for later use.

The plant has been reported to contain neo-clerodane diterpenoids (Ajugarin I, II, III, IV, & V, ajugapitin, bromoajugarin I, clerodin, deacetylajugarin IV, dihydroajugapitin, dihydroclerodin, 14-hydro-5-hydroxy, ajugapitin); iridoid glycosides (8-O-acetylharpagide, 8-acetylharpagide-6-O-β-glucoside, 6,7-dehydro-8-acetylharpagide, 7,8-dehydroharpagide, 6,8-diacetylharpagide, 6-keto-8-acetyl-harpagide, harpagide-6-O-β-glucoside); phytoecdysteroids (ajugalactone, ajugasterone C, cyasterone, ecdysterone, 20-hydroxyecdysone); flavonol glycosides (myricetin 3-O-rutinoside-4'-O-rutinoside, myricetin 3-O-rutinoside-3'-O-rutinoside, isorhamnetin 3-O-rutinoside-7-O-rutinoside-4'-O-β-glucoside, myricetin 3-O-rutinoside, 3-O-α-rhamnoside-4'-O-rutinoside); ergosterol 5,8-endoperoxide, kaempferol 3-O-α-rhamnoside, quercetin 3-O-β-glucoside, and quercetin 3-O-rutinoside (Kubo et al., 1976; Kubo et al., 1980; Kubo et al., 1981; Kubo et al., 1982; Kubo et al., 1983; Manguro et al., 2006; Cantrell et al., 1999; Kuria et al., 2001; Kuria et al., 2002; Coll et al., 2005; Manguro et al., 2006; Lemmen, 2007; Manguro et al., 2007).

18.5 ETHNOBOTANICAL ASPECTS OF *AJUGA* SP.

Ajuga species are very well distributed in India. Many tribal communities of India use *Ajuga* species for the treatment of different ailments, especially diabetes. Ethnobotanical uses of 23 *Ajuga* species have been listed in Table 18.2.

TABLE 18.2
Ethnobotanical Uses of 23 *Ajuga* Species

S. No.	Botanical name	Common name	Ethnobotanical importance	References
1.	*Ajuga macrosperma* Wallich ex Bentham	-	Fever, phlegm	Yunnan (1977)
2.	*A. spectabilis* Nakai	Korean bugle	Stimulates smooth and cardiac muscles	Chung et al. (1980)
3.	*A. taiwanensis* Nakai ex Murata (L.)	-	Hepatitis and hepatoma	Hou (1996)

S. No.	Botanical name	Common name	Ethnobotanical importance	References
4.	*A. bracteosa* Wallace ex Benth.	Bracted bugleweed and Nilkanthi	Used as an astringent, tonic blood purifier, anti-inflammatory, depurative, diuretic, and treatment of bronchitis, pneumonia, typhoid fever, agues, diarrhea, dysentery, palsy, jaundice, amenorrhea, and diabetes	Johnson (1999)
5.	*A. australis* R. Br.	Southern bugleweed, Australian bugle, austral bugle	External boils, sores, and wounds	Johnson (1999)
6.	*A. decumbens* Thunb.	Kiransou; creeping buglewood	Expectorant, depurative, analgesic, anticoagulant, anti-inflammatory, antipyretic, antitussive, for bladder ailments, diarrhea, eye troubles, febrifuge, hemostatic, joint pain, sore throat, and stomach ache	Johnson (1999)
7.	*A. forrestii* Diels	Forest bugle	Dysentery and ascariasis	Johnson (1999)
8.	*A. ophrydris* Burch. ex Benth.	-	Dysmenorrhea, sterility	Johnson (1999)
9.	*A. multiflora* var. multiflora	-	Diuretic	Johnson (1999)
10.	*A. nipponensis* Makino	-	Traumatic injury; inflammation	Johnson (1999)
11.	*A. panthanta* Hand.-Mazz.	-	Fever and phlegm used medicinally	Johnson (1999)
12.	*A. ciliata* (Bung.) var. villosior A. Gray	-	Hemolysis, tonsillitis, and sore throat	Johnson (1999)
13.	*A. orientalis* L.	Oriental bugle and Eastern bugle	External use: skin disorders; wound healing	Ali-Shtayeh et al. (2000)
14.	*A. pseudoiva* Robill& Castagne ex DC.	Yellow Southern bugle, French ground pine, and XantkuraSafra	hypoglycemic, insect antifeedant, antibacterial, antipyretic, anthelminthic, and wound healer	Ali-Shtayeh et al. (2000)
15.	*A. chamaepitys* (L.) Schreb	Ground pine; yellow bugle	Diuretic and emmenagogue (menstrual flow stimulant)	Lassak and McCarthy (2001)
16.	*A. turkestanica* (Rgl.) Brig	-	Heart disease, muscle aches, and stomach problems	Abdukadirov et al. (2004)
17.	*A. parviflora* Benth.	-	For fever, ulcer, colic, jaundice, stomach disorder, throat diseases, and as antidiabetic	Islam et al. (2006)
18.	*A. remota* Wall. ex Benth.	-	For fever, infection, malaria, and as an antimycobacterial agent	Muregi et al. (2007)
19.	*A. chia* L.	-	Wounds	Oran and Al- Eisawi (2015)
20.	*A. integrifolia* L.	Anamuro	Antimalarial	Asnake et al. (2016)
21.	*A. genevensis* L.	Blue bugle, Geneva bugleweed, blue bugleweed, and Suliman	Sedative, antihemorrhagic, anti-inflammatory, wound healing, epithelization capacity, precipitate the proteins from the digestive tract, antidiarrheal, and anticancer	Graham et al. (2000), Toiu et al. (2016)
22.	*A. reptans* L.	Bugleweed, bugle, common bugle and Vineriţa	Antioxidant and antimicrobial properties	Toiu et al. (2017)
23.	*A. iva*(L.) Schreb	Chondgoura and Chendghoura	Diabetes, rheumatism, allergy, cancer, and renal, metabolic, cardiovascular, digestive, and respiratory disorders	Bouyahya et al. (2020)

18.6 PHARMACOLOGICAL STUDIES OF *AJUGA* SP.

Antidiabetic activity of crude extracts of *Ajuga remota* Benth. of Ethiopia was determined by using alloxan-induced diabetic mice by Bekele et al. (2008). *A. remota* doses (300 mg/kg and 500 mg/kg body weight for 14 days) lowered blood glucose levels by 28.09% and 28.25% respectively as compared with diabetic untreated mice. Antidiabetic activity of *Ajuga remota* Benth. leaves in streptozotocin-induced diabetic rats were also determined by Assefa (2013).

Hypoglycemic effects of *Ajuga* sp. were also reported by Hsieh et al. (2014). Five *Ajuga* sp., namely *A. decumbens*, *A. taiwanensis*, *A. nipponensis*, *A. pygmaea*, and *A. dictyocarpa* of Taiwan also tests *in vitro* and *in vivo* for their hypoglycemic effect on streptozotocin-induced diabetic mice (Hsieh et al., 2014). The antidiabetic and antioxidant activity of ethanolic whole plant extract of *Ajuga parviflora* (Benth.) were also determined and results indicate the promising antidiabetic and antioxidant potential of the plant which can neutralize the harmful effects of diabetes in the human body. *In vitro* studies show that the significant percentage of inhibition in the α-amylase assay is even better than the positive control acarbose. 60 mg/kg of *A. parviflora* extract shows impressive results in Wistar male rats in terms of body weight, lipid profile, blood glucose level, and creatinine (Nirja et al., 2016).

The hydroalcoholic extract of leaves of *Ajuga bracteosa* (Ratpatiya) was reported to be antidiabetic, having hypoglycemic activity in alloxan-induced diabetic chicks. *Ajuga bracteosa* leaves were used at the rate of 10 mg/100 gm body weight of alloxan-induced diabetic chicks and blood glucose levels were estimated. *Ajuga bracteosa* leaves were also administered orally for 30 days daily to the diabetic chicks, and a significant reduction in the blood glucose levels was observed (Kumar and Bisht, 2017).

Ajuga integrifolia (100 mg/kg, 200 mg/kg, and 400 mg/kg) crude root extract and solvent fractions were tested for their hypoglycemic and antihyperglycemic activities using streptozotocin-induced diabetic mice. At the concentration of 200 mg/kg and 400 mg/kg, the plant shows significant body-weight improvement on the 14th day of treatment (Alene et al., 2020). The aqueous and 70% ethanol extract of *Ajuga remota* was tested using Swiss albino mice that were induced with alloxan to get diabetes. 70% ethanol extract caused a reduction of 27.94 ± 1.92% (300 mg/kg) and 28.26 ± 1.82% (500 mg/kg), and aqueous extract showed that a significant reduction in blood glucose levels may be due to the presence of flavonoids (Tafesse et al., 2017). *In vitro* antidiabetic activity of *Ajuga parviflora* Benth. shoot was also reported by Suryavanshi et al. (2022). The extract showed significant inhibition of key enzymes linked with hyperglycemia that has α-amylase enzyme at 132.38±1.18 μg/mL and α-glucosidase enzyme at 22.66±0.11 μg/mL concentration.

Several other therapeutic activities besides antidiabetic were also reported for different *Ajuga* sp. The antiplasmodial activity of *Ajuga remota* was determined by Kuria et al. (2002). The insecticidal activity of *Ajuga* sp. was investigated by Fekete et al. (2004). The larvicidal activity against anopheline and culicine mosquitos of *Ajuga remota* was determined by Sharma et al. (2004). Antiplasmodial activity of *Ajuga bracteosa* against *Plasmodium berghei*-infected mice was reported in the literature by Chandel and Bagai (2010). Antioxidant and antimicrobial effects of *Ajuga genevensis* and *Ajuga reptans* extracts were also examined by Toiu et al. (2016) and Toiu et al. (2017) respectively. The antiviral activity of *Ajuga bracteosa* and *Ajuga parviflora* against the hepatitis C virus was reported by Yousaf et al. (2018).

18.7 *IN VITRO* CULTURE OF *AJUGA* SP.

Some species of the genus *Ajuga* were also cultured *in vitro* through cell suspension cultures from long-cultured calluses, and growth patterns, morphophysiological characteristics, and percentage of ecdysteroid of *Ajuga reptans* L. were also monitored. The content of 20E (20-hydroxyecdysone), the main ecdysteroid in cell suspension of *A. reptans* was four to eight times higher than in the intact plant (Filippova et al., 2003). *In vitro* culture of shoot tips of *Ajuga reptans* L. "Burgundy Glow" on a Murashige-Skoog medium containing 1.0 mg/liter benzyladenine or 0.1 mg/liter naphthaleneacetic acid also reported in the literature which results in the multiple shoot proliferation (Lineberger and Wanstreet). Cell suspension and hairy root cultures were of *Ajuga turkestanica*, a medicinal plant indigenous to Uzbekistan, to enhance the metabolism for the production of phytoecdysteroids also reported in the literature (Cheng et al., 2008). *In vitro* micropropagation for conservation and genetic transformation of *Ajuga bracteosa* was also reported using MS medium supplemented with IAA, and an 82% survival rate was achieved after the hardening of plantlets (Kaul et al., 2013).

18.8 CONCLUSION AND FUTURE REMARKS

The *Ajuga* sp. has immense medicinal value and has the potential to be used as a future herbal drug. Different species have different medicinal values, including antidiabetic, antibacterial, diuretic, and larvicidal antioxidant. The most common compounds from *Ajuga* species include withanolides, neo-clerodane, phytoecdysteroids, iridoid glycosides, di- and triterpenoids, sterols, and several flavonoids and phenolic compounds. Further study on *in vivo* potential is required and clinical trials need to be conducted on crude plant extract. Furthermore, it can be used as a valuable component of polyherbal formulation (PHF) having medicinal values, especially for diabetes.

REFERENCES

Abdukadirov IT, Khodzhaeva MA, Turakhozhaev MT and Mamatkhanov AU, 2004. Carbohydrates from *Ajuga turkestanica. Chemistry of Natural Compounds(KhimPrirSoedin)* 40: 85–86.

Alene M, Abdelwuhab M, Belay A and Yazie TS, 2020. Evaluation of antidiabetic activity of *Ajuga integrifolia* (Lamiaceae) root extract and solvent fractions in Mice. *In Evidence-Based Complementary and Alternative Medicine* 2020: 1–11.

Ali-Shtayeh MS, Yaniv Z and Mahajna J, 2000. Ethnobotanical survey in the Palestinian area: A classification of the healing potential of medicinal plants. *Journal of Ethnopharmacology* 73: 221–232.

Arya D, Goel S, Joshi GC, Sharma OR and Sk S, 2018. Medico-ethno botanical practices among Bhotia tribe of Kumaon Himalaya: A case study from Bageshwar district, Uttarakhand, India. *Journal of Drug Research and Ayurvedic Sciences* 3: 43–47.

Asnake S, Teklehaymanot T, Hymete A, Erko B and Giday M, 2016. Survey of medicinal plants used to treat malaria by Sidama People of Boricha district, Sidama Zone, South Region of Ethiopia. *Evidence-Based Complementary and Alternative Medicine* 2016: 9690164.

Assefa F, 2013. Antidiabetic activity of *Ajuga remota* Benth (harmegusa) leaves in streptozotocin induced diabetic rats. PhD Thesis. Available at: http://thesisbank.jhia.ac.ke/7338/

Beauchamp PS, Bottini AT, Caselles MC, Dev V, Hope H, Larter M, Lee G, Mathela CS, Melkani AB, Millar PD and Miyatake M, 1996. Neo-clerodane diterpinoids from *Ajuga parviflora. Phytochemistry* 43: 827–834.

Bekele T, Hymete A, Tadesse M and Mekonnen Y, 2008. Antidiabetic activity and phytochemical screening of crude extracts of *Stevia rebaudiana* bertoni and *Ajuga remota* benth grown in ethiopia on alloxan-induced diabetic mice. PhD Thesis. Available at: 10.13140/2.1.3124.2564.

Bekeri D, Adane L and Mamo F, 2018. Phytochemical investigation and isolation of compounds from *Ajuga integrifolia* root extract. *World Journal of Chemistry* 13: 1–13.

Bhakuni RS, Shukla YN and Thakur RS, 1987. Chemical constituents of *Ajuga bracteosa. Indian Journal of Pharmaceutical Sciences* 49: 225–226.

Bhakuni RS, Shukla YN and Thakur RS, 1990. Heptacos-3–4n-25-one -A new unsaturated ketone from *Ajuga bracteosa. Journal of the Indian Chemical Society* 67: 355.

Bouyahya A, El Omari N, Elmenyiy N, Guaouguaou FE, Balahbib A, El-Shazly M and Chamkhi I, 2020. Ethnomedicinal use, phytochemistry, pharmacology, and toxicology of *Ajuga iva* (L.,) schreb. *Journal of Ethnopharmacology* 258: 112875.

Cantrell CL, Rajab MS, Franzblau SG, Fronczek FR and Fischer NH, 1999. Anti mycobacterial ergosterol-5,8-endoperoxide from *Ajuga remota. Planta Medica* 65: 732–734.

Chandel S and Bagai U, 2010. Antiplasmodial activity of *Ajuga bracteosa* against *Plasmodium berghei* infected BALB/c mice. *Indian Journal of Medical Research* 131: 440–444.

Chauhan NS, 1999. *Medicinal and Aromatic Plants of Himachal Pradesh*. Indus Publishing Company, New Delhi.

Chekole G, 2017. Ethnobotanical study of medicinal plants used against human ailments in Gubalafto District, Northern Ethiopia. *Journal of Ethnobiology and Ethnomedicine* 13: 55.

Cheng DM, Yousef GG, Grace MH, Rogers RB, Gorelick-Feldman J, Raskin I and Lila MA, 2008. In vitro production of metabolism-enhancing phytoecdysteroids from *Ajuga turkestanica. Plant Cell Tissue and Organ Culture* 93: 73–83.

Chopra RN, Nayar SL and Chopra IC, 1956. *Glossary Indian Medicinal Plants*. Council of Scientific and Industrial Research, New Delhi, 10.

Chopra RN, Nayar SL and Chopra IC, 1986. *Glossary of Indian Medicinal Plants (Including the Supplement)*. Council of Scientific and Industrial Research, New Delhi, 330.

Choudhary MI, Nawaz SA, Lodhi MA, Ghayur MN, Jalil S, Riaz N, Yousuf S, Malik A and Gilani AH, 2005. Withanolides, a new class of natural cholinesterase inhibitors with calcium antagonistic properties. *Biochemical and Biophysical Research Communications* 334: 276–287.

Chung B-S, Lee H-K and Kim J-W, 1980. Iridoid glycoside (I): Studies of the iridoid glycoside of *Ajuga spectabilis* Nakai. *Korean Journal of Pharmacognosy* 11: 15–23.

Coll J and Tandrón YA, 2005. Isolation and identification of neo-clerodane diterpenes from *Ajuga remota* by high-performance liquid chromatography. *Phytochemical Analysis* 16: 61–67.

DeFronzo RA, Eldor R and Abdul-Ghani M, 2013. Pathophysiologic approach to therapy in patients with newly diagnosed type 2 diabetes. *American Diabetes Association*. Available at: http://care.diabetesjournals.org/content/36/Supplement_2/S127

Eddouks M, Ouahidi ML, Farid O, Moufid A, Khalidi A and Lemhadri A, 2007. The use of medicinal plants in the treatment of diabetes in Morocco. *Phytotherapie* 5: 194–203.

Fekete G, Polgár L, Báthori M, Coll J and Darvas B, 2004. Peros efficacy of *Ajuga* extracts against sucking insects. *Pest Management Science* 60: 1099–1104.

Filippova VN, Zorinyants SE, Volodina SO and Smolenskaya IN, 2003. Cell cultures of ecdysteroid-containing *Ajuga reptans* and *Serratulacoronata* plants. *Russian Journal of Plant Physiology* 50(4).

Gaonkar VP and Hullatti K, 2020. Indian traditional medicinal plants as a source of potent anti-diabetic agents: A review. *Journal of Diabetes and Metabolic Disorders* 19: 1895–1908.

Graham JG, Quinn ML, Fabricant DS and Farnsworth NR, 2000. Plants used against cancer – an extension of the work of Jonathan Hartwell. *Journal of Ethnopharmacology* 73: 347–377.

Hedberg I, Kelbessa E, Edwards S, Demissew S and Persson E, 2006. Flora of ethiopia and eritrea. Plantaginaceae. The national herbarium, addis ababa, ethiopia, and department of Systematic Botany, Uppsala. *Sweden* 5: 527–528.

Hou D, 1996. *Flora of Taiwan, Editorial Committee of the Flora of Taiwan*, Taipei, Taiwan, 4: 442.

Hsieh C-W, Ju-Yun Cheng J-Y, Wang T-H, Wang H-J and Ho W-J, 2014. Hypoglycaemic effects of *Ajuga* extract in vitro and in vivo. *Journal of Functional Foods* 6: 224–230.

Husen A, 2021. *Traditional Herbal Therapy for the Human Immune System*, CRC Press, Boca Raton, FL. https://doi.org/10.1201/9781003137955

Husen A, 2022. *Herbs, Shrubs and Trees of Potential Medicinal Benefits*, Boca Raton and London: CRC Press. https://doi.org/10.1201/9781003205067

Islam M, Ahmad H, Rashid A, Razzaq A, Akhtar N and Khan I, 2006. Weeds and medicinal plants of Shawar valley, district Swat. *Pakistan Journal of Weed Science Research* 12: 83–88.

Johnson T, ed., 1999. *CRC Ethnobotany Desk Reference*, CRC Press, Boca Raton, FL, 1211.

Journal Storage. Available at: https://plants.jstor.org/compilation/ajuga.bracteosa.

Kampmann U, Madsen LR, Skajaa GO, Iversen DS, Moeller N and Ovesen P, 2015. Gestational diabetes: A clinical update. *World Journal of Diabetes* 6: 1065–1072.

Kaul S, Das S and Srivastava PS, 2013. Micropropagation of *Ajuga bracteosa*, a medicinal herb. *Physiology and Molecular Biology of Plants* 19: 289–296.

Kew. https://powo.science.kew.org/taxon/urn:lsid:ipni.org:names:20646-1

Khan PM, Ahmad S, Nawaz HR, Ullah N and Malik A, 1999a. New acetylated quinols from *Ajuga parviflora*. *Fitoterapia* 70: 229–232.

Khan PM, Malik A, Ahmad S and Nawaz HR, 1999b. Withanolides from *Ajuga parviflora*. *Journal of Natural Products* 62: 1290–1292.

Khan PM, Malik A, Ahmad S, Nawaz HR and Malik A, 1999d. The first report of a withanolide from the family of Labiatae. *Phytochemistry* 51: 669–671.

Khan PM, Nawaz HR, Ahmad S and Malik A, 1999c. Ajugins C and D, new withanolides from *Ajuga parviflora*. *Helvetica Chimica Acta* 82: 1423–1426.

Khare CP, 2007. Indian medicinal plants – An illustrated dictionary. 1st Indian Reprint Springer (India) Pvt. Ltd., New Delhi, India: 28. *International Journal of Advanced Research* 2: 389–394.

Kharroubi AT and Darwish HM, 2015. Diabetes mellitus: The epidemic of the century. *World Journal of Diabetes* 6: 850–867.

Kirtikar KR and Basu BD, 1918. *Indian Medicinal Plants*. Indian Press, Allahabad.

Kousar S, 2019. Type 1 diabetes: Causes, symptoms and treatments, review with personal experience. *Current Research in Diabetes and Obesity Journal* 11: 555817.

Kubo I, Kido M and Fukuyama Y, 1980. X-ray crystal structure of 12-bromoajugarin-I and conclusion on the absolute configuration of ajugarins. *Journal of the Chemical Society, Chemical Communications*: 897–898.

Kubo I, Klocke JA and Asano S, 1981. Insect ecdysis inhibitors from the East African medicinal plant *Ajuga remota* (Labiatae). *Agricultural and Biological Chemistry* 45: 1925–1927.

Kubo I, Klocke JA, Ganjian I, Ichkawa N and Matsumoto T, 1983. Efficient isolation of phytoecdysones from *Ajuga* plants by high-performance liquid chromatography and droplet counter-current chromatography. *Journal of Chromatography* 257: 157–161.

Kubo I, Klocke JA, Miura I and Fukuyama Y, 1982. Structure of ajugarin-IV. *Journal of the Chemical Society, Chemical Communications*: 618–619.

Kubo I, Lee Y-W, Balogh-Nair V, Nakanishi K and Chapya A, 1976. Structure of ajugarins. *Journal of the Chemical Society, Chemical Communications*: 949–950.

Kumar A, Aswal S, Chauhan A, Semwal RB, Kumar A and Semwal DK, 2019. Ethnomedicinal investigation of medicinal plants of Chakrata region (Uttarakhand) used in the traditional medicine for diabetes by Jaunsari tribe. *Natural Products Bioprospection* 9: 175–200.

Kumar D and Bisht R, 2017. Antidiabetic potency of leaves of *Ajuga bracteosa* on hypoglycemic activity in alloxan induced diabetic chicks. *International Journal of Advanced Research* 5: 836–841.

Kuria KAM, Chepkwony H, Govaerts C, Roets E, Busson R, de Witte P, Zupko I, Hoornaert G, Quirynen L, Maes L, Janssens L, Hoogmartens J and Laekeman G, 2002. The antiplasmodial activity of isolates from *Ajuga remota*. *Journal of Natural Products* 65: 789–793.

Kuria KAM, De Proft M, Masengo W, Kibwage I, Quirynen L, Maes L, Janssen L, Hoogmartens J and Laekeman, 2001. Cultivating the African plant *Ajuga remota* in Belgium and confirming its biological activity against *Plasmodium falciparum*. *Z Arzn Gew Pfl* 6: 69–72.

Lassak EV and McCarthy TM, 2001. *Australian Medicinal Plants*. New Holland, Sydney, 240.

Lemmen P, 2007. Further flavonol and iridoid glycosides from *Ajuga remota* aerial parts. *Journal of Asian Natural Products Research* 9: 617–629.

Lineberger RD and Wanstreet A, *Micropropagation ofAjuga reptans 'Burgundy Glow'*. https://aggie-horticulture.tamu.edu/tisscult/Microprop/ajuga/ajuga.html

Manandhar NP, 2002. *Plants and People of Nepal*. Timber Press, Oregon, 636.

Manguro LOA, Ogur JA, Okora DM, Wagai SO and Lemmen P, 2007. Further flavonol and iridoid glycosides from *Ajuga remota* aerial parts. *Journal of Asian Natural Products Research* 9: 617–629.

Manguro LOA and Wagai SO, 2006. Lemmen P. Flavonol and iridoid glycosides of *Ajuga remota* aerial parts. *Phytochemistry* 67: 830–837.

Manjunath BL, 1948. *The Wealth of India*, Vol. 1. Council of Scientific and Industrial Research, Delhi, 254.

McIntyre HD, Kapur A, Divakar H and Hod M, 2020. Gestational diabetes mellitus – innovative approach to prediction, diagnosis, management, and prevention of future NCD–mother and offspring. *Frontiers in Endocrinology* 11: 614533.

Muhammad P, Ahmad S, Nawaz HR, Ullah N and Malik A, 1999. New acetylated quinols from *Ajuga parviflora*. *Fitoterapia* 70: 229–232.

Muregi FW, Ishih A, Miyase T, Suzuki T, Kino H, Amano T, Mkoji GM and Terada M, 2007. Antimalarial activity of methanolic extracts from plants used in Kenyan ethnomedicine and their interactions with chloroquine (CQ) against a CQ-tolerant rodent parasite, in mice. *Journal ofEthnopharmacology* 111: 190–195.

Nadkarni KM, 1976. *Indian Materia Medica*. Popular Prakashan Private Ltd., Mumbai, 58.

Nawaz HR, Malik A, Khan PM and Ahmed S, 1999. Ajugin E and F: Two withanolides from *Ajuga parviflora*. *Phytochemistry* 52: 1357–1360.

Nawaz HR, Malik A, Muhammad P, Ahmed S and Riaz M, 2000. Chemical constituents of *Ajuga parviflora*. *Z Naturforsch* 55b: 100–103.

Nirja* R and Sharma ML, 2016. Antidiabetic and antioxidant activity of ethanolic extract of *Ajuga parviflora* benth. (Lamiaceae) vern. Neelkanthi, Neelbati. *International Journal of Pharmaceutical Sciences Review and Research* 41: 232–238.

Odek-Ogunde M, Rajab MS, Migwi GJ and Ndegwa JM, 1993. Blood pressure responses to an extract of *Ajuga remota* in experimentally hypertensive rats. [Letter]. *Planta Medica* 59: 573–574.

Olokoba AB, Obateru OA and Olokoba LB, 2012. Type 2 diabetes mellitus: A review of current trends. *Oman Medical Journal* 27: 269–273.

Oran SAS* and Al-Eisawi DMH, 2015. Ethnobotanical survey of the medicinal plants in the centralmountains (North-South) in Jordan. *Journal of Biodiversity and Environmental Sciences* 6: 381–400.

Pawar RS, Pal A, Jadon M, Katare YK, Singour PK, Rajak H and Patil UK, 2011. *Ajuga bracteosa* – A review on its ethnopharmacological and phytochemical studies. *Der pharmaciasinica* 2: 1–10.

The Plant List. Available at: www.theplantlist.org/browse/A/Lamiaceae/Ajuga/

Quattrocchi U, 2012. *CRC World Dictionary of Medicinal and Poisonous Plants: Common Names, Scientific Names, Eponyms, Synonyms, and Etymology*. CRC Press, New York, Taylor & Francis Group, Boca Raton, London, 137.

Rastogi PR and Mehrotra BN, 1996. *Central Drug Research*. Institute of Sciences Communication, New Delhi, 1: 19.

Reed J and Bain S, 2021. Venkateswarlu and Kanamarlapudi. A review of current trends with type 2 diabetes epidemiology, aetiology, pathogenesis, treatments and future perspectives diabetes. *Metabolic Syndrome and Obesity: Targets and Therapy* 14: 3567–3602.

Riaz N, Malik A, Aziz-ur-Rehman, Nawaz SA, Muhammad P and Choudhary MI, 2004. Cholinesterase-inhibiting withanolides from *Ajuga bracteosa*. *Chemistry and Biodiversity* 1: 1289–1295.

Riaz N, Nawaz SA, Mukhtar N, Malik A, Afza N, Ali S, Ullah S, Muhammad P and Choudhary MI, 2007. Isolation and enzyme-inhibition studies of the chemical constituents from *Ajuga bracteosa*. *Chemistry and Biodiversity* 4: 72–83.

Salehi et al., 2019. Antidiabetic potential of medicinal plants and their active components. *Biomolecules* 9: 551.

Seifu A, 2017. *Bioprospecting Potential of Ajuga Integrifolia for Access and Benefit Sharing*. FAO, Rome, Italy.

Shafi N, Khan GA, Arfan M, Ahmad KU and Gilani ND, 2004. Isolation and pharmacological screening of 8-O-acetyl harpagide from *Ajuga bracteosa* Wall. *Pakistan Journal of Scientific and Industrial Research* 47: 176–179.

Sharma P, Mohan L and Srivastava CN, 2004. Larval susceptibility of *Ajuga remota* against anopheline and culicine mosquitos. *Southeast Asian Journal of Tropical Medicine and Public Health* 35: 608–610.

Shen X, Isogai A, Furihata K, Sun H and Suzuki A, 1993b. Neo-clerodane diterpinoids from *Ajuga macrosperma* and *Ajuga pantantha*. *Phytochemistry* 34: 1091–1094.

Shimomura H, Sashida Y and Ogawa K, 1987. Iridoid glucosides and phenylpropanoid glycosides in *Ajuga* species of Japan. *Phytochemistry* 26: 1981–1983.

Singh N, Mahmood U, Kaul VK and Jirovetz L, 1996. A new phthalic acid ester from *Ajuga bracteosa*. *Natural Product Research* 20: 593–597.

Suryavanshi A, Kumar S, Kain D, Arya A and Vandana, 2022. In vitro antidiabetic, antioxidant activities and chemical composition of *Ajuga parviflora* Benth. shoot. *Journal of Herbmed Pharmacology* 11: 131–139.

Tafesse TB, Hymete A, Mekonnen Y and Tadesse M, 2017. Antidiabetic activity and phytochemical screening of extracts of the leaves of *Ajuga remota* Benth on alloxan-induced diabetic mice. *BMC Complementary and Alternative Medicine* 17: 243.

Toiu A, Laurian V, Gheldiu AM, Vodnar D and Oniga I, 2017. Evaluation of the antioxidant and antibacterial potential of bioactive compounds from *Ajuga Reptans* extracts. *Farmacia*65: 351–355.

Toiu A, Vlase L, Arsene AL, Vodnar DC and Oniga I, 2016. LC/UV/MS profile of polyphenols, antioxidant and antimicrobial effects of *Ajuga genevensis* Extracts. *Farmacia*64: 53–57.

Upadhyay SU, Patel VB, Patel AA, Upadhyay UM and Patel NM, 2012. *Ajuga bracteosa-* A promising herb. *Pharma Science Monitor* 20: 80–88.

Verma V, Mahmood U and Singh B, 2002. Clerodane diterpenoids from *Ajuga bracteosa* wall. *Natural Product Letters* 16: 255–259.

WHO. Available at: www.who.int/news-room/fact-sheets/detail/diabetes#:~:text=The%20number%20of%20people%20with,stroke%20and%20lower%20limb%20amputation

Yousaf T, Rafique S, Wahid F, Rehman S, Nazir A, Rafique J, Aslam K, Shabir G and Shah SM, 2018. Phytochemical profiling and antiviral activity of *Ajuga bracteosa, Ajuga parviflora, Berberis lyceum* and *Citrus lemon* against hepatitis C virus. *Microbial Pathogenesis* 118: 154–158.

Yunnan-Yunnan Institute of Botany, ed. 1977. *Flora Yunnanica; Tomus 1.* Science Press, Peking, 1: 850.

Ziyyat A, Legssyer A, Mekhfi H, Dassouli A, Serhrouchni M and Benjelloun W, 1997. Phytotherapy of hypertension and diabetes in Oriental Morocco. *Journal of Ethnopharmacology* 58: 45–54.

19 *Tinospora cordifolia* (Giloy) *An Important Antidiabetic Medicinal Plant and Herbal Drug*

Kolagani Chandramohan, Rohini Ganorkar, and Salman Khan

CONTENTS

19.1 INTRODUCTION

Plants were acknowledged for medicinal purpose before prehistoric times (Husen 2021, 2022). The term medicinal plant involves varied types of plants used in herbalism/herbology. Ancient Unani manuscripts, Egyptian papyrus, and Chinese writings described the use of herbs. The word herb is derived from the Latin word *herba*, also from the old French word *herbe*. Evidence exist that Unani hakims, Indian veds, and European and Mediterranean cultures were using herbs for over 4000 years as medicine. Indigenous cultures in Rome, Egypt, Iran, Africa, and America used herbs in their healing rituals, while others developed traditional medical systems such as Unani, Ayurveda, and Chinese medicine in which herbal therapies were used systematically. Indian traditional medicinal systems – Ayurveda, Yoga, Sidha, and Unani – are well recognized worldwide. Ayurveda knowledge was described and developed between 2500 and 500 BC in India (Subhose et al. 2005). India has 15 agro climatic zones resulting in a wide variety of medicinal plants.

Diabetes mellitus is a chronic metabolic disorder and one of the major risk factors for cardiovascular disease caused by defective insulin secretion, resistance to insulin action, or a combination of both. Around 300 million people across the globe are expected to suffer from diabetes by the year 2025. Several antidiabetic drugs already on the market have proven to be beneficial to these patients, but with the recent trends of overuse of hypoglycaemic drugs, the chances of resistance and complications arise. Moreover, the drugs' dosage needs to be increased in patients due to its sensitivity and effectiveness when one type of drug is used in the long term. So people are now switching to Ayurvedic and herbal medicines, which have fewer side effects and are also cost effective with minimal contraindications (Sharma et al. 2015).

DOI: 10.1201/b23347-19

T. cordifolia is widely recognized as Guduchi, Amrita, Chinnaruha, Vatsadaani, Tantrika, and Kundalini and is a traditional medicinal plant belonging to the Menispermaceae family of moonseeds. Menispermaceae is widespread in tropical lowland regions composed mainly of 70 genera and 450 species. *Tinospora* genus is among the prevailing genera in the family, containing around 15 distinct species. This family is a rich source of terpenes and alkaloids. They generally climb or twin, frequently with shrubs. *T. cordifolia* is indeed one of the plants of tremendous promise regarding its ability to tackle diabetes. It is an anti-allergic, anti-inflammatory, immunosuppressive, immunomodulatory, anticancer, hypoglycaemic plant. Besides these, it is also known for its antibacterial and antioxidant properties (Modi et al. 2020).

19.2 BOTANICAL DESCRIPTION

Tinospora cordifolia, commonly named Giloy, Guruch Amritvali, Vatsadani, Amara and Chinnodebha, Guruchi in Sanskrit (Figure 19.1). It belongs to the family Menispermaceae. Giloy is a gregarious glabrous, twiner, climbing shrub. Stem and branches are specked with white vertical lenticels. Older stems are up to 2 cm in diameter and have corky bark. Dried stem is cylindrical, slender, and slightly twisted in shape. Outer bark is thin and papery and brown to greyish in colour. The stem when sectioned transversely shows a wheel-like structure.

FIGURE 19.1 *Tinospora cordifolia.*

Lenticels are circular and prominent. The stem powder is creamish brown to dark brown in colour with characteristic odour and bitter taste. The stem is used for dyspepsia, fever, and urinary diseases. The starch obtained from the stem, known as "Guduchi-satva", is highly nutritive and digestive and used for many diseases (Bishayi et al. 2002). Aerial roots arise from nodal scars of branches, thread like, long filiform, threadlike, squarish. Dried aerial root is grey-brown or creamy white, warty, papery thin, and peels off easily. Microscopic observations of aerial roots are characterized by tetra- to penta-arch primary structure. However, cortex of root is divided into outer thick walled and inner parenchymatous zone (Spandana et al. 2013; Aiyer et al. 1963).

Leaves are greenish-yellow, heart-shaped 5–15 cm, ovate, and acute. They are membranous when young but become more or less leathery with age. Leaves are bitter and have an indistinct odour. Lamina is ovatecordate, 10–20 cm long, 8–15 cm broad. Leaves are rich in protein, calcium, and phosphorus (Ninama et al. 2022; Nasreen et al. 2010). Flowers are yellow, unisexual, minute, and less than 2 mm in size. Male flowers are grouped in axillary racemes, while female flowers are solitary. Sepals are six in two series of three each. Outer ones are smaller than the inner sepals. Petals are also six, smaller than sepals, free and membranous. Flowering occurs in May–June. Fruit is an ovoid and succulent drupe, lustrous, red in colour, and of the size of a large pea, having a single seed. Seed is fleshy and curved. Fruiting is witnessed in September–October. The variety of active components derived from the herb is alkaloids, steroids, diterpenoid, lactones, aliphatics, and glycosides, and these can be isolated from various parts of the body like root, stem, and whole plant (Rekha 2017).

19.3 DISTRIBUTION

The plant is distributed throughout the tropical region of India up to 1200 m above sea level from Kumaon to Assam and in the north extending through West Bengal, Bihar, Deccan, Karnataka, and Kerala. It is common throughout tropical and subtropical zones at an altitude of 600 m. It is indigenous to areas of India, Myanmar, Sri Lanka, China, Thailand, Philippines, Indonesia, Malaysia, Borneo, Vietnam, Bangladesh, North Africa, and South Africa. It is typically grown in deciduous and dry forests at elevations up to 1000 feet. Light medium sandy loam soil rich in organic matter, and with adequate drainage, is suitable for its cultivation. It does not tolerate high rainfall or waterlogged conditions (Ashish Kumar et al. 2019).

19.4 PHYTOCHEMICAL CONSTITUTION

Several biologically active components, including alkaloids, diterpenoid lactones, glycosides, steroids, sesquiterpenoids, phenolics, aliphatic compounds, and polysaccharides were isolated from stem, roots, shoots, leaves, and whole plants (refer to Figure 19.2). These compounds have been reported to have different biological roles in disease conditions, thus enabling potential application in clinical research (Deepika et al. 2015; Bisset et al. 1983; Sharma et al. 2012; Singh et al. 2017; Tiwari et al. 2018).

Whole plant: *T. cordifolia* contain terpenoids, lactones, cleodrane derivatives (5R, 10R)-4R-8Rdihydroxy-cleroda-13(16), 14-dieno-17, 12S:18, 1S-dilactone, tinosporin etc. showing vasorelaxant, anti-inflammatory, antimicrobial, anti-hypertensive, and antiviral activities (Sriramaneni et al. 2010; Dhanasekaran et al. 2009). Aliphatic compounds extracted from the whole plant constitute active parts like octacosanol, nanocosan- 15-one dichloromethane, heptacosanol showing anti-nociceptive and anti-inflammatory activity. They also inhibit tumour necrosis factor alpha (TNF-α) from binding to the DNA and provide protection against 6-hydroxydopamine induced parkinsonism in rats (De Oliveira et al. 2012; Wang et al. 2010; Thippeswamy et al. 2008).

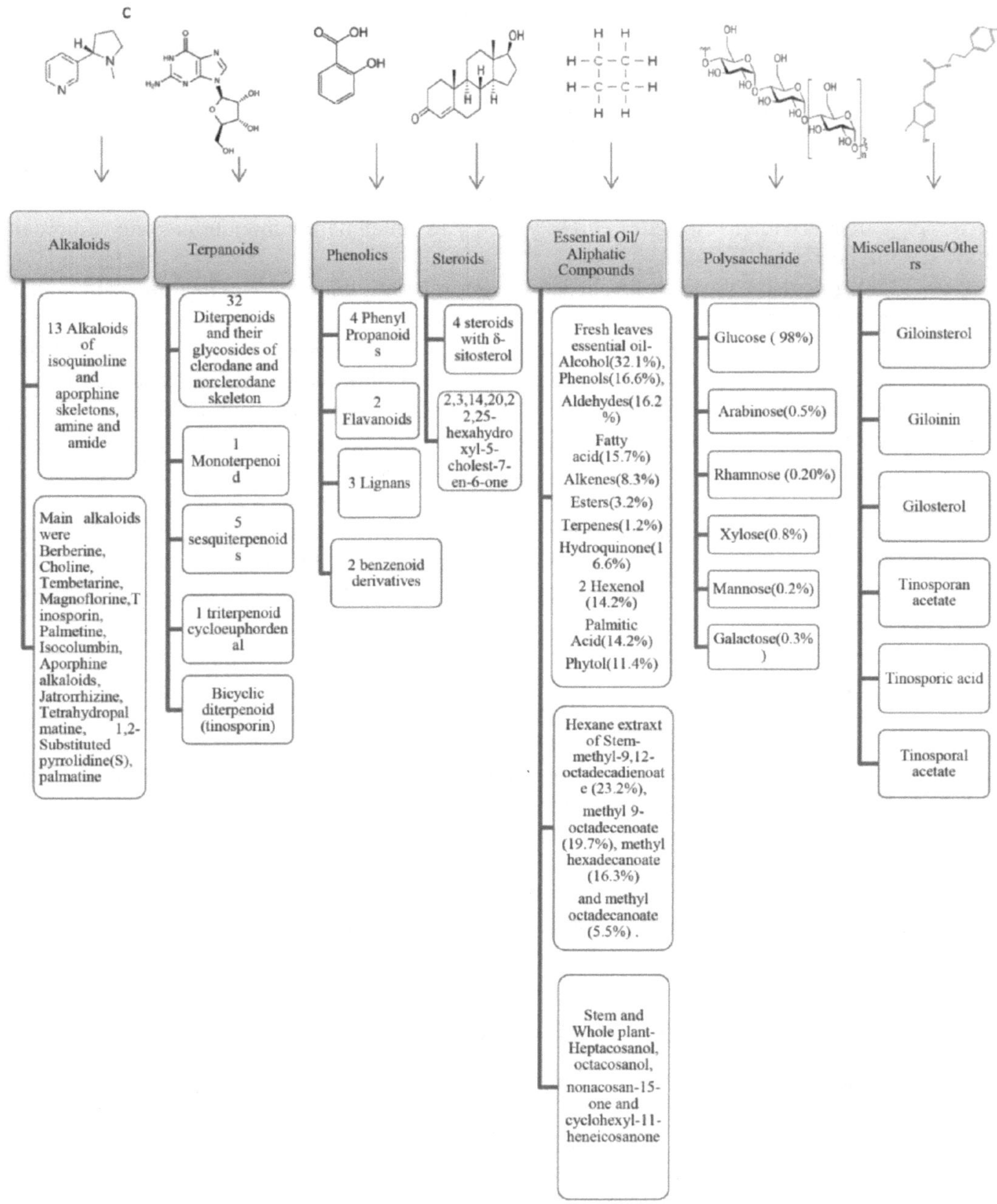

FIGURE 19.2 Phytoconstituents of *Tinospora cordifolia*.

Shoot: Steroids like B-sitosterol, δ-sitosterol, 20 β-hydroxyecdysone, giloinsterol, makisterone A, and ecdysterone were extracted from the shoot. They are effective in glucocorticoid-induced osteoporosis in early inflammatory arthritis. They encourage cell cycle arrest in G2/M phase and inhibit TNF-α, interleukin (IL) -1 β, IL-6 and cyclooxygenase type 2 (COX-2) and apoptosis through cellular myelocytomatosis oncogene (c-Myc) suppression (Lv et al. 2012; McKeown et al. 2012).

Stem: Stem and root part of *T. cordifolia* contain alkaloids as active constituents showing anticancer, antidiabetic, antiviral, anti-inflammatory, anti-psychotic, and immunomodulatory action (Upadhyay et al. 2010; Rout 2006). Glycosides' active constituents extracted from stem shows immunomodulation in Parkinson's disease, dementia, motor and cognitive disorder, and neurological disorders like amyotrophic lateral sclerosis (ALS). They inhibit nuclear factor kappa (NF-k) band to show anticancer properties (Ly et al. 2007; Kapil et al. 1997), whereas sesquiterpenoids and tinocordifolin exhibit an antiseptic activity (Maurya et al. 1998).

Leaves: Leaves are rich in protein, calcium, and phosphorus (Sinha et al. 2004). Methanol extract of leaves are rich in flavonoids, alkaloids, and glycosides (Soni et al. 2011).

19.5 PHARMACOLOGICAL IMPORTANCE

Tinospora cordifolia extracts have been extensively used for ages in traditional systems of medicine to treat different ailments. Various herbal preparations have been used for their anti-periodic, anti-spasmodic, antimicrobial, anti-osteoporotic, anti-inflammatory, anti-arthritic, anti-allergic, and antidiabetic properties, as shown in Table 19.1. This plant has many biological roles in treating disease and conditions, proving its role in clinical research. Its root part is known for its stress-relieving and antimalarial properties, while its stem is used as bitter stomachic and diuretic. It stimulates biliary secretion, enriches the blood, and cures jaundice (Tiwari et al. 2018).

TABLE 19.1
Pharmacological Importance of *Tinospora cordifolia*

Chemical constituent	Plant area	Biological impact in animal models, humans, and cell lines	References
Diterpenoid lactones	Whole plant	In rats, chemopreventive potential in diethylnitrosamine (DEN) induced hepatocellular carcinoma (HCC). In humans and cell lines, vasorelaxant: relaxes norepinephrine-induced contractions. Inhibits Ca++ influx. Anti-inflammatory, antimicrobial, anti-hypertensive, and antiviral. Induces apoptosis in leukaemia by activating caspase-3 and Bax, inhibits B-cell leukaemia (Bcl)-2.	Sriramaneni et al. 2010; Khuda et al. 1964; Yang et al. 2010; Zhao et al. 2008; Kohno et al. 2002, Dhanashekaran et al. 2009
Sesquiterpenoid	Stem	Antiseptic	Maurya et al. 1998
Alkaloids	Stem, root	Isoquinoline alkaloids have anti-cataract potential in rats. Antioxidant activity in mice, anticancer in Ehrlich ascites carcinoma (EAC) mice, hypoglycaemic activity in RINm5F rat insulinoma cell line. Antiviral, anticancer, antidiabetic, anti-inflammatory, and immunomodulatory.	Choudhary et al. 2013; Bisset et al. 1983; Upadhyay et al. 2010, Patel et al. 2011
Glycosides	Stem	In Swiss albino mice, cytotoxic action, protection against iron-mediated lipid peroxidation of rat brain homogenate, antioxidant and hydroxyl radical scavenging activities were observed. Treats neurological disorders like ALS, Parkinson's, dementia, motor and cognitive deficits, and neuron loss in spine and hypothalamus. Immunomodulation: immunoglobulin G (IgG) increase and macrophage activation. Inhibits NF-κB and acts as nitric oxide scavengers to show anticancer activities.	Ly et al. 2007; Kim et al. 2008

(Continued)

TABLE 19.1
(Continued)

Chemical constituent	Plant area	Biological impact in animal models, humans, and cell lines	References
Steroids	Shoot	Beta-ecdysone shows anabolic and anti-osteoporotic effects in mammals. IgA neuropathy, glucocorticoid-induced osteoporosis in early inflammatory arthritis, induces cell cycle arrest in G2/M phase and apoptosis through c-Myc suppression. Inhibits TNF-α, IL-1 β, IL-6 and COX-2. Activates NF-κB.	LV et al. 2012; Sundarraj et al. 2012
Aliphatic compounds	Whole plant	Radiosensitizing activity in Ehrlich ascites carcinoma mice. Modulating the pro-inflammatory cytokines. Inhibits proliferation of endothelial cells and Ehrlich ascites tumour cells. Anti-nociceptive and anti-inflammatory. Protection against 6-hydroxydopamine-induced parkinsonism in rats. Downregulates vascular endothelial growth factor (VEGF) and inhibits TNF-α from binding to the DNA.	De-oliviria et al. 2012; Thippeswamy et al. 2008
Miscellaneous/ others	Root, whole plant	Insulin-mimicking and insulin-releasing effect. Enhanced phagocytic activity of milk polymorphonuclear cells in bovine subclinical mastitis. Protease inhibitors for HIV and drug-resistant HIV. Tyramine is a neuromodulator. Used to treat anxiety and depression by inactivating neurotransmitters.	Ghosh et al. 2008; Mukherjee et al. 2010

The major biological activities of *T. cordifolia* include the following.

Anti-toxin activity: *T. cordifolia* lowers the concentration of thiobarbituric acid reactive substances (TBARS) and enhances the level of glutathione, ascorbic acid, and proteins along with elevated activities of antioxidant enzymes such as superoxide dismutase, catalase, glutathione peroxidase, glutathione S-transferase, and glutathione reductase in kidneys. This plant reverses the toxicity caused by aflatoxin in kidney (Swiss albino mice), where it substantially elevates the hormone level and enzyme activities and decreases the reactive oxygen species (ROS). And this anti-toxin activity is primarily brought by the alkaloids of this plant. The presence of alkaloids such as isocolumbin, palmetin, tetrahydropalmatine, and magnoflorine in *T. cordifolia* protects against aflatoxin-induced nephrotoxicity (Gupta et al. 2011; Sharma et al. 2010). Herbal extract from leaf and stem works against lead toxicity when oral doses were given to study lead nitrate toxicity in Swiss albino mice. The study shows a decrease in the level of the enzymes like glutamic pyruvic transaminase (GPT) or alanine aminotransferase (ALT) and aspartate aminotransferase (AST) and a rise in the enzyme responsible for scavenging free radicals such as catalase (Hamsa et al. 2012). Plant shows importance in overcoming cyclophosphamide-induced toxicity by substantially elevating the level of lowered GSH content and cytokines and gradually declining inflammatory cytokines (tumour necrosis factor) level in urinary-bladder and hepatic cell preventing the damage which confirms its anti-toxin activity (Chopra et al. 2012).

Anticancer activity: The first report on the anticancer activity of *T. cordifolia* was published in 1998, where HeLa, the human cervical cancer cells, were treated with 0, 5, 10, 25, 50, and 100 mg/ml of methanol, dichloromethane, and aqueous extracts of *T. cordifolia*. The most effective extract was dichloromethane extract, which killed HeLa cells more effectively than the other two extracts and almost equal to that of doxorubicin (Jagetia et al. 1998). Different extract preparations of *Tinospora* have been tested for their anticancer activity in various cancer model systems. In Ehrlich ascites carcinoma (EAC)-bearing mice, dichloromethane extract of *T. cordifolia* (TCE) has shown anticancer activity (Jagetia et al. 2006). A single dose of root extract of *T cordifolia* (RTc, 200 mg/kg b/w) was preirradiation administered and resulted in 76% survival in mice exposed to 10 Gy lethal gamma-irradiation, but without RTc treatment irradiated mice suffer from 100% mortality within 10 to 15 d. Most radioprotective agents show the maximum radioprotective effect at maximum tolerated level doses (MTD), but RTc shows significant protective effect at about 50% concentration of its MTD (Devi et al. 1999; Goel et al. 1998). *T. cordifolia* possesses the ability to scavenge free radicals and prevents radiation-induced damages (Kapil et al. 1997). *T. cordifolia* was found to possess antiangiogenic activity in melanoma B16F10 cell-induced capillary formation *in vivo* and *in vitro*. Many proinflammatory cytokines such as IL-1a, IL-6, TNF-a, granulocyte-monocyte colony-stimulating factor (GM-CSF), and vascular endothelial cell growth factor (VEGF) are upregulated by melanoma B16F10 cells. After intraperitoneal administration of *T. cordifolia* extract, capillary formation and the level of these cytokines are decreased while antiangiogenic agents IL-2 and tissue inhibitor of metalloprotease-1 (TIMP-1) were increased. Chloroform (Chl-TCE) and hexane (Hex-TCE) extracts significantly reduced the rate of proliferation and induced cell differentiation in glioblastoma and neuroblastoma with increased expression of stress markers HSP70 and mortalin and induced senescence. Chloroform and Hexane extracts also inhibit the migration of U87MG glioblastoma and neuroblastomas and thus may act as potential phytotherapeutic intervention in the treatment of neural cancers (Sharma et al. 2019). *T. cordifolia*'s ability to stimulate free radical formation and DNA damage in the tumour cells reduces the antioxidant status of tumour cells by increasing lipid peroxidation and lactate dehydrogenase. At the molecular level, inhibition of topoisomerases trigger DNA damage causing cytotoxic effects in tumour cells. Suppression of NF-κB, COX-II, Nrf2, STAT3, Bcl-2, Ca2+ release, cyclin-dependent kinase (CDK) 2, CDK4, cyclin B, cyclin D, and cyclin E may have induced cytotoxicity in tumour cells. The activation of p53, Weel and CDk1 proteins, Bax, P27, procaspase-9, caspase-9, caspase-3, and poly (ADP-ribose) polymerase (PARP) would have led to an increase in apoptotic death of tumour cells (Jagetia et al. 2019).

Anti-allergic activity: The anti-allergic and bronchodilator properties of an aqueous extract of the stem evaluated on histamine-induced bronchospasm in guinea pigs, capillary permeability in mice, and mast cell disruption in rats showed that it significantly decreased bronchospasm induced by 5% histamine aerosol, decreased capillary permeability, and reduced the number of disrupted mast cells (Nayampalli et al. 1986; Nayampalli et al. 1982). *Tinospora cordifolia* is mainly used for the treatment of asthma, and the juice is also employed for the treatment of chronic coughs (Spelman 2001). Of patients treated with *T. cordifolia* in a clinical study of sneezing, 83% reported reduced symptoms. Similarly, 69% experienced alleviation from nasal discharge, 61% from nasal blockages, and 71% from nasal pruritus. Only 21% of patients in the placebo group reported alleviation from sneezing, 16.2% from nasal discharge, 17% from nasal blockage, and 12% from nasal pruritus. As a result, *T. cordifolia* significantly reduced all allergic rhinitis symptoms while also being well tolerated (Badar et al. 2005).

Anti-HIV activity: A diterpenoid, tinosporin, showed activity against HIV, HTLV, and other viral diseases for its immunomodulatory and selective inhibition of the virus to target T helper cells. Anti-HIV effects of TCE was revealed by reduction in eosinophil count and stimulation of B lymphocytes, macrophages, polymorphonuclear leucocytes, and haemoglobin percentage, thus revealing its promising role in management of the disease (Kalikar et al. 2008; Akhtar et al. 2010).

Anti-arthritic and anti-osteoporotic activity: In traditional medicine, individual or combined formulations of *Tinospora cordifolia* and *Zingiber officinale* were used to treat rheumatoid arthritis. In an *in vitro* osteoblast model system, *T. cordifolia* inhibit the proliferation, differentiation, and mineralization of bone-like matrix showing used as an anti-osteoporotic agent. Alcoholic extract of *Tinospora cordifolia* shows increased rise of osteoblasts, as well as the differentiation of cells into the osteoblastic lineage and the mineralization of bone-like matrix (Abhiramasundari et al. 2012). In mammals, ecdysteroids extracted from the plant are shown to have protein anabolic and anti-osteoporotic properties. The compound beta-ecdysone (ECD) derived from *Tinospora cordifolia* extracts shows improved joint cartilage thickness, stimulates osteogenic differentiation in mouse mesenchymal stem cells, and alleviates osteoporosis in osteoporotic animal models. Another compound, 20-OH-Ecd, is also involved in osteoporosis and osteoarthritis treatment (Kapur et al. 2010).

Antimicrobial activity: Antimicrobial activity of *T. cordifolia* with different solvents on different microorganisms showed good antifungal and antibacterial activity (Duraipandiyan et al. 2012). Methanol extracts of *Tinospora cordifolia* assayed for antibacterial activity against *Escherichia coli*, *Staphylococcus aureus*, *Klebsiella pneumoniae*, *Proteus vulgaris*, *Salmonella typhi*, *Shigella flexneri*, *Salmonella paratyphi*, *Salmonella typhimurium*, *Pseudomonas aeruginosa*, *Enterobacter aerogene*, and *Serratia marcesenses* (gram-positive bacteria) are effective against microbiological infections. The aqueous, ethanol, and acetone extract of *T. cordifolia* inhibited the activity on clinical isolates of urinary pathogens *Klebsiella pneumoniae* and *Pseudomonas aeruginosa* (Shanthi et al. 2013). The aqueous extract of *T. cordifolia* showed potent activity against *A. fumigatus, Aspergillus flavus*, and *Aspergilles niger* (fungus) in the study (Alrumaihi et al. 2019). In bovine preclinical mastitis, intramammary infusion of hydro-methanolic extracts of *Tinospora cordifolia* therapy increased polymorphonuclear cell phagocytic activity (Sengupta et al. 2011). The ethanolic extract showed maximum 87.2% and 91.0% free radical scavenging activity concerning H_2O_2 scavenging and hydroxyl free radical scavenging assay (Prasad et al. 2019).

Immunomodulatory activity: Immunomodulators are those extrinsic or intrinsic substances which regulate or alter the scope, type, duration, or competency of the immune response. Active compounds of *Tinospora cordifolia*11-hydroxymustakone, N-methyl-2-pyrrolidone, N-formylannonain, cordifolioside A, magnoflorine, tinocordiside, and syringing have been reported to have potential immunomodulatory and cytotoxic effects (Tripathi et al. 1997; Bishayi et al. 2002; Subramanian et al. 2002). *Tinospora cordifolia* extract improves phagocytic and intracellular bacterial capacities of neutrophils in mice. Its natural compounds have been reported to improve the phagocytic activity of macrophages, enhance nitric acid production by stimulation of splenocyte, and produce reactive oxygen species (ROS) in human neutrophil cells (Mittal et al. 2014).

Antioxidant activity: Methanol extracts of *T. cordifolia* stem exhibited excellent antioxidant activity in scavenging superoxide anion radicals and inhibited deoxyribose degradation induced by hydroxyl radicals, scavenging them directly rather than via chelating iron ion. In addition, phytochemicals attribute a strong free radical scavenging activity. Antioxidant-rich plant extracts serve as a source of nutraceuticals that alleviate oxidative stress and therefore prevent or reduce the onset of degenerative diseases. Plant polyphenols act as reducing agents and antioxidants by the hydrogen-donating property of their hydroxyl groups (Aberoumand et al. 2008). (5R, 10R)-4R, 8R-dihydroxy-2S, 3R: 15, 16-diepoxycleroda-13 (16), 17, 12S: 18,1S-dilactone (ECD), a diterpenoid from *Tinospora cordifolia*, has been shown to possess chemo-preventive potential in diethylnitrosamine (DEN)-induced HCC rats. Treatment of ECD in both preventive and curative DEN-induced animals increased the level of antioxidants and detoxification enzymes (Dhanasekaran et al. 2009).

Role in COVID treatment: COVID-19 enters into a host cell by binding to angiotensin converting enzyme-2 (ACE2) via its spike protein receptor-binding domain (RBD). Infection rate will be significantly reduced if ACE2-RBD interaction could be disrupted and virus accession could be avoided. "Tinocordiside", a phytochemical compound, binds to the complex ACE2-RBD discouraging the entry of the virus (Gayatri et al. 2020; Rajput 2020). They may revive lung health by reducing oxidative stress and enhancing endothelial dysfunction (Dimri et al. 2020). Molecular docking findings showed that tinocordiside exhibited binding affinity as predicted to act as probable SARS-CoV-2 (severe acute respiratory syndrome coronavirus-2) main proteases (Mpro) inhibitor. Such phytoconstituents inhibit the transmission and propagation of viral protein into the host cell within the human body. Additionally, they are also safer to repurpose against COVID-19 without any toxicity (Shree et al. 2022; Modi et al. 2020).

19.6 ANTIDIABETIC ACTIVITY

Phytochemicals isolated from different parts of plant, including alkaloids, tannins, cardiac-glycosides, flavonoids, saponins, and steroids, exhibit antidiabetic potential. Isoquinoline alkaloid-rich fraction from stem, including palmatine, jatrorrhizine, and magnoflorine, are reported for insulin mimicking and insulin releasing effect both *in vitro* (using rat pancreatic β-cell line, RINm5F) and *in vivo* (Sudha et al. 2011; Rout 2006; Patel et al. 2011). "Berberine", an isoquinoline alkaloid tested *in vitro* and *in vivo*, lowers elevated glucose level, inhibits FOXO1, and integrates insulin signalling with mitochondrial function, thus improving hepatic metabolism during insulin resistance and metabolic syndrome. By adenosine monophosphate-activated protein kinase activation, it decreases the blood sugar and cholesterol level and maintains the blood pressure (Zhang et al. 2008; Yin et al. 2008; Cheng et al. 2009; Sharma et al. 2013). Pharmacological studies of various extracts of *T. cordifolia* have proven its potential *in vivo* antidiabetic properties, as is shown in Table 19.2.

In vitro studies: *In vitro* studies report initiation and restoration of cellular defence antioxidant markers including SOD, glutathione peroxidase, and glutathione, inhibition of glucose 6-phosphatase and fructose 1, 6-diphosphatase, and restoration of glycogen content in liver (Chougale et al. 2009). Water, ethanol, and methanol extracts of the herb were tested on Ehrlich ascites tumour cells model and showed glucose uptake-stimulatory activity (Joladarashi et al. 2014). Leaf powder effectively absorbs glucose and retards its diffusion across dialysis membrane. All these mechanisms together create a combined effect on lowering glucose absorption for decreasing postpardinal blood glucose (Ahmed et al. 2011).

TABLE 19.2
Pharmacological Profile of *Tinospora cordifolia* for Antidiabetic Activity

Plant part/ whole plant	Formulation	Animal model (drug-induced diabetes)	Dose (mg/kg)	Treatment (*p.o.* = per os, oral administration of dose)	After effects	Proposed mechanism/remarks	References
Whole plant	Aqueous	Alloxan-rats and rabbits	400	*p.o.*	Hypoglycaemic	Regulation of glucose metabolism	Raghunathan et al. 1969
Leaves	Aqueous extracted saponarin, (alpha-glucosidase inhibitor)	Maltose-fed rats	20.0–80.0	*p.o.*	Hypoglycaemic	Competitive inhibition on activities of alpha-glucosidase and sucrase of different origins and to show saponarin (apigenin-6-C-glucosyl-7-O-glucoside)	Sengupta et al. 2009
	Alcoholic and aqueous	Streptozotocin- mice	400	*p.o.* for 50 d		Amelioration of diabetic neuropathy and gastropathy	Grover et al. 2002; Sharma et al. 2011
	Aqueous		200	*p.o.* for 40 d		Increase in glucose metabolism results in reduced plasma glucose concentration by 7.45% through; to prevent polyuria, rise in urinary albumin levels and renal hypertrophy as well	Grover et al. 2001
Root	Aqueous	Alloxan rats	400	*p.o.* for 21–120 d	Antihyperglycaemic	Effects on key metabolic enzymes involved in carbohydrate metabolism, significant glycaemic control in mild and moderate type diabetes	Grover et al. 2000
				p.o.	Hypoglycaemic	Increases in body weight, total haemoglobin and hepatic hexokinase; decreases in hepatic glucose-6-phosphatase, serum acid phosphatase, alkaline phosphatase, and lactate dehydrogenase	Stanely et al. 2000
			400			Its effect equivalent to only 1 IU/kg of insulin	Dhaliwal 1999
			2.5, 5.0	*p.o.* for 6 weeks	Significant hypoglycaemic action	Reduction in serum and tissue cholesterol, phospholipids and free fatty acids	Prince et al. 1999
		Streptozotocin-rats		*p.o.* for 6 weeks	Significant antihyperglycaemic	Significant reduction in blood and urine glucose	Archana et al. 2010

	Alloxan-induced diabetic cataract rats	400	*p.o.* for 2 months	Hypoglycaemic	Decreases of 38.01% and 40.41% in serum glucose levels after 1 and 2 month treatment, respectively	Rathi et al. 2002
	Adrenaline-induced hyperglycaemia in rabbits	10		Antihyperglycaemic effect	Considerably inhibits hyperglycaemia	Raghunathan 1969
	Alloxan rats	100	*p.o.* for 6 weeks	Hypoglycaemic	To reduce blood and urine glucose levels and prevent weight loss	Prince et al. 2003
				Significant antihyperglycaemic	Normalized the antioxidant status of heart, brain, liver, and kidney, restores the antioxidant defence	Prince et al. 2004; Prince et al. 2004
Aqueous, alcoholic	Streptozotocin albino rats	200.0, 400.0	*p.o.* for 30 d	Antihyperglycaemic	To modulate renal tissue morphology and ameliorate activity of key gluconeogenic enzymes and to improve renal functions	Nagaraja 2007
	Fasted albino rats			Hypoglycaemic	To reduce fasting blood glucose by initiating endogenous insulin secretion, glucose uptake, inhibition of peripheral glucose release	George et al. 1949
	Fasting and adrenaline-induced hyper-glycaemia rabbits		*p.o.*		Decreases the blood glucose level and increases glucose tolerance	Zhang et al. 2008
Methanol	Alloxan and streptozotocin rats	150	*p.o.*	Significant hypoglycaemic	The extract without showing toxicity in acute toxicity study	Dhulia et al. 2011
	Normal and alloxan rats	500	*p.o.* for 6 weeks	Significant hypoglycaemic	Significant decreases in blood glucose, glycosylated haemoglobin, and cholesterol ($P < 0.05$); increases in body weight and protein ($P < 0.01$), hepatic enzyme hexokinase activity increased, glucose-6-phosphatase and significant decrease in fructose 1, 6-biphosphatase	Sivakumar et al. 2011
Hexane, ethyl acetate, methanol	Streptozotocin rats	250	*p.o.* for 100 d	Significant antihyperglycaemic	Decrease in glycosylated haemoglobin level and glucokinase, increased glucose-6-phosphatase activity, improvement in insulin secretagogue effect, insulin, and C-peptide levels which shows P-cell regeneration capacity of extracts	Rajalakshmi et al. 2009
Ethanolic	Fasted albino rats	250	*p.o.* for 1 d	Hypoglycaemic action	About 30% reduction in blood sugar	Dhar et al. 1968

(*Continued*)

TABLE 19.2 (Continued)

Plant part/ whole plant	Formulation	Animal model (drug-induced diabetes)	Dose (mg/kg)	Treatment (*p.o.* = per os, oral administration of dose)	After effects	Proposed mechanism/remarks	References
Stem	Ethanolic	Alloxan rats	250	Single dose, *p.o.*	Hypoglycaemic activity	Noteworthy effect within 1 week	Kar et al. 2003
	Aqueous, alcoholic	Streptozotocin albino rats	200.0, 400.0	*p.o.* for 30 d	Antihyperglycaemic	Increases hepatic glycogen synthase and decreases glycogen phosphorylase activity	Puranik et al. 2010
	Ethyl acetate, dichloromethane, chloroform and hexane extracts	Normal and glucose-loaded Wistar rats	15	*p.o.*		Reduction in postprandial glucose level; alpha glucosidase inhibitor, to inhibit the salivary and pancreatic amylase	Chougale et al. 2009
	Isoquinoline alkaloid-rich fraction	Normal and glucose-loaded Wistar rats	50.0, 100.0, 200.0	*p.o.*		Insulin-mimicking and insulin-releasing effect *in vitro* and *in vivo*	Patel et al. 2011
	Hydroalcoholic extraction (70% ethanol, 30% water)	High-fat diet fed and streptozotocin Sprague-Dawley rats	100.0, 200.0	*p.o.* for 14 d		To mitigate oxidative stress, promote insulin secretion, inhibit gluconeogenesis and glycogenolysis	Sangeetha et al. 2011
	Aqueous	Alloxan-induced diabetic rats	500	*p.o.* for 40 d		Noteworthy decreases in blood glucose, glycosylated haemoglobin, urea, cholesterol ($P < 0.05$), and escalations in protein and glycogen ($P < 0.01$), extract with nontoxic and well tolerated	Sivakumar et al. 2009; Nagaraja 2007
		High-fructose diet (66% fructose) induced diabetic Wistar rats	400	*p.o.* for 60 d		Improvement of glucose and lipid metabolism, to alleviate insulin resistance and oxidative stress; to prevent rise in glucose levels by 21.3%, insulin by 51.5%, triglycerides by 54.12%, and glucose-insulin index by 59.8%	Reddy et al. 2009
	Aqueous, alcoholic, chloroform	Normal and alloxan-induced diabetes in rabbits	50.0, 100.0, 200.0	*p.o.* for 1 d	Dose dependent hypoglyce mic	Results matches to glibenclamide and insulin	Wadood et al. 1992

19.7 TRADITIONAL AND OTHER POTENTIAL USES

Over the decades, tribals and other ethnic groups have established their own beliefs, knowledge, and tradition related to forest and usage of plants. Almost all parts of *T. cordifolia* are documented for its usage. Some important traditional and other noted usages are discussed in Table 19.3 and Table 19.4 respectively (Singh et al. 2003).

TABLE 19.3
Traditional and Other Potential Uses of *Tinospora cordifolia*

Name of tribe	State/district	Traditional uses	Usage
Baiga	Naugarh and Chakia Block Varanasi, Uttar Pradesh	Fever	Paste of stem of the Guduchi (*T. cordifolia*) and the roots of Bhatkatiaya (*Solanum surattense*) were used to prepare pills.
Tribals and Fishermen	Mumbai and its neighbouring area along the sea	Fever, jaundice, chronic diarrhoea, periodic fever	The whole plant is used.
Tribals of Khedbrahma	North Gujarat	Cancer, dysentery, diarrhoea, and periodic fever	Powdered root and steam bark of *T. cordifolia* with milk for cancer. Decoction of root for dysentery and diarrhoea. Decoction of old stems for periodic fever.
Tribals of Jammu and Bigwada	Jammu & Kashmir, Rajashthan respectively	Fever	Decoction of stem is administered orally.
Inhabitants of Bhubaneshwar	Odisha	Fever	The warm juice of root of *T. cordifolia* orally.
Inhabitants of Banka	Bihar	Daha (burning)	Paste or juice of *Amrita* (*T. cordifolia*) leaves and *Sarsapa beeja* churna (seed powder of *Brassica campestris*) are used for daha.
Local people	Patiala, Punjab	Fever	Juice or decoction of leaves is administered orally with honey.
Agaris, Bhils, Dhodias, Dublas, Khakaris, Rimoshis, Thakurs, Vandaries, Vagharis, and Varlis	Dhanu forest division of Maharashtra	General debility	Decoction of stem with cold and hot water (about 3–4 g) in morning in an empty stomach, as a tonic.
People of Dhurala	Haryana	Kasa (cough)	Powder of *Terminalia chebula* (Haritiki), *Tinospora cordifolia* (Amrita) and *Trachyspermum ammi* (Ajwain) in equal quantity is administered orally once daily early morning with salt.
Local women of Arjunpura	Rajasthan	Raktapradara (leukorrhoea)	Paste of Guduchi (*T. cordifolia*) and five seeds of Krishnamarich (*Piper nigrum*) is administered orally daily in morning.
Inhabitants of Badala	Uttar Pradesh	Swasa (asthma)	Juice of stem orally with honey.
Dehrabara, Kolaras tribes and people	Shivpuri, Madhya Pradesh	Twak-roga (skin disease)	Decoction of stem is administered orally.

TABLE 19.4
Other Potential Uses of *T. cordifolia*

Name of tribe	State/district	Other uses	Usage
Muslim tribals comprising Gujjars and Backwals	Rajouri, Jammu (Tawi)	Fracture	Whole part is used.
Mundas	Chhota Nagpur	Fracture	Paste of whole plant used as plaster.
In certain parts of India		Bites of poisonous insects and venomous snake, eye disorders.	The paste of Guduchi is applied to the part bitten and administered internally through mouth at intervals of half an hour.
Inhabitants of Banka	Bihar	Balashosha (emaciation in children)	Dyed shirt soaked in juice of Guduchi worn by children for Balashosha.

19.8 SAFETY ISSUES

Safety is a fundamental principle in the provision of herbal medicines for health care and a critical component of quality control. The drug, which is going to be administered into the human body, should be standard and should not produce any kind of untoward effects after its administration (Nesari 2022). Ministry of AYUSH has published an advisory to confirm Guduchi (*Tinospora cordifolia*) is safe to use and only *Tinospora cordifolia* should be used in therapeutics (PIB 2021). Chandrasekaran et al. evaluated the genotoxic risk of the aqueous extract of *T. cordifolia* (TC) in a battery of four different genotoxicity tests, viz., Ames, *in vitro* chromosome aberration (CA), rodent bone marrow micronucleus (MN), and Comet assay. Experimental results confirmed that in the Ames test, up to 5000 microg/plate of TC did not exhibit any mutagenic effect in *Salmonella typhimurium* mutant strains (TA97a, TA98, TA100, TA102, and TA1535). In the CA assay, TC was not clastogenic to human peripheral blood lymphocytes up to a concentration of 3000 microg/ml. In MN and Comet assays, TC was pre-treated for seven days at three dose levels (150, 200, and 250 mg/kg body weight) orally to male Balb/c mice. The results showed that TC treatment did not display clastogenicity and DNA damaging effect in bone marrow erythrocytes and peripheral blood lymphocytes respectively (Chandrasekaran et al. 2009).

19.9 CULTIVATION PRACTICE

Soil rich in organic matter ranging from light medium sandy loam, with adequate drainage, and tropical and subtropical climate are the basic requirements for *T. cordifolia* plant growth (Ashish et al. 2017). For raising commercial crop, stem cuttings are the best planting material. The cuttings of the small finger thickness with a 6- to 8-inch long stem having two or three nodes are used and can be obtained from mother plants in May–June. The healthy plant stem cuttings are sown directly in the ready field. Promotion of rooting of shoot cuttings by exogenous auxins application in several species has been reported. Plant propagation can be done by seeds and vegetative cuttings (Hartmann et al. 1997). For nursery cultivation, poly bags of 4-inch × 6-inch size are filled with mud, sand, and dry cow dung or vermin compost in the ratio 1:1:1 and planted in February–March. The rooting of the cuttings takes almost 15 to 25 days. The cuttings of *T. cordifolia* will be ready for planting into the main field by this time in the month of May–June. The plant requires support to grow, which can be provided by raising wooden stakes or trellis. Already-growing shrubs or trees, like neem and mango trees, can also support the plants. Such plants are supposed to possess better medicinal values. The stem is harvested during autumn when it develops to a diameter of more than 2.5 cm. Basal part is left for further growth (Sohani et al. 2022).

19.10 FUTURE PROSPECTS

Knowledge of collection and medication of traditional plants is passed down through generations. *T. cordifolia* has huge therapeutic value, leading to its overexploitation. Hence, for conservation, fulfilling increasing demand, and rapid propagation, plant tissue culture is an alternative method for cultivation and enhancement of secondary products. Asymptomatic fungal endophytes and highly diverse microorganisms are extremely common and reside in plant tissues involved in a variety of relationships from symbiotic to slightly pathogenic. They occupy a wide range of different habitats due to their vast nutritional diversity as heterotrophs and symbionts. Endophytic fungi were found associated with *T. cordifolia*. A biotechnological approach is now implemented for studying the role of endophytes in production of several metabolites (primary and secondary) with other groups of organisms. The endophytic fungi may produce a plethora of substances of potential use to modern medicine, agriculture, and industry (Gaiero et al. 2013; Tan RX et al. 2001; Sonaimuthu et al. 2010). Green synthesis technology has been used for the synthesis of noble metallic nanoparticles using plant extracts. It's a cheap, environmental friendly, and easy method. These nanoparticles play a vital role in drug delivery, diagnostics, imaging, sensing, gene delivery, artificial implant, tissue engineering, and more recently medical textiles for its antimicrobial activity (Singh et al. 2013).

T. cordifolia is a plant with multiple benefits that need to be explored with advanced techniques. Further studies are required in plant tissue culture to raise demand and success rate; studies at molecular level to understand genetic changes comes through an indirect regeneration process. Studies on fungal endophytes would reveal pathways of natural product synthesis and relationship with other microorganisms. The use of nanoparticles with phytoconstituents from plant extracts on various incurable diseases through drug delivery system should also be studied.

ACKNOWLEDGEMENT

The authors are greatly indebted to Shri. Anoop Singh IFS, Director General, Forest Survey of India, Dehradun and Shri. Venu Prasad IFS, Regional Director, Forest Survey of India Central Zone, Nagpur, for their support and facilities.

REFERENCES

Aberoumand, A. & S.S. Deokule. 2008. Comparison of phenolic compounds of some edible plants of Iran and India. *Pakistan Journal of Nutrition* 7(4): 582–585.

Abiramasundari, G., K.R. Sumalatha & M. Sreepriya. 2012. Effects of *Tinospora cordifolia* (Menispermaceae) on the proliferation, osteogenic differentiation and mineralization of osteoblast model systems in vitro. *Journal of Ethnopharmacology* 141(1): 474–480.

Ahmed, F., N. Siddaraju & A. Urooj. 2011. In vitro hypoglycemic effects of *Gymnema sylvestre, Tinospora cordifolia, Eugenia jambolana* and *Aegle marmelos*. *Journal of Natural Pharmaceuticals* 2(2): 52–52.

Aiyer, K.N. & M. Kolammal. 1963. *Pharmacognosy of ayurvedic drugs*, Series 1, Number 7. The Central Research Institute, Trivandrum, Dept of Pharmacognosy, University of Kerala, pp. 102–105.

Akhtar, S. 2010. Use of *Tinospora cordifolia* in HIV infection. *Indian Journal of Pharmacology* 42(1): 57.

Alrumaihi, F., K.S. Allemailem, A. Almatroudi, M.A. Alsahli, A. Khan & M.A. Khan. 2019. *Tinospora cordifolia* aqueous extract alleviates cyclophosphamide-induced immune suppression, toxicity and systemic candidiasis in immunosuppressed mice: In vivo study in comparison to antifungal drug fluconazole. *Current Pharmaceutical Biotechnology* 20(12): 1055–1063.

Archana, T., S. Ayodhya, T. Gaurav, A.K. Agarwal & V.K. Singh. 2010. In vitro and in vivo study on *Tinospora cordifolia* extract as an antidiabetic agent in rat. *Biochemical and Cellular Archives* 10(2): 175–177.

Badar, V.A., V.R. Thawani, P.T. Wakode, M.P. Shrivastava, K.J. Gharpure, L.L. Hingorani & R.M. Khiyani. 2005. Efficacy of *Tinospora cordifolia* in allergic rhinitis. *Journal of Ethnopharmacology* 96(3): 445–449.

Bishayi, B., S. Roychowdhury, S. Ghosh & M. Sengupta. 2002. Hepatoprotective and immunomodulatory properties of *Tinospora cordifolia* in CCl_4 intoxicated mature albino rats. *The Journal of Toxicological Sciences* 27(3): 139–146.

Bisset, N.G. & J. Nwaiwu. 1983. Quaternary alkaloids of *Tinospora* species. *Planta Medica* 48(8): 275–279.

Chandrasekaran, C.V., L.N. Mathuram, P. Daivasigamani & U. Bhatnagar. 2009. *Tinospora cordifolia*, a safety evaluation. *Toxicology in Vitro* 23(7): 1220–1226.

Cheng, Z., S. Guo, K. Copps, X. Dong, R. Kollipara, J.T. Rodgers, R.A. Depinho, P. Puigserver & M.F. White. 2009. Foxol integrates insulin signaling with mitochondrial function in the liver. *Nature Medicine* 15(11): 1307–1311.

Chopra, A., M. Saluja, G. Tillu, A. Venugopalan, G. Narsimulu, R. Handa, L. Bichile, A. Raut, S. Sarmukaddam & B. Patwardhan. 2012. Comparable efficacy of standardized Ayurveda formulation and hydroxychloroquine sulfate (HCQS) in the treatment of rheumatoid arthritis (RA): A randomized investigator-blind controlled study. *Clinical Rheumatology* 31(2): 259–269.

Choudhary, N., M.B. Siddiqui, S. Azmat & S. Khatoon. 2013. *Tinospora cordifolia*: Ethnobotany, phytopharmacology and phytochemistry aspects. *International Journal of Pharmaceutical Sciences and Research* 4(3): 891.

Chougale, A.D., V.A. Ghadyale, S.N. Panaskar & A.U. Arvindekar. 2009. Alpha glucosidase inhibition by stem extract of *Tinospora cordifolia. Journal of Enzyme Inhibition and Medicinal Chemistry* 24(4): 998–1001.

Deepika, S. & P.K. Chaudhuri. 2015. (+) Corydine from the stems of *Tinospora cordifolia. Asian Journal of Chemistry* 27(4): 1567–1568.

De Oliveira, A.M., L.M. Conserva, J.N. de Souza Ferro, F. de Almeida Brito, R.P. Lyra Lemos & E. Barreto. 2012. Antinociceptive and anti-inflammatory effects of octacosanol from the leaves of *Sabicea grisea* var. grisea in mice. *International Journal of Molecular Sciences* 13(2): 1598–1611.

Devi, P.U., A. Ganasoundari, B.S.S. Rao & K.K. Srinivasan. 1999. In vivo radioprotection by ocimum flavonoids: Survival of mice. *Radiation Research* 151(1): 74–78.

Dhaliwal, K.S. 1999. *Patent No. US 5886029.* United States Patent.

Dhanasekaran, M., A.A. Baskar, S. Ignacimuthu, P. Agastian & V. Duraipandiyan. 2009. Chemopreventive potential of Epoxy clerodane diterpene from *Tinospora cordifolia* against diethylnitrosamine-induced hepatocellular carcinoma. *Investigational New Drugs* 27(4): 347–355.

Dhar, M.L., M.M. Dhar, B.N. Dhawan, B.N. Mehrotra & C. Ray. 1968. Screening of Indian plants for biological activity: Part I. *Indian Journal of Experimental Biology* 6: 232–247.

Dhulia, I., V. Parcha, G. Pant, D. Kumar & A. Maithani. 2011. Antihyperglycemic effect of Methanolic extract of *Tinospora cordifolia* (Willd.) stem on experimentally induced diabetic rats. *Journal of Pharmaceutical Research* 4(8): 2828–2830.

Dimri, M., V.S. Rajwar & L. Kush. 2020. Rasayana drugs promise better anti-Covid-19 medications. *Asian Journal of Pharmaceutical Research and Development* 8(4): 148–149.

Duraipandiyan, V., S. Ignacimuthu, K. Balakrishna & N.A. AL-Harbi. 2012. Antimicrobial activity of *Tinospora cordifolia*: An ethnomedicinal plant. *Asean Journal of Traditional Knowledge* 7(2): 59–65.

Gaiero, J.R., C.A. McCall, K.A. Thompson, N.J. Day, A.S. Best & K.E. Dunfield. 2013. Inside the root microbiome: Bacterial root endophytes and plant growth promotion. *American Journal of Botany* 100(9): 1738–1750.

Gayatri, R., S. Lavanya, M. Hussain & J. Veslin. 2020. The new pandemic Covid-19: Treatment options and developments. *Asian Journal of Biology* 9(3): 1–13.

George, M. & K.M. Pandalai. 1949. Investigations on plant antibiotics. Part IV. Further search for antibiotic substances in Indian medicinal plants. *Indian Journal of Medical Research* 37(2): 169–181.

Ghosh, A.K., B.D. Chapsal, I.T. Weber & H. Mitsuya. 2008. Design of HIV protease inhibitors targeting protein backbone: An effective strategy for combating drug resistance. *Accounts of Chemical Research* 41(1): 78–86.

Goel, H.C., J. Prasad, A. Sharma & B. Singh. 1998. Antitumour and radioprotective action of *Podophyllum hexandrum. Indian Journal of Experimental Biology* 36(6): 583–587.

Grover, J.K., S.S. Rathi & V. Vats. 2000. Anti-hyperglycemic effect of *Eugenia jambolana* and *Tinospora cordifolia* in experimental diabetes and their effects on key metabolic enzymes involved in carbohydrate metabolism. *Journal of Ethnopharmacology* 73(3): 461–470.

Grover, J.K., S.S. Rathi & V. Vats. 2002. Amelioration of experimental diabetic neuropathy and gastropathy in rats following oral administration of plant (*Eugenia jambolana, Mucurna pruriens* and *Tinospora cordifolia*) extracts. *Indian Journal of Experimental Biology* 40: 273–276.

Grover, J.K., V. Vats, S.S. Rathi & R. Dawar. 2001. Traditional Indian anti-diabetic plants attenuate progression of renal damage in streptozotocin induced diabetic mice. *Journal of Ethnopharmacology* 76(3): 233–238.

Gupta, R. & V. Sharma. 2011. Ameliorative effects of *Tinospora cordifolia* root extract on histopathological and biochemical changes induced by aflatoxin-B1 in mice kidney. *Toxicology International* 18(2): 94.

Hamsa, T.P. & G. Kuttan. 2012. *Tinospora cordifolia* ameliorates urotoxic effect of cyclophosphamide by modulating GSH and cytokine levels. *Experimental and Toxicologic Pathology* 64(4): 307–314.

Hartmann, H.T., D.E. Kester, F.T. Davies & R.L. Geneve. 1997. *Plant propagation principles and practices.* 6th edition. Prentice-Hall of India Pvt. Ltd., New Delhi, pp. 276–238.

Husen, A. 2021. *Traditional herbal therapy for the human immune system.* CRC Press, Boca Raton, FL. https://doi.org/10.1201/9781003137955

Husen, A. 2022. *Herbs, shrubs and trees of potential medicinal benefits.* Taylor & Francis Group. https://doi.org/10.1201/9781003205067

Jagetia, G.C. 2019. Anticancer activity of Giloe, *Tinospora cordifolia* (Willd.) Miers Ex Hook F & Thoms. *International Journal of Complementary and Alternative Medicine* 12(2): 79–85.

Jagetia, G.C., V. Nayak & M.S. Vidyasagar. 1998. Evaluation of the antineoplastic activity of guduchi (*Tinospora cordifolia*) in cultured HeLa cells. *Cancer Letters* 127(1–2): 71–82.

Jagetia, G.C. & S.K. Rao. 2006. Evaluation of the antineoplastic activity of guduchi (*Tinospora cordifolia*) in Ehrlich ascites carcinoma bearing mice. *Biological and Pharmaceutical Bulletin* 29(3): 460–466.

Joladarashi, D., N.D. Chilkunda & P.V. Salimath. 2014. Glucose uptake-stimulatory activity of *Tinospora cordifolia* stem extracts in Ehrlich ascites tumor cell model system. *Journal of Food science and Technology* 51(1): 178–182.

Kalikar, M.V., V.R. Thawani, U.K. Varadpande, S.D. Sontakke, R.P. Singh & R.K. Khiyani. 2008. Immunomodulatory effect of *Tinospora cordifolia* extract in human immuno-deficiency virus positive patients. *Indian Journal of Pharmacology* 40(3): 107.

Kapil, A. & S. Sharma. 1997. Immunopotentiating compounds from *Tinospora cordifolia. Journal of Ethnopharmacology* 58(2): 89–95.

Kapur, P., W. Wuttke, H. Jarry & D. Seidlova-Wuttke. 2010. Beneficial effects of β-Ecdysone on the joint, epiphyseal cartilage tissue and trabecular bone in ovariectomized rats. *Phytomedicine* 17(5): 350–355.

Kar, A., B.K. Choudhary & N.G. Bandyopadhyay. 2003. Comparative evaluation of hypoglycaemic activity of some Indian medicinal plants in alloxan diabetic rats. *Journal of Ethnopharmacology* 84(1): 105–108.

Khuda, M.Q.I., A. Khaleque & N. Ray. 1964. *Tinospora cordifolia* constituents of plants fresh from the field. *Scientific Research* 1: 177–183.

Kim, S.K., H.J. Kim, S.E. Choi, K.H. Park, H.K. Choi & M.W. Lee. 2008. Anti-oxidative and inhibitory activities on nitric oxide (NO) and prostaglandin E2 (COX-2) production of flavonoids from seeds of *Prunus tomentosa* Thunberg. *Archives of Pharmacal Research* 31(4): 424–428.

Kohno, H., M. Maeda, M. Tanino, Y. Tsukio, N. Ueda, K. Wada, S. Sugie, H. Mori & T. Tanaka. 2002. A bitter diterpenoid furanolactone columbin from *Calumbae Radix* inhibits azoxymethane-induced rat colon carcinogenesis. *Cancer Letters* 183(2): 131–139.

Kumar, A. & A.C. Jnanesha. 2017. Cultivation, utilization and role of medicinal plants in tradition medicine in deccan eco-climate. *International Journal on Agriultural Science* 8(1): 98–103.

Kumar, A., A.C. Jnanesha & S. Nagaraju. 2019. Cultivation and medicinal properties of giloy [*Tinospora cordifolia* (Thunb.) miers. *Agrobios Newsletter*: 57–58.

Lv, J., D. Xu, V. Perkovic, X. Ma, D.W. Johnson, M. Woodward, A. Levin, H. Zhang, H. Wang & TESTING Study Group. 2012. Corticosteroid therapy in IgA nephropathy. *Journal of the American Society of Nephrology* 23(6): 1108–1116.

Ly, P.T.T., S. Singh & C.A. Shaw. 2007. Novel environmental toxins: Steryl glycosides as a potential etiological factor for age-related neurodegenerative diseases. *Journal of Neuroscience Research* 85(2): 231–237.

Maurya, R. & S.S. Handa. 1998. Tinocordifolin, a sesquiterpene from *Tinospora cordifolia. Phytochemistry* 49(5): 1343–1345.

McKeown, E., V.P. Bykerk, F. De Leon, A. Bonner, C. Thorne, C.A. Hitchon, G. Boire, B. Haraoui, D.S. Ferland, E.C. Keystone & J.E. Pope. 2012. Quality assurance study of the use of preventative therapies in glucocorticoid-induced osteoporosis in early inflammatory arthritis: Results from the CATCH cohort. *Rheumatology* 51(9): 1662–1669.

Mittal, J., M.M. Sharma & A. Batra. 2014. *Tinospora cordifolia*: A multipurpose medicinal plant-A. *Journal of Medicinal Plants Studies* 2(2): 32–47.

Modi, B., K. Kumari Shah, J. Shrestha, P. Shrestha, A. Basnet, I. Tiwari & S. Prasad Aryal. 2020. Morphology, biological activity, chemical composition, and medicinal value of *Tinospora cordifolia* (willd.) miers. *Advanced Journal of Chemistry-Section B*: 36–54.

Mukherjee, R., U.K. De & G.C. Ram. 2010. Evaluation of mammary gland immunity and therapeutic potential of *Tinospora cordifolia* against bovine subclinical mastitis. *Tropical Animal Health and Production* 42(4): 645–651.

Nagaraja, P.K. 2007. Modulation of morphology and some gluconeogenic enzymes activity by *Tinospora cordifolia* (Willd.) in diabetic rat kidney. *Biomed Res*earch 18: 179–183.

Nasreen, S., R. Radha, N. Jayashree, B. Selvaraj & A. Rajendran. 2010. Assessment of quality of *Tinospora cordifolia* (willd.) Mier.(Menispermacea): Pharacognostical and phytophysicolochemical profile. *International Journal of Comprehensive Pharmacy* 5(3): 1–4.

Nayampalli, S., S.S. Ainapure & P.M. Nadkarni. 1982. Study of antiallergic acid Bronchodilator effects of *Tinospora cordifolia. Indian Journal of Pharmacology* 14: 64–66.

Nayampalli, S., N.K. Desai & S.S. Ainapure. 1986. Antiallergic properties of *Tinospora cordifolia* in animal models. *Indian Journal of Pharmacology* 18(4): 250.

Nesari, T.M. 2022. Quality and safety of Ayurvedic medicinal plants is it the key to the success of therapeutic management? *Journal of Ayurveda Case Reports* 5(1): 1–5.

Ninama, R., A. Verma, M. Mishra, A. Nagle, R.K. Pati & R. Meshram. 2022. An exploration of physiological, medicinal and safety aspects of Guduchi (*Tinospora cordifolia*): A complete Ayurvedic and modern review. *Journal of Ayurveda and Integrated Medical Sciences* 7(4): 62–74.

Patel, M.B. & S. Mishra. 2011. Hypoglycemic activity of alkaloidal fraction of *Tinospora cordifolia. Phytomedicine* 18(12): 1045–1052.

PIB. 2021. Accessed from https://pib.gov.in/PressReleasePage.aspx?PRID=1761086 on 05/09/2022.

Prasad, B. & A. Chauhan. 2019. Anti-Oxidant and antimicrobial studies of *Tinospora cordifolia* (Guduchi/Giloy) stems and roots under in-vitro condition. *International Journal of Advanced Microbiology and Health Research* 3(1): 1–10.

Prince, P.S.M. & V.P. Menon. 2003. Hypoglycaemic and hypolipidaemic action of alcoholic extract of *Tinospora cordifolia* Roots in chemical induced diabetes in rats. *Phytotherapy Research* 17(4): 410–413.

Prince, P.S.M., N. Kamalakkannan & V.P. Menon. 2004. Restoration of antioxidants by ethanolic *Tinospora cordifolia* in alloxan-induced diabetic Wistar rats. *Acta Poloniae Pharmaceutica* 61(4): 283–287.

Prince, P.S.M., V.P. Menon & G. Gunasekaran. 1999. Hypolipidaemic action of *Tinospora cordifolia* roots in alloxan diabetic rats. *Journal of Ethnopharmacology* 64(1): 53–57.

Prince, P.S.M., M. Padmanabhan & V.P. Menon. 2004. Restoration of antioxidant defence by ethanolic *Tinospora cordifolia* root extract in alloxan-induced diabetic liver and kidney. *Phytotherapy Research* 18(9): 785–787.

Puranik, N., K.F. Kammar & S. Devi. 2010. Anti-diabetic activity of *Tinospora cordifolia* (Willd.) in streptozotocin diabetic rats; does it act like sulfonylureas? *Turkish Journal of Medical Sciences* 40(2): 265–270.

Raghunathan, K. 1969. The aqueous extract of *T. cordifolia* caused reduction of blood sugar in alloxan induced hyperglycemic rats and rabbits. *The Journal of Research and Education in Indian Medicine* 3: 203–211.

Raghunathan, K. & P.V. Sharma. 1969. Effect of *Tinospora cordifolia* Miers (guduchi) on adrenaline induced hyperglycaemia. *The Journal of Research and Education in Indian Medicine* 4(1): 59–62.

Rajalakshmi, M., J. Eliza, C.E. Priya, A. Nirmala & P. Daisy. 2009. Anti-diabetic properties of *Tinospora cordifolia* stem extracts on streptozotocin-induced diabetic rats. *African Journal Pharmacy and Pharmacology* 3(5): 171–180.

Rajput, D.S. 2020. Evolution, Ayurveda, immunity, and preventive aspects for emerging infectious diseases such as COVID-19. *International Journal of Research in Pharmaceutical Sciences* 1(11): 86–93.

Rathi, S.S., J.K. Grover, V. Vikrant & N.R. Biswas. 2002. Prevention of experimental diabetic cataract by Indian Ayurvedic plant extracts. *Phytotherapy Research* 16(8): 774–777.

Reddy, S.S., P. Ramatholisamma, R. Karuna & D. Saralakumari. 2009. Preventive effect of *Tinospora cordifolia* against high-fructose diet-induced insulin resistance and oxidative stress in male Wistar rats. *Food and Chemical Toxicology* 47(9): 2224–2229.

Rekha. 2017. Accessed from Vikaspedia: https://vikaspedia.in/agriculture/crop-production/package-of-practices/medicinal-and-aromatic-plants/tinospora-cordifolia on 05/09/2022.

Rout, G.R. 2006. Identification of *Tinospora cordifolia* (Willd.) Miers ex Hook F. & Thomas using RAPD markers. *Zeitschrift für Naturforschung C* 61(1–2): 118–122.

Sangeetha, M.K., H.R. Balaji Raghavendran, V. Gayathri & H.R. Vasanthi. 2011. *Tinospora cordifolia* attenuates oxidative stress and distorted carbohydrate metabolism in experimentally induced type 2 diabetes in rats. *Journal of Natural Medicines* 65(3): 544–550.

Sengupta, M., G.D. Sharma & B. Chakraborty. 2011. Effect of aqueous extract of *Tinospora cordifolia* on functions of peritoneal macrophages isolated from CCl4 intoxicated male albino mice. *BMC Complementary and Alternative Medicine* 11(1): 1–9.

Sengupta, S., A. Mukherjee, R. Goswami & S. Basu. 2009. Hypoglycemic activity of the antioxidant saponarin, characterized as α-glucosidase inhibitor present in *Tinospora cordifolia*. *Journal of Enzyme Inhibition and Medicinal Chemistry* 24(3): 684–690.

Shanthi, V. & R. Nelson. 2013. Anitbacterial activity of *Tinospora cordifolia* (Willd) Hook. F. Thoms on urinary tract pathogens. *International Journal of Current Microbiology and Applied Sciences* 2(6): 190–194.

Sharma, A. & A. Batra. 2013. Berberine a novel antidiabetic drug. *International Journal of Research and Reviews in Applied Sciences* 3(2): 216–230.

Sharma, A., S.K. Saggu, R. Mishra & G. Kaur. 2019. Anti-brain cancer activity of chloroform and hexane extracts of *Tinospora cordifolia* Miers: An in vitro perspective. *Annals of Neurosciences* 26(1): 10–20.

Sharma, P., J. Parmar, P. Verma & P.K. Goyal. 2011. Radiation-induced testicular injury and its amelioration by *Tinospora cordifolia* (an Indian medicinal plant) extract. *Evidence-Based Complementary and Alternative Medicine* 2011(643847): 1–9.

Sharma, R., H. Amin & P.K. Prajapati. 2015. Antidiabetic claims of *Tinospora cordifolia* (Willd.) Miers: Critical appraisal and role in therapy. *Asian Pacific Journal of Tropical Biomedicine* 5(1): 68–78.

Sharma, U., M. Bala, N. Kumar, N., B. Singh, R.K. Munshi & S. Bhalerao. 2012. Immunomodulatory active compounds from *Tinospora cordifolia*. *Journal of Ethnopharmacology* 141(3): 918–926.

Sharma, V. & D. Pandey. 2010. Beneficial effects of *Tinospora cordifolia* on blood profiles in male mice exposed to lead. *Toxicology International* 17(1): 8–11.

Shree, P., P. Mishra, C. Selvaraj, S.K. Singh, R. Chaube, N. Garg & Y.B. Tripathi. 2022. Targeting COVID-19 (SARS-CoV-2) main protease through active phytochemicals of ayurvedic medicinal plants – *Withania somnifera* (Ashwagandha), *Tinospora cordifolia* (Giloy) and *Ocimum sanctum* (Tulsi) – a molecular docking study. *Journal of Biomolecular Structure and Dynamics* 40(1): 190–203.

Singh, A., M.M. Sharma & A. Batra. 2013. Synthesis of gold nanoparticles using chick pea leaf extract using green chemistry. *Journal of Optoelectronics and Biomedical Materials* 5(2): 27–32.

Singh, D. & P.K. Chaudhuri. 2017. Chemistry and pharmacology of. *Tinospora cordifolia Natural Product Communications* 12(2): 299–308.

Singh, J., K. Sinha, A. Sharma, N.P. Mishra & S.P. Khanuja. 2003. Traditional uses of *Tinospora cordifolia* (Guduchi). *Journal of Medicinal and Aromatic Plant Sciences* 25: 748–751.

Sinha, K., N.P. Mishra, J. Singh & S.P.S. Khanuja. 2004. *Tinospora cordifolia* (Guduchi), a reservoir plant for therapeutic applications: A review. *Indian Journal of Traditional Knowledge* 3(3): 257–270.

Sivakumar, V. & M.D. Rajan. 2011. Hypoglycemic and antioxidant activity of *Tinospora cordifolia* in experimental diabetes. *International Journal of Pharmaceutical Sciences and Research* 2(3): 608–613.

Sivakumar, V. & A.M. Sadiq. 2009. Hypoglycemic activity of *Tinospora cordifolia* in Alloxan induced diabetic rats. *The Bioscan* 4(1): 75–78.

Sohani, S., L. Sawle, H. Surage & R.S. Chouhan. 2022. *Tinospora cordifolia* cultivation practices and medicinal properties: A review. *International Journal of Botany Studies* 7(2): 533–535.

Sonaimuthu, V., S. Krishnamoorthy & M. Johnpaul. 2010. Taxol producing endophytic fungus *Fusarium culmorum* SVJM072 from medicinal plant of *Tinospora cordifolia*-a first report. *Journal of Biotechnology* 150: 409–425.

Soni, H., G. Nayak, S.S. Patel, K. Mishra, R.V. Singh & A.K. Singhai. 2011. Pharmacognostic studies of the leaves of *Tinospora cordifolia*. *IJPI'S Journal of Pharmacognosy and Herbal Formuations* 1(3): 1–6.

Spandana, U., S.L. Ali, T. Nirmala, M. Santhi & S.S. Babu. 2013. A review on *Tinospora cordifolia*. *International Journal of Current Pharmaceutical Review and Research* 4(2): 61–68.

Spelman, K. 2001. Traditional and clinical use of *Tinospora cordifolia*, Guduchi. *Australian Journal of Medical Herbalism* 13(2): 49–57.

Sriramaneni, R.N., A.Z. Omar, S.M. Ibrahim, S. Amirin & A.M. Zaini. 2010. Vasorelaxant effect of diterpenoid lactones from *Andrographis paniculata* chloroform extract on rat aortic rings. *Pharmacognosy Research* 2(4): 242–246.

Stanely, P., M. Prince & V.P. Menon. 2000. Hypoglycaemic and other related actions of *Tinospora cordifolia* roots in alloxan-induced diabetic rats. *Journal of Ethnopharmacology* 70(1): 9–15.

Subhose, V., P. Srinivas & A. Narayana. 2005. Basic principles of pharmaceutical sciences in ayurveda. *Bulletin of the Indian Institute of History of Medicine* 35(2): 83–92.

Subramanian, M., G.J. Chintalwar & S. Chattopadhyay. 2002. Antioxidant properties of a *Tinospora cordifolia* polysaccharide against iron-mediated lipid damage and γ-ray induced protein damage. *Redox Report* 7(3): 137–143.

Sudha, P., S.S. Zinjarde, S.Y. Bhargava, S.Y. & A.R. Kumar. 2011. Potent α-amylase inhibitory activity of Indian Ayurvedic medicinal plants. *BMC Complementary and Alternative Medicine* 11(1): 1–10.

Sundarraj, S., R. Thangam, V. Sreevani, K. Kaveri, P. Gunasekaran, S. Achiraman & S. Kannan. 2012. γ-Sitosterol from *Acacia nilotica* L. induces G2/M cell cycle arrest and apoptosis through c-Myc suppression in MCF-7 and A549 cells. *Journal of Ethnopharmacology* 141(3): 803–809.

Tan, R.X. & W.X. Zou. 2001. Endophytes: A rich source of functional metabolites. *Natural Product Reports* 18(4): 448–459.

Thippeswamy, G., M.L. Sheela & B.P. Salimath. 2008. Octacosanol isolated from *Tinospora cordifolia* downregulates VEGF gene expression by inhibiting nuclear translocation of NF-< kappa> B and its DNA binding activity. *European Journal of Pharmacology*, 588(2–3): 141–150.

Tiwari, P., P. Nayak, S.K. Prusty & P.K. Sahu 2018. Phytochemistry and pharmacology of *Tinospora cordifolia*: A review. *Systematic Reviews in Pharmacy* 9(1): 70–78.

Tripathi, Y.B., M. Sharma & M. Manickam. 1997. Rubiadin, a new antioxidant from *Rubia cordifolia*. *Indian Journal of Biochemistry & Biophysics* 34(3): 302–306.

Upadhyay, A.K., K. Kumar, A. Kumar & H.S. Mishra. 2010. *Tinospora cordifolia* (Willd.) Hook. f. and Thoms.(Guduchi) – validation of the Ayurvedic pharmacology through experimental and clinical studies. *International Journal of Ayurveda Research* 1(2): 112–121.

Wadood, N., A. Wadood & S.A.W. Shah. 1992. Effect of *Tinospora cordifolia* on blood glucose and total lipid levels of normal and alloxan-diabetic rabbits. *Planta Medica* 58(2): 131–136.

Wang, T., Y.Y. Liu, X. Wang, N. Yang, H.B. Zhu & P.P. Zuo. 2010. Protective effects of octacosanol on 6-hydroxydopamine-induced parkinsonism in rats via regulation of ProNGF and NGF signaling. *Acta Pharmacologica Sinica* 31(7): 765–774.

Yang, S., A.M. Evens, S. Prachand, A.T. Singh, S. Bhalla, K. David & L.I. Gordon. 2010. Mitochondrial-mediated apoptosis in lymphoma cells by the diterpenoid lactone andrographolide, the active component of *Andrographis paniculata* andrographolide-induced apoptosis in Lymphoma. *Clinical Cancer Research* 16(19): 4755–4768.

Yin, J., H. Xing & J. Ye. 2008. Efficacy of berberine in patients with type 2 diabetes mellitus. *Metabolism* 57(5): 712–717.

Zhang, Y., X. Li, D. Zou, W. Liu, J. Yang, N. Zhu, L. Huo, M. Wang, J. Hong, P. Wu & G. Ren. 2008. Treatment of type 2 diabetes and dyslipidemia with the natural plant alkaloid berberine. *The Journal of Clinical Endocrinology & Metabolism* 93(7): 2559–2565.

Zhao, F., E.Q. He, L. Wang & K. Liu. 2008. Anti-tumor activities of andrographolide, a diterpene from *Andrographis paniculata*, by inducing apoptosis and inhibiting VEGF level. *Journal of Asian Natural Products Research* 10(5): 467–473.

20 Traditional Knowledge, Bioactive Compounds, and Antidiabetic and Other Potential Response of Holy Basil (*Ocimum sanctum*)

Yashashree Pradhan, Mohammad Abul Farah, Khalid Mashay Al-Anazi, and Ahmad Ali

CONTENTS

DOI: 10.1201/b23347-20

20.1 INTRODUCTION

In nature there are many known as well as unknown herbs and trees. These plants produce many compounds which are useful for the plant as well as for humans in many ways (Husen 2021, 2022). From ancient times plants have been used for many purposes such as food, ornaments, and medicines. In traditional medicine decoctions, pastes, and powders of many herbs are used to treat numerous diseases. In the modern world there are many chemical compounds present which act as medicines to cure many severe diseases. Apart from their positive effect against particular diseases, they show side effects on other systems (Vidyarthi et al. 2013). To overcome this problem, people are turning towards herbal medicine, which was used by our ancestors. From the report published by WHO, nearly 80% of the world population has turned towards herbal medicine (Ekor 2014). These herbal medicines show activities against many major ailments such as diabetes which affect major populations in the world. In Ayurveda as well as Unani medicine, many herbs are used for treatment of diabetes and for controlling blood sugar level. One of the herbs used is holy basil or tulsi, i.e., *Ocimum sanctum*.

O. sanctum is one of the herbs with spiritual importance in the Hindu religion (Siva et al. 2016) and has a wide range of applications in various industries like medicine and cosmetics. It has many phytochemicals which act against various diseases like throat infections, skin infections, stomach infections, respiratory tract infections, and many more (Cohen 2014).

20.2 BOTANICAL DESCRIPTION

Holy basil, also known as *O. sanctum* (Figure 20.1), is member of Lamiaceae family. It is an aromatic and a perennial herb. It is a straight and highly branched plant of 30 to 60 cm height and covered with fine hairs. Holy basil leaves have toothed margins as well as an elliptic and simple shape, opposite to each other and are of 5 cm in length (Pandey and Madhuri 2010). Flowers of holy basil are 5 mm in size and are purplish in color with cylindrical spikes. The plant has small yellow fruit containing reddish seeds. Seed is roundish to oval and 0.1 cm in diameter with no odor, and it is pungent in taste (Joseph and Nair 2013).

FIGURE 20.1 *Ocimum sanctum.*

20.3 DISTRIBUTION

O. sanctum grows in tropical and warm climatic areas. It is native to India, USA, Italy, Iran, France, and Egypt (Bano et al. 2017). It is cultivated widely in India. Wild varieties of holy basil grow in Africa, Asia, Malaysia, Arab countries, and Australia (Joseph and Nair 2013).

20.4 PHYTOCHEMICAL CONSTITUENTS

O. sanctum leaves contain many aromatic compounds present in leaf extracts as well as in essential oils. A major compound present in the essential oil of *O. sanctum* is methyleugenol, which is approximately 92.4% (Table 20.1). Other compounds include β-caryophyllene 1.3%, eugenol 2.4%, α-ylangene 0.1%, epi-cubebol 0.4%, trans-β-guaiene 0.5%, and β-elemene 0.4% (Joshi 2013). Phytochemicals mainly include ursolic acid, sesquiterpene alcohols, anthocyanins, eugenol, linoleic acid, and flavonoids like apigenin, orientin, vicenin, polyphenols, luteolin, and thymol (Table 20.2) (Joshi et al. 2011).

TABLE 20.1
Essential Oil Composition of *Ocimum sanctum* (Joshi 2013)

Compound	RI	Percentage
α-Pinene	943	t
Camphene	459	t
1,8-Cineole	1036	0.1
(Z)-β-Ocimene	1045	0.1
(E)-β-Ocimene	1058	0.2
Terpinolene	1092	t
Borneol	1173	t
α-Terpineol	1193	0.1
Methyl chavicol	1202	0.1
α-Cubebene	1353	t
Eugenol	1361	2.4
α-Ylangene	1379	0.1
β-Cubebene	1393	0.1
β-Elemene	1395	0.4
Methyleugenol	1407	92.4
β-Caryophyllene	1423	1.3
α-Humulene	1457	t
γ-Muurolene	1485	0.1
epi-Cubebol	1506	0.4
trans-β-Guaiene	1513	0.5
Cubebol	1527	0.1
δ-Cadinene	1533	0.1
10-*epi*-Cubebol	1546	0.2
1,10-di-*epi*-Cubenol	1629	0.1
epi-α-Cadinol	1651	0.1
Total identified (%)		98.9

RI – Retention index; t – Trace

TABLE 20.2
Major Bioactive Compounds of *Ocimum sanctum*

Structure	Name and reference
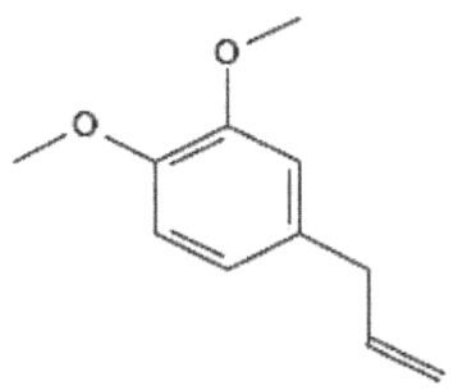	**Methyleugenol CID 7127** https://pubchem.ncbi.nlm.nih.gov/compound/7127. Retrieved July 23, 2022
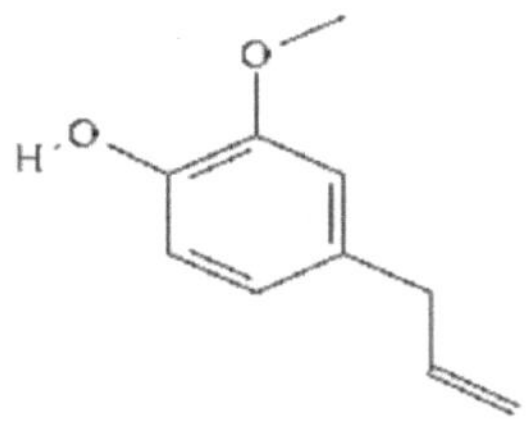	**Eugenol CID 3314** https://pubchem.ncbi.nlm.nih.gov/compound/3314. Retrieved July 23, 2022
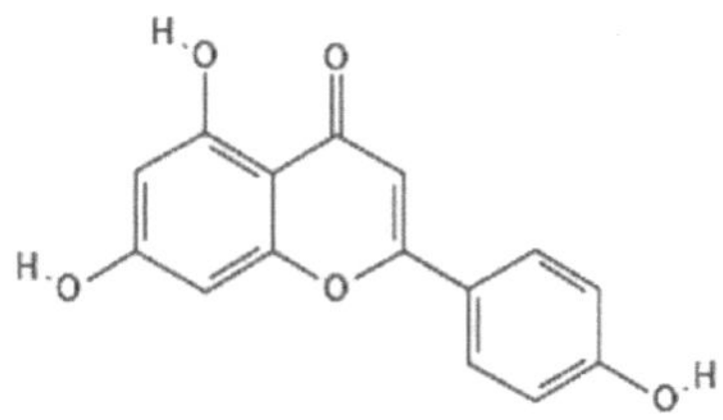	**Apigenin CID 5280443** https://pubchem.ncbi.nlm.nih.gov/compound/5280443. Retrieved July 23, 2022
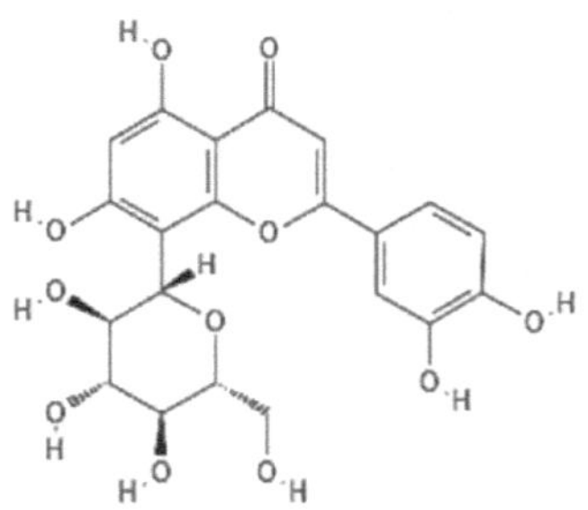	**Orientin CID 5281675**https://pubchem.ncbi.nlm.nih.gov/compound/5281675. Retrieved July 23, 2022
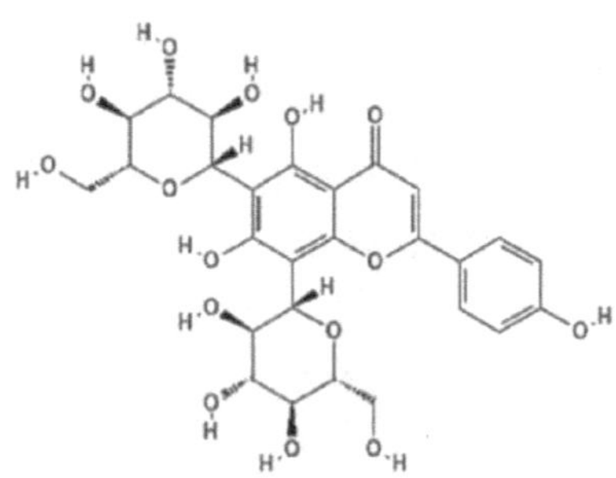	**Vicenin CID 442664**https://pubchem.ncbi.nlm.nih.gov/compound/442664. Retrieved July 23, 2022
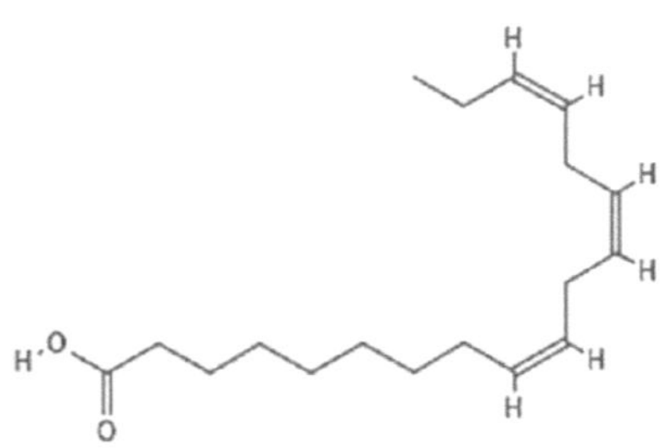	**Linolenic acid CID 5280934**https://pubchem.ncbi.nlm.nih.gov/compound/5280934. Retrieved July 23, 2022

Structure	Name and reference
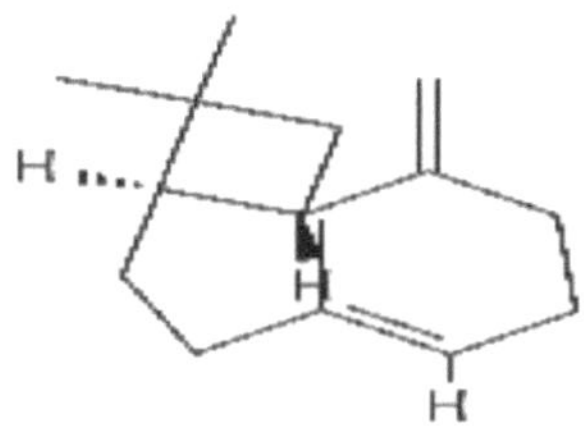	**β-Caryophyllene SID 134971569**https://pubchem.ncbi.nlm.nih.gov/substance/134971569. Retrieved July 23, 2022

20.5 PHARMACOLOGICAL STUDIES

O. sanctum is one of the traditional herbs and contains many phytochemicals that help to protect against many diseases and ailments. It has many pharmacological activities such as anticancer, antioxidant, antibacterial, antifungal, insecticidal, antiviral, antidiabetic, hyperlipidemic, anti-fertility, anti-fatigue, radioprotective, anti-inflammatory, immunomodulatory, cardioprotective, hepatoprotective, analgesic, antiulcer, and antiarthritic.

20.5.1 Anticancer Activity

Alcoholic extract of *O. sanctum* leaves show anticancer activity by enhancing the carcinogen metabolizing enzymes like cytochrome b_5, cytochrome P450, glutathione S-transferase, and aryl hydrocarbon hydroxylase (Pandey and Madhuri 2010). Leaf extract tested on DMBA (7,12-dimethylbenz(a) anthracene)-induced papilloma genesis in mice showed reduction of tumor formation. When the paste of leaves is given to DMBA-induced buccal pouch carcinogenesis mice, it showed prevention from formation of tumor (Prashar et al. 1994).

20.5.2 Antioxidant Activity

Essential oils and eugenol extracted from *O. sanctum* show antioxidant activity when tested using DPPH and ABTS assay (Joshi 2013). Orientin and vicenin, which are flavonoids present in *O. sanctum*, show a reduction in radiation-induced lipid peroxidation when tested in mouse liver (Pandey and Madhuri 2010). Phenolic compounds present in *O. sanctum* increase the activity of superoxide dismutase, glutathione, and catalase enzymes when tested in albino rats. This increase in activity of enzymes helps in the reduction of oxidative stress (Mahajan et al. 2013).

20.5.3 Antibacterial Activity

O. sanctum contains essential oils which show antibacterial activity against *Staphylococcus aureus*, *Bacillus subtilis*, *Streptococcus faecalis*, *Staphylococcus epidermidis*, *Micrococcus luteus*, *Micrococcus flavus*, *Salmonella typhimurium*, *Escherichia coli*, *Pseudomonas aeruginosa*, *Klebsiella pneumoniae*, *Proteus vulgaris*, and *Serratia marcescens* (Joshi 2013). Alcoholic extract of *O. sanctum* shows inhibition of *vibrio cholerae* as well as showing activity against antibiotic-resistant *Staphylococcus aureus* and *Neisseria gonorrhea* when tested using the agar diffusion method and MIC technique. These effects are mainly due to the presence of linolenic acid in the extract (Aqil et al. 2005).

20.5.4 Antifungal Activity

O. sanctum alcoholic extract shows antifungal activity against *Candida albicans*, whereas essential oils present in *O. sanctum* show antifungal activity against *Penicillium chrysogenum*, *Aspergillus niger*, and *Aspergillus fumigatus* (Joshi 2013).

20.5.5 Insecticidal Activity

Eugenol present in the extract of *O. sanctum* shows LD_{100} at 6.25 μg/ml and 200 μg/ml when tested against *Aedes aegyptii* larvae (Singh and Chaudhuri 2018). *O. sanctum* extract also showed activity against larvae of *Culex quinquefasciatus* and *Anopheles stephensi*. It also showed mosquito repellent activity (Mahajan et al. 2013). Leaf extract of *O. sanctum* shows repellent activity against *Culex quinquefasciatus* when tested in a concentration range of 150 to 900 ml (Joseph and Nair 2013).

20.5.6 Antiviral Activity

Eugenol, linalool, apigenin, and ursolic acid present in essential oils of *O. sanctum* show effects against viruses like HIV, HSV, new castle virus, hematopoietic necrosis virus, hepatitis B virus, adenovirus, and polio virus (Jayati et al. 2013).

20.5.7 Anti-Hyperlipidemic Activity

In triton-induced rats, hypolipidemic activity was observed when they are administered with *O. sanctum* aqueous extract. Fixed oil showed anti-hyperlipidemic effect due to suppression of liver lipid synthesis. Linolenic acid and linolic acid act as lipid-lowering compounds (Mahajan et al. 2013). Methyleugenol, β-caryophyllene, and eugenol reduces triglycerides and cholesterol in hyperlipidemic rats. Lipid peroxidation and an increase in antioxidant enzymes are observed in brain, liver, kidney, and lung when treated with aqueous extract of *O. sanctum* (Singh and Chaudhuri 2018).

20.5.8 Anti-Fertility Activity

Benzene extract of *O. sanctum* leaves when tested on albino rats with concentration of 250 mg/kg of body weight showed sperm mortality and decreased motility and forward velocity of sperm and total sperm count. This is the result of androgen reduction, and the effect is reversed after two weeks when the treatment is stopped. It also reduced sperm count in rabbits. FSH and LH levels are decreased, whereas testosterone levels are increased when tested in rabbits (Joseph and Nair 2013). When rats are treated with fresh leaves of *O. sanctum* they showed reduced weight of ventral prostate, epididymis, seminal vesicle, and testes (Mahajan et al. 2013). A reduction in fertility of 60% to 80% in female rats was observed when treated with benzene and petroleum extract (Siva et al. 2016).

20.5.9 Radioprotective Activity

When lethality testing was carried out on Swiss albino mice, the aqueous extract of *O. sanctum* showed greater radioprotective effect and also showed an increase in survival rate. Vicenin and orientin showed protective effects against γ radiation-induced lipid peroxidation in mouse liver. This activity is due to the Fenton reaction-induced OH radical *in vitro* (Siva et al. 2016).

20.5.10 Anti-Inflammatory Activity

Acute and chronic inflammation induced due to carrageenan and croton oil induced granuloma were inhibited by *O. sanctum*. Anti-inflammatory effect of essential oils is due to its inhibitory action on lipoxygenase and cyclooxygenase. Seed oil showed major inhibition of leukotriene-induced paw edema (Mahajan et al. 2013). Fixed oil shows anti-inflammatory effect by double inhibition of antihistaminic activity and arachidonate metabolism (Singh et al. 2007).

20.5.11 Immunomodulatory Activity

Aqueous extract administration of *O. sanctum* leaves in rats showed modification of humoral immune response like antibody production and release of hypersensitivity mediators in specific organs. Subclinical trials in bovine shows the immunomodulatory effect by reducing the total bacterial count and increase in lymphocyte and neutrophil count as well as increase in phagocytic activity. *O. sanctum* seed oil mediates GABAergic pathway which leads to an immunomodulatory effect (Bano et al. 2017).

20.5.12 Cardioprotective Activity

Hydroalcoholic extract of O. *sanctum* shows prevention of chronic resistant stress-induced rise in plasma cyclic AMP level as well as catalase and myocardial superoxide dismutase activities. It also shows changes in myocardium when observed under light microscope. Ursolic acid present in essential oils of *O. sanctum* shows protection against Adriamycin-induced lipid peroxidation in liver and heart microsomes (Sharma et al. 2001).

20.5.13 Hepatoprotective Activity

Alcoholic extract of *O. sanctum* showed hepatoprotective activity in anti-tuberculosis drug-, paracetamol-, and carbon tetrachloride-induced liver injury in albino rats (Bano et al. 2017). A hepatotonic effect against paracetamol and carbon tetrachloride was observed in albino rats when fed with cold water extract of *O. sanctum*. It also reduces serum alanine aminotransferase levels which are the markers of liver dysfunction (Mahajan et al. 2013).

20.5.14 Analgesic Activity

Oils present in *O. sanctum* inhibit histamine, prostaglandins, and acetylcholine, which leads to inhibition of writhing. This study was carried out in acetic acid-induced writhing model (Pandey and Madhuri 2010).

20.5.15 Antiulcer Activity

In pylorus ligated mice, oil of *O. sanctum*˚showed an effect of inhibition of gastric secretion. It also showed lipoxygenase inhibition and antihistaminic property, which leads to the antiulcer property. *O. sanctum* oil also reduced histamine-induced vasospastic effect and gastric secretion. Due to its anticholinergic, antisecretory, and histamine antagonist effects, it reduces stress-induced ulceration (Dharmani et al. 2004).

20.5.16 Antiarthritic Activity

Reduction of diameter of inflamed paw in formaldehyde-induced arthritic rats was observed after administration of fixed oil of *O. sanctum*. This effect is due to the reduction of carrageenan and other inflammatory mediators like bradykinin, serotonin, PGE 2, and histamine (Prakash and Gupta 2005).

20.6 ANTIDIABETIC RESPONSE

Diabetes mellitus represents a series of metabolic conditions caused by insufficient secretion of insulin or improper utilization of insulin. Insufficient secretion of insulin occurs in type I diabetes, whereas improper use of insulin occurs in type II diabetes. This leads to inefficient movement of glucose from blood to other cells for metabolism and starves the cells. This malfunctioning results in high blood glucose level as well as harm to certain tissues and organs (Siddiqui et al. 2013). There are treatments available for diabetes like insulin supplements and reduction of intake of sugars, and many artificial as well as herbal sweeteners are available (Ali 2020). Many herbs used in traditional medicine show antidiabetic potential which leads to treating of diabetes.

One of the reasons for diabetes is glycation. Glycation is a biochemical process in which spontaneous nonenzymatic amino-carbonyl reaction occurs between lipids, proteins, nucleic acid, and sugar molecules. In this reaction, formation of Schiff bases and Amadori product occurs that further leads to formation of advanced glycation end products, which can alter the functioning and structure of other biomolecules after accumulation. Complications that occur in diabetes are due to glycation-mediated cross-linking and structural alteration of nucleic acid, proteins, and lipids. These advanced glycation end products can be inhibited by using various synthetic inhibitors, but there are side effects associated with them. To avoid these side effects, many natural products are tested for similar effect (Tariq and Ali 2022).

One such herb is *O. sanctum*. Streptozotocin induces diabetes mellitus by reduction of β-cells which further reduces secretion of insulin in rats. When these models are treated with *O. sanctum* fixed oil, they showed increased level of serum insulin. α-Linolenic acid present in fixed oil is the major component responsible for antidiabetic effect. It also improves the insulin sensitivity by increasing GLUT4 protein in gastrocnemius muscle membranes (Suanarunsawat et al. 2016). Alloxan-induced diabetes mellitus rabbits are tested for fasting blood glucose after administration of seed powder of *O. sanctum* showed decrease in fasting blood glucose levels (Gupta et al. 2006). This effect is due to the presence of terpenoid in the extract. It increases the secretion of insulin from β-cells of the pancreas and also releases bounded insulin, which further leads to the decrease in blood glucose level (Patil et al. 2011).

Aqueous extract of *O. sanctum* facilitates the insulin secretion from isolated pancreatic β-islets cells which results in managing of type II diabetes mellitus (Nyarko et al. 2002). Soluble fibers from *O. sanctum* show effects against hyperglycemia and hyperlipidemia due to its stimulation action on phosphatidylinositol or adenyl cyclase and also promoting the entry of calcium in β-cells, further leading to secretion of insulin. A clinical trial of non-insulin-dependent diabetes mellitus patients' administration of *O. sanctum* extract showed reduction of fasting glucose level (Agrawal et al. 1996).

There was a study conducted on 60 patients with type 2 diabetes in which 30 were administered *O. sanctum* tablet (250 mg) plus glibenclamide (5 mg) and the other 30 patients were administered only glibenclamide (5 mg) for 90 days. Results of this study showed that after the course, fasting glucose level was decreased from 171.53 g/dl to 103.50 g/dl in patients administered *O. sanctum* tablet (250 mg) plus glibenclamide (5 mg), and in patients administered with only glibenclamide (5 mg), it was reduced from 174.35 g/dl to 114.50 g/dl (Somasundaram et al. 2012).

When an antidiabetic study was carried out on streptozotocin-induced diabetes in Wister rats, it was observed that the rats administered with insulin showed a 41.68% decrease in blood glucose level, whereas rats administered with *O. sanctum* root extract showed a 24.44% decrease in blood glucose level in 6 hours. When the same concentration of *O. sanctum* root extract was given for 21 days, it showed a 38.05% decrease when compared with rats administered with glimepiride for 21 days, which showed decrease of 66.12% (Mohd and Showkat 2012). The effect of *O. sanctum* leaf extract of glucose-fed hyperglycemic rats was observed to be 91.55%, whereas on streptozotocin-induced diabetic rats it was 70.43% (Chattopadhyay 1993).

O. sanctum extract showed a reduction of blood sugar level from 332.16 mg/dl to 263.5 mg/dl in 6 hours when tested in alloxan-induced diabetic rats (Bihari et al. 2011). Alloxan-induced diabetic

Wister rats were treated with *O. sanctum* leaf aqueous extract 300 µg/g body weight and compared with rats given treatment with pioglitazone. The results showed a reduction of blood glucose level from 345 mg/dl to 263 mg/dl, whereas in pioglitazone-treated rats it was reduced to 220 mg/dl (Raja et al. 2016).

Hydroalcoholic extracts of *O. sanctum* show inhibitory activity on enzymes, which plays a major role in glucose metabolism, *viz.* α-glucosidase and α-amylase. When these enzymes are inhibited, the formation of glucose is reduced, which further results in a reduction of blood glucose level. α-Glucosidase is inhibited 34.17%–71.45% when compared with acarbose, which showed result of 43.06%–97.10% inhibition (Mehta et al. 2016), whereas α-amylase was inhibited 61.08% when compared with acarbose, which showed result of 76.26% (Lokhande and Yadav 2018).

The alloxan-induced diabetic rats show a three-fold increase in blood glucose level, whereas the rats administered with *O. sanctum* extracts orally show a 56% decrease in blood glucose level when treated for 14 days. Administration of *O. sanctum* extract also reduces SGOT, SGPT, bilirubin, and ALP levels to 38.6%, 42%, 46%, and 78% respectively in diabetic rats. Increased levels of total cholesterol (78.5%), triglycerides (136.3%), and creatinine (186%) in diabetic rats was decreased to 35.3%, 44.4%, and 51% respectively (Jayant and Srivastava 2016).

O. sanctum ethanolic extracts show reduced plasma glucose levels by 26.4% after 30 days of administration to streptozotocin-induced diabetic rats, whereas the amount of glucokinase, hexokinase, and phosphofructokinase were reduced to 35%, 50%, and 60% respectively in diabetic rats, which is increased to 45%, 64%, and 76% respectively after administration of *O. sanctum* ethanolic extracts for 30 days (Vats et al. 2004). Butanol extract, ethyl acetate extract, and aqueous extract of *O. sanctum* showed increase in insulin secretion in perfused pancreas isolated from Long–Evans rats. The basal level of insulin was 0.3 ng/ml, which is increased to 1.2 ng/ml, 1.4 ng/ml, and 1.7 ng/ml respectively (Hannan et al. 2006).

20.7 TRADITIONAL AND OTHER POTENTIAL USES

In traditional Indian medicine like Ayurveda, *O. sanctum* is used for the treatment of many diseases. According to Ayurveda, it has properties as follows: rasa-katu and tikta, i.e., pungent and bitter, veerya-ushna, i.e., worm, vipaka-Katu, i.e., pungent, guna-Laghu, Snigdha, and Tikshna, i.e., light, oily, and strong, and dosha-kaphaghna and vataghna calms kapha and vata (Deshpande et al. 2018).

It is used for deworming, treating cough, reducing hiccups, asthma, treating tuberculosis, and pleural effusion. It is also used as an antiseptic and anti-inflammatory agent and as a digestive. For treatment of colonic ulcers caused in dysentery, decoction of seeds is used. The major function of *O. sanctum* is to treat respiratory diseases. Extract of *O. sanctum* mixed with honey is used for treatment of cough as well as hiccups. It also acts as a mucolytic. Paste and extract are effective against pain in pleural effusion. It is also used as a blood purifier and is cardioprotective. Extract of *O. sanctum*, honey, and black pepper is mixed and used in treatment of many types of fever including malaria. It acts as an insecticide and hence is used as mosquito repellant. The extract is applied locally as well as administered orally in many skin infections. It is also used as diuretic. For treatment of poison, the leaves are used (Deshpande et al. 2018). Eye drops containing *O. sanctum* are used to treat glaucoma and conjunctivitis. Decoction of leaves are used for treatment of sore throat (Gulhane et al. 2021).

20.8 SAFETY ISSUES

When *O. sanctum* hydroalcoholic extract was administered to rats up to 2000 mg/kg and tested for acute oral toxicity, it did not show any mortality. When tested in diabetic rats, it also did not show any toxicity or carcinogenicity. In treatment of many diseases, it is used without any side effects (Parasuraman et al. 2015).

20.9 CULTIVATION PRACTICES

O. sanctum is one of the ancient medicinal herbs. It contains essential oils that show activities against many ailments. Apart from medicine, it is also used in the food industry for flavoring. It is largely a household species in India. It is occasionally affected by root-rot under waterlogged conditions. In Madhya Pradesh, *O. sanctum* is commercially cultivated for bulk consumption of fresh leaves. Approximately 25,000 ha of land is under cultivation of *O. sanctum* in India (Smitha et al. 2019). It is one of the cash crops in India due to its easy cultivation methods. It requires unirrigated land; it has no depredation by wild animals and low pest infestations. In Uttarakhand, *O. sanctum* is cultivated as an alternative crop (Shah et al. 2019).

20.10 CONCLUSION AND FUTURE REMARKS

O. sanctum is one of the medicinal herbs which contains essential oils in its leaves, seeds, and flowers. These oils as well as aqueous and alcoholic extracts contain bioactive compounds which show activities against many diseases and disorders. The major compounds, such as methyleugenol, eugenol, β-caryophyllene, linolenic acid, vicenin, orientin, and apigenin, show many activities like anticancer, antioxidant, antibacterial, antifungal, insecticidal, antiviral, antidiabetic, hyperlipidemic, anti-fertility, anti-fatigue, radioprotective, anti-inflammatory, immunomodulatory, cardioprotective, hepatoprotective, analgesic, antiulcer, and antiarthritic. The antidiabetic potential of *O. sanctum* is due to the presence of α-linolenic acid, linolic acid, and terpenoids in its essential oils. These compounds reduce blood glucose level by increasing insulin secretion from the β-cells of the pancreas. It also unbinds insulin molecules, resulting in its availability to metabolize glucose. *O. sanctum* also shows antidiabetic activity by inhibiting the activity of α-glucosidase and α-amylase as well as increasing the amount of glucokinase, hexokinase, and phosphofructokinase (Figure 20.2). Due to these effects, *O. sanctum* can be used as a potential antidiabetic agent in many herbal products. It can also be used in many medicines for treatment of diseases.

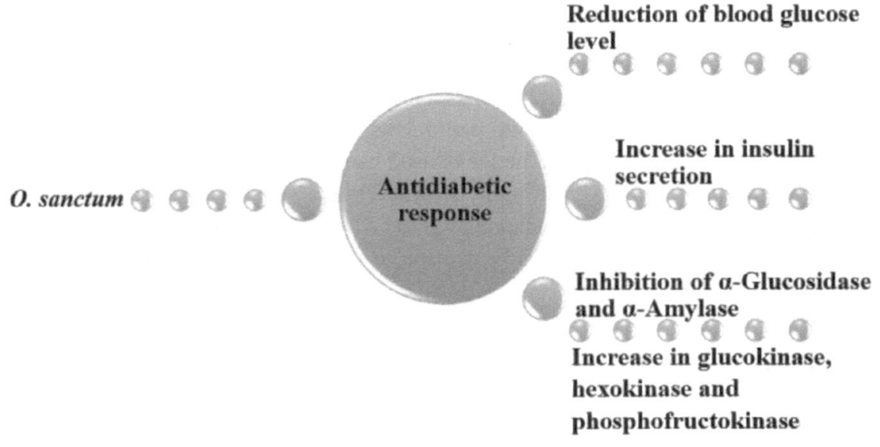

FIGURE 20.2 Overall antidiabetic response of *Ocimum sanctum*.

REFERENCES

Agrawal, P., V. Rai, & R. Singh. 1996. Randomized placebo-controlled, single blind trial of holy basil leaves in patients with noninsulin-dependent diabetes mellitus. *International Journal of Clinical Pharmacology and Therapeutics* 34(9): 406–409.

Ali, A. 2020. Herbs that heal: The philanthropic behavior of nature. *Annals of Phytomedicine* 9(1): 7–17.

Aqil, F., M. Khan, M. Owais & I. Ahmad. 2005. Effect of certain bioactive plant extracts on clinical isolates of β-lactamase producing methicillin resistant Staphylococcus aureus. *Journal of Basic Microbiology: An International Journal on Biochemistry, Physiology, Genetics, Morphology, and Ecology of Microorganisms* 45(2): 106–114.

Bano, N., A. Ahmed, M. Tanveer, G. Khan & M. Ansari. 2017. Pharmacological evaluation of *Ocimum sanctum*. *Journal of Bioequivalence and Bioavailability* 9(3): 387–392.

Bihari, C., B. Manaswini, P. Keshari & T. Kumar. 2011. Phytochemical investigation & evaluation for anti-diabetic activity of leafy extracts of various Ocimum (Tulsi) species by alloxan induced diabetic model. *Journal of Pharmacy Research* 4(1): 28–29.

Chattopadhyay, R. 1993. Hypoglycemic effect of *Ocimum sanctum* leaf extract in normal and streptozotocin diabetic rats. *Indian Journal of Experimental Biology* 31(11): 891–893.

Cohen, M. 2014. Tulsi- *Ocimum sanctum*: A herb for all reasons. *Journal of Ayurveda and Integrative Medicine* 5(4): 251.

Deshpande, A., R. Javalgekar & S. Ranade. 2018. *Dravyaguna vidnyan*, 4th ed. Proficient Publishing House, Pune.

Dharmani, P., V. Kuchibhotla, R. Maurya, S. Srivastava, S. Sharma & G. Palit. 2004. Evaluation of anti-ulcerogenic and ulcer-healing properties of Ocimum sanctum Linn. *Journal of Ethnopharmacology* 93(2–3): 197–206.

Ekor, M. 2014. The growing use of herbal medicines: Issues relating to adverse reactions and challenges in monitoring safety. *Frontiers in Pharmacology* 4: 177.

Gulhane, N., C. Ghode, A. Jadhao & P. Patil. 2021. Study of medicinal uses of *Ocimum sanctum* (Tulsi). *Journal of Pharmacognosy and Phytochemistry* 10(2): 1427–1431.

Gupta, S., P. Mediratta, S. Singh, K. Sharma & R. Shukla. 2006. Antidiabetic, antihypercholesterolaemic and antioxidant effect of *Ocimum sanctum* (Linn) seed oil. *Indian Journal of Experimental Biology* 44(4): 300–304.

Hannan, J., L. Marenah, L. Ali, B. Rokeya, P. Flatt & Y. Abdel-Wahab. 2006. *Ocimum sanctum* leaf extracts stimulate insulin secretion from perfused pancreas, isolated islets and clonal pancreatic β-cells. *Journal of Endocrinology* 189(1): 127–136.

Husen, A. 2021. *Traditional herbal therapy for the human immune system*. CRC Press, Boca Raton, FL. https://doi.org/10.1201/9781003137955

Husen, A. 2022. *Herbs, shrubs and trees of potential medicinal benefits*. Taylor & Francis Group, USA. https://doi.org/10.1201/9781003205067

Jayant, S. & N. Srivastava. 2016. Effect of *Ocimum sanctum* against alloxan induced diabetes and biochemical alterations in rats. *Integr Obesity Diabetes* 2(5): 1–4.

Jayati, B., A. Kumar, A. Goel, S. Gupta & A. Rahal. 2013. In vitro antiviral potential of *Ocimum sanctum* leaves extract against New Castle Disease Virus of poultry. *International Journal of Microbiology and Immunology Research* 2(7): 51–55.

Joseph, B. & V. Nair. 2013. Ethanopharmacological and phytochemical aspects of *Ocimum sanctum* Linn-the elixir of life. *British Journal of Pharmaceutical Research* 3(2): 273–292.

Joshi, B., G. Sah, B. Basnet, M. Bhatt, D. Sharma, K. Subedi, P. Janardhan & R. Malla. 2011. Phytochemical extraction and antimicrobial properties of different medicinal plants: *Ocimum sanctum* (Tulsi), *Eugenia caryophyllata* (Clove), *Achyranthes bidentata* (Datiwan) and *Azadirachta indica* (Neem). *Journal of Microbiology and Antimicrobials* 3(1): 1–7.

Joshi, R. 2013. Chemical composition, in vitro antimicrobial and antioxidant activities of the essential oils of *Ocimum gratissimum*, *O. sanctum* and their major constituents. *Indian Journal of Pharmaceutical Sciences* 75(4): 457.

Lokhande, V. & A. Yadav. 2018. In vitro antioxidant and antidiabetic activity of supercritical fluid extract of leaves *Ocimum sanctum* L. *Research Journal of Pharmacy and Technology* 11(12): 5376–5378.

Mahajan, N., S. Rawal, M. Verma, M. Poddar & S. Alok. 2013. A phytopharmacological overview on Ocimum species with special emphasis on *Ocimum sanctum*. *Biomedicine & Preventive Nutrition* 3(2): 185–192.

Mehta, V., A. Sharma, P. Kailkhura & U. Malairaman. 2016. Antioxidant, anti-inflammatory, and antidiabetic activity of hydroalcoholic extract of *Ocimum sanctum*: An in-vitro and in-silico study. *Asian Journal of Pharmaceutical and Clinical Research* 9(5): 44–49.

Mohd, A. & R. Showkat. 2012. Anti-diabetic activity of *Ocimum sanctum* L. roots and isolation of new phytoconstituents using two-dimensional nuclear magnetic resonance spectroscopy. *Journal of Pharmacognosy and Phytotherapy* 4(6): 75–85.

Nyarko, A., H. Asare-Anane, M. Ofosuhene & M. Addy. 2002. Extract of *Ocimum canum* lowers blood glucose and facilitates insulin release by isolated pancreatic β-islet cells. *Phytomedicine* 9(4): 346–351.

Pandey, G. & S. Madhuri. 2010. Pharmacological activities of *Ocimum sanctum* (tulsi): A review. *International Journal of Pharmacological Science Review and Research* 5(1): 61–66.

Parasuraman, S., S. Balamurugan, P. Christapher, R. Petchi, W. Yeng, J. Sujithra & C. Vijaya. 2015. Evaluation of antidiabetic and antihyperlipidemic effects of hydroalcoholic Extract of leaves of *Ocimum tenuiflorum* (Lamiaceae) and prediction of biological activity of its phytoconstituents. *Pharmacognosy Research* 7(2): 156–165.

Patil, R., R. Patil, B. Ahirwar & D. Ahirwar. 2011. Isolation and characterization of anti-diabetic component (bioactivity-guided fractionation) from *Ocimum sanctum* L.(Lamiaceae) aerial part. *Asian Pacific Journal of Tropical Medicine* 4(4): 278–282.

Prakash, P. & N. Gupta. 2005. Therapeutic uses of *Ocimum sanctum* Linn (Tulsi) with a note on eugenol and its pharmacological actions: A short review. *Indian Journal of Physiology and Pharmacology* 49(2): 125.

Prashar, R., A. Kumar, S. Banerjee & A. Rao. 1994. Chemopreventive action by an extract from Ocimum sanctum on mouse skin papillomagenesis and its enhancement of skin glutathione S-transferase activity and acid soluble sulfydryl level. *Anti-cancer Drugs* 5(5): 567–572.

Raja, T., R. Reddy & M. Buchineni. 2016. An evaluation of anti-hyperglycemic activity of *Ocimum sanctum* Linn (leaves) in Wister rats. *The Pharma Innovation* 5(1): 1.

Shah, G., F. Ahmad, S. Panwar, M. Khadka, A. Ali & S. Bisht. 2019. *Ocimum sanctum* [Tulsi] – an alternative additional livelihood option for the poor and smallholder farmers. *Sustainability* 11(1): 227.

Sharma, M., K. Kishore, S. Gupta, S. Joshi & D. Arya. 2001. Cardioprotective potential of *Ocimum sanctum* in isoproterenol induced myocardial infarction in rats. *Molecular and Cellular Biochemistry* 225(1): 75–83.

Siddiqui, A., S. Siddiqui, S. Ahmad, S. Siddiqui, I. Ahsan & K. Sahu. 2013. Diabetes: Mechanism, pathophysiology and management-A review. *International Journal of Drug Development and Research* 5(2).

Singh, D. & P. Chaudhuri. 2018. A review on phytochemical and pharmacological properties of Holy basil (*Ocimum sanctum* L.). *Industrial Crops and Products* 118: 367–382.

Singh, S., M. Taneja & D. Majumdar. 2007. Biological activities of *Ocimum sanctum* L. fixed oil – An overview. *Indian Journal of Experimental Biology* 45(5): 403–412.

Siva, M., K. Shanmugam, B. Shanmugam, S. Venkata, S. Ravi, R. Sathyavelu & K. Mallikarjuna. 2016. *Ocimum sanctum*: A review on the pharmacological properties. *International Journal of Basic and Clinical Pharmacology* 5: 558–565.

Smitha, G., B. Basak, V. Thondaiman & A. Saha. 2019. Nutrient management through organics, bio-fertilizers and crop residues improves growth, yield and quality of sacred basil (*Ocimum sanctum* Linn). *Industrial Crops and Products* 128: 599–606.

Somasundaram, G., K. Manimekalai, K. Salwe & J. Pandiamunian. 2012. Evaluation of the antidiabetic effect of *Ocimum sanctum* in type 2 diabetic patients. *International Journal of Life Science and Pharma Research* 5: 75–81.

Suanarunsawat, T., G. Anantasomboon & C. Piewbang. 2016. Anti-diabetic and anti-oxidative activity of fixed oil extracted from *Ocimum sanctum* L. leaves in diabetic rats. *Experimental and Therapeutic Medicine* 11(3): 832–840.

Tariq, M. & A. Ali. 2022. Plant derived natural products as anti-glycating agents. In: *Pharmacological Perspectives of Low Molecular Weight Ligands* (Eds. Debarshi Kar Mahapatra, & Sanjay Kumar Bharti). Apple Academic Press, Ontario, Canada, pp. 17–44.

Vats, V., S. Yadav & J. Grover. 2004. Ethanolic extract of *Ocimum sanctum* leaves partially attenuates streptozotocin-induced alterations in glycogen content and carbohydrate metabolism in rats. *Journal of Ethnopharmacology* 90(1): 155–160.

Vidyarthi, S., S. Samant & P. Sharma. 2013. Traditional and indigenous uses of medicinal plants by local residents in Himachal Pradesh, North Western Himalaya, India. *International Journal of Biodiversity Science and Management* 9: 185–200.

21 *Annona squamosa* L. (Sugar Apple) as a Potential Antidiabetic Agent

Safety Issues Related to the Presence of Acetogenins-Mediated Neurodegenerative Effect

M.H. Mohd. Sani, P. Palasuberniam, C.I. Che Noh, M.S. Rahman, and Z.A. Zakaria

CONTENTS

ABBREVIATIONS

$EEAS_L$: alcoholic extract of *A. squamosa* leaves
$W\text{-}EAS_L$: hydro-alcoholic extract of *A. squamosa* leaves
$EEAS_{YL}$: ethanolic extract of *A. squamosa* young leaves

DOI: 10.1201/b23347-21

STZ: streptozotocin
$MEAS_L$: methanol extract of *A. squamosa* leaves
$AEAS_L$: aqueous extract of *A. squamosa* leaf
$AcEAS_L$: acetone extract of *A. squamosa* leaves
$HAEAS_S$: hydroalcoholic extract of *A. squamosa* seeds
$PEEAS_S$: petroleum ether extract of *A. squamosa* seeds
$AWEAS_L$: acetone: water extract of *A. squamosa* leaf
$EEAS_L$: ethanol extract of *A. squamosa* leaf
$95EEAS_S$: 95% ethanol extract of *A. squamosa* seeds
$PEEAS_P$: petroleum ether extract of *A. squamosa* pericarps
$BEAS_P$: n-butanol extract of *A. squamosa* pericarps
$EEEAS_P$: ethyl acetate extract of *A. squamosa* pericarps
$EOAS_P$: essential oil of *A. squamosa* pericarps
$EEAS_S$: ethanol extract of *A. squamosa* seeds
$AEAS_S$: aqueous extract of *A. squamosa* seed
$AEAS_{PE}$: aqueous extract of *A. squamosa* peel
$AEAS_{PU}$: aqueous extract of *A. squamosa* pulp
$CEAS_S$: chloroform extract form *A. squamosa* seeds
$EAEAS_S$: ethyl acetate extract form *A. squamosa* seeds
$MEAS_S$: methanol extract of *A. squamosa* seeds
$PEEAS_L$: petroleum ether extract of *A. squamosa* leaf
$DEAS_L$: dichloromethane extracts of *A. squamosa* leaf
$HFAS_S$: n-hexane fraction of *A. squamosa* seeds
$MEAS_F$: methanol extract of *A. squamosa* fruits
$MEAS_{PU}$: methanol extract of *A. squamosa* pulp
$HAEAS_L$: hydroalcoholic extract of *A. squamosa* leaf
$HAEAS_L$: hot water extract of *A. squamosa* leaf
$PEEAS_{PE}$: petroleum ether extract of *A. squamosa* peel
$EAEAS_{PE}$: ethyl acetate extract of *A. squamosa* peel
$EEAS_{PE}$: alcoholic extract of *A. squamosa* peel
$EAEAS_{PU}$: ethyl acetate extract of *A. squamosa* pulp
NIDDM: non-insulin-dependent diabetes mellitus
OGTT: oral glucose tolerance test
ACGs: annonaceous acetogenins
α-MSH: α-melanocyte-stimulating hormone
TPC: total phenolic count
GAE: gallic acid equivalent
DPPH: 1,1-diphenyl-2-picrylhydrazyl
FRAP: ferric-reducing antioxidant power assay
SORS: superoxide radical scavenging assay
NORS: nitric oxide radical scavenging assay
TBARS: thiobarbituric acid reactive substance
TC: total cholesterol
HDL: High density lipoprotein
LDL: low density lipoprotein
TG: triglycerides
SOD: superoxide dismutase
CAT: catalase
GSH: reduced glutathione
ALP: alkaline phosphatase
AST: aspartate aminotransferase
ALT: alanine aminotransferase

DLA:	Dalton's lymphoma ascites
EAC:	Ehrlich ascites carcinoma
MIC:	minimal inhibitory concentration
MBC	minimal bactericidal concentration
TRP1:	tyrosinase related protein-1
TRP2:	tyrosinase related protein-2
MITF:	melanocyte inducing transcription factor
ERK:	extracellular signal-regulated kinase
JNK:	c-Jun N-terminal kinase
MDA:	*malondialdehyde*
NOx:	NADPH Oxidase
TFC:	total flavonoids content
GAE:	gallic acid equivalent
FBG:	fasting blood glucose
PPG:	postprandial glucose levels
HDL-C:	high-density lipoprotein-C
Hb:	total haemoglobin
HbA_1c:	glycosylated haemoglobin

21.1 INTRODUCTION

Diabetes mellitus is one of the common endocrine and metabolic disorders acquiring around 2.8% of the world's population and are anticipated to cross 5.4% by the year 2025 (Patel et al., 2011). This disease has caused significant morbidity and mortality due to complications of microvascular (retinopathy, neuropathy, and nephropathy) and macrovascular (heart attack, stroke, and peripheral vascular disease) networks. Currently, available therapies for diabetes include insulin and various oral antidiabetic agents such as sulfonylureas, biguanides, and glinides. Many have several serious adverse effects; therefore, searching for more effective and safer hypoglycaemic agents is one of the critical areas of investigation (Feingold, 2000).

For a long time, herbal medicines have been a highly esteemed source of medicine to help humans sustain health. Over the past century, plant-based natural products (i.e. phytochemicals and active constituents) have become a growing part of modern, high-tech medicine and have played a pivotal role in pharmaceutical discovery (Moghadamtousi et al., 2013). Plants from the family Annonaceae, which comprises 135 genera and 2300 species (Srivastava et al., 2012), have been traditionally used to treat various ailments worldwide. The plants have also been studied extensively for their medicinal values, including antidiabetic and hypoglycaemic effects.

Annona squamosa L. is one of the plants from the family Annonaceae that have been extensively used in traditional medicine to treat various ailments. *A. squamosa* is a shrub widely cultivated in the tropical and subtropical regions of the world and is thought to have originated in the Caribbean and/or in other parts of tropical America. The plant is considered an endemic species of the West Indies, South and Central America, Ecuador, Peru, Brazil, India, Mexico, Bahamas, Bermuda, Australia, Indomalaya, and Egypt, to name a few (Anaya-Esparza et al., 2020). The species is widely grown as a commercial fruit tree both within its native range and in tropical regions worldwide. Due to its wild distribution, *A. squamosa* is known by different common names (Table 21.1) (Morton, 1987). This review discusses current findings related to the antidiabetic potential of *A. squamosa*.

21.2 BOTANICAL DESCRIPTION

Annona squamosa is a well-branched shrub reaching 3–8 m in height with a broad, open crown or irregularly spreading branches and branched tap root (Figure 21.1A). This evergreen tree with thin, grey bark is a semi-deciduous tree with an oblong-lanceolate or lanceolate leaf. Dormancy is induced by fluctuations in rainfall and temperature/light. But not all trees respond simultaneously. In addition, the leaves, which are 6–17 cm long with 2–5 cm wide, are green, sharp or blunt at the tip, round or widely wedge-shaped

TABLE 21.1
Some Common Names of *A. squamosa* Based on Country

No.	Country	Common name
1	Australia	custard apple; sweet apple; sweet sop
2	Bolivia	cherimoya
3	Brazil	ata; fruta do conde; pinha do conde
4	Cambodia	tiep baay; tiep srôk
5	Cook Islands	katara apa Maori; katara'apa Maori; naponapo; tapotapo; tapotapo Maori
6	India	aatoa; amritaphala; ata; athichakku; luna; seethapalam; seethaphala; sharifa; shariffa; sitaphal
7	Indonesia	sirkaja; atis sarikaja
8	Malaysia	buah nona; nona sri kaya; sri kaya
9	Pakistan	sharifa; sitaphal
10	Philippines	ates; atis; atit; atti; yates
11	Thailand	lanang; makkhiap; noina
12	Vietnam	mang cầu ta; na

FIGURE 21.1 Various parts of *A. squamosa*: (A) tree; (B) leaves; (C) flowers; (D) fruits; and (E) seeds. (All images were obtained from www.google.com.)

at the base, and arranged alternately on short petioles (Figure 21.1B). Young leaves are hairy, and solitary and clustered crystals occur in epidermal cells. Furthermore, the flowers are bisexual, yellowish green, fleshy, drooping, extra-axillary, bracteate, pedicellate, actinomorphic, protogynous, spirocyclic, more on a leafy shoot than on the older wood, and tending to open as the shoot elongates (Figure 21.1C).

Moreover, the flowers are also fragrant and hang downward either singly or in clusters of two or four on each flower, which is about 2–4 cm long, contains three degenerated sepals and six petals, and the outer petals are oblong with greenish and purplish colours at the base. Inner petals reduce to minute scales or are absent. The six petals are organized into two whorls with three petals each, and the petals of the inconspicuous inner whorl degenerate into small scales or completely disappear. The multiple pistils grow on the conical receptacle, in the centre of the flower with several at the periphery. Flowering occurs during spring to early summer, and flowering occurs throughout the year in permanently humid regions. Inflorescences are supra-axillary.

On the other hand, the round, heart-shaped fruits can be 5–10 cm in diameter, ovate or conical, and contain many rounded protuberances (Figure 21.1D). The fruits also contain edible white pulp that is juicy and has a sweetly aromatic taste, which occurs in elongated segments – each segment surrounding a dark brown to blackish, oblong seed. The rind is thick with knobby segments but will turn soft and crack open, releasing a sweet aroma when it ripens. Trees start to bear fruit when they reach the age of 3–4 years old.

The seeds, which are blackish or dark brown, range in length from 1.3–1.6 cm, with an oblong shape that is smooth and shiny (Figure 21.1E). Each carpel holds the seed with a pale swelling at the hilum or albumen filled with numerous transverse, brown lines of clefts. There are 20–40 seeds in an average fruit (Dinesh Kumar et al., 2005; Chen et al., 2011; Ma et al., 2017).

21.3 DISTRIBUTION

Although the original home of *A. squamosa* is unknown, it is believed to be a native semi-deciduous tree in tropical America and the West Indies since it is widely distributed in these regions. According to Tien et al. (2005), the plant is nowadays cultivated in almost all tropical and subtropical countries. Several historical facts worth mentioning include the trees' introduction to southern India by the Portuguese and to the Philippines by the Spanish before 1590 (Morton et al., 1987). Moreover, the trees were also planted as fruit trees in Puerto Rico in 1626, which then spread from cultivated areas to roadsides and valleys (Chen et al., 2011).

21.4 PHYTOCHEMICAL CONSTITUENTS

The therapeutic properties of medicinal plants are possibly due to the presence of various phytoconstituents. Extensive approaches have been taken to evaluate the phytoconstituents of different parts of *A. squamosa*. Reports on the isolation and identification of bioactive compounds from other parts of *A. squamosa* have been published as early as 1972, and until 2016, approximately 153 compounds were identified (Ma et al., 2017), which consisted of 33 diterpenes, 19 alkaloids, 88 annonaceous acetogenins (ACGs), and 13 cyclopeptides from different parts of the plant.

Recent literature search on the progress of research related to *A. squamosa* phytoconstituents has revealed several new findings, as illustrated in Tables 21.2A and 21.2B.

21.5 PHARMACOLOGICAL STUDIES

Various pharmacological investigations have been carried out on *A. squamosa*, and most of the activities have been reported by Ma et al. (2017). Among them, antithyroid, vasorelaxant, antinociceptive, anti-inflammatory, wound healing, antiulcer, antihyperthyroidism, antifertility, vasorelaxant, cardiotonic, and immune modifier activities did not show any additional reports based on the literature search. The following pharmacological activities were reported, taking into account only papers published between 2017 and 2022.

TABLE 21.2A
Bioactive Compounds Isolated from the Fruits of *A. squamosa*

No.	Part of plant	Bioactive compounds	Reference
1.	Fruits	3,4-dihydroxybenzoic acid	De Moraes et al. (2020)
2.		caffeic acid	
3.		catechin	
4.		chlorogenic acid	
5.		epicatechin	
6.		gallic acid	
7.		p-coumaric acid	
8.		quercetin	
9.		rutin	
10.		ferulic acid	

TABLE 21.2B
Bioactive Compounds Isolated from the Fruits and Pericarp of *A. squamosa*

No.	Part of plant	Bioactive compounds	Reference
1.	Pericarps	4α-hydroxy-17,19-*dinor-ent*-kaurane-16-one	Chen et al. (2020)
2.		4β-hydroxy-16β-H-18-*nor-ent*-kaurane-17-oic acid	
3.		4β,17-dihydroxy-16α-acetoxy-18-nor-*ent*-kaurane	
4.		Annosquamosin Z	
5.		16α-H-*ent*-kaurane-17,18-dioic acid, 17-methy ester	

21.5.1 Antimicrobial Activity

The methanol ($MEAS_L$), but not aqueous ($AEAS_L$), extract of *A. squamosa* leaf exerts significant antimicrobial activity against both gram-positive (i.e. *Bacillus cereus* [medical isolate], *Staphylococcus aureus* [MTCC 1144]) and gram-negative (i.e. *Shigella sonnei*, *Shigella dysentriae*, and *Vibrio cholerae* MTCC 3904) bacteria with the recorded zone of inhibition ranging between 12 and 16 mm when assessed using the disc diffusion method. Further analysis revealed that $MEAS_L$ produced minimum inhibitory concentration ≤ 1 mg/mL only against *S. aureus*, *S. sonnei*, and *V. cholerae* (Aftabuddin et al., 2017).

Al-Nemari et al. (2020) investigated the antibacterial potential of $MEAS_L$, $AEAS_L$, and acetone extract of *A. squamosa* leaf ($AcEAS_L$) against several gram-positive (*Bacillus subtilis*, S. *aureus*, and *Enterococcus faecalis*) and gram-negative (*Escherichia coli*, *Psuedomonas aeruginosa*, *Klebsiella pneumoniae*, and *Salmonella typhimurium* LT2) bacteria using the agar well diffusion assay. At 50 mg/mL, $AcEAS_L$ inhibited the growth of both gram-positive and gram-negative bacterial strains, except *S. aureus*. $AcEAS_L$ was the most effective against *P. aeruginosa* (zone of inhibition = 24.6 mm), whereas the rest of the bacteria recorded the zone of inhibition ranging between 8.0 and 12.6 mm. $MEAS_L$, on the other hand, inhibited the growth of all the tested bacteria with the most effective inhibition recorded against *P. aeruginosa*, *S. aureus*, *B. subtitlis*, and *E. coli* (zone of inhibition ranged between 15.5 and 18.3 mm) followed by *S. typhimurium*, *K. pneumonia*, and *E. faecalis* (zone of inhibition ranged between 7.9 and 12.3 mm). Moreover, it was the only extract with marked antibacterial activity against *S. aureus*, which was more efficient than the standard antibiotic, tetracycline (zone of inhibition = 14.8 mm). Except for *B. subtilis* and *S. aureus*, $AEAS_L$ was more sensitive to gram-negative bacteria (zone of inhibition ranged between 10.5 and 17.0 mm) than gram-positive bacteria (only against *E. faecalis* with zone of inhibition of 9.5 mm). Tetracycline inhibited the growth of all bacteria with the zone of inhibition in the range of 12.7–16.2 mm and 14.3–19.3 mm against gram-positive and gram-negative bacteria, respectively.

Pinto et al. (2017) also determined the antibacterial activity of $MEAS_L$ and methanol extract of *A. squamosa* seeds ($MEAS_S$) by measuring the minimal inhibitory concentration (MIC) and minimal bactericidal concentration (MBC) using the broth microdilution techniques. The extracts were tested against a panel of microorganisms (i.e. *S. aureus* [ATCC 6538], *Enterobacter aerogenes* [ATCC 13048], *Enterobacter cloacae* [ATCC 23355], *K. pneumoniae* [ATCC 4552], *P. aeruginosa* [ATCC 9027], *B. cereus* [ATCC 14579], *E. faecalis* [ATCC 19433], *Shigella dysenteriae* [ATCC 13313], *E. coli* [ATCC 10536], *S. typhimurium* [ATCC 13311], and *Salmonella choleraesuis* [ATCC 10708]). Effects of $MEAS_L$ on established biofilms, time-kill kinetics, outer membrane permeability, and nucleotide leakage were also studied. From the MIC study, $MEAS_L$ exerted effective antibacterial activity only against *S. aureus*, *K. pneumonia*, and *E. faecalis* with MIC value of 78, 78, and 39 μg/mL, respectively. Generally, $MEAS_L$ recorded the MIC value in the range of 39–1250 μg/mL, whereas $MEAS_S$ recorded the MIC value in the range of 313–5000 μg/mL. Moreover, only $MEAS_L$ produced antibacterial activity against *S. aureus* and *E. faecalis* with the recorded MBC value of less than 100 μg/mL. $MEAS_L$ was also found to exert significant biofilm disruption, rapid time-dependent kinetics of bacterial killing, increased membrane permeability, and reduction in the cell numbers and viability when tested against the selected bacteria (*S. aureus*, *K. pneumoniae*, and *E. faecalis*).

21.5.2 Antiviral Activity

The hydroalcoholic extract of *A. squamosa* seeds ($HAEAS_S$), which was relatively nontoxic to TZM-bl cells at a CC_{50} value of 51.0 μg/ml, exerts preliminary inhibitory action against HIV-1_{IIIB} and HIV-1_{Ada5} laboratory-adapted strains only in the cell-associated assays with the recorded IC_{80} value of 27 and 20 μg/ml (Palshetkar et al., 2020). Further screening for anti-HIV in both cell-free and cell-associated assays performed against primary isolates HIV-1_{UG070} and HIV-1_{VB59} in TZM-bl shows that $HAEAS_S$ possessed a very good activity with lowest estimated IC_{80} of 25–27 μg/ml. On the other hand, similar screening against the same isolates in PM1 cell lines revealed that $HAEAS_S$ also exerts good activity at lower IC_{80} concentration of 0.8 μg/ml only in cell-associated assay.

21.5.3 Antipsoriatic Activity

The petroleum ether extract of *A. squamosa* seeds ($PEEAS_S$) exerts *in vitro* and *in vivo* antipsoriatic activity when assessed using the sulforhodamine B colorimetric assay in HaCaT cell lines and the oxazolone-induced ear psoriasis in Balb-C mice model. $PEEAS_S$ effectively inhibited the growth of HaCaT cells and markedly reduced the erythema and oedema formation in the mice model of psoriasis (Bhoir et al., 2019).

21.5.4 Antimelanogenesis Activity

Ko et al. (2020) reported that $MEAS_L$ (0.625 to 5 ng/mL) had no significant cytotoxic effects on B16F10 and human dermal fibroblasts cells. Additionally, no cytotoxic effect was observed in the latter at higher concentration 10 ng/mL. $MEAS_L$ (1.25, 2.5, and 5.0 ng/ml) treatment of α-MSH-stimulated B16F10 cells reduced the α-melanocyte-stimulating hormone (α-MSH)-induced over-production of melanin (91.0%, 76.8%, and 70.4%) and decreased intracellular tyrosinase activity

(85.1%, 79.6%, and 70.3%) in a dose-dependent manner, thus suggesting the extract to possess anti-melanogenesis and anti-tyrosinase activities. With regards to mechanisms of antimelanogenesis activity, $MEAS_L$ co-treatment with α-MSH significantly suppressed the expression of tyrosinase, TRP2, and α-MSH-induced MITF, TRP1, and TRP2 regulation, but caused only a slight decrease in TRP1 expression. $MEAS_L$ was also found to significantly increase the phosphorylation of p38 dose-dependently without altering the expression of ERK and JNK. Furthermore, GC-MS analysis of $MEAS_L$ identified major phytoconstituents such as ent-kaur-16-en-19-ol, 18-oxokauran-17-yl acetate, and β-sitosterol.

21.5.5 Antiparasitic Activity

Castañeda-Ramírez et al. (2019) reported on the *in vitro* anthelmintic activity of $MEAS_L$ and acetone:water extract of *A. squamosa* leaf ($AWEAS_L$) against *Haemonchus contortus* eggs. $MEAS_L$ caused the death of eggs mainly at the morula stage (ovicidal activity), whereas the $AWEAS_L$ caused egg-hatching failure of developed larvae (larvae failing eclosion activity). The EC_{50} value for $MEAS_L$ and $AWEAS_L$ against *H. contortus* eggs was approximately 870.0 and 1042.0 μg/ml, respectively.

21.5.6 Duretic and Antiurolithiatic Activity

Korah et al. (2020) reported on the diuretic and antiurolithiatic activities of ethanol extract of *A. squamosa* leaf ($EEAS_L$), at the concentrations of 250 and 500 mg/kg, when assessed using the rats' Lipschitz or ethylene glycol-induced urolithiasis model, respectively. The highest dose of $EEAS_L$ demonstrated both activities that were comparable to standard controls (furosemide and cystone). $EEAS_L$ affected several diuretic parameters (i.e. urine volume and urine pH, the concentration of sodium and potassium, and diuretic index, natriuretic index, and Lipschitz value) and several urolithiatic parameters (i.e. urine volume, urine pH, body weight, and biochemical parameters such as calcium, urea, uric acid, and creatine both from serum and urine).

21.5.7 Antinephritic Activity

Abd-Elrazek et al. (2021) reported on the protective effect of 95% ethanol extract of the seeds of *A. squamosa* ($95EEAS_S$) against ifosfamide-induced renal toxicity in rats. Ifosfamide (intraperitoneal; 80 mg/kg; 5 days) caused significant (i) elevation of creatinine, sodium, magnesium, and urea, along with depleted serum potassium and albumin levels; (ii) increases in renal malondialdehyde (MDA), NADPH oxide (NOx), and upregulated iNOS; (iii) increase mRNA and protein expression of NF-kB; and (iv) immunohistological expression of caspase-3 as well as Bax in kidney tissues, while it significantly (v) reduced renal glutathione (GSH) and (vi) downregulated mRNA and protein expression of Bcl-2. Pre-treatment with 50 mg/kg $95EEAS_S$ orally for two weeks reversed the toxic effect of ifosfamide by causing downregulation of iNOS and NF-kB expressions, downregulation of caspase-3 and Bax immunohistological expressions, and upregulation of GSH and Bcl-2 expression.

21.5.8 Anti-Hepatoma Activity

Chen et al. (2017) reported on the anti-hepatoma activity of petroleum ether ($PEEAS_P$), n-butanol ($BEAS_P$), and ethyl acetate ($EAEAS_P$) of *A. squamosa* pericarps against SMMC-7721 cells and verified using H22 xenografts bearing mice. $EAEAS_P$ (IC_{50} = 7.46 μg/mL) demonstrated the most effective *in vitro* anti-hepatoma activity in comparison to $PEEAS_P$ (IC_{50} = 15.96 μg/mL) and $BEAS_P$ (IC_{50} = 31.98 μg/mL). In the *in vivo* assay, only $EAEAS_P$ (5 and 50 mg/kg) reduced tumour weight (0.36 and 0.27 g, respectively) in comparison to the control untreated mice (0.66 g). At 50 mg/kg,

$EAEAS_P$ caused more than 50% tumour inhibition rate (58.5%). Ent-kauran-16-en-19-oic acid and ent-kauran-15-en-19-oic acid was isolated from $EAEAS_P$ and demonstrated significant inhibition of SMMC7721 cells (IC_{50} = 67.4 and 45.7μM, respectively) and triggered apoptosis in a dose-dependent manner by inducing G0/G1 phase arrest. In addition, ent-kauran-16-en-19-oic acid activated caspase-3, -8 and -9 in a dose-dependent manner, and also increased the protein level of Bax and decreased the level of Bcl-2 expression, thus suggesting to induce apoptosis via a mitochondria-mediated pathway.

In another study, Chen et al. (2016) reported on the *in vitro* anti-hepatoma potential of the essential oil of *A. squamosa* pericarps ($EOAS_P$) when assessed using the SMMC-7721 hepatoma cell line. Interestingly, at the concentration of 3.1 μg/mL, $EOAS_P$ caused more than 40% (43.3%) inhibition of cell viability with anti-hepatoma activity observed at IC_{50} = 54.8 μg/ml. $EOAS_P$ might act possibly by causing nucleus shrinkage or breaking as seen through fluorescent microscope and through modulation of the pro-apoptosis and cell cycle arrest effects as confirmed by flow cytometry analysis. (−)-S pathulenol (32.51%) was the main compound in the oil based on the GC-MS analysis.

Later, Chen et al. (2020) reported on the *in vitro* anti-hepatoma activity of the oil obtained from *A. squamosa* pericarps via the response surface model against SMMC-7721 cell line in *vitro* (IC_{50} = 16.0 μg/ml) and H_{22} cell line *in vivo* (50 mg/kg = 54.0% tumour inhibition rate). Western blot analysis revealed that the oil triggered apoptosis via upregulation of Bax, downregulation of Bcl-2, and cleavage of caspase-3, -8, and -9 proteins. Overall, the mechanisms of apoptosis involve modulation of Bcl-2/Bax/caspase-9/caspase-3 and Fas/FasL/caspase-8/caspase-3 pathways.

21.5.9 Hepatoprotective Activity

Zahid et al. (2020) reported on the hepatoprotective activity of ethanol extract of *A. squamosa* seeds ($EEAS_S$) (200 and 400 mg/kg; orally) when assessed using the 50% alcohol-induced rat's liver toxicity model. $EEAS_S$ dose-dependently prevented the alcohol-induced increase in serum levels of hepatic enzymes (i.e. alanine aminotransferase [ALT], aspartate aminotransferase [AST], and alkaline phosphatase [ALP]) and reduced the level of MDA, but significantly increased the levels of superoxide dismutase (SOD), catalase (CAT), and GSH. Histopathologically, it is suggested that the extract diminished the hepatocellular necrosis leading to regeneration and repair of cells toward normal. Overall, the hepatoprotective and antioxidant potentials of the extract contribute to the protective effect observed against alcohol-induced liver injury.

21.5.10 Anti-Tumour Activity

Shehata et al. (2021) reported on the non-cytotoxic property of the aqueous extract of the seed ($AEAS_S$), peel ($AEAS_{PE}$), and pulp ($AEAS_{PU}$) of *A. squamosa* (Balady) cultivar against normal human lung fibroblast WI-38 cells indicated by the high EC_{100} value of 654.0, 551.4, and 552.5 μg/ml, respectively. All extracts also exerted *in vitro* anti-tumour activity against Caco-2, PC-3, HepG-2, and MCF-7 cell lines with $AEAS_S$ being the most effective extract (IC_{50} = 7.3–14.3 μg/ml) compared to $AEAS_{PE}$ (IC_{50} = 14.4–30.7 μg/ml) and $AEAS_{PU}$ (IC_{50} = 22.7–45.1 μg/ml). $AEAS_S$ induced apoptosis by causing a collapse in spindle shape with cellular shrinking with the highest total apoptosis percentages recorded for PC-3 (51.3%) followed by HepG-2 (50.1%), MCF-7 (49.3%), and Caco-2 (48.8%). $AEAS_S$ also downregulated mRNA expression of Bcl-2 and upregulated p53 in all treated cells while reducing the expression of proliferation marker (Ki-67) in HepG-2 cell.

Al-Nemari et al. (2020) also reported on the anti-tumour potential of $MEAS_L$, $AcEAS_L$, and $AEAS_L$ against Lovo and HCT-116 cell lines. The extracts (1–100 μg/mL) affect the number but not the cell morphology of those cancer cells, and at 100 μg/ml, inhibited the proliferation of Lovo cell (> 80%) and HCT-116 cell (94, 91, and 58%, respectively, possibly by causing dose-dependent notable damage to the nucleus and inducing cytotoxicity, which is more apparent in HCT-116 cells, in a concentration-dependent manner in both cells.

21.5.11 Antioxidant Activity

$PEEAS_S$, chloroform ($CEAS_S$), ethyl acetate ($EAEAS_S$), and methanol ($MEAS_S$) extracts from *A. squamosa* seeds possessed significant free radical scavenging power when assessed using the 1,1-diphenyl-2-picrylhydrazyl (DPPH) radical scavenging effect, evaluation of ABTS cation decolourization capacity (ABTS), ferric-reducing antioxidant power (FRAP) assay, hydroxyl radical scavenging assay, superoxide radical scavenging (SORS) assay, and nitric oxide radical scavenging (NORS) assay (Vikas et al., 2017). The total phenolic count (TPC) value recorded for each extract was 0.64, 0.54, 0.49, and 0.57 mg gallic acid equivalent (GAE)/g dry material, respectively. At 800 µg/ml, each extract produced (i) the EC_{50} value of 34.4, 43.8, 34.7, and 28.8 µg/ml, respectively, in DPPH assay; (ii) 63.0%–70.0% inhibition in SORS assay; (iii) 73.3–80.0 µg/ml concentration in ABTS assay; (iv) 62.5%–73.3% inhibition in H_2O_2 assay; and (v) 66.8%–91.0% inhibition in NORS assay.

Chatatikun and Chiabchalard (2017) reported that 100 µg/ml of petroleum ether ($PEEAS_L$) and dichloromethane ($DEAS_L$) extracts of *A. squamosa* leaf, together with $EEAS_L$, produced the percentage of scavenging activity of 29.9, 36.3, and 99.1, respectively, in the ABTS assay with only $DEAS_L$ (17.9%) and EEASL (94.0%) showing scavenging activity in the DPPH assay. Each extract (1.0 mg/ml) produced TPC and total flavonoids content (TFC) values of 4.1, 9.3, and 62.7 mg GAE/g dry material and 8.9, 9.7, and 13.0 mg quercetin equivalent (QE)/g dry material, respectively, with only $EEAS_L$ exerted anti-tyrosinase activity (21.9±1.5% tyrosine inhibition) and anti-collagenase activity (55.1±3.2% collagenase inhibition).

When assessed using the DPPH assay, $95EEAS_S$ as well as its n-hexane fraction ($HFAS_S$) (100–1000 µg/ml) exerted a concentration-dependent antioxidant activity (Zahid et al., 2018a). At 1,000 µg/mL, $95EEAS_S$ and $HFAS_S$ exerted approximately 66.0% and 92.0% antioxidant activity with the recorded IC_{50} value of approximately 193.0 and 180.0 µg/ml, respectively. GC-MS analysis identified n-hexadecanoic acid (10.1%), heptadecene-(8)-carbonic acid-(1) (29.7%), y-sitosterol (8.3%), and stigmasta-5,22-dien-3-ol (5.9%) as the major compounds in $95EEAS_S$, while in $HFAS_S$, the major compounds identified were n-hexadecanoic acid (14.42%), 9-octadecenoic acid (Z)-, 2,3-dihydroxypropyl ester (13.33%), y-sitosterol (10.87%), and cis-vaccenic acid (10.39%).

Silva and Sirasa (2018) reported that the methanol extract of *A. squamosa* fruits ($MEAS_F$) exhibited a free radical scavenging activity against DPPH assay (27.4 mg ascorbic acid equivalent antioxidant capacity (AEAC)/g) and FRAP assay (46.8 µmol $FeSO_4$/g) and produced the recorded TPC and TFC values of 747.0 mg gallic acid equivalents (GAE)/100 g and 170.6 mg catechin equivalents (CE)/100 g, respectively.

$MEAS_L$, $AcEAS_L$, and $AEAS_L$ were reported to exert antioxidant activity when assessed using the DPPH, H_2O_2, and NORS assays (Al-Nemari et al., 2020). In DPPH assay, $AcEAS_L$ (10–100 µg/mL) exhibited the highest activity followed by $MEAS_L$ and $AEAS_L$ with the recorded IC_{50} value of 33.9, 51.0, and 98.3 µg/ml; in H_2O_2 assay, $AEAS_L$ (0.1–1.0 µg/mL) exerted the highest activity followed by $MEAS_L$ and $AcEAS_L$ with the recorded IC_{50} value of 110.0, 516.7, and 735.0 µg/mL; and in NORS assay, $MEAS_L$ (10–100 µg/mL) showed the highest activity followed by $AcEAS_L$ and $AEAS_L$ with the recorded IC_{50} value of 12.0, 44.0, and 81.0 µg/mL, respectively. At 0.75 µg/ml, all extracts exerted reducing power with the highest to lowest reducing power based on the recorded absorbance values was listed as follows: $AEAS_L$ (0.984) > $MEAS_L$ (0.975) > $AcEAS_L$ (0.95). In addition, the TPC and TFC values recorded for $MEAS_L$, $AcEAS_L$, and $AEAS_L$ (80 µg/mL) were 282.1, 256.3, and 16.9 mg GAE/g, and 1.8, 7.1, and 0. 1 mg quercetin/g, respectively. GC-MS analysis of $MEAS_L$ revealed the presence of germacrene D (22.0%), palmitone (16.9%), and trans-caryophyllene (12.1%).

Leite et al. (2021) reported that $MEAS_{PU}$ has better activity in comparison to methanol extract of *A. squamosa* seeds ($MEAS_S$) when measured using the DPPH, ABTS, FRAP, oxidative degradation of 2-deoxyribose (2-DR) protection, and co-oxidation of β-carotene/linoleic acid method (β-carotene) assays with the recorded IC_{50} value of 0.14–0.57 mg/ml for $MEAS_S$ and 0.38–1.36

mg/ml for $MEAS_{PU}$, respectively. Moreover, $MEAS_S$ have the highest TPC and TFC values (32.5 µg GAE/mg Ext and 893.3 µg QE/g Ext) than $MEAS_{PU}$ (2.2 µg GAE/mg Ext and 246.6 µg QE/g Ext), respectively. Additionally, $MEAS_{PU}$ (6.25–200.0 µg/ml; IC_{50} = 18.8 µg/ml) exerted better anti-acetylcholine esterase activity than did $MEAS_S$ (IC_{50} = 22.3 µg/ml).

21.5.12 Cytotoxicity Activity

Pinto et al. (2017) reported that $MEAS_L$ and $MEAS_S$ (non-serially diluted from 50 to 1.5 µg/mL) exerted antiproliferative activity against Jurkat, HL60, MCF-7, and HCT-116 cell lines by increasing DNA fragmentation in all cells, indicating the involvement of apoptosis pathway with $MEAS_S$ being more active (IC_{50} value of 1.1, 2.1, and 5.6 µg/mL) compared to $MEAS_L$ (IC_{50} value of 4.2, 6.4, 11.0, and 15.2 µg/mL), respectively. Both extracts were safe as they exhibited lower activity against Vero cells (IC_{50} value higher than 30.0 µg/ml). UPLC-ESI-MS/MS analysis identified the presence of anonaine, asimilobine, corypalmine, liriodenine, nornuciferine, and reticuline.

At 25–800 µg/ml, $PEEAS_S$, $CEAS_S$, $EAEAS_S$, and $MEAS_S$ exerted cytotoxic effect at 12 h against the Dalton's lymphoma ascites (DLA) and K-562 cells in the tryphan blue assay (Vikas et al., 2019) with $PEEAS_S$ (800 µg/ml) followed by $MEAS_S$ exerted maximum cytotoxicity of 57.0% and 50.0% against DLA and 60.8% and 51.3% against K-562 cells, respectively. At 24 h, all extracts (400 µg/ml) produced cytotoxic effect against DLA, K-562, KB, A-549, MCF-7, K-562, and Ehrlich ascites carcinoma (EAC) cells in the MTT assay with $PEEAS_S$ still producing the highest percentage of cytotoxic effect (KB [68.0%], A-549 [62.0%], MCF-7 [61.0%], K-562 [59.8%], DLA [60.0%], and EAC [62.0%]). $PEEAS_S$ (400 µg/ml) treated for 12, 24, and 48 h produced percentage of cytotoxic effect of 63.0%–76.0% against KB, 56.3%–64.5% against A-549, 53.5%–60.5% against MCF-7, 53.8%–63.8% against K-562, 56.5%–62.3% against DLA cells, and 38.0%–64.0% against EAC cells. The IC_{50} value calculated at 24 h against these cells was 22.4, 154.0, 136.0, 90.0, 149.0, and 98.0 µg/ml, respectively.

21.6 ANTIDIABETIC RESPONSE

Alcoholic extract of *A. squamosa* leaves ($EEAS_L$) exerted antidiabetic activity when assessed using the streptozotocin (STZ)-induced non-insulin-dependent diabetes mellitus (NIDDM) model (Shirwaikar et al., 2004). At 500 mg/kg, $EEAS_L$ reduced plasma glucose level (66.33 mg/dl) after 120 min compared to normal control group (97.83 mg/dl) in the oral glucose tolerance test (OGTT) and significantly lowered fasting blood glucose (FBG) level in STZ-induced diabetic groups from 187.93 mg/dl at Day 5 and 202.74 mg/dl and Day 12 to 134.53 and 119.51 mg/dl, respectively. Serum cholesterol (TC) level was reduced after treatment with 250 and 500 mg/kg extract with the latter also reduced the triglycerides (TG) level when compared to diabetic control group. $EEAS_L$, at both concentrations, also improved the level of liver glycogen (in the range of 10.0–11.60 mg/g) and pancreatic thiobarbituric acid reactive substance (TBARS) (range of 2.10–2.30 mg/g) in comparison to the diabetic control rats (range of 10.0–11.56 mg/g), but did not affect the serum insulin level.

$EEAS_L$ (350mg/kg) significantly reduced the FBG level by 26.8% at 0 h, 38.5% at 1 h and 40.6% at 2 h, and OGTT level by 37.2% and 60.6% during 1 and 2 h of OGTT observation in STZ-induced diabetic rats and alloxan-induced diabetic rabbits (Gupta et al., 2005a). The extract also reduced the level of TC, TG, and low-density lipoprotein (LDL), but improved the high-density lipoprotein (HDL) level in the treated diabetic rats.

Hydroalcoholic extract of *A. squamosa* leaf ($HAEAS_L$) (350 and 700 mg/kg) exerted antidiabetic effect against alloxan-induced diabetic rats by reducing the FBG level by 30.9% and 40.6%, respectively, compared to diabetic control group (278.00 mg/dl), decreasing the level of serum TC and TG while improving the level of serum HDL (Tomar and Sisodia, 2014).

Rabintossaporn et al. (2009) explored the antidiabetic activity of ethanolic extract of *A. squamosa* young leaves ($EEAS_{YL}$) (125–500 mg/kg) against STZ-induced diabetic rats. The extract, at

250 and 500 mg/kg, significantly reduced fasting blood glucose level 60 min onwards. However, post-withdrawal of the extract for three consecutive days led to the blood glucose levels at both doses returned to the earlier levels before the diabetic rats were treated with the extract. Diabetic rats that received 250 and 500 mg/kg of $EEAS_{YL}$ showed significant decrease in blood glucose levels reaching the pre-glucose load level at 150 and 180 minutes when compared to diabetic rats receiving vehicle (distilled water). Histological examination of the pancreas of diabetic rats showed the irregular shape of islets of Langerhans and pretreatment with $EEAS_{YL}$ dose-dependently reversed the shape of pancreas towards rounder islets of Langerhans.

Mittal et al. (2010) also reported on the $EEAS_L$ potential to exert antidiabetic activity against alloxan-induced diabetic rats. $EEAS_L$ (100–300 mg/kg) caused a dose-dependent reduction in FBG after Day 20 (134.4–156.2 mg/dl) and 30 (118.6–148.7 mg/dl) of treatment in comparison to the diabetic control (241.0 mg/dl) and positive control (97.05 mg/dl) rats. At the same time, $EEAS_L$ caused a significant and dose-dependent reduction in the level of TC, TG, and blood urea in comparison to the diabetic control rats.

Basha and Subramanian (2011) further reported on the antidiabetic potential of $EEAS_L$ (100 mg/kg) against STZ-induced diabetic rats. The extract significantly reduced the plasma glucose level of treated diabetic rats (91.54 ± 2.74 mg/dl) after daily treatment for 30 days in comparison to the diabetic control rats (292.72 ± 12.33 mg/dl). The extract also significantly decreased the plasma protein, urea, and creatinine level in treated diabetic rats in comparison to the diabetic control rats. Moreover, the extract at 100 mg/kg was also claimed to show no significant toxicity.

Sengupta et al. (2011) reported for the first time on the antidiabetic potential of $MEAS_L$. At the concentration of 250 mg/kg, the extract given orally against alloxan-induced hyperglycaemia in mice significantly reduced FBG level after 20 and 40 min by 68.5% and 42.6%, respectively, compared to the diabetic control group when measured using the OGTT. The antihyperglycaemic activity of $MEAS_L$ was observed following 60-, 90-, 120-, 240-, and 360-min post treatment with a reduction of blood glucose level by 17.44%, 21.8%, 23.88%, 39.51%, and 30.36%, respectively.

$MEAS_L$, $AEAS_L$, and hexane extract of *A. squamosa* leaves ($HEAS_L$), at 200 mg/kg, produced hypoglycaemic activity in STZ-induced diabetic rats after 2 h of oral treatment with $HEAS_L$ showed the highest potency (Ranjana and Tripathi, 2014). $HEAS_L$ followed by $AEAS_L$ and $MEAS_L$ lowered the blood glucose level by 14.5%, 10.5%, and 8.0% compared to the diabetic control group (295.2 mg/dl). Moreover, seven days of treatment with $HEAS_L$ (100 and 400 mg/kg) increased the blood insulin (11.6 and 16.3 μU/ml) and decreased the blood glucose level (41.2% and 78.0%) in comparison to diabetic control group (8.6 μU/ml and 386.9 mg/dl), respectively. $HEAS_L$ (100 and 400 mg/kg) reduced the peak glucose value at 30 min (GP_{30}) to about 49.9% and 38.3% in comparison to 368.1 mg/dl for diabetic control group when measured using OGTT, and reduced the α-glucosidase activity by 57.1% and 75.7% when measured in the small intestine of STZ-induced rats after seven days of treatment. Moreover, $HEAS_L$ (1.0–15.0 mg/ml) also inhibited α-glucosidase activity (IC_{50} = 9.2 mg/ml) in a concentration-dependent manner (12.4%–66.8%).

$AEAS_L$, $HEAS_L$, $MEAS_L$, and $PEEAS_L$ were earlier screened in 2-DG uptake bioassay whereby $HEAS_L$, at 10 μg/ml, was found to show no cytotoxic effect on differentiated L6 myotubes and stimulated glucose uptake in a dose-dependent manner (EC_{50} = 39.0 ng/ml) (Davis et al., 2012). $HEAS_L$ (10 and 100 ng/ml) *in vitro* antihyperglycaemic effect involved stimulation of insulin signalling events such as IR-β and IRS-1 phosphorylation and upregulation of GLUT4 mRNA expression while at 100 ng/ml $HEAS_L$ augmented PI3 kinase expression. Moreover, $HEAS_L$ also inhibited human PTP1B in a dose-dependent manner (IC_{50} = 17.4 μg/ml), inhibited DPP-IV even at 30 μg/ml but did not activate GPR40 or SIRT1. $HEAS_L$ reduced random glucose (27.7%) and plasma triglyceride (30.5%), but had no effect on total cholesterol level when compared to diabetic control group after 21 days treatment orally. $HEAS_L$ (500 mg/kg; given orally for 21 days) lowered blood glucose level in OGTT at all intervals (30, 60, and 120 min), but did not result in body weight gain when compared to the diabetic control group.

Following the OGTT test on $HAEAS_L$ (200, 300, 350, and 400 mg/kg), 350 mg/kg of $HAEAS_L$ was found to be the effective dose (Gupta et al., 2005b). In the 2 h treatment regime, $HAEAS_L$ reduced FBG levels in STZ-induced diabetic rats from 350 mg/dl after 1 h of treatment and 325 mg/dl after 2 h of treatment to 266 and 106 mg/dl, respectively. $HAEAS_L$ (350 mg/kg) also reduced FBG reading to 190.0 mg/dl after 1 h and 103.0 mg/dl after 2 h of treatment in comparison to the diabetic control group (301 and 173 mg/dl, respectively) in alloxan-induced diabetic rabbits. In the 10-day treatment regimen, $HAEAS_L$ still reduced the FBG (75%) and postprandial glucose levels (PPG) (51%) when compared to the initial value of 246 and 261 mg/dl, respectively. Moreover, $HAEAS_L$ (350 mg/kg) also reduced the urine sugar level to the undetected level at the end of the experiment. In diabetic rabbits, $HAEAS_L$ significantly reduced the urine sugar, FBG, and PPG levels, but significantly improved the serum lipid profile in diabetic rabbits with the recorded value of 125 mg/dl (TC), 56 mg/dl (LDL-C), and 110 mg/dl (TG) compared to diabetic rabbits (213, 157.4, and 148 mg/dl, respectively). $HAEAS_L$ also improved the level of high-density lipoprotein-C (HDL-C), total haemoglobin (Hb), and glycosylated haemoglobin (HbA_1c) levels by 46 mg/dl, 13.3 g/dl, and 7.4% when compared to diabetic control (26 mg/dl, 12.0 g/dl, and 10.6%, respectively). At 100 and 150 ug/kg, $HAEAS_L$ significantly inhibited glucose-6-phophatase activity by 32.2% and 34.3%, respectively.

Gupta et al. (2005c) reported that the $AEAS_{PU}$ (2.5, 5, and 10 g/kg) improved various nutritional indicators in alloxan-induced diabetic rabbits in comparison to the diabetic control group. Only the PPG, but not the FBG, of diabetic rabbits significantly lowered by up to 41.9% following treatment with $AEAS_{PU}$ with 5 g/kg exerting the highest activity. Other than that, $AEAS_{PU}$ also significantly reduced the level of TC (23.6%–32.5%) and TG (18%–36%), but caused significant increase in HDL-C level by 36.6% in diabetic group when treated with 5 g/kg $AEAS_{PU}$.

Additionally, $AEAS_L$ also significantly improved the serum lipid profile, namely TC, LDL-C, and TG, by reducing the value to 40.0%, 57.6%, and 41.7% when compared to the diabetic control group (168.2 mg/100 ml for TC, 78.7 mg/100 ml for LDL-C, and 109.2 mg/100 ml TG, respectively) (Kaleem et al., 2006). On the other hand, $AEAS_L$ significantly improved the level of HDL-C while lowering the levels of TBARS and hydroperoxides in both the liver and kidney of diabetic treated rats.

$AEAS_L$ (300 mg/kg) was reported to significantly decrease the level of blood glucose (97.5 mg/dl) while increasing the plasma insulin (11.9 μU/ml) and C-peptide (242.1 pmol/l) of STZ-induced diabetic rats, which almost reach the untreated control level (91.0 mg/dl, 16.0 μU/ml, and 260.2 μU/ml, respectively (Kaleem et al., 2008). Moreover, $AEAS_L$ significantly increased the level of protein, albumin, and A/G ratio, while significantly reducing the level of urea, uric acid, and creatinine compared to the diabetic control group. Furthermore, $AEAS_L$ also affected the activities of AST, ALT, ALP, and γ-GT in the plasma, liver, and kidney of diabetic control rats and caused their levels to return to near normal value as observed in the control untreated group.

Kaur et al. (2013) investigated the antidiabetic activity of $AEAS_L$ against STZ-induced diabetic rats. Results showed that $AEAS_L$, at the dose of 300 mg/kg, produced significant improvement in the blood glucose level by approximately 39% from the initial reading of 348.2 ± 2.85 mg/dl in the diabetic control group. In comparison, 5 mg/kg glipizide reduced the blood glucose level of diabetic rats by approximately 55%. In addition, no significant changes were observed to the weight of the treated experimental rats. It is important to highlight that the effect of a combination between $AEAS_L$ and glipizide as a polypharmacy is not further discussed to focus only *A. squamosa* potential as an antidiabetic agent.

Petroleum ether ($PEEAS_{PE}$), ethyl acetate ($EAEAS_{PE}$), and alcoholic ($EEAS_{PE}$) extracts of *A. squamosa* peel exerted antidiabetic activity against STZ-induced diabetic rats by significantly (i) improving the body weight of diabetic treated rats compared to the diabetic control rats with $EEAS_{PE}$, exhibiting a more pronounced effect on the 14th and 21st days of study; (ii) diminishing blood glucose level on Day 0, 7, 14, and 21, which was increased throughout the entire study period, with $EEAS_{PE}$ (at 250 mg/kg) producing maximum effect over $PEEAS_{PE}$ and $EEAS_{PE}$ (Sharma et al., 2013). In addition, all extracts significantly decreased the liver content of TC, TG, LDL, and VLDL levels, but significantly increased in HDL levels, and significantly decreased the ALP, SGOT, SGPT, and bilirubin levels when compared to the diabetic control rats. Again, $EEAS_{PE}$ was more effective than $PEEAS_{PE}$ or $EAEAS_{PE}$.

Wong et al. (2020) have reported that the recorded TPC value for 70% ethanol extract of *A. squamosa* leaf ($70EEAS_L$) was 199.62±7.40 mg GAE/g crude extract. Furthermore, $70EEAS_L$ was found to exert antioxidant potential when assessed using the DPPH and NORS assays with the recorded IC_{50} value of 5.00 ± 0.20 and 109.02 ± 3.18 µg/ml, respectively. The extract also exerted a potential to inhibit α-glucosidase activity with IC_{50} value of 3.59 ± 0.18 µg/ml.

Ansari et al. (2020), when studying the *in vitro* insulin-releasing effects of hot water extract of *A. squamosa* leaf ($HAEAS_L$) on BRIN-BD11 and isolated mouse islets cells, reported among others that $HAEAS_L$ (i) increased the insulin secretion from BRIN-BD11 cells in the presence of 5.6 mM and 16.7 mM glucose in a concentration-dependent manner (40–5000 µg/mL); (ii) did not affect cell viability, but increased release of LDH levels (by 16% to 60%) at higher concentrations (up to 200 µg/mL); (iii) triggered concentration-dependent (25–200 µg/mL) activation of glucose-induced insulin secretion (1.4–25-folds) in comparison to the diabetic control (16.7 mM glucose alone); (iv) caused approximately 12%–50% reduction in insulin glycation in a concentration-dependent manner (50–200 µg/mL); and (v) inhibited starch digestion by 18%–28% at the higher concentrations (500–1000 µg/mL). At 200 µg/mL, $HAEAS_L$ enhanced insulin release activity in the presence of various modulators of insulin secretion (i.e. 16.7 mM glucose, 200 µM isobutylmethylxanthine, or 200 µM tolbutamide), which was impeded in the presence of a K_{ATP} channel opener (300 µM diazoxide) and a voltage-gated calcium channel blocker (50 µM verapamil) and inhibited DPP-IV activity (8%–22%) at concentrations between 200 and 5000 µg/mL. In the *in vivo* study, 50 mg/ml/kg $HAEAS_L$ (i) improved glucose tolerance in high-fat fed rats compared to glucose alone due to a significant increased plasma insulin response at 30 min; (ii) caused a reduction in DPP-IV enzyme activity from 30 min onwards with 20%–24% decrease in high-fat fed rats receiving oral glucose (18 mmol/kg) compared to the high-fat fed control rats (glucose alone); (iii) reduced food intake, energy intake, fluid intake, and blood glucose and enhanced the plasma insulin level accompanied by a significant decrease of DPP-IV enzyme activity from Day 6 onwards following the extract administration (orally twice daily for nine days to high-fat fed rats; and (iv) produced a significant reduction of the islet area, as well as the beta and alpha cell areas. The isolation processes using RP-HPLC led to the identification of rutin and its metabolites (isoquercitrin and quercetin).

21.7 TRADITIONAL AND OTHER POTENTIAL USES

All parts of the *A. squamosa* tree possess various medicinal values and are widely used as ethnic medicine by various communities in different parts of the world for the treatment of different acute and chronic diseases (Kalidindi et al., 2015; Anaya-Esparza et al., 2020). The fruits of *A. squamosa*, which are reported to be a good tonic and are also used as apophlegmatisant, are consumed as a sedative to the heart, to enrich blood, to increase muscle strength, to alleviate vomiting, and to help cool and to relieve burning perception and the tendency to biliousness according to Indian Ayurvedic medicine (Ma et al., 2017). In other reports, the unripe fruits are also used as astringent (Saha, 2011) and for treating diarrhoea and dysentery (Zahid et al., 2018b). Zahid et al. (2018b) also cited that the pulp juice is used for chills and fever in Mexico, whereas in Brazil the juice is used as vermifuge. Moreover, the fruits or fruit juice are also consumed in Brazil to treat worm and parasite infections, to cool fevers, to increase mother's milk after childbirth, and as an astringent for diarrhoea and dysentery.

The seeds, on the other hand, are used as abortient and antiparasitic agent to eliminate lice in hair as well as fleas according to Yunani medicine. In addition, the oil and resin of the seed mixed with gram-flour are claimed to be good for hair wash (Al-Nemari et al., 2020). Moreover, the seeds were utilized as a folkloric medication in the south of China to treat "malignant sores" (cancer) (Ma et al., 2017). In addition, the seeds are also cited to have an antifertility activity (Saha, 2011) and are used in traditional medicine to prevent insect and parasite activity (Zahid et al., 2018a, 2018b).

With regards to the leaves of *A. squamosa*, the leaves are prepared as poultice in India to appease boils and ulcers, whereas the leaf infusion is applied in the treatment of prolapse in children, to fight against heart failure and palpitations, for proper digestion, and as an antispasmodic agent (Ma et al., 2017). Moreover, a mixture of four to five newly grown young leaves with black pepper (*Piper nigrum*) are consumed for management of diabetes by tribes in the Aligarh district of Northern India (Gajalakshmi et al., 2011). Other than that, the bruised leaves are mixed with salt to form a cataplasm, which is applied to extract guinea-worms while the crushed leaves are applied on ulcers and to treat gastritis and internal and external wounds. Zahid et al. (2018a, 2018b) also added that the leaf decoction is consumed throughout tropical America as an emmenagogue, febrifuge, or tonic and to alleviate dysentery, colds, and digestive and urinary tract ailments, in addition to being utilized in baths to alleviate rheumatic pain. In Brazil, in particular, the leaves are also utilized to treat hysteria and fainting spells with its decoction applied in the treatment of cough, intestinal infections, and acidity, condition while in Cuba, the leaves are also consumed to reduce uric acid level (Al-Nemari et al., 2020). Additionally, Saha (2011) reported that the leaves can be used as a vermicide and are applied to abscesses, insect bites, and other skin complaints and for treating cancerous tumours.

With regards to the root and bark, Saha (2011) reported that the root can be used as a drastic purgative, whereas the bark can be used to treat cancer (Suresh et al., 2008) and consumed as a tonic to stop diarrhoea (Zahid et al., 2018a, 2018b). In addition, the scrapped root bark can also be applied in the treatment of toothache. According to Zahid et al. (2018a, 2018b), the root, bark, and leaves also are considered as a sedative and are used in ulcer treatment and as a nervine tonic.

21.8 SAFETY ISSUES

Plants from the Annonaceae family have uses in folk remedies and traditional medicine with different plant parts (i.e. fruit, leaves, bark, roots) scientifically proven to possess medicinal values (Ma et al., 2017). Hence, preparations made from different plant parts of Annonaceae trees have been utilized as herbal medicine and are also sold internationally as over-the-counter food supplements as they are believed to support general health and to treat a wide range of health conditions. Despite the traditional medicinal claims and scientifically proven medicinal values of *A. squamosa* as described earlier, there are possible health risks associated with the consumption of Annonaceae plants, including *A. squamosa*, and their derived food supplements (Höllerhage et al., 2015).

The issue of safety of Annonaceae-based products and their dietary supplements have been triggered by the earlier report of an incidence of atypical parkinsonism in Guadeloupe, an island group in the southern Caribbean Sea, that has been associated with the consumption of fruits from Annonaceae family (Caparros-Lefebvre and Elbaz, 1999). Further study revealed that the neurodegenerative tauopathy endemic in Guadeloupe Island is strongly linked with the consumption of ACGs-contained fruits (Escobar-Khondiker et al., 2007). Later, Derbré et al. (2008) reported on the neurotoxic potential of squamocin, one of the ACGs, later isolated from the fruits of *A. squamosa* (Bonneau et al., 2017). These findings were concurrent with an earlier report made by Bonneau et al. (2012), who proposed the fruits of *A. squamosa* as a source of exposure to ACGs, the natural lipophilic inhibitors of mitochondrial complex I that act as environmental neurotoxins to cause neurological disorders (i.e. atypical parkinsonism in Guadeloupe) (Höllerhage et al., 2015). Among the ACGs in *A. squamosa*, squamocin was one of the abundant phytoconstituents of *A. squamosa* and have been reported to possess neurotoxicity effect (Derbré et al., 2008).

Among the attempt to prove the neurotoxicity effect of *A. squamosa*, a study by Höllerhage et al. (2015) is worth mentioning. In this study, the fruit pulps of *A. squamosa*, collected from Belém, Pará, Brazil, was subjected to hot, pressurized ethyl acetate extraction to obtain the ethyl acetate extract of *A. squamosa* pulp ($EAEAS_{PU}$). The extract was reported to partly contain annonacin and squamocin, which are neurodegenerative in nature, in the amount of 2.2 and 0.7 µg/g (dry weight), respectively. A neurotoxicity study using Lund human mesencephalic neurons cells revealed that $EAEAS_{PU}$ (1.0 and 10 µg/mL) reduced cell viability with the recorded percentages of 14.4% + 5.4% ($P < 0.0009$) and 4.6% + 2.5% ($P < 0.0003$) in comparison to the negative control. Moreover, 10 µg/mL $EAEAS_{PU}$ caused cell death by increasing ($P < .0001$) the LDH level to 68.7% + 2.1% compared to 15.9% + 0.8% in negative controls. Based on the results obtained it is plausible to suggest that the intake of dietary supplements containing plant material particularly from *A. squamosa* may be hazardous to health. Although the amount of annonacin and squamocin quantified were low in *A. squamosa* pulp extract, the neurotoxicity effect observed by Caparros-Lefebvre and Elbaz (1999) could be due to continuous consumption of the fruits of *A. squamosa* and the presence of other ACGs that work synergistically, but were not detected in this study. Taking the neurotoxic effect of ACGs, in general, Ma et al. (2017) did suggest the monitoring of Annonaceae products' consumption to ensure that the amount taken should be below a certain value.

Regarding the use of *A. squamosa* in food supplements, scientific information from human intervention studies, which might be used for safety evaluation of *A. squamosa*, is currently lacking, thus impeding risk assessment. This could be attributed to several factors such as that the composition of such preparations may differ considerably, depending, among others, on conditions of plant growth, conditions of harvest, part of plant used, method of extraction, and further processing of the preparation. Reliable toxicological data, human studies, or toxicokinetic *in vivo* data, however, are currently lacking for specific food supplement products based on *A. squamosa* preparations. There is also a lack of reliable *in vitro* and *in vivo* studies with respect to potential interactions between *A. squamosa* preparations and conventional drugs.

21.9 CULTIVATION PRACTICES

The trees of *A. squamosa* grow well in subtropical to tropical areas on an array of soil types (sandy soil to clay loams) provided that they have good drainage. In addition, the trees succeed in soil with optimal pH of 6.0–6.5, areas up to 2000 m above sea level with a mean annual rainfall of greater than 700 mm and a mean annual temperature up to 41°C (Dinesh Kumar et al., 2005; Ma et al., 2017). Although the plant is considered a hardy, drought-resistant crop as the tree grows well in hot and relatively dry climates of many tropical countries, tolerates occasional light frosts, and succeeds on rocky, alkaline soils with a pH up to 8, it requires adequate moisture during the growing season. The plant is most commonly propagated by seed, but it can also be propagated using grafting or budding techniques. To allow maximum germination, it is suggested that the seeds be removed from the flesh and allowed to dry for one week before propagating. Moreover, grafting has been acknowledged as the technique of propagation that can yield high-quality fruit in a shorter time period in comparison to seed (Chen et al., 2011; Ma et al., 2017). Other than that, poor pollination is frequently observed with *A. squamosa* under high temperatures (>300°C) and low humidity (<60% relative humidity) even when the hand pollination method was implemented. In contrast, lower temperature (25°C) and high humidity (>70% relative humidity) improves pollination significantly. The mature trees of *A. squamosa*, which are approximately 5 m high, start to produce fruits (200–300g each) in a season when they reach three to four years old. The fruit has a sweet sugar-liked taste whereby the ripened fruit is indicated by the sweet aroma of the fruit. In addition, the fruit is said to be borne on old and new wood, but is more commonly on new wood making pruning an advantage. The trees of *A. squamosa* required a spacing of 6 m apart to enable them to grow to their maximal size and require fertilizer that is rich in three main nutrients, namely nitrogen, phosphorus, and potassium, to maximize fruit production. The trees are susceptible to seed borers, scale insects, beetles, mealybugs, and fungal diseases (Dinesh Kumar et al., 2005; Chen et al., 2011; Ma et al., 2017).

21.10 CONCLUSION AND FUTURE REMARKS

One of the promising medicinal values attributed to *A. squamosa* is antidiabetic activity, which was scientifically studied and proven via standardized *in vitro* and *in vivo* models as described in detail in this review chapter. Hence, attempts have been made in many parts of the world to utilize the preparations made from different parts of *A. squamosa* tree as herbal medicine, which are marketed as over-the-counter dietary supplements. However, the potential antidiabetic activity as well as other medicinal values of *A. squamosa* have been overshadowed by findings that *A. squamosa*, like other Annonaceae plants, exhibits a neurotoxicity effect and that this observation is linked to the presence of ACGs, including annonacin and squamocin. This claim has been supported by the epidemiology findings that associated the development of atypical parkinsonism in Guadeloupe Island to the consumption of *A. squamosa* fruits. Moreover, there are substantial uncertainties regarding the safe use of *A. squamosa*-based food supplements despite the strong indications of neurotoxic potential of certain *A. squamosa* preparations. Unfortunately, the paucity of adequate human intervention studies, particularly related to long-term use of *A. squamosa* supplements, does not allow the establishment of a safe intake level.

Hence, to fully exploit the medicinal benefits of *A. squamosa* in the future, further investigations utilizing the right scientific and systematic approach should be planned, which include establishing the threshold at which the beneficial effect is observed by those phytoconstituents of *A. squamosa*, elucidating the mechanisms of action that is a vital hinge for exploiting it in pharmaceutical and agricultural productions, and exploring the neurodegenerative effect due to its medicinal usages at clinical level. Overall, it is hoped that this review can be used as the foundation and motivation for researchers directly working on the bioactivities of *A. squamosa* to further explore ways of discarding its toxic effect while increasing its usage in the pharmaceutical areas.

REFERENCES

Abd-Elrazek A, Shapana H, Shukry W, Galilah D. Comparison between *Annona squamosa*, *Annona cherimolia* and *Annona atemoya* ethanolic extracts extenuative impact against oxidative stress, inflammation and apoptosis in rat kidney induced by Ifosfamid. *Toxicol Res.* 2021;10(4):947–958.

AftabUddin S, Siddique MAM, Romkey SS, Shelton WL. Antibacterial function of herbal extracts on growth, survival and immunoprotection in the black tiger shrimp Penaeus monodon. *Fish Shellfish Immunol.* 2017;65:52–58.

Al-Nemari R, Al-Senaidy A, Semlali A, Ismael M, Badjah-Hadj-Ahmed AY, Ben Bacha A. GC-MS profiling and assessment of antioxidant, antibacterial, and anticancer properties of extracts of Annona squamosa L. leaves. *BMC Complement Med Ther.* 2020;20(1):296.

Anaya-Esparza LM, García-Magaña ML, Abraham Domínguez-Ávila J, et al. Annonas: Underutilized species as a potential source of bioactive compounds. *Food Res Int.* 2020;138(Pt A):109775.

Ansari P, Flatt PR, Harriott P, Abdel-Wahab YH. Evaluation of the antidiabetic and insulin releasing effects of *A. squamosa*, including isolation and characterization of active phytochemicals. *Plants.* 2020;9(10):1348.

Basha SH, Subramanian S. Biochemical evaluation of antidiabetic and antioxidant potentials of Annona squamosa leaves extracts studied in STZ induced diabetic rats. *Int J Pharm Sci Res.* 2011;2(3):643.

Bhoir SS, Vishwapathi V, Singh KK. Antipsoriatic potential of Annona squamosa seed oil: An in vitro and in vivo evaluation. *Phytomedicine.* 2019;54:265–277.

Bonneau N, Baloul L, Bajin Ba Ndob I, Sénéjoux F, Champy P. The fruit of Annona squamosa L. as a source of environmental neurotoxins: From quantification of squamocin to annotation of Annonaceous acetogenins by LC-MS/MS analysis. *Food Chem.* 2017;226:32–40.

Bonneau N, Le Ven J, Schmitz-Afonso I, Guérineau V, Bajin ba Ndob I, Baloul L, et al. Annonaceous acetogenins as environmental neurotoxins: Human exposure from edible Annona fruits. *Planta Med.* 2012;78:PH25.

Caparros-Lefebvre D, Elbaz A. Possible relation of atypical parkinsonism in the French West Indies with consumption of tropical plants: A case-control study. Caribbean Parkinsonism Study Group: A case-control study. Caribbean Parkinsonism Study Group. *Lancet.* 1999;354:281–286.

Castañeda-Ramírez GS, Torres-Acosta JFJ, Mendoza-de-Gives P, Tun-Garrido J, Rosado-Aguilar JA, Chan-Pérez JI, Hernández-Bolio GI, Ventura-Cordero J, Acosta-Viana KY, Jímenez-Coello M. Effects of different extracts of three *Annona* species on egg-hatching processes of *Haemonchus contortus. J Helminthol.* 2019;94:e77.

Chatatikun M, Chiabchalard A. Thai plants with high antioxidant levels, free radical scavenging activity, anti-tyrosinase and anti-collagenase activity. *BMC Complement Altern Med.* 2017;17(1):487.

Chen JW, Chen Y, Li X. Beneficial aspects of custard apple (Annona squamosa L.) seeds. In: VR Preedy, RR Watson and VB Patel (eds.), *Nuts and Seeds in Health and Disease Prevention*, Academic Press, London, 2011, pp. 439–445.

Chen YY, Cao YZ, Li FQ, Zhu XL, Peng CX, Lu JH, Chen JW, Li X, Chen Y. Studies on anti-hepatoma activity of annona squamosa L. pericarp extract. *Bioorg Med Chem Lett.* 2017;27(9):1907–1910.

Chen YY, Li FQ, Zhu XL, Chen JW, Li X. Chemical composition and anti-hepatoma effect of *Annona squamosa* L. pericarp oil. *Nat Prod Res.* 2020;36(1):401–404.

Chen YY, Peng CX, Hu Y, Bu C, Guo SC, Li X, Chen Y, Chen JW. Studies on chemical constituents and anti-hepatoma effects of essential oil from *Annona squamosa* L. pericarps. *Nat Prod Res.* 2016;31(11):1305–1308.

Davis JA, Sharma S, Mittra S, Sujatha S, Kanaujia A, Shukla G, Katiyar C, Lakshmi BS, Bansal VS, Bhatnagar PK. Antihyperglycemic effect of Annona squamosa hexane extract in type 2 diabetes animal model: PTP1B inhibition, a possible mechanism of action? *Indian J Pharmacol.* 2012;44(3):326.

De Moraes MR, Ryan SM, Godoy HT, Thomas AL, Maia JGS, Richards KM, Tran K, Smith RE. Phenolic compounds and metals in some edible annonaceae fruits. *Biol Trace Elem Res.* 2020;197(2):676–682.

Derbré S, Gil S, Taverna M, Boursier C, Nicolas V, Demey-Thomas E, Vinh J, Susin SA, Hocquemiller R, Poupon E. Highly cytotoxic and neurotoxic acetogenins of the Annonaceae: New putative biological targets of squamocin detected by activity-based protein profiling. *Bioorg Med Chem Lett.* 2008;18(21):5741–5744.

Escobar-Khondiker M, Höllerhage M, Muriel M, et al. Annonacin, a natural mitochondrial complex I inhibitor, causes tau pathology in cultured neurons. *J Neurosci.* 2007;27(29):7827–7837.

Feingold KR. Oral and injectable (non-insulin) pharmacological agents for the treatment of type 2 diabetes. [Updated 2022 Aug 26]. In: KR Feingold, B Anawalt, A Boyce, et al. (eds.), *Endotext* [Internet]. South Dartmouth, MA: MDText.com, Inc., 2000. Available from: www.ncbi.nlm.nih.gov/books/NBK279141/

Gajalakshmi S, Divya R, Deepika VD, Mythili S, Sathiavelu A. Pharmacological activities of Annona squamosa: A review. *Int J Pharm Sci Rev Res.* 2011;10:24–29.

Gupta RK, Kesari AN, Murthy PS, Chandra R, Tandon V, Watal G. Hypoglycemic and antidiabetic effect of ethanolic extract of leaves of Annona squamosa L. in experimental animals. *J Ethnopharmacol.* 2005a;13;99(1):75–81.

Gupta, RK, Kesari AN, Watal G, Murthy PS, Chandra R, Maithal K, Tandon V. Hypoglycaemic and antidiabetic effect of aqueous extract of leaves of Annona squamosa (L.) in experimental animal. *Curr Sci.* 2005b;88:1244–1254.

Gupta RK, Kesari AN, Watal G, Murthy PS, Chandra R, Tandon V. Nutritional and hypoglycemic effect of fruit pulp of Annona squamosa in normal healthy and alloxan-induced diabetic rabbits. *Ann Nutr Metab.* 2005c;49(6):407–413.

Höllerhage M, Rösler TW, Berjas M, Luo R, Tran K, Richards KM, Sabaa-Srur AU, Maia JG, Moraes MR, Godoy HT, Höglinger GU, Smith RE. Neurotoxicity of dietary supplements from annonaceae species. *Int J Toxicol.* 2015;34(6):543–550.

Kaleem M, Asif M, Ahmed QU, Bano B. Antidiabetic and antioxidant activity of Annona squamosa extract in streptozotocin-induced diabetic rats. *Singap Med J.* 2006;47(8):670–675.

Kaleem M, Medha P, Ahmed QU, Asif M, Bano B. Beneficial effects of Annona squamosa extract in streptozotocin-induced diabetic rats. *Singap Med J.* 2008;49(10):800–804.

Kalidindi N, Thimmaiah NV, Jagadeesh NV, Nandeep R, Swetha S, Kalidindi B. Antifungal and antioxidant activities of organic and aqueous extracts of Annona squamosa Linn. leaves. *J Food Drug Anal.* 2015;23(4):795–802.

Kaur R, Afzal M, Kazmi I, Ahamd I, Ahmed Z, Ali B, Ahmad S, Anwar F. Polypharmacy (herbal and synthetic drug combination): A novel approach in the treatment of type-2 diabetes and its complications in rats. *J Nat Med.* 2013;67(3):662–671.

Ko GA, Kang HR, Moon JY, Ediriweera MK, Eum S, Bach TT, Cho SK. Annona squamosa L. leaves inhibit alpha-melanocyte-stimulating hormone (α-MSH) stimulated melanogenesis via p38 signaling pathway in B16F10 melanoma cells. *J Cosmet Dermatol.* 2020;19(7):1785–1792.

Korah MC, Rahman J, Rajeswari R, Sherief H, Lalitha V, Sengottavelu S, Sivakumar T. Evaluation of diuretic efficacy and antiurolithiatic potential of ethanolic leaf extract of *Annona squamosa* Linn. in experimental animal models. *Indian J Pharmacol.* 2020;52(3):196–202.

Kumar CD, Rajendran K, Lobo R, Annie S. An identity based pharmacognostical profile of folium annona squamosa. *Nat Prod Sci.* 2005;11(4):213–219.

Leite DOD, Camilo CJ, Nonato CFA, Carvalho NKG, Salazar GJT, de Morais SM, Costa JGMD. Chemical profile and evaluation of the antioxidant and anti-acetylcholinesterase activities of *Annona squamosa* L. (Annonaceae) extracts. *Foods.* 2021;10(10):2343.

Ma C, Chen Y, Chen J, Li X, Chen Y. A Review on Annona squamosa L.: Phytochemicals and biological activities. *Am J Chin Med.* 2017;45(5):933–964.

Mittal M, Juval V, Singh A. Antidiabetic and cryoprotective effect of Annona squamosa leaves on diabetic Wistar albino rat. *Pharmacol Online.* 2010;2:908–913.

Moghadamtousi SZ, Goh BH, Chan CK, Shabab T, Kadir HA. Biological activities and phytochemicals of Swietenia macrophylla King. *Molecules (Basel, Switzerland).* 2013;18(9):10465–10483.

Morton JF. Sugar apples. In: JF Morton (ed.), *Fruits of Warm Climates.* Creative Resource System Inc., Winterville, NC, 1987, pp 69–72.

Palshetkar A, Pathare N, Jadhav N, Pawar M, Wadhwani A, Kulkarni S, Singh KK. In vitro anti-HIV activity of some Indian medicinal plant extracts. *BMC Complement Med Ther.* 2020;20(1):69.

Patel DK, Kumar R, Prasad SK, Sairam K, Hemalatha S. Antidiabetic and in vitro antioxidant potential of Hybanthusenneaspermus (Linn) F. Muell in streptozotocin-induced diabeticrats. *Asian Pac J Trop Biomed.* 2011;1(4):316–322.

Pinto NCC, Silva JB, Menegati LM, Guedes MCMR, Marques LB, Silva TPD, Melo RCN, Souza-Fagundes EM, Salvador MJ, Scio E, Fabri RL.Cytotoxicity and bacterial membrane destabilization induced by Annona squamosa L. extracts. *An Acad Bras Cienc.* 2017;89(3 Suppl):2053–2073.

Rabintossaporn P, Saenthaweesuk S, Thuppia A, Ingkaninan K, Sireeratawong S. Antihyperglycemic and histological effects on the pancreas of the aqueous leaves extract of Annona squamosa L. in normal and diabetic rats. *Songklanakarin J Sci Technol.* 2009;31(1).

Ranjana, Tripathi YB. Insulin secreting and alpha-glucosidase inhibitory activity of hexane extract of Annona squamosa Linn. in streptozotocin (STZ) induced diabetic rats. *Indian J Exp Biol.* 2014;52(6):623–629.

Saha R. Pharmacognosy and pharmacology of Annona squamosa: A review. *Int J Pharm Life Sci.* 2011;2(10):1183–1189.

Sengupta R, Sawant CS, Karmarkar SM, Bhagwat AM. Antihyperglycaemic effects of herbal extracts on alloxan induced hyperglycaemic mice. *Int J Pharm Pharm Sci.* 2011;3(4):307–310.

Sharma A, Chand T, Khardiya M, Yadaav KC, Mangal R, Sharma AK. Antidiabetic and antihyperlipidemic activity of Annona squamosa fruit peel in streptozotocin induced diabetic rats. *Int J Toxicol Pharmacol Res.* 2013;5(1):15–21.

Shehata MG, Abu-Serie MM, Abd El-Aziz NM, El-Sohaimy SA. Nutritional, phytochemical, and in vitro anticancer potential of sugar apple (Annona squamosa) fruits. *Sci Rep.* 2021;11(1):6224.

Shirwaikar A, Rajendran K, Kumar C. Oral antidiabetic activity of annona squamosa leaf alcohol extract in NIDDM rats. *Pharm Biol.* 2004;42(1):30–35.

Silva KDRR, Sirasa MSF. Antioxidant properties of selected fruit cultivars grown in Sri Lanka. *Food Chem.* 2018;238:203–208.

Srivastava S, Lal VK, Pant KK. Preliminary screening of some of the emerging antidiabetic plants for quality control. *Int J Pharm Pharm Sci.* 2012;4(3):331–333.

Suresh K, Manoharn S, Blessy D. Protective role of Annona squamosa linn bark extracts in DMBA-induced genotoxicity. *Kathmandu Univ Med J.* 2008;6(23):364–369.

Tien YY, Ng CC, Chang CC, Tseng WS, Kotwal S, Shyu YT. Studies on the lactic-fermentation of sugar apple (Annona squamosa L.) puree. *J Food Drug Anal.* 2005;13(4):377–381.

Tomar RS, Sisodia SS. Antidiabetic activity of Annona squamosa Linn. in alloxan-induced diabetic rats. *Int J Green Pharm.* 2014;8(4):237–241.

Vikas B, Akhil BS, Remani P, Sujathan K. Free radical scavenging properties of annona squamosa. *Asian Pac J Cancer Prev.* 2017;18(10):2725–2731.

Vikas B, Anil S, Remani P. Cytotoxicity profiling of *Annona squamosa* in cancer cell lines. *Asian Pac J Cancer Prev.* 2019;20(9):2831–2840.

Wong PL, Fauzi NA, Mohamed Yunus SN, Abdul Hamid NA, Abd Ghafar SZ, Azizan A, Zolkeflee NKZ, Abas F. Biological activities of selected plants and detection of bioactive compounds from *Ardisia elliptica* using UHPLC-Q-exactive orbitrap mass spectrometry. *Molecules.* 2020;25(13):3067.

Zahid M, Arif M, Rahman MA, Mujahid M. Hepatoprotective and antioxidant activities of *Annona squamosa* seed extract against alcohol-induced liver injury in Sprague Dawley rats. *Drug Chem Toxicol.* 2020;43(6):588–594.

Zahid M, Arif M, Rahman MA, Singh K, Mujahid M. Solvent extraction and gas chromatography-mass spectrometry analysis of *Annona squamosa* L. Seeds for determination of bioactives, fatty acid/fatty oil composition, and antioxidant activity. *J Diet Suppl.* 2018a;15(5):613–623.

Zahid M, Mujahid M, Singh PK, Farooqui S, Singh K, Parveen S, Arif M. Annona squamosa Linn. (Custard Apple): An aromatic medicinal plant fruit with immense nutraceutical and therapeutics potentials. *Int J Pharm Sci Res.* 2018b;9(5):1745–1759.

22 Guava (*Psidium guajava* L.) *Chemical Diversity, Antidiabetic, and Other Pharmacological Properties*

Keddagoda Gamage Piyumi Wasana and Anoja Priyadarshani Attanayake

CONTENTS

ABBREVIATIONS

ALP: Alkaline phosphatase
ALT: Alanine aminotransferase
AST: Aspartate aminotransferase
LD_{50}: Median lethal dose

22.1 INTRODUCTION

Phytomedicine refers to the utilization of medicinal plants or herbs in order to treat and relieve human diseases (Abate et al., 2021; Bachheti et al., 2021; Husen and Iqbal, 2021; Husen, 2021, 2022; Husen et al., 2021; Sonkar et al., 2022; Rahman and Husen, 2021, 2022). The history of use of medicinal flora for therapeutic uses is as long as the history of human civilization. In the past decades, the practice of phytomedicine has increasingly become one of the important aspects of flourishing global health care systems, since it is believed to be safer than allopathic medicines with no or fewer adverse effects. Extraction, isolation, and purification of drug leads from medicinal flora have resulted in novel compounds with potent activity that protect humankind from debilitation.

Among the widely used medicinal plants in the field of phytomedicine, *Psidium guajava* L., belonging to the Myrtaceae family, is a prominent plant. It is frequently called guava or the poor man's apple of the tropics. From this point onward, the eminence of *P. guajava* is accentuated deeply.

DOI: 10.1201/b23347-22

FIGURE 22.1 (a) *Psidium guajava* in natural habitat, (b) bunch of *Psidium guajava* leaves with dorsal view on the left and ventral view on the right, (c) *Psidium guajava* leaf with dorsal view on the left and ventral view on the right, and (d) *Psidium guajava* fruit.

22.2 BOTANICAL DESCRIPTION

P. guajava is a tropical shrub or small tree with a maximum height of 7 m and wide spreading branches. *P. guajava* has a small trunk and scaly bark with an open canopy (Figure 22.1). The leaves are coriaceous, elliptical, oval, and characterized by an obtuse-type apex with prominent veins. *P. guajava* has white flowers and round, ovoid, or pear-shaped fruit types with firm pulp, seeds, and thin or thick peels with sugary to acetic flavor (Rishika and Sharma, 2012). The *P. guajava* fruit matures approximately 120 days after the flowering stage. The weight of the fruit varies up to 500 g and is governed by the variety of *P. guajava* and the environment (Omayio et al., 2019).

22.3 DISTRIBUTION

Even though *P. guajava* is native to tropical areas from southern Mexico to northern South America, globally the distribution of *P. guajava* is extended to tropical and subtropical climates except the areas with low drainage capacity, high clay content, and saline or acid soil (Morais-Braga et al., 2016). Current habitats of *P. guajava* include waysides, savannahs, and orchards from near sea level to 1600 m (Landrum, 2017). In general, *P. guajava* is introduced for areas having a temperature between 15 and 30 °C and average rainfall range from 1000 to 2000 m^3/ha per year. Seeds of the *P. guajava* fruit play a vital role in distribution, as seeds are disseminated effectually by birds and mammals. Domestic ungulates are able to diffuse seeds between 18,000 and 49,000 per day during the fruiting climax (Sugahara and Takaki, 2004). In order to attain > 90% germination of *P. guajava* seeds, it is important to have a long photoperiod (>10 h) under temperatures between 20 and 30 °C

(Sugahara and Takaki, 2004). When exposed to direct sunlight, the mature *P. guajava* trees produce abundant fruits with spreading branches (Arévalo-Marín et al., 2021).

22.4 PHYTOCHEMICAL CONSTITUENTS

Since time immemorial, *P. guajava* has been used frequently by the natives living in different regions of the world as complementary medicine. With increasing interest in the importance of *P. guajava*, the analysis of its phytochemical constituents has been conducted. *P. guajava* contains a wide range of phytochemicals, comprising alkaloids, flavonoids, tannins, polysaccharides, minerals, vitamins, essential oils, and saponins (Kumar et al., 2021). These phytochemicals have brought substantial attentiveness because of their extensive therapeutic potential. According to the variety of *P. guajava*, the composition of the phytochemicals may be different. Since the pigments, polyphenols and carotenoids, are responsible for the color of the peel and pulp, red-orange *P. guajava* contains more pigments than yellow-green *P. guajava* (Joseph and Priya, 2011). Furthermore, the profile of phenolic compounds of *P. guajava* may vary depending on the environmental condition and the removal process (Schulz et al., 2020). Polyphenols and carotenoids of *P. guajava* have been reported to exert powerful antioxidant activity (Lim et al., 2006). In addition to antioxidant activity, *P. guajava* polyphenols are believed to have an imperative function in controlling diabetes mellitus by inhibiting the absorption of glucose from the small intestine, enhancing the secretion of insulin, and facilitating the controlled release of glucose from the liver (Zhu et al., 2020). Furthermore, the phenolic compounds in *P. guajava* modulate several physiological processes, such as cell proliferation, cellular redox potential, and enzymatic activities. Polyphenols are present in *P. guajava* in free form or bound to sugar or other compounds resulting in these imperative bioactivities. A hybrid technique of chromatography and spectrometry resulted in identifying 72 different phenolic compounds from *P. guajava* leaves (Díaz-de-Cerio et al., 2016). Flavonoids, as one of the members of the polyphenol family in *P. guajava*, are reported to exert antioxidant activity and depend on the substitution pattern of hydroxyl groups, the availability of phenolic hydrogens, and the possibility of stabilization of the resulting phenoxyl radicals (Furlan et al., 2010). According to Sobral-Souza et al. (2019), flavonoids and other phenolic compounds are the most represented phytochemical groups in *P. guajava*. Due to the pleasant taste of *P. guajava* polyphenols, there is an increasing interest in the utilization of *P. guajava* leaves as functional tea or as a raw material in other herbal beverages. Phenolic compounds of *P. guajava* have been claimed as food for specific chronic diseases including diabetes mellitus (Arai et al., 2008) and, in fact, several complementary products including *P. guajava* leaf tea are available in the market, especially in Japan.

The typical smell of *P. guajava* leaves is mainly due to its essential oil. The essential oil in *P. guajava* consists of volatile, lipophilic, aromatic, and liquid substances. The prominent essential oil components of *P. guajava* are α-pinene, β-caryophyllene, limonene, veridiflorol, and nerolidol (Silva et al., 2016). Essential oils are currently attracting interest in the scientific community due to their medicinal properties such as anticancer, antimicrobial, antiviral, antioxidant, and antiproliferative activities (Sepahvand et al., 2014; Satyal et al., 2015). The chemical composition of the essential oils of *P. guajava* varies due to several factors such as variety, temperature, precipitation, humidity, photoperiod, and soil condition. An investigation conducted by Arain et al. (2019) identified 50 compounds from the essential oil profile of *P. guajava* leaves after gas chromatography and mass spectroscopy analysis, and among the compounds found the most was β-caryophyllene followed by globulol and trans-nerolidol. The essential oil profile of *P. guajava* leaves in the Philippine variety consisted of limonene, α-pinene, β-caryophyllene, and longicyclene as prolific compounds (Smith and Oliveros-Belardo, 1977). Essential oil of Ecuadorian *P. guajava* leaves was found to contain diverse monoterpenes such as limonene and α-pinene, whereas Tunisian *P. guajava* leaves resulted in veridiflorol and trans-caryophyllene (Sacchetti et al., 2005; Khadhri et al., 2014).

Accordingly, a wide range of phytoconstituents are present in *P. guajava* and, importantly, these phytoconstituents are responsible for diverse biological activities.

22.5 PHARMACOLOGICAL STUDIES

It is important to scientifically evaluate the biological activities of *P. guajava*. To date, several biological activities of *P. guajava*, including antidiabetic, antiobesity, antihyperglycemic, antimicrobial, antidiarrheal, antioxidant, and anticancer, have been precisely investigated. Table 22.1 represents the pharmacological studies carried out on different extracts/fractions of *P. guajava* during 2019–2021. The reported bioactivities, except for the antidiabetic activity of *P. guajava*, are mentioned in Table 22.1. The antidiabetic activity of *P. guajava* is elaborated separately.

TABLE 22.1
Pharmacological Studies Carried Out on *Psidium guajava* during 2019–2021

Part of the plant	Type of extract/ fraction	Study model	Reported activity/ activities	Key reference
Leaves	Ethanol	CCl_4-induced hepatotoxic rats	Hepatoprotective	Vijayakumar et al., 2020
	Quercetin fraction	CCl_4-induced hepatotoxic rats	Hepatoprotective	Vijayakumar et al., 2020
	Aqueous	Streptozotocin-induced diabetic rats	Antidyslipidemic	Tella et al., 2019
	Polysaccharides	DPPH assay, and ABTS assay	Antioxidant	Luo et al., 2019
	Lycopene-rich extract	Hypercholesterolemia-induced hamsters	Hypotriglyceridemic and antioxidant	Brito et al., 2019
	-	High-fat fed obese rats	Antiobesity, antioxidant, and anti-inflammatory	Mamun et al., 2019
	Ethanol	Diarrhea-induced rats	Antidiarrheal	Hirudkar et al., 2020
	Ethanol	MTT assay, colony formation assay, cell migration assay, tube formation assay, hanging drop assay, and rat aortic ring assay	Antiangiogenic and anticancer	Lok et al., 2020
	Methanol	Hole-plate diffusion method	Antimicrobial	Naili et al., 2020
	Methanol	DPPH assay	Antioxidant	Jani et al., 2020
	Aqueous	ABTS assay and DPPH assay	Antioxidant	Cheon et al., 2019
	Aqueous/ethanol	Agar dilution method	Antibacterial	Ngene et al., 2019
	Ethanol	PC-3, A549, and BT549 human breast cancer cells	Antitumor	Alhamdi et al., 2019
	70% ethanol	Cell viability assay, quantification of nitric oxide, and proinflammatory cytokine production	Antioxidant and anti-inflammatory	Phromnoi et al., 2019
	Ethanol/methanol	Human red blood cell membrane stabilization method	Anti-inflammatory	Hotta et al., 2020
Fruits (with seeds)	Acetone and water (70:30)	MDA-MB-435 and MCF-7 human breast cancer cells	Antiproliferative	Correa et al., 2020

ABTS: 2,2'-azino-bis(3-ethylbenzothiazoline-6-sulfonic acid), CCl_4: carbon tetrachloride, MTT: 3-(4,5-dimethylthiazol-2-yl)-2,5-diphenyl-2H-tetrazolium bromide, DPPH: 2,2-diphenyl-1-picrylhydrazyl, A549: human alveolar epithelial cells, MCF-7: michigan cancer foundation-7, PC-3: prostate cancer cell line

22.6 ANTIDIABETIC RESPONSE

Since ancient times, several traditional claims have been made concerning the antidiabetic activity of *P. guajava*. According to the folklore survey carried out among tribal and non-tribal people in Andhra Pradesh, India, the intake of remedy prepared from the dry powder of *P. guajava* fruits with buttermilk twice daily for 15 days was highly accepted in the treatment of diabetes mellitus (Nagaraju and Rao, 1989). Similarly, the ethnobotanical survey conducted in Tamil Nadu, India, and the central region of Togo cited the use of *P. guajava* leaves and fruits for the management of diabetes mellitus (Karou et al., 2011; Makheswari and Sudarsanam, 2012).

In the present era, investigations on the antidiabetic activity of *P. guajava* have been reported through preclinical and clinical trials. Furthermore, several studies have demonstrated and/or suggested the mechanisms of actions related to the antidiabetic response of *P. guajava*. Imperatively, butanol and ethyl acetate fractions collected from the ethanol extract of *P. guajava* leaves and aqueous, ethanol, and methanol leaf extracts of *P. guajava* were reported to exert *in vitro* antidiabetic activity through inhibitory activities of α-amylase, α-glucosidase, and dipeptidyl peptidase-IV enzymes, glucose diffusion, and inhibition of the formation of advanced glycation end products (Soman et al., 2010; Wang et al., 2010; Basha and Kumari, 2012; Manikandan et al., 2016; Wang et al., 2018). Remarkably, the polysaccharides extracted from the fresh *P. guajava* fruits exerted higher α-glucosidase enzyme inhibitory activity than the standard α-glucosidase inhibitor, acarbose (Zhang et al., 2016). Recently, 3T3-L1 cells treated with the methanol leaf extract of *P. guajava* showed inhibition of adipogenesis, improvement of adipocyte function, and insulin-stimulated glucose uptake in adipocytes (Choi et al., 2021).

Based on the imperative results of *in vitro* studies, several *in vivo* studies have been carried out to evaluate the efficacy of the antidiabetic potential of *P. guajava*. A study carried out in streptozotocin-induced diabetic rats revealed that the ethanol leaf extract of *P. guajava* at a dose of 1 g/L per day for eight weeks significantly reduced plasma glucose concentration (Bahrani et al., 2012). Supplementation of the ethyl acetate fraction collected from the sequential extraction of *P. guajava* leaf extract (25 mg/kg per day) for 30 days in streptozotocin-induced diabetic rats has led to a significant reduction in blood glucose levels and formation of glycation end products such as glycated hemoglobin and fructosamine (Soman et al., 2010). According to Luo et al. (2019), treatment of the mixture of polysaccharides isolated from *P. guajava* leaves at doses of 100 and 200 mg/kg for 30 days in streptozotocin-induced diabetic mice showed a significant decrease in fasting blood glucose concentration. It is well known that the insulin signaling pathway is the key regulator of diabetes mellitus and insulin receptors play a vital role in the initiation of signal transduction. Supplementing the aqueous leaf extract of *P. guajava* (200 mg/kg per day) for 45 days in streptozotocin-induced diabetic rats elevated the expressions of genes as insulin receptor substrate-1 and -2 and protein kinase B, and thereby activated the insulin signaling pathway (Jayachandran et al., 2020). Hexokinase enzymes are the key limiting factor during glucose phosphorylation. Administration of the aqueous leaf extract of *P. guajava* (200 mg/kg per day) for 45 days in streptozotocin-induced diabetic rats revealed a significant increment of insulin, glycogen, hexokinase, glucose-6-phosphatase dehydrogenase, and significant belittled hepatic markers and gluconeogenic enzymes (Vinayagam et al., 2018). The investigation undertaken by Tella et al. (2019) revealed that the aqueous leaf extract of *P. guajava* leaves at the dose of 400 mg/kg per day for 14 days in streptozotocin-induced diabetic rats significantly decreased hormone-sensitive lipase activity in the adipose tissue and significantly increased glycogen level in the liver. Supplementation of aqueous leaf extract of *P. guajava* at a dose of 800 mg/kg per day for 20 days in streptozotocin-induced diabetic rats showed a significant reduction in blood glucose concentration (Xu et al., 2020). Furthermore, a study revealed significant metabolic pathways underlining the antihyperglycemic potential of *P. guajava* leaves as cysteine and methionine metabolism, valine, leucine, and isoleucine biosynthesis, phenylalanine metabolism, glycine, serine, and threonine metabolism, and histidine metabolism (Xu et al., 2020).

Based on the lines of clinical science, the performance of clinical trials is mandatory to assess the benefits and risks of medicinal plant preparations. To date, very few clinical trials have been performed to assess the effectiveness of *P. guajava* preparations. The glucose-lowering potency of *P. guajava* leaf tea was demonstrated through a placebo-controlled clinical trial (Jayasudha et al., 2017). In that study, patients with type 2 diabetes mellitus ($n = 40$) were randomly assigned into two groups. One group received *P. guajava* leaf tea after breakfast every day for 28 days, while the other group was considered a control. On day 28 of the intervention, the mean fasting blood glucose concentration and the postprandial blood glucose concentration of the test group were significantly lower compared with the control group (Jayasudha et al., 2017). A crossover clinical trial involving 20 hospitalized patients with type 2 diabetes mellitus revealed that single administration of *P. guajava* leaf tea significantly reduced postprandial glucose levels without developing hypoglycemic status (Ishibashi et al., 2004). However, the glucose-reducing potential was lower than that of voglibose, which is an α-glucosidase inhibitor that has been used by patients with diabetes mellitus.

Persistent hyperglycemia for chronic periods results in diabetic complications in uncontrolled diabetic patients. The mechanisms underlying the development of complications of diabetes mellitus include increased oxidative stress, chronic inflammation, and dyslipidemia. Interestingly, several lines of studies have reported potent antioxidant, anti-inflammatory, and lipid-lowering activities of *P. guajava*, and therefore *P. guajava* could serve as a potential therapeutic agent for combating diabetic complications and associated commodities (Soman et al., 2010; Bahrani et al., 2012; Vinayagam et al., 2018; Luo et al., 2019; Tella et al., 2019; Jayachandran et al., 2020).

Based on the positive feedback of the reported investigations, several active compounds have been isolated from different extracts/fractions of *P. guajava* that could serve as prospective drug leads for the development of novel antidiabetic agents to ameliorate diabetes mellitus (Table 22.2).

TABLE 22.2
Isolated Active Phytoconstituents of *Psidium guajava*

Parts of the plant	Extract/ fraction	Isolated compound/s	Key references
Leaves	Ethyl acetate/ butanol/ aqueous	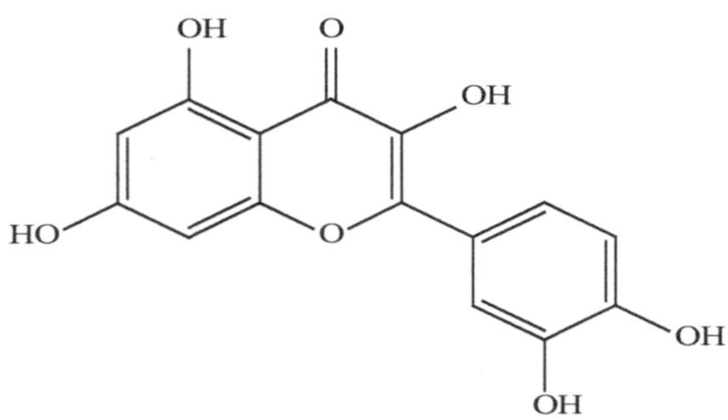 Quercetin	Wu et al., 2009; Wang et al., 2010; Sampath Kumar et al., 2021
Leaves	Ethyl acetate/ butanol	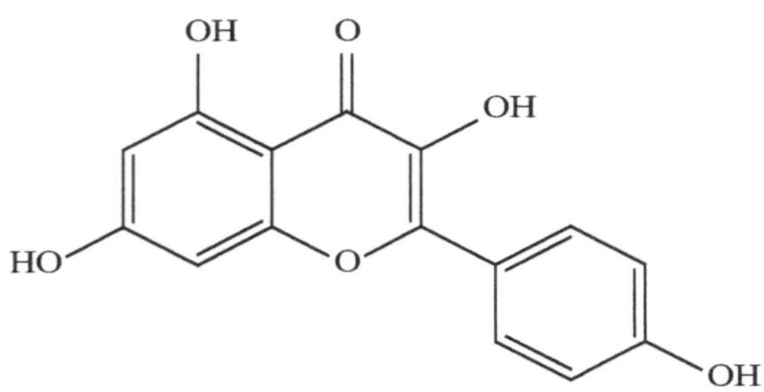 Kaempferol	Wang et al., 2010

Parts of the plant	Extract/ fraction	Isolated compound/s	Key references
Leaves	Ethyl acetate/ butanol	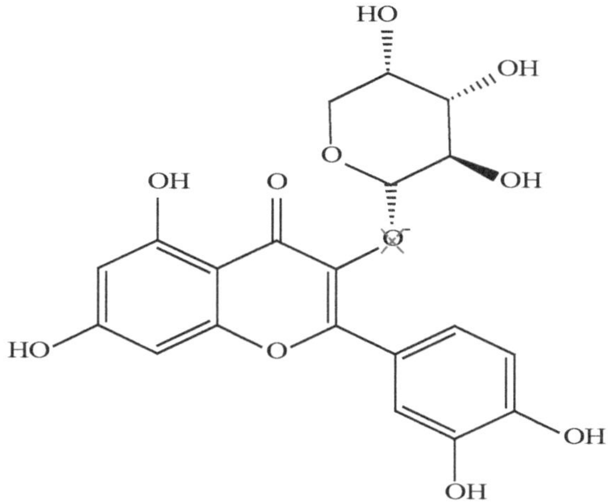 Guaijaverin	Wang et al., 2010
Leaves	Ethyl acetate/ butanol	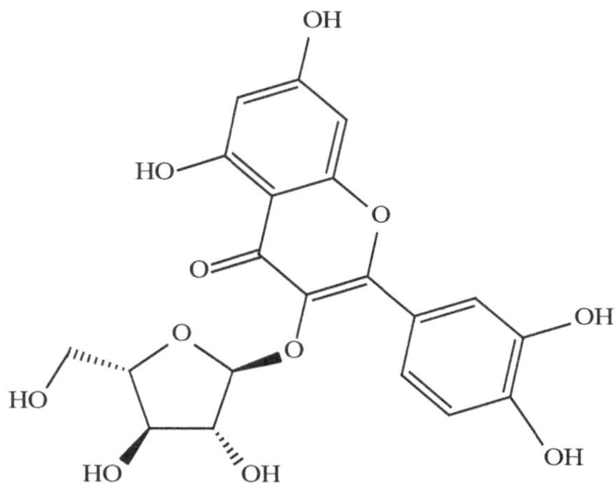 Avicularin	Wang et al., 2010; Ouyang et al., 2016
Leaves	Ethyl acetate	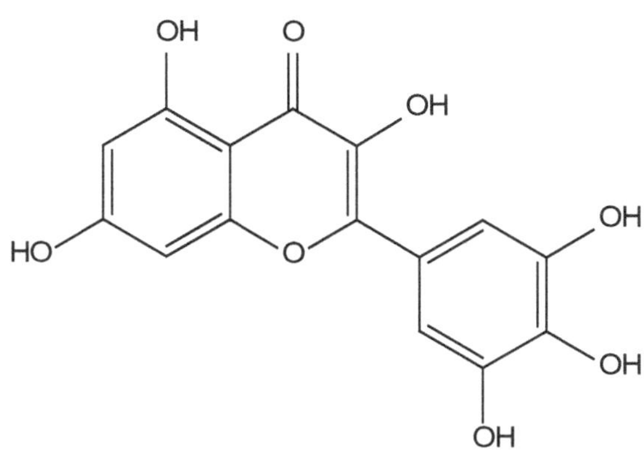 Myricetin	Wang et al., 2010

(*Continued*)

TABLE 22.2
(Continued)

Parts of the plant	Extract/ fraction	Isolated compound/s	Key references
Leaves	Ethyl acetate/ butanol	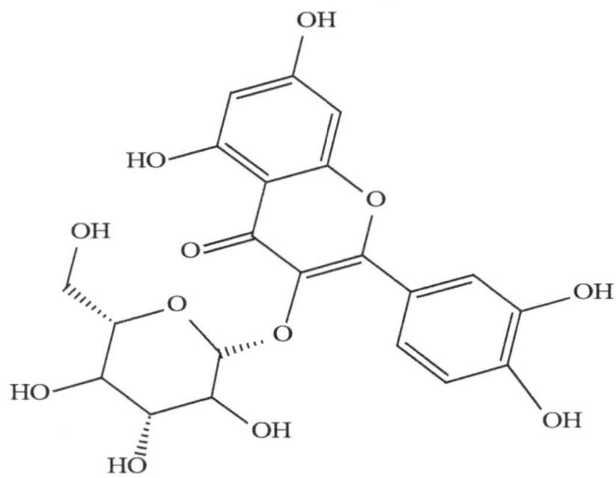 Hyperin	Wang et al., 2010
Leaves	Aqueous	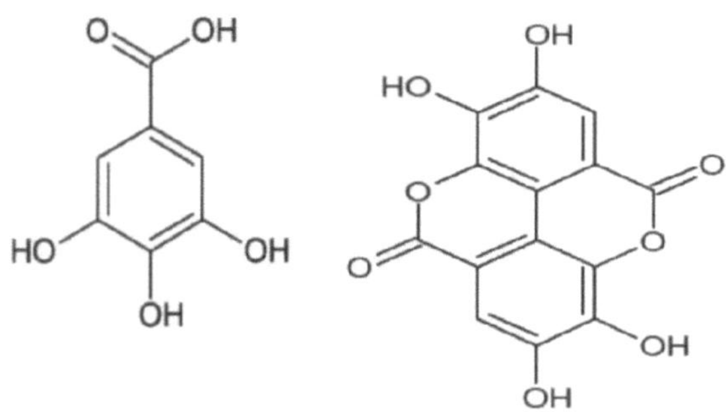 Gallic acid, ellagic acid 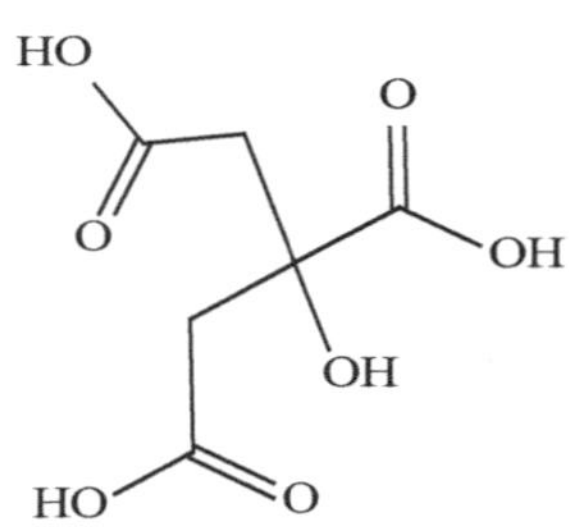 Citric acid 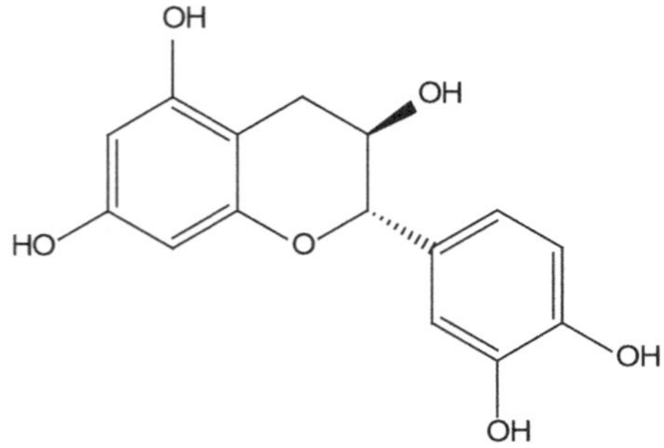 Catechin	Wu et al., 2009; Sampath Kumar et al., 2021

Parts of the plant	Extract/ fraction	Isolated compound/s	Key references
Leaves	Ethanol	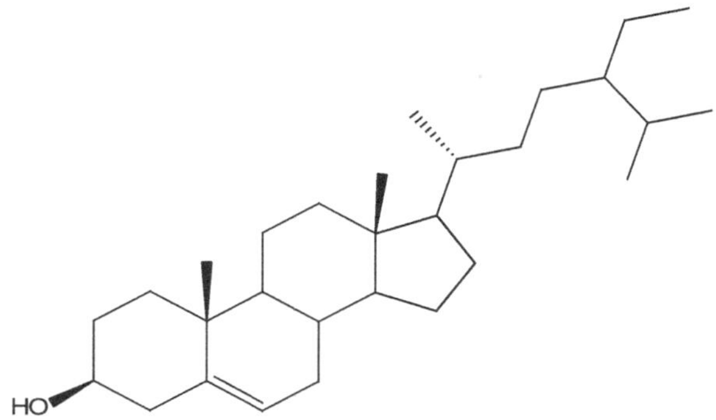 Beta-sitosterol 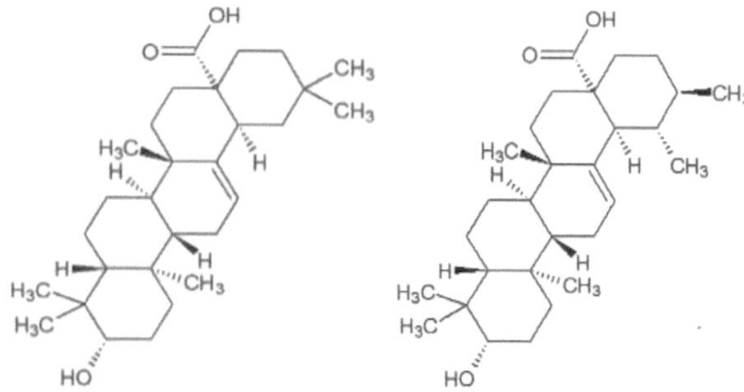 Oleanolic acid, ursolic acid	Begum et al., 2004

22.7 TRADITIONAL AND OTHER POTENTIAL USES

In addition to the use of *P. guajava* in antidiabetic remedies, traditional healers in different settings have recommended the use of *P. guajava* for the management of several other ailments (Díaz-de-Cerio et al., 2017). Depending on the condition of aliments, the application of *P. guajava* remedies is varied, usually either oral or topical. Several modes of preparation of *P. guajava* as decoction, infusion, maceration, juice, and so on are commonly used in the management of diseases (Díaz-de-Cerio et al., 2017). It is well reputed that *P. guajava* leaf decoction is used as a gargle for mouth ulcers by Southeast Asians (Morais-Braga et al., 2016). In Nigeria, the leaf decoction of *P. guajava* is applied in the treatment of bacterial infections (Gutiérrez et al., 2008). The poultice made from *P. guajava* is used externally in Mexico, Brazil, Philippines, and Nigeria for wound healing (Gutiérrez et al., 2008). Guajava leaf chewing sticks are one of the prominent therapies among the Nigerian population for oral care. In addition, different natives in the world exploit *P. guajava* to manage several conditions such as gestational diseases, antibacterial pain, blood pressure, fever, malaria, diarrhea, tuberculosis, and rheumatism (Morais-Braga et al., 2016; Díaz-de-Cerio et al., 2017).

Apart from the therapeutic value, the *P. guajava* fruit exerts high economic value mainly due to its wide acceptance in nature and industries. The reason emphasized for the wide acceptance is the high nutritional value of the *P. guajava* fruit. The fruit is rich in vitamin C (100–260 mg/100 g), vitamin A, vitamin B_1, vitamin B_2, calcium, phosphorus, dietary fiber, protein (1%), and energy (67.78 cal/100 g) (Chavan et al., 2015; Chauhan et al., 2015). The dehydrated fruit is widely commercialized in the form of sweets, jam, pasta, canned fruit, purees, soft drinks, syrups, ice cream, sauces, chutney, and cheese and in fermented dairy products (Chauhan et al., 2015; Chavan et al., 2015; Shukla et al., 2018).

22.8 SAFETY ISSUES

Safe ingestion of *P. guajava* has been addressed through the performance of several toxicity studies. In fact, safety assessment is important to ensure the safety of potential applications of *P. guajava* as a therapeutic agent or as a functional food. Administration of aqueous leaf extract of *P. guajava* in gel form (200 mg/kg per day) for 30 days to male and female albino rats resulted in nontoxic and adverse effects on the rat liver (Uboh et al., 2010). It was evident through the absence of significant changes in biochemical parameters as serum concentrations of total protein, albumin, aspartate aminotransferase (AST), alanine aminotransferase (ALT), and alkaline phosphatase (ALP). Histopathological study revealed that there were no changes in the morphological architecture of liver tissues in both sexes of rats. However, red blood cell counts and hemoglobin concentration were increased upon administration (Uboh et al., 2010).

According to Atik et al. (2019), administration of the ethanol extract of *P. guajava* fruit at doses of 2000 and 5000 mg/kg to female albino mice as a single dose resulted in its safe consumption. It was evident through the assessment of body weight changes, accounting for mortality and based on morphological changes in the liver and kidney. No significant changes in body weight were observed. The H and E stained paraffin sections showed no significant changes in hepatocytes and podocytes (Atik et al., 2019).

The phenolic-rich extract collected from the fermented *P. guajava* leaves was screened for toxicity assessment (Huang et al., 2021). The extract was safe with a median lethal dose (LD_{50}) greater than 5000 mg/kg (Huang et al., 2021). An acute toxicity study conducted through the supplementation of the phenolic-rich extract in male and female Kunming mice showed no significant differences in food and water consumption, behavioral changes, excreta production, or body weight (Huang et al., 2021). Histopathological analysis revealed no pathological changes in the heart, liver, kidney, lung, or spleen upon the treatment of the extract. Furthermore, the extract was nontoxic according to subchronic administration up to the dose limit of 2000 mg/kg in male and female Wistar rats for 30 days (Huang et al., 2021). No significant body weight changes were observed, and the hematological evaluation showed a significant decrement in red cell distribution width – corpuscular volume, mean platelet volume, and monocytes in male Wistar rats, and red blood cell counts, platelet counts, hematocrit, and plateletcrit in female Wistar rats (Huang et al., 2021).

In an acute toxicity study of 14 days, it was observed that all mice treated with *P. guajava* seed oil, which was obtained by extracting *P. guajava* seeds, at a single dose of 10 mg/kg survived throughout the study period (Prommaban et al., 2019). No abnormal clinical signs or adverse effects were observed for the 14-day post-treatment period. The subchronic study carried out in male and female Wistar rats for 90 days revealed that repeated daily oral administration of *P. guajava* seed oil was safe based on terms of nonsignificant changes in body weight, the concentration of liver enzymes, including AST, ALT, and ALP, hematological parameters, and with no gross abnormalities of the internal organs such as liver, kidney, spleen, heart, lungs, intestine, pancreas, or ovaries (Prommaban et al., 2019).

A study carried out by Manekeng et al. (2019) revealed that the supplementation of a single dose of crude methanol bark extract of *P. guajava* (5000 mg/kg body weight) to female Wistar rats did not cause mortality and signs of toxicity. Therefore, the bark extract of LD_{50} of the *P. guajava* was suggested to have a value greater than 5000 mg/kg body weight. Similarly, single-dose administration (5000 mg/kg) of the *P. guajava* methanol leaf extract did not show evidence of adverse effects in female rats (Yamssi et al., 2020). Subacute toxicity was performed through the prolonged oral administration of *P. guajava* methanol bark extract at daily doses of 250 mg/kg, 500 mg/kg, and 1000 mg/kg body weight for 28 days in male and female Wistar rats (Manekeng et al., 2019). It was reported that there were no significant changes in relative organ weight, body weight, and biochemical parameters, including ALT, AST, urea, creatinine, and total protein upon administration of the extract (Manekeng et al., 2019). Furthermore, histopathological examination of H and E stained liver and kidney sections showed minor liver inflammation in Wistar female rats at the dose of 1000 mg/kg (Manekeng et al., 2019). Consequently, several studies were conducted to assess the safety *P. guajava*, and the majority of the studies revealed its safety.

22.9 CULTIVATION PRACTICES

Cultivation of *P. guajava* has become an important consideration among the minor tropical fruits. Annual world production of *P. guajava* fruit between 2015 and 2017 has been estimated at 2.3 million tons (Arévalo-Marín et al., 2021). The highest production of *P. guajava* could be seen in India, followed by China, Mexico, Egypt, and Brazil (Chavan et al., 2015). In India, *P. guajava* is cultivated primarily in Maharashtra, Uttar Pradesh, Maharashtra, Bihar, and Andra Pradesh (Chavan et al., 2015). In particular, these countries where the cultivation of *P. guajava* is prominent show the plant to be quite invasive. The success of *P. guajava* as an invasive cultivated species is its ability to combine clonal and sexual propagation during invasive events (Richardson and Rejmánek, 2011). Furthermore, this fact is supported by the production of a large number of *P. guajava* seeds that can persist as sustainable for a long duration (Adhiambo et al., 2019). These cultivation practices have led to the production of numerous horticultural products that could be exported throughout the world. The main circumstance associated with the cultivation practices of *P. guajava* is the considerable loss of fruits during the post-harvest linkage, as the fruits are highly perishable and therefore cannot be stored for longer periods (Daswani et al., 2017). The high moisture content of the fruit causes accelerated growth of contaminant microbes and thus decreases quality and safety. In fact, it is important to develop suitable technologies and preservative methods associated with surplus cultivation practices.

22.10 FUTURE REMARKS

Future remarks should aim to further investigate the antidiabetic activity of different extracts/fractions of *P. guajava* through preclinical and clinical studies. It is important to isolate active compounds, investigate their structure-activity relationship by means of antidiabetic activity, and perform clinical studies on the isolated active compounds to elaborate on efficacy and safety. The fact will merit commenting on the underlined molecular mechanisms of actions of the different extracts, fractions, and isolated active compounds of *P. guajava* in terms of antidiabetic activity. The development of new antidiabetic drug leads and food cum medicines with improved efficacy and safety for the management of type 2 diabetes mellitus will be a key research area. In parallel, it will be beneficial to develop and commercialize the herbal medicines and nutraceutical products of *P. guajava*.

REFERENCES

Abate, L., Bachheti, A., Bachheti, R.K. and Husen, A., 2021. Antibacterial properties of medicinal plants: Recent trends, progress, and challenges. In: *Traditional Herbal Therapy for the Human Immune System* (Ed. Husen A.). CRC Press, Boca Raton, FL, pp. 13–54.

Adhiambo, R., Muyekho, F., Creed, I.F., Enanga, E., Shivoga, W., Trick, C.G. and Obiri, J., 2019. Managing the invasion of guava trees to enhance carbon storage in tropical forests. *Forest Ecology and Management*, *432*, pp. 623–630.

Alhamdi, H., Boadi, W. and Tiriveedhi, V., 2019. Investigating the anti-tumor potential of *Psidium guajava* extracts on PC-3, A549, and BT-549. *MedCrave Online Journal of Cell Science & Report*, *6*(1), pp. 10–15.

Arai, S., Yasuoka, A. and Abe, K., 2008. Functional food science and food for specified health use policy in Japan: State of the art. *Current Opinion in Lipidology*, *19*(1), pp. 69–73.

Arain, A., Hussain Sherazi, S.T., Mahesar, S.A. and Sirajuddin, 2019. Essential oil from *Psidium guajava* leaves: An excellent source of β-caryophyllene. *Natural Product Communications*, *14*(5). https://doi.org/10.1177/1934578X19843007.

Arévalo-Marín, E., Casas, A., Landrum, L., Shock, M.P., Alvarado-Sizzo, H., Ruiz-Sanchez, E. and Clement, C.R., 2021. The taming of *Psidium guajava*: Natural and cultural history of a Neotropical fruit. *Frontiers in Plant Science*, *12*. https://doi.org/10.3389/fpls.2021.714763.

Atik, N., Muda, I., Rahmadi, A.R., Achadiyani, A. and Djunaedi, D.D., 2019. Phytochemical screening and histology appearance of acute oral toxicity study on ethanol extract of *Psidium guajava* Linn. fruit in mice. *Asian Journal of Pharmaceutical and Clinical Research*, *12*(1), pp. 351–355.

Bachheti, A., Deepti, Bachheti, R.K. and Husen, A., 2021. Medicinal plants and their pharmaceutical properties under adverse environmental conditions. In: *Harsh Environment and Plant Resilience* (Ed. Husen A.). Springer International Publishing, Cham, pp. 457–502. https://doi.org/10.1007/978-3-030-65912-7_19.

Bahrani, A.H.M., Zaheri, H., Soltani, N., Kharazmi, F., Keshavarz, M. and Kamalinajad, M., 2012. Effect of the administration of *Psidium guava* leaves on blood glucose, lipid profiles and sensitivity of the vascular mesenteric bed to Phenylephrine in streptozotocin-induced diabetic rats. *Journal of Diabetes Mellitus*, *2*(1), pp. 138–145.

Basha, S.K. and Kumari, V.S., 2012. *In vitro* antidiabetic activity of *Psidium guajava* leaves extracts. *Asian Pacific Journal of Tropical Disease*, *2*, pp. 98–100.

Begum, S., Hassan, S.I., Ali, S.N. and Siddiqui, B.S., 2004. Chemical constituents from the leaves of *Psidium guajava*. *Natural Product Research*, *18*(2), pp. 135–140.

Brito, A.K.D.S., Lima, G.D.M., Farias, L.M.D., Rodrigues, L.A.R.L., Carvalho, V.B.L.D., Pereira, C.F.D.C., Frota, K.D.M.G., Conde-Júnior, A.M., Silva, A.M.O., Rizzo, M.D.S. and Fonseca, C.M.B., 2019. Lycopene-rich extract from red guava (*Psidium guajava* L.) decreases plasma triglycerides and improves oxidative stress biomarkers on experimentally-induced dyslipidemia in hamsters. *Nutrients*, *11*(2). https://doi.org/10.3390/nu11020393.

Chauhan, A.K., Singh, S., Singh, R.P. and Singh, S.P., 2015. Guava-enriched dairy products: A review. *Indian Journal of Dairy Science*, *68*(1), pp. 1–5.

Chavan, U.D., Pawar, U.B. and Pawar, G.H., 2015. Studies on preparation of mixed toffee from guava and strawberry. *Journal of Food Science and Technology*, *52*(10), pp. 6791–6797.

Cheon, W., Seo, D. and Kim, Y., 2019. Antioxidative and hepatocyte protective effects of guava (*Psidium guajava* L.) leaves cultivated in Korea. *The Korean Journal of Food & Nutrition*, *32*(1), pp. 33–40.

Choi, E., Baek, S., Baek, K. and Kim, H.K., 2021. *Psidium guajava* L. leaf extract inhibits adipocyte differentiation and improves insulin sensitivity in 3T3-L1 cells. *Nutrition Research and Practice*, *15*(5), pp. 568–578.

Correa, M.G., Couto, J.S., Trindade, B.B., Abreu, J.P., Nakajima, V.M., Oliveira, F.L., Farah, A. and Teodoro, A.J., 2020. Antiproliferative effect of guava fruit extracts in MDA-MB-435 and MCF-7 human breast cancer cell lines. *Anais da Academia Brasileira de Ciências*, *92*(2), p. e20191500.

Daswani, P.G., Gholkar, M.S. and Birdi, T.J., 2017. *Psidium guajava*: A single plant for multiple health problems of rural Indian population. *Pharmacognosy Reviews*, *11*(22), pp. 167–174.

Díaz-de-Cerio, E., Gómez-Caravaca, A.M., Verardo, V., Fernández-Gutiérrez, A. and Segura-Carretero, A., 2016. Determination of guava (*Psidium guajava* L.) leaf phenolic compounds using HPLC-DAD-QTOF-MS. *Journal of Functional Foods*, *22*, pp. 376–388.

Díaz-de-Cerio, E., Verardo, V., Gómez-Caravaca, A.M., Fernández-Gutiérrez, A. and Segura-Carretero, A., 2017. Health effects of *Psidium guajava* L. leaves: An overview of the last decade. *International Journal of Molecular Sciences*, *18*(4), p. 897. https://doi.org/doi/10.3390/ijms18040897.

Furlan, C.M., Santos, D.Y., Motta, L.B., Domingos, M. and Salatino, A., 2010. Guava flavonoids and the effects of industrial air pollutants. *Atmospheric Pollution Research*, *1*(1), pp. 30–35.

Gutiérrez, R.M.P., Mitchell, S. and Solis, R.V., 2008. *Psidium guajava*: A review of its traditional uses, phytochemistry and pharmacology. *Journal of Ethnopharmacology*, *117*(1). https://doi.org/doi:10.1016/j.jep.2008.01.025.

Hirudkar, J.R., Parmar, K.M., Prasad, R.S., Sinha, S.K., Lomte, A.D., Itankar, P.R. and Prasad, S.K., 2020. The antidiarrhoeal evaluation of *Psidium guajava* L. against enteropathogenic *Escherichia coli* induced infectious diarrhoea. *Journal of Ethnopharmacology*, *251*, p. 112561. https://doi.org/doi: 10.1016/j.jep.2020.112561.

Hotta, S.K. and Neelima, N., 2020. *In vitro* evaluation of anti-inflammatory activity of methanolic and ethanolic leaf extract of *Psidium guajava*. *World Journal of Current Medical and Pharmaceutical Research*, *2*(2), pp. 159–165.

Huang, Z., Luo, Y., Xia, X., Wu, A. and Wu, Z., 2021. Bioaccessibility, safety, and antidiabetic effect of phenolic-rich extract from fermented *Psidium guajava* Linn. leaves. *Journal of Functional Foods*, *86*, p. 104723. https://doi.org/10.1016/j.jff.2021.104723.

Husen, A., 2021. *Traditional Herbal Therapy for the Human Immune System*. CRC Press, Boca Raton, FL. https://doi.org/10.1201/9781003137955.

Husen, A., 2022. *Herbs, Shrubs and Trees of Potential Medicinal Benefits*. CRC Press, USA. https://doi.org/10.1201/9781003205067.

Husen, A., Bachheti, R.K. and Bachheti, A., 2021. *Non-Timber Forest Products: Food, Healthcare and Industrial Applications*. Springer, Cham, Switzerland. https://doi.org/10.1007/978-3-030-73077-2.

Husen, A. and Iqbal, M. 2021. Plant-based potential nutraceuticals for improving the human immune system. In: *Traditional Herbal Therapy for the Human Immune System* (Ed. Husen A.). CRC Press, Boca Raton, FL, pp. 1–11.

Ishibashi, K., Oka, M., Hachiya, M. and Maeda, T., 2004. Tajima N: Comparison of voglibose and Guava tea (Bansoureicha®) on postprandial blood glucose level. *Journal of Practice Diabetes*, *21*, pp. 455–458, In Japanese.

Jani, N.A., Azizi, N.A.A. and Aminudin, N.I., 2020. Phytochemical screening and anti-oxidant activity of *Psidium guajava*. *Malaysian Journal of Analytical Sciences*, *24*(2), pp. 173–178.

Jayachandran, M., Vinayagam, R. and Xu, B., 2020. Guava leaves extract ameliorates STZ induced diabetes mellitus via activation of PI3K/AKT signaling in skeletal muscle of rats. *Molecular Biology Reports*, *47*(4), pp. 2793–2799.

Jayasudha, P., Chitra, W. and Nirmala, T., 2017. Effect of guava leaf tea on blood sugar level among clients with type ii diabetes mellitus at a selected community, Coimbatore. *International Journal of Applied Research*, *3*(11), pp. 218–220.

Joseph, B. and Priya, M., 2011. Review on nutritional, medicinal and pharmacological properties of guava (*Psidium guajava* Linn.). *International Journal of Pharma and Bio Sciences*, *2*, pp. 53–69.

Karou, S.D., Tchacondo, T., Djikpo Tchibozo, M.A., Abdoul-Rahaman, S., Anani, K., Koudouvo, K., Batawila, K., Agbonon, A., Simpore, J. and de Souza, C., 2011. Ethnobotanical study of medicinal plants used in the management of diabetes mellitus and hypertension in the Central Region of Togo. *Pharmaceutical Biology*, *49*(12), pp. 1286–1297.

Khadhri, A., El Mokni, R., Almeida, C., Nogueira, J.M.F. and Araújo, M.E.M., 2014. Chemical composition of essential oil of *Psidium guajava* L. growing in Tunisia. *Industrial Crops and Products*, *52*, pp. 29–31.

Kumar, M., Tomar, M., Amarowicz, R., Saurabh, V., Nair, M.S., Maheshwari, C., Sasi, M., Prajapati, U., Hasan, M., Singh, S. and Changan, S., 2021. Guava (*Psidium guajava* L.) leaves: Nutritional composition, phytochemical profile, and health-promoting bioactivities. *Foods*, *10*(4), p. 752.

Landrum, L.R., 2017. *The Genus Psidium (Myrtaceae) in the State of Bahia, Brazil*. Herbarium: Natural History Collections, School of Life Sciences, Arizona State University.

Lim, Y.Y., Lim, T.T. and Tee, J.J., 2006. Antioxidant properties of guava fruit: Comparison with some local fruits. *Sunway Academic Journal*, *3*, pp. 9–20.

Lok, B., Sandai, D., Baharetha, H.M., Nazari, V.M., Asif, M., Tan, C.S. and Majid, A.A., 2020. Anticancer effect of *Psidium guajava* (Guava) leaf extracts against colorectal cancer through inhibition of angiogenesis. *Asian Pacific Journal of Tropical Biomedicine*, *10*(7), p. 293. https://doi.org/10.4103/2221-1691.284944.

Luo, Y., Peng, B., Wei, W., Tian, X. and Wu, Z., 2019. Antioxidant and anti-diabetic activities of polysaccharides from guava leaves. *Molecules*, *24*(7), p. 1343. https://doi.org/10.3390/molecules24071343.

Makheswari, M.U. and Sudarsanam, D., 2012. Database on antidiabetic indigenous plants of Tamil Nadhu, India. *International Journal of Pharmaceutical Sciences and Research*, *3*(2), pp. 287–293.

Mamun, M., Al, A., Faruk, M., Rahman, M., Nahar, K., Kabir, F., Alam, M.A. and Subhan, N., 2019. High carbohydrate high fat diet induced hepatic steatosis and dyslipidemia were ameliorated by *Psidium guajava* leaf powder supplementation in rats. *Evidence-Based Complementary and Alternative Medicine*, p. 1897237. https://doi.org/10.1155/2019/1897237.

Manekeng, H.T., Mbaveng, A.T., Ntyam Mendo, S.A., Agokeng, A.J.D. and Kuete, V., 2019. Evaluation of acute and subacute toxicities of *Psidium guajava* methanolic bark extract: A botanical with *in vitro* antiproliferative potential. *Evidence-Based Complementary and Alternative Medicine*, p. 8306986. http://doi: 10.1155/2019/830698.

Manikandan, R., Anand, A.V. and Kumar, S., 2016. Phytochemical and *in vitro* antidiabetic activity of *Psidium guajava* leaves. *Pharmacognosy Journal*, *8*(4), pp. 392–394.

Morais-Braga, M.F.B., Carneiro, J.N.P., Machado, A.J.T., Dos Santos, A.T.L., Sales, D.L., Lima, L.F., Figueredo, F.G. and Coutinho, H.D.M., 2016. *Psidium guajava* L., from ethnobiology to scientific evaluation: Elucidating bioactivity against pathogenic microorganisms. *Journal of Ethnopharmacology, 194*, pp. 1140–1152.

Nagaraju, N. and Rao, K.N., 1989. Folk medicine for diabetes from Rayalaseema of Andhra Pradesh. *Ancient Science of Life*, *9*(1), pp. 31–35.

Naili, M., Errayes, A., Alghazeer, R., Mohammed, W.A. and Darwish, M., 2020. Evaluation of antimicrobial and antioxidant activities of *Psidium guajava* L growing in Libya. *International Journal of Advanced Biological and Biomedical Research*, *8*(4), pp. 419–428.

Ngene, A.C., Aguiyi, J.C., Chibuike, C.J., Ifeanyi, V.O., Ukaegbu-Obi, K.M., Kim, E.G., Ohaeri, U.C. and Onyemegbulem, B.O., 2019. Antibacterial activity of *Psidium guajava* leaf extract against selected pathogenic bacteria. *Advances in Microbiology*, *9*(12), pp. 1012–1022.

Omayio, D.G., Abong, G.O., Okoth, M.W., Gachuiri, C.K. and Mwang'ombe, A.W., 2019. Current status of guava (*Psidium guajava* L.) production, utilization, processing and preservation in Kenya: A review. *Current Agriculture Research Journal*, *7*(3), pp. 318–331.

Ouyang, W., Zhu, X.A., Shao, X.H. and Cao, Y., 2016. Avicularin as the major hypoglycemic compound isolated from Guava leaves. *Planta Medica*, *82*(5). http://doi: 10.1055/s-0036-1578761.

Phromnoi, K., Sinchaiyakij, P., Khanaree, C., Nuntaboon, P., Chanwikrai, Y., Chaiwangsri, T. and Suttajit, M., 2019. Anti-inflammatory and antioxidant activities of medicinal plants used by traditional healers for antiulcer treatment. *Scientia Pharmaceutica*, *87*(3). https://doi.org/10.3390/scipharm87030022.

Prommaban, A., Utama-ang, N., Chaikitwattana, A., Uthaipibull, C. and Srichairatanakool, S., 2019. Linoleic acid-rich guava seed oil: Safety and bioactivity. *Phytotherapy Research*, *33*(10), pp. 2749–2764.

Rahman, S. and Husen, A., 2021. Potential role of medicinal plants in the cure of liver and kidney diseases. In: *Non-Timber Forest Products* (Eds. Husen A., Bachheti R.K., Bachheti A.). Springer International Publishing, Cham. https://doi.org/10.1007/978-3-030-73077-2_10.

Rahman, S. and Husen, A., 2022. *Asparagus racemosus* (Shatavari) and *Dioscorea villosa* (Wild yams). In: *Herbs, Shrubs, and Trees of Potential Medicinal Benefits* (Ed. Husen A.). CRC Press, Boca Raton, FL, pp. 95–112.

Richardson, D.M. and Rejmánek, M., 2011. Trees and shrubs as invasive alien species – a global review. *Diversity and Distributions*, *17*(5), pp. 788–809.

Rishika, D. and Sharma, R., 2012. An update of pharmacological activity of *Psidium guajava* in the management of various disorders. *International Journal of Pharmaceutical Sciences and Research*, *3*(10), pp. 3577–3584.

Sacchetti, G., Maietti, S., Muzzoli, M., Scaglianti, M., Manfredini, S., Radice, M. and Bruni, R., 2005. Comparative evaluation of 11 essential oils of different origin as functional antioxidants, antiradicals and antimicrobials in foods. *Food Chemistry*, *91*(4), pp. 621–632.

Sampath Kumar, N.S., Sarbon, N.M., Rana, S.S., Chintagunta, A.D., Prathibha, S., Ingilala, S.K., Jeevan Kumar, S.P., Sai Anvesh, B. and Dirisala, V.R., 2021. Extraction of bioactive compounds from *Psidium guajava* leaves and its utilization in preparation of jellies. *AMB Express*, *11*(1), pp. 1–9.

Satyal, P., Paudel, P., Lamichhane, B. and Setzer, W.N., 2015. Leaf essential oil composition and bioactivity of *Psidium guajava* from Kathmandu, Nepal. *American Journal of Essential Oils and Natural Products*, *3*(2), pp. 11–14.

Schulz, M., Seraglio, S.K.T., Della Betta, F., Nehring, P., Valese, A.C., Daguer, H., Gonzaga, L.V., Costa, A.C.O. and Fett, R., 2020. Determination of phenolic compounds in three edible ripening stages of yellow guava (*Psidium cattleianum* Sabine) after acidic hydrolysis by LC-MS/MS. *Plant Foods for Human Nutrition*, *75*(1), pp. 110–115.

Sepahvand, R., Delfan, B., Ghanbarzadeh, S., Rashidipour, M., Veiskarami, G.H. and Ghasemian-Yadegari, J., 2014. Chemical composition, antioxidant activity and antibacterial effect of essential oil of the aerial parts of Salvia sclareoides. *Asian Pacific Journal of Tropical Medicine*, *7*, pp. 491–496.

Shukla, R., Shukla, Y.K. and Pathak, S., 2018. Standardization and preparation of Guava cheese from different cultivars. *International Journal of Current Microbiology and Applied Sciences*, *7*(3), pp. 583–595.

Silva, E.A.J.D., Silva, V.P.D., Alves, C.C.F., Alves, J.M., Souchie, E.L. and Barbosa, L.C.D.A., 2016. Harvest time on the content and chemical composition of essential oil from leaves of guava. *Ciência Rural*, *46*, pp. 1771–1776.

Smith, R.M. and Oliveros-Belardo, L., 1977. The composition of leaf essential oils of *Psidium guajava* L. from Manila, Philippines. *Asian Journal of Pharmaceutics*, *3*(3), pp. 5–9.

Sobral-Souza, C.E., Silva, A.R., Leite, N.F., Rocha, J.E., Sousa, A.K., Costa, J.G., Menezes, I.R., Cunha, F.A., Rolim, L.A. and Coutinho, H.D., 2019. *Psidium guajava* bioactive product chemical analysis and heavy metal toxicity reduction. *Chemosphere*, *216*, pp. 785–793.

Soman, S., Rauf, A.A., Indira, M. and Rajamanickam, C., 2010. Antioxidant and antiglycative potential of ethyl acetate fraction of *Psidium guajava* leaf extract in streptozotocin-induced diabetic rats. *Plant Foods for Human Nutrition*, *65*(4), pp. 386–391.

Sonkar, S., Singh, A.K. and Husen, A., 2022. *Convolvulus pluricaulis* (Shankhpushpi) and *Erythroxylum coca* (Coca plant). In: *Herbs, Shrubs, and Trees of Potential Medicinal Benefits* (Ed. Husen A.). CRC Press, Boca Raton, FL, pp. 83–93.

Sugahara, V.Y. and Takaki, M., 2004. Effect of light and temperature on seed germination in guava (*Psidium guajava* L.-Myrtaceae). *Seed Science and Technology*, *32*(3), pp. 759–764.

Tella, T., Masola, B. and Mukaratirwa, S., 2019. The effect of *Psidium guajava* aqueous leaf extract on liver glycogen enzymes, hormone sensitive lipase and serum lipid profile in diabetic rats. *Biomedicine & Pharmacotherapy*, *109*, pp. 2441–2446.

Uboh, F.E., Okon, I.E. and Ekong, M.B., 2010. Effect of aqueous extract of *Psidium guajava* leaves on liver enzymes, histological integrity and hematological indices in rats. *Gastroenterology Research*, *3*, pp. 32–38.

Vijayakumar, K., Arumugam, V.A., Ramasamy, M., Natesan, M., Palanisamy, S., Thajuddin, N.B., Balasubramanian, B. and Meyyazhagan, A., 2020. Hepatoprotective effects of *Psidium guajava* on mitochondrial enzymes and inflammatory markers in carbon tetrachloride-induced hepatotoxicity. *Drug Development and Industrial Pharmacy*, *46*(12), pp. 2041–2050.

Vinayagam, R., Jayachandran, M., Chung, S.S.M. and Xu, B., 2018. Guava leaf inhibits hepatic gluconeogenesis and increases glycogen synthesis via AMPK/ACC signaling pathways in streptozotocin-induced diabetic rats. *Biomedicine & Pharmacotherapy*, *103*, pp. 1012–1017.

Wang, H., Du, Y.J. and Song, H.C., 2010. α-Glucosidase and α-amylase inhibitory activities of guava leaves. *Food Chemistry*, *123*(1), pp. 6–13.

Wang, L., Luo, Y., Wu, Y., Liu, Y. and Wu, Z., 2018. Fermentation and complex enzyme hydrolysis for improving the total soluble phenolic contents, flavonoid aglycones contents and bio-activities of guava leaves tea. *Food Chemistry*, *264*, pp. 189–198.

Wu, J.W., Hsieh, C.L., Wang, H.Y. and Chen, H.Y., 2009. Inhibitory effects of guava (*Psidium guajava* L.) leaf extracts and its active compounds on the glycation process of protein. *Food Chemistry*, *113*(1), pp. 78–84.

Xu, C., Li, X., Zeng, D., Liu, Y., Gao, Y., Tsunoda, M., Deng, S., Xie, X., Wang, R., Li, L.S. and Song, Y., 2020. Amino acid profiling study of *Psidium guajava* L. leaves as an effective treatment for type 2 diabetic rats. *Evidence-Based Complementary and Alternative Medicine*, *2020*, p. 9784382. https://doi.org/10.1155/2020/9784382.

Yamssi, C., Payne, V.K., Noumedem Anangmo, C.N., Tateng Ngouateu, A., Megwi, L. and Kuiate, J.R., 2020. Acute toxicity of *Pentaclethra macrophylla* and *Psidium guajava* use as antiprotozoan medicinal plants. *Journal of Drug Discovery, Development and Delivery*, *6*(1), p. 1037.

Zhang, Z., Kong, F., Ni, H., Mo, Z., Wan, J.B., Hua, D. and Yan, C., 2016. Structural characterization, α-glucosidase inhibitory and DPPH scavenging activities of polysaccharides from guava. *Carbohydrate Polymers*, *144*, pp. 106–114.

Zhu, X., Ouyang, W., Lan, Y., Xiao, H., Tang, L., Liu, G., Feng, K., Zhang, L., Song, M. and Cao, Y. 2020. Anti-hyperglycemic and liver protective effects of flavonoids from *Psidium guajava* L. (guava) leaf in diabetic mice. *Food Bioscience*, *35*, p. 100574. https://doi.org/10.1016/j.fbio.2020.100574.

23 Phytochemical Constituents, Antidiabetic Potential and Other Pharmacological Activities of Mulberry (*Morus alba* L.)

Phaniendra Alugoju and Tewin Tencomnao

CONTENTS

23.1 INTRODUCTION

Diabetes has become a global health issue in both developing and developed countries. The global diabetic population is expected to increase to about 360 million by year 2030. Intriguingly, about 90% of the global diabetic cases are of type 2 diabetes. Therapeutic approaches such as the use of oral antidiabetic drugs/products may be the right solution for the proper management of diabetes-related hyperglycemia and associated microvascular complications. In this context, the use of herbal medicine could be the possible alternative therapeutic strategy for the treatment of diabetes. *Morus alba* L. (commonly known as white mulberry or mulberry) is one such medicinal plant that has been widely used in traditional medicine in Asia for the management of diabetes. This chapter deals with the botanical description, phytochemical constituents, antidiabetic potential and other pharmacological activities, and other traditional uses of mulberry.

23.2 BOTANICAL DESCRIPTION

Morus alba L. is a small to medium size deciduous fast-growing tree belonging to the class Magnoliopsida, order Rosales, and family Moraceae (Figure 23.1). It can grow up to 20–30 feet in height. The leaves are 5–6 inches long and are generally known as folium mori. They are ovate in

DOI: 10.1201/b23347-23

FIGURE 23.1 *Morus alba.*

shape with a cordate-shaped base, and the surface of the leaf is glabrous and shiny. The fruit of mulberry is a fleshy, succulent berry composed of a cluster of drupes, and each drupe is composed of a single seed. The fruit of mulberry is 1–1.5 cm long and occurs in a variety of colors depending on the ripening stage. Fruit occurs in white, pink, and red to dark purple. It has unisexual flowers (0.5–1.5 inches long), and the sexes are borne in separate catkin clusters which are tiny and cylindrical.

23.3 DISTRIBUTION

M. alba is widely distributed in warm, temperate, and subtropical regions around the globe including Asia, Africa, and the Americas. It is naturalized and largely cultivated in various other regions across the world due to its extensive use as a feed in sericulture.

23.4 PHYTOCHEMICAL CONSTITUENTS

Mulberry leaves are rich sources of different iminosugars (also known as azasugars) including 1-deoxynojirimycin (DNJ), N-methyl-DNJ, 2-O-α-D-galactopyranosyl-DNJ, and fagomine (Asano et al. 2001). Several other bioactive compounds such as flavonoids (e.g., rutin, isoquercitrin, astragalin), phenolic acid (e.g., chlorogenic acid), and terpene lactones (e.g., loliolide) were also identified in the *M. alba* leaf (Hunyadi et al. 2013; Kim et al. 2014). Zhang et al. (2009) have isolated different phenolic compounds such as moracin M, steppogenin-4'-O-beta-D-glucosiade, and mullberroside A from the root bark of *M. alba*. Structures of different phytochemicals identified in *M. alba* are illustrated in Figure 23.2.

23.5 PHARMACOLOGICAL ACTIVITIES OF *M. ALBA*

Several *in vitro* and *in vivo* research studies have demonstrated various pharmacological properties of *M. alba*, including antioxidant (Martins et al. 2021; Polumackanycz et al. 2021; Suriyaprom et al. 2021), anti-microbial (Z. Li et al. 2021; Suriyaprom et al. 2021), anti-inflammatory (Umeyama et al. 2021; Wu et al. 2020; Yu et al. 2021), anti-cancer (Chan et al. 2020; Eo et al. 2014; Wang et al. 2020), anti-skin aging (Lim et al. 2014; Woo et al. 2017), cardioprotective (Carrizzo et al. 2016; Chaiwong

Iminosugars

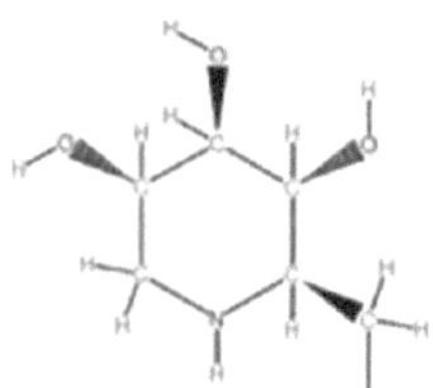

1-deoxynojirimycin (DNJ)

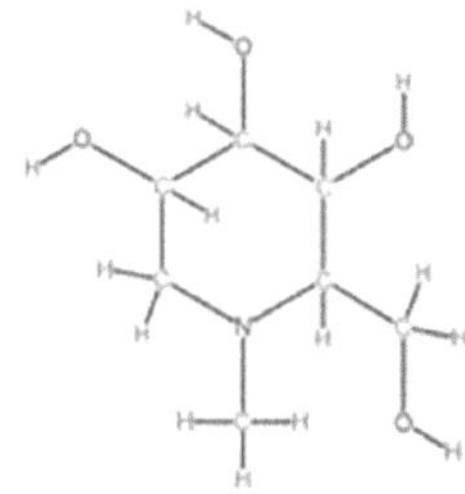

N-Methyl-1-deoxynojirimycin

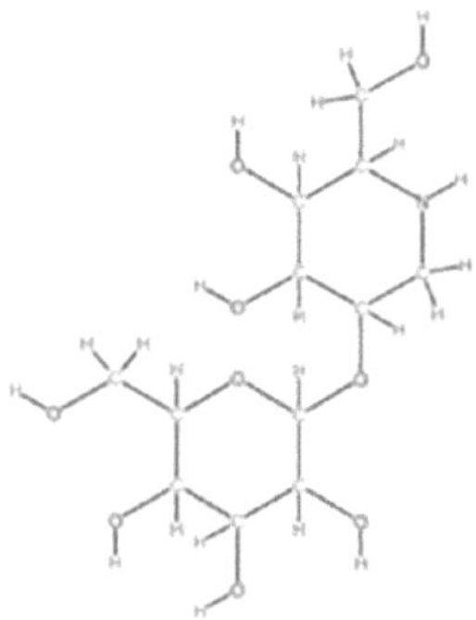

2-O-alpha-D-Galactopyranosyl-1-deoxynojirimycin

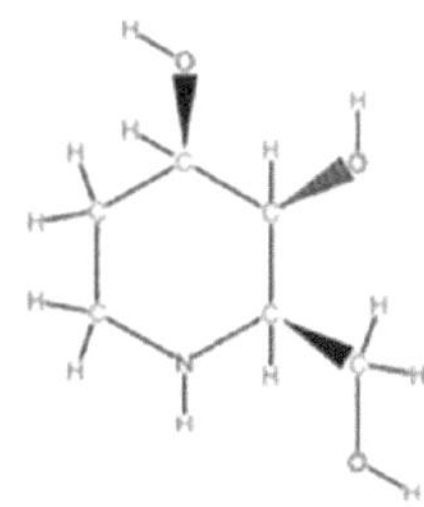

Fagomine

Flavonoids

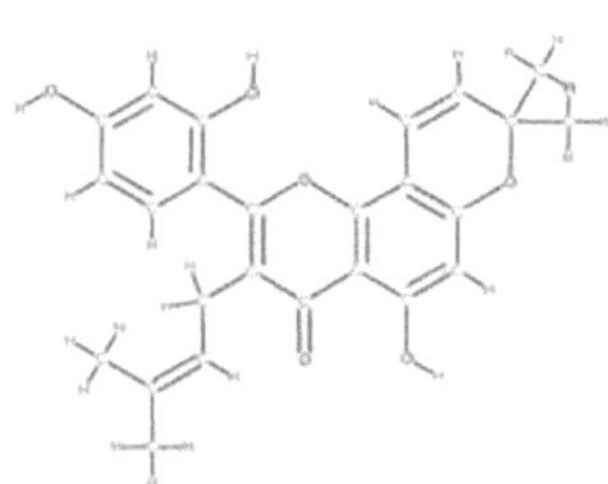

Morusin

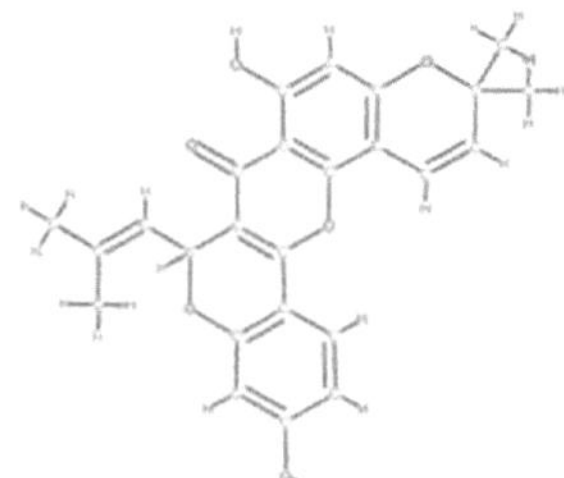

Cyclomorusin

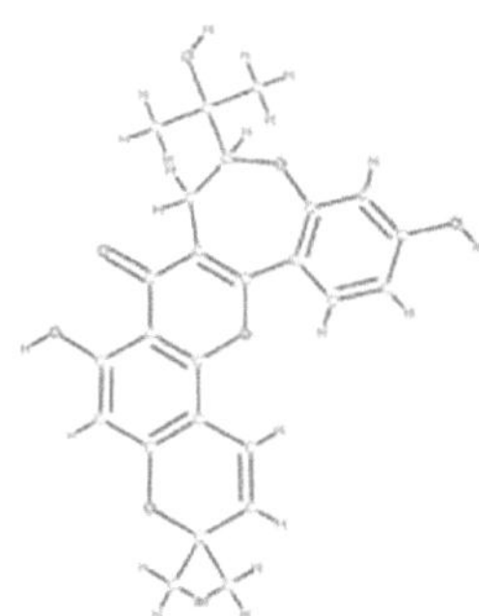

Neocyclomorusin

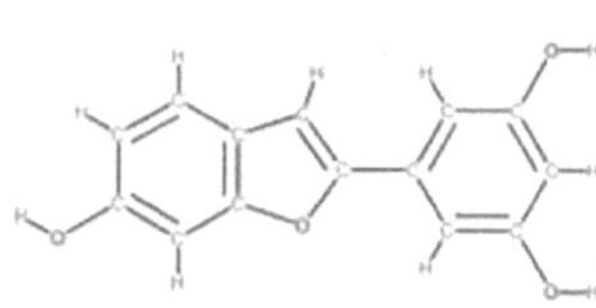

Moracin M

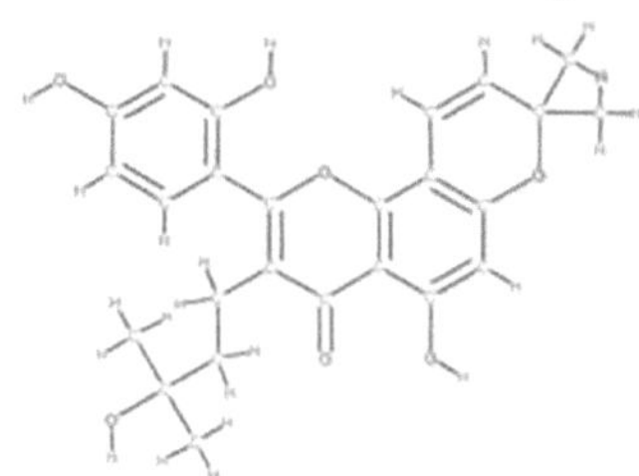

Morusinol

FIGURE 23.2 Phytochemicals reported in different parts of *M. alba*.

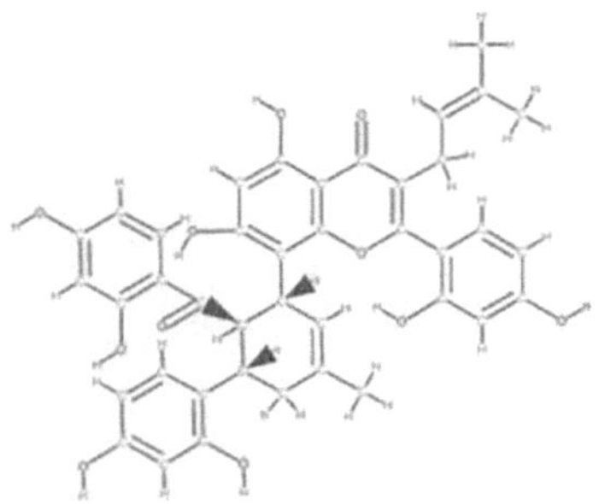

Kuwanon

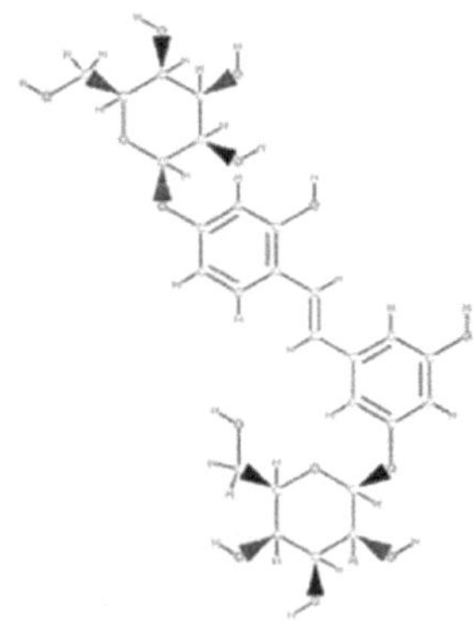

Mulberroside A

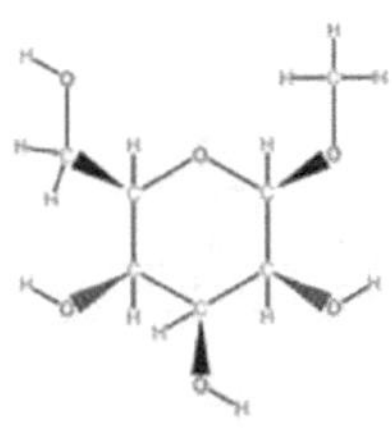

Steppogeninn-4'-O-beta-D-glucoside

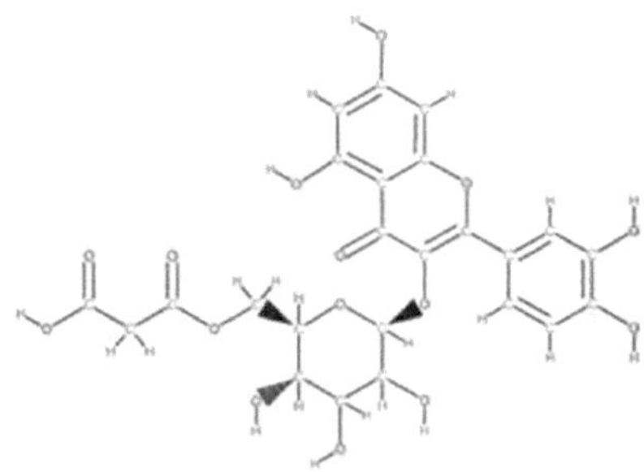

Quercetin-3-(6-malonylglucoside)

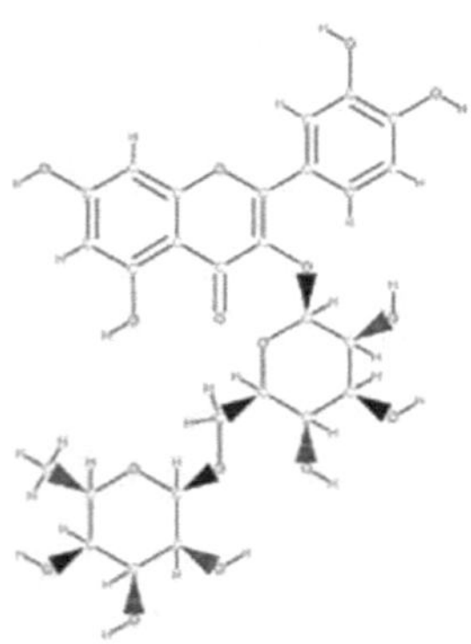

Rutin

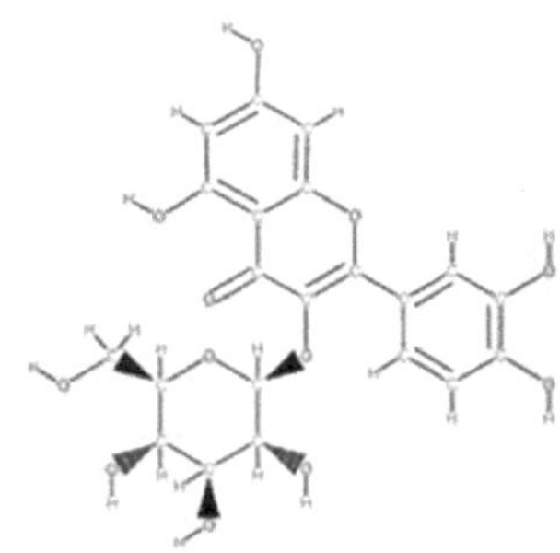

Isoquercitrin

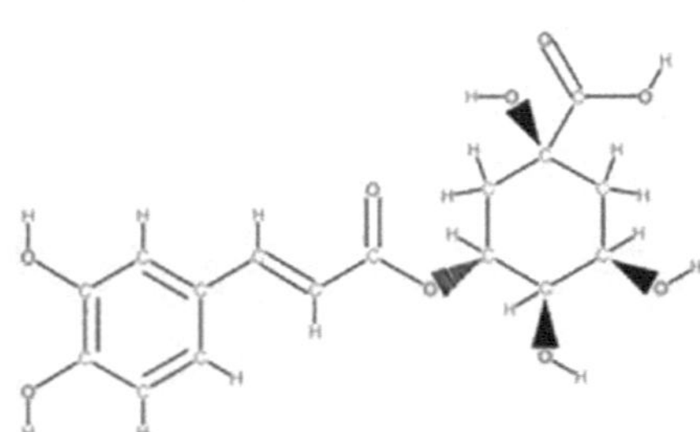

Chlorogenic Acid

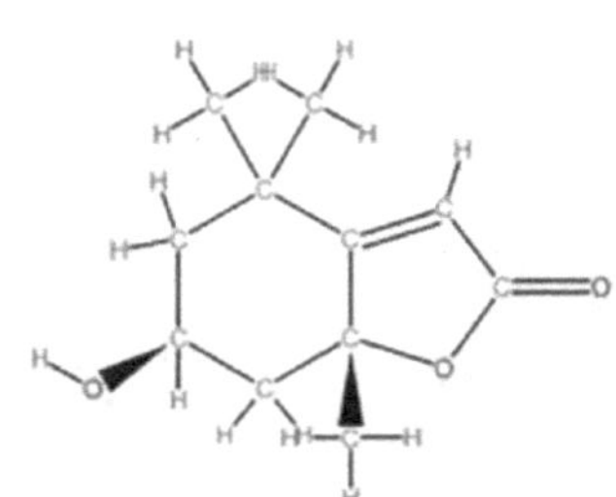

Loliolide

FIGURE 23.2 (Continued)

et al. 2021; Lee et al. 2012), hepatoprotective (Lee et al. 2020; Shendge et al. 2021), renoprotective (Lee et al. 2021; Ullah et al. 2016), neuroprotective (Chang et al. 2021; Liu and Du 2020; Rebai et al. 2017), and antidiabetic activities (Ji et al. 2021; Ramos et al. 2021). In the following section, we discuss the antidiabetic potential of different parts of the *M. alba* including leaf, fruit, and roots. A list of studies on the antidiabetic effect of different extracts or phytochemical constituents of *M. alba* are given in Table 23.1.

TABLE 23.1
List of Studies on the Antidiabetic Effects of Different Solvent Extracts or Phytochemical Constituents of *M. alba*

Name of the extract	Concentration used	Antidiabetic mechanism	Key reference
Leaf			
50% ethanolic leaf extract	1, 5, 15, and 45 μg/mL	Increased GLUT4 translocation to the plasma membrane and PI3K signaling pathway activation result in increased glucose absorption by adipocytes (isolated from diabetic rats)	Naowaboot et al. (2012)
Leaf flavonoids	180 mg/kg b.w/day	Enhanced expression of GLUT4, PGC-1, and p-Akt and p-AMPK phosphorylation in the skeletal muscle, which improved glucose sensitivity in db/db diabetic mice model	Meng et al. (2020)
Aqueous leaf extract	2 g/kg b.w/day	Activated IRS-1/PI3K/GLUT4 signaling pathway, thereby lowered blood glucose, total cholesterol, triglyceride, and low-density lipoprotein levels as well as improved insulin resistance in STZ-diabetic rats	Cai et al. (2016)
Hot water leaf extract	200 mg/kg b.w/day	Reduced glucose levels in STZ-induced diabetic mice	Chen et al. (1995a)
Acetone-water & ethanol-water leaf extracts	6 g/kg diet & 6 g/kg diet	Exhibited hypoglycemic effects and antioxidant effects high-fat diet/STZ-induced diabetic rats	Król et al. (2016)
Aqueous leaf extract	10, 50, and 100 mg/kg b.w/day	Inhibited STZ-induced rise in hypothalamic NOS expression and NO production, thereby reduced diabetic associated hyperphagia in diabetic mice	Jang et al. (2002).
1-DNJ	0.4–520 μM	Significantly inhibited α-glucosidase enzyme and subsequently reduced the glucose levels	Asano et al. (2001)
Hot water extract	0.01–0.04 mg	Significantly inhibited rat intestinal α-glucosidases	Hansawasdi and Kawabata (2006)
50% Ethanolic leaf extracts	-	Inhibited human as well as rat disaccharidases *in vitro* and also prevented elevation of serum glucose levels *in vivo*.	Oku et al. (2006)
Aqueous leaf extracts	20 mg/100 g b.w	Reduced food intake and serum glucose levels in STZ-induced diabetic rats	Musabayane et al. (2006)
90% ethanolic leaf extract	400 mg and 600 mg/kg b.w	Prevented STZ-induced elevation in the levels of blood glucose, triglycerides, LDL, VLDL, blood urea, and cholesterol, except HDL	Mohammadi and Naik (2008)
Leaf powder mixed with rat chow	3% (w/w)	Increased adiponectin expression, decreased TNF- and MCP-1 expression, and decreased levels of oxidative stress indicators in the liver and white adipose tissue (WAT) of db/db mice	Sugimoto et al. (2009)

(Continued)

TABLE 23.1
(Continued)

Name of the extract	Concentration used	Antidiabetic mechanism	Key reference
95% ethanolic leaf extract	10, 30, and 100 μg/kg	Inhibited MCH1, thereby reduced the body weight and hepatic lipid accumulation in diet induced obese mice	Oh et al. (2009)
50% ethanolic leaf extract	0.25, 0.5, and 1 g/kg b.w	Reduced STZ-induced lipid peroxidation in liver, kidney, heart, and aorta, restored the arterial pressure, and prevented AGEs accumulation in STZ-induced diabetic rats	Naowaboot et al. (2009)
Leaf powder/ quercetin 3–6-malonylglucoside	10 g/kg b.w/ 1g/kg b.w	Both diets raised the expression of genes involved in glycolysis, which decreased blood glucose levels in diabetic mice	Katsube et al. (2010)
Leaf powder	3%–6% powder mixed with high-sucrose diets	A significant decrease in blood glucose and insulin levels KK-Ay mice	Tanabe et al. (2011)
Leaf extract enriched with DNJ	3, 6, or 9 mg DNJ	Improved postprandial glucose levels in humans with impaired glucose metabolism.	Asai et al. (2011)
70% aqueous ethanol leaf extract	250 or 750 mg/kg or equal amounts of chlorogenic acid or rutin	Significantly and dose-dependently decreased the non-fasting blood glucose levels in the type 2 diabetic rat model	Hunyadi et al. (2012)
Leaf extract or 1-DNJ	1, 3, 10, and 30 mg/kg/day or 30 mg/kg/day	Blood sugar levels diminished as a result of elevated serum insulin levels. Diabetes-related oxidative stress is less severe in the kidneys of STZ-related nephropathy.	Huang et al. (2014)
Ethanolic leaf extract	100 mg/kg	Lowering of serum glucose, sorbitol, and fructose levels. Downregulation of pro-inflammatory cytokines, VEGF, oxidative stress, and apoptotic markers (caspase-3 and Bax), upregulation of Bcl-2 in the retinas of STZ-rats.	Mahmoud et al. (2017)
Root			
70% alcohol extract root bark extract	600 mg/kg b.w	Reduced lipid peroxidation in pancreatic beta cells, therefore enhanced insulin levels and subsequently decreased the glucose levels in STZ-induced diabetic rats	Singab et al. (2005)
Moracin M, steppogenin-4'-O-beta-D-glucosiade, & mullberroside A	50 and 100 mg/kg b.w	Significantly exerted hypoglycemic effect in alloxan-induced diabetic mice	Zhang et al. (2009)
Morusinol from the root bark	5, 10, and 30 μg/mL (*in vitro*) 20 mg/kg (*in vivo*)	Prevented platelet aggregation and TXB(2) formation in platelets *in vitro* and FeCl(3)-induced thrombosis rodent model *in vivo*	Lee et al. (2012)
Root bark powder	10 g/kg	Reduced the depressed behavior caused by STZ in diabetic rats by upregulating BDNF and phosphorylating ERK and Akt in the prefrontal cortex.	Ye et al. (2017).
Morusalfuran B, mulberrofuran D2, H, & sanggenofuran A from the root bark	Up to 50 μM	Strongly inhibited PTP1B, the negative regulator of insulin signaling	Shrestha et al. (2019)

Name of the extract	Concentration used	Antidiabetic mechanism	Key reference
Moran 20K from root methanolic extract		Showed hypoglycemic effect in STZ-induced diabetic mice	(E. S. Kim et al., 1999)
Fruit			
Ethyl acetate soluble fraction	100 and 200 mg/kg b.w	Reduced fasting blood glucose levels, glycosylated serum protein levels, and increased enzymatic antioxidant activities STZ induced diabetic mice	Wang et al. (2013)
Anthocyanin fraction of fruit	125 and 250 mg/kg b.w	High dose but not the low dose reduced glucose levels in leptin receptor-deficient Zucker diabetic fatty rats	Sarikaphuti et al. (2013)
Anthocyanin fraction of fruit	0.1, 0.3, and 0.5 mg/mL	Exerted hypolipidemic effects via activation of AMPK and subsequent inhibition of acetyl coenzyme A carboxylase and fatty acid synthase, resulting in the inhibition of FA biosynthesis. Additionally increased the expression of lipolytic enzymes (PPAR α and CPT1), resulting in the elevated FA oxidation	Chang et al. (2013)
80% methanolic fruit extract	Chow supplied with 0.5% w/w fruit extract	Enhanced GLUT4 expression and p-Akt and p-AMPK phosphorylation in the muscle of db/db diabetic mice, which increased glucose sensitivity	Choi et al. (2016).
Fruit polysaccharide portion (MFP50 and MFP90)	400 mg/kg	Increased expression levels of insulin signaling proteins (INSR, IRS-2, Akt and GLUT4), thereby reduced fasting blood glucose levels, decreased triglycerides but increased HDL cholesterol in high-fat diet and STZ-induced diabetic rats	Jiao et al. (2017)

23.6 ANTIDIABETIC EFFECTS OF *M. ALBA* LEAVES

Diabetes is characterized by elevated plasma glucose levels, reduced insulin levels, and changes in insulin signaling proteins, including the insulin receptor (IR), insulin receptor substrate (IRS)-1/2, Akt, phosphoinositide 3-kinase (PI3K), and glucose transporter 4 (GLUT4). *In vitro* mulberry leaf extract has been shown to exert its antidiabetic mechanism mainly through its ability to increase the glucose uptake via activation of PI3K signaling pathway and enhanced translocation of GLUT4 to the plasma membrane, thus maintaining glucose homeostasis in diabetes (Naowaboot et al. 2012). Skeletal muscle insulin resistance plays a significant role in the pathogenesis of type 2 diabetes. The leaf flavonoids have also been shown to exert antidiabetic effects through increasing glucose sensitivity via increasing the expression of GLUT4, PGC-1α, and phosphorylation of p-IRS1, p-PI3K, p-Akt, and p-AMPK in skeletal muscle of db/db diabetic mice model. Thus, mulberry leaf extract exerts its antidiabetic effects through the activation of AMPK-PGC1α signaling pathway as well as through the activation of IRS1/PI3/Akt pathway (Meng et al. 2020). Leaf extract was also shown to reduce the body weight, blood glucose, total cholesterol, triglyceride, and low-density lipoprotein (LDL) levels and improve insulin resistance in diabetic rats. This indicates that leaf extract exerts significant antihyperglycemic and antihyperlipidemic activities via the IRS-1/PI3K/Akt/GLUT4 signaling pathway (Cai et al. 2016). Furthermore, the ethanolic leaf extracts of *M. alba* has been reported to induce an increase in glucose uptake, thereby preventing STZ-induced increase in blood glucose levels in mice model of type 2 diabetes (Chen et al. 1995a, 1995b). Besides, the aqueous leaf extract and its active compounds were shown to exert hypoglycemic effects by elevating serum insulin levels and reducing serum glucose levels in STZ-induced diabetic mice (Chen et al. 1995a; Nazari et al. 2013). Short-term oral supplementation of acetone-water and ethanol-water extracts of mulberry leaf have also been reported to exhibit hypoglycemic effects and antioxidant effects in high-fat diet/STZ-induced diabetic rats (Król et al. 2016).

Diabetes is usually associated with an increase in carbohydrate metabolism. In humans, there are different classes of disaccharidases, namely alpha-glucosidases (such as sucrase, isomaltase, maltase, and glucoamylase) and beta-galactosidase (lactase), which are located in the intestinal brush border and are involved in the breakdown of the carbohydrates, especially disaccharides such as sucrose, maltose, and lactose, respectively, and the monosaccharide glucose, resulting in the hyperglycemic condition. Therefore, inhibition of alpha glucosidases is considered a potential strategy for the management of diabetes (Tao et al. 2013). Ethanolic leaf extracts of *M. alba* was reported to significantly inhibit human as well as rat disaccharidase activities *in vitro*. This suggest that *M. alba* exerts its antidiabetic effects via inhibiting the disaccharidases that play a crucial role in the elevation of serum glucose levels in diabetes (Hansawasdi and Kawabata 2006; Oku et al. 2006). Previous studies have investigated the hypoglycemic effects of confections containing mulberry leaf extract in human subjects. It was found that mulberry leaf extract added to confections was able to lower the postprandial glucose and insulin levels in humans. Moreover, they showed an inhibitory effect on intestinal sucrase. From this study's results, it is obvious that mulberry leaf extract embedded confectionaries can be given to diabetic patients to regulate glucose levels (Nakamura et al. 2009).

Mulberry leaves are abundant in various iminosugars, also known as azasugars, such as 1-deoxynojirimycin (DNJ), N-methyl-DNJ, 2-O-α-D-galactopyranosyl-DNJ, and fagomine (Figure 23.2). These iminosugars are critically known for their ability to act as potent inhibitors of glycosidase enzyme, therefore they can exert hypoglycemic effects. As a result, there is an increase in interest in the discovery of naturally derived iminosugars to ameliorate diabetes-associated hyperglycemic conditions. Among these different iminosugars, DNJ is the most predominant iminosugar, constituting about 50% of the total iminosugars. Thus, the antidiabetic efficacy of mulberry leaves is attributed to the presence of DNJ, which is a potent alpha-glucosidase inhibitor (Asano et al. 2001). The efficacy of DNJ-enriched mulberry powder was tested in a human clinical trial and the results revealed that consumption of DNJ-rich mulberry leaf powder significantly reduced the postprandial blood glucose levels in humans, suggesting the beneficial role of mulberry leaf powder as a food supplement to regulate glucose levels in diabetic patients (Kimura et al. 2007). A randomized, double-blind clinical trial was conducted to check the efficacy of long-term (12 weeks) supplementation of mulberry leaf extract enriched with DNJ or the single injection of mulberry leaf extract enriched with DNJ on the postprandial glucose levels in humans with impaired glucose metabolism. These studies found that long-term supplementation of mulberry leaf extract enriched with DNJ has profoundly improved the postprandial glucose levels in humans with impaired glucose metabolism. These trials further warrant the substantial beneficial effects of mulberry on diabetes-associated pathology in humans (Asai et al. 2011). Several other bioactive compounds such as flavonoids (e.g., rutin, isoquercitrin, astragalin), terpene lactone (e.g., loliolide), and iminosugar (e.g., fagomine) were also identified in *M. alba* leaf (Figure 23.2) (Hunyadi et al. 2013; Kim et al. 2014). Previous studies also demonstrated that other mulberry leaf bioactive flavonoids (e.g., chlorogenic acid and rutin but not isoquercitrin) could significantly reduce the non-fasting blood glucose levels in type 2 diabetic rat model in a dose-dependent manner. This study further confirms that mulberry leaf flavonoids (chlorogenic acid and rutin) contribute to the antidiabetic activity of mulberry leaves (Hunyadi et al. 2012; Salemi et al. 2016).

Diabetes is also associated with severe microvascular complications such as retinopathy, neuropathy, and nephropathy. Several studies have investigated the protective effects of *M. alba* leaf extract against diabetic microvascular complications. For example, short-term (one week) supplementation of diabetic rats with leaf extract (1, 3, 10, and 30 mg/kg/day) or 1-DNJ (30 mg/kg/day) resulted in increased serum insulin levels and concomitant reduction in blood glucose levels. Furthermore, diabetic-associated oxidative stress burden was also attenuated in the kidney of rats, indicating that *M. alba* can have significant beneficial effect on the diabetic complications such as diabetic nephropathy (Huang et al. 2014). In another study, ethanolic leaf extract was tested for its beneficial efficacy against diabetic retinopathy. In this study, STZ-induced diabetic rats were orally supplemented with leaf ethanolic extracts (100 mg/kg) for more than 90 days, and the results showed a significant reduction in glucose levels and loss of body weight, diminished levels of sorbitol, fructose, pro-inflammatory cytokines, and oxidative stress markers in retinas. Additionally, it was discovered that treatment with leaf ethanolic extracts raised expression of anti-apoptotic Bcl-2 and lowered expression of apoptotic markers (caspase-3 and Bax), vesicular endothelial growth factor, and other apoptotic markers in diabetic rats' retinas. These studies collectively demonstrate that folium mori has powerful protective effects against various macrovascular problems, such as diabetic retinopathy, by reducing the retina's oxidative stress, apoptosis, inflammation, activation of the polyol pathway, and expression of the growth factor VEGF (Mahmoud et al. 2017). Recent *in vitro* studies also suggest that 1-DNJ and isoquercitrin isolated from the leaf extract possess antidiabetic effects via increasing glucose uptake and inhibiting AGEs formation and p38MAPK/Nf-kB pathway (Li et al. 2022).

Diabetes is characterized by hypopituitarism (undersecretion of pituitary hormones) and hyperphagia (a strong desire to eat food more often, i.e., overeating). The hypothalamus is known to play a critical role in the regulation of food intake and energy imbalance. Therefore, undersecretion of pituitary hormones is usually associated with hyperphagia. Nitric oxide synthase (NOS) is an enzyme which catalyzes the synthesis of nitric oxide (NO), a neurotransmitter in the brain and other tissues. NO in turn is implicated in the regulation of food intake, therefore, inhibition of NOS and subsequent NO levels play a crucial role in the modulation of food intake in diabetics. Research studies have demonstrated that aqueous extract of folium mori significantly inhibited STZ-induced upsurge in hypothalamic NOS expression. This suggests that *M. alba* exerts antidiabetic effects through the modulation of hypopituitarism and hyperphagia via inhibiting expression of NOS in the hypothalamus (Jang et al. 2002). In a short-term study, it was shown that when rats were administered mulberry leaf extracts for five weeks, a reduction in food intake and subsequently a reduction in serum glucose levels was noted, indicating the hypoglycemic ability of mulberry leaves in rats (Musabayane et al. 2006).

Diabetes and associated cardiovascular complications are usually characterized by an altered carbohydrate utilization and enhanced lipolysis, resulting in hyperlipidemia (O'Brien et al. 1998). In a short-term *in vivo* study, oral administration of mulberry leaf ethanolic extract at dose of 600 mg/kg but not 400 mg/kg for five weeks significantly prevented STZ-induced elevation in the levels of blood glucose, triglycerides, LDL, VLDL, blood urea, and cholesterol, except HDL. Notably, treatment with leaf ethanolic extract could also reduce STZ-induced increase in number of pancreatic beta cells and diameter of the islets of Langerhans in rats (Mohammadi and Naik 2008). It was also demonstrated that dietary supplementation of mulberry leaf maintains pancreatic β-cell function and ameliorates insulin resistance in db/db mice (Suthamwong et al. 2020).

Accumulating data suggest that the adipose tissue plays a crucial role in the maintenance of glucose levels (i.e., glucose homeostasis) in healthy as well as diseases conditions. Moreover, adipose tissue functions as an endocrine organ due to its ability to synthesize a number of peptide hormones as well as adipokines (the cytokines released by adipocytes). In fact, there is a direct neural connection between the adipose tissue and the brain. White adipose tissues release different pro-inflammatory adipokines such as TNF-alpha and monocyte chemoattractant protein-1 (MCP-1). As an endocrine organ, adipose tissue secretes a peptide hormone adiponectin which exhibits antidiabetic and anti-atherogenic properties (Guilherme et al. 2008). Dysregulation of adipokines leads to the development of atherosclerosis, a cardiovascular disease. Experimental studies revealed that in db/db mice fed on mulberry leaf powder–containing chow for 12 weeks, a decrease in both blood glucose levels and plasm triglycerides were observed. Moreover, co-treatment with mulberry leaf powder and pioglitazone (an antidiabetic drug) exerted additive effects when compared to pioglitazone treatment alone. In addition, co-treatment attenuated pioglitazone-induced increase in body weight. Mice fed on mulberry leaf powder–containing diets showed increased levels of adiponectin and decreased levels of inflammatory markers such as TNF-alpha and MCP-1 in white adipose tissue. Furthermore, mulberry leaf power attenuated lipid peroxidation and oxidative stress in white adipose tissue and liver, indicating that mulberry leaf exerts protective effects by ameliorating the adipocytokine dysregulation via inhibiting the oxidative stress. This study also suggests that mulberry leaf might be also useful for reducing any adverse effects of pioglitazone (Sugimoto et al. 2009). The KK-Ay mice exhibit diabetic features such as insulin resistance, dyslipidemia, altered adipokine expression, and obesity, therefore these mice serve as another suitable animal model to study the various aspects of type 2 diabetes. In another study, KK-Ay mice were fed with diets containing mulberry leaf powder which resulted in a significant decrease in blood glucose and insulin levels, indicating potent antihyperglycemic and antihyperinsulinemic effects of mulberry in diabetes. Thus, it can also be suggested that dietary intake of mulberry leaf might be beneficial in reducing insulin resistance in diabetic patients (Tanabe et al. 2011).

Melanin concentrating hormone (MCH), a hormone produced in the hypothalamus, is connected to the metabolism of food and energy. Melanin concentrating hormone receptor subtype 1 is a receptor linked to the G-protein coupled receptor family, and it interacts with MCH to produce these actions (MCH1). MCH1 has therefore been thought of as a therapeutic target for metabolic disease (Pissios and Maratos-Flier 2003). A study was done utilizing the europium-ligand binding test, in which the MCH peptide was labeled with europium at the N1 position to determine the antagonistic effect of mulberry ethanolic leaf extract on MCH-1. The results showed that treatment with ethanolic extract substantially and concentration-dependently blocked the binding of Eu-MCH to MCH1 receptor. Additional *in vivo* experiments showed that oral treatment of mulberry leaf ethanolic extract decreased body weight and hepatic fat buildup in mice that had been made obese by diet. These *in vitro* and *in vivo* experiments indicate that MCH1 inhibition is the mechanism via which mulberry leaf extract exerts its antidiabetic benefits (Oh et al. 2009).

Diabetes-related hyperglycemia results in increased glucose oxidation, the production of advanced glycation end products (AGEs), lipid peroxidation, and a variety of reactive oxygen species (ROS). All of these things increase the burden of oxidative stress, which in turn causes diabetic vascular problems. The STZ-induced lipid peroxidation in the liver as well as other extra hepatic organs like the kidney, heart, and aorta was significantly decreased when ethanolic mulberry leaf extracts (0.5 and 1 g/kg body weight) were administered orally. Additionally, diabetic rats' arterial pressure was also restored (Naowaboot et al. 2009). These authors have also reported that ethanolic mulberry leaf extract could also reduce the levels of AGEs in STZ-induced diabetic rats, indicating the potent antiglycation effects of mulberry (Naowaboot et al. 2009). In an eight-week experiment, diabetic mice were orally supplemented with either mulberry leaf powder (10 g/kg b.w) or mulberry leaf flavonoid glucoside quercetin 3–6-malonylglucoside (1g/kg b.w), and this study found that both diets increased the expression of glycolysis-related genes, thereby reducing the blood glucose levels in diabetic mice. This further suggests that antihyperglycemic effects of mulberry are in part associated with quercetin 3–6-malonylglucoside, a major flavonoid compound of the mulberry leaves (Katsube et al. 2010).

A randomized, single-blind, placebo-controlled, cross-over study was carried out by Jóozefczuk et al. (2017) to examine the impact of mulberry leaf extract added to cornflakes on postprandial glycemic levels. According to the study's findings, participants' starch digestion and absorption were lowered by consuming a single dosage of cornflakes (50 g cornflakes plus 100 ml low-fat milk) enhanced with mulberry leaf extract (36 mg of active ingredient 1-deoxynojirimycin). These results imply that mulberry leaf extract may help diabetic individuals manage their postprandial glucose levels. In order to determine the safe dosage of mulberry leaf extract on the glycemic profiles in obese people with borderline diabetes, Thaipitakwong et al. (2020) did a randomized controlled trial. The results of this study showed that taking leaf extract supplements for 12 weeks reduced insulin resistance and fasting plasma glucose, and that 12 mg of mulberry leaf DNJ was the lowest effective dose for reducing postprandial hyperglycemia. Although bloating, gas, and loose stools are a few unfavorable effects of mulberry leaf consumption, no major side effects were seen in this study. Patients with type 2 diabetes received 300 mg of *M. alba* leaf extract twice daily for 12 weeks in another randomized, double-blind, placebo-controlled experiment. According to the study's findings, compared to a placebo, *M. alba* leaf extract significantly decreased insulin and lipid peroxidation levels while raising HDL cholesterol levels. These clinical studies indicate that using *M. alba* leaf extract for the treatment of type 2 diabetes is safe (Taghizadeh et al. 2022).

23.7 ANTIDIABETIC EFFECTS OF *M. ALBA* ROOTS

The roots of *M. alba* have been widely used as antidiabetic remedy in traditional Chinese medicine. A new glycoprotein called Moran 20k isolated from the root methanolic extracts of M. *alba* was shown to possess antihyperglycemic effects in STZ-induced diabetes in mice. Similar to insulin, this Moran 20k was found to be composed of more than 20% of amino acids like cysteine and serine (Kim et al. 1999). Hydroalcoholic root bark extract of *M. alba* significantly prevented lipid peroxidation in pancreatic beta cells, subsequently enhanced insulin levels, and reduced glucose levels in STZ-treated diabetic rats. This potent hypoglycemic effect of root extract is attributed to the presence of flavonoids such as morusin, cyclomorusin, neocyclomorusin, and kuwanon E (Singab et al. 2005).

Zhang et al. have isolated different phenolic compounds such as moracin M, steppogenin-4'-O-beta-D-glucosiade, and mullberroside A (Figure 23.2) from the root bark of *M. alba*. All these phenolic compounds were found to exhibit hypoglycemic effects on alloxan-induced diabetic mice model. Among these phenolics, steppogenin-4'-O-beta-D-glucosiade showed significant hypoglycemic effect at a dose of 50 mg/kg while the other two compounds at a dose of 100 mg/kg exerted remarkable hypoglycemic effect on diabetic mice (Zhang et al. 2009). Type 2 diabetes (T2D) increases the risk for developing cardiovascular diseases. Diabetic patients usually show increased incidence of acute thrombotic events (e.g., myocardial infarction (MI) and stroke). Morusinol, a flavonoid isolated from the root bark of *M. alba*, was investigated for its antithrombotic effect. *In vitro* assays such as rabbit platelet aggregation and thromboxane B(2) formation assays were generally used to determine the anti-platelet potential of natural compounds. Morusinol significantly inhibited prevented platelet aggregation and TXB (2) formation *in vitro* in a concentration-dependent manner. Ferric chloride (FeCl(3))-induced thrombosis rodent model is used to investigate arterial thrombus formation *in vivo*. Authors have also reported that short-term (three days) oral administration of morusinol (20 mg/kg) substantially augmented the time to occlusion *in vivo* when compared to control group. This indicates that morusinol's anti-platelet therapeutic activity may be beneficial for the management of diabetic-associated cardiovascular complications (Lee et al. 2012).

Diabetic patients usually suffer from depression. *In vivo* studies employ various tests such as forced swimming test, locomotor activity test, and open field test to investigate depressive-like behaviors. Previous studies have demonstrated that treatment with root bark of *M. alba* attenuated the STZ-induced depressive behavior in diabetic rats. Additionally, administration of root bark extract to diabetic rats resulted in an increase in the expression of brain-derived neurotrophic factor (BDNF) and the phosphorylation of extracellular signal-regulated kinase (ERK) and Akt in the prefrontal cortex. This study's results further points to the fact that root bark of *M. alba* can be used as therapeutic for the treatment of depression seen in diabetics (Ye et al. 2017).

Protein-tyrosine phosphatase 1B (PTP1B) enzyme belongs to the family of protein tyrosine phosphatases which can dephosphorylate tyrosine phosphorylated proteins. PTP1B is widely expressed on muscle, brain, adipose tissue, and liver and is involved in the insulin signaling pathway, where it removes phosphate groups from insulin receptor or insulin receptor substrate 1. This leads to reduction in insulin sensitivity or shutting down of insulin signaling pathway. Thus, PTB1 acts as a negative regulator of the insulin signaling pathway. Therefore, natural inhibitors of PTP1B may serve as potent therapeutic compounds for the management of type 2 diabetes (Zhao et al. 2018). Arylbenzofuran analogs such as morusalfuran B, mulberrofuran D2, mulberrofuran H, and sanggenofuran A that were isolated from the root bark of *M. alba* have been reported to possess potent inhibitory action on the PTP1B. Therefore, these mulberry compounds may be developed as potent antidiabetic therapeutics to ameliorate the insulin signaling in type 2 diabetes (Shrestha et al. 2019; Su et al. 2020; Ha et al. 2020).

23.8 ANTIDIABETIC EFFECTS OF *M. ALBA* FRUITS

Both *in vitro* and *in vivo* studies showed that ethyl acetate soluble fraction of mulberry fruit can exhibit potent alpha glucosidase inhibitory effect as well as antioxidant activity in STZ-induced diabetic rat model. Ethyl acetate soluble fraction significantly reduced the blood glucose levels and also enhanced the endogenous enzymatic antioxidant levels. This indicates that mulberry fruit's antioxidant activity may be beneficial for the amelioration of hyperglycemia and associated oxidative stress burden (Wang et al. 2013). Mulberry fruit is rich in anthocyanins, and the reported antioxidant activity of mulberry fruit could be possibly linked to its potent antidiabetic activity. Moreover, the anthocyanin fraction reduced the blood glucose levels in diabetic rats. Therefore, consumption of mulberry fruit may have beneficial effects against diabetes associated oxidative stress burden (Sarikaphuti et al. 2013). Obesity is also a risk factor for the development of diabetes and hyperlipidemia. The fruit of *M. alba* is known for its ability to modulate lipid levels in hyperlipemia. *In vitro* studies have suggested that anthocyanin extract of mulberry improve hyperlipidemic condition in diabetes. The anthocyanin portion was shown to activate adenosine monophosphate-activated protein kinase (AMPK) and subsequent inhibition of acetyl coenzyme A carboxylase and fatty acid synthase, therefore the inhibition of fatty acid (FA) biosynthesis. Furthermore, the anthocyanin fraction was also shown to increase the expression of lipolytic enzymes such as peroxisome proliferator activated receptor α and carnitinepalmitol- transferase-1 (CPT1), therefore promoting fatty acid oxidation. This hypolipidemic action of anthocyanin portion of mulberry fruit could be the other possible reason for the antidiabetic efficacy of mulberry fruit (Chang et al. 2013). Skeletal muscle insulin resistance plays a significant role in the pathogenesis of type 2 diabetes. Similar to leaf extracts, the fruit extract has also been shown to exert its antidiabetic effects through increasing the glucose sensitivity via increasing the expression of GLUT4, PGC-1α, and phosphorylation of p-Akt and p-AMPK in skeletal muscle of db/db diabetic mice model (Choi et al. 2016). Likewise, the polysaccharide portion of mulberry fruit has been demonstrated to possess antidiabetic effects by inducing the expression levels of insulin signaling proteins such as InsR, IRS-2, Akt, and GLUT4, thereby reducing the fasting blood glucose levels and triglyceride levels and enhancing the HDL cholesterol levels in high-fat diet- and STZ-induced diabetic rat model (Jiao et al. 2017).

23.9 ANTIDIABETIC EFFECTS OF TWIGS OF *M. ALBA*

Dried twigs of *Morus alba* are known as ramulus mori. Polysaccharides isolated from ramulus mori (RMP) were shown to decrease the expression of pro-inflammatory markers (tumor necrosis factor-α (TNF-α), interleukin-8 (IL-8), interleukin-6 (IL-6), and cyclo-oxygenase-2 (COX-2)), increase the levels of endogenous enzymatic antioxidant levels (SOD and glutathione reductase), increase insulin levels, and reduce oxidative stress in the pancreas of STZ-induced diabetic mice. This further points to the fact that mulberry prevents diabetes-associated inflammation and oxidative stress (Guo et al. 2013).

23.10 TRADITIONAL AND OTHER POTENTIAL USES

The leaves of *M. alba*, generally known as folium mori, have been widely used in the traditional medicine systems of various regions across the world including India, China, Chile, Korea, Jordan, Thailand (Chichioco-Hernandez et al. 2011; Hansawasdi and Kawabata 2006; Lemus et al. 1999; Oh et al. 2009; P et al. 2011), and the Caribbean (Lans 2006) as a therapeutic remedy for diabetes.

Mulberry leaves have long been used as a feed in sericulture. In addition, mulberry leaves and fruits have been largely used as ethnomedicine due to the presence of a variety of health promoting bioactive nutrients such as polyphenols, flavonoids, alkaloids, anthocyanins, vitamins, and minerals (Wang et al. 2013). Several other functional foods such as teas, beverages, and noodles are usually prepared from the mulberry in Asian countries including China, Thailand, Japan, and Korea. In Korea, folium mori is used as herbal remedy for the treatment of body swelling, dropsy, and vitamin deficiency (beriberi) (Kim et al. 2014). The mulberry plant has been used for centuries in Chinese medicine to control blood pressure, strengthen joints, protect the liver, improve eyesight, and lower body temperature when a person has a fever (Katsube et al. 2010). In Thailand, people drink mulberry tea popularly known as Mon Tea as a therapeutic remedy for diabetes (Hansawasdi and Kawabata 2006).

23.11 TOXICITY AND SAFETY

Assessment of the safety and toxicity of mulberry is crucial for the development of various functional foods or potential drugs of mulberry for the implementation in evidence-based practice and also to benefit consumers. *In vivo* research studies demonstrated that *M. alba* is not associated with either cytotoxicity or genotoxicity. Chang et al. evaluated the subchronic oral toxicity and genotoxicity of mulberry fruit in rodents. In this study, rats were orally administered with (up to 1000 mg/kg) of mulberry fruit for 90 days. This subchronic oral toxic investigation has documented no animal deaths or adverse effects, such as changes in appetite, water and food intake, body weight, or hematological parameters. Furthermore, mulberry fruit did not show any genotoxicity against all tested different strains of *Salmonella typhimurium* (Chang et al. 2016). In a 12-week clinical prospective study, obese persons with borderline diabetes were given 12 mg of mulberry leaf powder, and the results revealed that 12 mg leaf powder is enough to decrease the postprandial glucose levels and glycated hemoglobin. However, human subjects showed mild gastrointestinal adverse events such as bloating, flatulence, and loose stools but no serious side effects (Thaipitakwong et al. 2020). Mehmood et al. (2016) conducted a door-to-door survey to assess the health risk associated with the consumption of mulberry fruit. The authors reported that consumption of mulberry fruits by adults did not show any noncarcinogenic or carcinogenic effects in the consumers. Therefore, it can be suggested that mulberry fruit and folium tea may be developed as functional foods for the safe use in the management of diabetes.

23.12 CONCLUSION AND FUTURE REMARKS

Morus alba has long been used in various traditional medicine systems across the world. Various parts of the plant including leaves, fruits, root bark, and twigs of the plant have been reported to possess a variety of bioactive flavonoids, alkaloids, and polyphenols which are responsible for possible health-promoting effects. Several *in vitro* and *in vivo* studies have demonstrated that different solvent extracts of various parts of the mulberry exhibited antidiabetic activity. Mulberry exerts antidiabetic effects by lowering blood glucose levels (hypoglycemic effect), increasing insulin sensitivity, and ameliorating insulin resistance, through the activation of different signaling pathways such as AMPK-PGC1α signaling pathway, IRS1/PI3/Akt pathway, and IRS-1/PI3K/Akt/GLUT4 signaling pathway, and also through the inhibition of AGEs formation and p38MAPK/Nf-kB pathway. The bioactive compounds of mulberry and different solvent extracts exhibit potent inhibitory action on alpha glucosidase and PTB1, thereby lowering the diabetic-associated enhanced glucose levels. Even though the chemical makeup of the mulberry's many components has been documented, further research is still needed to pinpoint any unidentified substances and the biological processes by which they might contribute to the fruit's health benefits. Therefore, it is essential to unravel complete phytochemical composition of mulberry for its use in the clinical trials.

REFERENCES

Asai, A., K. Nakagawa, O. Higuchi, T. Kimura, Y. Kojima, J. Kariya, T. Miyazawa & S. Oikawa. 2011. Effect of mulberry leaf extract with enriched 1-deoxynojirimycin content on postprandial glycemic control in subjects with impaired glucose metabolism. *Journal of Diabetes Investigation* 2(4), 318–323.

Asano, N., T. Yamashita, K. Yasuda, K. Ikeda, H. Kizu, Y. Kameda, A. Kato, R. J. Nash, H. S. Lee & K. S. Ryu. 2001. Polyhydroxylated alkaloids isolated from mulberry trees (Morusalba L.) and silkworms (Bombyx mori L.). *Journal of Agricultural and Food Chemistry* 49(9), 4208–4213.

Cai, S., W. Sun, Y. Fan, X. Guo, G. Xu, T. Xu, Y. Hou, B. Zhao, X. Feng & T. Liu. 2016. Effect of mulberry leaf (Folium Mori) on insulin resistance via IRS-1/PI3K/Glut-4 signalling pathway in type 2 diabetes mellitus rats. *Pharmaceutical Biology* 54(11), 2685–2691.

Carrizzo, A., M. Ambrosio, A. Damato, M. Madonna, M. Storto, L. Capocci, P. Campiglia, E. Sommella, V. Trimarco, F. Rozza, R. Izzo, A. A. Puca & C. Vecchione. 2016. Morus alba extract modulates blood pressure homeostasis through eNOS signaling. *Molecular Nutrition & Food Research* 60(10), 2304–2311.

Chaiwong, S., U. Chatturong, R. Chanasong, W. Deetud, K. To-On, S. Puntheeranurak, E. Chulikorn, T. Kajsongkram, V. Raksanoh, K. Chinda, N. Limpeanchob, K. Trisat, J. Somran, N. Nuengchamnong, P. Prajumwong & K. Chootip. 2021. Dried mulberry fruit ameliorates cardiovascular and liver histopathological changes in high-fat diet-induced hyperlipidemic mice. *Journal of Traditional and Complementary Medicine* 11(4), 356–368.

Chan, E. W. C., S. K. Wong, J. Tangah, T. Inoue & H. T. Chan. 2020. Phenolic constituents and anticancer properties of Morus alba (white mulberry) leaves. *Journal of Integrative Medicine* 18(3), 189–195.

Chang, B. Y., S. B. Kim, M. K. Lee, H. Park & S. Y. Kim. 2016. Nonclinical safety assessment of morus alba L. fruits: Study of 90-D toxicity in Sprague Dawley rats and genotoxicity in Salmonella. *Journal of Food Science* 81(5), T1328–1335.

Chang, J. J., M. J. Hsu, H. P. Huang, D. J. Chung, Y. C. Chang & C. J. Wang. 2013. Mulberry anthocyanins inhibit oleic acid induced lipid accumulation by reduction of lipogenesis and promotion of hepatic lipid clearance. *Journal of Agricultural and Food Chemistry* 61(25), 6069–6076.

Chang, Y., C. F. Hung, H. H. Ko & S. J. Wang. 2021. Albanin A, derived from the root bark of morus alba L., depresses glutamate release in the rat cerebrocortical nerve terminals via Ca(2+)/calmodulin/adenylate cyclase 1 suppression. *Journal of Medicinal Food* 24(3), 209–217.

Chen, F., N. Nakashima, I. Kimura & M. Kimura. 1995a. Hypoglycemic activity and mechanisms of extracts from mulberry leaves (folium mori) and cortex mori radicis in streptozotocin-induced diabetic mice. *Yakugaku Zasshi* 115(6), 476–482.

Chen, F., N. Nakashima, I. Kimura, M. Kimura, N. Asano & S. Koya. 1995b. Potentiating effects on pilocarpine-induced saliva secretion, by extracts and N-containing sugars derived from mulberry leaves, in streptozocin-diabetic mice. *Biological and Pharmaceutical Bulletin* 18(12), 1676–1680.

Chichioco-Hernandez, C., J. Wudarski, L. Gevaert & L. Verschaeve. 2011. Evaluation of cytotoxicity and genotoxicity of some Philippine medicinal plants. *Pharmacognosy Magazine* 7(26), 171–175.

Choi, K. H., H. A. Lee, M. H. Park & J. S. Han. 2016. Mulberry (Morus alba L.) fruit extract containing anthocyanins improves glycemic control and insulin sensitivity via activation of AMP-activated protein kinase in diabetic C57BL/Ksj-db/db mice. *Journal of Medicinal Food* 19(8), 737–745.

Eo, H. J., J. H. Park, G. H. Park, M. H. Lee, J. R. Lee, J. S. Koo & J. B. Jeong. 2014. Anti-inflammatory and anti-cancer activity of mulberry (Morus alba L.) root bark. *BMC Complementary and Alternative Medicine* 14, 200.

Guilherme, A., J. V. Virbasius, V. Puri & M. P. Czech. 2008. Adipocyte dysfunctions linking obesity to insulin resistance and type 2 diabetes. *Nature Reviews. Molecular Cell Biology* 9(5), 367–377.

Guo, C., R. Li, N. Zheng, L. Xu, T. Liang & Q. He. 2013. Anti-diabetic effect of ramulus mori polysaccharides, isolated from Morus alba L., on STZ-diabetic mice through blocking inflammatory response and attenuating oxidative stress. *International Immunopharmacology* 16(1), 93–99.

Ha, M. T., S. Shresth, T. H. Tran, J. A. Kim, M. H. Woo, J. S. Choi & B. S. Min. 2020. Inhibition of PTP1B by farnesylated 2-arylbenzofurans isolated from Morus alba root bark: Unraveling the mechanism of inhibition based on in vitro and in silico studies. *Archieves of Pharmacol Reserach* 43(9), 961–975.

Hansawasdi, C. & J. Kawabata. 2006. Alpha-glucosidase inhibitory effect of mulberry (Morus alba) leaves on Caco-2. *Fitoterapia* 77(7–8), 568–573.

Huang, S. S., Y. H. Yan, C. H. Ko, K. M. Chen, S. C. Lee & C. T. Liu. 2014. A comparison of food-grade Folium mori (Sāng Yè) extract and 1-deoxynojirimycin for glycemic control and renal function in streptozotocin-induced diabetic rats. *Journal of Traditional and Complementary Medicine* 4(3), 162–170.

Hunyadi, A., E. Liktor-Busa, A. Márki, A. Martins, N. Jedlinszki, T. J. Hsieh, M. Báthori, J. Hohmann & I. Zupkó. 2013. Metabolic effects of mulberry leaves: Exploring potential benefits in type 2 diabetes and hyperuricemia. *Evidence Based Complementary and Alternative Medicine* 2013, 948627.

Hunyadi, A., A. Martins, T. J. Hsieh, A. Seres & I. Zupkó. 2012. Chlorogenic acid and rutin play a major role in the in vivo anti-diabetic activity of Morus alba leaf extract on type II diabetic rats. *PLoS ONE* 7(11), e50619.

Jang, M. H., H. Kim, M. C. Shin, B. V. Lim, T. H. Lee, S. B. Jung, C. J. Kim & E. H. Kim. 2002. Administration of Folium mori extract decreases nitric oxide synthase expression in the hypothalamus of streptozotocin-induced diabetic rats. *Japanese Journal of Pharmacology* 90(2), 189–192.

Ji, S., C. Zhu, S. Gao, X. Shao, X. Che, H. Zhang & D. Tang. 2021. Morus alba leaves ethanol extract protects pancreatic islet cells against dysfunction and death by inducing autophagy in type 2 diabetes. *Phytomedicine* 83, 153478.

Jiao, Y., X. Wang, C. Jiang, F. Kong, S. Wang & C. Yan. 2017. Antidiabetic effects of Morus alba fruit polysaccharides on high-fat diet- and streptozotocin-induced type 2 diabetes in rats. *Journal of Ethnopharmacology* 199, 119–127.

Józefczuk, J., K. Malikowska, A. Glapa, B. Stawińska-Witoszyńska, J. K. Nowak, J. Bajerska, A. Lisowska & J. Walkowiak. 2017. Mulberry leaf extract decreases digestion and absorption of starch in healthy subjects-A randomized, placebo-controlled, crossover study. *Advances in Medical Sciences* 62(2), 302–306.

Katsube, T., M. Yamasaki, K. Shiwaku, T. Ishijima, I. Matsumoto, K. Abe & Y. Yamasaki. 2010. Effect of flavonol glycoside in mulberry (*Morus alba* L.) leaf on glucose metabolism and oxidative stress in liver in diet-induced obese mice. *Journal of the Science of Food and Agriculture* 90(14), 2386–2392.

Kim, D. S., Y. M. Kang, W. Y. Jin, Y. Y. Sung, G. Choi & H. K. Kim. 2014. Antioxidant activities and polyphenol content of Morus alba leaf extracts collected from varying regions. *Biomedical Reports* 2(5), 675–680.

Kim, E. S., S. J. Park, E. J. Lee, B. K. Kim, H. Huh & B. J. Lee. 1999. Purification and characterization of Moran 20K from Morus alba. *Archieves of Pharmacol Research* 22(1), 9–12.

Kimura, T., K. Nakagawa, H. Kubota, Y. Kojima, Y. Goto, K. Yamagishi, S. Oita, S. Oikawa & T. Miyazawa. 2007. Food-grade mulberry powder enriched with 1-deoxynojirimycin suppresses the elevation of postprandial blood glucose in Humans. *Journal of Agricultural and Food Chemistry* 55(14), 5869–5874.

Król, E., M. Jeszka-Skowron, Z. Krejpcio, E. Flaczyk & R. W. Wójciak. 2016. The effects of supplementary mulberry leaf (Morus alba) extracts on the trace element status (Fe, Zn and Cu) in relation to diabetes management and antioxidant indices in diabetic rats. *Biological Trace Element Research* 174(1), 158–165.

Lans, C. A. 2006. Ethnomedicines used in Trinidad and Tobago for urinary problems and diabetes mellitus. *Journal of Ethnobiology and Ethnomedicine* 2, 45–45.

Lee, D., S. R. Lee, B. J. Park, J. H. Song, J. K. Kim, Y. Ko, K. S. Kang & K. H Kim. 2021. Identification of renoprotective phytosterols from mulberry (Morus alba) Fruit against cisplatin-induced cytotoxicity in LLC-PK1 kidney cells. *Plants (Basel)* 10(11).

Lee, J. J., H. Yang, Y. M. Yoo, S. S. Hon, D. Lee, H. J. Lee, H. J. Lee, C. S. Myung, K. C. Choi & E. B. Jeung. 2012. Morusinol extracted from Morus alba inhibits arterial thrombosis and modulates platelet activation for the treatment of cardiovascular disease. *Journal of Atherosclerosis and Thrombosis* 19(6), 516–522.

Lee, M. R., J. E. Kim, J. W. Park, M. J. Kang, H. J. Choi, S. J. Bae, Y. W. Choi, K. M. Kim, J. T. Hong & D. Y. Hwang. 2020. Fermented mulberry (Morus alba) leaves suppress high fat diet-induced hepatic steatosis through amelioration of the inflammatory response and autophagy pathway. *BMC Complementary Medicine and Therapies* 20(1), 283.

Lemus, I., R. García, E. Delvillar & G. Knop. 1999. Hypoglycaemic activity of four plants used in Chilean popular medicine. *Phytotherapy Research* 13(2), 91–94.

Li, J. S., T. Ji, S. L. Su, Y. Zhu, X. L. Chen, E. X. Shang, S. Guo, D. W. Qian & J. A. Duan. 2022. Mulberry leaves ameliorate diabetes via regulating metabolic profiling and AGEs/RAGE and p38 MAPK/NF-κB pathway. *Journal of Ethnopharmacology* 283, 114713.

Li, Z., X. Chen, G. Liu, J. Li, J. Zhang, Y. Cao & J. Miao. 2021. Antioxidant activity and mechanism of resveratrol and polydatin isolated from mulberry (Morus alba L.). *Molecules* 26(24).

Lim, H. S., H. Ha, H Lee, J. K. Lee, M. Y. Lee & H. K. Shin. 2014. Morus alba L. suppresses the development of atopic dermatitis induced by the house dust mite in NC/Nga mice. *BMC Complementary and Alternative Medicine* 14, 139.

Liu, D. & D. Du. 2020. Mulberry fruit extract alleviates cognitive impairment by promoting the clearance of amyloid-β and inhibiting neuroinflammation in Alzheimer's disease mice. *Neurochemical Research* 45(9), 2009–2019.

Mahmoud, A. M., S. M. Abd El-Twab & E. S. Abdel-Reheim. 2017. Consumption of polyphenol-rich Morus alba leaves extract attenuates early diabetic retinopathy: The underlying mechanism. *European Journal of Nutrition* 56(4), 1671–1684.

Martins, B. A., D. Sande, M. D. Solares & J. A. Takahashi. 2021. Antioxidant role of morusin and mulberrofuran B in ethanol extract of Morus alba roots. *Natural Products Research* 35(24), 5993–5996.

Mehmood, A. A., M. H. Shah, X. Guo & N. Khan. 2016. Comparison of nutritional value, antioxidant potential, and risk assessment of the mulberry (Morus) fruits. *International Journal of Fruit Science* 16(2), 113–134.

Meng, Q., X. Qi, Y. Chao, Q. Chen, P. Cheng, X. Yu, M. Kuai, J. Wu, W. Li, Q. Zhang, Y. Li & H. Bian. 2020. IRS1/PI3K/AKT pathway signal involved in the regulation of glycolipid metabolic abnormalities by Mulberry (Morus alba L.) leaf extracts in 3T3-L1 adipocytes. *Chinese Medicine* 15, 1.

Meng, Q., X. Qi, Y. Fu, Q. Chen, P. Cheng, X. Yu, X. Sun, J. Wu, W. Li, Q. Zhang, Y. Li, A. Wang & H. Bian. 2020. Flavonoids extracted from mulberry (Morus alba L.) leaf improve skeletal muscle mitochondrial function by activating AMPK in type 2 diabetes. *Journal of Ethnopharmacology* 248, 112326.

Mohammadi, J. & P. R. Naik. 2008. Evaluation of hypoglycemic effect of Morus alba in an animal model. *Indian Journal of Pharmacology* 40(1), 15–18.

Ramos, J. G.M., A. T. Esteves Pairazamán, M. E. S. Mocarro Willis, S. Collantes Santisteban & E. Caldas Herrera. 2021. Medicinal properties of Morus alba for the control of type 2 diabetes mellitus: A systematic review. *F1000Research* 10, 1022.

Musabayane, C. T., P. T. Bwititi & J. A. Ojewole. 2006. Effects of oral administration of some herbal extracts on food consumption and blood glucose levels in normal and streptozotocin-treated diabetic rats. *Methods and Findings in Experimental and Clinical Pharmacology* 28(4), 223–228.

Nakamura, M., S. Nakamura & T. Oku. 2009. Suppressive response of confections containing the extractive from leaves of *Morus alba* on postprandial blood glucose and insulin in healthy human subjects. *Nutrition and Metabolism (Lond)* 6, 29.

Naowaboot, J., P. Pannangpetch, V. Kukongviriyapan, B. Kongyingyoes & U. Kukongviriyapa. 2009. Antihyperglycemic, antioxidant and antiglycation activities of mulberry leaf extract in streptozotocin-induced chronic diabetic rats. *Plant Foods for Human Nutrition* 64(2), 116–121.

Naowaboot, J., P. Pannangpetch, V. Kukongviriyapan, U. Kukongviriyapan, S. Nakmareong & A. Itharat. 2009. Mulberry leaf extract restores arterial pressure in streptozotocin-induced chronic diabetic rats. *Nutrition Research* 29(8), 602–608.

Naowaboot, J., P. Pannangpetch, V. Kukongviriyapan, A. Prawan, U. Kukongviriyapan & A. Itharat. 2012. Mulberry leaf extract stimulates glucose uptake and GLUT4 translocation in rat adipocytes. *The American Journal of Chinese Medicine* 40(1), 163–175.

Nazari, M., M. R. Hajizadeh, M. Mahmoodi, M. R. Mirzaei & G. Hassanshahi. 2013. The regulatory impacts of *Morus alba* leaf extract on some enzymes involved in glucose metabolism pathways in diabetic rat liver. *Clinical Laboratory* 59(5–6), 497–504.

O'Brien, T., T. T. Nguyen & B. R. Zimmerman. 1998. Hyperlipidemia and diabetes mellitus. *Mayo Clinic Proceedings* 73(10), 969–976.

Oh, K. S., S. Y. Ryu, S. Lee, H. W. Seo, B. K. Oh, Y. S. Kim & B. H Lee. 2009. Melanin-concentrating hormone-1 receptor antagonism and anti-obesity effects of ethanolic extract from Morus alba leaves in diet-induced obese mice. *Journal of Ethnopharmacology* 122(2), 216–220.

Oku, T., M. Yamada, M. Nakamura, N. Sadamori & S. Nakamura. 2006. Inhibitory effects of extractives from leaves of Morus alba on human and rat small intestinal disaccharidase activity. *British Journal of Nutrition* 95(5), 933–938.

P, S., Zinjarde, S. Y. Bhargava & A. R. Kumar. 2011. Potent α-amylase inhibitory activity of Indian Ayurvedic medicinal plants. *BMC Complementary and Alternative Medicine* 11, 5.

Pissios, P. & E. Maratos-Flier. 2003. Melanin-concentrating hormone: From fish skin to skinny mammals. *Trends in Endocrinology & Metabolism* 14(5), 243–248.

Polumackanycz, M., M. Wesolowski & A. Viapiana. 2021. Morus alba L. and Morus nigra L. Leaves as a promising food source of phenolic compounds with antioxidant activity. *Plant Foods for Human Nutrition* 76(4), 458–465.

Rebai, O., M. Belkhir, A. Boujelben, S. Fattouch & M. Amri. 2017. Morus alba leaf extract mediates neuroprotection against glyphosate-induced toxicity and biochemical alterations in the brain. *Environmental Science and Pollution Research* 24(10), 9605–9613.

Salemi, Z., S. Barzin Tond, S. Fallah, A. Shojaii & M. Seifi. 2016. The effect of Morus alba leaves extract and powder on resistin levels and liver transaminase enzymes activities in diabetes. *Cellular and Molecular Biology (Noisy-le-grand)* 62(6), 112–118.

Sarikaphuti, A., T. Nararatwanchai, T. Hashiguchi, T. Ito, S. Thaworanunta, K. Kikuchi, Y. Oyama, I. Maruyama & S. Tancharoen. 2013. Preventive effects of Morus alba L. anthocyanins on diabetes in Zucker diabetic fatty rats. *Experimental and Therapeutic Medicine* 6(3), 689–695.

Shendge, A. K., S. Panja, T. Basu, N. B. Ghate & N. Mandal. 2021. Ameliorating effects of white mulberry on iron-overload-induced oxidative stress and liver fibrosis in Swiss albino mice. *Food Chemical Toxicology* 156, 112520.

Shrestha, S., S. H. Seong, S. G. Park, B. S. Min, H. A. Jun & J. S. Choi. 2019. Insight into the PTP1B Inhibitory Activity of Arylbenzofurans: An in vitro and in silico study. *Molecules* 24(16).

Singab, A. N., H. A. El-Beshbishy, M. Yonekawa, T. Nomura & T. Fukai. 2005. Hypoglycemic effect of Egyptian Morus alba root bark extract: Effect on diabetes and lipid peroxidation of streptozotocin-induced diabetic rats. *Journal of Ethnopharmacology* 100(3), 333–338.

Su, C., Y. Duan, J. Tian, J. Liu, K. Xie, D. Chen, F. Ye, R. Chen & J. Dai. 2020. Morusalisins A-F, six new Diels-Alder type adducts, as potential PTP1B inhibitors from cell cultures of Morus alba. *Fitoterapia* 146, 104682.

Sugimoto, M., H. Arai, Y. Tamura, T. Murayama, P. Khaengkhan, T. Nishio, K. Ono, H. Ariyasu, T. Akamizu, Y. Ueda, T. Kita, S. Harad, K.Kamei & M. Yokode. 2009. Mulberry leaf ameliorates the expression profile of adipocytokines by inhibiting oxidative stress in white adipose tissue in db/db mice. *Atherosclerosis* 204(2), 388–394.

Suriyaprom, S., T. Kaewkod, I. Promputtha, M. Desvaux & Y. Tragoolpua. 2021. Evaluation of antioxidant and antibacterial activities of White mulberry (Morus alba L.) fruit extracts. *Plants (Basel)* 10(12).

Suthamwong, P., M. Minami, T. Okada, N. Shiwaku, M. Uesugi, M. Yokode & K. Kamei. 2020. Administration of mulberry leaves maintains pancreatic β-cell mass in obese/type 2 diabetes mellitus mouse model. *BMC Complementary Medicine and Therapies* 20(1), 136.

Taghizadeh, M., A. Mohammad Zadeh, Z. Asemi, A. H. Farrokhnezhad, M. R. Memarzadeh, Z. Banikazemi, M. Shariat & R. Shafabakhsh. 2022. *Morus alba* leaf extract affects metabolic profiles, biomarkers inflammation and oxidative stress in patients with type 2 diabetes mellitus: A double-blind clinical trial. *Clinical Nutrition ESPEN* 49, 68–73.

Tanabe, K., S. Nakamura, K. Omagari & T. Ok. 2011. Repeated ingestion of the leaf extract from Morus alba reduces insulin resistance in KK-Ay mice. *Nutrition Research* 31(11), 848–854.

Tao, Y., Y. Zhang, Y. Cheng & Y. Wang. 2013. Rapid screening and identification of α-glucosidase inhibitors from mulberry leaves using enzyme-immobilized magnetic beads coupled with HPLC/MS and NMR. *Biomedical Chromatography* 27(2), 148–155.

Thaipitakwong, T., O. Supasyndh, Y. Rasmi & P. Aramwit. 2020. A randomized controlled study of dose-finding, efficacy, and safety of mulberry leaves on glycemic profiles in obese persons with borderline diabetes. *BMC Complementary Therapies in Medicine* 49, 102292.

Ullah, N., M. A. Khan, S. Khan, H. Ahmad, A. H. Asif & T. Khan. 2016. Nephro-protective potential of Morus alba, a prospective experimental study on animal models. *Pharmaceutical Biolology* 54(3), 530–535.

Umeyama, L., B. Hardianti, S. Kasahara, D. F. Dibwe, S. Awale, S. Yokoyama & Y. Hayakawa. 2021. Anti-inflammatory effects of Morus alba Linne bark on the activation of toll-like receptors and imiquimod-induced ear edema in mice. *BMC Complementary Medicine and Therapies* 21(1), 115.

Wang, J., X. Liu, H. Zheng, Q. Liu, H. Zhang, X. Wang, T. Shen, S. Wang & D. Ren. 2020. Morusin induces apoptosis and autophagy via JNK, ERK and PI3K/Akt signaling in human lung carcinoma cells. *Chemico Biological Interactions* 331, 109279.

Wang, Y., L. Xiang, C. Wang, C. Tang & X. He. 2013. Antidiabetic and antioxidant effects and phytochemicals of mulberry fruit (Morus alba L.) polyphenol enhanced extract. *PLoS ONE*, 8(7), e71144.

Woo, H., J. Lee, D. Park & E. Jung. 2017. Protective effect of mulberry (Morus alba L.) extract against Benzo[a] pyrene induced skin damage through inhibition of aryl hydrocarbon receptor signaling. *Journal of Agricculture and Food Chemistry* 65(50), 10925–10932.

Wu, Y. X., Y. J. Kim, T. H. Kwon, C. P. Tan, K. H. Son & T. Kim. 2020. Anti-inflammatory effects of mulberry (Morus alba L.) root bark and its active compounds. *Natural Products Research* 34(12), 1786–1790.

Ye, M., Y. Ke, B. Liu, Y. Yuan, F. Wang, S. Bu & Y. Zhang. 2017. Root bark of Morus alba ameliorates the depressive-like behaviors in diabetic rats. *Neuroscience Letters* 637, 136–141.

Yu, J. S., S. H. Lim, S. R. Lee, C. I. Choi & K. H Kim. 2021. Antioxidant and anti-inflammatory effects of White mulberry (Morus alba L.) fruits on lipopolysaccharide-stimulated RAW 264.7 macrophages. *Molecules* 26(4).

Zhang, M., M. Chen, H. Q. Zhang, S. Sun, B. Xia & F. H. Wu. 2009. In vivo hypoglycemic effects of phenolics from the root bark of Morus alba. *Fitoterapia* 80(8), 475–477.

Zhao, B. T., D. H. Nguyen, D. D. Le, J. S. Choi, B. S. Min & M. H. Woo. 2018. Protein tyrosine phosphatase 1B inhibitors from natural sources. *Archives of Pharmacal Research* 41(2), 130–161.

24 Antidiabetic and Other Pharmacological Activities of *Syzygium cumini* (Black Plum or Jamun)

Jamal Akhtar and Fouzia Bashir

CONTENTS

DOI: 10.1201/b23347-24

24.1 INTRODUCTION

One of the most prevalent health issues in the globe is diabetes. Figures from 2017 show that 425 million individuals worldwide had diabetes. There are 114.4 million diabetics in China, making them the nation with the greatest percentage of the condition. In 2015, India has 72.9 million diabetic patients, making it the country with the second-highest number. Between 2005 and 2015, the number of diabetic patients in India increased by over 50%. The third-largest number of diabetes patients (32.2 million) lives in the USA, where this illness is likewise a serious health concern. Diabetes treatment is extremely expensive on a global scale, and alternative medical systems aid in giving diabetic patients more treatment options. The precise therapeutic applications of various medication sources, particularly those with a plant origin, are supplemented in medicinal plants based on conventional medical systems like Unani and Ayurveda. Due to the fact that they tend to cause fewer side effects and are effective and safe at the same time, people are now placing more attention on herbal medicines.

Syzygium cumini (syn. *Eugenia jambolana* Lam.) is one of the important medicinal plants with the potential to treat several ailments successfully (Rastogi 2004). It belongs to the kingdom Plantae; class Magnoliopsida; order Myrtales; and family Myrtaceae. Vernacular names include black plum, black berry; jaman and phalenda in Urdu; jam, jamun, jaman, phalanda, and phalinda in Hindi; jambula and jambu in Sanskrit; and neereedu, neredu, jambuvu, naeraedu, and nesedu in Telugu (Kirtikar 1987; Nadkarni 1976; Khare 2007; Anonymous 2006a; Anonymous 2006b; Anonymous 2014; Pullaiah 2006; Khan 2013). *Syzygium cumini* plays an important role in prevention as well as in managing non-communicable diseases such as diabetes mellitus, cancer, gout, and ischaemic heart disease (IHD).

24.2 BOTANICAL DESCRIPTION

Syzygium cumini is a sizable, glabrous, evergreen tree with a height of up to 30 metres and a girth of 3.6 metres. It is also regarded as a tree that grows quickly because it reaches its full size in 40 years. In India, it can grow to a height of 100 feet (30 metres), with a spread of 36 feet (11 metres) and a trunk diameter of 2 or 3 feet (0.6–0.9 m) (Anonymous 2006b).

24.3 MACROSCOPIC

Old stems have pale brown, slightly rough bark that has tiny fractures (Kirtikar 1975). The bark's upper section is smooth and light grey to grey or greyish-brown in hue, while the lower portion is rough, cracked, flaking and stained (Nair 2017). The typical size and shape of a leaf is 7.5–15 by 3.8–6.3 cm. They are lanceolate, elliptic-oblong or widely ovate-elliptic, acute, acuminate or subobtuse, coriaceous, smooth and shiny, with numerous close parallel fine secondary nerves joining to form an intra-marginal vein (Kirtikar 1987), as shown in Figure 24.1. The blooms have different colour shades. They could be white, creamy white or sporadically a little bit green. They grow on terminal or axillary panicled racemes in groups of up to 40 (Nair 2017). Flowers are sessile, 7.5–13 mm diameter, fragrant and typically clustered in groups of three in trichotomous panicles that are 3.8–10 cm long and typically emerge from the scars of dead leaves (Kirtikar 1987). March marks the beginning of blossoming, which lasts until May. Full blossoming takes place between April and May (Anonymous 1997; Nair 2017).

24.4 DISTRIBUTION

Ibn Batuta, who travelled to India in 1332 and encountered this shrub, characterised it as one of the fruits of Delhi (Dymock 1890). Since ancient times, it has been found all across India, as well as in other nations like Nepal, Myanmar, Sri Lanka, Indonesia, Pakistan, Bangladesh, Malaysia,

Fig. 1a. Tree of *Syzygium cumini*

Fig. 1b. Seeds of *Syzygium cumini*

Fig. 1c. Fruit of *Syzygium cumini*

Fig. 1d. Flower of *Syzygium cumini*

FIGURE 24.1 *Syzygium cumini.*

Thailand, the Philippines, Australia, and other tropical regions of the world like South America and Madagascar (Anonymous 2014).

24.5 PHYTOCHEMICAL CONSTITUENTS

The native Indian plant known as jamun, or black plum, has a significant phytochemical composition. It is a plentiful source of numerous phytoconstituents, including sterols, alkaloids, flavonoids and anthrocyanins (Baliga et al. 2011).

All of *Syzygium cumini*'s plant components have important phytochemical and pharmacological properties (Figure 24.2). According to reports, the stem bark contains myricetine, gallic acid, ellagic acid, galllotannin, friedelan-3-ol, betulinic acid, kaempferol and friedelan-3-ol-D-glucoside (Ayyanar and Babu 2012). It has been shown that the plant's leaves include betulinic acid, quercetin, -sitosterol, myricitrin, myricetin, myricetin 3-O-(4″-acetyl)–L-rhamnopyranosides, n-nonacosane, mycaminose, n-hepatcosane, crategolic (maslinic) acid (Mahmoud et al. 2001). Oleanolic acid, ellagic acid, isoquercetin, quercetin, kampferol and myricetin have all been discovered to be present in the flowers. According to studies, the leaf's essential oil contains the phytochemicals pinocarveol, -terpeneol, cineole, myrtenol, eucarvone, muurolol, -myrtenal, geranyl acetone, -cadinol and pinocarvone (Shafi et al. 2002). The most researched plant component is the seeds, which are said to

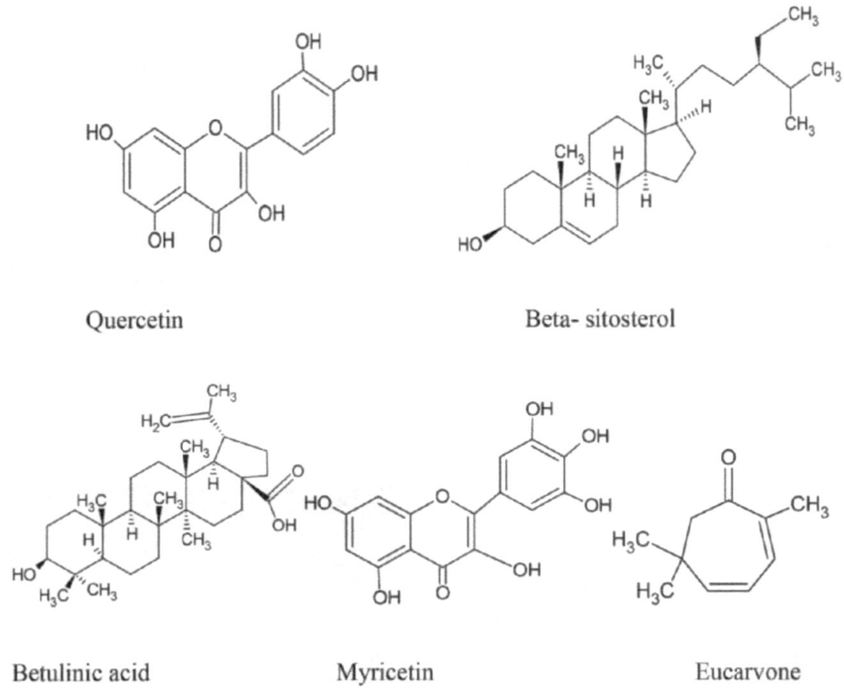

FIGURE 24.2 Chemical derivatives of *Syzygium cumini.*

include jambosine, gallic acid, ellagic acid, corilagin, 3,6-hexahydroxydiphenoylglucose, 4,6-hexahydroxydiphenoylglucose, 1-galloylglucose, 3-galloylglucose, quercetin, and -sitosterol (Ramya et al. 2012). Studies have shown that the pulp of *Syzygium cumini* contains anthocyanins, delphinidin, petunidin and malvidin-diglucosides, and these compounds are responsible for their bright purple colour.

24.6 PHARMACOLOGICAL STUDIES

S. cumini possesses various properties like those mentioned in the Figure 24.3.

24.6.1 Anti-Inflammatory Activity

A 70% ethanolic extract of *S. cumini* bark demonstrated strong anti-inflammatory action in an experimental trial, with activity comparable to that of acetylsalicylic acid (300 mg/kg/oral). It demonstrates that it has a good anti-inflammatory effect against different phases of inflammation without any side effect on gastric mucosa (Muruganandan et al. 2001).

24.6.2 Cardio-Protective Activity

An investigation was made on the antihypertensive and vasorelaxant properties of the hydro-alcoholic extract from the fruits of *S. cumini*. According to the findings, the extract had antihypertensive and hypotensive effects. This finding demonstrates that this plant's fruits can be used as a form of cardioprotection (Herculano et al. 2014).

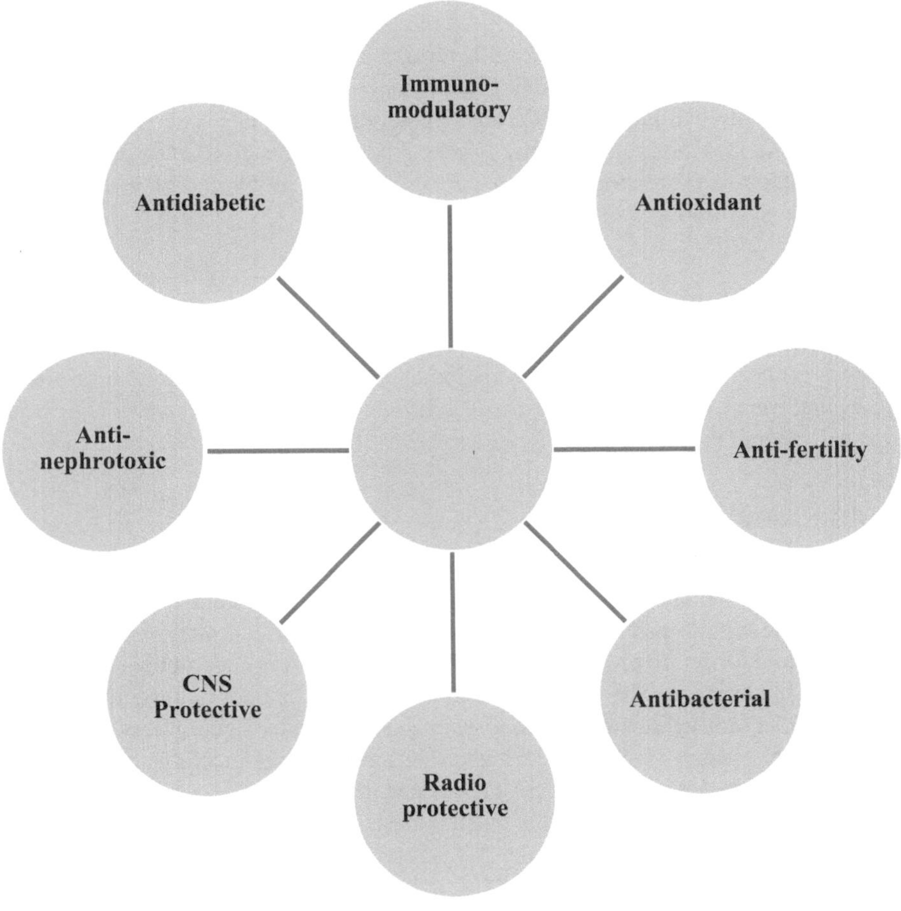

FIGURE.24.3 Pharmacological activities of *Syzygium cumini*.

24.6.3 Anti-Retinitis

Five anthocyanins from the *S. cumini* fruit peel were tested for their ability to bind to the X-linked retinitis pigmentosa (RP2) gene (a mutant of this gene causes loss of vision in humans). It was discovered that the cyanidin 3, 5 diglucoside with the lowest G score (–12.62 kcal/mol) functions as an inhibitor and may be used to treat human retinitis pigmentosa (Priya et al. 2013).

24.6.4 Antipyretic Activity

S. cumini seed extract was found to have anti-inflammatory and antipyretic properties when tested on adult male Charles-Foster rats (120–160 g). Significant antipyretic efficacy was achieved by the seed extract (50, 100 and 200 mg/kg, i.p.) against yeast-induced pyrexia in rats (Mahapatra et al.1986; Chaudhuri et al. 1990).

24.6.5 Anti-Diarrhoeal Activity

According to a study, *Eugenia jambolana* bark extract significantly inhibited the diarrhoea that was induced by castor oil in experimental animal models (Mukherjee et al. 1998).

24.6.6 Hepatoprotective Activity

The ethanolic extract of *Eugenia jambolana* L. or *Syzygium cumini* pulp was studied at doses of 100 mg/kg and 200 mg/kg. When rats were given hepatotoxic paracetamol to induce liver damage, the extract significantly exhibited hepatoprotective action (Das and Sarma 2009). Another study showed that *Eugenia jambolana* methanolic extract, when administered orally at a dose of 400 mg/kg/day, was effective in preventing the liver damage brought on by carbon tetrachloride (CCL_4) (Sisodia and Bhatnagar 2009).

24.6.7 Anti-Cancer Activity

For clinical application in the United States, a study was conducted to look into *Syzygium cumini*'s anti-cancerous properties. Vinblastine, vincristine and paclitaxel were among nine substances of plant origins that were given approval. The research found that *Syzygium cumini* (L.) fruits may be able to prevent cancer (Pradhan 2016).

24.6.8 Anti-Leishmania Activity

The effects of *Syzygium cumini* (L.) essential oil (ScEO) and its main component -pinene on *Leishmania* (Leishmania) *amazonensis* were investigated in a study. Tetrazolium salt (MTT) assay was used to evaluate the anti-proliferative effect on *Leishmania*, effects on promastigote, and effects on axenic amastigote forms. To measure the survival index, the intra-macrophagic amastigotes were exposed to ScEO and pinene. The results showed that -pinene had a 50% inhibitory concentration (IC_{50}) value of 19.7 g/mL against *Leishmania amazonensis* promastigote forms. In comparison to ScEO, which had IC_{50} values of 43.9 and 38.1 g/mL for axenic and intracellular amastigotes, respectively, -pinene was more active, with values of 16.1 and 15.6 g/mL. The study's findings suggested that it might be utilised to treat and prevent leishmaniasis (Rodrigues et al. 2015).

24.6.9 Antihyperlipidaemic Activity

In a study, streptozotocin-induced diabetic rats received an oral dose of *Syzygium cumini* ethanolic extract (100 mg/kg body weight), which demonstrated antihyperlipidaemic action (Ravi et al. 2005).

24.6.10 Anti-Anaemic Activity

Aqueous seed extract of *S. cumini* has been said to have anti-anaemic properties. The *S. cumini* seed extract was found to improve total haemoglobin in a study (Prince 1998).

24.6.11 Antibacterial Activity

In a study, extracts of *Syzygium cumini* seeds in methanol and ethyl acetate at a concentration of 200 g/disc demonstrated antibacterial action against *Bacillus creus*, *B. subtalis*, *B. megateriun*, *Steptococcus β-haemolyticus*, *S. aureus*, *Shigella dysenteriae*, *Sh. shiga*, *Sh. boydii* and *Sh. flexneriae* (Bijauliya et al. 2017). An extract from *Syzygium cumini* leaves also demonstrated some antibiotic action against *Micrococcus pyogenes* var. aureus and mild antibacterial activity against *Escherichia coli* (Anonymous 2014).

24.6.12 Anti-Fertility Activity

Oleanolic acid, which was extracted from *Syzygium cumini* flowers, has been shown in a study to have anti-fertility properties. Without significantly changing the body or reproductive organ weights, it reduced the male albino rats' ability to reproduce. It resulted in a considerable reduction in spermatocyte conversion to spermatids and an early meiotic stoppage of spermatogenesis, both of which led to a drop in sperm count without any abnormalities in spermatogenic cells, Leydig interstitial cells or Sertoli cells (Rajasekaran et al. 1988).

24.6.13 CNS Protective Activity

On albino mice, 200 and 400 mg/kg doses of ethyl acetate and methanol extracts of *S. cumini* seeds were tested for CNS action. Significant CNS protective action was shown by both extracts (Kumar et al. 2007).

24.6.14 Anti-Nephrotoxic Activity

An ethanol extract of the fruits of *S. cumini* was tested for its ability to protect the kidneys against cisplatin-induced nephrotoxicity in albino rats (6 mg/kg intraperitoneally), which was determined by calculating the kidney's levels of urine protein, blood urea nitrogen, serum total proteins, creatinine and lipid peroxidation. Cisplatin increased the amount of a serum marker, boosted protein excretion in the urine, decreased creatinine clearance and raised the level of renal MDA. Animals given an ethanol extract of *S. cumini* fruit effectively and dose-dependently reversed the effects of cisplatin (Adikay et al. 2010).

24.6.15 Radioprotective Effect

Using a micronucleus assay, the leaves of *S. cumini* were evaluated for their radioprotective properties. It was discovered that *S. cumini* inhibits the development of micronuclei in lymphocytes (Jagetia 2002). Arun et al. (2011) additionally confirmed that *S. cumini* seed extract prevented the development of micronuclei in mouse bone marrow cells brought on by genotoxic stress. Similar to earlier findings, *S. cumini* extract decreased the frequency of micronuclei, and mice brain extract prepared after irradiation likewise revealed a concentration-dependent suppression of lipid peroxides.

24.6.16 Immunomodulatory Activity

The ability of *S. cumini* seed extract to modulate the immune system was investigated. It was revealed that it had a dose-dependent effect on rats' humoral antibody titres and delayed-type hypersensitivity (DTH) reactions. Similar to this, the therapy also led to an increase in the quantity of neutrophils, lymphocytes and overall white blood cells in rats. This study revealed that *S. cumini* seed extract might boost the body's hematopoietic system, which suggests that the plant may be used to treat immune-deficient illnesses that develop during radiation therapy or chemotherapy (Mastan et al. 2008).

24.6.17 Antioxidant Activity

Using the 2,2-diphenyl-1-picrylhydrazyl (DPPH) free radical scavenging and ferric reducing antioxidant power (FRAP) tests, the antioxidant activity of *S. cumini* leaf extracts was examined. The findings of this investigation indicated that the ethyl acetate fraction had greater antioxidant activity than the other fractions. Data from high-performance liquid chromatography (HPLC) showed that phenolic components including ferulic acid and catechin, which are responsible for its antioxidant action, were present in *S. cumini* leaf extracts (Zhi-Ping et al. 2008; Ahmad et al. 2019).

24.7 *SYZYGIUM CUMINI* AS ANTIDIABETIC AGENT

A metabolic illness of a complex nature, diabetes mellitus (DM) causes either insulin deficiency or insulin malfunction. There are two varieties of it: type 1 diabetes, which requires insulin, and type 2 diabetes, which is not insulin-dependent and affects 90% of people with diabetes (Jagetia 2018; Makheswari and Sudarsanam 2011). In most developed nations, it ranks as the fourth most common cause of death. Diabetes is a factor in at least one out of every ten fatalities of adults aged 35 to 64. According to a research by the International Diabetes Federation, diabetes claims the lives of 3.2 million people annually and one person every seven seconds. In 2011, there were 366 million cases of diabetes worldwide, and by 2030, there will likely be 552 million cases (Wild et al. 2004). Incidence rates of diabetes for all age categories were projected to reach 2.8% in 2000 and 4.4% in 2030. The overall number of diabetic patients in India was 31.7 million in 2000 and is anticipated to reach 79.4 million by 2030. Between 2000 and 2030, it is expected that the number of urban diabetics in developing nations would double (Whiting et al. 2011).

In classical Unani literature as well as in Ayurvedic pharmacopoeia, *Syzygium cumini* has been noted to have antidiabetic properties. *Syzygium cumini* seed powder has been shown to be useful in lowering blood sugar levels. For more than 130 years, Western nations have used it to control blood sugar levels. Numerous researchers have looked at the hypoglycaemic effects of various *Syzygium cumini* components to manage diabetes in preclinical animals, and they have found that it lowers blood sugar levels. The details are summarised in Table 24.1.

In a study, extract from *Syzygium cumini* seed powder reduced blood sugar levels in diabetic rabbits, demonstrating positive outcomes (Brahmachari 1961). It has been noted that diabetic rats administered jamun aqueous seed extract at a dose of 1 g/kg body weight experienced a hypoglycaemic effect. (Kedar 1983).

Aqueous extract of *Syzygium cumini* has also been shown in numerous additional experiments to lower blood sugar levels in diabetic rats (Achrekar et al. 1991; Bhaskaran 1986; Prince 1998).

TABLE 24.1
The Hypoglycaemic Effect of Different Parts/Extracts of *Syzygium cumini* in Preclinical Models and Clinical Patients

S. No.	Parts used	Extract type	Species	References
1.	Seed	Aqueous	Rabbit	Brahmachari (1961)
			Rat	Kedar (1983); Pandey and Khan (2002)
			Mice	Prince (1998); Sharma et al. (2003); Sharma et al. (2011a); Deb et al. (2013)
		Extract	Mice	Grover et al. (2001); Siddiqui et al. (2014)
		Powder	Humans	Ayya et al. (2015); Banu (2016); Sidana et al. (2017)
		Ethanol and powder	Rat & Rabbit	Ravi et al. (2004, 2005); Sridhar et al. (2005); Chaturvedi et al. (2009); Raza et al. (2017); Jonnalagadda et al. (2012); Sharma et al. (2011b)
		Methanol	Rat and mice	Kumar et al. (2008); Mastan et al. (2009) Deb et al. (2013); Yadav et al. (2013)
2.	Fruit	Lyophilized, Aqueous	Rats	Pepato et al. (2005); Sharma et al. (2006); Rekha et al. (2010)
		Phenol		Gajera et al. (2017)
3.	Roots	Methanol	Rat	Deb et al. (2013)
4.	Leaf	Aqueous	Humans	Bopp et al. (2009)
		Ethanol	Rat	Chattu (2016)

Another study found that diabetic mice and rats' blood glucose levels decreased when lyophilised powdered aqueous seed extract was given (Grover et al. 2001; Vikrant et al. 2001).

Likewise, the aqueous extract of jamun seed, which contains gummy fibres, was also found to be useful in managing diabetes in rats with alloxan-induced diabetes. Contrarily, the aqueous extract lacking gummy fibre had no impact on blood sugar levels (Pandey and Khan 2002).

In a previous study, ethanol extract of jamun seeds was given to alloxan-induced diabetes rabbits, and the animals' fasting blood glucose levels decreased (Sharma et al. 2003). Similarly, the ethanol-extracted jamun seed kernel reduced blood glucose levels in streptozotocin-induced diabetic rats and also restored catalase, glutathione peroxidase and superoxide dismutase enzyme activities along with restoration of glutathione levels in the liver and kidney of diabetic rats (Ravi et al. 2004).

Various doses of jamun seed powder were administered to diabetic rats produced with streptozotocin, and the results revealed attrition in the fasting glucose level (Sridhar et al. 2005).

Similar to this, administering 100 mg/kg body weight of seed kernel ethanol extract to streptozotocin-induced diabetic rats decreased blood sugar levels, urea and cholesterol, increased glucose tolerance, and decreased glutamate oxaloacetate transaminase and glutamate pyruvate transaminase activities (Ravi et al. 2005).

It has been noted that jamun seed methanol extract can lower serum glucose levels in alloxan-induced diabetic mice and rats (Chaturvedi et al. 2009; Jonnalagadda et al. 2012).

In some other investigations, the administration of an methanolic extract of jamun seeds reduced the levels of blood serum glucose in diabetic rats produced by streptozotocin (Mastan et al. 2009; Yadav et al. 2013).

Sephadex gel, which contains the active ingredients extracted from the ethanol portion of jamun seed extract, has been shown to reduce serum glucose levels in rats with both mild and severe alloxan-induced diabetes (Sharma et al. 2011b).

400 mg/kg of aqueous jamun seed extract was administered to type II diabetic rats produced by streptozotocin, and it was discovered to lower blood sugar levels and boost PPAR and PPAR protein expression in the rat liver (Sharma et al. 2011a).

In mice with alloxan-induced diabetes, the aqueous jamun seed extract has been shown to lower serum glucose levels (Siddiqui et al. 2014).

Jamun seed powder supplementation for 30 days in human diabetes subjects decreased fasting and postprandial blood glucose levels (Ayya et al. 2015).

The root, stem bark, leaf and seed extracts of jamun have been shown to reduce serum glucose levels in selected diabetic subjects of 51–60 years of age (Deb et al. 2013).

Recently, hyperglycaemic/diabetic rats were given an ethanol extract of jamun seeds and fruits for 60 days. The results of instant research depicted that both seed and fruit extracts reduce the blood glucose level significantly and also regulate the insulin levels in hyperglycemic rats (Raza et al. 2017).

A significant drop in serum blood glucose and alleviation from symptoms including exhaustion, polyurea and tiredness were seen in a trial on 60 type II diabetics who were given jamun seed powder in various forms for 60 days (Banu 2016).

In a scientific research with a double-blind control group, diabetic patients who consumed 10 g of jamun seed powder for up to 90 days saw reductions in their fasting blood sugar levels of 9%, 18%, and 30% and their postprandial glucose levels of 8%, 15%, and 22% after 30, 60, and 90 days, respectively (Sidana et al. 2017).

In a study, streptozotocin (STZ) induced significant rise in blood glucose levels in the rats. SC100 mg/kg and 200 mg/kg caused a reduction in blood sugar levels (BSL) (192.50 ± 6.189 and 175.00 ± 6.782 respectively). SC 200 mg/kg alone and in combination with metformin caused a significant reduction in HbA_1C levels at the end of the eighth week (8.84 ± 0.65 and 6.86 ± 0.40 respectively) as compared to their baseline levels. Increase in dose led to more significant reduction in BSL at the end of the eighth week in groups A and B ($p < 0.05$). However, reduction of BSL was superior with metformin alone (159.17 ± 13.060) compared to *Syzygium cumini* administered alone. Also, antihyperglycaemic effect of *Syzygium cumini* administered along with metformin was significant ($p < 0.05$) compared to either doses of *Syzygium cumini* given alone (Mulkalwar et al. 2021).

In a study, it was discovered that feeding diets containing 15% unextracted (intact), 15% defatted *S. cumini* seeds, and 6% water soluble gummy fibre for 21 days significantly improved glucose tolerance in both normal and diabetic rats compared to their respective control groups and appreciably lowered (26%–28%) blood glucose levels (Pandey and Khan 2002).

24.8 TRADITIONAL AND OTHER POTENTIAL USES OF *SYZYGIUM CUMINI*

The entire *Syzygium cumini* plant, including the seed, fruit, leaves, flower and so on, is utilised in folk medicine. Its bark is said to be bitter, pleasant, diuretic, astringent to the bowels, anthelmintic and beneficial for treating ulcers, bronchitis, asthma, thirst, dysentery and blood impurities (Kirtikar 1975). *Syzygium cumini* seeds, leaves and fruits were utilised by renowned Vaidya Charaka in concoctions to treat diarrhoea, and the bark was used as an astringent. In cases of obesity, vaginal discharges and monthly abnormalities, another Vaidya Sushruta recommended taking the fruit internally; cases of intrinsic haemorrhage required cold infusion (Nair and Subramanian 1962). The bark's juice is administered in quantities to treat chronic diarrhoea, dysentery and menorrhagia because it is astringent. The bark's decoction works well as a mouthwash and gargle to cure conditions like sore throats, stomatitis and spongy gums. It is also used to treat skin inflammation as well as for tanning, dyeing and colouring fishnets. It is recommended to treat bloody diarrhoea with a mixture of goat milk, honey and juice from jamun, amla and amalaka leaves. To treat diabetes, *S. cumini* leaf juice is used orally. Every morning, the juice is consumed diluted with milk. For stomach ache, fresh *S. cumini* leaf juice is administered orally (Bhandary et al. 1995). The juice of ripe fruit can be used to make syrup, which is a delicious beverage. Astringent properties of the syrup or vinegar made from the ripe fruit are effective in treating chronic diarrhoea and spleen enlargement, respectively. For diabetes, acid reflux and stomach ulcers, hot water extract of dried fruits is utilised (Katiyar et al. 2016).

Additionally, if we look at the classical literature of the Unani system of medicine, it is discovered that *S. cumini* has been utilised for many years to treat a variety of illnesses. Its traditional actions and uses are described in Table 24.2.

TABLE 24.2
Traditional Actions and Uses of *Syzygium cumini*

Actions and uses	References
Antidiabetic	(Dymock 1890; Kirtikar 1987; Nadkarni 1976; Pullaiah 2006; Khare 2007; Khān 2013; Ghani 1998; Bhattacharya 2008)
Antipyretic	(Khān 2013; Tarique 2010; Kabeeruddin 2000; Ghani 1998)
Astringent	(Dymock 1890; Kirtikar 1987; Nadkarni 1976; Khān 2013; Tarique 2010; Kabeeruddin 2000; Bahadur 2012; Ghani 1998)
Blood Purifier	(Khān 2013; Bahadur 2012; Ghani 1998)
Bronchial Asthma, Bronchitis	(Kirtikar 1987; Khān 2013; Ghani 1998)
Carminative	(Dymock 1890; Kirtikar 1987; Nadkarni 1976; Khare 2007; Khān 2013; Ghani 1998)
Diuretic	(Dymock 1890; Kirtikar 1987; Nadkarni 1976; Khare 2007)
Flatulence	(Khān 2013; Ghani 1998)
Hepato-tonic	(Pullaiah 2006; Khān 2013; Kabeeruddin 2000; Nabi 2007; Ghani 1998)
Infertility	(Pullaiah 2006)
Palpitation	(Khān 2013; Ghani 1998)
Splenomegaly	(Tarique 2010; Nabi 2007)
Stomachic	(Dymock 1890; Nadkarni 1976; Khare 2007; Khān 2013; Anonymous 2007; Tarique 2010; Kabeeruddin 2000; Bahadur 2012; Nabi 2007; Ghani 1998; Ghani 2003)

Its bark has therapeutic potential and is used as a bowel anthelmintic, astringent and digestive (Kirtikar 1987; Nadkarni 1976). It is recommended for the treatment of bronchitis, bronchial asthma, diarrhoea and ulcers (Kirtikar 1987).

Seed: The seeds are reputed to be effective intestinal astringents (Kirtikar 1987) as well as anti-inflammatory, anti-arthritic, antipyretic and analgesic properties (Pullaiah 2006). In diabetes mellitus, it is recommended as an antidiabetic medication (Kirtikar 1987). Additionally, the seed is employed for liver and stomach weakness (Anonymous 2007; Nadkarni 1976).

Leaves: Its leaves are also employed as astringents for dysentery (Kirtikar 1987) and as an anti-inflammatory for gingivitis and stomatitis (Anonymous 2006b). Both vomiting and haemorrhoids have been treated by using the leaves. It is also used as anti-venom. When burns occur, leaves are ground into a paste to reduce the burning. For the purpose of strengthening the teeth and gums, leaf ash is applied (Kirtikar 1987). Decoction of an equal amount of its stem barks and leaves cave in the smell of armpit.

Fruits: They are used to energise the liver and improve blood quality. Because of its astringent qualities, they help to strengthen gums and teeth. They are also applied on ringworm and sore throats (Kirtikar 1987). The fruit's vinegar has tonic, stomachic, astringent and carminative properties (Kirtikar 1987; Nadkarni 1976). It is useful in diseases of the spleen (Nadkarni1976).

It also functions as a diuretic and an antidiabetic (Kirtikar 1987; Nadkarni 1976). The fruits are a strong source of iron and are used as a powerful remedy for asthma, liver and heart conditions, as well as other conditions (Pullaiah 2006).

24.9 CULTIVATION PRACTICES

Due to urbanisation and intensive agriculture, the jamun crop is experiencing serious genetic erosion and the extinction of several of its species. In the search for sources of resistance to biotic and abiotic challenges, the genetic diversity of the associated wild species is particularly valuable (Singh et al. 2011).

24.9.1 Plant Propagation

Since jamun has a long gestation period, utmost care is needed in selecting the genuine planting material at the time of orchard establishment. Jamun, being heterozygous in nature, does not produce true to the type from seeds; however, polyembryony has been reported to the extent of 20%–50%, hence nucellar seedlings may be utilised to produce uniform rootstocks for grafting and budding (Singh et al. 2011).

Due to the prolonged gestation period of jamun, extreme care must be taken when establishing an orchard to choose genuine planting material. Although polyembryony has been observed to occur to a level of 20%–50% in jamun, which is heterozygous in nature, nucellar seedlings may be used to produce uniform rootstocks for grafting and budding (Singh et al. 2011).

24.9.2 SEED PROPAGATION

Fresh seeds can be placed 4–5 cm deep in the nursery because jamun seeds do not undergo dormancy. Two weeks after seeding, seeds begin to sprout. In the spring, the seedlings are prepared for grafting (Singh et al. 2007, 2010). Polythene bags can also be used to sow seeds since they make rootstocks and grafted plants easier to handle (Singh et al. 2007).

24.10 PROCESSING FOR VALUE-ADDED PRODUCTS

Due to its very perishable nature, jamun sustains significant losses following harvest, especially during postharvest processing. It is required to use various fruit preservation techniques to turn the fruits into various value-added goods. The fruits may be used to make pickles, jam, jelly, drinks, wine and vinegar (Das 2009; Mishra 2018). Nectar, squash, syrup, vinegar and cider are significant jamun postharvest products (Singh et al. 2010).

24.11 MARKETING, ECONOMICS, AND SOCIOECONOMIC CONSIDERATIONS

Due to its high level of perishability and availability only during certain seasons, jamun has greater marketing issues. Since small and marginal farmers cannot afford to transport their goods to distant markets, around 75% of farmers sell their produce locally to village merchants, retailers or preharvest contractors. The marketing of product requires knowledge of the consumer's preferences as well as information on supply, demand, pricing, market forecast and distribution networks. Jamun fruits are taken from trees that have grown naturally or from a few plants planted in a small area. It is impossible to assemble trustworthy statistical data to determine the crop's economics because the majority of the fruit originates from trees spread around backyards and field boundaries. The global market for jamun seed powder is quite promising as well.

24.12 FUTURE REMARKS

This chapter explored *Syzygium cumini*, a traditional medicinal herb, and its ethnobotanical usage, short phytochemistry, pharmacological actions, including antidiabetic activity, and clinical studies (jamun). *S. cumini* is a plant with a long history in ethnobotany. India, along with nations like Nepal, Myanmar, Sri Lanka, Indonesia, Pakistan, Bangladesh, and other tropical parts of the planet including South America and Madagascar, are all home to it. Due to the presence of active chemical components including gallotannin, ellagitannin and myricetine, among others, it has a broad spectrum of pharmacological activities including cardio-protective, anti-inflammatory, hepatoprotective, anti-diarrhoeal, antioxidant and especially antidiabetic.

This plant has been used for medicinal purposes from the beginning of time, and it is referenced in conventional pharmacopoeias. This plant is rapidly vanishing because of overuse and overexploitation to fulfil demand. As a result, the manufacture of medicines should be legalised and carefully regulated. It is necessary to do further in-depth scientific research on the mechanism of action of *Syzygium cumini*, its phytochemical components, medication interactions, effective dose estimation and adverse effects.

REFERENCES

Achrekar, A., G. S. Kaklij, M. S. Pote, et al. 1991. Hypoglycemic activity of *Eugenia jambolana* and *Ficus bengalensis*: Mechanism of action. *In Vivo* 5(2):143–147.

Adikay, S., P. Belide, and B. Koganti. 2010. Protective effect of fruits of *Syzygium cumini* against Cisplatin-induced acute renal failure in rats. *Journal of Pharmacy Research* 3(11): 2756–2758.

Ahmad, N., M. Nawab, and M. H. Kazmi. 2019. Medicinal potential of Jamun (*Syzygium cumini* Linn): A review. *Journal of Drug Delivery and Therapeutics* 9(5): 175–180.

Anonymous. 1997. *Standardization of Single Drugs of Unani Medicine*. Central Council for Research in Unani Medicine: Ministry of Health & Family Welfare, Government of India. Part III, pp. 20–25.

Anonymous. 2006a. *Qarabadeen Sarkari. Central Council for Research in Unani Medicine: Ministry of AYUSH*, Government of India, 2, p. 49.

Anonymous. 2006b. *Standardization of Single Drugs of Unani Medicine. Central Council for Research in Unani Medicine: Department of AYUSH*, Government of India. Part V, pp. 113–124.

Anonymous. 2007. *The Unani Pharmacopeia of India*. Ministry of Health and Family Welfare, New Delhi, vol. IV (1): 55–57.

Anonymous. 2014. *The Wealth of India (Raw Materials and Industrial Products)*. Council of Scientific and Industrial Research, New Delhi, 10, pp. 100–104.

Arun, R., M. V. Prakash, S. K. Abraham, and K. Premkumar. 2011. Role of *Syzygium cumini* seed extract in the chemoprevention of in vivo genomic damage and oxidative stress. *Journal of Ethnopharmacology* 134(2): 329–333.

Ayya, N., V. Nalwade, and T. N. Khan 2015. Effect of Jamun (*Syzygium cumini* L.) seed powder supplementation on blood glucose level of type-II diabetic subject. *Journal of Food Science Research* 6(2): 353–356.

Ayyanar, M., and P. S. Babu. 2012. *Syzygium cumini* (L.) Skeels: A review of its phytochemical constituents and traditional uses. *Asian Pacific Journal of Tropical Biomedicine* 2(3): 240–246. https://doi.org/10.1016/S2221-1691(12)60050-1.

Bahadur, M. 2012. *Mufradat-i- Maseehahi*. Masood Publishing House, Deoband, p. 39.

Baliga, M. S., H. P. Bhat, B. R. V. Baliga, R. Wilson, and P. L. Palatty. 2011. Phytochemistry, traditional uses and pharmacology of *Eugenia jambolana* Lam. (black plum): A review. *Food Research International* 44(2011): 1776–1789.

Banu, H., and A. Jyothi. 2016. Hypoglycemic and hypocholesterolemic effect of *Eugenia jambolana* (kala jamun) spicy mix on type II diabetic subjects. *Imperial Journal of Interdisciplinary Research* 2(4): 850–857.

Bhandary, M. J., K. R. Chandrashekar, and K. M. Kaveriappa. 1995. Medical ethnobotany of the siddis of Uttara Kannada district, Karnataka, India. *Journal of Ethnopharmacology* 47(3):149–158.

Bhaskaran, N. R., and G. Santhakumari. 1986. Anti-diabetic activity of the seed kernel of *Syzygium cumini* Linn. *Ancient Science Life* 6(2): 80–84.

Bhattacharya, S. K. 2008. *Hand Book of Medicinal Plants*. 5th ed. Pointer Publication, Jaipur, p. 338.

Bijauliya, R. K., S. Alok, M. Singh, and S. B. Mishra. 2017. Morphology, phytochemistry and pharmacology of *Syzygium cumini* (Linn.) – an overview. *International Journal of Pharmaceutical Sciences and Research* 8(6): 2360–2371. https://doi.org/10.13040/IJPSR.0975-8232.8(6).2360-71.

Bopp, A., K. S. De Bona, and L. P. Bellé. 2009. *Syzygium cumini* inhibits adenosine deaminase activity and reduces glucose levels in hyperglycemic patients. *Fundamental & Clinical Pharmacology* 23(4): 501–507. https://doi.org/10.1111/j.1472-8206.2009.00700.x.

Brahmachari, H. D., and K. T. Augusti. 1961. Hypoglycemic agents from Indian indigenous plants. *Journal of Pharmacy and Pharmacology* 13: 381–182.

Chattu, M. 2016. Evaluation of antidiabetic activity for ethanolic extract of *Syzygium cumini* leaf in dexamethasone induced diabetic rats. *Research in Pharmacy and Health Sciences* 2(2): 85–90.

Chaturvedi, A., G. Bhawani, P. K. Agarwal, et al. 2009. Antidiabetic and antiulcer effects of extract of *Eugenia jambolana* seed in mild diabetic rats: Study on gastric mucosal offensive acid–pepsin secretion. *Indian Journal of Physiology and Pharmacology* 53(2): 137–146.

Chaudhuri, A. K., S. Pal, A. Gomes, and S. Bhattacharya. 1990. Anti-inflammatory and related actions of *Syzygium cuminii* seed extract. *Phytotherapy Research* 4(1): 5–10.

Das, J. N. 2009. Studies on storage stability of jamun beverages. *Indian Journal of Horticulture* 66(4): 508–510.

Das, S., and G. Sarma. 2009. Study of the hepatoprotective activity of the ethanolic extract of the pulp of *Eugenia jambolana* (jamun) in albino rats. *Journal of Clinical and Diagnostic Research* 3(2): 1466–1474.

Deb, L., C. Bhattacharjee, and S. R. Shetty. 2013. Evaluation of anti-diabetic potential of the Syzygium cumini (Linn) skeels by reverse pharmacological approaches. *Bulletin of Pharmaceutical Research* 3(3): 135–145.

Dymock, W. 1890. *Pharmacographia Indica: A History of the Principal Drugs of Vegetable Origin*. Vol. 2, pp. 24–26.

Gajera, H. P., S. N. Gevariya, D. G. Hirpara, et al. 2017. Antidiabetic and antioxidant functionality associated with phenolic constituents from fruit parts of indigenous black jamun (*Syzygium cumini* L.) landraces. *Journal of Food Science and Technology* 54(10): 3180–3191. https://doi.org/10.1007/s13197-017-2756-8.

Ghani, A. 2003. *Medicinal Plants of Bangladesh with Chemical Constituents and Uses*. 2nd ed. Asiatic Society of Bangladesh, Dhaka, p. 184.

Ghani, N. M. 1998. *Khazain ul Advia*. Vol. 2. Idara Matbuat-i- Sulemani, Lahore, Pakistan, pp. 240–241.

Grover, J. K., V. Vats, S. S. Rathi, et al. 2001. Traditional Indian anti-diabetic plants attenuate progression of renal damage in streptozotocin induced diabetic mice. *Journal of Ethnopharmacology* 76(3): 233–238.

Herculano, E. D., C. da Costa, A. K. Rodriques, et al. 2014. Evaluation of cardiovascular effects of edible fruits of *Syzygium cumini* Myrtaceae (L) skeels in rats. *Tropical Journal of Pharmaceutical Research* 13(11): 1853–1861.

Jagetia, G. C. 2018. A review on the role of jamun, *Syzygium cumini* skeels in the treatment of diabetes. *International Journal of Complementary and Alternative Medicine* 11(2): 91–95.

Jagetia, G. C., and M. S. Baliga. 2002. *Syzygium cumini* (jamun) reduces the radiation induced DNA damage in the cultured human peripheral blood lymphocytes, A preliminary study. *Toxicology Letters* 132(1): 19–25.

Jonnalagadda, A., K. K. Maharaja, and N. Prem Kumar. 2012. Combined effect of *Syzygium cumini* seed kernel extract with oral hypoglycemics in diabetes induced increase in susceptibility to ulcerogenic stimuli. *Journal of Diabetes and Metabolism* 4(1): 236.

Kabeeruddin, M. 2000. *Makhzan ul Mufradat Al-Maroof Khawas ul Advia.* Faisal Publication, Deoband, p. 215–216.

Katiyar, D., V. Singh, and M. Ali. 2016. Recent advances in pharmacological potential of *Syzygium cumini*: A review. *Advances in Applied Science Research* 7(3): 1–12.

Kedar, P., and C. H. Chakrabarti. 1983. Effects of Jambolan seed treatment on blood sugar, lipids and urea in streptozotocin induced diabetes in rabbits. *Indian Journal of Physiology and Pharmacology* 27(2): 135–140.

Khān, M. A. 2013. *Muḥiṭ-i-A'ẓam.* Vol. II. Central Council for Research in Unani Medicine, New Delhi, pp. 132–134.

Khare, C. P. 2007. *Indian Medicinal Plants.* 1st ed. Springer Science, New Delhi, p. 639.

Kirtikar, K. R., and B. D. Basu. 1975. *Indian Medicinal Plants.* Vol. II. Periodical Experts, New Delhi, pp. 1052–1053.

Kirtikar, K. R., and B. D. Basu. 1987/1999. *Indian Medicinal Plants.* Vol. I. International Book Distributors, Dehradun, pp. 1052–1054.

Kumar, A., N. Padmanabhan, and M. R. V. Krishnan. 2007. Central nervous system activity of *Syzygium cumini* seed. *Pakistan Journal of Nutrition* 6(6): 698–700.

Kumar, E. K., S. K. Mastan, K. R. Reddy, et al. 2008. Anti-arthritic property of methanolic extract of *Syzygium cumini* seeds. *International Journal of Integrative Biology* 4(1): 55–61.

Mahapatra, P. K., D. Chakraborty, and A. N. Chaudhuri. 1986. Anti-inflammatory and antipyretic activities of *Syzygium cumini. Planta Medica* 52(6): 540.

Mahmoud, I. I., M. S. A. Marzouk, F. A. Moharram, M. R. El-Gindi, and A. M. K. Hassan. 2001. Acylated flavonol glycosides from *Eugenia jambolana* leaves. *Phytochemistry* 58(8): 1239–1244. https://doi.org/10.1016/s0031-9422(01)00365-x.

Makheswari, M. U., and D. Sudarsanam. 2011. Phytomedicine for diabetes mellitus: An overview. *Research in Pharmacy* 1(4): 28–37.

Mastan, S. K., T. B. Latha, T. S. Latha, et al. 2009. Influence of methanolic extract of *Syzygium cumini* seeds on the activity of gliclazide in normal and alloxan-induced diabetic rats. *Pharmacology Online* 3: 845–850.

Mastan, S. K., A. Saraseeruha, V. Gourishankar, et al. 2008. Immunomodulatory activity of methanolic extract of *Syzygium cumini* seeds. *Pharmacology Online* 3: 895–903.

Mishra, D. S. 2018. Enhancing income through value-addition. *Indian Horticulture* 63(5): 107–109.

Mukherjee, P. K., K. Saha, T. Murugesan, M. Pal, and B. P. Saha. 1998. Screening for antidiarrhoeal profile some plant extract of specific region of West Bengal, India. *Journal of Ethnopharmacology* 60(1): 85–89. https://doi.org/10.1016/s0378-8741(97)00130-x.

Mulkalwar, S., V. Kulkarni, T. Deshpande, et al. 2021. Antihyperglycemic activity of syzygium cumini (Jamun) in diabetic rats. *Journal of Pharmaceutical Research International* 33(35A): 12–19.

Muruganandan, S., K. Srinivasan, S. Chandra, S. K. Tandan, J. Lal, and V. Raviprakash. 2001. Anti-inflammatory activity of Syzygium cumini bark. *Fitoterapia* 72(4): 369–375. https://doi.org/10.1016/s0367-326x(00)00325-7.

Nabi, G. N. 2007. *Makhzan- i-Mufradat-wa- Murakkabat.* CCRUM, New Delhi, p. 103.

Nadkarni, K. M. 1976. *Indian Materia Medica, Bombay Popular Parkashan.* Vol. I, 3rd ed., pp. 516–518.

Nair, A. G., and S. Subramanian. 1962. Chemical examination of flowers of Eugenia-jambolana. *Journal of Scientific & Industrial Research* 21(9): 457.

Nair, K. N. 2017. *The Genus* Syzygium*:* Syzygium cumini *and Other Underutilized Species*. Boca Raton and London: CRC Press, pp. 57, 58, 60, 61.

Pandey, M., and A. Khan. 2002. Hypoglycaemic effect of defatted seeds and water soluble fibre from the seeds of *Syzygium cumini* (Linn.) skeels in alloxan diabetic rats. *Indian Journal of Experimental Biology* 40(10): 1178–1182.

Pepato, M. T., D. M. Mori, A. M. Baviera, et al. 2005. Fruit of the jambolan tree (*Eugenia jambolana* Lam.) and experimental diabetes. *Journal of Ethnopharmacology* 96(1–2): 43–48. https://doi.org/10.1016/j.jep.2004.07.029.

Pradhan, M. 2016. Phytochemistry, pharmacology and novel delivery applications of *Syzygium cumini* (L.). *International Journal of Pharmacy and Pharmaceutical Research* 7(1): 659–675.

Prince, P. S., V. P. Menon, and L. Pari. 1998. Hypoglycaemic activity of *Syzygium cumini* seeds: Effect on lipid peroxidation in alloxan diabetic rats. *Journal of Ethnopharmacology* 61(1): 1–7.

Priya, S. S. L., P. R. Devi, and A. Madeswaran. 2013. In silico docking studies of RP2 (X-linked retinitis pigmentosa) protein using anthocyanins as potential inhibitors. *Bangladesh Journal of Pharmacology* 8(3): 292–299.

Pullaiah, T. 2006. *Encyclopaedia of World Medicinal Plants*. Vol. 4. Regency Publication, New Delhi, pp. 1892.

Rajasekaran, M., J. S. Bapna, S. Lakshmanan, A. R. Nair, A. J. Veliath, and M. Panchanadam. 1988. Antifertility effect in male rats of oleanolic acid, a triterpene from *Eugenia jambolana* flowers. *Journal of Ethnopharmacology* 24(1): 115–121. https://doi.org/10.1016/0378-8741(88)90142-0.

Ramya, S., et al. 2012. Profile of bioactive compounds in Syzygium cumini – a review. *Journal of Pharmacy Research* 5(8): 4548–4553.

Rastogi, R. P., and B. N. Mehrotra. 2004. *Compendium of Medicinal Plants*. Central Drug Research Institute (CDRI), Lucknow and National institute of communication and information resources, New Delhi, p. 406.

Ravi, K., S. Rajasekaran, and S. Subramanian. 2005. Antihyperlipidemic effect of *Eugenia jambolana* seed kernel on streptozotocin-induced diabetes in rats. *Food and Chemical Toxicology* 43(9):1433–439. https://doi.org/10.1016/j.fct.2005.04.004.

Ravi, K., B. Ramachandran, and S. Subramanian. 2004. Protective effect of *Eugenia jambolana* seed kernel on tissue antioxidants in streptozotocin-induced diabetic rats. *Biological and Pharmaceutical Bulletin* 27(8): 1212–1227. https://doi.org/10.1248/bpb.27.1212.

Raza, A., M. S. Butt, and H. A. Suleria. 2017. Jamun (*Syzygium cumini*) seed and fruit extract attenuate hyperglycemia in diabetic rats. *Asian Pacific Journal of Tropical Biomedicine* 7(8): 750–754.

Rekha, N., R. Balaji, and M. Deecaraman. 2010. Antihyperglycemic and antihyperlipidemic effects of extracts of the pulp of *Syzygium cumini* and bark of *Cinnamon zeylanicum* in streptozotocin-induced diabetic rats. *Journal of Applied Biosciences* 28: 1718–1730.

Rodrigues, K. A., L. V. Amorim, C. N. Dias, D. F. Moraes, and S. M. Carneiro. 2015. *Syzygium cumini* (L.) Skeels essential oil and its major constituent α-pinene exhibit anti-Leishmania activity through-immunomodulation in vitro. *Journal of Ethnopharmacology* 3(160): 32–40. https://doi.org/10.1016/j.jep.2014.11.024.

Shafi, P. M., M. K. Rosamma, K. Jamil, and P. S. Reddy. 2002. Antibacterial activity of *Syzygium cumini* and *Syzygium travancoricum* leaf essential oils. *Fitoterapia* 73(5): 414–416. https://doi.org/10.1016/s0367-326x(02)00131-4.

Sharma, A. K., S. Bharti, R. Kumar, et al. 2011a. Syzygium cumini ameliorates insulin resistance and β-cell dysfunction via modulation of PPAR, dyslipidemia, oxidative stress, and TNF-α in type 2 diabetic rats. *Journal of Pharmacological Sciences* 119(3): 205–213.

Sharma, S. B., A. Nasir, K. M. Prabhu, et al. 2003. Hypoglycaemic and hypolipidemic effect of ethanolic extract of seeds of *Eugenia jambolana* in alloxan-induced diabetic rabbits. *Journal of Ethnopharmacology* 85(2–3): 201–206. https://doi.org/10.1016/s0378-8741(02)00366-5.

Sharma, S. B., A. Nasir, K. M. Prabhu, et al. 2006. Antihyperglycemic effect of the fruit-pulp of *Eugenia jambolana* in experimental diabetes mellitus. *Journal of Ethnopharmacology* 104(3): 367–373.

Sharma, S. B., R. S. Tanwar, A. Nasir, et al. 2011b. Antihyperlipidemic effect of active principle isolated from seed of *Eugenia jambolana* on alloxan-induced diabetic rabbits. *Journal of Medicinal Food* 14(4): 353–359. https://doi.org/10.1089/jmf.2010.1227.

Sidana, S, V. B. Singh, B. L. Meena, et al. 2017. Effect of *Syzygium cumini* (jamun) seed powder on glycemic control: A double-blind randomized controlled trial. *Journal of Medical Society* 31(3): 185–189.

Siddiqui, M. S., B. Sharma, and G. Ram. 2014. Anti-hyperglycemic and anti-hyperlipemia effects of *Syzygium cumini* seed in alloxan induced diabetes mellitus in Swiss albino mice (Mus musculus). *Medicinal & Aromatic Plants* 3(4): 166. https://doi.org/10.4172/2167-0412.1000166.

Singh, S., A. K. Singh, B. G. Bagle, and H. K. Joshi. 2007. Scientific cultivation of jamun (Syzygium cumini Skeels). *Asian Science* 2(2): 37–41.

Singh, S., A. K. Singh, B. G. Bagle, and T. A. More. 2010. *The Jamun*. ICAR-DKMA, New Delhi, pp. 1–48.

Singh, S., H. P. Singh, A. K. Singh, and P. S. Sisodia. 2011. *The Jamun (Fruit for Future)*. Agro-tech Publishing Agency Udaipur, Rajasthan, pp. 1–100.

Sisodia, S. S., and M. Bhatnagar. 2009. Hepatoprotective activity of *Eugenia jambolana* Lam. in carbon tetrachloride treated rats. *Indian Journal of Pharmacology* 41(1): 23–27. https://doi.org/10.4103/0253-7613.48888.

Sridhar, S. B., U. D. Sheetal, M. R. Pai, et al. 2005. Preclinical evaluation of the antidiabetic effect of *Eugenia jambolana* seed powder in streptozotocin-diabetic rats. *Brazilian Journal of Medical and Biological Research* 38(3): 463–468. https://doi.org/10.1590/s0100-879x2005000300018.

Tarique, N. A. 2010. *Taj-al-Mufradat*. 1st ed. Idara Kitab-al -Shifa, New Delhi, p. 262.

Vikrant, V., J. K. Grover, and N. Tandon, et al. 2001. Treatment with extracts of Momordica charantia and Eugenia jambolana prevents hyperglycemia and hyperinsulinemia in fructose fed rats. *Journal of Ethnopharmacology* 76(2): 139–143. https://doi.org/10.1016/s0378-8741(01)00218-5.

Whiting, D., L. Guariguata, C. Weil, and J. Shaw. 2011. IDF diabetes atlas: Global estimates of the prevalence of diabetes for 2011 and 2030. *Diabetes Research and Clinical Practice* 94(3): 311–321.

Wild, S., G. Roglic, A. Green, R. Sicree, and H. King. 2004. Global prevalence of diabetes-estimates for the year 2000 and projections for 2030. *Diabetes Care* 27(5): 1047–1053.

Yadav, D., A. Lalit, S. Singh, et al. 2013. Evaluation of antidiabetic and phytochemical activity of 50% methanolic extract of jamun seed (*Syzygium cumini*). *Search and Research* 4(3): 13–16.

Zhi-Ping, R., Z. Liang-Liang, and L. Yi-Ming. 2008. Evaluation of the antioxidant activity of *Syzygium cumini* leaves. *Molecules* 13(10): 2545–2556. https://doi.org/10.3390/molecules1310254.

25 Antidiabetic and Other Potential Features of *Azadirachta indica* A. Juss. (Neem)

Rohini Ganorkar, Salman Khan, and Kolagani Chandramohan

CONTENTS

25.1 INTRODUCTION

Neem (*Azadirachta indica* A. Juss.) is a deciduous tree species belonging to the family Meliaceae (Vietmeyer 1992) and native to India and Burma (Myanmar). The name *Azadirachta indica* is derived from the Persian name "Azad-darakht-e-hind" which means "Free Tree of India". Other names of neem are margosa (in Spanish), azidarac or azadira d'inde (in French), Niembaum (In German), Vembu/Veppan (in Tamil), tamar/tamarkha (in Burmese) mindi (in Indonesia) and dongoyaro (in Nigeria). The tree is considered as a divine tree or village pharmacy tree (Islas et al. 2020). The Indian traditional medical system, Ayurveda, the Charak Samhita, Sushrut Samhita and other ancient books all mention neem for medical use (Chaguthi et al. 2018). The IUCN status of neem is "Least Concern". The tree is tall (~30 meters), has a girth of around 2.5 meters and has a large crown with bright to dark green colour leaves. Pollarding is common practice in *A. indica* trees due to the gregarious emergence of the leaves. The aroma produced by monoecious neem flowers attracts a large number of pollinators. Neem is a multipurpose tree species having varied uses from afforestation, agriculture (as fertilizer, agrochemicals and others), medicines and insecticides to other uses (Koul 2004). The wood obtained through neem trees is useful in making agricultural implements and other furniture. Seeds are being used for extracting neem oil, which has tremendous uses in different industries from product ingredients to machinery greasing (Axtell and Fairman 1992; Koul 2004). In rural areas, neem provides shade to farmers and cattle on sunny days.

These days, due to increased pollution and adulteration, the risk of disease and disorder is prevalent. Also, people are developing trust in herbal or plant-based medicines due to their nil/negligible

DOI: 10.1201/b23347-25

side effects in comparison to other medical systems (Eid et al. 2017). Also, they are the cost-effective and traditional medical system of India. Neem is considered a tree of high value in terms of medicinal and therapeutic uses. Bark, root, flower, fruit, seed and leaves of neem have considerate medicinal properties and are of tremendous use. They exhibit properties like antimicrobial, analgesic, antipyretic, anti-ulcerous, anti-cancerous, antimalarial, antioxidant, antidiabetic and anti-inflammatory (Patel et al. 2016; Shareef and Akhtar 2018; Islas et al. 2020).

25.2 BOTANICAL DESCRIPTION

Azadirachta indica (Juss 1830) is a large tree 40–50 ft high, with a straight trunk. Leaves simply pinnate, 8–15 in. long, crowded near the ends of the branches; leaflets 9–12, subopposite, 1–3 by 1/2–1-1/2 in., obliquely lanceolate, sometimes falcate, acuminate, serrate, glabrous on both surfaces, base inequilateral, acute; petiolules very short. Flowers white, fragrant, in branched glabrous panicles shorter than the leaves; bracts minute, lanceolate, caduceus. Calyx puberulous outside, divided almost to the base; lobes rotund-ovate, minutely ciliolate. Petals 1/4-inch long, obovate-oblong, faintly puberulous outside, ciliolate. Staminal-tube glabrous, a little shorter than the petals, obconic, the laciniae truncate and toothed at the apex; anthers ten, opposite the laciniae and a little shorter than them, apiculate. Disk 0, ovary glabrous, three-celled, the cells opposite to the petals; ovules two in each cell, collateral; stigma three-toothed, included in the tube. Drupes the shape of an olive, 1/2–3/4-inch long, glabrous, one-seeded. A photograph of the different plant parts is shown in Figure 25.1. Flowering in neem is mostly seen from March to April followed by

FIGURE 25.1 A neem tree and its different parts in India.

fruiting is April to June and can be extended up to August. The seed yield obtained from a neem tree of five years of age is about 5–10 kg, whereas it can go up to 10–15 kg for a tree of ten or more years of age (TNAU 2021).

25.3 DISTRIBUTION

India: Neem is commonly available in almost all parts of India. It is available from the Himalayas and Shivalik hills to Kerala under multiple climatic conditions and varying altitudes (up to 700 m). It favours mostly drier areas with considerate water and salt conditions. It is available in the state of Andhra Pradesh, Assam, Bihar, Chhattisgarh, Delhi, Gujarat, Haryana, Karnataka, Madhya Pradesh, Maharashtra, Meghalaya, Odisha, Punjab, Rajasthan, Tamilnadu, Uttar Pradesh, Uttarakhand and West Bengal in India (Makheswari and Sudarsanam 2012; Shukla et al. 2020).

Outside India: *Azadirachta indica* has a large geographical range in almost more than 30 countries. It is available in Australia, Bangladesh, the Caribbean, China, Cuba, Eritrea, Fiji, Haiti, Indonesia, Iran, Iraq, Kenya, Malaysia, Mauritius, Mexico, Mozambique, Nepal, Oman, Pakistan, Sri Lanka, Somalia, Saudi Arabia, Tanzania, Thailand, United States of America, West Indies, Central and South America and Vietnam (Vietmeyer 1992; Koul 2004; Islas et al. 2020).

25.4 PHYTOCHEMICAL CONSTITUENTS

Neem contains a natural compound group known as triterpenes. These include azadirachtin, nimbin, nimoins, epoxy-azadiradione, nimbidin, miliantriol and salannin (Vietmeyer 1992; Naik et al. 2014; Islas et al. 2020). The structure of important bioactive compounds extracted from neem is depicted in Figure 25.2.

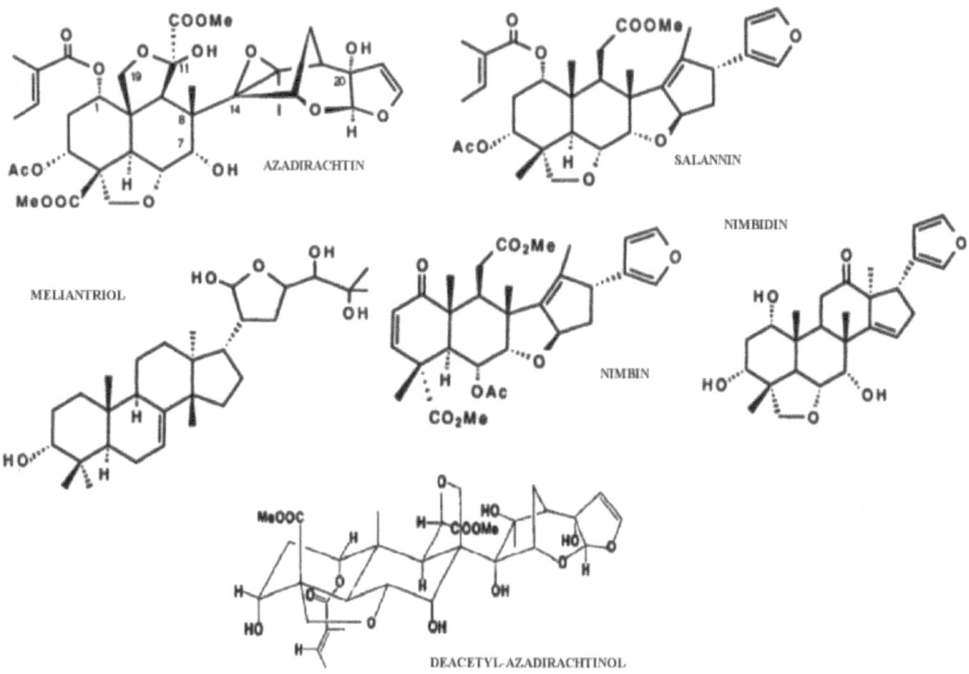

FIGURE 25.2 Structures of neem extracted bioactive compound with high medicinal value.

Azadirachtin: This chemical compound extracted from the neem tree has multi-functional properties which include antifeedant, insect repellent, growth regulator, pesticides, manipulating insect hormones and controlling insect metamorphosis (Saleem et al. 2018). Vietmeyer (1992) mentioned that structural similarity is observed in both azadirachtin and ecdysones (insect growth hormone). Azadirachtin can be extracted from various parts of the neem tree such as leaves, seeds and fruits. Nearly two to four milligrams of azadirachtin is present in the neem kernel. It is an established compound being used as a herbal insecticide and therapeutic agent (Chattopadhyay et al. 2004). The structural formula of azadirachtin is shown in Figure 25.2.

Meliantriol: This is another important chemical compound extracted from neem. Meliantriol affects the feeding process and behaviour of the insect pest. It will save the tree from extreme defoliation by manipulating the insect's feed. Meliantriol is also being used as a mosquito repellent and botanical pesticide (Shoaib et al. 2010).

Salannin: In insects, salannin shows antifeedant property against various insects due to its inhibitory action (Vietmeyer 1992; Tembe-Fokunang et al. 2019). The isolation protocol for salannin is most developed after azadirachtin (Koul 2004). The efficacy of salannin group compound is well known against most insects such as the beetle, *Epilachna varivestis* (Koul et al. 2004). Apart from antifeedant activity, it also regulates the growth of insects.

Nimbin: Nimbin is a triterpene obtained from the neem extract. This biochemical compound possesses antiviral, antioxidant, antipyretic, antihistamine and antimicrobial qualities (Islas et al. 2020) (Figure 25.2). Nimbin is also useful as an antifeedant and herbal insecticide.

Nimbidin: This compound extracted from neem seeds shows antiviral properties. In the neem kernel, nimbidin content is around 2% (Islas et al. 2020). Nimbidin also possesses anti-arthritic and anti-inflammation properties. Apart from this, nimbidin is a good moderator for neutrophils and macrophages (Kaur et al. 2004). It also helps in the treatment of gastro-ulcer (Suganthan et al. 1978).

Nimbidiol: This molecule is collected from the neem stem, bark and roots. It is useful in controlling diabetes and has a high potential to be used in industry (Mukherjee and Sengupta 2013). It has antipyretic, anti-cancerous and anti-protozoan properties (Tembe-Fokunang et al. 2019).

Gedunin: The bioactive compound gedunin has properties like antimalarial and antifungal and it helps in blood vessel dilation (Tembe-Fokunang et al. 2019). Gedunin also inhibits the growth of tumour cells.

Quercetin: This compound extracted from the neem tree possesses anti-protozoal properties (Tembe-Fokunang et al. 2019). Quercetin is a well-established bioflavonoid for consumption purposes (Kelly 2011). It exhibits properties like anti-inflammation, anti-cancerous, antioxidant and immunity enhancer (Islas et al. 2020).

Other constituents: Various other compounds which are extracted from the neem tree show anti-hormonal activity. Compounds such as deacetyl-azadirachtinol is equally important from an insecticidal point of view. Also, 3-deacetylesalannin and salannol are two other biochemical compounds extracted from the neem tree that make the insect cease to feed (Vietmeyer 1992). Neem seed kernels possess biochemical compounds which are more concentrated and obtainable than any other plant part. Extraction of such bioactive compounds can be done by water and organic solvents. There are a few flavonoids which are also extracted from the neem tree that bear medicinal properties (Islas et al. 2020). They are capable of inhibiting processes like alprostadil/dinoprostone (eicosanoids lipid) pathways and work as an anti-inflammatory by regulating kinase and phosphodiesterases (PDE) enzymes (Naik et al. 2014; Islas et al. 2020).

25.5 PHARMACOLOGICAL STUDIES

A. indica shows enormous pharmacological properties. In Ayurveda and Unani medical systems, neem is considered for treating multiple diseases and disorders. Various parts of the plant like flower, leaf, fruit, seeds, bark, wood and roots are being used for such treatment. Neem oil is the most widely used ingredient for the preparation of different medicines as well as products like toothpaste, face wash and so on, and thus it bears high economic value (Patel et al. 2016). Worldwide, in plant-based medicines, neem is a major ingredient. *A. indica* leaves are largely being used in various ailments and skin disease treatment. Combining neem extract with different additive materials enhances its effectiveness by 10–20 times. Neem exhibits properties like antifungal, antiviral, antibacterial, anti-inflammatory, antipyretic, analgesic and anti-ulcer (Biswas et al. 2002). Also, neem has shown a significant effect against malaria and Chagas disease. The following pharmacological uses of neem have been recorded in various pieces of literature:

Antifungal: Neem has always been used for the treatment of fungal infections in humans. Khan and Wassilew (1987) mentioned that neem has been potentially toxic to the following fungi: *Trichophyton*, *Epidermophyton*, *Microsporum*, *Trichosporon*, *Geotrichum* and *Candida*. These fungi are causal organisms for infection of hair, skin, nails, hand, intestinal tract, lungs and mucous membrane. Thus, neem helps get rid of various skin and fungal diseases due to its antifungal property. Various researchers have also reported that *Aspergillus flavus*, *Alternaria solani* and *Rhizopus* can successfully show suppression when neem extract is applied (Lessa et al. 2010; Chhavi et al. 2014).

Antibacterial: Neem has also shown antibacterial properties due to the presence of active bioactive compounds. Two bacteria, namely *Staphylococcus aureus*, which causes food poisoning and pus-forming disorder, and *Salmonella typhi* (Patel and Trivedi 1962), which causes typhoid, food poisoning and intestinal inflammation, can be controlled by using neem oil. Also, Shareef and Akhtar (2018) reported that neem emulsion is effective against the following disease-causing bacteria: *Aeromonas salmonicida*, *Vibrio* sp., *Pseudomonas aeroginosa* and others (Taylor and Webster 2011). Moreover, there are few bacterial species and strain that show no effect from neem oil or extract (Vietmeyer 1992).

Analgesic: Neem oil is obtained from the neem seed through a solvent extraction procedure. Neem oil contains an important active compound, nimbidin. This nimbidin has analgesic, anti-arthritic and anti-inflammatory properties which help in reducing joint pain (Kaur et al. 2004).

Immunomodulatory activity: Neem, being an important medicinal plant, has a variety of uses in the Ayurveda medical system and immunomodulation is one of the uses (Patel et al. 2012). Neem flavonoids such as catechin and epicatechin extracted from the neem bark exhibit immunomodulatory properties.

Anti-cancerous: Neem trees contain different biochemicals, and a few of them are responsible for the control of cell-signalling path. They help to regulate the genes by removing and deactivating tumour cells. Also, they moderate different biological functioning such as removing undesirable/unwanted cells, providing the protein required for transcription from DNA to RNA and helping in the formation of blood vessels (Shareef and Akhtar 2018). It was also reported that neem with cisplatin exhibits anti-cancerous properties against cervical and breast cells in humans and also reduces cell toxicity. Also, neem leaf-based treatment can help from nephron and hepatotoxicity caused by cisplatin (Sivakumar et al. 2014). Efferth and Koch (2011) informed that neem is expressive for decreasing the oncogene expression in the cancerous cell of mice. Chen et al. (2018) reported that azadiramide-A, a biochemical compound obtained from neem leaves ethanolic extract, shows anti-cancerous properties in breast cells. Neem leaf glycoprotein compounds maintain the protected micro-environment of cancerous cells by keeping the cytokines and chemotactic cytokines in balance.

Antioxidant: Neem has a greater concentration of antioxidants and is thus useful as a free-radical scavenger due to the presence of azadirachtin and other limonoids (Shareef and Akhtar 2018). As per them, free radicals are responsible for plenty of diseases in humans; thus antioxidants are important as they help in defusing and neutralizing free radicals/reactive oxygen species. Almost all parts of the neem tree exhibit antioxidant properties. The use of neem tea and oil can fulfil the antioxidant requirements of our body (Page and Hawes 2013). The antioxidant property of trichloromethane crude extract is superior and at par with raw antioxidant (Biswas et al. 2002). Shori and Baba (2013) reported that neem-supplemented yoghurt enhances the natural scavenging of free radicals.

Antidiabetic: Neem contains antidiabetic properties due to the presence of various bioactive compounds (Patil et al. 2022). Various parts of the neem trees such as leaves, roots, bark, leaves, fruits and flowers contain antidiabetic properties (Rafe 2017). Aqueous leaf extract doses help in combating diabetes. The antidiabetic properties are discussed in detail later.

Anti-ulcerous: Neem is an essential plant for the treatment of ulcers. Researchers have also demonstrated the efficiency of neem bark extract against ulcers. Neem extract can control gastric acid secretion along with gastro-oesophagus and gastro-duodenal ulcers (Srivastava et al. 2020).

Hepatoprotective activity: Azadirachtin extracted from neem in combination with CCl_4 shows hepatoprotective activity against necrosis of hepatic cells in mice (Singh and Sastry 1997). The leaf extract in methanol, aqueous and alcohol is suitable for protection against hepatic necrosis.

Neuroprotective activity: Neem extract shows a significant effect on the central nervous system. It shows depressant and sedative activity in mice (Singh et al. 1980), and it also depicts anti-anxiety properties in rats (Jaiswal et al. 1994).

Antiviral: Vietmeyer (1992) in his work mentioned that the paste of the *A. indica* foliage is being applied on pox virus-infected areas, though this was mentioned as unreliable practice. However, it was also reported that further transmission of infection to the healthy cell is restricted when neem paste is applied. A few researchers have also reported that neem extract can restrict the essential enzyme for DNA replication belonging to the hepatitis B virus. Studies in Germany suggested that neem extract has a significant effect on the herpes virus. Also, plant viruses can be tracked down by using neem extract. Govindachari et al. (1998) mentioned the antiviral property of neem bark and leaf extract against enteroviruses.

Anti-inflammatory: Various researchers have reported that neem extract has a significant impact on inflammation. The biochemical compounds of neem control the activity of inflammatory enzymes (Shareef and Akhtar 2018; Rupani and Chavez 2018). Nimbidin has proved to be useful in managing the white blood cells and polymorphonuclear leukocyte activity which eventually reduces inflammation (Govindachari et al. 1998). Naik et al. (2014) stated that neem can act as an anti-inflammatory moderator due to the presence of limonoids. Soares et al. (2014) experimented and concluded that neem can inhibit swelling and fibrovascular muscle development in rats. It was also claimed that neem inhibits (macrophage) white blood cell movement which causes enhancement in inflammation and the favourable reaction of multiple diseases. IL- (1,2,6,8,12,18) and TNF-α are commonly known inflammatory mediators which are largely affected by NF-kβ. Ephoxyazadiradione (a bioactive compound derived from neem) regulates and uplifts the level of NF-kβ (Alam et al. 2012). Thus, NF-kβ is important for inflammation inhibition in blood cells, B cells and mast cells (Alam et al. 2012; Naik et al. 2014).

Skin insect: The compounds collected from the neem plant have a notable impact on skin insects. For example, head lice can be controlled by using neem-based hair oil. This is the traditional knowledge of most of the tribes in India and other countries (Vietmeyer 1992).

Dental remedies: As per the Ayurveda medical system of India, neem is considered for most of the uses and of which utilization for dental use has prevailed. In India, specifically, rural peoples largely depend on neem twigs for daily brushing. This helps them to remove dental and periodontal problems (Henkes 1986). Also, many toothpaste brands have formulated neem-based products. In almost all toothpaste brands' products, neem is an essential element. The bark of the neem tree helps maintain good gum health (Vietmeyer 1992).

Antipyretic: Significant results are obtained when neem-based formulation or its extract is used against body fever (Rafe 2017). The bioactive compound, nimbidine, exhibits antipyretic potential (David 1969).

Antimalarial: Neem is equally important for treating various diseases such as malaria (Srivastava et al. 2020). As per Ayurveda, neem leaf has antimalarial properties. The medication is orally in the form of leaf tea or aqueous extract (Khalid et al. 1989). Similarly, neem seeds are equally effective against malarial parasites. Vietmeyer (1992) mentioned that gedunin, a compound extracted from neem seeds, has been proven as an antimalarial agent.

Other medical uses: Extracted oil from seeds and leaves of neem contains bioactive chemicals which perform as a spermicide. It was reported that neem oil has great potential as a contraceptive in females. Vietmeyer (1992) mentioned in his report that the neem oil product called "Sensal" is being sold in India for birth control. Apart from this, neem-based extracts have been continuously used for making veterinary medicines.

25.6 ANTIDIABETIC RESPONSE

Rising sugar (glucose) level in the blood causes a group of a chronic disease called diabetes or diabetes mellitus. Whenever we eat food, the body breaks the food into glucose and other molecules. This glucose is further transmitted to the blood system. Once the sugar level in the blood rises, the pancreas releases insulin to maintain the sugar level by converting sugar into energy (CDC 2022). In diabetes, the body's insulin production got affected or it may not be used efficiently. This causes a rise in the blood glucose/sugar that keeps on increasing. Diabetes can be of various types: type 1, type 2, pre-diabetes and gestational diabetes (Balwan et al. 2022). In type 1, the pancreas insulin production is affected badly; in type 2, the body's blood sugar level cannot be managed efficiently (Islas et al. 2020); pre-diabetes is an early condition of diabetes which implies that the glucose is elevating; gestational diabetes is a condition for the surge of blood sugar in pregnant women. Diabetes causes kidney failure, cardiovascular disease and other problems (Shailey and Basir 2012; Islas et al. 2020). Ogurtsova et al. (2022) reported that a large population of the world is undiagnosed for this chronic disease (diabetes). As per them, more than 50% of the population of Africa, the southeast region and the western Pacific has undiagnosed diabetes. Also, International Diabetes Federation in Belgium mentioned that approximately 0.536 billion people (in 2021) worldwide are living with diabetes (both detected and undetected), and this figure is continuously rising.

Shinkafi et al. (2015) reported that different plant parts of *Azadirachta indica* such as leaves, bark and root are used as an antidiabetic in Nigeria. In India, the antidiabetic potential of neem is reported by various researchers such as Tarak et al. (2011) and Tag et al. (2012). Neem causes a hypoglycaemic effect, inhibits epinephrine activity and glycogenolysis and also drops marginal consumption of sugar (Balwan et al. 2022). In the diabetic rat (streptozotocin-induced), neem extract treatment recovers glucose-6-phosphate dehydrogenase (Islas et al. 2020). Patil et al. (2022) investigated the potential of neem for lipid profile, reactive oxygen species, hyperglycaemia, enzyme ingestion and so on. Neem leaf and bark extracts exhibit similar glucose homeostasis to that obtained after using insulin. Also, superoxide dismutase, nitric oxide dioxygenase and gluthione functioning can be reactivated through these neem extracts (Shailey and Basir 2012). Patil et al. (2013) demonstrate that epoxy-azadiradione present in neem seed extract drops the sugar level in streptozotocin-induced diabetic mice by more than one third of its level and reaches half level after several applications.

Chloroform extract of neem rescues glucose-6-phosphate dehydrogenase and elevates the release of insulin.

McCalla et al. (2016) reported that neem extract can enhance insulin secretion and can restore β-cells. Nimbidin extracted from the neem tree contains antidiabetic potential when taken orally. Also, the chloroform extract of neem possesses the potential for glucose tolerance. Akter et al. (2013) revealed that leaf extract shows hypoglycaemic activity in animals (alloxan-induced diabetic). Christian et al. (2019) studied the phytochemical profile of neem leaves and found that leaf extract contains saponins, phenols, alkaloids, tannins and cardiac glycosides of which saponin was adequately available. They also tested the neem extract against diabetic rats and found that the blood sugar level decreased with the application of the extract. Bisht and Sisodia (2010) identified the effect of neem on diabetes and found that the alcoholic extract of neem can regulate hyperglycaemic conditions in diabetic-induced rats.

Akter et al. (2014) reported the reduction of high glucose levels in diabetic mice using an aqueous extract of *A. indica*. Researchers have found that the consumption of dried leaves of neem is potent enough to restore and manage the pancreas β-cells, which is an essential component for elevating natural insulin secretion (Patel et al. 2012). Bhat et al. (2011) identified that methanolic and aqueous extract of neem can be proven to have glucosidase inhibition activity but not in chloroform extract. Atangwho et al. (2012) suggested the use of neem along with *Vernonia amygdalina* for a better antidiabetic result. Mukherjee and Sengupta (2013) mentioned the antidiabetic importance of nimbidiol, a compound extracted from the stem, root and bark of neem tree.

25.7 TRADITIONAL AND OTHER POTENTIAL USES

Entomologists throughout the world have reported that neem bears a significant effect on insect control as the seeds and leaf possess insecticidal properties (Chattopadhyay et al. 2004; Shoaib et al. 2010). The metamorphosis process of insect larvae can be disturbed due to the effect of azadirachtin (Vietmeyer 1992; Campos et al. 2016). The major effect of the neem-based product on insects involves the following mode of action: disturbing the developmental process, affecting the reproduction process, resisting egg laying or ovipositioning, reducing reproduction potential or sterilization, poisoning individuals, ceasing feeding and restricting chitin growth. The effectiveness of neem-based insecticides is considered equivalent to commercial pesticides (organo-chlorine or organo-phosphate) (Vietmeyer 1992). The neem kernel exhibits properties to alter the normal life cycle of more than 200 insect pests belonging to different orders such as Lepidoptera, Diptera, Coleoptera, Hemiptera, Homoptera, Orthoptera and others (Vietmeyer 1992; Nisbet and Mordue 2000). Apart from this, medically important insects such as mosquitoes (*Aedes* and *Anopheles*), aphids, fruit flies and blowflies are also majorly affected by bioactive compounds extracted from neem. Azatin, Turplex, Align, and Margosan-O are neem-based pesticides which are formulated and accepted in the United States (Vietmeyer 1992), although the desired results using neem products can be delayed when compared with commercial pesticides (Schmutterer 1985). Description of different bioactive compounds collected from various parts of the neem plant is given in Table 25.1. Also, various other organisms such as nematodes, snails, crustaceans, fungi, aflatoxin and plant viruses can be controlled by using neem-based products due to the presence of various bioactive compounds.

Neem is being used in cosmetics as a beauty enhancer. Neem oil has been a great lubricant which is used mostly in greasing (Idris et al. 2018). Also, the product of neem leaves and neem cake is of high importance as fertilizer in agriculture industries. A variety of products are made up of neem parts such as neem shampoo, neem soap, neem hair oil, neem-based washing bar and toothpaste. A few common products available in the market that use neem as a major ingredient are shown in Figure 25.3. In India, paddy fields are covered by neem foliage as a preparation followed by sprouts transplant. The timber obtained from neem trees has great economic value as it is primarily used in building structures, rural agricultural gears, furniture, idols, boats and other farm instruments

TABLE 25.1
Bioactive Compounds of Neem and Importance of Its Different Parts

Part of neem tree	Bioactive compound	Traditional, medicinal and other uses	Key references
Leaves	Alkaloids, saponins, tannins, flavonoids, glycosides, nimbidin, nimbiol, nimbidiol nimbandiol, nimbolide, proline, epoxy-azadiradione, azadiramide-A, cyclic trisulphide, quercetin, nimonol, nimicinol	Skin infections, antifungal, helpful in acne, injuries and boils, antioxidant, anti-cancerous, antibacterial, antiviral, used for treating Alzheimer's and Parkinson's disease, anti-inflammatory, wound healing, hepatoprotective, neuroprotective, nephroprotective, immunomodulatory, reduced intestinal glucosidase activity, hyperglycaemic, leprosy	Dash et al. 2017; Shareef and Akhtar 2018; Efferth and Koch 2011; Page and Hawes 2013; Sivakumar et al. 2014; Zong et al. 2012; Govindachari et al. 1998; Islas et al. 2020; Chen et al. 2018; Jaiswal et al. 1994; Noor et al. 2013; Chaguthi et al. 2018; Mukherjee and Sengupta 2013; Saleem et al. 2018
Twigs	Azadirachtin, nimbin, nimbidin	As toothbrush, for gum health, maintain saliva pH, oral deodorant, toothache, antibacterial, anti-candidial, remove bad odour and plaque, anti-cariogenic, make teeth whiter, overall dental care	Shareef and Akhtar 2018; Sengupta 2018; Gupta 2021; Srivastava et al. 2020; Lakshmi et al. 2015; Saleem et al. 2018; Drabu et al. 2012
Fruits	Resin, azadiradione, nomolin, nimolicin, 5-hydroxymethylfurfural, phytosterol, limonoids (gedunin and azadiradione), mahmoodin, azadirachtol, limocin-A & B, limocinol	Fruits/seeds provide neem oil, oil is used in making face wash, mosquito repellent, shampoo, soap, hand wash, hair oil, free radical scavenging, anti-inflammatory, analgesic, piles pain reliever, helps in eye irritation, anti-ulcerous, antidiabetic	Shareef and Akhtar 2018; Nahak and Sahu 2011; Ilango et al. 2013; Naik et al. 2014; Trivedi et al. 2019; Ponnusamy et al. 2015; Saleem et al. 2018; Siddiqui et al. 1991
Flower	Flavonoids, quercetin, flowerine, kaemferol, flowerone O-methylazadironolide, diepoxyazadirol, sitosterol, triterpenoids, nimbaflavone, sesquiterpenes, azadirone	Aromatherapy, nausea, anorexia, antioxidant, anti-infertility, stomach ulcer, jaundice, intestinal worms, bile reducer, phlegm	Shareef and Akhtar 2018; Saleem et al. 2018; Islas et al. 2020; Nahak and Sahu 2011; Gbotolorun et al. 2008; Eid et al. 2017; Srivastava et al. 2020; Trivedi et al. 2019; Tembe-Fokunang et al. 2019; Siddiqui et al. 1998
Bark	Azadirachtin, phenols, triterpenes, sugiol, saponins, unsaturated sterols, nimbin, nimbinin, nimbiol, gallic acid, catechin, epicatechin, margolone, margolonone, glycosides, nimbidiol	Antibacterial, antiviral, antidiabetic, antioxidant, antimicrobial, hypotensive, diuretic, spasmolytic, antimalarial, anti-ulcer, antipyretic, analgesic, anti-cancerous, anti-inflammatory, immunomodulatory	Govindachari et al. 1998; Shailey and Basir 2012; Patil et al. 2013; Srivastava et al. 2020; Subramanian and Lakshmanan 1996; Trivedi et al. 2019; Chaguthi et al. 2018; Mukherjee and Sengupta 2013
Root	Nimbin, nimbidin, nimbilin, nimolinin, nimbidiol	Antidiabetic, antihyperglycaemic and hypoglycaemic properties	Shinkafi et al. 2015; Patil et al. 2013; Mukherjee and Sengupta 2013; Ara et al. 1989; Saleem et al. 2018
Seed	Nimbidin, epoxy-, azadirachtin, nimbin, nimbolide, mahmoodin, nimbidiol, limonoids (gedunin and azadiradione), salannin, tetranortriterpene alcohol	Anti-inflammatory, antibacterial, antimicrobial, insecticide, antidiabetic, antidiuretic, immunomodulatory, antimalarial, antipyretic, analgesic, leprosy	Govindachari et al. 1998; Zong et al. 2012; Gupta et al. 2019; Mukherjee and Sengupta 2013; Quelemes et al. 2015; Patil et al. 2013; Patel et al. 2012; Chaguthi et al. 2018; Elakovich 1996; Tembe-Fokunang et al. 2019; Ponnusamy et al. 2015; Saleem et al. 2018

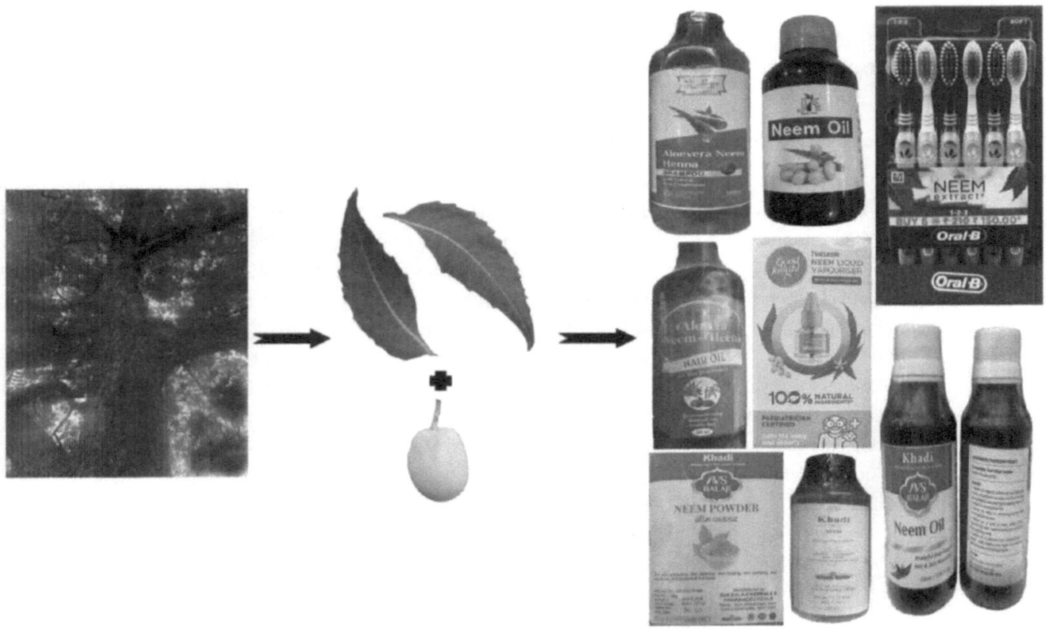

FIGURE 25.3 Uses of neem in manufacturing of different commercial products.

(Koul 2004). Neem oil can be utilized as fuel for igniting lamps, and small branches are used as fuel wood.

Several other potential products derived or collected from the neem plant parts are resin, tannins, neem-linked honey, fruit and other food products. It was also seen that Giloy (*Tinospora cordifolia*), a medicinal climber which grows over neem trees, has more medicinal and economical values when compared with other than neem (Jain and Dhupper 2021).

25.8 SAFETY ISSUES

Neem trees have a large geographical extent and thus can be easily found in most countries worldwide. This may bring genetic variability among the individuals growing at different locations (Kaushik et al. 2007). So, the testing, experiment results and formulation of the different products may not have similar or equal effects on the target (Jessinta et al. 2014).

The product formulated using a different part of neem may showcase different potential and activity which may be based on the age of the tree, habitat and environmental condition, geographical location, phenology, biotic stress and so on. (Jessinta et al. 2014). Loke et al. (1990) reported that *A. indica* products can have a negative effect on plant growth and its properties, e.g. retarded growth in cabbage and depletion of leaf waxy surface of onion crop. Similarly, neem oil can have a significant phytotoxic effect if the purity is not met (Vietmeyer 1992).

Even though the tree has high insecticidal properties, it has natural enemies such as oriental yellow scale (*Aonidella orientalis*), scale insect (*Pinnaspis strachani*), tortricid moth (*Adoxophyes aurata*) and tea mosquito bug (*Helopeltis antonii*) (Vietmeyer 1992; TNAU 2015). Similarly, TNAU (2020) documented a few pests of neem, i.e. scale insect (*Pulvinaria maxima*), tip borer (*Laspeyresia* spp.) and black tea thrips (*Heliothrips haemorrhoidalis*). Vietmeyer (1992) mentioned that an *A. orientalis* outbreak leading to the death of neem trees was observed in 1986 in West Africa. Areas such as Nigeria, Cameroon and Sahel show significant defoliation and other impacts on neem tree health. In Nigeria, Schmutterer (1990) reported more than 14 insect pest attacks on neem. DTE (2012) reported the severe defoliation of neem trees in Uttar Pradesh, India, due to a lepidopteron pest. This suggests that in the future, similar infestations or outbreaks could also be seen by this or other insect pests.

Apart from the insect pests, fungus, bacteria and viruses do attack neem trees. *Ganoderma lucidium* causes root rot, *Corticium salmonicolor* and *Pseudomonas azadirachtae* cause stem rot and blight, *Cercospora subsessilis* causes leaf spot, *Oidium* sp. causes powdery mildew, *Fusarium* and *Rhizoctonia* cause seedling rot and blight (Bakshi 1976; Vietmeyer 1992). Also, some fungi belonging to the genus *Alternaria*, *Colletotrichum* and *Pseudocercospora* are reported as a pest of neem trees and cause leaf blight and spot (TNAU 2020).

Del Serrone et al. (2015) identified that neem has a significant negative impact on meat quality. Neem, being an antibacterial, kills bacteria including those which are healthy for the human body. It was also reported that neem has a remarkable effect on young children causing severe disorders like organ swelling and Reye's syndrome (Sinniah and Baskaran 1981). Also, Mishra and Dave (2013) highlighted the toxicity caused by neem oil in humans. They mentioned the various bioactive complexes such as azadirachtin present in the neem may cause encephalopathy in aged men. As per Vietmeyer (1992), *Azadirachta indica* tea may have possibly caused kidney malfunction in West Africa, though the actual research or validity of this statement is pending. Sadre et al. (1984) mentioned that the neem compound has a toxicological effect on goats and guinea pigs. Similarly, Shori and Baba (2013) and Kato-Noguchi et al. (2014) reported that neem seed and foliage extract (aqueous) can cause mild toxicity in rats due to higher concentration of nimbolide B and nimbic acid B. Neem extract collected from solvents other than water causes severe allergy to the rat (Patel et al. 2016; Deng et al. 2013). Page and Hawes (2013) reported that in a few cases of the clinical trial of higher doses (neem tea), haemolytic anaemia, icterus, diarrhoea and dizziness were seen. It was also stated that the clinical studies of neem oil should be carried out with proper caution and a dose of 1600 mg/kg should be given in a day, and this process should not go beyond 90 days (Deng et al. 2013).

25.9 CULTIVATION PRACTICES

In the United States, neem propagation and harvesting benefitted rural people by providing them with a source of income and continuous employment (Vietmeyer 1992). The propagation of neem requires following a predefined protocol. The neem is frost resistant and under warm climatic conditions becomes susceptible to microbial and pest attacks.

Singh et al. (2009) mentioned that the cultivation of *Azadirachta indica* can be done by both sexual and vegetative propagation, but propagation through seeds is a common practice. However, researchers have also reported the propagation of neem through cuttings, leaf discs, forced axillary branching, and root and shoot tip explant methods (Gehlot et al. 2014). For degraded areas, propagation through seed is recommended, whereas, for large-scale production or forestry plantation, propagation using the cutting method is suggested (TNAU 2020).

- *Seed preparation*: The fresh seeds are collected from the healthy neem trees (TNAU 2020). Further, the pulpy portion including the fruit coat is removed by rubbing it on a coarse surface followed by washing and cleaning with water (Vietmeyer 1992). In African countries, seeds whose pulp is consumed by fauna are taken up for propagation. The viability of neem seeds decreases with time. So, the seed should be stored in an earthen pot having wet sandy soil with good moisture content to extend the seed viability (TNAU 2020).
- *Site conditions*: The optimum annual precipitation required for the growth of neem seedlings is 400 to 1200 mm (TNAU 2020). Neem can bear large temperature and climatic variations but the growth is severely affected under frost conditions (Vietmeyer 1992; TNAU 2020). Also, neem can grow under various soil type conditions like acidic, sandy

and dry or on barren and rocky sites. The pH should be around 5–6.2 for better growth (Phogat et al. 2009). Waterlogged condition is not suitable for neem propagation (Singh et al. 2009).

- *Planting technique and nursery management*: In the nursery, seedlings emerge after almost a week of seed sowing (TNAU 2021). Also, the twigs can be used for vegetative propagation. The seedlings should be grown in polybags of dimensions 16 × 20 cm. No insecticide is recommended, but bio-fertilizer such as *Azospirillum*, phospho-bacteria and vermicompost can help the seedlings to grow healthy (TNAU 2021). Further, after 5–6 months of germination, followed by hardening, seedlings can be shifted to the soil bed for growth. The dose of the previously mentioned bio-fertilizer can be provided for better growth. TNAU (2020) reported that the germination potential of seeds is 85% and 15% in recently collected and stocked seeds respectively.
- *Growth*: Neem is a fast-growing species with a large crown and shows gregarious flowering and fruiting patterns. The rotation period of the neem trees is around eight years yielding about 169 m^3 of wood. TNAU (2020) documented that a neem tree of 25 years of age may reach almost ten meters in height. The watering of neem trees is generally required during the initial days, but as the trees get older, watering can be skipped (Phogat et al. 2009). The herbs and other unwanted plants are naturally suppressed due to the phytochemical properties of neem plant parts (Vietmeyer 1992). The oil content is directly proportional to the precipitation in the neem tree locality (TNAU 2020).

There are various other factors which regulate the growth of the neem plant. Chlorosis and necrosis of the neem leaf are caused by Zn and K deficiency. So, fertilizer containing Zn and K should be provided for better growth. Fire, cyclones or high winds cause tree death (Singh et al. 2009). The grazing and browsing of neem leaves by cattle reduce the growth rate and in a few cases the tree dies.

25.10 FUTURE REMARKS

The health of people depends on good food and a good environment, but due to the increase in population, adulteration, depleting natural resources, industrialization, developmental activities and many others, it is getting deteriorated every day. A few chemicals have a negative impact on our bodies and are not good for the environment. So now people are inclining more toward herbal medicines and products due to no or negligible side effects. Neem is found in most countries of the world and exhibits tremendous quality such as antioxidant, antiviral, antibacterial, antifungal, analgesic, antidiabetic, anti-cancerous, insecticide, insect repellent and antimalarial. Neem oil contains triterpenes and other bioactive compounds with medicinal properties that are largely extracted from seeds and leaves. Due to its wide uses and potential, industries are formulating many neem-based products related to cosmetics, insecticides, oil and others. In agriculture, neem-based fertilizers and insecticides are economical, are easily available and have less/no soil persistency. More research should be carried out on the utilization of neem in the agricultural industry for farmers' welfare. Similarly, industries should take up neem as the primary major ingredient for the preparation of different products like anti-smell socks and cloth with a neem-based coating to avoid odour. Government should also promote local products made up of trees including neem. The ancient medical system of India mentioned neem and its importance for treating different diseases and disorders. Very few of them we know, and we should explore more about their uses as there is a great possibility that neem can help in fighting and treating future chronic/deadly diseases. Whenever any product is being formulated, its safety should be completely checked and noted for any adverse effects. There should also be research on pathogens and insect pests that attack neem trees. Also, the effect of climate change on neem tree phenology, bioactive compound content, health and growth should be studied to save this magnificent tree. The relationship of neem trees and local microclimate, habitat, ecology and biodiversity should also be studied to know about their ecological importance. Trees are very important for humankind, and thus efforts should be made to plant more trees of many species to combat global issues like climate change and global warming.

REFERENCES

Akter, F., M.M. Rahman, M. Mostofa & E.H. Chowdhury. 2014. Anti-diabetic effect of neem and spirulina in alloxan induced diabetic mice. *International Journal of Current Research and Academic Review* 2(4): 124–134.

Akter, R., M. Mahabub-Uz-Zaman, M.S. Rahman, M.A. Khatun, A.M. Abdullah, N.U. Ahmed & F. Islam. 2013. Comparative studies on antidiabetic effect with phytochemical screening of *Azadirachta indicia* and *Andrographis paniculata*. *IOSR Journal of Pharmacy and Biological Sciences* 5(2): 122–128.

Alam, A., S. Haldar, H.V. Thulasiram, R. Kumar, M. Goyal, M.S. Iqbal, C. Pal, S. Dey, S., Bindu, S., Sarkar & U. Pal. 2012. Novel anti-inflammatory activity of epoxyazadiradione against macrophage migration inhibitory factor: Inhibition of tautomerase and proinflammatory activities of macrophage migration inhibitory factor. *Journal of Biological Chemistry* 287(29): 24844–24861.

Ara, I., B.S. Siddiqui, S. Faizi & S. Siddiqui. 1989. Two new terpenoids from root bark of *Azadirachta indica*. *Journal of Natural Products* 52(6): 1209–1213.

Atangwho, I.J., P.E. Ebong, E.U. Eyong, M.Z. Asmawi & M. Ahmad. 2012. Synergistic antidiabetic activity of *Vernonia amygdalina* and *Azadirachta indica*: Biochemical effects and possible mechanism. *Journal of Ethnopharmacology* 141(3): 878–887.

Axtell, B.L. & R.M. Fairman. 1992. Minor oil crops, Part I Edible oil, Part II Non-edible oil, Part III Essential oil. *FAO Agriculture Science Bulletin*, No. 94.

Bakshi, B.K. 1976. *Forest pathology: Principles and practice in forestry*. Controller of Publication, Govt. of India, Delhi, India, p. 400.

Balwan, W.K., N. Saba & J.I. Zargar. 2022. Burden of diabetes and role of medicinal plants in its treatment. *Saudi Journal of Medical and Pharmaceutical Sciences* 8(7): 355–361.

Bhat, M., S.S. Zinjarde, S.Y. Bhargava, A.R. Kumar & B.N. Joshi. 2011. Antidiabetic Indian plants: A good source of potent amylase inhibitors. *Evidence-Based Complementary and Alternative Medicine*: 1–6.

Bisht, S. & S.S. Sisodia. 2010. Anti-hyperglycemic and antidyslipidemic potential of *Azadirachta indica* leaf extract in STZ-induced diabetes mellitus. *Journal of Pharmaceutical Sciences and Research* 2(10): 622–627.

Biswas, K., I. Chattopadhyay, R.K. Banerjee & U. Bandyopadhyay. 2002. Biological activities and medicinal properties of neem (*Azadirachta indica*). *Current Science* 82(11): 1336–1345.

Campos, E.V., J.L. De Oliveira, M. Pascoli, R. De Lima & L.F. Fraceto. 2016. Neem oil and crop protection: From now to the future. *Frontiers in Plant Science* 7(1494): 1–8.

CDC. 2022. *What is diabetes*? Accessed from www.cdc.gov/diabetes/basics/diabetes.html#:~:text=Diabetes%20is%20a%20chronic%20(long,your%20pancreas%20to%20release%20insulin.

Chaguthi, P., A. Mittal, S. Singh & D.S. Baghel. 2018. An overview on therapeutic potential of *Azadirachta indica*. *Journal of Emerging Technologies and Innovative Research* 5(12): 5–17.

Chattopadhyay, I., B. Nandi, R. Chatterjee, K. Biswas, U. Bandyopadhyay & R.K. Banerjee. 2004. Mechanism of antiulcer effect of Neem (*Azadirachta indica*) leaf extract: Effect on H+-K+-ATPase, oxidative damage and apoptosis. *Inflammopharmacology* 12(2): 153–176.

Chen, J., X. Fan, J. Zhu, L. Song, Z. Li, F. Lin, R. Yu, H. Xu & J. Zi. 2018. Limonoids from seeds of *Azadirachta indica* A. Juss and their cytotoxic activity. *Acta Pharmaceutica Sinica B* 8(4): 639–644.

Chhavi, S., J.V. Andrea, G. Payal, M.G. Taher, A.R. Tahir & H. Arif. 2014. Ethanolic neem (*Azadirachta indica*) leaf extract prevents growth of MCF-7 and HeLa cells and potentiates the therapeutic index of cisplatin. *Journal of Oncology*: 1–10.

Christian, E.O., E.C. Felicia, I.N. Helen, S.V. Nneka, C.U. Vivian & A.P. Ogochukwu. 2019. Antidiabetic property and antioxidant potentials of aqueous extract of *Azadirachta indica* leaves in streptozotocin-induced diabetic rats. *Journal of Medicinal Plant Studies* 7(6): 18–23.

Dash, S.P., S. Dixit & S. Sahoo. 2017. Phytochemical and biochemical characterizations from leaf extracts from *Azadirachta indica*: An important medicinal plant. *Biochemistry & Analytical Biochemistry* 6(2): 2–5.

David, S.N. 1969. Anti-pyretic of neem oil and its constituents. *Mediscope* 12: 25–27.

Del Serrone, P., C. Toniolo & M. Nicoletti. 2015. Neem (*Azadirachta indica* A. Juss) oil: A natural preservative to control meat spoilage. *Foods* 4(1): 3–14.

Deng, Y.X., M. Cao, D.X. Shi, Z.Q. Yin, R.Y. Jia, J. Xu, C. Wang, C. Lv, X.X. Liang, C.L. He & Z.R. Yang. 2013. Toxicological evaluation of neem (*Azadirachta indica*) oil: Acute and subacute toxicity. *Environmental Toxicology and Pharmacology* 35(2): 240–246.

Drabu, S., S. Khatri & S. Babu. 2012. Neem: Healer of all ailments. *Research Journal of Pharmaceutical, Biological and Chemical Sciences* 3(1): 120–126.

DTE. 2012. *Neem, the wonder tree, under attack: A new major pest.* Accessed from www.indiaenvironmentportal.org.in/content/351839/neem-the-wonder-tree-under-attack-a-new-major-pest/ on 08/09/2022.

Efferth, T. & E. Koch. 2011. Complex interactions between Phytochemicals. The multi-target therapeutic concept of phytotherapy. *Current Drug Targets* 12(1): 122–132.

Eid, A., N. Jaradat & N. Elmarzugi. 2017. A review of chemical constitutents and traditional usage of Neem plant (*Azadirachta indica*). *Palestinian Medical and Pharmaceutical Journal* 2(2): 75–81.

Elakovich, S.D. 1996. *The neem tree: Source of unique natural products for integrated pest management, Medicinal, Industrial, & Other Purposes Edited by Heinrich Schmutterer (Giessen U., FGR).* VCH, New York, p. 680.

Gbotolorun, S.C., A.A. Osinubi, C.C. Noronha & A.O. Okanlawon. 2008. Antifertility potential of Neem flower extract on adult female Sprague-Dawley rats. *African Health Sciences* 8(3): 168–173.

Gehlot, A., S. Arya & I.D. Arya. 2014. Vegetative propagation of *Azadirachta indica* A. Juss (Neem) through cuttings: A review. *Nativa* 2(4): 239–246.

Govindachari, T.R., G. Suresh, G. Gopalakrishnan, B. Banumathy & S. Masilamani. 1998. Identification of antifungal compounds from the seed oil of *Azadirachta indica*. *Phytoparasitica* 26(2): 109–116.

Gupta, A. 2021. *Should you brush your teeth with a neem stick instead of a toothbrush? Here's the answer.* Accessed from www.healthshots.com/preventive-care/self-care/benefits-of-using-neem-stick-for-teeth/ on 12/09/2022.

Gupta, A., S. Ansari, S. Gupta & M. Narwani. 2019. Therapeutics role of neem and its bioactive constituents in disease prevention and treatment. *Journal of Pharmacognosy and Phytochemistry* 8(3): 680–691.

Henkes, R. 1986. The neem tree: A farmer's friend. *The Furrow*, 16.

Idris, M.N., M. Usman & I.A. Igbafe. 2018. Experimental studies on neem seed (*Azadirachta indica*) as a possible engineering lubricating fluid. *International Journal of Agriculture and Earth Science* 4(2): 25–34.

Ilango, K., G. Maharajan & S. Narasimhan. 2013. Anti-nociceptive and anti-inflammatory activities of *Azadirachta indica* fruit skin extract and its isolated constituent azadiradione. *Natural Product Research* 27(16): 1463–1467.

Islas, J.F., E. Acosta, G. Zuca, J.L. Delgado-Gallegos, M.G. Moreno-Treviño, B. Escalante & J.E. Moreno-Cuevas. 2020. An overview of Neem (*Azadirachta indica*) and its potential impact on health. *Journal of Functional Foods* 74(104171): 1–13.

Jain, H. & R. Dhupper. 2021. A review on healing properties of *Tinospora cordifolia* (Indian Giloy). *International Journal for Research in Applied Science & Engineering Technology*, 9(5): 1114–1118.

Jaiswal, A.K., S.K. Bhattacharya & S.B. Acharya. 1994. Anxiolytic activity of *Azadirachta indica* leaf extract in rats. *Indian Journal of Experimental Biology* 32(7): 489–491.

Jessinta, S., H.N. Azhari, N.T. Saiful & H.N. Abdurahman. 2014. Impact of geographic variation on physicochemical properties of neem (*Azadirachta indica*) seed oil. *International Journal of Pharmaceutical Sciences and Research* 5(10): 4406–4413.

Juss, A. 1830. Pars secunda genera: *Quibus Specierum* enumeration accredit-meliaceae. *Memories Museum of Paris* t-19: 221.

Kato-Noguchi, H., M.A. Salam, O. Ohno & K. Suenaga. 2014. Nimbolide B and nimbic acid B, phytotoxic substances in neem leaves with allelopathic activity. *Molecules* 19(6): 6929–6940.

Kaur, G., M. Sarwar Alam & M. Athar. 2004. Nimbidin suppresses functions of macrophages and neutrophils: Relevance to its antiinflammatory mechanisms. *Phytotherapy Research* 18(5): 419–424.

Kaushik, N., B.G. Singh, U.K. Tomar, S.N. Naik, S. Vir, S.S. Bisla, K.K. Sharma, S.K. Banerjee & P. Thakkar. 2007. Regional and habitat variability in azadirachtin content of Indian neem (*Azadirachta indica* A. Jusieu). *Current Science* 92(10): 1400–1406.

Kelly, G.S. 2011. Quercetin. *Alternative Medicine Review* 16(2): 172–195.

Khalid, S.A., H. Duddeck & M. Gonzalez-Sierra. 1989. Isolation and characterization of an antimalarial agent of the neem tree *Azadirachta indica*. *Journal of Natural Products* 52(2): 922–926.

Khan, M. & S.W. Wassilew. 1987. The effect of raw material from the neem tree, neem oil, and neem extracts on fungi pathogenic to humans. In *Natural pesticides from the neem tree and other tropical plants.* GTZ, Eschborn, pp. 645–650.

Koul, O. 2004. Neem: A global perspective. In *Neem: Today and in the new millennium.* Springer, Dordrecht, pp. 1–19.

Koul, O., G. Singh, R. Singh, J. Singh, W.M. Daniewski & S. Berlozecki. 2004. Bioefficacy and mode-of-action of some limonoids of salannin group from *Azadirachta indica* A. Juss and their role in a multi-component system against lepidopteran larvae. *Journal of Biosciences* 29(4): 409–416.

Lakshmi, T., V. Krishnan, R. Rajendran & N. Madhusudhan. 2015. *Azadirachta indica*: A heral panacea in dentistry-An update. *Pharmacognosy Reviews* 9(17): 41–44.

Lessa, F.C., I. Nogueira, S. Vargas Fda, D.M. Spolidorio, J. Hebling & F. García-Godoy. 2010. Direct and transdentinal antibacterial activity of chlorhexidine. *American Journal of Dentistry* 23: 255–259.

Loke, W.H., C.K. Heng, N. Basirun & A. Rejab. 1990. Non-target effects of neem (*Azadirachta indica* A. Juss) on *Apanteles plutellae* Kurdj., cabbage, sawi and padi. In *3 International Conference on Plant Protection in the Tropics*, Genting Highlands, Pahang (Malaysia). Malaysian Plant Protection Society, pp. 108–110.

Makheswari, M.U. & D. Sudarsanam. 2012. Database on antidiabetic indigenous plants of Tamil Nadhu, India. *International Journal of Pharma Science and Research* 3(2): 287–293.

McCalla, G., O. Parshad, P.D. Brown & M.T. Gardner. 2016. Beta cell regenerating potential of *Azadirachta indica* (Neem) extract in diabetic rats. *West Indian Medical Journal* 65(1): 13–17.

Mishra, A. & N. Dave. 2013. Neem oil poisoning: Case report of an adult with toxic encephalopathy. *Indian Journal of Critical Care Medicine* 17(5): 321–322.

Mukherjee, A. & S. Sengupta. 2013. Characterization of nimbidiol as a potent intestinal disaccharidase and glucoamylase inhibitor present in *Azadirachta indica* (neem) useful for the treatment of diabetes. *Journal of Enzyme Inhibition and Medicinal Chemistry* 28(5): 900–910.

Nahak, G. & R.K. Sahu. 2011. Evaluation of antioxidant activity of flower and seed oil of *Azadirachta indica* A. juss. *Journal of Applied and Natural Science* 3(1): 78–81.

Naik, M., D. Agrawal, R. Behera, A. Bhattacharya, S. Dehury & S. Kumar. 2014. Study of anti-inflammatory effect of neem seed oil (*Azadirachta indica*) on infected albino rats. *Journal of Health Research and Reviews* 1(3): 66–69.

Nisbet, A.J. & A.J. Mordue 2000. Azadirachtin from the neem tree *Azadirachta indica*: Its action against insects. *Anais da Sociedade Entomológica do Brasil* 29(4): 615–632.

Noor, A., V.S. Bansal & M.A. Vijayalakshmi. 2013. Current update on anti-diabetic biomolecules from key traditional Indian medicinal plants. *Current Science* 104(6): 721–727.

Ogurtsova, K., L. Guariguata, N.C. Barengo, P.L.D. Ruiz, J.W. Sacre, S. Karuranga, H. Sun, E.J. Boyko & D.J. Magliano. 2022. IDF diabetes Atlas: Global estimates of undiagnosed diabetes in adults for 2021. *Diabetes Research and Clinical Practice* 183(109118): 1–9.

Page, C. & E.M. Hawes. 2013. Haemolytic anaemia after ingestion of Neem (*Azadirachta indica*) tea. *BMJ Case Reports*: 1–3.

Patel, R.P., P. Harde, J. Pillai, N. Darji & B. Patel. 2012. Antidiabetic herbal drugs a review. *Pharmacophore* 3(1): 18–29.

Patel, R.P. & B.M. Trivedi. 1962. The in-vitro antibacterial activity of some medicinal oils. *Indian Journal of Medical Research* 50: 218–222.

Patel, S.M., K.C.N. Venkata, P. Bhattacharyya, G. Sethi & A. Bishayee 2016. Potential of neem (*Azadirachta indica* L.) for prevention and treatment of oncologic diseases. *Seminars in Cancer Biology* 40–41: 100–115.

Patil, P., S. Patil, A. Mane & S. Verma. 2013. Antidiabetic activity of alcoholic extract of Neem (*Azadirachta indica*) root bark. *National Journal of Physiology, Pharmacy and Pharmacology* 3(2): 142–146.

Patil, S.M., P.S. Shirahatti & R. Ramu. 2022. *Azadirachta indica* A. Juss (neem) against diabetes mellitus: A critical review on its phytochemistry, pharmacology, and toxicology. *Journal of Pharmacy and Pharmacology* 74(5): 681–710.

Phogat, V., S. Kumar, K.K. Singh & Satyavan. 2009. Techniques of neem plantation and its economic benefits. In *Neem, a treatise.* I.K. International Pvt Ltd., New Delhi, India, pp. 67–84.

Ponnusamy, S., S. Haldar, F. Mulani, S. Zinjarde, H. Thulasiram & A. RaviKumar. 2015. Gedunin and azadiradione: Human pancreatic alpha-amylase inhibiting limonoids from neem (*Azadirachta indica*) as anti-diabetic agents. *PLoS ONE* 10(10): 1–19.

Quelemes, P.V., M.L. Perfeito, M.A. Guimarães, R.C. dos Santos, D.F. Lima, C. Nascimento, M.P. Silva, M.J.D.S. Soares, C.D. Ropke, P. Eaton & J. de Moraes. 2015. Effect of neem (*Azadirachta indica* A. Juss) leaf extract on resistant *Staphylococcus aureus* biofilm formation and *Schistosoma mansoni* worms. *Journal of Ethnopharmacology* 175: 287–294.

Rafe, M.R. 2017. A review of five traditionally used anti-diabetic plants of Bangladesh and their pharmacological activities. *Asian Pacific Journal of Tropical Medicine* 10(10): 933–939.

Rupani, R. & A. Chavez. 2018. Medicinal plants with traditional use: Ethnobotany in the Indian subcontinent. *Clinics in Dermatology* 36(3): 306–309.

Sadre, N.L., V.Y. Deshpande, K.N. Mendulkar & D.H. Nandal. 1984. Male anti-fertility activity of *Azadirachta indica* A. Juss (Neem) in different species. *Paper Presented at the Proceedings of the 2nd International Neem Conference (INCR'83)*, Rauischholzhausen, Germany, pp. 473–482.

Saleem, S., G. Muhammad, M.A. Hussain & S.N.A. Bukhari. 2018. A comprehensive review of phytochemical profile, bioactives for pharmaceuticals, and pharmacological attributes of *Azadirachta indica*. *Phytotherapy Research* 32(7): 1241–1272.

Schmutterer, H. 1985. Which insect pests can be controlled by application of neem seed kernel extracts under field conditions? *Zeitschrift für angewandte Entomologie* 100(1–5): 468–475.

Schmutterer, H. 1990. Observations on pests of *Azadirachta indica* and some species of *Melia*. *Journal of Applied Entomology* 109(4): 390–400.

Schmutterer, H. and K.R.S. Ascher, eds. 1984. *Natural pesticides from the neem tree (Azadirachta indica A. Juss.) and other tropical plants.* Deutsche Gesellschaft für Technische Zusammenarbeit (GTZ), Eschborn, Germany pp. 473–482.

Sengupta, S. 2018. *Does brushing with neem twig help give you stronger teeth and gums? We find out.* Accessed from https://food.ndtv.com/food-drinks/does-brushing-with-neem-twig-help-give-you-stronger-teeth-and-gums-we-find-out-1897901 on 12/09/2022.

Shailey, S. & S.F. Basir. 2012. Strengthening of antioxidant defense by *Azadirachta indica* in alloxan-diabetic rat tissues. *Journal of Ayurveda and Integrative Medicine* 3(3): 130–135.

Shareef, M. & M.S. Akhtar. 2018. Neem (*Azadirachta indica*) and its potential for safeguarding health, prevention and treatment of diseases. *Matrix Science Medica* 2(1): 4–8.

Shinkafi, T.S., L. Bello, S.W. Hassan & S. Ali. 2015. An ethnobotanical survey of antidiabetic plants used by Hausa-Fulani tribes in Sokoto, Northwest Nigeria. *Journal of Ethnopharmacology* 172: 91–99.

Shoaib, M.A., M.F. Mahmoud, N. Loutfy, M.A. Tawfic & M. Barta. 2010. Effect of botanical insecticide Nimbecidine® on food consumption and egg hatchability of the terrestrial snail *Monacha obstructa*. *Journal of Pest Science* 83(1): 27–32.

Shori, A.B. & A.S. Baba. 2013. Antioxidant activity and inhibition of key enzymes linked to type-2 diabetes and hypertension by *Azadirachta indica*-yogurt. *Journal of Saudi Chemical Society* 17(3): 295–301.

Shukla, V., M.D. Khurshid & B. Kumar. 2020. A review on phytochemistry and pharmacological activity of *Azadirachta indica* (Neem). *International Journal of Pharmacy and Biological Sciences* 10: 172–180.

Siddiqui, B.S., M. Ghaiussdin & S. Faizi. 1998. Tetracyclic triterpenoids of the fruit coats of *Azadirachta indica*. *Phytochemistry* 47(8): 1631–1636.

Siddiqui, B.S., B.S. Siddiqui, M. Ghiasuddin & S. Faizi. 1991. Terpenoids from fruit coatings of *Azadirachta indica*. *Phytochemistry* 30(5): 1615–1619.

Singh, K.K., S. Phogat, A. Tomar & R.S. Dhillon. 2009. *Neem, a treatise*. I.K. International Pvt Ltd., New Delhi, India, pp. 1–546.

Singh, N. & M.S. Sastry. 1997. Antimicrobial activity of Neem oil. *Indian Journal of Pharmacology* 13: 102–106.

Singh, P.P., A.Y. Junnarkar, G.P. Thomas, R.M. Tripathi & R.K. Varma. 1980. Antidepressant activity of. *Azadirachta indica*. *Indian Journal of Pharmacology* 61: 164–168.

Sinniah, D. & G. Baskaran. 1981. Margosa oil poisoning as a cause of Reye's syndrome. *The Lancet* 317(8218): 487–489.

Sivakumar, M., S.Y. Tang & K.W. Tan. 2014. Cavitation technology – a greener processing technique for the generation of pharmaceutical Nano emulsions. *Ultrasonic Sonochemistry* 21(6): 2069–2083.

Soares, D.G., A.M. Godin, R.R. Menezes, R.D. Nogueira, A.M.S. Brito, I.S. Melo, G.M.E. Coura, D.G. Souza, F.A. Amaral, T.P. Paulino & M.M. Coelho. 2014. Anti-inflammatory and antinociceptive activities of azadirachtin in mice. *Planta Medica* 80(8/9): 630–636.

Srivastava, S.K., B. Agrawal, A. Kumar & A. Pandey. 2020. Phytochemicals of *Azadirachta indica* source of active medicinal constituent used for cure of various diseases: A review. *Journal of Scientific Research* 64(1): 385–390.

Subramanian, M.S. & K.K. Lakshmanan. 1996. *Azadirachta indica* A. Juss. Stem bark as an anti-leprosy source. *Neem and Environment* 2: 1143–1150.

Suganthan, D., C. Seshadri & G. Santhakumari. 1978. Anti-gastric ulcer activity of nimbidin. *The Indian Journal of Medical Research* 68: 169–175.

Tag, H., P. Kalita, P. Dwivedi, A.K. Das & N.D. Namsa. 2012. Herbal medicines used in the treatment of diabetes mellitus in Arunachal Himalaya, northeast, India. *Journal of Ethnopharmacolgy* 141: 786–795.

Tarak, D., N.D. Namsa, S. Tangjang, S.C. Arya, B. Rajbonshi, P.K. Samal & M. Mandal. 2011. An inventory of the ethnobotanicals used as anti-diabetic by a rural community of Dhemaji district of Assam, Northeast India. *Journal of Ethnopharmacology* 138: 345–350.

Taylor, E. & T.J. Webster. 2011. Reducing infections through nanotechnology and nanoparticles. *International Journal of Nanomedicine* 6: 1463–1473.

Tembe-Fokunang, E.A., F. Charles, N. Kaba, G. Donatien, A. Michael & N. Bonaventure. 2019. The potential pharmacological and medicinal properties of neem (*Azadirachta indica* A. Juss) in the drug development of phytomedicine. *Journal of Complementary and Alternative Medical Research* 7(1): 1–18.

TNAU. 2015. *Crop protection: Pest of neem.* Accessed from https://agritech.tnau.ac.in/crop_protection/crop_prot_crop_insect_medi_neem.html on 25/08/222.

TNAU. 2020. *Neem – A versatile tree.* Accessed from https://agritech.tnau.ac.in/banking/pdf/Forestry%20 Wasteland_Neem.pdf on 09/09/2022.

TNAU. 2021. *Forestry technologies: Neem.* Accessed from https://agritech.tnau.ac.in/forestry/forestry_Neem.html on 09/09/2022.

Trivedi, A., N. Fatima, I. Husain & A. Misra. 2019. An update on the therapeutic potential of neem and its active constituents: A panacea for all diseases. *Era's Journal of Medical Research* 6(1): 110–117.

Vietmeyer, N.D. 1992. Neem: A tree for solving global problems. *Report of an Ad Hoc Panel of the Board on Science and Technology for International Development, National Research Council*, National Academy Press, Washington, DC, pp. 1–141.

Zong, A., H. Cao & F. Wang. 2012. Anticancer polysaccharides from natural resources: A review of recent research. *Carbohydrate Polymers* 90(4): 1395–1410.

26 Antidiabetic and Other Medicinal Properties of *Aegel marmelos* L. Correa (Golden Apple)

Shakeelur Rahman and Azamal Husen

CONTENTS

26.1 INTRODUCTION

Some of the trees and their parts are not only used for timber or sources of food but are also a valuable resource of medicines (Husen 2021, 2022; Husen et al. 2021). This is because of the availability of phytochemicals in different parts of the plant at its different stages. These phytochemicals are very important due to their efficacy and side effects and being synthesized easily and economically viable. A majority of tree species are associated with the traditional knowledge system of medicines that provides clues for their specific use. *Aegle marmelos* (L.) Correa is an important tree species of family Rutaceae, which is also called a medicinally treasured tree species (Chanda 2008). The tree species is also known as golden apple, bilva, bael, begal-quince and wood or stone apple. There are some common cultivars of *A. marmelos*, these are Kagzi Etawah, Kagzi Gonda, Baghel and Mirzapuri (Sharma et al. 2007). *A. marmelos* is one of the popular medicinal trees used since ancient times in traditional systems of medicines such as Ayurvedic, Unani and Siddha (Jagetia and Baliga 2004). The historical records shows that *A. marmelos* has been used as a traditional medicine for various ailments and food since 5000 B.C. The scientific research also proves the presence of nutritional elements and phytochemicals including flavonoids, alkaloids, tannins, coumarins and phenolic acids as well as useful compounds like carbohydrates, fatty acids, amino acids, minerals, vitamins and fibres. Vitamins present in the tree species are vitamin A, riboflavin, ascorbic acid, niacin and thiamine, which make the fruits of *A. marmelos* highly nutritious (Bhardwaj 2014; Bhattacherjee et al. 2015). All the parts of this aromatic tree including flowers, fruits, seeds, leaves, bark and roots are rich in medicinal properties (Kintzios 2006; Baliga et al. 2011) such as for treating vomiting, dysentery, dyspepsia, neurological diseases, oedema, rheumatism and malabsorption (Chanda 2008). The root of the tree is one of the important ingredients used in ten type of root recipes or *Dasmula* recipe (Chopra 1982). Root and root bark after decoction is used to

DOI: 10.1201/b23347-26

treat palpitation of the heart, intermittent fever, melancholia and hypochondriasis (Nadkarni 1954). The leaves and stem bark are given for diabetes and medicated enema. The unripe fruit is suggested for chronic cases of diarrhoea and irritable bowel syndrome due to the presence of a high amount of mucilage (Chopra 1982; Nadkarni 1954; Satyavati et al. 1976; Dhuley 2003). *A. marmelos* was described as a Rasaya by Charaka (Pandeya 1983). The most important rationale of the review of the plant is to understand the role of its phytochemicals in diabetes and other diseases as well as the nutritional and economic value of the plant species. It will also help researchers in further study.

26.2 BOTANICAL DESCRIPTION

Aegle marmelos is a deciduous tree whose growth is not fast. The structure of the tree is variable depending on climatic condition and nature of the soil (Figure 26.1). However, the important botanical characters remain uniform. The tree is woody hard and extensively adaptable to adverse environments. The tree crown is highly dense with strong branches. The height of the tree reaches to 13–15 m with many branches and has a maximum girth of 1 m. The trunk is thick and short with tapered oval shape ends. There is central pith in young wood. *A. marmelos* has non-spiny as well as small spiny branches originating from the axis of the leaves of maximum 3 cm in length. The bark of the tree species are peeling, flexible, greyish blue in colour with asymmetrical linings on the young branches. The tree has dimorphic branches with 3–5 cm long internodes. The branches grown in the first year have glabrous surface, whereas branches of the second and third year have striated surface. The suckers develop from the trunk after maturity of the tree. The sucker is protected from herbivores by the spines present on it. The wounded tree secretes resins which dry up into solid long crystals. The leaves are deciduous, trifoliate, single or compound, alternate with one or sometimes two pairs of small stalked opposite leaflets, petioles are glabrous and long. There are 2–5 leaflets in oval or ovate shaped found in the compound leaf. The leaflet is toothed leaflets, 4–10 cm long and 2–5 cm wide, thin with a visible midrib from beneath. The petiole of terminal leaflet is longer in size. The colour of new foliage is light pink and glossy in look. The colour of leaves is different in various stages; in early stage the leaf is deep green and bright, while the mature leaves appears deep green

FIGURE 26.1 *Aegel marmelos* (golden apple).

(Bhar et al. 2019). The flowers found in 4–7 clusters, having sweet aromatic properties. Generally flowering comes in April to May in India when new leaves grow out. The flower is bisexual, hypogynous, collectively held by some lateral panicles of leaf axils and greenish-white in appearance (Patel et al. 2012). The flower consists of four or five curved, 2 cm wide and fleshy petals with greenish-yellow colour having 50 or more light green or yellow stamens, erect, stalked, lax and aroused terminal cymes. The calyx is thin, pubescent, having five short and broad teeth. The stigma is capitate, ovary is oblong-ovoid and slightly tapering into the thick short style. The fruits are mostly found in the periphery of the canopy. The fruits are slow ripening and can take about one year to ripen. The nature of the fruit is smooth, hard with woody shell. The crust is grey-green at early phase, turns orange or yellowish at the ripening stage and transforms into very hard and orange-red when dried. The fruit is found in different shapes ranging from oval, round, pyriform to oblong. The size of fruit is 5–20 cm in diameter, about 77.2 g in weight, 73.7 mL in volume, and 93.72 ± 2.78% in sphere (Sonawane et al. 2020). The scented minute oil glands are present on the fruit shell. There is a hard central core with 8 to 20 hazily triangular segments and dark-orange walls inside the fruit. The pulp of the ripe fruit is sweet-smelling, light-orange, mucilaginous, resinous with astringent pulp and fibrous. The seeds embed in a sticky transparent mucilage adhesive pulp, 15 to 50 in number, mucilage, and dried seeds look solidified glassy crystal. The size of seed is 1 cm long with woolly hairs, flattened-oblong shape and white testa. The embryo has a short superior radicle with large cotyledons having no endosperm. Some of the seeds can get terminated during the development. There is no seed dormancy as it germinates within 2–3 weeks. The seedlings of the species become ready to transplant within 2–3 months (Sing et al. 2014). Table 26.1 shows the botanical classification of *A. marmelos.*

TABLE 26.1
***Aegel marmelos* – Phytochemical Constituents and Their Medicinal Properties**

Name	Compound	Plant part	Medicinal property	Key reference
Alkaloids	Aegelenine Aegeline Aegelinosides A Aegelinosides B Dictamine Ethyl cinnamamide Ethyl cinnamate Fragrine Halfordinol	Fruit, leaves	Antidiabetic, antibacterial, anti-inflammatory, andanticancerous	Shoeb et al. (1973); Chatterjee et al. (1977); Bhar et al. (2019).
Coumarins	Alloimperatorin Imperatorin Isoimperatorin Marmelide Marmelosin Marmesin Marmin Methyl ether Psoralen Psoralen-a Scoparone Scopolentin Scopoletin Umbelliferone Xanthotoxol Zanthotoxol	All parts	Antidiabetic, antioxidant, anti-inflammatory, andantianalgesic	Bhar et al. (2019); Roy et al. (1976)

(*Continued*)

TABLE 26.1
(Continued)

Name	Compound	Plant part	Medicinal property	Key reference
Terpenoids	Caryophyllene Cineol *cis*-Limonene oxide *cis*-Linalool oxide Cubedol Elemol Epi-cubebal Hexanylhexanoate Humulene Isosylvestrene Limonene Linalool Methyl perilate Myrcene P-cymene Terpinolene Valencene Caryophyllene Cineol *cis*-Limonene oxide *cis*-Linalool oxide Cubedol Elemol Epi-cubebal Hexanylhexanoate Humulene Isosylvestrene Limonene Linalool Methyl perilate Myrcene P-cymene Terpinolene Valencene	Fruit, leaf, andbark	Anticancer	Bhar et al. (2019); Basak et al. (1981)

26.3 DISTRIBUTION

India is the origin place of *A. marmelos*. According to historical findings, the tree has existed in India since 800 B.C. (Nagar et al. 2017). It has also been reported that *A. marmelos* has an evolutionary relation with the African genus *A. fraaegle*. The scientific argument is that the genus *Aegle* would have originated in Africa and later spread to South Asia (Pathirana 2020). The tree grows well in mixed deciduous and dry dipterocarp forests and also on uplands and plains. It is extended to southern and central India and Burma, Sri Lanka, Pakistan and Bangladesh, Thailand, Vietnam, the Philippines, Cambodia, Malaysia, Java, Egypt, Surinam and Trinidad. *A. marmelos* is a subtropical species that grows up to an altitude of 1200 m where the temperature rises to 48.89°C in summer and descends to −6.67°C in the winter. The tree cannot fruit well where there is no long dry season as in the southern region of Malaysia. *A. marmelos* usually needs soil of pH 5–8, but it can equally flourish in shallow, alkaline, tough soils (Roy and Singh1979).

26.4 PHYTOCHEMICAL CONSTITUENTS

A. marmelos is rich in phytochemicals such as marmesin, scoparone, scopoletin, umbelliferone and coumarins (Chatterjee et al. 1977; Shoeb et al. 1973). Fruits of the tree contain alloimperatorin, imperatorin, β-sitosterol, xanthotoxol tannins and alkaloids such as aegeline and marmeline (Saha and Chatterjee 1957). The increased level of tannin and riboflavin was found in fully ripe fruits, while the content of ascorbic acid decreases significantly in ripened fruits (Yadav et al. 2011). The antioxidant activity reduces with the fruit's maturation. Mandal and Mukherjee (1981) have given a partial structure of glycoprotein discovered in the fruit. The tree contains anticancerous phenyl proponoids (Kintzios 2006). Marmelin is found as a key bioactive compound in the unripe fruit (Sharma et al. 1981). Monoterpenes and sesquiterpenes are present as main volatile compounds (Charoensiddhi and Anprung 2009). A newly discovered compound shahidine is present, which has very high oxazoline activity against some of the gram-positive bacteria (Faizi et al. 2009). The gum adhered with the seeds also contains phytochemicals (Roy et al. 1976). Leaves of the tree contain mermesinin, rutin, phenylethyl, cinnamides, anhydromarmeline, aegelinosides, sterols, essential oils and alkaloids (Shoeb et al. 1973). Stem bark and roots contain a coumarin as aegelinol. Roots also contain xanthotoxin, coumarins, tembamide, mermin skimmianine and psoralen (Basak et al. 1981). The details of phytochemicals constituents are given in Table 26.1.

26.5 PHARMACOLOGICAL STUDIES

A. marmelos is one of the important Indian trees which have both nutritional and medicinal properties (Satyavati et al. 1976). All parts of the tree are generally considered safe as a medicine and dietary item. The medicinal preparations of *A. marmelos* are used to cure dysentery, chronic diarrhoea, peptic ulcers and respiratory ailments and as a laxative for astringency (Baliga et al. 2011). Effective tonics that are good for the brain are prepared from the plant parts (Sood and Katoch 2014). One of the important bioactive compounds is marmelosin, which acquires restorative, astringent, laxative and toxic properties as well as is a good constituent for human heart and brain (Shailajan et al. 2011; Kirtikar and Basu 1935). Apart from that, the tree root contains anti-inflammatory properties (Benni et al. 2011). The aqueous extract of *A. marmelos* fruit has been studied to be nonmutagenic to *Salmonella typhimurium* strain (Kruawan and Kangsadalampai 2006). The hydroalcoholic extract of the fruit is toxic-free up to a dose of 6 g/kg body weight in mice (Jagetia et al. 2004). It was found in a study that *A. marmelos* stops the colonization by *E. coli* E134, *E. coli* B170 and *S. flexneri*. This is because of its effect on the metabolism of HEp-2 cells or modification of cell receptors to prevent adherence or bacterial infection as seen on the pre-incubation of HEp-2 with the decoction. This shows that the decrease in attack may not be due to the inhibition of initial attachment of the bacteria to the epithelial cells by the plant decoction but due to its effect on the engulfment process of the bacteria at a post adherence stage. Therefore, the result shows that *A. marmelos* does not permit the pathogens to establish themselves in the intestinal epithelium, which could be a very important feature in the antidiarrhoeal activity of the plant. The decoction also decreased the binding of both cholera toxin (CT) and labile toxin (LT) to the ganglioside monosialic acid receptor (GM1), and the CT and LT are known to be antigenically similar (Ganguly and Kaur 1996). Thus, the effect of the decoction on their binding suggests that it may contain some compound(s) which may directly block the GM1 on the cell membrane, thereby inhibiting their binding to the receptor (Das and Das 1995; Roy and Singh 1979).

Anticancerous activity: There are more than 187 plant species which are effective against cancer and tumour (Kintzios 2006). The anticancerous properties of the tree has been studied by using sea urchin eggs assay, brine shrimp lethality assay, haemolysis assay and 3-(4, 5-dimethylthiazol-2-yl)-2, 5-diphenyl tetrazolium bromide (MTT) assay applying on tumour cells where important toxic effects were found (Costa-Lotufo et al. 2005).

The compounds extracted from the tree showed *in vitro* antitumor activity in cancer cell lines, Ehrlich ascites carcinoma (EAC) (LK562), body cavity-based lymphoma cell line (BC) (MCF7) and mast cell line (MC) (Colo38) (Jagetia and Baliga 2004). The bioactive molecules available in the tree parts extracts were later recognized by gas chromatography-mass spectrometry (GC-MS) as chromomycin, cisplatin, cytosine arabinoside, 5-fluorouracil, butyl-p-tolyl sulfide, 5-methoxypsolaren and 6-methyl-4-chromanone (Lampronti 2003). The tree extracts can adjust reactive oxygen species, minimize inflammation and angiogenesis and increase immunity. High-tech techniques such as single photon emission computed tomography (SPECT), positron emission tomography (PET) and nuclear magnetic resonance (NMR) were applied to verify the anti-inflammatory and antiangiogenesis activities of *A. marmelos* (Sagar and Wong 2010). The phytochemicals of the tree hold protective effects against chemo- and radiotherapy treatments of cancer. For example, the fruit extract was reported to inhibit the doxorubicin-induced genetic damage (Venkatesh 1990) and is less toxic and acts as an immune enhancer (Sagar and Wong 2010), therefore, the tree extract is radioprotective. Hydroalcoholic extracts of the fruit were also effective for those cancer patients who have undergone gamma radiation (Jagetia and Baliga 2004).

Antimicrobial activity: It has been reported that the phytochemical extracts of *A. marmelos* have antibacterial, antifungal and antiviral activities (Rani and Khullar 2004; Rana et al. 1997; Balasubramanian et al. 2007). The coumarin compound was found effective against *Shigella dysenteriae* (Raja et al. 2008). The combination of both phytochemicals of the tree and popular antibiotic β-lactum has inhibitory effect on *S. flexneri* and *S. dysenteriae*. It was observed that generally these bacteria are resistant to β-lactum, but the fruit extract restores the inhibitory activity of β-lactam by changing the dynamics of porin channel (Raja et al. 2008).

The essential oil of *A. marmelos* leaf has an antifungal effect as it inhibits the spore germination of *F. udum* (Rana et al. 1997). Moreover, the seed extracts of the tree provide anthraquinone, which is effective against some fungi such as *Candida* spp. and *Aspergillus*. The antimicrobial effect of the phytochemicals of the tree helps to control skin diseases and food contamination (Mishra et al. 2010). It also has an antimicrofilarial effect and insecticidal activity on Japanese encephalitis vector *Culex tritaeniorhynchus* (Sharma et al. 2010; Elango et al. 2012). The antiviral effect was also noticed against the white spot syndrome virus in shrimp (Balasubramanian et al. 2007).

General ailments: The phytochemicals of *A. marmelos* are reported to reduce fertility in male rats by reducing the motility and density of sperms. It indicates the potential use of the phytochemicals as a future birth-controlling herbal medicine (Chauhan et al. 2007). The antihistamine activities of the phytochemicals are efficient against respiratory ailments such as asthma. Similarly anti-inflammatory, antipyretic and analgesic effects of the phytochemicals have also been reported (Arul et al. 2005). The herbal preparations of *A. marmelos* control the expression of the I-18 gene in brachial epithelial cells; therefore, it has the potential to apply in treating cystic fibrosis (Nicolis et al. 2009). The phytochemicals, including flavonoids, are effective in hepatotoxicity (Singh and Rao 2008), cirrhosis, repairing structure and function of the liver and contributing as hepatoprotective (Singh and Rao 2008). The decoction of dried leaves of the tree is supposed to be effective in hyperthyroidism. However, the thyroxin level did not go down by the leaf extract (Kar et al. 2002). It was found in an experiment on rabbit that the fruit extract was effective in intraocular pressure. The controlling of intraocular pressure could be one of the treatment strategies for glaucoma (Agarwal et al. 2009). The antivenom effect of the tree roots–based formulation has also been reported by some researchers (Pitre and Srivastava 1988). The methanol extracts of *A. marmelos* was used in the formulations of ointments and injections for healing wounds (Jaswant et al. 2001). The marmelosin compound of the tree has potential to cure digestive problems (Lampronti 2003). The fruit extracts help to enhance the immunity upon exposure to infections (Patel et al. 2010). The aqueous extract of the tree works as cardiac and uterine stimulant and muscle relaxant (Hema and Lalithakumari 1999).

26.6 ANTIDIABETIC RESPONSE

A. marmelos can be effective in controlling diabetes mellitus, which is very common in modern society because of stressful lifestyles and some genetic problems (Maity et al. 2009). It was reported that the presence of amino acid in fruit extract enhances the utilization of glucose, dilating the blood vessels, reducing blood clot and stimulating insulin sensitivity (Upadhya et al. 2004; Bhardwaj 2014). It was reported that the fruit juice enhances the plasma insulin and glycogen in diabetic rats (Kamalakkannan and Prince 2003). The mechanism of action of *A. marmelos* is mentioned in Figure 26.2.

The tree extracts result in significantly higher antidiabetic activities when experimented with in animals (Anandharajan et al. 2006). It protects the pancreatic tissues of diabetic rats and elevates vitamin C (Kamalakkannan and Prince 2005). Aqueous and alcoholic extracts of the fruits given at a dose of 500 mg per kg of body weight produced hyperglycaemia in rabbits (Hema and Lalithakumari 1999). Moreover, the antihypoglycaemic and hyperglycaemic activities of the aqueous extracts were found effective in rat (Kesari et al. 2006). Choudhary et al. (2017) observed the same effect by applying aqueous extracts of the tree seeds. The tree leaves have also good impact on hyperglycaemia (Arul et al. 2002) as it decreases the mi-receptor gene expression and inhibits aldose reductase activity and free radicals (Gireesh et al. 2008). According to Narender (2007), aegeline-2 of the tree leaves is an effective compound in antihyperglycaemic activity.

26.7 TRADITIONAL AND OTHER POTENTIAL USES

A. marmelos has been used in Ayurveda, Unani and traditional systems of medicine since ancient times. According to the Ayurvedic system, the ripe fruit helps to control vatha and kapha and keepsthe heart healthy. The different parts of the tree are preserved by sun drying for a long period and are used as processed food and medicines. In the Unani system, the fully ripened fruit is used as a cardiac and brain tonic, astringent, restorative and laxative (Kirtikar and Basu 1939). The unripe fruit is an appetizer and antidysenteric and reduces pain. According to Siddha, the fruit reduces vatha, pitha, and kapha dosha. Gastrointestinal problems, rectum inflammation, piles and

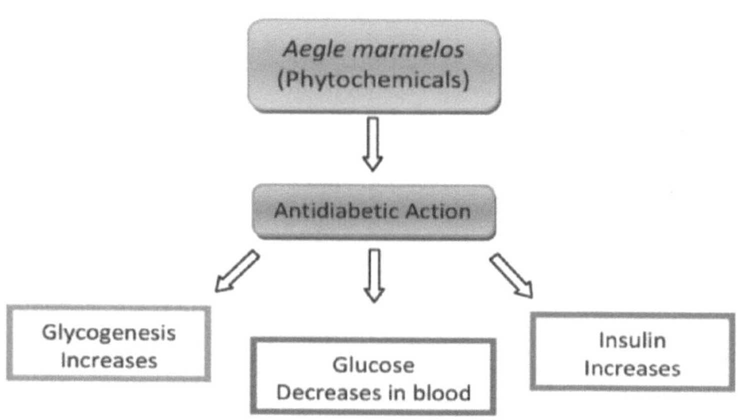

FIGURE 26.2 Antidiabetic action of *Aegel marmelos.*

blood sugar level is treated with the extract of ripe fruit (Dhankhar et al. 2011). Fruit extract is also effective in thyroid problems. It is helpful to control vomiting tendency during pregnancy. The pregnant woman drinks the juice mixed with boiling rice water. The unripe fruit pulp powder is traditionally used to heal wounds (Kala 2006). Urinogenital problems are prevented by giving fruit pulp mixed with milk. Other uses of the fruit include as stomachic agent and antiscorbutic and to treat chronic constipation, indigestion, intestinal ulcer and gonorrhoea (Sampath et al. 2012). The unripe fruit powder mixed with mustard oil is applied in skin burn (Sharma et al. 2007). The people of Thailand use the dried leaf and fruit pulp of *A. marmelos* as a syrup and making of cake, squash and jam (Chauhan et al. 2016). The mixture of dried fruit and Indian sweet fennel seeds both in a quantity of 2-1/2 drops with one drop gum of silk cotton tree, two drops honey, 1/2 drop dried ginger, and sugar is used in chronic diarrhoea and dysentery. The effective traditional formula of use of the fruit powder in chronic dysentery is mixing of 1–3 drachms of the powder Indian sweet fennel seeds one part, two parts of *Holarrhena antidysentrica*, ispagula two parts, sugar, *Chebulic myrobalan* and plantago. Another effective traditional formula to treat both diarrhoea and chronic dysentery is formulated by mixing one drop *Andropogon muricatus*, four drops fruit extract, one drop *Symplocos racemosa* and one drop *Scindapsus officinalis* (Nadkarni 2009). The fruit juice is popular as a traditional drink in summer and in the preparation of fruit jam (Satyavati, et al. 1976). In the modern age, the use of fruit juice has been found as the best way to get micronutrients and essential antioxidants for good health (Tarwadi and Agte 2005). The fruit is a rich source of nontoxic natural food additives (Okwu and Emenike 2006). Activated carbon is produced from the fruit shell which is used as an effective, low-cost adsorbent to eliminate heavy metals such as chromium from drinking water (Anandkumar and Mandal 2009). The seed oil of *A. marmelos* is rich in fatty acid and 12-hydroxyoctadec-cis-9-enoic acid; these phytochemicals have the potential to produce biodiesel (Katagi et al. 2011). The seed gum and resin is generally used as glue, an adhesive by jewellers and a soap substitute. The gum is mixed with lime plaster to use to waterproof wells and cement walls. The gum is used as water colours and applied as a protective coating on paintings. The leaf and unripe fruit tannin is applied in tanning and the manufacturing of yellow dye. Leaf extract has insecticidal properties against the brown plant hopper (*Nilaparvata lugens*), a significant pest of rice crop. The leaf extract is also used in abortion and as a birth control (Pushpakumara et al. 2007). Traditionally the tree leaves are used in hepatitis, asthma, febrifuge and hypoglycaemia. The leaf powder is given in the treatment of beriberi and bowel syndrome (Sampath et al. 2012; Atul et al. 2012). Young and fresh leaves and shoots are traditionally used as food material in Thailand and Indonesia (Lakht-e-Zehra et al. 2015). The flower is an astringent used asantiseptic as has wound-healing properties. The flower is also used in epilepsy and conjunctivitis (Gautam et al. 2014; Mani et al. 2017). The roots are given to cure dysentery, fever, colitis and flatulence (Kala 2006). The root and bark decoction is effective in heart palpitation, melancholia and intermittent fever (Kakiuchi et al. 1991). Traditionally, the root bark is used as a fish poison (Lock and Gaur 2001).

26.8 NUTRITIONAL PROPERTIES

A. marmelos is rich in nutrients with the important compositions of carbohydrate (9%–21%), water (60%–65%), sugars (11%–17%), and fibres (5%) along with minerals, vitamins, glucose, amino acids and fatty acids. The key parts of nutrients are found in seeds and leaf. *A. marmelos* seeds consist of 34.4% oil, 8.8% stearic acid, 8.1% linolenic acid, 16.6% palmitic acid and 30.5% oleic acid (Kumar et al. 2012). The seeds contain vitamin A (55 mg), vitamin B and vitamin C (8 mg), which act as an antioxidant (Bhardwaj and Nandal 2015). The fruit contains 1.7% mineral with potassium (610 mg), copper (0.21 mg), calcium (80 mg), iron (0.60 mg/100 g) and phosphorous (52 mg) (Singh et al. 2014). Its calorific value is 88 cal/100 g (Bhardwaj 2014) from xylose, threose, glucose, galactose, fructose, arabinose, sucrose, galacturonic acid, polysaccharides, ricinoleic, linolenic, oleic, linoleic, palmitic and myristic (Bhattacherjee et al. 2020). The seed oil is used in cosmetics and aromatherapy and has antiseptic, antimicrobial, antioxidant, astringent, disinfectant, carminative, cytophylactic and anti-inflammatory effects (Bajaniya et al. 2015; Gopalan et al. 1971).

26.9 CULTIVATION PRACTICES

The seeds of *A. marmelos* are the primary source of planting material. The mature seeds are used to generate seedlings in specially prepared nurseries. *A. marmelos* seed germination rate is low, and the seedlings usually express segregation of fruit traits. Therefore, the superior tree is propagated by grafting, budding and micropropagation of vegetative plant parts. The one-month-old buds are collected and budded onto two-year matured rootstocks to obtain successful regeneration (Ghosh and Roy 2012). The agronomic and genetic characteristics are focused on improving traits (Sharma et al. 2007; Pathirana 2020). The grafting method is followed to propagate the best mother plants (Roy and Singh 1979). Tissues culture technique is used for mass propagation of *A. marmelos* using 8–10 mm sized explants of juvenile auxiliary and terminal buds from the well-performed tree in tissue culture (Arya and Shekhawat 1986). The Murashige and Skoog (MS) medium with 5 mg/l, 2–4D, and NAA are used for the induction of callus. Kinetin (1 mg/l) and BAP (0.5 mg/l) are applied for the induction of rooting. High-frequency plantlet regeneration can be achieved by using cotyledonary nodes of the tree (Nayak et al. 2007).

26.10 CONCLUSION AND FUTURE REMARKS

A. marmelos is one of the important indigenous fruit trees which is significantly contributing to solving problems of malnutrition, hunger, health and environmental pollution. It is a valuable tree species that provides pharmacological and nutritional supports. Therefore, it is an important cash crop to improve the livelihood and health of rural and urban communities. The fruit can play an important role in the economy through promoting value-added food and pharmaceutical industries. Many scientific studies have validated the ethnomedicinal importance of *A. marmelos*. The tree has huge medicinal and nutritional properties to replace a large number of allopathic drugs. The available phytochemical constituents in *A. marmelos* are very useful in different therapeutic actions. This review concentrated on the medicinal uses of the tree. Therefore, more research and investigation are needed to know the efficacy, optimum doses and optimum ways of administering the phytochemicals at a molecular level that can provide more competent medicine in future for various diseases, including diabetes.

REFERENCES

Agarwal, R., S. R. Gupta, S. Srivastava, R. Saxena & S. S. Agrawal. 2009. Intraocular pressure-lowering activity of topical application of *Aegle marmelos* fruit extract in experimental animal models. *Ophthalmic Research* 42:112–116.

Anandharajan, R., S. Jaiganesh, N. P. Shankernarayanan, R. A. Viswakarma & A. Balakrishnan. 2006. *In vitro* glucose uptake activity of *Aegle marmelos* and *Syzygium cumini* by activation of Glut-4, PI3 kinase and PPAR*c* in L6 myotubes. *Phytomedicine* 13:434–441.

Anandkumar, J. & B. Mandal. 2009. Removal of Cr (VI) from aqueous solution using Bael fruit (*Aegle marmelos* correa) shell as an adsorbent. *Journal of Hazardous Materials* 168:2–3:633–640.

Arul, V., S. Miyazaki & R. Dhananjayan. 2002. Mechanisms of the contractile effect of the alcoholic extract of *Aegle marmelos* Corr. on isolated Guinea pig ileum and tracheal chain. *Phytomedicine* 11:679–683.

Arul, V., S. Miyazaki & R. Dhananjayan. 2005. Studies on the anti-inflammatory, antipyretic and analgesic properties of the leaves of *Aegle marmelos* Corr. *Journal of Ethnopharmacology* 96:1–2:159–163.

Arya, H. C. & N. S. Shekhawat. 1986. Clonal multiplication of tree species in the Thar Desert through tissue culture. *Forest Ecology and Management* 16:1–4:201–208.

Atul, N. P., V. D. Nilesh, A. R. Akkatai & S. K. Kamlakar. 2012. A review on *Aegle marmelos*: A potential medicinal tree. *International Research Journal of Pharmacy* 3:86–91.

Bajaniya, V. K., U. K. Kandoliya, N. H. Bodar, N. V. Bhadja & B. A. Golakiya. 2015. Fatty acid profile and phytochemical characterization of bael seed (*Aegle marmelos* L.) oil. *International Journal of Current Microbiology and Applied Sciences* 4:2:97–102.

Balasubramanian, G., M. Sarathi, R. S. Kumar & A. S. S. Hameed. 2007. Screening the antiviral activity of Indian medicinal plants against white spot syndrome virus in shrimp. *Aquaculture* 263:15–19.

Baliga, M. S., H. P. Bhat, N. Joseph & F. Fazal. 2011. Phytochemistry and medicinal uses of the bael fruit (*Aegle marmelos* Correa): A concise review. *Food Research International* 44:7: 1768–177.

Basak, R. K., P. K. Mandal & A. K. Mukherjee. 1981. Studies on a neutral polysaccharide isolated from bael (*Aegle marmelos*) fruit pulp. *Carbohydrate Research* 97:315–321.

Benni, J. M., M. K. Jayanthi & R. N. Suresha. 2011. Evaluation of the anti-inflammatory activity of *Aegle marmelos* (Bilwa) root. *Indian Journal of Pharmacology* 43:393–397.

Bhar, K, S. Mondal & P. Suresh. 2019. An eye-catching review of *Aegle marmelos* L. (golden apple). *Pharmacognosy Journal* 11:207–224.

Bhardwaj, R. L. 2014. Role of Bael fruit juice in nutritional security of Sirohi tribals. *Benchmark Survey Report of Sirohi Tribals*, Krishi Vigyan Kendra, AU, Jodhpur 11–37.

Bhardwaj, R. L. & U. Nandal. 2015. Nutritional and therapeutic potential of Bael (*Aegle marmelos* Corr.) fruit juice: A review. *Nutrition & Food Science* 45:895–919.

Bhattacherjee, A. K., A. Dikshit, D. Pandey & D. K. Tandon. 2015. High performance liquid chromatographic determination of marmelosin and psoralen in bael (*Aegle marmelos* (L.) Correa) fruit. *Journal of Food Science and Technology* 52:1:597–600.

Bhattacherjee, A. K., A. Dikshit & D. K. Tandon. 2020. Nutraceutical changes during ripening of bael [*Aegle marmelos* (L.) Correa] fruits harvested at different maturity periods. *Indian Journal of Traditional Knowledge* 19:2:416–422.

Chanda, R. 2008. Phytochemical & pharmacological activity of *Aegle marmelos* as a potential medicinal plant: An overview. *The Internet Journal of Pharmacology* 6:1:3.

Charoensiddhi, S. & P. Anprung. 2009. Characterization of bael fruit (*Aegle marmelos* [L.] correa) hydrolysate as affected by enzyme treatment. *Journal of Food Biochemistry* 34:1249–1267.

Chatterjee, A., R. Sen & G. D. Aegelinol. 1977. Amonorlactonic constituent of *Aegle marmelos*. *Phytochemistry* 17:328–329.

Chauhan, A., M. Agarwal, S. Kushwaha & A. Mutreja. 2007. Suppression of fertility in male albino rats following the administration of 50% ethanolic extract of *Aegle marmelos*. *Contraception* 76:474–481.

Chauhan, H., R. K. Kaul, N. Ahmed, P. Gupta & A. Anjum. 2016. Preparation and evaluation of Bael (*Aegle marmelos*) vermouth. *International Journal of Food and Fermentation Technology*. 6:1:41–47.

Chopra, R. 1982. *Indigenous drugs of India*. Academic Publishers, Calcutta.

Choudhary, Y., A. Saxena, Y. Kumar, S. Kumar & V. Pratap. 2017. Phytochemistry, pharmacological and traditional uses of *Aegle marmelos*. *UHJPB* 5:27–33.

Costa-Lotufo, L. V., M. T. H. Khan & A. Ather. 2005. Studies of the anticancer potential of plants used in Bangladeshi folk medicine. *Journal of Ethnopharmacology* 99:21–30.

Das, B. & R. Das. 1995. Medicinal properties and chemical constituents of *Aegle marmelos* Correa. *Indian Drugs* 32:93–99.

Dhankhar, S., S. Ruhil, M. Balhara, S. Dhankhar & A. K. Chhillar. 2011. *Aegle marmelos* (Linn.) Correa: A potential source of phytomedicine. *Journal of Medicinal Plants Research* 5:9:1497–1507.

Dhuley, J. N. 2003. Investigation on the gastroprotective and antidiarrhoeal properties of *Aegle marmelos* unripe fruit extract. *Hind Antibiot Bull*: 41–46.

Elango, G., A. A. Rahuman, C. Kamaraj, A. Bagavan & A. A. Zahir. 2012. Adult emergence inhibition and adulticidal activity of leaf crude extracts against Japanese encephalitis vector, *Culex tritaeniorhynchus*. *Journal of King Saud University-Science* 24:1:73–80.

Faizi, S., F. Farooqi & S. Zikr-Ur-Rehman. 2009. Shahidine, a novel and highly labile oxazoline from *Aegle marmelos*: The parent compound of aegeline and related amides. *Tetrahedron* 65:5:998–1004.

Ganguly, N. K. & T. Kaur. 1996. Mechanism of action of cholera toxin and other toxins. *Indian Journal of Medical Research* 104:28–37.

Gautam, M. K., V. Purohit, M. Agarwal, A. Singh & R. K. Goel. 2014. *In vivo* healing potential of *Aegle marmelos* in excision, incision, and dead space wound models. *Scientific World Journal* 9:459–486.

Ghosh, S. N. & S. Roy. 2012. Studies on propagation of bael (*Aegle marmelos* L.) under Jhargram conditions. *The Journal of Horticultural Science and Biotechnology* 7:214–216.

Gireesh, G., S. K. Reas, M. Jobin & C. S. Paulose. 2008. Decreased muscarinic M1 receptor gene expression in the cerebral cortex of streptozotocin-induced diabetic rats and *Aegle marmelos* leaf extract's therapeutic function. *Journal of Ethnopharmacology* 116:296–304.

Gopalan, C., B. V. B. Sastri & S. C. Balasubramanian. 1971. *Nutritive value of Indian food*. National Institute of Nutrition, Indian Council of Medical Research, Hyderabad, India.

Hema, C. G. & K. Lalithakumari. 1999. Screening of pharmacological actions of *Aegle marmelos. Indian Journal of Pharmacology* 20:80–85.

Husen, A. 2021. *Traditional herbal therapy for the human immune system*. CRC Press, Boca Raton, FL. https://doi.org/10.1201/9781003137955

Husen, A. 2022. *Herbs, shrubs and trees of potential medicinal benefits*. Taylor & Francis Group, Boca Raton. https://doi.org/10.1201/9781003205067

Husen, A., R. K. Bachheti & A. Bachheti. 2021. *Non-timber forest products: Food, healthcare and industrial applications*. Springer, Cham, Switzerland. https://doi.org/10.1007/978-3-030-73077-2

Jagetia, G. C. & M. S. Baliga 2004. The evaluation of nitric oxide scavenging activity of certain Indian medicinal plants in vitro: A preliminary study. *Journal of Medicinal Food* 7:3:343–348.

Jagetia, G. C., P. Venkatesh & M. S. Baliga. 2004. Fruit extract of *Aegle marmelos* protects mice against radiation-induced lethality. *Integrative Cancer Therapies* 3:323–332.

Jaswant, A., V. Loganathan & V. Manimaran. 2001. Wound healing activity of *Aegle marmelos. Indian Journal of Pharmaceutical Sciences*. 63:41–44.

Kakiuchi, N., L. R. E. Senaratne, S. L. Huang, X. W. Yang, M. Hattori & U. Pilapitiya. 1991. Effects of constituents of Beli (*Aegle marmelos*) on spontaneous beating and calcium-paradox of myocardial cells. *Planta Medica* 57:43–47.

Kala, C. P. 2006. Ethnobotany & ethnoconservation of *Aegle marmelos* (L.) correa. *Indian Journal of Traditional Knowledge* 5:4:537–540.

Kamalakkannan, N. & P. S. M. Prince. 2003. Hypoglycaemic effect of water extract of *Aegle marmelos* Corr: Fruit in streptozotocin diabetic rats. *Journal of Ethnopharmacology* 87:2/3:207–210.

Kamalakkannan, N. & P. S. M. Prince. 2005. Antihyperlipidaemic effect of *Aegle marmelos* fruit extract in streptozotocin-induced diabetes in rats. *Journal of the Science of Food and Agriculture* 85:569–573.

Kar, A., S. Panda & S. Bharti. 2002. Relative efficacy of three medicinal plant extracts in the alteration of thyroid hormone concentrations in male mice. *Journal of Ethnopharmacology* 81:281–285.

Katagi, K. S., R. S. Munnoli & K. M. Hosamani. 2011. Unique occurrence of unusual fatty acid in the seed oil of *Aegle marmelos* Corre: Screening the rich source of seed oil for bioenergy production. *Applied Energy* 88:5:1797–1802.

Kesari, A. N., R. K. Gupta, S. K. Singh, S. Diwakar & G. Watal. 2006. Hypoglycemic and anti hyperglycemic activity of *Aegle marmelos* seed extract in normal and diabetic rats. *Journal of Ethnopharmacology* 107:374–379.

Kintzios, S. E. 2006. Terrestrial plant-derived anticancer agents and plant species used in anticancer research. *Critical Reviews in Plant Sciences* 25:79–113.

Kirtikar, K. R. & B. D. Basu. 1935. *Indian medicinal plants*. Lalit Mohan Basu, Allahabad, India.

Kirtikar, K. R. & B. D. Basu. 1939. *Indian medicinal plants*. 2nd ed. Lalit Mohan Basu, Allahabad, India, pp. 499–502.

Kruawan, K. & K. Kangsadalampai. 2006. Antioxidant activity, phenolic compound contents and antimutagenic activity of some water extract of herbs. *Thai Journal of Pharmaceutical Sciences* 30:1–47.

Kumar, K. S., M. Umadevi, D. Bhowmik, D. M. Singh & A. S. Dutta. 2012. Recent trends in medicinal uses and health benefits of Indian traditional herbs *Aegle marmelos. The Pharma Innovation* 1:4:57–65.

Lakht-e-Zehra, A., N. G. Dar, N. Saleem, U. A. Soomro, W. Afzal & B. Naqvi. 2015. Nutritional exploration of leaves, seed and fruit of bael (*Aegle marmelos* L.) grown in Karachi region. *Pakistan Journal of Biochemistry & Molecular Biology* 48:3:61–65.

Lampronti, I. 2003. In Vitro antiproliferative effects on human tumor cell lines of extracts from the Bangladeshi medicinal plant *Aegle marmelos* Correa. *Phytomedicine*10:300–308.

Lock, J. M. & R. D. Gaur. 2001. Flora of the district garhwal north west Himalaya (with ethnobotanical notes). *Kew Bulletin* 56:251.

Maity, P., D. Hansda, U. Bandyopadhyay & D. K. Mishra. 2009. Biological activities of crude extracts and chemical constituents of *Aegle marmelos* L. Corr. *Indian Journal of Experimental Biology* 47:11:849–861.

Mandal, P. K. & A. K. Mukherjee. 1981. Studies on bael (*Aegle marmelos*) seed glycoproteins. *Carbohydrate Research* 90:233–241.

Mani, A., A. Singh, N. Jain & S. Misra. 2017. Flowering, fruiting and physiochemical characteristics of bael (*Aegle marmelos* correa.) grown in northern districts of West Bengal. *Current Journal of Applied Science and Technology* 23:1–8.

Mishra, B. B., N. Kishore, V. K. Tiwari, D. D. Singh & V. Tripathi. 2010. A novel antifungal anthraquinone from seeds of *Aegle marmelos* Correa (family Rutaceae). *Fitoterapia* 81: 104–107.

Nadkarni, A. K. 1954. *Indian materia medica.* Volume 1. 3rd ed., Popular Prakashan, Mumbai.

Nadkarni, M. K. 2009. *Indian materia medica.* 3rd ed. Popular Prakashan, Mumbai, India, pp. 45–49.

Nagar, S., M. Kumar, R. B. Kumatkar, J. R. Sharma & S. Sing. 2017. Evaluation of bael (*Aegle marmelos* Corr.) germplasms for seed and qualitative characters under semi-arid conditions of Haryana. *International Journal of Pure & Applied Bioscience* 5:436–442.

Narender, T. 2007. Antihyperglycemic and antidyslipidemic agent from *Aegle marmelos. ChemInform* 38:31:1808–1811.

Nayak, P., P. R. Behera & T. Manikkannan. 2007. High frequency plantlet regeneration from cotyledonary node cultures of *Aegle marmelos* (L.) Corr. *In Vitro Cellular & Developmental Biology-Plant* 43:231–236.

Nicolis, E., I. Lampronti, M. Dechecchi. 2009. Modulation of expression of *IL-8* gene in bronchial epithelial cells by 5-methoxypsoralen. *International Immunopharmacology* 9:1411–1422.

Okwu, D. E. & I. N. Emenike. 2006. Evaluation of the phytonutrients and vitamins content of *Citrus* fruits. *International Journal of Molecular Medicine and Advance Sciences* 2:1–6.

Pandeya, G. S. 1983. *Charaka-Samhita.* Chowkhamba Sanskrit Sansthan, Varanasi.

Patel, A. R., D. Garach, M. Chakraborty & J. Kamath. 2012. *Aegle marmelos* (Linn.): A therapeutic boon for human health. *International Journal of Research in Ayurveda and Pharmacy* 3:2:159–163.

Patel, P., S. Mohammed & B. Asdaq. 2010. Immunomodulatory activity of methanolic fruit extract of *Aegle marmelos* in experimental anima. *Saudi Pharmaceutical Journal* 18:3:161–165.

Pathirana, C. K. 2020. Assessment of the elite accessions of bael (*Aegle marmelos* L. Corr.) in Sri Lanka based on morphometric, organoleptic, and elemental properties of the fruits and phylogenetic relationships. *PLoS ONE* 15:Article ID e0233609.

Pitre, S. & S. K. Srivastava. 1988. Pharmacological, microbiological and phytochemical studies on roots of *Aegle marmelos. Journal of Ethnopharmacology* 23:2–3:356.

Pushpakumara, D. K. N. G., H. P. M. Gunasena & V. P. Singh. 2007. *Beli Aegle marmelos L. Correa: Underutilized fruit trees in Sri Lanka.* World Agroforestry Centre, South Asia Office, New Delhi, India, 8, pp. 249–276.

Raja, S. B., M. R. Murali & S. N. Devaraj. 2008. Differential expression of *ompC* and *ompF* in multidrug-resistant *Shigella dysenteriae* and *Shigella flexneri* by aqueous extract of *Aegle marmelos*, altering its susceptibility toward β-lactam antibiotics. *Diagnostic Microbiology and Infectious Disease* 61:321–328.

Rana, B. K., U. P. Singh & V. Taneja 1997. Antifungal activity and kinetics of inhibition by essential oil isolated from leaves of *Aegle marmelos. Journal of Ethnopharmacology* 57:29–34.

Rani, P. & N. Khullar. 2004. Antimicrobial evaluation of some medicinal plants for their anti-enteric potential against multi-drug resistant *Salmonella Typhi. Phytotherapy Research* 18:670–673.

Roy, A., S. B. Bhattacharya, A. K. Mukherjee & C. V. N. Rao. 1976. The Structure of degraded Bael (*Aegle marmelos*) gum. *Carbohydrate Research* 50:87–96.

Roy, S. K. & R. N. Singh. 1979. Bael fruit (*Aegle marmelos*) – A potential fruit for processing. *Economic Botany* 33:203–212.

Sagar, S. M. & R. K. Wong. 2010. Increasing therapeutic gain and controlling radiation-induced injuries with Asian botanicals and acupuncture. In *Supportive cancer care with Chinese medicine*, W. C. S. Cho, Ed., Springer, Dordrecht, Netherlands, pp. 109–139.

Saha S. K. and A. Chatterjee. 1957. Isolation of alloimperatorin and 3-sitosterol from the fruits of *Aegle marmelos. Journal of Indian Chemical Society* 34:228–230.

Sampath, K. P., M. Umadevi, D. Bhowmik, D. M. Singh & A. S. Dutta. 2012. Recent trends in medicinal uses and health benefits of Indian traditional herbs *Aegle marmelos. Pharma Innovation* 1:57–65.

Satyavati, G. V., A. K. Gupta & N. Tandon. 1976. *Medicinal plants of India Volume 1*, Indian Council of Medical Research, New Delhi.

Shailajan, S., S. Menon & H. Hande. 2011. Method validation of Marmelosin from fruit pulp of *Aegle marmelos* (L.) Correa using HPTLC Technique. *Journal of Pharmacy Research* 4:1353–1355.

Sharma, B. R., R. K. Rattan & P. M. Sharma. 1981. An alkaloid, and other components of unripe fruits of *Aegle marmelos. Phytochemistry* 20:2606–2607.

Sharma, P. C., V. Bhatia, N. Bansal & A. Sharma. 2007. A review on bael tree. *Indian Journal of Natural Products and Resources* 6:171–178.

Sharma, R. D., S. Petare, G. B. Shinde, K. Goswami & M. V. R. Reddy. 2010. Novel drug designing rationale against-*Brugiamalayi* microfilariae using herbal extracts. *Asian Pacific Journal of Tropical Medicine* 3:11:846–850.

Shoeb, A., R. S. Kapil & S. P. Popli. 1973. Coumarins and Alkaloids of *Aegle marmelos. Phytochemistry* 12:2071–2072.

Sing, A. K., S. Singh, R. P. Singh, H. K. Joshi & S. K. Sharma. 2014. Characterization of bael (*Aegle marmelos*) varieties under rainfed hot semi-arid environment of western India. *Indian Journal of Agricultural Sciences* 84:1236–1242.

Singh, K., I. Chakraborty & A. K. Chaurasiya. 2014. Bael preserve syrup as booster of human health as a health drink. *The Bioscan* 9:2:565–569.

Singh, R. & H. S. Rao. 2008. Hepatoprotective effect of the pulp/seed of *Aegle marmelos* correa ex Roxb against carbon tetrachloride induced liver damage in rats. *International Journal of Green Pharmacy* 2:232–234.

Sonawane, A., S. S. Pathak & R. C. Pradhan. 2020. Physical, thermal, and mechanical properties of Bael fruit. *Journal of Food Process Engineering* 43:e13393.

Sood, S. & S. Katoch. 2014. Development and evaluation of dietetic products prepared from bael (*Aegle marmelos*) fruit. *Journal of Krishi Vigyan* 2:22–24.

Tarwadi, K. & V. Agte. 2005. Antioxidant and micronutrient quality of fruit and root vegetables from the Indian subcontinent and their comparative performance with green leafy vegetables and fruits. *Journal of the Science of Food and Agriculture* 85:9:1469–1476.

Upadhya, S., K. K. Shanbhag, G. Suneetha, B. M. Naidu & S. Upadhya. 2004. A study of hypoglycemic and antioxidant activity of *Aegle marmelos* in alloxan induced diabetic rats. *Indian Journal Physiology and Pharmacology* 48:4:476–480.

Venkatesh, Y. S. S. 1990. Anaplastic carcinoma of the thyroid. A clinicopathologic study of 121 cases. *Cancer* 66:321–330.

Yadav, N., P. Singe & R. Mehrotra. 2011. Determination of some ethnomedicinally important constituents of *Aegle marmelos* fruit during different stages of ripening. *Chinese Journal of Natural Medicines* 9:204–209.

27 Antidiabetic Properties of *Cinnamomum verum* J. Presl (True Cinnamon Tree or Ceylon Cinnamon Tree)

Recent Trends, Progress, and Challenges

Weria Weisany, Shima Yousefi, Katayoon Samiakalantari, and Maryam Tahmasebi

CONTENTS

ABBREVIATIONS

CA:	Trans-cinnamaldehyde
C. verum:	*Cinnamomum verum*
C. zeylanicum:	*Cinnamomum zeylanicum*
EO:	Essential oil
FBS:	Fasting blood sugar
GC:	gas chromatography
GC–MS:	gas chromatography–mass spectrometry
HDL:	High-density lipoprotein
HPLC:	High-performance liquid chromatography
LDL:	Low-density lipoprotein
ROS:	Reactive oxygen species

DOI: 10.1201/b23347-27

STZ:	Streptozotocin
T2DM:	Type 2 diabetes mellitus
TC:	total cholesterol

27.1 INTRODUCTION

Spices are common dietary additives that improve the taste and flavor of foods. Spices are pungent or aromatic substances derived from dried parts of plants, typically seeds, fruits, leaves, roots, bark, and other tropic-derived plant parts (Sangal 2011). Polyphenols are found in plant seeds, fruits, leaves, and bark and have long been used as food additives to add flavor and taste to foods and improve food quality (Aparna, Narayanan et al. 2014). A variety of spices also serve as excellent preservatives, extending the shelf life of the food by delaying the decaying process (Asimi, Sahu et al. 2013). They contain a number of bioactive compounds that can affect digestion and metabolism. Many studies have shown that spices have a variety of beneficial physiological effects, including antioxidant, antimicrobial, anti-inflammatory, antidiabetic, and anticancer properties, as well as insulin potentiating activity in both normal and experimentally induced diabetic animal models, as well as humans (Abeysekera, Premakumara et al. 2013). Cinnamon is one spice that is emerging as a potential therapeutic agent for diabetes management. *Cinnamomum* is one of the world's oldest and most widely used spices, as well as an herbal remedy. Ceylon cinnamon (*Cinnamomum zeylanicum*) is a member of the Lauraceae family that originated in Ceylon but is now grown throughout southern Asia and North America. The raw material is bark (Cinnamomi cortex) stripped of its internal layer, known as primary bark. Depending on the origin of the raw material, Ceylon cinnamon bark contains 0.5% to 4.0% oil. Cinnamaldehyde (65%–75%), cinnamyl acetate, eugenol (approximately 5% in total), and caryophyllene (up to 4%) are the main components of the oils. Furthermore, polysaccharides (mucilage), phenolic acids (cinnamic acid and its derivatives), oligomeric proanthocyanidins, diterpenes, and other compounds are found in the bark (Momtaz, Hassani et al. 2018). It's an evergreen tree in the Lauraceae family. *Cinnamomum cassia* (L.) J. Presl, *Cinnamomum camphora*, and *Cinnamomum zeylanicum* are the most well-known species. For centuries, several cultures have used this plant as a flavor modifier to make food more palatable, primarily in culinary applications (Rao and Gan 2014). Though many species in this genus are marketed as cinnamon, the inner dried bark of *Cinnamomum verum* J. Presl (family Lauraceae) has long been regarded as the true cinnamon (Avula, Smillie et al. 2015). Its medicinal and culinary uses have been well documented in ancient literature dating back 4000 years (Rao and Gan 2014). The various parts of the plant and its essential oil (EO) are widely used as spices and flavoring agents all over the world. In the industry, composites of this spice are used. The addition of cinnamon essential oil extracts to poultry diets improves immunity and microbiological aspects (El-Hack, Mohamed et al. 2020). Cinnamon oils have also been discovered in the development of new active packaging films based on whey protein and containing chitosan nanofibers. The results of Fourier-transform infrared spectroscopy (FTIR) and scanning electron microscopy (SEM) tests show a homogeneous distribution of chitosan nanofibers in the film, as well as the effect of chitosan nanofibers on mechanical properties and water vapor permeability (Mohammadi, Mirabzadeh et al. 2020). Cinnamon is made up of many compounds that have a biologically active formula and are responsible for its properties.

It is also used as a seasoning, sauce, bakery, confectionery, and beverage condiment to preserve food products and enhance their pharmacological activities (Pittman 2011). This means they have distinct properties. The same monoterpene hydrocarbon spectrum is found in volatile oils extracted from the cinnamon plant's leaves, bark, and root bark. Their main compounds, however, differ. The basic compound found in cinnamon bark is cinnamon aldehyde. The primary compound in the leaf oil is eugenol, whereas camphor is the primary compound in the root bark oil (Gruenwald, Freder et al. 2010). In ancient times, cinnamon was utilized for the preservation of mummies by the Egyptian and Roman empires, owing to its alluring fragrance and flavoring properties (Ribeiro-Santos, Andrade et al. 2017). Cinnamaldehyde, the main bioactive component in cinnamon, is primarily responsible for the characteristic taste, scent, and flavor that cinnamon adds to food (Isaac-Renton, Li et al. 2015). It also provides resistance to oxidative stress, microbial infection, and other chronic diseases (Ribeiro-Santos, Andrade et al. 2017). Biologically active compounds are a diverse group. These compounds can be classified into two groups based on their effects on human and animal organisms. Because of their novel properties, both of these groups are of particular interest today. Cinnamon's biologically active compounds are of interest in terms of human health (Gruenwald, Freder et al. 2010). Cinnamon is a popular spice because of its flavor, but it may also be useful in pharmaceutical applications (Gharibzahedi 2018). (Gharibzahedi 2018). Secondary metabolites account for a significant portion of the synthetic compounds with health-promoting properties. The majority of them are dietary neutral, but they have a positive effect on human health. Plant oils are thought to be among the safest compounds used in medicine (Gharibzahedi 2018). Polyphenolic constituents such as phenolic acids, coumarin, and proanthocyanidin, as well as volatile essential oils, contribute to the pharmacological properties. Cinnamon was found to be beneficial for cardiovascular-related comorbidities such as diabetes and other metabolic disorders, as well as for its antioxidant and anti-inflammatory properties (Gruenwald, Freder et al. 2010).

27.2 HISTORICAL PERSPECTIVES

Cinnamon, which has been prized for centuries for its distinct flavor and aroma, is the dried inner bark of *Cinnamomum verum* (syn. *C. zeylanicum* Blume), an evergreen tree native to Sri Lanka, India's Malabar Coast, and Myanmar and also grown in South America and the West Indies. As a result, it is also known as Mexican cinnamon. The spice is brown in color and made up of dried inner bark. It has a delicately fragrant aroma and a warm, sweet flavor. It is known as "dalchini" in Urdu. Chinese cinnamon (*C. aromaticum* or Cassia), Vietnamese or Saigon cinnamon (*C. loureiroi*), Indonesian cinnamon (*C. burmannii*), camphor laurel (*C. camphora*), and Malabar cinnamon (*C. citriodorum* or *C. tamala*, also known as tejpata, tejpat, or Indian bay leaf) are all cultivated as sources of cinnamon spice (Braudel 1984). Canella is an Italian word that means "little tube," which perfectly describes cinnamon quills. Ceylon cinnamon, also known as true cinnamon, is distinguished from the other two primary types by the appearance of its quill. It is made up of soft and light-colored layers of roll, whereas the others are dark, hard, hollow, and rolled into one layer. It is a preferred species due to its low coumarin levels and delicate taste (Corn 1998).

27.3 BOTANICAL DESCRIPTION

Cinnamon is a tropical tree whose inner bark is mostly used as a spice. It belongs to the Lauraceae family and contains about 250 species, four of which are commercially important and traded globally: Ceylon cinnamon (*Cinnamomum verum* or *Cinnamomum zeylanicum*) from India and Sri Lanka, Chinese cinnamon (*Cinnamomum cassia* or *Cinnamomum aromaticum*), and Indonesian cinnamon (*Cinnamomum burmannii*; *Cinnamomum loureiroi*). *Cinnamomum verum* J. Presl (syn. *Cinnamomum zeylanicum* Blume) is a valuable evergreen tree in the Lauraceae family. True cinnamon, Ceylon cinnamon, Mexican cinnamon, Dalchini, Qirfa, and Darchini are some of the other names for it (Ribeiro-Santos, Andrade et al. 2017). *C. verum* has trees (up to 50 feet tall) with long lance-shaped leaves, small yellow flowers organized in a cluster, and ovoid-shaped fruits. *C. verum* grows to a height of 10–15 m. It has opposite leaves that are ovate to oblong in shape, glabrous, and usually have a leathery texture on both surfaces. The dark green leaves measure 7–18 cm in length and have pointed tips. Abaxially, the lamina is greenish-white, and adaxially, it is shiny green (Chakraborty, Sankaran et al. 2015). The flowers, which are greenish or yellowish-white, are arranged in five panicles and have a distinct odor. Its fruit is a one-centimeter-long berry with a single seed (Ribeiro-Santos, Andrade et al. 2017). Its bark is known as milled bark and is extremely valuable economically (Cardoso-Ugarte, López-Malo et al. 2016). When the red flush of the young leaves turns green, it is time to peel off the bark (Balasubramanian, Singh et al. 2012). The dried bark, which has been formed into a tubular form known as a quill or cinnamon stick, is primarily used in food preparation (Hamidpour, Hamidpour et al. 2015). The different parts of *C. verum* are depicted in Figure 27.1.

FIGURE 27.1 Different parts of *Cinnamomum verum* J. Presl.

27.4 DISTRIBUTION

C. verum is indigenous to Sri Lanka and southern India, but it can also be found in Southeast Asia, China, Burma, Indonesia, Madagascar, the Caribbean, Australia, and Africa (Goyal, Kaur et al. 2018). It is most commonly found in tropical rain forests, marshy areas, and well-drained soils. *C. verum* is the most commonly harvested cinnamon species, and it can be harvested after three years of cultivation (Ribeiro-Santos, Andrade et al. 2017). During the 17th century, Java began to cultivate cinnamon, and the East India Company became the largest exporter of cinnamon to European nations. Nowadays, the leading cinnamon exporters are Sri Lanka, China, Vietnam, and Madagascar.

27.5 BIOACTIVE COMPOUNDS DETERMINED IN CINNAMON

Cinnamomum zeylanicum antioxidant compounds are found in many of parts of the plant such as leaves, buds, flowers, fruits, bark, root bark, and oils. Additionally, *C. zeylanicum* is also rich with volatile compounds, most of which act as antioxidants. *C. zeylanicum* contains cinnamyl acetate, eugenol, trans-cinnamaldehyde (the main component of cinnamon flavor), cymene, cinnacassiol, cineol, camphene, catechins, coumarin cinnamic acid and gamma-terpinene, terpinolene, and α-thujene, α-terpineol, linalool, l-borneol, E-nerolidol, pinene, phyllandrene, proanthocyanidins, safrole, tannins constituting polymeric 5,7,3,4-tetrahydroxy-tetrahydroxy flavan-3–4-diol units, α-cubene, and resins (Figure 27.2) (Abdelgadir, Hassan et al. 2020). In addition, most of the compounds are mainly derived from cinnamyl, hydrolyzed phenol, tannins, phenylpropanoids, and terpenoid compounds (Tang, Chen et al. 2020). However, common antioxidants of *C. zeylanicum* species include eugenol, benzyl benzoate, linalool, and eugenyl acetate (Schmidt, Jirovetz et al. 2006). Among the bioactive constituents of *C. zeylanicum*, cinnamaldehyde and transcinnamaldehyde are thought to be the most important, particularly in terms of anti-tyrosinase activity (Arisha, Sakr et al. 2020). Cinnamaldehyde is primarily responsible for *C. zeylanicum*'s spicy and fragrant characteristics (Rao and Gan 2014). *C. zeylanicum* is traditionally used to provide aroma and essence compounds due to its high concentration of bioactive compounds and medicinal properties. It is also used medicinally as an antioxidant, anti-inflammatory, antihyperglycemic, antilipidemic, antidiabetic, anticancer, antitumor, anthelmintic, anti-aflatoxigenic, antifungal, and antimicrobial (Abdelgadir, Hassan et al. 2020). There are numerous examples of its use in medicine, such as the proanthocyanidin metabolite of cinnamon, which reduces inflammatory compounds and supports pathways like insulin signaling. Additionally, bioactive compounds found in essential oils are used to combat a variety of pathogenic (including food-borne) and spoilage bacteria (Arisha, Sakr et al. 2020). The type of plant, the tree section, and the maturity level all affect the concentration of cinnamon compounds. The essential characteristics of cinnamon extract may be impacted by these elements. *Cinnamomum cassia* (L.) J. Presl, for instance, contains more coumarin. Depending on the part of the tree, the ingredient content also varies. The mature tree has the highest yield value of cinnamon trans-aldehyde, and the upper and middle segments of the cinnamon bark are more

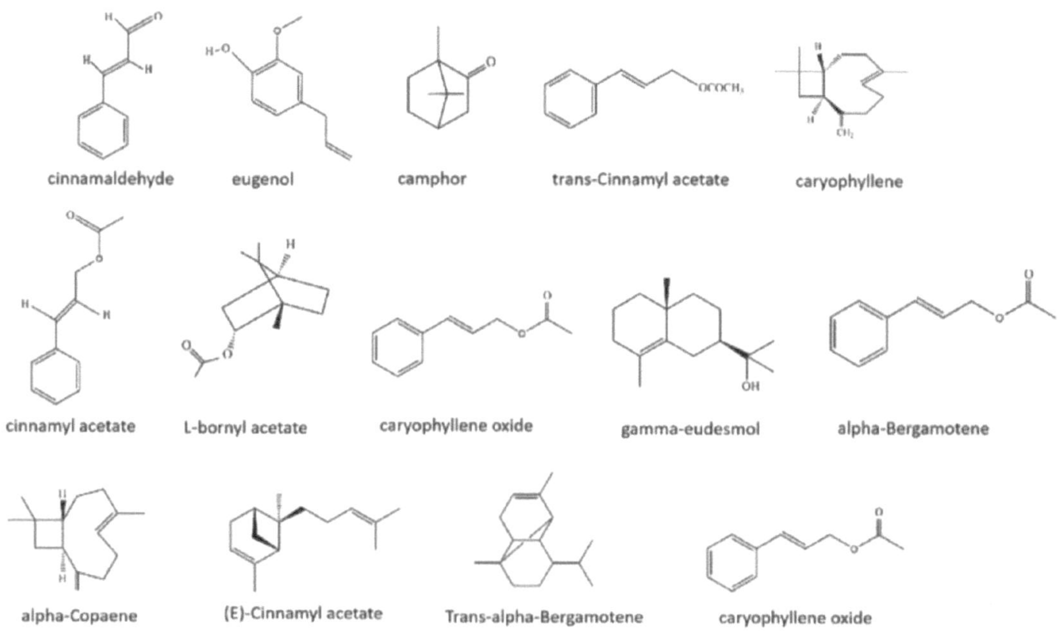

FIGURE 27.2 The chemical structures of some important compounds isolated from cinnamon.

effective at extracting cinnamon oil. However, the separation techniques, solvents, and all other factors related to the extraction and separation process, such as time, temperature, and pressure, affect how effectively the compounds in the extracted cinnamon oil are separated. Optimizing the extraction process is a significant factor that affects the composition of oil (Adarsh, Chettiyar et al. 2020). Cinnamon's compounds are separated using a variety of techniques. Organic solvent extraction, hydrodistillation, and steam distillation are common techniques for obtaining essential oils. The ones that employ solvents, like methanol, ethanol, and chloroform, are the most prevalent (Adarsh, Chettiyar et al. 2020). New extraction techniques have recently been developed that use supercritical fluid and are aided by microwaves or ultrasounds (Chen, Sun et al. 2021). Additionally, supercritical CO_2 extraction is employed. Cinnamic aldehyde, β-caryophyllene, longifolene, and β-amyrin are the primary components of the extract obtained using this method, according to gas chromatography–mass spectrometry (GC–MS) analysis (Li, Guan et al. 2018). Water extraction, however, is the most dependable and secure method for protecting human health. By using ultraperformance liquid chromatography-high-resolution mass spectrometry (UPLC–HRMS) to profile the aqueous cinnamon extract, it was possible to identify compounds with biologically active properties like camphor, L (-)-carnitine, and rosavin (Tang, Chen et al. 2020).

27.5.1 Determination Methods for Compounds Naturally Occurring in Cinnamon

Cinnamon is a plant that contains a variety of bioactive substances. The most popular technique for making cinnamon extract oil is hydrodistillation, which involves steam distillation in a device similar to a Clevenger (Muhammad, Lemarcq et al. 2020). Extraction is used on tiny pieces of fruit, leaves, or bark. After several hours, this process produces an oil that is full of bioactive compounds. Extraction of oils from leaves (Khasanah, Prasetyawan et al. 2017) and bark (Othman, Maskat et al. 2020) can also be carried out using maceration extraction. For bark extraction, it is also possible to use a wide range of other extraction techniques, including distillation (Gotmare and Tambe 2019), steam distillation (Wong, Ahmad-Mudzaqqir et al. 2014), supercritical CO_2 extraction (Masghati and Ghoreishi 2018), subcritical extraction (Liang, Li et al. 2019), extraction in

Soxhlet apparatus (Wong, Ahmad-Mudzaqqir et al. 2014), and water extraction (Tang, Chen et al. 2020). Chromatographic analysis of the obtained extract is usually carried out with the use of gas chromatography techniques, mainly GC–MS (Farias, Monteiro et al. 2020), but also GC (Masghati and Ghoreishi 2018), GC–FID (Farias, Monteiro et al. 2020), or GC×GC–TOFMS (Silva, Silva et al. 2020). Additionally, liquid chromatography is used for the analysis of cinnamon extracts, including HPLC (Wong, Ahmad-Mudzaqqir et al. 2014), reversed-phase high-performance liquid chromatography (RP–HPLC) (Othman, Maskat et al. 2020), liquid chromatography with tandem mass spectrometry (LC–MS/MS) (Gulcin, Kaya et al. 2019), and ultraperformance liquid chromatography–high-resolution mass spectrometry (UPLC–HRMS) (Muhammad, Tuenter et al. 2021). Other examples of methods used in extract analyses are the Folin–Ciocalteu colorimetric method (Othman, Maskat et al. 2020).

27.6 ETHNOMEDICINAL USES

Throughout its history, *C. verum* has been utilized in numerous food preparations and ethnomedicinal formulations. It was the first aromatic plant whose potential for treating bronchitis had been reported (Ribeiro-Santos, Andrade et al. 2017). For centuries, rural people have used the herb to treat ailments such as bronchitis, asthma, cephalalgia, odontalgia, cardiac diseases, diarrhea, uropathy, nausea, vomiting, flatulence, fever, halitosis, arthritis, coughing, hoarseness, impotence, frigidity, eye inflammation, cramps, intestinal spasms, vaginitis, neuralgia, rheumatism, and sore throats (Gulcin, Kaya et al. 2019). It also improves blood circulation in the uterus. It is used in the production of candy, chewing gum, mouthwash, lip sunscreen, fragrance, cinnamon toast, volatile oils, and toothpaste (Isaac-Renton, Li et al. 2015). It also possesses tissue regenerative, coagulant, and wound-healing properties (Rao and Gan 2014). Cinnamon essential oil (CEO) is effective for treating biliousness, headache, piles, parched mouth, bronchitis, diarrhea, itching, and heart and urinary disease (Farahpour and Habibi 2012).

27.7 PHYTOCHEMICAL CONSTITUENTS OF CINNAMON

A qualitative phytochemical screening of a methanolic extract of *C. verum* bark revealed the presence of all four secondary metabolite categories. The botanical part of the tree used to produce cinnamon extract also influences its phytoprofile. At the same time, essential oils from the *C. verum* bark primarily contain cinnamaldehyde and linalool, while flower and fruit extracts contain (E)-cinnamyl acetate, and leaf extracts primarily contain eugenol (Alizadeh Behbahani, Falah et al. 2020). The bark of the cinnamon tree has also been reported to contain coumarin, a benzenoid lactone. *C. cassia* is particularly rich in coumarin (3462.0 mg/kg in *C. cassia* vs. 12.3 to 143.0 mg/kg for *C. verum*) (Ananthakrishnan, Chandra et al. 2018). The solvent and temperature should also be carefully selected according to the molecule one wishes to extract; for example, water is a better solvent for extracting the phenols from *C. verum* than polar organic solvents at 200 °C (Pramote, Nucha et al. 2012). During extraction, the state of the cinnamon cell structures is primarily responsible for efficient extraction (Klejdus and Kováčik 2016). O-methoxycinnamaldehyde has been previously isolated from powder cinnamon. Cinnamon is a rich source of vital oils as well as other derivatives including cinnamic acid, cinnamaldehyde, and cinnamate. Various chemical constitutes such as cinnamaldehyde, cinnamyl acetate, α-thujen, terpineol, α-cubebene, eugenol, and coumarin has been documented from cinnamon, and some of these compounds have been reported for antimicrobial activity (Muhammad and Dewettinck 2017). Cinnamaldehyde is of great importance as the majority of scientific studies claimed that it has a powerful antibacterial activity (Doyle and Stephens 2019).

27.8 PHARMACOLOGICAL ACTIVITY

Cinnamon plant parts and essential oil are primarily used as a spice and condiment to flavor seasonings, sauces, bakery, confectionery, and beverages. Because of their ability to preserve food, they are also used as a food additive in a variety of foods (Pittman 2011). *C. verum* is a potent medicinal plant that is used for healthcare purposes in addition to being used as a food additive. It has antioxidant, antibacterial, antifungal, antiviral, antiulcer, antilipidemic, anticancer, antipyretic, antiplatelet, antiallergic, antihypertensive, insecticidal, nematicidal, antidiabetic, and anesthetic properties (Gulcin, Kaya et al. 2019). It also possesses gastroprotective, cardioprotective, immunomodulatory, and cognitive function boosting activities (Gopinath, Vidya et al. 2014). Its effective usefulness was also reported during flatulence, diarrhea, amenorrhea, toothache, fever, leucorrhea, common cold, headache, and blood pressure control (Hajimonfarednejad, Ostovar et al. 2019). According to Gulcin, Kaya et al. (2019), enzymatic inhibition is the most studied therapeutic approach for combating various health problems such as obesity, diabetes, and Alzheimer's. Cinnamon's ability to inhibit amylase, -glycosidase, acetylcholinesterase (AChE), and butyrylcholinesterase (BChE) is being investigated in the pharmaceutical, cosmeceutical, and food industries.

27.9 CINNAMON IN THE TREATMENT OF DISEASES AND DISORDERS

Cinnamon is well known in Western natural medicine as a natural species with healing properties. Nowadays, it is also reported as a plant with unique properties that can be used to treat a variety of ailments. Cinnamon's bioactive compounds have a wide range of effects on the human body and can be used to treat a variety of diseases and metabolic disorders (Table 27.1). There is enough evidence to suggest that cinnamon can be used as a spice in everyday life and has a positive impact on

TABLE 27.1
Examples of Potential Application of Cinnamon in Human Health

Potential effect	Description	References
Autoimmune disorders	Cinnamon and its metabolite sodium benzoate (NaB) upregulate anti-autoimmune Tregs and Th2, suppress autoimmune Th17 and Th1, inhibit inflammatory infiltration, and reduce the expression of pro-inflammatory molecules.	(Pahan and Pahan 2020)
Antidiabetic	Cinnamon extract lowers blood glucose and cholesterol levels.	(Sivaranjani, Zachariah et al. 2021)
Neuroprotective	Results of research suggest that cotreatment of rats with cinnamon oil alleviated the adverse effects of deltamethrin pesticide in rats. Cinnamon oil may prevent deltamethrin's neurotoxic effects through an enhanced antioxidant defense system, decreased stress levels, and regulation of gene expression.	(Ahmed, Abdel-Azeem et al. 2021)
Anticancer	Cinnamon essential oil and cinnamon aldehyde exhibit antimicrobial properties, in patients with suspected colon cancer and infections caused by *Escherichia coli*.	(Kosari, Taheri et al. 2020)
Lipid lowering	Cinnamon extract exhibits hypolipidemic activity in hypercholesterolemic albino rats. A daily dose of the extract reduced total cholesterol levels, triglyceride levels, and LDL cholesterol levels (LDL-C).	(Alsoodeeri, Alqabbani et al. 2020)
Antioxidant and antimicrobial	Herbs including, but not limited to, cinnamon have antibacterial, antiviral, and antioxidant properties with double the effectiveness of synthetic supplements.	(Parham, Kharazi et al. 2020)
COVID-19 treatment	Cinnamon extract has the potential to limit overshooting immune reactions in COVID-19.	(Lucas, Fröhlich-Nowoisky et al. 2021)
TiO2NPs protector	Cinnamon oil is a promising substance for the protection against the hazards of TiO2NPs.	(Salman, Al-Shaikh et al. 2021)

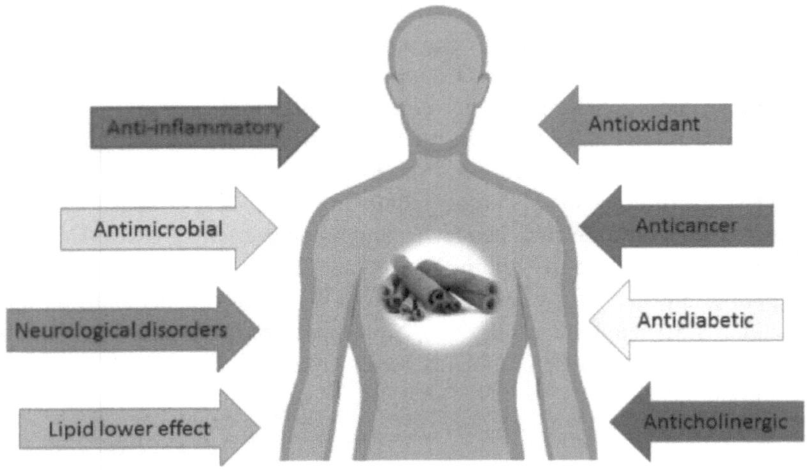

FIGURE 27.3 Cinnamon in the treatment of diseases and disorders.

human health (Figure 27.3). It is used to treat infections, diabetes, and cancer and offers hepatoprotection, neuroprotection, cardioprotection, and immunomodulation. It can be used to kill microbes due to its antioxidant and antimicrobial properties. These compounds' anti-inflammatory and antidiabetic properties have an indirect effect on the receptor-mediated mechanism (Othman, Maskat et al. 2020).

27.10 ANTIDIABETIC ACTIVITY

Diabetes is a metabolic disorder with numerous health consequences. Diabetes mellitus type 2 is the most common. Diabetes is a metabolic disorder characterized by insulin resistance and hyperglycemia that is typically caused by increased oxidative stress and inflammation. Herbal formulations, rather than conventional synthetic drugs, have become more popular in the treatment of diabetes in recent years. Cinnamon has long been regarded as a potentially useful treatment for type 2 diabetes (Gulcin, Kaya et al. 2019). However, the most recent scientific evidence points to limited and contradictory outcomes in humans. It's possible that the use of various cinnamon doses, extracts, and species, along with various administration methods, is what's causing the inconsistent results (food or capsule) (Deyno, Eneyew et al. 2019). Furthermore, other studies have looked into the impact of cinnamon and found that it could be influenced by the variability of intervention time or the oral antidiabetic drugs used. Furthermore, a number of studies have demonstrated cinnamon's efficacy as a potential therapeutic agent in lowering fasting blood glucose and improving insulin resistance in type 2 diabetes patients (Deyno, Eneyew et al. 2019). Although cinnamon has the ability to lower fasting blood glucose levels, its supplementation appears to change the hemoglobin A1c (HbA1c) level. This finding could be attributed to the short-term effects of cinnamon administration observed in various studies (Deyno, Eneyew et al. 2019). There are some differences between human studies and animal models because animal models showed significantly beneficial effects on lipid profiles. A greater number of studies, particularly well-controlled clinical trials with control groups for comparison, should be conducted. Cinnamon supplementation may improve the lipid profile of diabetic patients with dyslipidemia. Cinnamon appears to lower serum triglyceride (TG), total cholesterol (TC), and low-density lipoprotein (LDL) levels (Jamali, Kazemi et al. 2020). Due to the fact that dyslipidemia is one of the main risks for cardiovascular disease in diabetics, these properties of cinnamon are assumed to play a critical role in diabetic patients (Stein, Ferrari et al. 2019). Research conducted by Mathew and Abraham in 2006 suggests that Cinnamon extract exhibits *in vitro* anti-oxidant properties and demonstrates a considerable capacity for protecting against oxidant damage. However, few studies in *in vivo* models have been conducted to assess the total serum anti-oxidant status. According to the findings of this study, cinnamon has anti-inflammatory and antioxidant properties that are both beneficial and protective. Proanthocyanins, a polyphenol compound, could be the bioactive compound responsible for these properties (Yokozawa, Cho et al. 2012). This compound is found in aqueous cinnamon extract and has been shown to help prevent the formation of advanced glycation-end products (AGE) (Peng, Cheng et al. 2008), which originate by reactive oxygen species during the hyperglycemic status (Davari, Hashemi et al. 2020). Type 2 diabetes mellitus patients are treated mainly with antihyperglycemic drugs (Artasensi, Pedretti et al. 2020), combined with daily exercise, weight loss, and diet modification to obtain optimal metabolic outcomes (Chawla, Madhu et al. 2020). They not only lower blood glucose and lipid levels, but they also delay or prevent long-term complications associated with T2DM. As a result, as part of a dietary plan, cinnamon may be used as an adjunct to drug therapy for T2DM prevention and/or management (Chawla, Madhu et al. 2020). Moreover, it is important to emphasize that cinnamon ingestion should be carefully determined so that it does not exceed the recommended doses per day (L Gonçalves, Fernandes et al. 2019). In the scientific literature, various mechanisms of action for cinnamon and its bioactive compounds' beneficial properties have been proposed. Cinnamaldehyde, procyanidin type-A polymers, cinnamic acid, and coumarin are the main compounds found in cinnamon. Cinnamon, which is high in polyphenols, has been shown to be a potential source of natural antioxidants with strong free radical scavenging properties *in vitro*, which may help protect against oxidative stress (Rachid, Moncada et al. 2022). The results of *in vitro* and *in vivo* studies suggest that cinnamon may provide beneficial antihyperglycemic effects via insulin-mimetic action through signaling pathway inhibition (Zare, Nadjarzadeh et al. 2019). Cinnamon seems to promote the GLUT4 translocation and glucose uptake in insulin-dependent tissue (Shen, Fukushima et al. 2010). Furthermore, cinnamon can downregulate phosphoenolpyruvate carboxykinase (PEPCK) in the liver (Shen, Fukushima et al. 2010). Another possible mechanism of action for cinnamon is a decrease in intestinal enzyme activity, which affects glucose absorption and thus postprandial blood glucose concentration (Rekha, Balaji et al. 2010). Furthermore, cinnamon appears to be effective as an antihyperlipidemic agent. The proposed mechanism of action for this effect is related to the regulation of lipid metabolism in enterocytes. Cinnamon bioactive compounds can also inhibit cholesterol and fatty acid absorption in gut cells by inactivating the Niemann-Pick c1-like 1 and Cd36 mRNA receptors, respectively. Finally, cinnamon can inhibit chylomicron synthesis by lowering microsomal triglyceride transfer protein (MTTP) levels and apo B48 secretion from enterocytes (Qin, Dawson et al. 2012). Cinnamon's other main strategy is the inhibition of -amylase and -glucosidase enzymes, which are involved in the hydrolysis of carbohydrates in the digestive tract. Benzoic acid, (E)-cinnamaldehyde, trans-cinnamic acid, eugenol, and o-methoxy-cinnamaldehyde were discovered in a water extract of *Cinnamomum zeylanicum* prepared using high pressure and the decoction method. This method's bioactive compounds have a high potential for inhibiting α-glucosidase and controlling hyperglycemia (Wariyapperuma, Kannangara et al. 2020). Cinnamaldehyde is the most abundant compound in *Cinnamomum zeylanicum* bark oil. It appears to be more effective than metformin, which is commonly used in traditional medicine, in lowering plasma glucose levels. Cinnamon oil's bioactive compounds boost the expression of proteins involved in glucose transport, insulin signaling, and dyslipidemia regulation. Tissue damage can also result from disruptions in the balance of free radicals and oxidative stress. High levels of oxidative stress increase insulin resistance. This results in impaired glucose tolerance and T2DM. Azimi, Ghiasvand et al. (2014) investigated the antidiabetic effect of cinnamon and discovered significant results on total cholesterol, LDL, and high-density lipoprotein (HDL) levels, proving cinnamon's antidiabetic effect. Shokri, Fathi et al. (2015) demonstrated that methanol extracts of cinnamon combined with green tea had an antidiabetic effect on 50 STZ-induced diabetic Wistar rats. After six weeks of treatment, the rats' glucose levels and body weight were significantly lower. Both extracts have a synergistic effect on diabetes control. Joshi, Jain et al. (2017) studied the hypolipidemic and antiatherosclerotic effects of methanolic extract of cinnamon bark on male rabbits. Cinnamon administration significantly reduced high cholesterol levels and restored normal arteries with no signs of abnormality, according to biochemical and histological studies. Sangi and Elwahab (2017) investigated the antihyperglycemic, antihyperlipedemic, and renoprotective action of *C. verum* in combination with ginger on 40 adult albino male rats and discovered that *C. verum* was more effective than *Z. officinale* against the studied unusual albino health issues. Singh, Rao et al. (2021) studied the antidiabetic effect of cinnamon methanol extract and observed inhibition of α-amylase and α-glucosidase activity. Gaber, Mourad et al. (2012) tested the antidiabetic potential of cinnamon aqueous extract on alloxan-induced diabetic rats. For 30 days, the rats were given different doses of cinnamon aqueous extract (200, 400, 600, and 1200 mg/kg). After 30 days of cinnamon administration, the rats showed antidiabetic symptoms such as lower FBS, HDL cholesterol, LDL cholesterol, triglycerides, and serum total cholesterol levels. The rat supplemented with the lowest dose of cinnamon extract, 200 mg/kg, had the greatest antidiabetic effects. Beji, Khemir et al. (2018) investigated the antidiabetic effect of cinnamon powder in diabetic rats induced by alloxan. It was observed that cinnamon powder significantly reduced the FBS, serum levels of total cholesterol, and triglycerides and improved the hyperglycemia and lipid profile of diabetic rats. Beejmohun, Peytavy-Izard et al. (2014) investigated the antidiabetic effect of hydro-alcoholic extract of Ceylon cinnamon on post-meal glycemia. The rats were administrated with 1 g of Ceylon cinnamon hydro-alcoholic extract after a standardized meal. The rats were analyzed via *in vitro* enzymatic assay and *in vivo* starch tolerance tests. *In vitro* and *in vivo* studies suppressed α-amylase activity and acutely reduced glycemic response to starch, respectively. They also revealed that cinnamon hydro-alcoholic extract is a more potent antidiabetic agent than its aqueous extract. Cinnamon was found to have antidiabetic properties in diabetic mice induced with STZ and fed a high sugar diet (Li, Liang et al. 2013). For 14 days, diabetic mice were fed cinnamon. Diabetic mice were later tested using glucose oxidase (GOD) and radioimmunoassay (RIA). These tests revealed significantly lower levels of blood glucose and insulin. Enzyme-linked immunosorbent assay (ELISA) revealed a significantly lower level of oxidative stress markers. The researchers hypothesized that cinnamon polyphenols aid in the repair of pancreatic beta cells, which in turn has hypoglycemic and hypolipidemic effects. Couturier, Qin et al. (2011) investigated the effects of cinnamon on glycogen synthesis, associated gene expression, and protein levels in insulin-resistant male Wister rats and discovered that including cinnamon in a high-fat, high-fructose diet increased glycogen synthesis. According to Tulini, Souza et al. (2016), solid lipid microparticles (SLM) of proanthocyanidin-rich cinnamon extract boost the antioxidant and antidiabetic potential of the food. All of the aforementioned studies clearly show that cinnamon has an antidiabetic effect. Among the various extracts, methanol (precisely more) and aqueous extract were found to be extremely beneficial for their antidiabetic potential. They primarily work to reduce oxidative stress, TC, LDL, HDL, FBS, and triglyceride levels, as well as to help restore normal insulin levels, which results in normal glucose levels. Santos and da Silva (2018) examined the role of cinnamon in diabetes management. They also discovered that a dose of 1–6 g/d cinnamon powder can be used safely to treat T2D mellitus and hyperglycemia. They suggested that more clinical studies are needed to establish cinnamon's antidiabetic activity. As a result, using cinnamon as a functional food ingredient can boost antioxidant and antidiabetic properties. Cinnamon in daily meals should be studied further in people with type 2 diabetes to gain a better understanding of its long-term effects. Its ability to repair pancreatic beta cell mass, insulin synthesis and secretion, as well as its mimetic effects on L and K cells in the gut to activate the production and release of glucagon-like peptide-1 (GLP-1) are just a few examples.

27.11 ANTIOXIDANT PROPERTIES AND BENEFICIAL EFFECTS

Antioxidants are known as substances or compounds that delay/stop the oxidation by ceasing the damage caused by free radicals. Because of their ability to oxidize free radicals, most reactions occur in a single or multistep process. Antioxidants can also react by transferring a single electron, transferring a hydrogen atom, or chelating transitional metals. Furthermore, antioxidants in biological systems can be found in both enzymatic and nonenzymatic forms in both extracellular and intracellular environments (Nimse and Pal 2015). To minimize oxidative stress, it is crucial to maintain a balance between free radicals and antioxidant defense mechanisms. (Arisha, Sakr et al. 2020). Oxidative stress, which is induced by free radicals, is associated with many chronic diseases such as cancer, osteoporosis, diabetes, and coronary heart disease (Nimse and Pal 2015). Reactive oxygen species (ROS) induce oxidative stress and are responsible for the cumulative damage imparted on DNA, lipids, proteins, and other molecules, subsequently resulting in even permanent damage (Arisha, Sakr et al. 2020). Several spices, fruits, and vegetables have already been identified as being rich in antioxidant compounds such as polyphenols, vitamins, flavonoids, and carotenoids (Borzoei, Rafraf et al. 2018). Moreover, antioxidant-rich foodstuff are good sources to combat and prevent the incidence of many chronic diseases associated with oxidative stress (Rao and Gan 2014). *C. zeylanicum* is rich in phenolic compounds. These compounds and their activities are defined by their structure (reactive benzene rings), which is directly linked with quenching radicals in biological systems (Wilson, Pierre et al. 2020). Cinnamaldehyde, eugenol, carvacrol, cinnamic acetate, and thymol are the main phenolic compounds that can be found in essential oils of *C. zeylanicum* (Weerasekera, Samarasinghe et al. 2021). Characterization of phenolic compounds in *C. zeylanicum* revealed that it can improve hyperlipidemia, possibly by lowering cholesterol production and suppressing lipid peroxidation (Abdelgadir, Hassan et al. 2020). Among the parts used in the *C. zeylanicum* tree for various medicinal purposes, the bark demonstrated the highest antioxidant activity compared to the leaves and flowers (Abeysekera, Premakumara et al. 2013). However, essential oils appear to have the greatest antioxidant activity compared to leaves, bark, and extracts from other parts of the plant (Castro, Pante et al. 2020). Peroxynitrite (ONOO−) is a compound that can react with almost any type of biomolecule due to the formation of NO2• and OH• radicals during degradation. These radicals have the potential to cause oxidative damage to blood vessels, skin, the heart, lungs, kidneys, and the brain. Eugenol, a component of cinnamon's active oils, was found to be effective in preventing peroxynitrite-induced damage *in vitro*. However, the concentration of eugenol in active oil extracts varies depending on the cinnamon variety used, with *C. zeylanicum* activity being the highest. As a result, cinnamon oil extracts with a high eugenol content can be classified as a spice to inhibit the activity of radicals NO2• and OH• from a pure peroxynitrite inhibitory standpoint (Chericoni, Prieto et al. 2005). Besides, many studies have been conducted to assess the antioxidant properties of *C. zeylanicum* with extractions from different parts of the tree, under both *in vitro* and *in vivo* conditions (Abeysekera, Premakumara et al. 2013). Multiple studies have exposed the total antioxidant capacity and its beneficial results, such as a decrease in blood lipid peroxide levels through the improvement of hepatic antioxidant enzyme activities (Borzoei, Rafraf et al. 2018) and lowered risks of male infertility and inflammatory diseases (Arisha, Sakr et al. 2020). A study done with Swiss albino mice by using cinnamon 0.25% and cardamom 0.5%, orally administered at doses of 100 ml/mouse/day, observed that azoxymethane-induced colon carcinogenesis could be significantly controlled by inhibiting lipid peroxidation and enhancing glutathione-S-transferase (GST) activity in liver and colon (Bhattacharjee, Rana et al. 2007). As well as their health benefits, these antioxidants are used as primary additives or preservatives in the food industry to prevent or delay the spoilage of fat- and oil-rich foods (Rao and Gan 2014) and to enhance flavor (Ranjbar, Ghasmeinezhad et al. 2006). Nowadays, food industries are increasingly concerned about producing food that is less toxic, has fewer health risks, and contains fewer synthetic compounds. Therefore, plant-derived antioxidants, especially those coming from *C. zeylanicum*, have commanded the attention of manufacturers and consumers (Wilson, Pierre et al. 2020). Studies have also revealed that when *C. zeylanicum* is used as an antioxidant in food, it enhances antioxidant enzymes and removes ROS while decreasing malondialdehyde, which is naturally present in oxidative stress situations (Arisha, Sakr et al. 2020). Moreover, *C. zeylanicum* is used in the pharmaceutical industry as a nutraceutical. It is also used in the essence industries due to its fragrance to produce foods, perfumes, and drugs (Abeysekera, Premakumara et al. 2013). In terms of the bioactive antioxidant compounds present in *C. zeylanicum*, cinnamaldehyde and eugenol act as the main bioactive antioxidant compound because of their active functional groups in the structures (Adarsh, Chettiyar et al. 2020). Tauopathy neurodegeneration is a group of diseases characterized by neurofibrillary tangling. When the microtubular protein tau is hyperphosphorylated, it dissociates from the microtubules and forms insoluble aggregates. These tau neurofibrillary tangles are thought to be one of the central pathologies of Alzheimer's disease. Cinnamon extract was found to effectively inhibit human tau aggregation *in vitro*. A proanthocyanidin trimer and cinnamaldehyde were found to be responsible for the activity. According to the same study, while cinnamon extract inhibited tau aggregation, not all polyphenols in the extract were active in the inhibitory process. As a result, the inhibitory activity cannot be attributed to the extract's overall antioxidant properties. However, the studies were conducted *in vitro*, raising concerns about compound bioavailability. Regardless, this study has set the stage for further testing of cinnamon extract in clinical trials (Peterson, George et al. 2009). In a study with Wistar albino rats, cinnamon extract also demonstrated significant gastroprotective effects. An orally administered indomethacin solution was used to induce gastric lesions. A cinnamon suspension was given 30 minutes before the oral indomethacin, and the animals were sacrificed six hours later. Across a variety of models, the results revealed a significant decrease in basal gastric acid secretion and ulcer protective effects (Saleh 2012).

27.12 CONTRAINDICATIONS

While cinnamon has a wide range of medicinal properties, it can also have negative health effects if consumed on a regular basis. Cinnamon inhalation or dry consumption can trigger a hypersensitive airway and irritate mucous membranes in the lungs due to its cellulose fiber composition, which does not dissolve or biodegrade in the lungs (Grant-Alfieri, Schaechter et al. 2013). Due to the apoptotic effect of the cinnamon component cinnamaldehyde on B- and T-cells, the consumption of cinnamon is contraindicated in patients under an immunotherapy treatment (Roth-Walter, Moskovskich et al. 2014). The consumption of cinnamon supplements should be avoided during pregnancy since cinnamon can lead to uterine contractions, miscarriage, or premature labor (Ernst 2002). Importantly, studies conducted both *in vitro* and *in vivo* suggest that the toxic compound coumarin, found abundantly in *C. cassia* and less in Ceylon cinnamon (~250 times less), is a potential carcinogen to individuals with mutations of the cytochrome P450 2A6 (Hsieh, Sun et al. 2019).

27.13 CONCLUSION AND FUTURE PERSPECTIVE

C. verum is a popular medicinal herb that is used for a variety of pharmacological purposes. This herb is used in nearly every pharmaceutical system on the planet. *C. verum* has antidiabetic, antibacterial, antioxidant, anti-inflammatory, and anticancer properties. Each of these characteristics is essential for the advancement of human health. The main components of the *C. verum* plant are eugenol, cinnamaldehyde, cinnamyl acetate, copane, and camphor. Cinnamon, in addition to its antibacterial properties, shows great promise in treating a wide range of disorders when combined with extracts from medicinal plants. However, only a few combinations of phytochemicals from medicinal plants and nanoparticles have been studied, and the mode of action of various bioactive compounds is poorly understood. To identify less toxic and more effective drugs that will benefit the global population, much attention must be paid to evaluating and exploring molecular characterization of potential bioactive compounds found in plants.

REFERENCES

Abdelgadir, A., H. Hassan, A. Eltaher, G. Khnsaa Mohammed, A. Lamya Mohammed and T. Hago (2020). "Hypolipidemic effect of cinnamon (*Cinnamomum zeylanicum*) bark ethanolic extract on triton X-100 induced hyperlipidemia in albino rats." *Medicinal and Aromatic Plants (Los Angeles)* **9**(351): 2167–0412.2120.

Abeysekera, W., G. Premakumara and W. Ratnasooriya (2013). "In vitro antioxidant properties of leaf and bark extracts of ceylon cinnamon (*Cinnamomum zeylanicum* Blume)." *Tropical Agricultural Research* **24**(2): 128–138.

Adarsh, A., B. Chettiyar, B. Kanthesh and N. Raghu (2020). "Phytochemical screening and antimicrobial activity of 'Cinnamon zeylanicum'." *International Journal of Pharmaceutical Research and Innovation* **13**: 22–33.

Ahmed, W. M., N. M. Abdel-Azeem, M. A. Ibrahim, N. A. Helmy and A. M. Radi (2021). "Neuromodulatory effect of cinnamon oil on behavioural disturbance, CYP1A1, iNOStranscripts and neurochemical alterations induced by deltamethrin in rat brain." *Ecotoxicology and Environmental Safety* **209**: 111820.

Alizadeh Behbahani, B., F. Falah, F. Lavi Arab, M. Vasiee and F. Tabatabaee Yazdi (2020). "Chemical composition and antioxidant, antimicrobial, and antiproliferative activities of *Cinnamomum zeylanicum* bark essential oil." *Evidence-Based Complementary and Alternative Medicine* **2020**.

Alsoodeeri, F. N., H. M. Alqabbani and N. M. Aldossari (2020). "Effects of cinnamon (*Cinnamomum cassia*) consumption on serum lipid profiles in albino rats." *Journal of Lipids* **2020**.

Ananthakrishnan, R., P. Chandra, B. Kumar and K. Rameshkumar (2018). "Quantification of coumarin and related phenolics in cinnamon samples from south India using UHPLC-ESI-QqQLIT-MS/MS method." *International Journal of Food Properties* **21**(1): 50–57.

Aparna, Y., L. Narayanan and J. Sarada (2014). "Quorum quenching ability of dietary spice Cinnamomum verum on pathogenic bacteria." *International Journal of Pharmaceutical Sciences and Research* **5**(12): 5216.

Arisha, S. M., S. A. Sakr and F. R. Abd-Elhaseeb (2020). "*Cinnamomum zeylanicum* alleviate testicular damage induced by high fat diet in albino rats; histological and ultrastructural studies." *Heliyon* **6**(11): e05584.

Artasensi, A., A. Pedretti, G. Vistoli and L. Fumagalli (2020). "Type 2 diabetes mellitus: A review of multi-target drugs." *Molecules* **25**(8): 1987.

Asimi, O. A., N. Sahu and A. Pal (2013). "Antioxidant activity and antimicrobial property of some Indian spices." *International Journal of Scientific and Research Publications* **3**(3): 1–8.

Avula, B., T. J. Smillie, Y.-H. Wang, J. Zweigenbaum and I. A. Khan (2015). "Authentication of true cinnamon (*Cinnamon verum*) utilising direct analysis in real time (DART)-QToF-MS." *Food Additives & Contaminants: Part A* **32**(1): 1–8.

Azimi, P., R. Ghiasvand, A. Feizi, M. Hariri and B. Abbasi (2014). "Effects of cinnamon, cardamom, saffron, and ginger consumption on markers of glycemic control, lipid profile, oxidative stress, and inflammation in type 2 diabetes patients." *The Review of Diabetic Studies: RDS* **11**(3): 258.

Balasubramanian, S., K. Singh, A. M. Mohite and T. J. Zachariah (2012). "Physical properties of cinnamon bark." *Journal of Spices and Aromatic Crops* **21**: 161–163.

Beejmohun, V., M. Peytavy-Izard, C. Mignon, D. Muscente-Paque, X. Deplanque, C. Ripoll and N. Chapal (2014). "Acute effect of Ceylon cinnamon extract on postprandial glycemia: Alpha-amylase inhibition, starch tolerance test in rats, and randomized crossover clinical trial in healthy volunteers." *BMC Complementary and Alternative Medicine* **14**(1): 1–11.

Beji, R. S., S. Khemir, W. A. Wannes, K. Ayari and R. Ksouri (2018). "Antidiabetic, antihyperlipidemic and antioxidant influences of the spice cinnamon (*Cinnamomum zeylanicumon*) in experimental rats." *Brazilian Journal of Pharmaceutical Sciences* **54**.

Bhattacharjee, S., T. Rana and A. Sengupta (2007). "Inhibition of lipid peroxidation and enhancement of GST activity by cardamom and cinnamon during chemically induced colon carcinogenesis in Swiss albino mice." *Asian Pacific Journal of Cancer Prevention* **8**(4): 578–582.

Borzoei, A., M. Rafraf, S. Niromanesh, L. Farzadi, F. Narimani and F. Doostan (2018). "Effects of cinnamon supplementation on antioxidant status and serum lipids in women with polycystic ovary syndrome." *Journal of Traditional and Complementary Medicine* **8**(1): 128–133.

Braudel, F. (1984). *Perspective of the World; Civilization and Capitalism, 15th–18th Century*, Volume III. Harper & Row Publishers, New York.

Cardoso-Ugarte, G. A., A. López-Malo and M. E. Sosa-Morales (2016). "Cinnamon (*Cinnamomum zeylanicum*) essential oils. Essential oils in food preservation, flavor and safety." *Elsevier*: 339–347.

Castro, J. C., G. C. Pante, B. M. Centenaro, R. T. R. D. Almeida, E. J. Pilau, B. P. Dias Filho, S. A. G. Mossini, B. A. D. Abreu Filho, G. Matioli and M. Machinski Junior (2020). "Antifungal and antimycotoxigenic effects of Zingiber officinale, *Cinnamomum zeylanicum* and Cymbopogon martinii essential oils against Fusarium verticillioides." *Food Additives & Contaminants: Part A* **37**(9): 1531–1541.

Chakraborty, A., V. Sankaran, M. Ramar and D. R. Chellappan (2015). "Chemical analysis of leaf essential oil of Cinnamomum verum from Palni hills, Tamil Nadu." *Tic* **3**(e10): x10.

Chawla, R., S. Madhu, B. Makkar, S. Ghosh, B. Saboo and S. Kalra (2020). "RSSDI-ESI clinical practice recommendations for the management of type 2 diabetes mellitus 2020." *Indian Journal of Endocrinology and Metabolism* **24**(1): 1.

Chen, G., F. Sun, S. Wang, W. Wang, J. Dong and F. Gao (2021). "Enhanced extraction of essential oil from *Cinnamomum cassia* bark by ultrasound assisted hydrodistillation." *Chinese Journal of Chemical Engineering* **36**: 38–46.

Chericoni, S., J. M. Prieto, P. Iacopini, P. Cioni and I. Morelli (2005). "In vitro activity of the essential oil of *Cinnamomum zeylanicum* and eugenol in peroxynitrite-induced oxidative processes." *Journal of Agricultural and Food Chemistry* **53**(12): 4762–4765.

Corn, C. (1998). *The Scents of Eden: A Narrative of the Spice Trade*. Kodansha America Inc., p. 352. ISBN: 9781568362021.

Couturier, K., B. Qin, C. Batandier, M. Awada, I. Hininger-Favier, F. Canini, X. Leverve, A. M. Roussel and R. A. Anderson (2011). "Cinnamon increases liver glycogen in an animal model of insulin resistance." *Metabolism* **60**(11): 1590–1597.

Davari, M., R. Hashemi, P. Mirmiran, M. Hedayati, S. Sahranavard, S. Bahreini, R. Tavakoly and B. Talaei (2020). "Effects of cinnamon supplementation on expression of systemic inflammation factors, NF-kB and Sirtuin-1 (SIRT1) in type 2 diabetes: A randomized, double blind, and controlled clinical trial." *Nutrition Journal* **19**(1): 1–8.

Deyno, S., K. Eneyew, S. Seyfe, N. Tuyiringire, E. L. Peter, R. A. Muluye, C. U. Tolo and P. E. Ogwang (2019). "Efficacy and safety of cinnamon in type 2 diabetes mellitus and pre-diabetes patients: A meta-analysis and meta-regression." *Diabetes Research and Clinical Practice* **156**: 107815.

Doyle, A. A. and J. C. Stephens (2019). "A review of cinnamaldehyde and its derivatives as antibacterial agents." *Fitoterapia* **139**: 104405.

El-Hack, A., E. Mohamed, M. Alagawany, A.-M. E. Abdel-Moneim, N. G. Mohammed, A. F. Khafaga, M. Bin-Jumah, S. I. Othman, A. A. Allam and S. S. Elnesr (2020). "Cinnamon (*Cinnamomum zeylanicum*) oil as a potential alternative to antibiotics in poultry." *Antibiotics* **9**(5): 210.

Ernst, E. (2002). "Herbal medicinal products during pregnancy: Are they safe?" *BJOG: An International Journal of Obstetrics & Gynaecology* **109**(3): 227–235.

Farahpour, M. and M. Habibi (2012). "Evaluation of the wound healing activity of an ethanolic extract of Ceylon cinnamon in mice." *Veterinary Medicine* **57**(1): 53–57.

Farias, A. P. P., O. D. S. Monteiro, J. K. R. da Silva, P. L. B. Figueiredo, A. A. C. Rodrigues, I. N. Monteiro and J. G. S. Maia (2020). "Chemical composition and biological activities of two chemotype-oils from Cinnamomum verum J. Presl growing in North Brazil." *Journal of Food Science and Technology* **57**(9): 3176–3183.

Gaber, E. E.-D., A. A.-S. Mourad and S. A.-N. Khalid (2012). "Antidiabetic and hypolipidemic effects of Ceylon cinnamon (Cinnamomum verum) in alloxan-diabetic rats." *Journal of Medicinal Plants Research* **6**(9): 1685–1691.

Gharibzahedi, S. M. T. (2018). "The preparation, stability, functionality and food enrichment ability of cinnamon oil-loaded nanoemulsion-based delivery systems: A review." *International Journal on Nutraceuticals, Functional Foods and Novel Foods* **17**: 97–105.

Gonçalves, L., T. Fernandes, M. A. Bernardo and J. A Brito (2019). "Assessment of human health risk of toxic elements due to cinnamon ingestion in the diet." *Biological Trace Element Research* **189**(2): 313–324.

Gopinath, S., G. Vidya and T. Narasimha Murthy (2014). "Isolation, characterization and purification of bioactive compounds from andrographispaniculata and cinnamon verum for anti-hiv activity." *International Journal of Innovative Research in Science, Engineering and Technology* **3**(6): 13777–13783.

Gotmare, B. and E. Tambe (2019). "Identification of chemical constituents of cinnamon bark oil by GCMS and comparative study garnered from five different countries." *Global Journal of Science Frontier Research: C Biological Science* **19**(1).

Goyal, M., H. Kaur, M. Bhandari, A. A. Rizvanov, S. F. Khaiboullina and M. Baranwal (2018). "Antioxidant and immune effects of water soluble polysaccharides isolated from Cinnamomum verum bark." *BioNanoScience* **8**(3): 935–940.

Grant-Alfieri, A., J. Schaechter and S. E. Lipshultz (2013). "Ingesting and aspirating dry cinnamon by children and adolescents: The 'cinnamon challenge'." *Pediatrics* **131**(5): 833.

Gruenwald, J., J. Freder and N. Armbruester (2010). "Cinnamon and health." *Critical Reviews in Food Science and Nutrition* **50**(9): 822–834.

Gulcin, I., R. Kaya, A. C. Goren, H. Akincioglu, M. Topal, Z. Bingol, K. Cetin Çakmak, S. B. Ozturk Sarikaya, L. Durmaz and S. Alwasel (2019). "Anticholinergic, antidiabetic and antioxidant activities of cinnamon (Cinnamomum verum) bark extracts: Polyphenol contents analysis by LC-MS/MS." *International Journal of Food Properties* **22**(1): 1511–1526.

Hajimonfarednejad, M., M. Ostovar, M. J. Raee, M. H. Hashempur, J. G. Mayer and M. Heydari (2019). "Cinnamon: A systematic review of adverse events." *Clinical Nutrition* **38**(2): 594–602.

Hamidpour, R., M. Hamidpour, S. Hamidpour and M. Shahlari (2015). "Cinnamon from the selection of traditional applications to its novel effects on the inhibition of angiogenesis in cancer cells and prevention of Alzheimer's disease, and a series of functions such as antioxidant, anticholesterol, antidiabetes, antibacterial, antifungal, nematicidal, acaracidal, and repellent activities." *Journal of Traditional and Complementary Medicine* **5**(2): 66–70.

Hsieh, C. J., M. Sun, G. Osborne, K. Ricker, F. C. Tsai, K. Li, R. Tomar, J. Phuong, R. Schmitz and M. S. Sandy (2019). "Cancer hazard identification integrating human variability: The case of coumarin." *International Journal of Toxicology* **38**(6): 501–552.

Isaac-Renton, M., M. K. Li and L. M. Parsons (2015). "Cinnamon spice and everything not nice: Many features of intraoral allergy to cinnamic aldehyde." *Dermatitis* **26**(3): 116–121.

Jamali, N., A. Kazemi, J. Saffari-Chaleshtori, M. Samare-Najaf, V. Mohammadi and C. C. Clark (2020). "The effect of cinnamon supplementation on lipid profiles in patients with type 2 diabetes: A systematic review and meta-analysis of clinical trials." *Complementary Therapies in Medicine* **55**: 102571.

Joshi, S. C., P. K. Jain and P. Sharma (2017). "Antiatherosclerotic and lipid-lowering effects of Cinnamomum verum in cholesterol-fed rabbits." *International Journal of Current Pharmaceutical Research* **9**(1): 75–80.

Khasanah, L., P. Prasetyawan, R. Utami, W. Atmaka, G. Manuhara and A. Sanjaya (2017). Optimization and characterization of cinnamon leaves (Cinnamomum burmannii) oleoresin. *IOP Conference Series: Materials Science and Engineering*, IOP Publishing.

Klejdus, B. and J. Kováčik (2016). "Quantification of phenols in cinnamon: A special focus on 'total phenols' and phenolic acids including DESI-Orbitrap MS detection." *Industrial Crops and Products* **83**: 774–780.

Kosari, F., M. Taheri, A. Moradi, R. Hakimi Alni and M. Y. Alikhani (2020). "Evaluation of cinnamon extract effects on CLBB gene expression and biofilm formation in Escherichia coli strains isolated from colon cancer patients." *BMC Cancer* **20**(1): 1–8.

Li, J., E. Guan, L. Chen, X. Zhang, L. Yin, L. Dong, Q. Pan, X. Fu and L. Zhang (2018). "Optimization for extraction of an oil recipe consisting of white pepper, long pepper, cinnamon, saffron, and myrrh by supercritical carbon dioxide and the protective effects against oxygen-glucose deprivation in PC12 cells." *Revista Brasileira de Farmacognosia* **28**: 312–319.

Li, R., T. Liang, L. Xu, Y. Li, S. Zhang and X. Duan (2013). "Protective effect of cinnamon polyphenols against STZ-diabetic mice fed high-sugar, high-fat diet and its underlying mechanism." *Food and Chemical Toxicology* **51**: 419–425.

Liang, Y., Y. Li, A. Sun and X. Liu (2019). "Chemical compound identification and antibacterial activity evaluation of cinnamon extracts obtained by subcritical n-butane and ethanol extraction." *Food Science & Nutrition* **7**(6): 2186–2193.

Lucas, K., J. Fröhlich-Nowoisky, N. Oppitz and M. Ackermann (2021). "Cinnamon and hop extracts as potential immunomodulators for severe COVID-19 cases." *Frontiers in Plant Science* **12**: 589783.

Masghati, S. and S. M. Ghoreishi (2018). "Supercritical CO2 extraction of cinnamaldehyde and eugenol from cinnamon bark: Optimization of operating conditions via response surface methodology." *The Journal of Supercritical Fluids* **140**: 62–71.

Mathew, S. and T. E. Abraham (2006). "Studies on the antioxidant activities of cinnamon (Cinnamomum verum) bark extracts, through various in vitro models." *Food Chemistry* **94**(4): 520–528.

Mohammadi, M., S. Mirabzadeh, R. Shahvalizadeh and H. Hamishehkar (2020). "Development of novel active packaging films based on whey protein isolate incorporated with chitosan nanofiber and nano-formulated cinnamon oil." *International Journal of Biological Macromolecules* **149**: 11–20.

Momtaz, S., S. Hassani, F. Khan, M. Ziaee and M. Abdollahi (2018). "Cinnamon, a promising prospect towards Alzheimer's disease." *Pharmacological Research* **130**: 241–258.

Muhammad, D. R. A. and K. Dewettinck (2017). "Cinnamon and its derivatives as potential ingredient in functional food – A review." *International Journal of Food Properties* **20**(sup2): 2237–2263.

Muhammad, D. R. A., V. Lemarcq, E. Alderweireldt, P. Vanoverberghe, D. Praseptiangga, J. G. Juvinal and K. Dewettinck (2020). "Antioxidant activity and quality attributes of white chocolate incorporated with Cinnamomum burmannii Blume essential oil." *Journal of Food Science and Technology* **57**(5): 1731–1739.

Muhammad, D. R. A., E. Tuenter, G. D. Patria, K. Foubert, L. Pieters and K. Dewettinck (2021). "Phytochemical composition and antioxidant activity of Cinnamomum burmannii Blume extracts and their potential application in white chocolate." *Food Chemistry* **340**: 127983.

Nimse, S. B. and D. Pal (2015). "Free radicals, natural antioxidants, and their reaction mechanisms." *RSC Advances* **5**(35): 27986–28006.

Othman, Z. S., M. Y. Maskat and N. H. Hassan (2020). "Optimization of cinnamaldehyde extraction and antioxidant activity of Ceylon cinnamon extract." *Sains Malaysiana* **49**(5): 995–1002.

Pahan, S. and K. Pahan (2020). "Can cinnamon spice down autoimmune diseases?" *Journal of Clinical & Experimental Immunology* **5**(6): 252.

Parham, S., A. Z. Kharazi, H. R. Bakhsheshi-Rad, H. Nur, A. F. Ismail, S. Sharif, S. RamaKrishna and F. Berto (2020). "Antioxidant, antimicrobial and antiviral properties of herbal materials." *Antioxidants* **9**(12): 1309.

Peng, X., K.-W. Cheng, J. Ma, B. Chen, C.-T. Ho, C. Lo, F. Chen and M. Wang (2008). "Cinnamon bark proanthocyanidins as reactive carbonyl scavengers to prevent the formation of advanced glycation endproducts." *Journal of Agricultural and Food Chemistry* **56**(6): 1907–1911.

Peterson, D. W., R. C. George, F. Scaramozzino, N. E. LaPointe, R. A. Anderson, D. J. Graves and J. Lew (2009). "Cinnamon extract inhibits tau aggregation associated with Alzheimer's disease in vitro." *Journal of Alzheimer's Disease* **17**(3): 585–597.

Pittman, S. (2011). "Cinnamon: It's not just for making cinnamon rolls." *Ethnobotanical Leaflets* **2000**(2): 11.

Pramote, K., S. Nucha, S. Suched, P. Parinda, S. Prasong and A. Shuji (2012). "Subcritical water extraction of flavoring and phenolic compounds from cinnamon bark (*Cinnamomum zeylanicum*)." *Journal of Oleo Science* **61**(6): 349–355.

Qin, B., H. D. Dawson, N. W. Schoene, M. M. Polansky and R. A. Anderson (2012). "Cinnamon polyphenols regulate multiple metabolic pathways involved in insulin signaling and intestinal lipoprotein metabolism of small intestinal enterocytes." *Nutrition* **28**(11–12): 1172–1179.

Rachid, A. P., M. Moncada, M. F. d. Mesquita, J. Brito, M. A. Bernardo and M. L. Silva (2022). "Effect of aqueous cinnamon extract on the postprandial glycemia levels in patients with type 2 diabetes mellitus: A randomized controlled trial." *Nutrients* **14**(8): 1576.

Ranjbar, A., S. Ghasmeinezhad, H. Zamani, A. A. Malekirad, A. Baiaty, A. Mohammadirad and M. Abdollahi (2006). "Antioxidative stress potential of *Cinnamomum zeylanicum* in humans: A comparative cross-sectional clinical study." *Clinical Practice* **3**(1): 113.

Rao, P. V. and S. H. Gan (2014). "Cinnamon: A multifaceted medicinal plant." *Evidence-Based Complementary and Alternative Medicine* **2014**.

Rekha, N., R. Balaji and M. Deecaraman (2010). "Antihyperglycemic and antihyperlipidemic effects of extracts of the pulp of Syzygium cumini and bark of Cinnamon zeylanicum in streptozotocin-induced diabetic rats." *Journal of Applied Bioscience* **28**: 1718–1730.

Ribeiro-Santos, R., M. Andrade, D. Madella, A. P. Martinazzo, L. d. A. G. Moura, N. R. de Melo and A. Sanches-Silva (2017). "Revisiting an ancient spice with medicinal purposes: Cinnamon." *Trends in Food Science & Technology* **62**: 154–169.

Roth-Walter, F., A. Moskovskich, C. Gomez-Casado, A. Diaz-Perales, K. Oida, J. Singer, T. Kinaciyan, H. C. Fuchs and E. Jensen-Jarolim (2014). "Immune suppressive effect of cinnamaldehyde due to inhibition of proliferation and induction of apoptosis in immune cells: Implications in cancer." *PLoS ONE* **9**(10): e108402.

Saleh, A. (2012). "Anti-secretagogue and antiulcer effects of Cinnamon *Cinnamomum zeylanicum* in rats." *Journal of Pharmacognosy and Phytotherapy* **4**(4): 53–61.

Salman, A. S., T. M. Al-Shaikh, Z. K. Hamza, A. A. El-Nekeety, S. S. Bawazir, N. S. Hassan and M. A. Abdel-Wahhab (2021). "Matlodextrin-cinnamon essential oil nanoformulation as a potent protective against titanium nanoparticles-induced oxidative stress, genotoxicity, and reproductive disturbances in male mice." *Environmental Science and Pollution Research* **28**(29): 39035–39051.

Sangal, A. (2011). "Role of cinnamon as beneficial antidiabetic food adjunct: A review." *Advances in Applied Science Research* **2**(4): 440–450.

Sangi, S. and M. Elwahab (2017). "Experimental evaluations of the nephroprotective properties of ginger (Zingiber officinale), Cinnamomum verum and Nigella sativa in STZ induced diabetic rats." *International Journal of Biology, Pharmacy and Allied Sciences* **6**(6): 1195–1209.

Santos, H. O. and G. A. da Silva (2018). "To what extent does cinnamon administration improve the glycemic and lipid profiles?" *Clinical Nutrition ESPEN* **27**: 1–9.

Schmidt, E., L. Jirovetz, G. Buchbauer, G. A. Eller, I. Stoilova, A. Krastanov, A. Stoyanova and M. Geissler (2006). "Composition and antioxidant activities of the essential oil of cinnamon (*Cinnamomum zeylanicum* Blume) leaves from Sri Lanka." *Journal of Essential Oil Bearing Plants* **9**(2): 170–182.

Shen, Y., M. Fukushima, Y. Ito, E. Muraki, T. Hosono, T. Seki and T. Ariga (2010). "Verification of the antidiabetic effects of cinnamon (*Cinnamomum zeylanicum*) using insulin-uncontrolled type 1 diabetic rats and cultured adipocytes." *Bioscience, Biotechnology, and Biochemistry* **74**(12): 2418–2425.

Shokri, G., H. Fathi, M. Jafari Sabet, N. Nasri Nasrabadi and R. Ataee (2015). "Evaluation of anti-diabetic effects of hydroalcoholic extract of green tea and cinnamon on streptozotocin-induced diabetic rats." *Pharmaceutical and Biomedical Research* **1**(2): 20–29.

Silva, F. L., R. V. Silva, P. C. Branco, L. V. Costa-Lotufo, C. Murakami, M. C. Young, D. A. Azevedo and P. R. Moreno (2020). "Chemical composition of the Brazilian native Cinnamomum stenophyllum (Meisn.) Vattimo-Gil essential oil by GC-qMS and GC× GC-TOFMS, and its cytotoxic activity." *Arabian Journal of Chemistry* **13**(4): 4926–4935.

Singh, N., A. S. Rao, A. Nandal, S. Kumar, S. S. Yadav, S. A. Ganaie and B. Narasimhan (2021). "Phytochemical and pharmacological review of Cinnamomum verum J. Presl-a versatile spice used in food and nutrition." *Food Chemistry* **338**: 127773.

Sivaranjani, R., T. J. Zachariah and N. Leela (2021). "Phytotherapeutic potential of bi-herbal extract of cinnamon and turmeric: In vivo antidiabetic studies." *Clinical Phytoscience* **7**(1): 1–9.

Stein, R., F. Ferrari and F. Scolari (2019). "Genetics, dyslipidemia, and cardiovascular disease: New insights." *Current Cardiology Reports* **21**(8): 1–12.

Tang, P. L., Y. T. Chen, J. Qin, X. Hou and J. Deng (2020). "Effect of cinnamon bark and twig extracts on the chemical, physicochemical and antioxidant properties of fermented milk." *Journal of Food Measurement and Characterization* **14**(4): 2271–2281.

Tulini, F. L., V. B. Souza, M. A. Echalar-Barrientos, M. Thomazini, E. M. Pallone and C. S. Favaro-Trindade (2016). "Development of solid lipid microparticles loaded with a proanthocyanidin-rich cinnamon extract (*Cinnamomum zeylanicum*): Potential for increasing antioxidant content in functional foods for diabetic population." *Food Research International* **85**: 10–18.

Wariyapperuma, W. N. M., S. Kannangara, Y. S. Wijayasinghe, S. Subramanium and B. Jayawardena (2020). "In vitro anti-diabetic effects and phytochemical profiling of novel varieties of *Cinnamomum zeylanicum* (L.) extracts." *PeerJ* **8**: e10070.

Weerasekera, A. C., K. Samarasinghe, H. K. S. de Zoysa, T. C. Bamunuarachchige and V. Y. Waisundara (2021). *Cinnamomum zeylanicum*: Morphology, antioxidant properties and bioactive compounds. *Antioxidants-Benefits, Sources, Mechanisms of Action*, IntechOpen.

Wilson, A. A., M. D. J. Pierre, N. T. Leopold, P. Priya and P. Nisha (2020). "Stabilisation potentials of the essential oils of Thymus vulgaris L., *Cinnamomum zeylanicum* B. and Mentha piperita L. on palm olein at accelerated storage." *African Journal of Biotechnology* **19**(7): 464–477.

Wong, Y., M. Ahmad-Mudzaqqir and W. Wan-Nurdiyana (2014). "Extraction of essential oil from cinnamon (*Cinnamomum zeylanicum*)." *Oriental Journal of Chemistry* **30**(1): 37.

Yokozawa, T., E. J. Cho, C. H. Park and J. H. Kim (2012). "Protective effect of proanthocyanidin against diabetic oxidative stress." *Evidence-Based Complementary and Alternative Medicine* **2012**.

Zare, R., A. Nadjarzadeh, M. M. Zarshenas, M. Shams and M. Heydari (2019). "Efficacy of cinnamon in patients with type II diabetes mellitus: A randomized controlled clinical trial." *Clinical Nutrition* **38**(2): 549–556.

28 Ethnobotany, Phytochemistry, and Pharmacology of the Indian Banyan (*Ficus benghalensis* L.); Specific Focus on Antidiabetic Properties

Sujata Mandal, Tuyelee Das, Abhijoy Paul, Protha Biswas, Abdel Rahman Al-Tawaha, and Abhijit Dey

CONTENTS

28.1 INTRODUCTION

The banyan tree, also known as *Ficus bengalensis*, is a member of the order Urticales and the family Moraceae. Over 800 different *Ficus* species exist, the majority of which are indigenous to the old-world tropics (Rahman and Khanom 2013). *Ficus* plants are indigenous to Bangladesh, India, and Sri Lanka. In India, *F. benghalensis* is revered as the National Tree, symbolising spiritual knowledge and eternal life (Kala 2011). In addition to being one of the most diverse species, *Ficus* trees produce the highest amount of oxygen and also have the highest rate of photosynthesis. Several types of leaf minerals have been found in deeply lobed leaves, including calcium oxalates, amorphous calcium carbonate cystoliths, and silica phytoliths.

Natural remedies are used in Ayurveda and traditional Chinese medicine to treat diabetes since they work well in this regard. Elevated fasting blood sugar levels, a sign of diabetes mellitus, can be caused by reduced insulin sensitivity, insulin resistance, inadequate insulin, or any combination

DOI: 10.1201/b23347-28

of the three. *F. benghalensis* is used as a traditional folk remedy to treat "Madhumey" because it includes a variety of bioactives, such as triterpenes, flavonoids, phytosterols, and polyphenols (Prashanth 2017). The Ayurvedic pharmacopoeia of India lists it as "Nyagrodha" for treating Prameha, or metabolic diseases. Additionally, it has been claimed that the bark of *F. benghalensis* lowers high blood sugar levels in test animals (Gayathri and Kannabiran 2008). *F. benghalensis* will be highlighted in this review for its antidiabetic activity and bioactive compounds, as well as its traditional uses. Thus, the review aims to update the latest scientific findings of *F. benghalensis* and indicate any gaps in the research that may be addressed in future studies.

28.2 BOTANICAL DESCRIPTION

28.2.1 Geographical Distribution

The genus *Ficus* contains more than 800 species including trees, shrubs, creepers and epiphytes. It is mainly distributed in different parts of tropical and subtropical zones of Asia, Africa, America, and Australia. *F. benghalensis* is native to India, Nepal, East Himalaya (up to 170 m altitudes), Assam, Pakistan, Andaman Islands, Sri Lanka, Maldives, among other places (POWO, Tiwari et al. 2014).

28.2.2 Morphology

F. benghalensis belongs to the family Moraceae (the mulberry family). More than 800 species and 2000 varieties are included in the *Ficus* genus. It is generally an evergreen tree, but only in some dry areas the plant becomes leafless for a small period of time. It is popularly called Indian fig or Nyagrodha. After successful generation of the trunk, root branches are gradually converted into shoots and thus the tree attains a vigorous growth. This has made the tree very useful on roadsides and other places for shade. At young stage only, the plant is epiphytic. Upon maturation, the height of the tree becomes approximately 30 m. Leaf is stipulate (stout), ovate to orbicular in shape, apex obtuse, base rounded, petiolate, cuticularised, and pubescent and the margin of the lamina is entire. Leaf whorl is spiral. Anatomical study of the leaf reveals the presence of calcium carbonate or cystolith. Female flowers are sessile, ovoid globose ovary and the style is either erect or curved. Male and female flowers are born in different stalks. The Indian banyan tree generally lives more than a thousand years.

28.3 PHYTOCHEMICAL CONSTITUENTS

The importance of *F. benghalensis* from a biological point of view is remarkable. HPLC-MS analysis of *F. benghalensis* revealed the presence of compounds such as leucopelargonidin-3–0-α-L rhamnoside, leucocynidin 3–0-α-D galactosylcellobioside, β-glucoside, glucoside, 20-tetratriaconthene-2-one, 6-heptatriacontene-10-one, pentatriacontan-5-one, β-sitosterol-α–D glucose, meso-inositol, and tetratriaconthene-2-one from the bark of the plant. The compounds found in the latex are tirucalloleuphorbinol, α-amyrin acetate, β-sitosterol, palmitic acid, and bergapten (Ashraf et al. 2021). Compounds like 1,3,4,5-tetrahydroxy cyclohexane carboxylic acid, N, n-dimethyl-1-non decanamine, hexadecanoic acid, n-hexadecanoic acid, 14,17-octadecadienoic acid, 2-hexadecen-1-ol (diterpene), 14-methyl-8-hexadecynl-ol, 2-hydroxy-1-(hydroxymethyl) ethyl ester, 1H-benzimidazol-1 ylmethanol, (Z,Z)-6,9-cis-3,4-epoxy-nonadecadiene, alpha-selinene (sesquiterpenoids), cholestane-3,16-diol, Gorgost-5-en-3-ol present in the leaves of *F. benghalensis* were recognised using GC-MS. Compared to the other parts of the plant, *F. benghalensis* fruits possess the highest number of phytochemicals, such as 7-tetradecenal, hexadecanoic acid, octadecanoic acid, 4H-pyran-4-one 2,3-dihydro3,5-dihydroxy-6-methyl-9,12-octadecadienoic acid, methyl ester, 2-methyl-Z,Z-3,13-octadecadienol, 9,12,15-octadecatrienoic acid, methyl ester, hexadecanoic acid, 2-hydroxy-1-(hydromethyl) ethyl ester, formic acid, 2-propenyl ester, 1,2,3-propanetriol, 5-hydroxymethylfurfural, 1,2–15,16-diepoxyhexadecane, 2,4-dihydroxy-2,5-dimethyl-3(2H)-furan-3-one, azulene, 9-octadecenoic acid (Z), gamma-tocopherol, pentadecanoic acid, 14-methyl-methyl ester, tetrahydrocyclopenta [1,3] dioxin-4-one, 1,2-benzenediol, 2-methoxy-4-vinylphenol, alpha-tocopherol, 1,2-benzenedicarboxylic acid, dibutyl ester, and 9-hexadecenoic acid, found using GC-MS analysis (Radhakrishnan and Venkatachalam 2020; Murugesu et al. 2021). Coumarin compounds such as psoralen were found in the seed of *F. benghalensis*. Phytochemical analysis of *F. benghalensis* root via GC-MS reveals the presence of different compound classes such as sesterterpenoids, diterpenoids, triterpenoids, ketones, tetracyclic cyclitol, fatty acid, and phytosterols. These chemical classes include cycloartenol, taraxerone, euphorbol, ingenol, quinic acid, palmitic acid, eicosadienoic acid, methyl ester, sitosterol, cycloartanyl acetate, amyrin acetate, and lupenyl acetate (Figure 28.1). Table 28.1 compiles a list of phytochemicals reported from the plant.

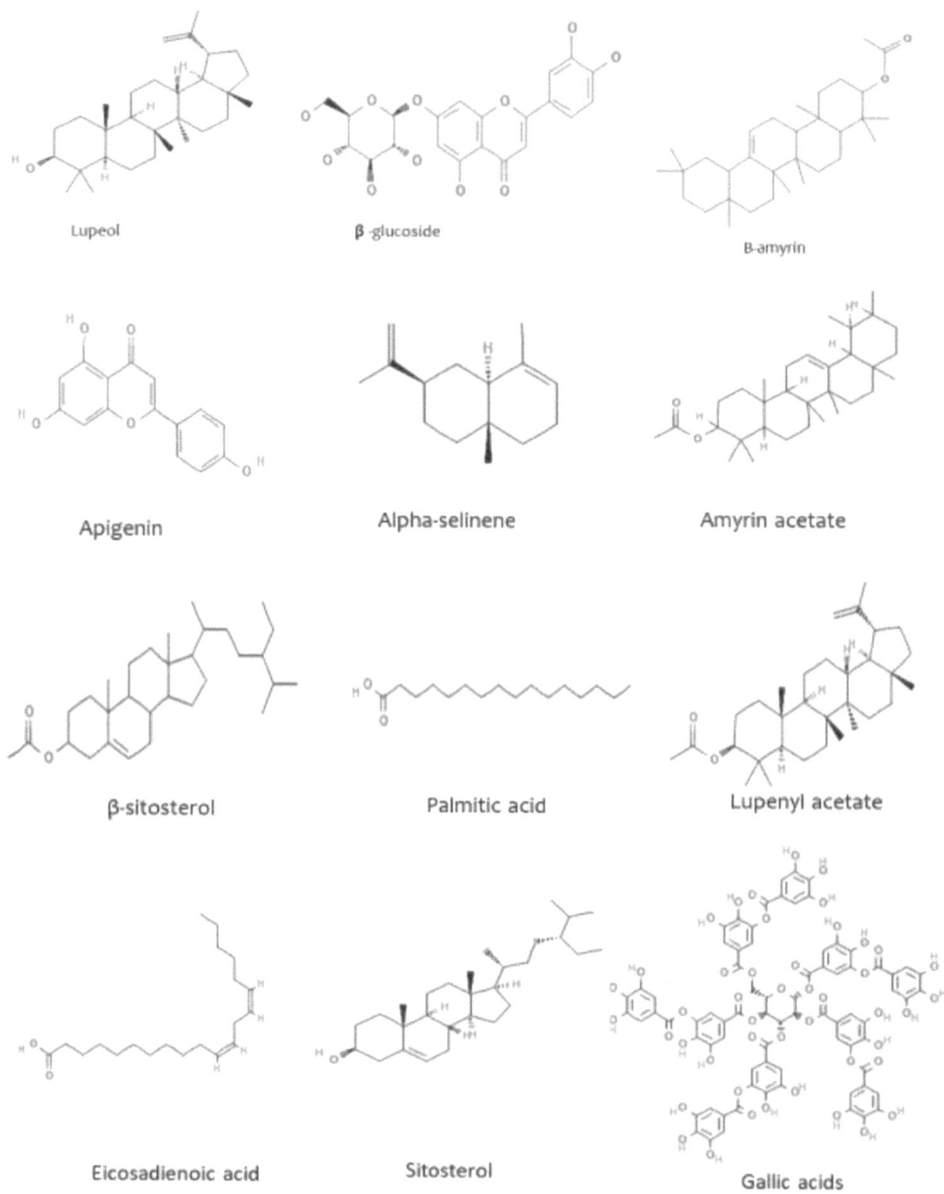

FIGURE 28.1 Chemical structures of phytochemicals obtained from *Ficus benghalensis*.

TABLE 28.1
List of Phytochemicals Found in Plant Parts

Plant parts	Phytochemicals	Compound classes	Reference
Bark	leucopelargonidin-3–0-α-L rhamnoside	Flavonoids	Joseph et al. (2010), Ashraf et al. (2021)
	Lupeol	Terpenoids	
	β-glucoside		
	glucoside		
	20-tetratriaconthene-2-one		
	6-heptatriacontene-10-one	Alkanes	
	pentatriacontan-5-one		
	B-sitosterol-α–D glucose	Phytosterols	
	meso-inositol	Miscellaneous	
	leucocynidin 3–0-α-D galactosylcellobioside		
Leaves	Gallic acids	Phenolics	Karthikeyan et al. (2019); Mudhafar et al. (2021)
	Apigenin		
	Hexadecanoic acid	Palmitic acid	
	n-Hexadecanoic acid		
	14,17-Octadecadienoic acid	Miscellaneous	Ahmad et al. (2011)
	2-Hexadecen-1-ol	Diterpene	
	14-Methyl-8-hexadecyn1-ol	Alcohol	
	1H-benzimidazol-1 ylmethanol	Organic Compound	
	Alpha-selinene	Sesquiterpenoids	Bhaskara Rao et al. (2014)
	B-amyrin	Triterpene	
	Lupeol		
	Cholestane-3,16-diol		
	Friedelin		
Latex	tirucalloleuphorbinol	Terpenoids	Ashraf et al. (2021)
	α -amyrin acetate		
	β-sitosterol		
	palmitic acid		
	Bergapten	Furanocoumarin	
Fruits	7-tetradecenal	Fatty aldehyde	Gopukumar et al. (2016)
	hexadecanoic acid	Terpenoids	
	octadecanoic acid		
	4H-Pyran-4-one 2,3-dihydro3,5-dihydroxy-6-methyl-	5-decenedioic acid	
	9,12-octadecadienoic acid, methyl ester	Linoleic acid	
	2-methyl-Z,Z-3,13-octadecadienol		
	9,12,15-octadecatrienoic acid, methyl ester		
	Hexadecanoic acid, 2-hydroxy-1-(hydromethyl)ethyl ester	Dodecanoate ester	
	Formic acid, 2-propenyl ester	Ester	Radhakrishnan and Venkatachalam (2020)
	1,2,3-propanetriol	Glycerol	
	5-hydroxymethylfurfural	Organic compound	
	2,4-Dihydroxy-2,5-dimethyl-3(2H)-furan-3-one	Ester	
	1,2-Benzenediol	Organic compound	
	Azulene	Mancudecarbobicyclic parent	
	9-octadecenoic acid (Z)	Oleic acid	
	Gamma-tocopherol	Fat soluble	
	Pentadecanoic acid, 14-methyl-methyl ester	Ester	
	2-methoxy-4-vinylphenol	Phenols	
	Alpha-tocopherol	Fat Soluble	

Plant parts	Phytochemicals	Compound classes	Reference
	hexadecanoic acid	Miscellaneous	
	undecanoic acid		
	1,2-Benzenedicarboxylic acid, dibutyl ester	Terpenoids	
	9-hexadecenoic acid		
	Hexadecanoic acid, ethyl ester		
	8-pentadecanone	Ketones	
Seeds	Psoralen	Coumarins	Salem et al. (2013)
Roots	Cycloartenol	Triterpenoid	Verma et al. (2015); Murti et al. (2012)
	Taraxerone	Sesterterpenoid	
	Euphorbol	Ketones	
	Ingenol	Tetracyclic Diterpenoid	
	Quinic acid	Cyclitol	
	Palmitic acid	Fatty acid	
	Eicosadienoic acid, Methyl ester		
	Sitosterol	Phytosterols	
	Cycloartanyl acetate	Triterpenoid	
	Amyrin acetate		
	Lupenyl acetate		

28.4 TRADITIONAL AND OTHER POTENTIAL USES

In accordance with Hindu mythology, *F. benghalensis* is referred to as "the wish-fulfilling tree" that symbolises eternal life. It is extensively used in traditional medicines. The milky sap of the banyan tree is externally used in the treatment of pains and bruises. This whitish sap is also utilised as a remedy for toothache. Charaka, a great contributor of ancient Indian Ayurveda, directed the watery extract of banyan leaf buds properly mixed with honey and sugar to check diarrhoea, milk prepared with leaf buds or aerial roots in bleeding piles and haemorrhages. He also prescribed aerial root and leaf decoction with honey to prevent thirst, vomiting, and burning sensation during fever (Wagner et al. 1999). The bark is utilised against burning sensation, dysentery, diarrhoea, haemoptysis, diabetes, skin diseases, hyperpiesia, and so on and the leaves are useful in leprosy, burning sensations, skin allergy, and abscesses. The latex is useful in neuralgia, gonorrhoea, rheumatism, cracks of sole, and inflammations.

Apart from traditional uses in medicine, *F. benghalensis* is utilised to produce shellac, a crucial constituent in French polish. Shellac is also used as an ingredient of hair lacquer. Because of its hard and water-durable nature, banyan wood is used for householding purposes. Woods from aerial roots are often utilised in cart yokes and poles. Banyan bark fibre is considered as a remarkable raw material for paper and rope making. Fruits are edible in both fresh and dried condition. Even young shoots and leaves are taken as famine food.

Recently interest has built up on using *F. benghalensis* in nanotechnology. Utilisation of plant extracts to synthesise nanoparticles is actively carried out in current times. Tripathi et al. (2018) reported the presence of sulfur nanoparticles in leaf extracts of *F. benghalensis*. Sulphur nanoparticles are widely used in fungicides and pesticides, cancer therapy, catalytic applications like battery, and more. Roots of banyan trees are also employed as a reducing agent in the process of green synthesis.

28.5 PHARMACOLOGICAL STUDIES

Pharmacological properties of *F. benghalensis* are tremendously important and their applications in modern medicines are increasing day by day (Figure 28.2). Constituents extracted from each part,

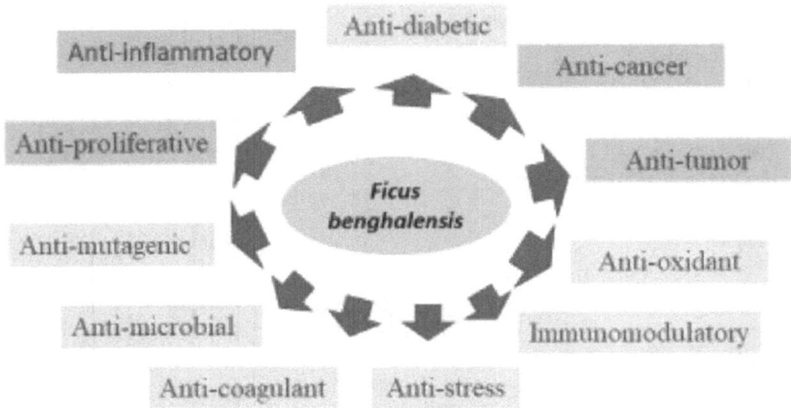

FIGURE 28.2 Pharmacological activities of *Ficus benghalensis*.

i.e. root, stem, bark, leaves, fruits, and latex of plant, possess healing and therapeutic potentials. Pharmacological activity of the plant includes anti-inflammatory, antidiabetic, anticancer, antiproliferative, antitumour, antimutagenic, antioxidants, anti-helminthic, antimicrobial, hepatoprotective, immunomodulatory activities, anticoagulant, wound healing, and antistress. The plant is also extensively used in toxicity studies and as insect repellents. Fresh leaf decoction of *F. benghalensis* is generally utilised in Ayurvedic practices to strengthened the immune system against a group of diseases (Chopra et al. 1958). Table 28.2 compiles the list of pharmacological properties of *F. benghalensis*.

28.5.1 Hypoglycaemic Properties

Hypoglycaemic properties of *F. benghalensis* have been studied extensively by the scientific community. The presence of antidiabetic constituents, for example leucocyanids, attracts the interest of researchers. Their antidiabetic effects have been explored by utilising *in vitro* assays and several animal models.

28.5.2 Antidiabetic Responses

Management of diabetes implies the regulation of enzymes that hydrolyse glucose to minimise postprandial glucose content. An *in vitro* study had been performed on carbohydrate hydrolysing enzyme inhibition by using powdered extract of *F. benghalensis* bark (Ahmed et al. 2011; Khanal and Patil 2020). A positive result was found along with IC_{50} value of 77 μg/mL and 141 μg/mL against α-glucosidase and also sucrose hydrolysing enzymes (Ahmed et al. 2011). Reduction in the level of blood glucose using bark extracts was reported in another study. Diabetic rats with streptozotocin (STZ) induction were allowed oral administration of bark extracts for the experiment. Those rats were also stimulated for insulin secretion from beta (β) cells (Kasireddy et al. 2021). There was more evidence of *F. benghalensis* bark assisting the uptake of glucose involving *in vitro* and *in silico* prospective (Madiwalar et al. 2022). Meanwhile, hypoglycaemic activity of ethanolic leaf extracts from *F. benghalensis* was observed in an *in vivo* study. Dosages were applied as 200 mg and 400 mg per body weight of diabetic albino rats with alloxan induction. A significant decrease in glucose level, cholesterol, and triglycerides was reported (Saraswathi et al. 2013). An *in silico* study demonstrated that the probable α-glucosidase inhibitors marked in *F. benghalensis* can prevent the activity of aldose reductase. Aldose reductase, as having a

TABLE 28.2
Pharmacological Activities of *F. benghalensis*

Activity	Plant extract and dosage	Root of administration	Model organism	Results	References
Anti-inflammation	Methanol extract of leaves (500 mg/kg)	Oral administration	Gastric ulcer rat model	Reductions in ulcer index	Saha et al. (2010)
	Methanol extract of leaves (250 mg/kg)	Oral administration	Gastric ulcer rat model	Reductions in ulcer index	Saha et al. (2010)
	Ethanol extract of leaves (200 mg/kg)	-	Excision and incision wound rat	Acceleration in wound healing with absence of mortality	Garg et al. (2011)
Anticancer	Ethyl acetate extract of latex	-	Lung cancer (A549), human breast cancer line (MDA-MB-231), HeLa cell line	Anticancer activity on lung cancer (A549), breast cancer (MDA-MB-231), and cervical cancer (Hela) cell lines	Tulasi et al. (2018)
	Ethanol extract of latex	-	Human breast cancer line (MCF-7)	Anticancer activity on colorectal and neuroblastoma cells	Gulecha et al. (2011)
Antiproliferative	Crude butanol extract of stem bark	-	*Allium cepa* root tip assay and yeast cell model respectively	Antimitotic, antiproliferative and antioxidant activities	Raheel et al. (2017)
Antitumour	Butanol extract of stem bark	-	*Allium cepa* root tip assay and yeast cell model respectively	Chromosomal and mitotic aberrations including accumulation of prophases, sticky chromosomes at metaphase, spindle disturbance at prophase and anaphase bridges	Raheel et al. (2017)
Antioxidants	Ethanol extract from seed	-	Assessed against DPPH, nitric oxide, lipid peroxidation, and reducing power assay	Prescribed as a diet for peptic ulcer	Govindan et al. (2015)
Antimicrobial	Ethanol extract from roots	Oral administration	*Staphylococcus aureus*, *Escherichia coli*, *Klebsiella pneumonia*	Used as antibiotics	Murti et al. (2011)
Hepatoprotective	Methanol extract from latex	Oral administration	CCl4-induced hepatotoxicity in albino rats	Pharmacological application against multitude of diseases like cancer, diabetes, liver disorders, and oxidative stress	Chandrasekaran et al. (2012)
Immunomodulatory activities	Butanol fraction extract from leaves	-	*Candida albicans*		Bhanwase et al. (2016)
Anticoagulant	Crude methanol extract from leaves	-	Used healthy human plasma, using activated partial thromboplastin time (APTT) and prothrombin time (PT) methods	To be used as anticoagulant	Ambreen et al. (2021)

role in the glucose metabolism pathway, is a crucial enzyme for operating diabetes mellitus. It had been predicted that this might involve the p53 pathway of cell signalling (Khanal et al. 2019; Khanal and Patil 2019). In fructose-induced insulin-resistant mice, *F. benghalensis* bark hydroalcoholic extract was found to reduce insulin resistance through reducing gluconeogenesis (Khanal and Patil 2022). In another study, hydroalcoholic extracts of *F. benghalensis* bark also decreased gluconeogenesis and enhanced glycolysis, acting as antidiabetic agents. Furthermore, bark extract promotes glucose uptake and insulin secretion via PI3K/Akt signalling and inhibits protein tyrosine phosphatase 1b activity (Khanal et al. 2021). Phytoconstituents showing the best results in molecular docking with aldolase reductase (target of diabetes mellitus type 2) were reported to be isolated from *F. benghalensis* (Singh et al. 2019).

28.6 SAFETY ISSUES

During the procedure of drug development, toxicological screening is an important parameter for new drugs as well as for the extension of existing drugs. According to the FDA (Food and Drug Administration), it has become necessary to evaluate any molecule for its probable therapeutic action and noxious prospective using animal models. Plant extracts may contain various synergistic and/or antagonistic constituents which may cause allergy or other hypersensitive reactions. Therefore, screening of plant extracts is crucial (Murugesu et al. 2021).

Aerial roots of the plant are safe as they do not possess any toxicity – 5000 mg/kg has proven safe in a severe toxicity study (Renata et al. 2019). In order to study the toxicity of leaves, 50% ethanolic extracts were tested in adult albino mice up to 2000 mg/kg, with no lethal effect or major effects observed. Application of *F. benghalensis* fruit extract on albino mice does not reveal any mortality or abnormalities up to 2000 mg/kg (Bhardwaj et al. 2016).

28.7 CULTIVATION PRACTICES

F. benghalensis is a common plant that often grows in roadsides in tropical countries. It normally prefers to grow in well-drained areas. Organic compost with balanced NPK is needed for the development of the tree. Propagation takes place by apical cuttings or internodal cuttings. A humid environment is required with free-draining compost. Latex is exuded from plants after cut. Fresh seed propagation also occurs. For successful fresh seed propagation, the seed needs to be soaked in warm water up to 12 hours. Pollination in *F. benghalensis* takes place by a single wasp species named *Eupristinamasoni*. Cultivation of the banyan tree in "bonsai" preparation has become quite popular nowadays.

28.8 CONCLUSION

The varied pharmacological activities of *Ficus* species have been influenced by the phytochemicals that are contained in them. Sesterterpenoid, diterpenoid, triterpenoid, ketones, tetracyclic cyclitol, fatty acids, and phytosterols are among the substances found in this plant. The species' antidiabetic effect has been verified by a number of studies that have been examined, giving traditional medicine a scientific justification for its usage. Additionally, the plant's components work well as insect repellents. As a result, using the plant can significantly enhance human health. *F. benghalensis* has received more attention from researchers than other *Ficus* species since it has a variety of medicinal uses. To quantify the chemicals' potential bioactivity, additional research on animal models would be beneficial. With the help of this compilation of chemical components and their pharmacological characteristics, it will be possible to identify future research gaps, as well as pharmacological issues that need more study and experimental implications.

REFERENCES

Ahmad, S., H. Rao, M. Akhtar, I. Ahmad & M. Munawar. 2011. Phytochemical composition and pharmacological prospectus of *Ficus bengalensis* Linn. (Moraceae)-A. *Journal of Medicinal Plants Research* 5(28): 6393–6400.

Ahmed, F., S. Chavan, A. Satish & K.R. Punith. 2011. Inhibitory activities of *Ficus benghalensis* bark against carbohydrate hydrolyzing enzymes-An in vitro study. *Pharmacognosy Journal* 3(20): 33–37.

Ambreen, S., M. Tariq, M.S. Masoud, I. Ali, M. Qasim, A. Mushtaq, M. Ahmed & R. Asghar. 2021. Anticoagulant potential and total phenolic content of six species of the genus *Ficus* from Azad Kashmir, Pakistan. *Tropical Journal of Pharmaceutical Research* 18(6): 1245–1251.

Ashraf, K., M.R. Haque, M. Amir, N. Ahmad, W. Ahmad, S. Sultan, S.A.A. Shah, A.M. Alafeefy, M. Mujeeb & M.F.B. Shafie. 2021. An overview of phytochemical and biological activities: *Ficus deltoidea* Jack and other *Ficus* spp. *Journal of Pharmacy and Bioallied Sciences* 13(1): 11.

Bhanwase, A.S. & K.R. Alagawadi. 2016. Antioxidant and immunomodulatory activity of hydroalcoholic extract and its fractions of leaves of *Ficus benghalensis*. Linn. *Pharmacognosy Research* 8(1): 50–55.

Bhardwaj, L.K., K.K. Chandrul & U. Sharma. 2016. Evaluation of anti-arthritic activity of *Ficus benghalensis* Linn. root extracts on freund's adjuvant induced arthritis in rats. *Journal of Phytopharmacology* 5: 10–14.

Bhaskara Rao, K.V., V. Ojha, Preeti, G. Kumar & L. Karthik. 2014. Phytochemical composition and antioxidant activity of Ficus benghalensis (Moraceae) leaf extract. *Journal of Biologically Active Products from Nature* 4(3): 236–248.

Chandrasekaran, C., S. Dethe, D. Mundkinajeddu, M.K. Pandre, J. Balachandran, A. Agarwal, B.D. Hiraganahalli & V.C. Chinampudur. 2012. Hepatoprotective and antioxidant activity of standardized herbal extracts. *Pharmacognosy Magazine* 8(30): 116–123.

Chopra, R.N., I.C. Chopra, K.L. Handa & L.D. Kapur. 1958. *Indigenous drugs of India*, U.N. Dhur and Sons Pvt. Ltd., Calcutta, 673–675.

Garg, V.K. & S.K. Paliwal. 2011. Wound-healing activity of ethanolic and aqueous extracts of *Ficus benghalensis*. *Journal of Advanced Pharmaceutical Technology and Research* 2(2): 110–114.

Gayathri, M. & K. Kannabiran. 2008. Antidiabetic and ameliorative potential of *Ficus bengalensis* bark extract in streptozotocin induced diabetic rats. *Indian Journal of Clinical Biochemistry* 23(4): 394–400.

Gopukumar, S., P. Alexander, M. Jainamboo & P. Praseetha. 2016. Phytochemical screening and FT-IR analysis of *Ficus benghalensis* fruits. *International Journal of Pharmacognosy and Phytochemical Research* 8(9): 1529–1534.

Govindan, V. & G.S. Francis. 2015. Qualitative and quantitative determination of secondary metabolites and antioxidant potential of *Ficus benghalensis* linn seed. *International Journal of Pharmacy and Pharmaceutical Sciences* 7: 118–124.

Gulecha, V. & T. Sivakuma. 2011. Anticancer activity of *Tephrosia purpurea* and *Ficus religiosa* using MCF 7 cell lines. *Asian Pacific Journal of Tropical Medicine* 4: 526–529.

Joseph, B. & S.J. Raj. 2010. Phytopharmacological and phytochemical properties of three *Ficus* species-an overview. *International Journal of Pharma and Bio Sciences* 1(4): 246–253.

Kala, C.P. 2011. Traditional ecological knowledge, sacred groves and conservation of biodiversity in the Pachmarhi Biosphere Reserve of India. *Journal of Environmental Protection* 2(7): 967.

Karthikeyan, M., S. Parthiban & S. Ramalingam. 2019. Phytochemical analysis in economically important *Ficus benghalensis* L. and *Ficus krishnae* C. DC. using GC-MS. *International Journal of Pharma and Bio Sciences* 10(4): 5–12.

Kasireddy, G.R., K.S. Kumar, L.R. Nadithe & P. Chinnam. 2021. Experimental evaluation of hypoglycemic effect of bark extract of *Ficus benghalensis* in streptozotocin-induced diabetic rats. *National Journal of Physiology, Pharmacy and Pharmacology* 11: 320–325.

Khanal, P. & B.M. Patil. 2019. Gene set enrichment analysis of alpha-glucosidase inhibitors from *Ficus benghalensis*. *Asian Pacific Journal of Tropical Biomedicine* 9(6): 263.

Khanal, P. & B.M. Patil. 2020. Integration of *in silico*, *in vitro* and *ex vivo* pharmacology to decode the antidiabetic action of *Ficus benghalensis* L. bark. *Journal of Diabetes and Metabolic Disorders* 19(2): 1325–1337.

Khanal, P. & B.M. Patil. 2021. Consolidation of network and experimental pharmacology to divulge the antidiabetic action of *Ficus benghalensis* L. bark. *3 Biotech* 11(5): 1–16.

Khanal, P. & B.M. Patil. 2022. Reversal of insulin resistance by *Ficus benghalensis* bark in fructose-induced insulin-resistan S. Venkatachalam. 2020. A holistic approach for microwave assisted solvent extraction of phenolic compounds from *Ficus benghalensis* fruits and its phytochemical profiling. *Journal of Food Process Engineering* 43(11): e13536.

Madiwalar, V.S., P.S. Dwivedi, A. Patil, S. Gaonkar, V.J. Kumbhar, P. Khanal & B.M. Patil. 2022. *Ficus benghalensis* promotes the glucose uptake-Evidence with *in silico* and *in vitro*. *Journal of Diabetes and Metabolic Disorders* 21: 429–438.

Mudhafar, M., H. Alsailawi, M. Abdulrasool, R.K.M. Jawad & A. Mays. 2021. Mini-review of phytochemicals of ten *Ficus* species. *International Journal of Chemistry Research* 5(4): 7–18.

Murti, K. & U. Kumar. 2011. Antimicrobial activity of *Ficus benghalensis* and *Ficus racemosa* roots L. *American Journal of Microbiological Research* 2: 21–24.

Murti, K., U. Kumar, M. Kaushik & A. Kaushik. 2012. Preliminary phytochemical investigation of roots of *Ficus benghalensis*. *Asian Journal of Traditional Medicines*, 7(1): 44–49.

Murugesu, S., J. Selamat & V. Perumal. 2021. Phytochemistry, pharmacological properties, and recent applications of *Ficus benghalensis* and *Ficus religiosa*. *Plants* (Basel) 10(12): 2749.

Prashanth, B.K. 2017. Banyan tree: *Ficus benghalensis*: Uses, research, renedies, side effects. *Easyayurveda*.

Radhakrishnan, A.J. & S. Venkatachalam. 2020. A holistic approach for microwave assisted solvent extraction of phenolic compounds from *Ficus benghalensis* fruits and its phytochemical profiling. *Journal of Food Process Engineering* 43(11): e13536.

Raheel, R., Z. Saddiqe, M. Iram & S. Afzal. 2017. *In vitro* antimitotic, antiproliferative and antioxidant activity of stem bark extracts of *Ficus benghalensis* L. *South African Journal of Botany* 111: 248–257.

Rahman, A.H.M.M. & A. Khanom. 2013. Taxonomic and ethno-medicinal study of species from Moraceae (Mulberry) family in Bangladesh flora. *Research in Plant Sciences* 1(3): 53–57.

Renata-Maria, V. 2019. Determination of acute and subacute toxicity of six plant extracts. *Medico Research Chronicles* 6: 64–67.

Saha, S. & G. Goswami. 2010. Study of antiulcer activity of *Ficus religiosa* L. on experimentally induced gastric ulcers in rats. *Asian Pacific Journal of Tropical Medicine* 3(10): 791–793.

Salem, M.Z., A.Z.M. Salem, L.M. Camacho & H.M. Ali. 2013. Antimicrobial activities and phytochemical composition of extracts of *Ficus* species: An over view. *African Journal of Microbiological Research* 7(33): 4207–4219.

Saraswathi, S., R. Senthamarai & S. Sundari. 2013. Antidiabetic activity of leaves extract of Ficus benghalensislinn on alloxan induced diabetic rats. *International Journal of Pharmacology & Biological Sciences* 7(3): 47–51.

Singh, P., V.K. Singh & A.K. Singh. 2019. Molecular docking analysis of candidate compounds derived from medicinal plants with type 2 diabetes mellitus targets. *Bioinformation* 15(3): 179.

Tiwari, A.K., A. Chaturvedi & S.K. Chaturvedi. 2014. Pharmacognostical and Phytochemical Investigation of Ficus benghalensis Linn. Fruit. *Research Journal of Pharmacognosy and Phytochemistry* 6(4): 147.

Tripathi R.M., R.P. Rao & T. Tsuzuki. 2018. Green synthesis of sulfur nanoparticles and evaluation of their catalytic detoxification of hexavalent chromium in water. *RSC Advances* 8(63): 36345–36352.

Tulasi, C.D.S.L.N., N.M. Lakshmi & L. Saida. 2018. Cell viability assay of *Ficus benghalensis* latex solvent extracts on different cell lines. *Asian Journal of Pharmaceutical and Clinical Research* 11: 335–339.

Verma, V.K., N. Sehgal & O. Prakash. 2015. Characterization and screening of bioactive compounds in the extract prepared from aerial roots of *Ficus benghalensis*. *International Journal of Pharmaceutical Sciences and Research* 6(5056): 5056–5069.

Wagner, W.L., D.R. Herbst & S.H. Sohmer. 1999. *Manual of the Flowering Plants of Hawai'i*. 2 vols. Bishop Museum Special Publication 83, University of Hawai'i and Bishop Museum Press, Honolulu, HI.

29 *Terminalia bellerica* (Bahera) *Antidiabetic and Other Potential Benefits*

Jamal Akhtar and Fouzia Bashir

CONTENTS

29.1 INTRODUCTION

Diabetes mellitus (DM), a dangerous, chronic, and complex metabolic disorder with various etiologies, has both immediate and long-term repercussions. Diabetes mellitus, also known as diabetes, affects people in both industrialized and developing countries and has a serious socio-economic impact (Malviya et al. 2010). According to estimates, 25% of people worldwide have this illness (Makheswari and Sudarsanam 2011). Environmental and genetic variables have a major role in the development of diabetes (Wild et al. 2004; Whiting et al. 2011). Insulin and several oral antidiabetic medications such as sulfonylureas, biguanides, and glinides are being

DOI: 10.1201/b23347-29

used to treat diabetes. One of the key areas of research is the search for more powerful and secure hypoglycemic agents because many of them have a number of serious side effects. Herbal medicines can be utilized as an alternative source of antidiabetic medications (Saxena and Vikram 2004; Wild et al. 2004). Recently, certain medicinal plants have been utilized empirically in antidiabetic and antihyperlipidemic therapies and reported to be helpful in treating diabetes globally. The primary mechanisms through which plants have antihyperglycemic activity are their capacity to boost insulin secretion, restrict glucose absorption from the intestine, or facilitate metabolites in insulin-dependent activities. Although there are more than 400 plant species with hypoglycemic activity documented in the literature, finding new antidiabetic medications derived from natural plants is still appealing since they contain compounds that have different and harmless effects on diabetes mellitus. Glycosides, alkaloids, terpenoids, flavonoids, carotenoids, and other compounds found in the majority of plants are usually thought to have antidiabetic effects (Singh 2011).

Terminalia bellirica is commonly known as bahera and is one of the major components of triphala. It has been a part of the Indian System of Medicine (ISM) incorporating Unani, Ayurveda, and Siddha systems of medicine (Zhang et al. 2019). It belongs to the kingdom Plantae; class Mangoliopsida; order Myrtales; and family Combretaceae (Swati et al. 2017). Commonly, it is known as belleric myrobalan in English, bahera, baheda, and bullain in Hindi, bayada and bahura in Bengali, bibhitaki in Sanskrit, and bibhitaki in Ayurveda (Sharma et al. 2021).

29.2 BOTANICAL DESCRIPTION

T. bellerica is a sizable deciduous tree with a rounded crown that is 50 meters in height and 30 meters in circumference (Figure 29.1). The first 20 meters are branchless. Since it demands a chilly temperature, it is perpetual (Kabeeruddin YNMb; Ghani 2008).

The inner bark is yellow in color, as opposed to the outer bark's bluish or ashy gray hue. There are several longitudinal fissures in the bark. The colorof young leaves is copper red, which eventually changes to parrot green and then dark green. Large, alternating, glabrous leaves measure 4–24 cm × 2–11 cm and are primarily grouped at the ends of the twigs. The leaf's base is cunate to rounded, and it has 6–9 pairs of secondary veins. On both surfaces, there are noticeable secondary and tertiary venations, which are typically grouped around the ends of branchlets. A pair of sessile glands, occasionally unclear, is located immediately above the middle of the 2.5–9 cm long, flattened petiole (Kabeeruddin YNMa; Hakeem 2009).

Lamina measures 8–20 × 4–14 cm and can be either broad elliptical or broad obovate. A rounded to sharply short acuminate apex is present. They typically appear with new leaves and are greenish white in color. They smell strongly of honey or have an unpleasant odor. Simple, solitary, and sessile flowers are found. The inflorescence consists of 3 to 15 cm long axillary spikes. The spike's upper flowers are male. Lower blooms have two sexes. Fruit is a pale-yellow hue. It is a drupe that is 2–4 × 1.8–2.2 cm, globose or ovoid, thickly velutinous or sericeous. It is 3 cm broad and slightly 5 ridged. It has one seed and is covered in little, light pubescence (Nadkarni 1976; Kirtikar 1987; Rastogi and Mehrotra 2004; Pullaiah 2006; Bhattacharya 2008; Anonymous 2014).

29.3 DISTRIBUTION

T. bellerica may grow up to 2000 meters above sea level in a wide range of ecologies. It is indigenous to China, Indonesia, Pakistan, Malaysia, Sri Lanka, India, Bangladesh, Bhutan, Bangladesh, Thailand, Cambodia, and Vietnam (Chopra et al. 1956; Brahmachari and Augusti 1961). It is often found in Madhya Pradesh, Uttar Pradesh, Punjab, and Maharashtra in India. It mainly originates alongside teak in monsoon forests, mixed forests, or dry deciduous forests (Anonymous 2007).

Fig 29.1 a. Whole plant of *T. bellerica*

Fig 29.1 b. Leaves of *T. bellerica*

Fig 29.1 c. Fruits of *T. bellerica*

Fig 29.1 d. Flowers of *T. bellerica*

FIGURE 29.1 *Terminalia bellerica*: (a) whole plant; (b) leaves; (c) fruit; (d) flowers.

29.4 PHYTOCHEMICAL CONSTITUENTS

T. bellerica has a wide variety of phytoconstituents including glucoside (bellericanin), gallo-tannic acid, ellagic acid, gallic acid, lignans (termilignan and thannilignan), 7-hydroxy 3', 4'(methylene-dioxy) flavone and anolignanB, tannins, ellagic acid, ethyl gallate, galloyl glucose and chebulagic acid, phyllemblin, β-sitosterol, mannitol, glucose, fructose and rhamnose, coloring matter, resins, and a greenish yellow oil (Deb et al. 2016). The phytoconstituents are as follows in Table 29.1 (Swati et al. 2017) and Figure 29.2.

29.5 PROVEN POTENTIAL HEALTH BENEFITS

29.5.1 Antioxidant Activity

There have been reports of strong antioxidant effects from *T. bellerica*. An imbalance in the production and metabolism of reactive oxygen species (ROS) leads to oxidative stress. It is linked to a variety of pathophysiological problems in people. Antioxidants are chemicals that block or lessen free radical reactions, preventing cellular damage as a result (Young 2001). Supplemental antioxidants aid in minimizing oxidative cell damage. Antioxidant defense mechanisms differ depending on the composition of phytochemicals. The intracellular and extracellular environments contain both enzymatic and non-enzymatic antioxidants (Fahmy et al. 2015).

TABLE 29.1
Phytoconstituents of *Terminalia bellerica*

Compounds	Chemical constituents
Flavone	7-hydroxy 3', 4'(methylenedioxy) flavone, luteoline
Steroids	β-sitosterol
Lignans	Termilignan, thannilignan, anolignanB
Tannins	Gallic acid, ellagic acid, methyl gallate, ethyl gallate (phenyllemblin), chebulagininc acid, chebulagic acid, hexahydroxydiphenic acid ester
Glycosides	Fructose, sucrose, galactose, D-glucose, mannose, rhamnose
Terpenoids	Belleric acid, chebulagic acid, arjungenin
Saponin	Bellericoside and bellericanin
Cardenolide	Cannogenol 3-O-β-galactopyranosyl-(1→4)-Oα-L-rhamopyranoside
Flavonol aglycones	Quercetine and kaempferol
Flavonol Glycosides	Quercetin-3-O-[6"-α-L-rhamnopyranosyl]- (1→6)-β-D-glucopyranoside (rutin), quercetin3-O-α-L-rhamnopyranoside, quercetin-3-O-βD-glucopyranoside and kaempferol-3-O-β-Dglucopyranoside
Fatty acids present in oil	Palmitic acid, linoleic acid, stearic acid, myristic acid, and oleic acid
Glycerides of fatty acids	Palmitooleolinolein, stearo-oleolinolein, palmitodiolein, stearodiolein, dioleolinolein, and triolein

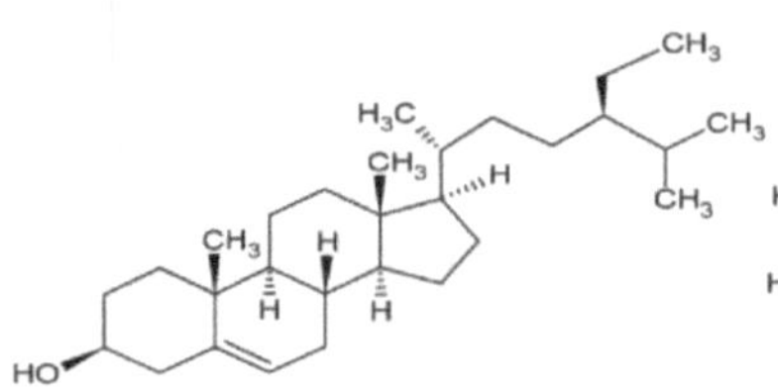

Beta-sitosterol

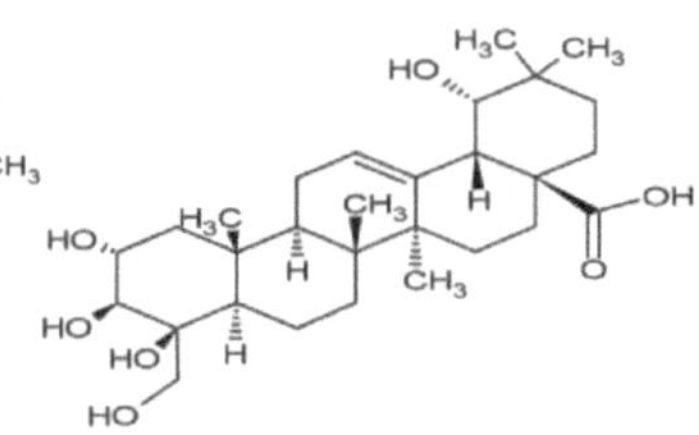

Arjungenin

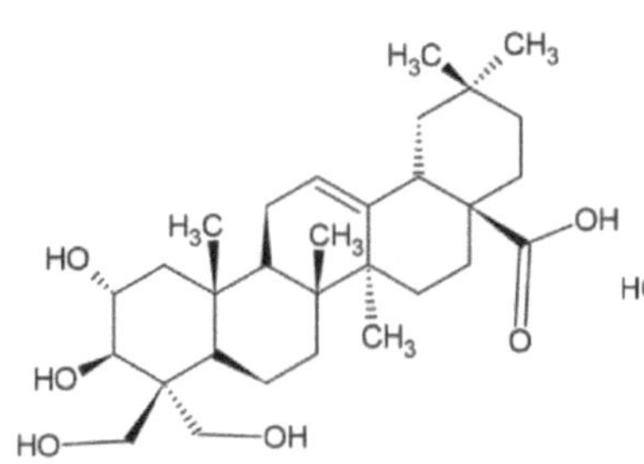

Belleric acid

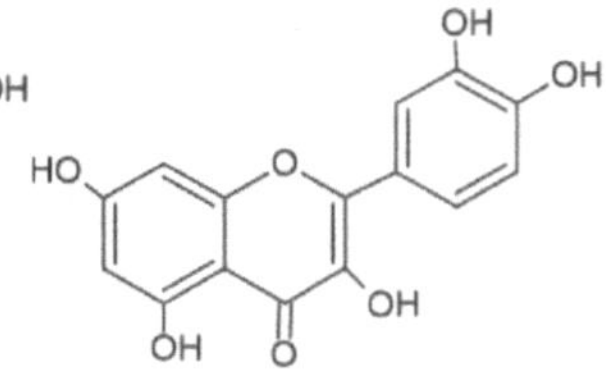

Quercetin

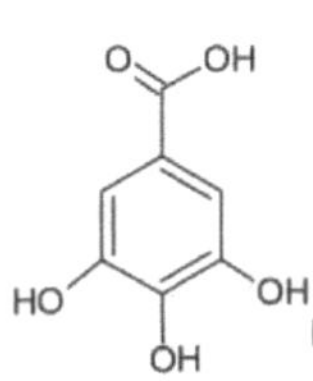

Gallic acid

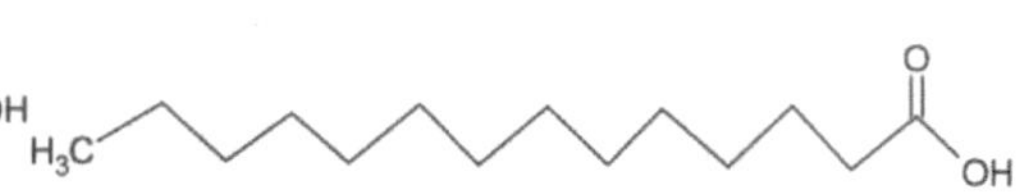

Myristic acid

FIGURE 29.2 Phytochemical constituents of *Terminalia bellerica*.

29.5.2 Anticancer Activity

The research on *T. bellerica*'s anticancer characteristics has proven to be helpful in developing a viable anticancer medication. In an *in vitro* study to determine the anticancer activity of 70% methanolic extract of *T. bellerica* (TBME) against human lung (A549) and breast (MCF-7) cancer cells and its potential mechanism, the extract is significantly cytotoxic to both A549 and MCF-7 cells, but not to non-malignant WI-38 cells. A549 and MCF-7 were subjected to a flow cytometric study using 100 g/mL of TBME as the effective concentration for triggering apoptosis in the cancer cell lines. This level of TBME was found to result in an apoptotic DNA fragmentation pattern. A study was carried out by using western blotting to determine the mechanism of apoptosis induction. They found that both A549 and MCF-7 had an elevated Bax/Bcl2 ratio, which in turn activated the caspase cascade and caused PARP to be cleaved. By altering the Bcl-2 family of proteins, these findings demonstrated that TBME had anticancer properties in both lung and breast cancer cell lines (Nishida et al. 2008).

29.5.3 Antifungal Activity

Five clinical and five environmental isolates of *Cryptococcus neoformans* were tested to see if an ethanolic extract of *Terminalia bellerica* fruit had any antifungal properties. The disc diffusion method was used to assess anticryptococcal activity. Comparing clinical isolates to environmental isolates, it was discovered that clinical isolates were more sensitive. To prevent drug-resistant fungus strains, an ethanol extract of *T. bellerica* fruit showed antifungal action (Valli 2013).

29.5.4 Antispasmodic and Bronchodilator Activity

To ascertain the mode of action for the therapeutic use of *T. bellerica* fruit in hyperactive gastrointestinal and respiratory illnesses, *in vivo* and *in vitro* research was carried out. It demonstrated a synergistic impact of Ca++ antagonist and anticholinergic. It relaxes spontaneous spasms in the jejunum of rabbits. While ethyl acetate, chloroform, and the aqueous fraction all had anticholinergic effects, only the aqueous fraction had calcium channel blocking effects (Gilani et al. 2008).

29.5.5 Antimicrobial Activity

A study was done to show *T. bellerica* methanol extract's antibacterial efficacy against the respiratory infections *Staphylococcus aureus* and *Klebsiella pneumonia*. Initial antimicrobial investigation revealed that it alters the biochemistry of both strains while inhibiting the coagulase activity of *S. aureus*. After 24 and 48 hours of treatment, it significantly alters the capsular shape in *Klebsiella pneumonia*. According to the findings, *T. bellerica* has antimicrobial action against respiratory pathogens and can thus be used to treat disorders brought on by pathogens (Sabnis 2014).

29.5.6 Anti-Ulcer Activity

Pylorus ligation and ethanol-induced ulcer models in Wistar rats were used to assess the anti-ulcer efficacy of the ethanolic extract of *T. bellerica* fruit. In addition to reducing free acidity and ulcer index as compared to control, ethanol extract indicated prevention of the gastric lesion brought on by pylorus ligation-induced ulcer and ethanol-induced gastric ulcer. As a result, it may have anti-ulcer properties (Choudhary 2012).

29.5.7 Anti-Alzheimer's Activity

The three main components of triphala, the fruits of *Terminalia chebula*, *Terminalia bellerica*, and *Emblica officinalis*, were tested for their acetylcholinesterase inhibitory activities using a methanol extract of triphala. Every extract demonstrated dose-dependent reduction of enzyme activity. Acetylcholinesterase was inhibited by phytoconstituents like gallic acid, ellagic acid, and phenolic acids found in the fruit of all three plants. Due to its ability to suppress acetylcholinesterase, it may be used to treat the symptoms of Alzheimer's disease (Nag and De 2011).

29.5.8 Antihypertensive Activity

In order to understand how *Terminalia bellerica* crude extract lowers blood pressure, a study was conducted. While it decreased the force and rate of arterial contraction in isolated guinea pig atria, it caused a dose-dependent decline in arterial BP in anaesthetized rats. It relaxes phenylephrine and K+-induced contractions in the rabbit thoracic aorta and suppresses PE peaks in Ca2+ free media. The endothelium-independent vasodilator effect was present in the endothelium-denuded tissues at comparable concentrations. Since it lowers BP by Ca2+ antagonist, it can therefore be utilized for hypertension (Ahmad and Mishra 2017; Khan 2008).

29.5.9 Wound-Healing Activity

The effectiveness of an ethanol extract of *Terminalia bellerica* fruit used as an ointment on excision and incision wound models was investigated. Compared to regular, it showed a stronger ability to contract while a wound heals (Choudhary 2008).

29.5.10 Hepatoprotective Activity

BALB/CN mice were given an ethyl acetate extract of *T. bellerica* aerial parts once daily for two days in a row, followed by carbon tetrachloride poisoning. In addition to reducing the expression of the oxidative stress indicators 4-hydroxynonenal (4-HNE) and 3-nitrotyrosine (3-NT) and restoring P450 2E1 (CYP2E1) expression, it greatly improved liver necrosis. Additionally, it revealed a decrease in the upregulation of tumor necrosis factor alpha (TNF-), cyclooxygenase-2 (COX-2), and nuclear factor-kappa B (NF-B) in wounded livers, indicating an improvement in the inflammatory response. Transforming growth factor-beta 1 (TGF-1) and alpha smooth muscle actin (-SMA) expression demonstrated a considerable reduction of hepatic fibrosis (Pingale 2011; Rashed et al. 2014).

29.5.11 Analgesic and Antipyretic Activity

In a study, the analgesic and antipyretic effects of ethanolic and aqueous extract from *Terminalia bellerica* fruits were examined in mouse and rat models of acetic acid-induced writhing, Eddy's hot plate method, and Brener's yeast-induced fever. Both the extract and the hot plate approach were found to increase the licking time to heat stimuli while significantly reducing the amount of writhes caused by acetic acid. When compared to the matching control group, both extracts significantly reduced high body temperature. The ethanolic and aqueous extracts were found to have considerable analgesic and antipyretic properties in rats and mice at the doses 200 mg/kg, respectively (Sharma et al. 2010).

29.5.12 Antidepressant Activity

The forced swim test (FST) and tail suspension test were used in a study to examine the antidepressant efficacy of aqueous and ethanolic extracts of *Terminalia bellerica* fruits in Swiss young male albino mice (TST). For ten days in a row, both extracts were given orally. The mice's mobility times in the FST and TST were significantly reduced by ethanol extract and aqueous extract in a dose-dependent manner, but the mice's locomotor activity was not significantly affected. Therefore, both extracts interacted with the adrenergic, dopaminergic, and serotonergic systems to have a strong antidepressant-like effect in mice (Dhingra and Valecha 2007).

29.6 *TERMINALIA BELLERICA* AS ANTIDIABETIC AGENT

A study was done to look for antidiabetic effect in the fruit rind of *Terminalia bellerica*. By using bioassay-guided fractionation, glycemic acid was extracted from *Terminalia bellerica*. Male Wistar rats that had been treated with diabetes using streptozotocin (STZ) were given isolated and synthetic gallic acid at varying concentrations for 28 days. There was a significant, dose-dependent decrease in plasma glucose level. When diabetic rats were not treated, gallic acid-treated rats displayed regeneration of -islets cells after histological analysis. Gallic acid was administered orally, and the serum levels of total cholesterol, triglycerides, LDL cholesterol, urea, uric acid, and creatinine were all significantly lower while plasma insulin, C-peptide, and glucose tolerance levels were noticeably higher. Additionally, it helped diabetic rats regain their body weight, albumin, and total protein. So, *Terminalia bellerica*'s extracted gallic acid could be used as a diabetic agent (Latha and Daisy 2011).

Hexane, ethylacetate, and methanolic extracts of the *T. bellirica* fruit significantly ($p < 0.05$) reduced the amount of diabetes that streptozotocin-induced diabetic rats developed over the course of 60 days. In addition to increasing body weight, serum total protein, plasma insulin, C-peptide, and glucose tolerance levels, it also markedly reduced the levels of urea, uric acid, creatinine, total cholesterol, triglycerides, and low-density lipoprotein cholesterol (Latha and Daisy 2010).

In a study conducted, Vero, L_6, and 3T3 cell lines' glucose absorption was increased by 27.15–1.19, 18.73–1.29, and 24.43–0.88 at 500 g/mL concentration, indicating that the ethanolic extract of *T. bellirica* was determined to be significant (Das 2015).

Treatment with *T. bellirica*fruits significantly reduced blood glucose levels and oxidative stress in rats when used as an antioxidant defense against alloxan-induced hyperglycemia. Increased amounts of antioxidant enzymes like catalase, glutathione reductase, and superoxide dismutase were seen in the blood and liver (Sabu 2009).

Following treatment with *T. bellerica* extract, diabetic rats showed striking reductions in free radicals as well as increased glutathione, superoxide dismutase, catalase, glutathione peroxidase, and glutathione reductase activities. Similar to glibenclamide, an aqueous extract of the *T. bellerica* fruit increased the production of insulin from a pancreatic α-cell line (Kasabri et al. 2010). Additionally, the aqueous extract exhibited insulin derivativeactivity, increasing glucose uptake into 3T3-L1 adipocytes while decreasing protein and starch glycation and starch absorption.

In another study, the fruit of *T. bellerica* was used to investigate the hypoglycemic effect, lipid profile, and safety profile. An ethanolic extract of *T. bellerica* fruit in alloxan-induced diabetic rats was administered. In rats, diabetes was induced by intraperitoneal alloxan injection at a dose of 150 mg/kg body weight, and *T. bellerica* fruit extract wasfed to the rats at a dose of 750mg/kg. Blood glucose levels, lipid profile, and safety profile were assessed by measuring serum cholesterol, free fatty acids, phospholipids and triglycerides, SGOT, SGPT, and creatinine levels in diabetic nondiabetic rats before and after extract administration. After evaluating the amount of blood glucose, the hypoglycemic efficacy was found to be equivalent to that of metformin ($p > 0.05$) given at a dose of 500 mg/kg (Tahsin et al. 2021).

29.6.1 POLYHERBAL FORMULATIONS

Itrifal-e-Sagheer
Itrifal-e-Muqil
Itrifal-e-Ustukhuddus
Majoon-e-Jograj Gugal
Majoon-e-Fanjnosh.

(Anonymous 2007; Kabeeruddin YNMa; Anonymous 1986)

29.7 TRADITIONAL AND OTHER POTENTIAL USES

Fruits of *T. bellerica* have therapeutic properties. They are utilized as laxatives, astringents, anthelmintics, and antipyretics. Fruits can be used to treat a variety of ailments, including hepatitis, bronchitis, asthma, dyspepsia, piles, diarrhea, coughs, hoarseness of voice, eye disorders, scorpion stings, and dyspepsia. They can also be used as a hair tonic (Singh 2011; Rastogi and Mehrotra 2004). The green fruit's decoction is used to treat coughs. Leprosy, dropsy, piles, and dysenteric diarrhea can all be treated with fruit pulp. Half-ripe fruit is used as a purgative. The fruit's kernel contains a narcotic. Fruits are used by Khagrachari residents in Bangladesh to treat menstruation irregularities. Rheumatism is treated with seed oil. The bark's demulcent and purgative gum is used. The triterpenoid found in the fruits has strong antibacterial properties. In mice, continuous usage of kernel oil produces a purgative effect and is well tolerated (Hakeem 2009). Details are mentioned in Table 29.2.

29.8 CULTIVATION PRACTICES

T. bellerica is a heavy demander. However, while young, it can tolerate some light shading. Heavy shadow is known to harm young seedlings after the first and second years, however it is known that young seedlings may endure it throughout the first two years. The plant is very vulnerable to frost. It may grow on a variety of soil types as long as there is enough moisture availability. On deep sandy loam soils with good drainage, the best growth is achieved (Kirtikar 1975; Gupta et al. 2003).

29.8.1 Propagation

Trees naturally repopulate through coppicing and seeds. *T. bellerica* makes decent coppice; May–June cuts make good coppice. Natural regeneration is aided by good seed yields, healthy seeds with a high capacity for germination, and seedlings that can survive for the first year or two in partial shade. The fruits ripen artificially between November and February. Freshly dropped fruits are gathered from the ground and swept beforehand. The seed is promptly separated from the pulp and dried in the sun before being stored. They continue to be viable for around a year. The ability of such stored seeds to germinate is between 40% and 50% less (Swati et al. 2017).

29.8.2 Direct Sowing/Transplanting

Direct sowing or transplanting nursery-raised seedlings that are a year old can both be used to grow the plant in the field. As the monsoon season begins in June and July, sowing is completed. For the

TABLE. 29.2
Traditional Actions and Uses of *Terminalia bellerica*

Actions and uses	References
Anthelmintic	Singh 2011; Rastogi and Mehrotra 2004; Chopra et al. 1956
Antipyretic	Ghani 2003
Astringent	Singh 2011; Kirtikar 1987; Nadkarni 1976; Chopra et al. 1956
Bronchial asthma, bronchitis	Kirtikar 1987; Rastogi and Mehrotra 2004
Diarrhea	Ghani 2003
Dyspepsia	Deb et al. 2016
Laxative	Khare 2007; Rastogi and Mehrotra 2004; Kirtikar et al. 1987
Piles	Rastogi and Mehrotra 2004; Ghani 2003
Scorpion sting	Ghani 2003

purpose of softening the seed coat, the seed is immersed in water for 24 hours. Sowing is done in lines spaced about 3 meters apart. Well-worked soil can be found along the lines that will be sowed. The seeds are lightly encased when the plants are roughly 3–4 months old. In June–July, planting takes place with 3m × 3m as the appropriate spacing (Bhandary et al. 1995).

29.8.3 Collection and Harvesting Time

The month of January is the best time to harvest fruits since it has the most fruit, and tannin quality declines after that. Collections made too early also display a comparable level of tannic quality. Following 7–8 years of growth, harvesting is completed. There is no information available regarding the ideal time to harvest, however it is between mid-December and mid-January (Anonymous 2014).

29.8.4 Processing and Storage

During gathering, fruits that have been bored or harmed by insects should be avoided. Before being stored, the fruits are depulped and dried. It takes two days of sun drying in a dry, well-ventilated area after deseeding. It can be kept in this state for a year in storage. Bahera that has been deseeded must be disposed of within a year of the processing day in order to retain its valuable characteristics. Bahera should be stored in a dry, clean, and airtight gunny bag. It must never be stored somewhere damp because it helps fungus to grow and cause discoloration (Hakeem 1999).

29.9 CONCLUSION

It has been determined that the medicinal plants mentioned in conventional medical systems play a significant role in both disease management and disease prevention. A thorough description of the ethnobotanical and phytochemical representation of *Terminalia bellerica* (bahera) as well as its possibly beneficial effects on health has been provided. *Terminalia bellerica* and its main chemical components include beta-sitosterol, gallic acid, ellagic acid, ethylgallate, galloyl glucose, and belleric acid. It also has antioxidant, immune-modulator, analgesic, antidiabetic, anti-diarrheal, anti-androgenic, antifungal, and anti-helminthic properties. The primary ingredient in itrifalatin Unani medicine and triphala in Ayurveda is *Terminalia bellerica*. In order to facilitate its use in the future, thorough scientific studies on the mechanism of action of this medicinal plant, its phytochemical compounds, drug interactions, effective dose determination, and side effects should be carried out.

REFERENCES

Ahmad, N., and P. Mishra. 2017. Effect of Terminalia bellerica against high fat diet induced hyperlipidemia and obesity. *Pharmacoeconomics* 2(1):62.

Anonymous. 1986. *Qarabadin e Majeedi*. Hamdard Waqf Laboratories, New Delhi, India, pp. 23–24, 88.

Anonymous. 2007. *The Unani Pharmacopeia of India*. Ministry of Health and Family Welfare, New Delhi. Vol. I (1), pp. 5–6, 17–18, 32–33.

Anonymous. 2014. *The Wealth of India (Raw materials and Industrial Products)*. Council of Scientific and Industrial Research, New Delhi. Vol. 10, pp. 100–104.

Bhandary, M. J., K. R. Chandrashekar, and K. M. Kaveriappa. 1995. Medical ethnobotany of the siddis of Uttara Kannada district, Karnataka, India. *Journal of Ethnopharmacology* 47(3):149–158. https://doi.org/10.1016/0378-8741(95)01274-h

Bhattacharya, S. K. 2008. *Hand Book of Medicinal Plants*. Pointer Publication, Jaipur. 5th ed., p. 338.

Brahmachari, H. D., and K. T. Augusti. 1961. Hypoglycemic agents from Indian indigenous plants. *Journal of Pharmacy and Pharmacology*13(1):381–182.

Chopra, R. N., S. L. Nayar, and I. C. Chopra. 1956. *Glossary of Indian Medicinal Plants*. Council of Scientific and Industrial Research, New Delhi. 1st ed., p. 238.

Choudhary, G. P. 2008. Wound healing activity of the ethanol extract of *Terminalia bellirica* Roxb. fruits. *Natural Product Radiance* 7(1):19–21.

Choudhary, G. P. 2012. Anti-ulcer activity of the ethanolic extract of *Terminalia bellirica* Roxb. *International Journal of Pharmacy and Biological Sciences* 1(4):1293–1297.

Das, M. S., and G. Devi. 2015. In vitro cytotoxicity and glucose uptake activity of fruits of *Terminalia bellirica* in Vero, L-6 and 3T3 cell lines. *Journal of Applied Pharmaceutical Science* 5(12):92–95.

Deb, A., G. Choudhury, S. Barua, and B. Das. 2016. Pharmacological activities of Baheda (*Terminalia bellerica*): A review. *Journal of Pharmacognosy and Phytochemistry* 5(1):194–197.

Dhingra, D., and R. Valecha. 2007. Evaluation of antidepressant-like activity of aqueous and ethanolic extracts of *Terminalia bellerica* Roxb. fruits in mice. *Indian Journal of Experimental Biology* 45(7):610–616.

Fahmy, N. M., E. Al-Sayed, and A. N. Singab. 2015. Genus *Terminalia*: A phytochemical and biological review. *Medicinal and Aromatic Plants* 4(5):1–21.

Ghani, A. 2003. *Medicinal Plants of Bangladesh with Chemical Constituents and Uses*. Asiatic Society of Bangladesh, Dhaka. 2nd ed., p. 184.

Ghani, N. 2008. *Khazain ul Advia*. Idara Kitab us Shifa, New Delhi, India. 2nd ed., pp. 187–189, 421–422, 1352–1354.

Gilani, A. H., A. U. Khan, T. Ali, and S. Ajmal. 2008. Mechanisms underlying the antispasmodic and bronchodilatory properties of *Terminalia bellerica* fruit. *Journal of Ethnopharmacology* 116(3):528–538. https://doi.org/10.1016/j.jep.2008.01.006. PMID: 18289813

Gupta, A. K., N. Tendon, and M. Sharma. 2003. *Quality Standard of Indian Medicinal Plants*. ICMR, New Delhi. Vol. 7, p. 295.

Hakeem, H. M. 2009. *Mufradat Azeezi*. CCRUM, New Delhi, India, pp, 14, 46, 103, 105, 115, 118.

Hakeem, M. A. 1999. *Bustanul Mufradat*. Idara e Kitabul Shifa, New Delhi, India, pp. 72, 100, 347–348.

Kabeeruddin, H. YNMa. *Biyaz Kabir*. Hikmat Book Depot, Hyderabad, India. Vol. 2, p. 5.

Kabeeruddin, H. YNMb. *Makhzan Al Mufradat*. Ejaz Publishing House, New Delhi, India, pp. 55, 145, 590–591.

Kasabri, V., P. R. Flatt, and Y. H. A. Abdel-Wahab. 2010. *Terminalia bellerica* stimulates the secretion and action of insulin and inhibits starch digestion and proteinglycation in vitro. *British Journal of Nutrition* 103:212–217.

Khan, A. U., and A. H. Gilani. 2008. Pharmacodynamic evaluation of *Terminalia bellerica* for its anti-hypertensive effect. *Journal of Food and Drug Analysis* 16(3):6–14.

Khare, C. P. 2007. *Indian Medicinal Plants*. Springer Science, New Delhi. 1st ed., p. 639.

Kirtikar, K. R., and B. D. Basu. 1975. *Indian Medicinal Plants*. Periodical Experts, New Delhi. Vol. II, pp. 1052–1053.

Kirtikar, K. R., and B. D. Basu. 1987. *Indian Medicinal Plants*. International Book Distributors, Dehradun. Vol. I, pp. 1052–1054, 1999.

Latha, R. C., and P. Daisy. 2010. Influence of *Terminalia bellerica* Roxb. fruit extract on biochemical parameters in streptozotocin diabetic rats. *International Journal of Pharmacology* 6(2):89–96.

Latha, R. C., and P. Daisy. 2011. Insulin-secretagogue, anti-hyperlipidemic and other protective effects of gallic acid isolated from *Terminalia bellerica* Roxb. in streptozotocin-induced diabetic rats. *Chemico Biological Interactions* 189(1–2):112–118.

Makheswari, M. U., and D. Sudarsanam. 2011. Phytomedicine for diabetes mellitus: An overview. *Research in Pharmacy* 1(4):28–37.

Malviya, N., S. Jain, and S. Malviya. 2010. Antidiabetic potential of medicinal plants. *Acta Poloniae Pharmaceutica* 67(2):113–118.

Nadkarni, K. M. 1976. *Indian Materia Medica*. Bombay Popular Parkashan. Vol. I. 3rd ed., pp. 516–518.

Nag, G., and B. R. De. 2011. Acetylcholinesterase inhibitory activity of Terminalia chebula, Terminalia bellerica and Emblica officinalis and some phenolic compounds. *International Journal of Pharmacy and Pharmaceutical Sciences* 3(3):121–124.

Nishida, K., O. Yamaguchi, and K. Otsu. 2008. Crosstalk between autophagy and apoptosis in heart disease. *Circulation Research* 103(4):343–351. https://doi.org/10.1161/CIRCRESAHA.108.175448. PMID: 18703786

Pingale, S. S. 2011. Hepatoprotective action of *Terminalia bellericaon* CCl_4 induced hepatic disorders. *Der Pharma Chemica* 3(1):42–48.

Pragya, M., A. Nesar, M. Tarique, S. Arshiya, P. Bagga, and S. Prashant. 2016. Effect of *Terminalia bellerica* against high fat diet induced hyperlipidemia and obesity. *World Journal of Pharmaceutical Sciences* 4(4):33–37.

Pullaiah, T. 2006. *Encyclopaedia of World Medicinal Plants*. Regency Publication, New Delhi. Vol. 4, p. 1892.

Rashed, K., I. Potočnjak, J. Giacometti, M. Škoda, and R. Domitrović. 2014. *Terminalia bellerica* aerial parts ethyl acetate extract exhibits antioxidant, anti-inflammatory and antifibrotic activity in carbon tetrachloride intoxicated mice. *Journal of Functional Foods* 8(1):319–330.

Rastogi, R. P., and B. N. Mehrotra. 2004. *Compendium of Medicinal Plants*. Central Drug Research Institute (CDRI), Lucknow and National Institute of Communication and Information Resources, New Delhi, p. 406.

Sabnis, S. 2014. Antimicrobial efficacy of *Terminalia bellerica* against virulence factors of respiratory pathogens. *International Journal of Current Microbiology and Applied Sciences* 3(5):215–221.

Sabu, M. C., and R. Kuttan. 2009. Antidiabetic and antioxidant activity of *Terminalia bellirica* Roxb. *Indian Journal of Experimental Biology* 47:270–275.

Saxena, A., and N. K. Vikram. 2004. Role of selected Indian plants in management of type 2 diabetes: A review. *Journal of Alternative and Complementary Medicine* 10(2):369–378.

Sharma, P., K. K. Verma, H. Raj, and N. Thakur. 2021. A review on ethnobotany, phytochemistry and pharmacology on Terminalia belerica (Bibhitaki). *Journal of Drug Delivery and Therapeutics* 11(1-s):173–181.

Sharma, U. S., U. K. Sharma, and A. Singh. 2010. Screening of *Terminalia bellerica* fruits extracts for its analgesic and antipyretic activities. *Jordan Journal of Biological Sciences* 3(3):121–124.

Singh, A. S. 2011. *Herbalism Phytochemistry and Ethanopharmacology*. CRC Press, Boca Raton. 1st ed., pp. 357–361.

Swati, K., M. J. Krishna, A. B. Joshi, et al. 2017. A pharmacognostic, phytochemical and pharmacological review of *Terminalia bellerica*. *Journal of Pharmacognosy and Phytochemistry* 6(5):368–376.

Tahsin, M. R., B. Jannat, A. Mostafa, et al. 2021. An evaluation of diabetes ameliorating capacity of *Terminalia bellerica* ethanolic extract on alloxan-induced diabetic rat including safety profile study. *Journal of Clinical and Molecular Endocrinology* 6(1):33.

Valli, S., and G. S. Shankar. 2013. *Terminalia bellerica*: A promising challenge to cryptococcosis. *International Journal of Pharmaceutical Research and Bioscience* 2(5):154–169.

Whiting, D., L. Guariguata, C. Weil, and J. Shaw. 2011. IDF Diabetes atlas: Global estimates of the prevalence of diabetes for 2011 and 2030. *Diabetes Research and Clinical Practice* 94(3):311–321.

Wild, S., G. Roglic, A. Green, R. Sicree, and H. King. 2004. Global prevalence of diabetes-estimates for the year 2000 and projections for 2030. *Diabetes Care* 27(5):1047–1053.

Young, I. S., and J. V. Woodside. 2001. Antioxidant in health and diseases. *Journal of Clinical Pathology* 54(3):176–186.

Zhang, X. R., J. S. Kaunda, H. T. Zhu, D. Wang, C. R. Yang, and Y. J. Zhang. 2019. The genus terminalia (Combretaceae): An ethnopharmacological, phytochemical and pharmacological review. *Natural Products and Bioprospecting* 9(6):357–392.

30 *Pterocarpus marsupium* Roxb. (Malabar Kino or Indian Kino)

Antidiabetic and Other Potential Features

Salman Khan, Kolagani Chandramohan, and Rohini Ganorkar

CONTENTS

DOI: 10.1201/b23347-30

30.1 INTRODUCTION

The catalogue of world plants provides information on useful plants, and some of them may hold biological compounds which cure irremediable diseases. Plants are always a common source of biologically active compounds which have traditional importance in the treatment of different ailments (Abate et al. 2021; Bachheti et al. 2021; Husen 2021, 2022; Husen et al. 2021; Sonkar et al. 2022; Rahman & Husen 2021, 2022). Due to their effectiveness, minimum side effects in therapy and low cost, the demand for herbal drugs has had a major share in the world's economy.

Pterocarpus marsupium Roxb. (Leguminosae) is a multipurpose and internationally demanded tree for its timber and pharmaceutical properties. Locally, it is also called bija, Indian kino tree, Malabar kino tree, bijayasal, and yegisa. It is native to India, Nepal and Sri Lanka, where it grows naturally and abundantly. In India, it is usually found in hilly regions throughout the Deccan Peninsula and extends to the states of Gujarat, Chhattisgarh, Madhya Pradesh, Uttar Pradesh, Bihar and Odisha (Krishnamurthi 1998), and it also occurs in parts of the Western Ghats (Gamble 1935; Mathew 1983). *Pterocarpus marsupium* is well known for its timber that ranks after teak and rosewood. There is no information on population size and conservation strategies in natural habitats for this species. Very few natural stands of the tree are noticed in peninsular India, based on the assessment of forest resources by the Forest Survey of India (ISFR 2021). For the last few decades, it is disappearing quickly from natural habitats due to land use changes, forest degradation, overexploitation and other reasons (Kadavul & Parthasarathy 1988; Anis et al. 2005). When compared with other timber species, this tree has been paid less attention due to lack of cultivation practices among growers and farmers. It is declining slowly from the wild because of habitat degradation, land use change, pressure from invasive alien species (pers.obs.) and forest fire. Hence, it has been placed in the red data book under the category Near Threatened. There is some concern of the world's foresters and researchers towards the reforestation of crop with desirable timber qualities.

Nowadays, the exploitation of medicinal plants from forest has increased for its demand and medicinal value. This tree is exploited for its less-known timber value and its bark, which is medicinally significant (Ramya et al. 2008).

30.2 TAXONOMY AND PHENOLOGY

Pterocarpus marsupium Roxb. is a large deciduous tree; stem stout, crooked and widely spreading branches; stout and straight cylindrical clean bole; bark thick, yellowish grey, the outer layer corky, falling off in flakes; inwardly red and fibrous. Leaves stipulate; stipules early deciduous; compound, 15–24 cm long; rachis glabrous, prolonged up to 3 cm long beyond the insertion of the upper lateral leaflet; leaflets 5–7, coriaceous, 6–14 cm long, oblong, obtuse or rounded, truncate or more or less retuse at the apex, glabrous on both surfaces, shining above, subacute to obtuse at the base; main nerves numerous, close, prominent; petiolules 0.6–0.9 cm long. Flowers in short lateral and terminal fusco pubescent paniculate racemes, usually shorter than leaves; pedicels short, a little downy; articulated beneath the flower; bracts small, falling one below each division and subdivision of the panicle. Flowers papilionaceous, numerous, white with a small tinge of yellow or golden yellow. Calyx 0.6 cm long, veined, brown pubescent; teeth very short, broadly triangular; the upper one is largest. Corolla 1.2 cm long, pale yellow or golden yellow with crisped margins; standard petal with a slender claw, broad; sides reflexed, waved, curled and veined; wing petal clawed, ovate, waved, similar to standard petal; keel petals adhering each other for a little way near the middle only. Stamens monadelphous or sometimes the vexillary stamen often nearly or quite free; anthers globose, two lobed. Style ascending; ovary shortly; ovules two. Pods 2.5–5 cm in diameter, nearly

FIGURE 30.1 Roxburgh's illustration of *Pterocarpus marsupium* Roxb.

orbicular, glabrous or nearly so, the upper side which extends from the pedicel to the remains of the style, straight; the whole surrounded with a membranous wing, waved, veined and downy. Seed small (Gamble 1935; Figures 30.1 and 30.3).

The wood is hard and durable in nature. The sapwood is pale yellowish white or white; the heartwood is golden yellowish brown with darker streaks, staining yellow when damp and turning darker on exposure, grains broadly interlocked, textures strong, tough, very hard and moderately heavy (Figure 30.2).

30.2.1 Phenology

Flowering usually appears during July–October but in Maharashtra during May–June and in Madhya Pradesh during October to November due to climatic factors. The fruiting takes place from December to March. The ripened pods hang in clusters on the tree for 3–5 months. The dried and ripened pods are disseminated through wind during the hot season. A peculiar phenomenon of gregarious flowering was reported from South Chandrapur, Maharasthtra, in 1942. All flowers opened simultaneously on one day only and after that no open flowers were noticed (Singh & Karthikeyan 2000).

FIGURE 30.2 Wood structure of *Pterocarpus marsupium* Roxb.

FIGURE 30.3 *Pterocarpus marsupium* Roxb. (A) habitat; (B) bark; (C) flowering twig; (D) inflorescence; (E) pod.

30.3 ECOLOGY, DISTRIBUTION AND ASSOCIATED SPECIES

It is native to Bangladesh, India, Nepal and Sri Lanka. In India, it is mainly distributed in greater parts of India from South India to western peninsular India. It is also found to a limited extent in sub-Himalayan tract from Haldwani eastwards to Gorakhpur. There is rare distribution towards north-eastern states. It attains 5 m in girth and 30 m height under favourable habitats in Karnataka, Tamil Nadu and Kerala. But in other parts of India, it does not exceed 2 m in girth. It has been introduced in some parts of world for various purposes (Gamble 1935).

It grows usually on hills or undulating lands between 150–900 m elevations at temperature varies from 0° to 48° on a wide variety of soils and geological formations. The tree is scarcely distributed in moist and dry deciduous forests and shows 1%–5% of the total crop where it is found. As per Champion and Seth's classification, it is restricted to the southern tropical semi-evergreen forests, south and north Indian tropical moist dry deciduous forests and rarely in dry deciduous forests (Gamble 1935).

It is chiefly associated with *Terminalia paniculata*, *Xylia xylocarpa*, *Dalbergia latifolia*, *Kydia calycina*, *Lagerstroemia* sp. and *Haldina cordifolia* in moist deciduous forests. *Anogeissus latifolia*, *Sterculia urens*, *Boswellia serrata*, *Diospyros melanoxylon*, *Lagerstroemia parviflora* and *Terminalia tomentosa* are major associates in dry deciduous forests.

30.4 PHYTOCHEMICAL CONSTITUENTS

Plant parts of the *P. marsupium* tree (gum, heartwood, bark, roots, leaves and flowers) have long been used for their medicinal properties in traditional systems of medicine. Earlier research on chemical investigations on the genus *Pterocarpus* has revealed that it is arich source of polyphenolic compounds (Seshadri 1972). All active principles of *P. marsupium* are thermostable (Ivorra et al. 1989; Kameswara et al. 2001). The plant parts contain pterostilbene 45%, alkaloids 0.4%, tannins 5%, protein, pentosan, pterosupin, pseudobaptigenin, liquiritigenin, isoliquiritigenin, garbanzol, 5-de-oxykaempferol, P-hydroxybenzaldehyde, beudesmol, erythrodirol-3-monoacetate, l-epicatechin, marsupol, carpusin, propterol, propterol B, marsupinol, irisolidone-7-O-A-L-rhamnopyranoside, which have been extracted chiefly from the heartwood and root (Mathew et al. 1977; Chakravarthy et al. 1981a, 1981b; Subbarao & Mathew 1982; Maurya et al. 1984; Maneesha et al. 2015; Table 30.1). *P. marsupium* contains terpenoids and phenolic compounds: β-sitosterol, lupenol, aurone glycosides, epicatechins and iso-flavonoids (Mitra & Joshi 1983; Kumar & Seshadri 1976).

The gum "kino" is known as an astringent drug which exudes from the phloem of the stem after wounding and contains red-coloured fluid and provides non-glucosidal tannins. Kino is odourless, has an astringent taste, sticks in the teeth, and colours the saliva red (Figure 30.4).

a. Kinotannic acid
b. Kinonin (C28H24O12)
c. Kinored (C28H22O11)
d. Pyrocatechin
e. Pyrocatechin acid and small quantities of resin, pectin and gallic acid.

The heartwood is used as an astringent and in the treatment of inflammation. The wood and bark of *P. marsupium* are known for their antidiabetic activity (Badkhane et al. 2010).

Aqueous extract of the heartwood of *P. marsupium* contains five new flavonoids: C-glucosides namely 6-hydroxyl-2-(4-hydroxybenzyl)-benzo-furan-7-C-a-D- glucopyranoside, 3-(a-methoxy-4-hydroxybenzylidene)-6-hydroxybenzo-2(3H)-furanone-7-C-â-D-glucopyranoside, 2-glucopyranoside, 8-(C-a-D-glucopyranosyl)-7,3,4-trihydroxyflavone and 1,2-bis (2,4-dihydroxy, 3-C-glucopyranosyl)- ethanedione (Maneesha et al. 2015).

Methanolic extract of heartwood contains an isoflavone -7-O-a-L-rhamnopyranosyloxy-4-methoxy-5-hydroxyisoflavone. Three new isoflavone glycosides *viz* retusin 7-glucoside, irisolidone

TABLE 30.1
Chemical Components and Uses of *Pterocarpus marsupium*

Plant part	Biological compounds	Biological activity	References
Kino gum	Kinotannic acid, kinonin, kinored, pyrocatechin, pyrocatechin acid	Diarrhoea, dysentery, leucorrhoea, passive haemorrhages	Badkhane et al. 2010; Rahman et al. 2008
Bark	Liguiritigenin, isoliguiritigenin, pteosupin, epicatechin, pterostilbene and marsupinol	Astringent, toothache, severe diarrhoea, elephantiasis, leucoderma, rectalgia, cough and greyness of hair, wounds, fever, stomach ache, diabetes, jaundice and antiulcer, cholera, dysentery, tongue diseases, urinary complaints, treatment of tumours of the gland, urethral discharges, chronic ulcers	Badkhane et al. 2010; Chawla et al. 2013; Rahman et al. 2008; Chakravarthy & Gode 1985; Maruthupandian & Mohan 2011; Jahromi et al. 1993; Mankani et al. 2005; Goel & Bhattacharya 1991; Takehara et al. 1997
Heartwood	Propterol, marsupin, liquiritigenin, pterostilbene, pterosupun and (-)epicatechin	Controls blood sugar level. Astringent, bitter acrid, anti-inflammatory, anthelmintic, anodyne	Rahman et al. 2008; Hugar & Londonkar 2017; Jahromi et al. 1993; Hougee et al. 2005; Salunkhe et al. 2005; Sashikanth et al. 2012; Banerji et al. 1999; Dudeja et al. 2001; Kubish et al. 1997; Naziroğlu & Cay 2001
Leaves	Propterol, liquiritigenin, pterostilbene and pterosupun	External application for boils, sores and skin diseases, stomach pain	Badkhane et al. 2010; Sapha 1956; Gayathri & Kannabiran 2009; Nair et al. 2005; Manisckam et al. 1997
Flowers	Pentosan, isoflavonoid glycol, liquiritigenin	Fever	Badkhane et al. 2010; Rahman et al. 2008

FIGURE 30.4 Extraction of kino gum from bija tree.

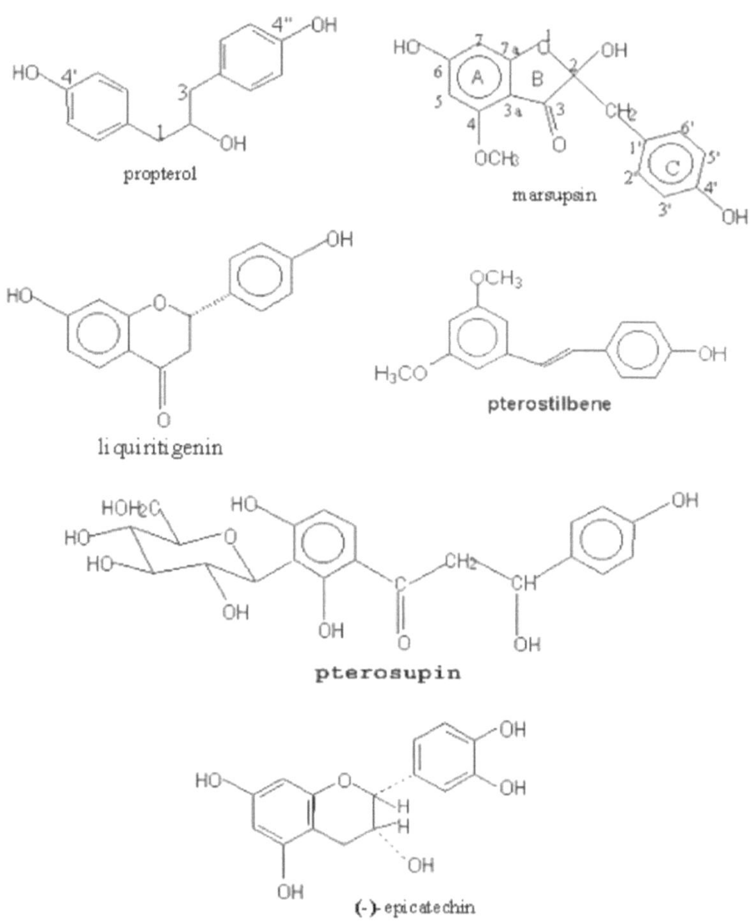

FIGURE 30.5 Chemical constituents of *Pterocarpus marsupium* Roxb.

7-rhamnoside and 5,7-dihydroxy-6-methoxyisoflavone7-rhamnoside have been isolated from the heartwood of *P. Marsupium* (Figure 30.5).

Ether extract of the roots of *P. marsupium* consists of a new flavonol glycoside 6-hydroxy-3,5,7,4-tetramethoxyflavone 6-O-rhamnopyranoside, 8-hydroxy-4-methoxy isoflavone-7-O-glucopyranoside. A benzofuranone derivative 2,4,6-trihydroxy-4-methoxy benzofuran-3(2H) one designated carpusin, 1,3 bis (4-hydroxyphenyl) propan-2-ol designated propterol, 1-(2,4-dihydroxyphenyl)-3-(4-hydroxyphenyl) propan-2-ol-designated propterol, 6-hydroxy-7-O-methyl-3-(3-hydroxy-4-O-methylbenzyl) chroman-4-one (Maneesha et al. 2015).

Ethyl acetate extract of the root contains benzofuranone, marsupin, dihydrochalcone, pterosupin, stilbene, pterostilbene, a homoisoflavonepteromarsupone, transstilbene, liquiritigenin and isoliquiritigenin.

The biological compounds liguiritigenin, isoliguiritigenin, pteosupin, epicatechin, pterostilbene and marsupinol have been isolated from the bark of *P. marsupium* (Badkhane et al. 2010).

30.5 PHARMACOLOGICAL STUDIES

Various parts of the *P. marsupium* tree have been used as traditional Ayurveda medicine in India since ancient times, and the utilization of leaf, wood, gum and bark have been described. The bark is used for the treatment of stomach ache, cholera, dysentery, urinary complaints, tongue diseases and toothache. The gum exude kino, derived from this tree, is used as an astringent (Maurya et al. 1984); it is also antipyretic, anthelmintic and tonic to liver, useful in all diseases of body and styptic vulnerant and good for griping and biliousness, opthalmiya, boils and urinary discharges. The flowers are bitter, improve the appetite and cause flatulence (Kitikar & Basu 1999). *P. marsupium* has a long history of its utilization in India as a treatment for diabetes. It is a drug that is believed to have some unique features such as beta cell protective and regenerative properties apart from blood glucose reduction (WHO 1980; Chakravarthy et al. 1981a). *P. marsupium* is reported here effective against in treatment of different diseases.

30.5.1 Activity on Central Nervous System (CNS)

The bioactive compound (-)epicatechin was the major component in the bark, and research on rats, mice and frogs revealed that it has no impacts on the CNS. A higher dose introduced in rats leads to hyperglycaemia (Chakravarthy & Gode 1985; Gupta & Gupta 2009).

30.5.2 Activity to Reduce Levels of Lipids and Lipoproteins in Blood (Antihyperlipidaemic Activity)

Shah (1967) and Singh et al. (2012) studied the effect of a combination drug (methanolic extract of *Ocimum sanctum* leaves and *P. marsupium* heartwood) on Wistar rats (female) and revealed that it has shown significant effect on levels of lipids and lipoproteins in the blood. It shows a significant role in maintaining endogenous antioxidant levels at the dosage of 500 mg/kg b.wt. in female rats. The extract of ethanolic wood and bark of *P. marsupium* (150+150 mg/kg b.wt.) significantly decreased blood glucose and lipid in albino Wistar rats (Maruthupandian & Mohan 2011; Jahromi et al. 1993).

30.5.3 Activity Against Inflammation

The extract of *P. marsupium* which contains pterostilbene exhibits potential for its anti-inflammatory activity. The COX-1/2 selective inhibitory activity was also investigated (Jahromi et al. 1993; Hougee et al. 2005; Salunkhe et al. 2005; Hugar & Londonkar 2017).

30.5.4 Hepatoprotective Activity

The methanol and aqueous stem bark extracts were evaluated for hepatoprotective effects in carbon tetrachloride (CCl4)-induced hepatotoxicity model after obtaining results on histology and liver biomarkers (serum protein, total bilirubin, alanine amino transaminase, alkaline phosphatase and aspartate amino transaminase) (Mankani et al. 2005; Rahman et al. 2008).

The reduction of levels of the enzymes (lactate dehydrogenase, aspartate transaminase, alkaline phosphatase, alanine transaminase and bilirubin) were noticed after treatment of plant extracts (100 mg/kg b.wt orally) in CCl4 induced hepatotoxicity model. Hence, the analysis reflects as a good protector on CCl4 induced hepatotoxicity (Devipriya et al. 2007).

30.5.5 Antidiarrheal Activity

The investigations on ethanolic extract of *P. marsupium* heartwood (250 and 500 mg/kg b.wt.) revealed that it potentially decreased the severity and frequency of charcoal and castor oil induced gastrointestinal motility or diarrhoea, as cited in traditional use of this plant since ancient days (Manisckam et al. 1997; Dilpesh et al. 2011).

30.5.6 Cardiotonic Activity

Cardiotonic activity was reported of the aqueous extract of heartwood of *P. marsupium* which contains 5, 7, 2–4 tetrahydroxy isoflavone 6–6 glucoside to prevent cardiovascular diseases due to the presence of antioxidants (Mohire et al. 2007).

30.5.7 Antibacterial Activity

The bark and leaf extracts of *Pterocarpus marsupium* were investigated for antibacterial activity in hexane, ethyl acetate and methanol extract methods against four selected gram-positive and gram-negative bacteria (Sapha 1956; Gayathri & Kannabiran 2009). *In vitro*, it inhibits activity on bacteria like *Pseudomonas aeruginosa*, *Streptococcus pyrogens* and *Staphylococcus aureus*. Ethyl and methanol extracts showed variation in their antibacterial activity upon concentration, whereas ethanol extracts of *P. marsupium* exhibited significant antiulcer and antioxidant properties in rats (Nair et al. 2005; Manisckam et al. 1997).

The growth of gram-positive bacteria *Bacillus coagulans* and *Escherichia coli* was significantly inhibited when methanolic extract was applied against 100 mg/ml by the paper disc diffusion method (Kachhawa et al. 2012). The result of acetone and isopropyl extract of stem against the gram-positivebacteria (*Staphylococcus aureus* and *Bacillus cereu*) is specific (Pant et al. 2017).

30.5.8 Anticancer Activity

Antitumor effect of pterostilbene, a chemical constituent of *P. marsupium*, was proved against Hep-G2 (liver) and HCT-116 (colon) cancers and able to inhibit cell proliferation and induce apoptosis in a specific concentration and time through mitochondrial and Fas/FasL pathway and GADD expression and by modifying cell cycle progress and changes in several cycle-regulating proteins (Pan et al. 2007).

30.5.9 Antimicrobial Activity

Several experiments were done on antimicrobial activity by using the disc diffusion method revealed that aqueous, ethanolic and methanolic bark extracts of *P. marsupium* showed significant antimicrobial action against microbes. Methanol extract showed significant effect by preventing *A. niger* at 25 µg/ml and *E. faecalis* and *S. typhi* at 12.5 µg/ml (Londonkar & Hugar 2017). Ethanolic extract of bark against *Candida albicans*, *Vibrio cholera* and *Bacillus polymyxa* was evaluated by using cyclic voltammetry at different dosages (Deepa et al. 2014).

30.5.10 Toxicity Test

The investigation on toxic effect in mice by aqueous extracts of *P. marsupium* bark revealed that no adverse effects were noticed while taken orally at single dose of 2.5–12.5 g/kg and examined general behaviour and mortality. The histopathological studies of kidney, lungs, liver and pancreas did not show any morphological alteration. Mortality was increased with the dose (LD50>10.75 g/kg b. wt) for the mouse where hepatic catalase activity has increased due to overdose of 600 mg/kg b.wt. (Tchamadeu et al. 2011; Mahnaz et al. 2010).

30.5.11 Antiulcer Activity

Methanolic heartwood extract of *P. marsupium* (750 mg/kg b.wt) decreases the blood glucose level in both normal rats and NIDDM rats. It did not indicate any antiulcer activity caused by aspirin, pylorus ligation, cold restraint stress or ethanol in normal rats (Goel & Bhattacharya 1991; Takehara et al. 1997; Rahman et al. 2008).

30.5.12 Analgesic Activity

The effect of ethyl acetate, petroleum ether and methanol leaf extracts of *P. marsupium* were evaluated for analgesic activity in Swiss albino mice. The methanolic extract at doses of 120 mg/kg b.wt was more effective than ethyl acetate and petroleum ether extracts (Sikdar et al. 2013), whereas bark extract at dose of 500 mg/ml significantly improved reaction in mice (Tippani et al. 2010).

30.5.13 Anticataract Activity

The effect of aqueous bark extract of *P. marsupium* (1 g/kg/day), alcoholic seed extract of *Trigonella foenum-graecum* (2 g/kg/day) and leaf extract of *Ocimum sanctum* (200 mg/kg/day) revealed the growth of cataract in diabetic rats is stable (Vats et al. 2004; Vats et al. 2002).

30.5.14 Antifungal Activity

The effect of alcoholic and aqueous extract was studied and showed that the ointment prepared from alcoholic extract was more effective than aqueous extract (Dhir et al. 1982).

30.5.15 Anthelmintic Activity

The leaf extracts of ethyl acetate, ethanol, *n*-butanol and petroleum ether were analysed on Indian earthworms at different concentrations to determine paralysis and death time of the worms. Of them, petroleum ether and ethanol extracts show more significant anthelmintic activity by using albendazole (10 mg/ml) (Panda et al. 2015).

30.5.16 Antioxidant Activity

Ethanol, isopropyl alcohol (IPA) and acetone stem wood extracts of *P. marsupium* stimulates antioxidant activity in 2,2-diphenyl-1-picrylhydrazyl (DPPH) radical scavenging method (Pant et al. 2017). Methanolic extract (100 μg/ml) of *P. marsupium* shows maximum 2,2-diphenyl-1-picrylhydrazyl free radical scavenging effect, followed by ethyl acetate and aqueous extracts Abirami et al. 2012). The effect of ethyl acetate leaf extract of *P. marsupium* was investigated *in vitro* using DPPH assay, hydroxyl radical scavenging activity, ABTS assay, FRAP assay, NO radical scavenging activity, TRAP assay, reducing power assay and hydrogen peroxide (H2O2) radical scavenging activity methods. The leaf extract showed higher free radical scavenging activity, i.e. 71% in FRAP assay method (Kumaravel et al. 2013). The effect of *P. marsupium* bark extract on antioxidant activity was significant in DPPH, superoxide, ABTS, hydroxyl radical, nitric oxide scavenging and suppression of *in vitro* lipid peroxidation (Abirami et al. 2012).

30.6 ANTIDIABETIC RESPONSE

Diabetes mellitus is a metabolic disorder and dysregulation primarily of carbohydrate metabolism due to hyperglycaemia and deficiency of secretion or action of endogenous insulin (Sashikanth et al. 2012). It is a common disease affecting over 124 million individuals throughout the world. If the situation of hyperglycaemia persistent during diabetic conditions which leads to production of free radicals or impaired antioxidant defences via several mechanisms (Valko et al. 2007). Indians show a significantly higher age-related prevalence of diabetes when compared with several other populations (Ali et al. 2010). Due to peripheral insulin resistance, insulin shows at a higher level in Asian Indians (Banerji et al. 1999; Dudeja et al. 2001). Excess body fat, typical abdominal deposition pattern, low muscle mass and racial predisposition may claim the prevalence of hyperinsulinaemia and leads to type 2 diabetes in Asian Indians. Uncontrolled diabetes leads to a plethora of complications affecting the vascular system, eyes, nerves and kidneys, leading to peripheral vascular disease, nephropathy, neuropathy, retinopathy and mortality.

The major chemical constituents' flavonoids, diphenylpropane derivatives and sesquiterpenes are present in plant parts of the tree and are suggested to be useful for therapy and prophylaxis of diabetes (Achari et al. 2012).

The plant parts of *P. marsupium* shows unique pharmacological properties, which include lowering activity of blood glucose and beta cell protective and regenerative properties (Manisckam et al. 1997; Ahmad et al. 1991). These effects have been examined in numerous animal species and in human beings over the past 50 years. Several animal studies have investigated *P. marsupium* in rats, dogs and rabbits with induced diabetes and subsequent treatment with the plant's bark, leaf and wood extracts. As a result of all of these studies, *P. marsupium* was found to be completely suitable to restore normal insulin secretion (Pandey & Sharma 1976; Chakravarthy et al. 1982a; Valko et al. 2007). Many investigations revealed that antioxidants capable of neutralizing free radicals are effective in preventing induced diabetes in animals (Kubish et al. 1997; Naziroğlu & Cay 2001) and also act as an agent in the reduction of diabetic complications (Lipinski 2001). Recent studies on antidiabetic and antioxidant activity of plant extracts suggested that their proper intake in the diet may help decrease the oxidative load in diabetes mellitus. The extracts of bark and wood of *P. marsupium* would have a better therapeutic value in reducing the oxidative load in diabetic patients.

Marsupsin and pterostilbene are the most important phenolic compounds of the heartwood of *P. marsupium*. The ethanolic extract of *P. marsupium* stem wood has proven to lower blood glucose in 180 min by using a dose of 200 mg/kg b. wt. (51.30%) and at 400 mg/kg b. wt. (55.13%) (Pant et al. 2017). The ethanolic extract of heartwood (1gm/kg *per oral* and 2 gm/kg *per oral*) significantly ($p<0.05$) reduced glucose levels in dexamethasone-induced hyperglycaemia and hyperinsulinaemia in Wistar male albino rats (Narender et al. 2016). The methanol extract of *P. marsupium* activates the glucose transport in a PPARγ mediated PI3 kinase dependent fashion on GLUT-4, PPARγ and PI3 kinase (Anandharajan et al. 2005). Aqueous extract of heartwood of *P. marsupium* stimulates the insulin secretion and glucose levels in mouse muscle and pancreatic tissues with a concentration-dependent manner (10 and 100 μL or 0.1 and 1 μL). Bioassay revealed that extracts had potent stimulation of insulin *in vitro* and *in vivo* (Mohankumar et al. 2012).

The blood sugar-lowering activity has been due to the presence of tannates in the different extracts of the plant. Antihyperlipidaemic activity is initiated due to the presence of marsupin, pterosupin and liquiritigenin in plant extracts (Jahromi et al. 1993). Epicatechin has insulinogenic property by enhancing insulin release where proinsulin to insulin converted (Ahmad et al. 1989; Rizvi & Zaid 2001). The epicatechin plays an important role in the insulin signalling by activating key proteins of that pathway and regulating glucose production through AKT and AMPK modulation in HepG2 cells (Cordero-Herrera et al. 2013).

The wood has made a significant mark in the traditional system of medicine. The wooden glasses and wooden vessels made up of heartwood of *Pterocarpus marsupium* are being used for drinking water to control glucose levels in blood and as strong a antidiabetic in the Ayurvedic system of medicine due to presence of compounds iso-flavonoids, terpenoids and related phenolic compounds, ß-sitosterol, lupenol, epicatechin, and aurone glycosides (Kumar & Seshadri 1976; Mitra & Joshi 1983; Ivorra et al. 1989; Kameswara et al. 2001; Gairola et al. 2010; Madhuri et al. 2019). Leaves, wood, stem bark, seed and flowers are used in African traditional medicine, especially in the Cameroonian pharmacopoeia, for treating various diseases including hypertension, diabetes, intestinal parasites and renal systems.

30.7 TRADITIONAL AND OTHER POTENTIAL USES

As per the Ayurveda system of medicine, *P. marsupium* is one of the most versatile medicinal plants with a wide range of biological activities, and each part has been acknowledged for its therapeutic potential value. Heartwood, leaves, flowers and gum are the major useful parts of *P. marsupium* used in the treatment of leucoderma, severe cough, elephantiasis, diarrhoea, hair fall and intestinal diseases (Mankani et al. 2005). It is used as a cooling external application for inflammations and headache, as antipyretic, anti-helminthic, aphrodisiac and alexeteic and in biliousness, mental aberrations and ulcers in traditional systems of medicines (Kumar et al. 2006).

It is nontoxic and useful in treatment of jaundice, fever, wounds, diabetes, stomach ache and ulcer (Jung et al. 2006). The genus is widely distributed on the earth, and the astringent drug from this genus is known as kino. The phloem of the stem contains red astringent fluid present in secretory cells, which exudes after incision. As an astringent, it is used in the treatment of diarrhoea, dysentery and more. Bruised leaves are applied on fractures, leprosy, leucoderma, skin diseases, sores and boils, constipation, depurative, rectalgia, ophthalmology, haemorrhages and rheumatoid arthritis. The chemical constituents marsupin and pterostilbene significantly lower the blood glucose levels in animals. The bark extract is used as a diuretic. The flowers are used to cure fever.

The extract of bark contains l-epicatechin and a reddish brown coloured dye which is used as an astringent and in toothache. The heartwood consists of liquiritigenin, isoliquiritigenin, a natural unidentified component and resin. The wood also contains an essential oil and semi-drying fixed oil. The therapeutic value of kino is due to kino tannic acid. Kino is a powerful astringent and has antipyretic, anthelmintic properties (Singh et al. 1965). It is locally applied in leucorrhoea and in passive haemorrhages. It is also used for toothache. Kino finds application in dyeing, tanning and printing. It is of potential use in the paper industry (Anonymous 2003). The extract of bark is used in treatment of different ailments like stomach ache, cholera, dysentery, urinary complaints, tongue diseases, toothache and mouth ulcers (Barmukh & Nikam 2008; Ravi 2013).

The flowers are used to cure fever. The flowers are bitter in taste, improve the appetite and cause flatulence (Kitikar & Basu 1999; Madhuri et al. 2019).

30.8 SAFETY ISSUES

Besides educing a strong antidiabetic property, *Pterocarpus marsupium* is reported to be effective in treatment against several diseases and does not cause any major side effects when administered in prescribed doses (Singh & Ali 2016).

30.9 CULTIVATION PRACTICES

30.9.1 Climate and Soil

It grows at elevations from 200 to 500 metres and in areas where the mean annual temperature is within the range of 1°C to 48°C. It is found in areas where there is a distinct dry season, preferring a mean annual rainfall of 1000 to 1500 mm, rarely to 2000 mm. It requires a sunny position in a well-drained soil for better growth. It grows finest in a deep, rich, light to medium soil. It prefers a pH in the range 6 to 7. This species fixes atmospheric nitrogen into the soil by symbiotic relationship with certain soil bacteria; these bacteria form nodules on the roots.

30.9.2 Seed Collection, Storage and Pre-Sowing Treatment

The ripened pods are collected in February–May by beating off the trees or collecting from the ground. 1550 to 1620 dry pods are estimated per kg weight depending upon geographical area. Seed extraction from pods is very difficult due to hardness of pods. Dried pods store well for 9–12 months upon proper maintenance of moisture content (up to 12%) and insect attack, otherwise seed viability is low. The percentage of germination is reported to be 70%–80% and plant percent is to be 60%–65%.

Owing to the hardness of the pods, germination takes place in 2–4 weeks eventually through proper germination techniques as follows (Madhuri et al. 2019):

a. Physical scarification, cutting of pod wings and soaking in water for few days prior to sowing.
b. Soaking the pods in slurry of cow dung for few days and keeping them in wet gunny bags and watering regularly before sowing.
c. By keeping alternate layers of pods and dead leaves in a pit. Flooding it with water until germination starts and sowing the pre-germinated seeds in root trainers.
d. 24 hrs soaked seeds tied up in gunny bag, the excess water to be drained off. After a few days, germinating seeds to be taken out and used for sowing and shifting to root trainers which are filled with coco pit.

The development of seedlings after sowing/spouting is slow, but it becomes faster once the tap root system is established. The root system is sensitive to drought and frost while in young stage of seedlings. Dibbling of pods can also be done in polythene bags or coco pit containers where root system develops vigorously. Watering and weeding is done regularly to enhance better growth. After attaining 1 ft height, the seedling can be shifted to bigger poly bags or planted in the field. The young seedlings are prone to high temperature, and maintaining the nursery under control situations is very important in nursery raising.

30.9.3 Planting in the Field

Farmland soil is loosened and dug by ploughing in April–May. Pits of appropriate size (50 cm × 50 cm) are dug at a spacing of 8 m × 8 m. About 25 kg FYM (farm yard manure), along with 200 g of nitrogen and 150 g of phosphorus, is mixed with soil of each pit as basal dose. The pits are refilled with this mixture after weathering of soil. Transplanting may be done in July–August (monsoon season) when the plants attain 1 ft height. A spacing of 8 m × 8 m is recommended, which accommodates about 160 plants per hectare. Gap filling in the field is done in September. When bijasal is planted at a spacing of 8 m × 8 m, intercropping can be done with a number of species such as medicinal plants and vegetable crops. The species can also be raised as a pure crop at smaller spacing.

30.9.4 Crop Maintenance

Fertilizer is applied in two split doses, the first in September and the second in January. Two manual weedings, the first one in August and the second in November, are recommended for good growth. Irrigation should be done six times in the first year (preferably once a month) through the check basin system or filling the basin of the pit with water.

30.9.5 Disease and Pest Control

No serious insect pests or diseases are noticed in mature stems and roots of the crop. However, young or ripened seeds are prone to seed borer, which decreases seed viability. This can be controlled by four sprays of endosulphan @ 0.003% at fortnightly intervals and application of phorate 10 g near the root zone, respectively.

30.9.6 Harvest Management and Crop Yield

The tree is ready to harvest after 10–15 years for production of heartwood. Kino gum is collected through wounding to the bark before logging of tree, and then it is dried well in shade. Each mature tree yields approximately 500 kg of dry heartwood after 10–15 years. Thus, an estimated yield of 750–800 quintals per hectare is obtained.

30.9.7 Economic Importance

Pterocarpus marsupium yields a high-quality timber with high percentages of strength, stiffness, shock-resisting ability, hardness, retention of shape and shear. The timber is mainly used for building purposes as doors, window frames, rafters, beams and posts and as a substitute for teak after suitable seasoning and treatment. It is used in the construction of railway carriages, wagons, carts, boats and occasionally for ship building. It is also used for railway sleepers, electric transmission poles and pit props in mines. It is employed for a variety of other purposes such as agricultural implements, drums, tool handles, camp furniture, mathematical instruments, picture frames, combs, cheap guns, sport rifles and parts of textile looms. It is found suitable for chip boards and carving joinery the building of timber bridges. Mixed with other woods, it can be utilized in the manufacture of pulp for wrapping paper.

30.10 CONCLUSION AND FUTURE REMARKS

After critical review, it was concluded that *Pterocarpus marsupium* can be used in the treatment of different ailments and has been proven therapeutically. *P. marsupium* is one of the most potent medicinal plants with a wide range of biological activities, and each part has been acknowledged for its therapeutic potential value. Bark, heartwood, leaves, flowers and kino gum are the major useful parts of *P. marsupium* used in the treatment of leucoderma, severe cough, elephantiasis, diarrhoea, hair fall and intestinal diseases, as cooling external application for inflammations and headache, as antipyretic, anti-helminthic, aphrodisiac, alexeteic and in biliousness, mental aberrations and ulcers in traditional systems of medicines The extracts of this plant now also play a major role in the treatment of diabetes. Various bioactive compounds of this plant's extracts are potentially active in reducing sugar levels in blood. The bioactive compounds are being used for their antihyperlipidaemic, anti-inflammatory, hepatoprotective, antidiarrheal, cardiotonic, antibacterial, anticancer, antimicrobial, antiulcer, analgesic, anticataract, antifungal, anthelmintic, antioxidant and antidiabetic effects. So, further research on the therapeutic potency of *P. marsupium* would be required for a better healthcare system.

ACKNOWLEDGMENTS

The authors are greatly indebted to Shri. Anoop Singh IFS, Director General, Forest Survey of India, Dehradun and Shri. Venu Prasad IFS, Regional Director, Forest Survey of India Central Zone, Nagpur, for their support and facilities.

REFERENCES

Abate, L., A. Bachheti, R.K. Bachheti & A. Husen. 2021. Antibacterial properties of medicinal plants: Recent trends, progress, and challenges. In*Traditional Herbal Therapy for the Human Immune System* (Ed. Husen A.). CRC Press, Boca Raton, FL, pp. 13–54.

Abirami, B., P. Gayathri & D. Uma. 2012. *In vitro* antioxidant potential of *Pterocarpus marsupium* bark. *International Journal of Chemistry and Pharmaceutical Sciences* 3: 17–24.

Achari, B., P.K. Dutta, S.K. Roy, P. Chakraborty, J. Sengupta, D. Bandyopadhyay, J.K. Maity, I.A. Khan, Y. Ding & D. Ferreira. 2012. Fluorescent pigment and phenol glucosides from the heartwood of *Pterocarpus marsupium*. *Journal of Natural Products* 75:655–660.

Ahmad, F., P. Khalid, M.M. Khan, M. Chaaubey, A.K. Rastogi & J.R. Kidwai. 1991. Hypoglycemic activity of *Pterocarpus marsupium* wood. *Journal of Ethnopharmacology* 35(1): 71–75.

Ahmad, F., P. Khalid, M.M. Khan, A.K. Rastogi & J.R. Kidwai. 1989. Insulin like activity in (-) epicatechin. *Acta Diabetologica Latina* 26(4): 291–300.

Ali, M.K., K.M.V. Narayan & N. Tandon. 2010. Diabetes & coronary heart disease: Current perspectives. *Indian Journal of Medical Research* 132(11): 584–597.

Anandharajan, R., K. Pathmanathan, N.P. Shankernarayanan, R.A. Vishwakarma & A. Balakrishnan. 2005. Upregulation of GLUT-4 and PPARγ by an isoflavone from *Pterocarpus marsupium* on L6 myotubes: A possible mechanism of action. *Journal of Ethnopharmacology* 97(2): 253–260.

Anis, M., M.K. Husain & A. Shahzad. 2005. In vitro plantlet regeneration of *Pterocarpus marsupium* Roxb., an endangered leguminous tree. *Current Science* 88(6): 861–863.

Anonymous. 2003. *The Wealth of India: Raw Materials*. Vol. VIII, Publications and Information Directorate, C.S.I.R., New Delhi, pp. 302–305.

Bachheti, A., Deepti, R.K. Bachheti & A. Husen. 2021. Medicinal plants and their pharmaceutical properties under adverse environmental conditions. In*Harsh Environment and Plant Resilience* (Ed. Husen A.). Springer International Publishing, Cham, pp. 457–502. https://doi.org/10.1007/978-3-030-65912-7_19

Badkhane, Y., A.S. Yadav, A.K. Sharma, D.K. Raghuwanshi, U. Shailendra Kumar, A.M. Firdous, A.L. Shabir & M. Tanuja. 2010. *Pterocarpus marsupium* Roxb.-Biological activities and medicinal properties. *International Journal of Advances in Pharmaceutical Sciences* 1: 350–357.

Banerji, M.A., N. Faridi, R. Atluri, R.L. Chaiken & H.E. Lebovitz. 1999. Body composition, visceral fat, leptin, and insulin resistance in Asian Indian men. *Journal of Clinical Endocrinology and Metabolism* 84(1): 137–144.

Barmukh, R.B. & T.D. Nikam. 2008. Promotion of seed germination in *Pterocarpus marsupium* Roxb. *Indian Journal of Plant Physiology* 13(2): 143–150.

Chakravarthy, B.K. & K.D. Gode. 1985. Isolation of (-)-epicatechin from *Pterocarpus marsupium* and its pharmacological actions. *Planta Medica* 51(1): 56–59.

Chakravarthy, B.K., S. Gupta, S.S. Gambhir & K.D. Gode. 1981a. Pancreatic beta-cell regeneration in rats by (-)-epicatechin. *The Lancet* 2: 759–760.

Chakravarthy, B.K., S. Gupta, S.S. Gambhir & K.D. Gode. 1981b. The prophylactic action of (-)-epicatechin against alloxan induced diabetes in rats. *Life Sciences* 29(20): 2043–2047.

Chakravarthy, B.K., S. Gupta & K.D. Gode. 1982a. Antidiabetic effect of (-)-epicatechin. *The Lancet* 2: 272–273.

Chakravarthy, B.K., S. Gupta & K.D. Gode. 1982b. Functional beta cell regeneration in the islets of pancreas in alloxan-induced diabetic rats by (-)-epicatechin. *Life Sciences* 31(24): 2693–2697.

Chawla, A., J. Kaur & A.K. Sharma. 2013. Systemic review: Pharmacognosy, phytochemistry, pharmacology and clinical applications of *Pterocarpus marsupium* Roxb. *International Journal of Pharmaceutical and Phytopharmacological Research* 2(5): 319–325.

Cordero-Herrera, I., M.A. Martin, L. Bravo, L. Goya & S. Ramos. 2013. Cocoa flavonoids improve insulin signalling and modulate glucose production via AKT and AMPK in HepG2 cells. *Molecular Nutrition and Food Research* 57(6): 974–985.

Deepa, R., H. Manjunatha, V. Krishna & B.E. Kumara Swamy. 2014. Evaluation of antimicrobial activity and antioxidant activity by electrochemical method of ethanolic extract of *Pterocarpus marsupium* Roxb. Bark. *Journal of Biotechnology & Biomaterials* 4(1): 166–170.

Devipriya, D., S. Gowri & T.R. Nideesh. 2007. Hepatoprotective effect of *Pterocarpus marsupium* against carbon tetrachloride induced damage in albino rats. *Ancient Science of Life* 27(1): 19–25.

Dhir, G.G., G. Mohan, R.B. Verma & S.S. Mishra. 1982. Studies on the antifungal activity of *Pterocarpus marsupium*: A clinical evaluation. *Clinical and Laboratory Investigations* 48(3): 154–156.

Dilpesh, J., I. Patel & R. Somani. 2011. Anti-diarrhoeal activity of ethanolic heartwood extract of *Pterocarpus marsupium*. *Pharmacologyonline* 1: 552–559.

Dudeja, V., A. Misra, R.M. Pandey, G. Devina, G. Kumar & N.K. Vikram. 2001. BMI does not accurately predict overweight in Asian Indians in northern India. *British Journal of Nutrition* 86(1): 105–112.

Gairola, S., V. Gupta, B. Singh, M. Maithani & P. Bansal. 2010. Phytochemistry and pharmacological activities of *Pterocarpus marsupium* – a review. *International Research Journal of Pharmacy* 1: 100–104.

Gamble, J.S. 1935. *Flora of the Presidency of Madras*. Adlard and Sons Ltd, London.

Gayathri, M. & K. Kannabiran. 2009. Antimicrobial activity of *Hemidesmus indicus, Ficus bengalensis* and *Pterocarpus marsupium* Roxb. *Indian Journal of Pharmaceutical Sciences* 71: 578–581.

Goel, R.K. & S.K. Bhattacharya. 1991. Gastroduodenal mucosal defence and mucosal protective agents. *Indian Journal of Experimental Biology* 29(8): 701–714.

Gupta, R. & R.S. Gupta. 2009. Effect of *Pterocarpus marsupium* in streptozotocin-induced hyperglycemic state in rats: Comparison with glibenclamide. *Diabetologia Croatica* 38(2): 39–45.

Hougee, S., J. Faber, A. Sanders, M.A. Hoijer & H.F. Smit. 2005. Selective COX-2 inhibition by a *Pterocarpus marsupium* extract characterized by pterostilbene and its activity in healthy human volunteers. *Planta Medica* 71(5): 387–392.

Hugar, A.L. & R.L. Londonkar. 2017. GC-MS profiling of bioactive components from aqueous extract of *Pterocarpus marsupium. International Journal of ChemTech Research* 10(9): 557–564.

Husen, A. 2021. *Traditional Herbal Therapy for the Human Immune System.* CRC Press, Boca Raton, FL. https://doi.org/10.1201/9781003137955

Husen, A. 2022. *Herbs, Shrubs and Trees of Potential Medicinal Benefits.* CRC Press, Boca Raton. https://doi.org/10.1201/9781003205067

Husen, A., R.K. Bachheti & A. Bachheti. 2021. *Non-Timber Forest Products: Food, Healthcare and Industrial Applications.* Springer, Cham, Switzerland. https://doi.org/10.1007/978-3-030-73077-2

ISFR. 2021. *India State of Forest Report 2021. Forest Survey of India.* Ministry of Environment, Forest and Climate Change, Government of India, DehraDun, India.

Ivorra, M.D., M. Payá & A. Villar. 1989. A review of natural products and plants as potential antidiabetic drugs. *Journal of Ethnopharmacology* 27: 243–275.

Jahromi, M.F., A.B. Ray & J.P. Chansouria. 1993. Antihyperlipidemic effect of flavonoids from *Pterocarpus marsupium. Journal of Natural Products* 56(7): 989–994.

Jung, M., M. Park, H.C. Lee, Y.H. Kang, E.S. Kang & S.K. Kim. 2006. Antidiabetic agents from medicinal plants. *Current Medicinal Chemistry* 13(10): 1203–1218.

Kachhawa, J.B., N. Sharma, S. Tyagi, R.S. Gupta & K.K. Sharma. 2012. In vitro of evaluation of antibacterial activity of Pterocarpus marsupium Roxb. *International Journal of Pharmacy and Pharmaceutical Sciences* 4(1): 67–68.

Kadavul, K. & N. Parthasarathy. 1998. Plant biodiversity and conservation of tropical semi-evergreen forest in the Shervarayan hills of Eastern Ghats, India. *Biodiversity and Conservation* 8(3): 421–439.

Kameswara, R.B., R. Guiri, M.M. Kesavulu & C.H. Apparao. 2001. Effect of oral administration of bark extracts of *Pterocarpus santalinus* L. on blood glucose level in experimental animals. *Journal of Ethnopharmacology* 74: 69–74.

Kitikar, K.R. & B.D. Basu. 1999. *Indian Medicinal Plants.* Vol. I. Scientific Publishers, DehraDun, India.

Krishnamurthi, A. 1998. *The Wealth of India – A Dictionary of Indian Raw Materials and Industrial Products.* Vol. VIII. CSIR-National Institute of Science Communication and Poly Research, New Delhi.

Kubish, H.M., J. Vang, T.M. Bray & J.P. Phillips. 1997. Targeted over expression of Cu/Zn superoxide dismutase protects pancreatic beta-cells against oxidative stress. *Diabetes* 46: 1563–1566.

Kumar, N. & T.R. Seshadri. 1976. A new triterpene from *Pterocarpus santalinus* bark. *Phytochemistry* 15(9): 1417–1418.

Kumar, S.R., T. Sivakumar, R.S. Sundram, P. Sivakumar, R. Nethaji, M. Gupta & U.K. Mazumdar. 2006. Antimicrobial and antioxidant activities of *Careya arborea* Roxb. Stem bark. *Iranian Journal of Pharmacology and Therapeutics*5: 35–41.

Kumaravel, R.S., S. Begum, H. Parvathi & M.S. Kumar. 2013. Phytochemical screening and *in vitro* antioxidant activity of ethyl acetate leaf extracts of *Pterocarpus marsupium* Roxb. (Fabaceae). *International Journal of Current Science and Research* 9: 46–55.

Lipinski, B. 2001. Pathophysiology of oxidative stress in diabetes mellitus. *Journal of Diabetes and Its Complications* 15(4): 203–210.

Londonkar, R.L. & A.L. Hugar. 2017. Physicochemical, phytochemical profiling and anti-microbial activity of *Pterocarpus marsupium. International Journal of Pharmaceutical Sciences and Research* 8(5): 2177–2183.

Madhuri, S., C. Dholariya, L.K. Behera, A.A. Mehta, S.A. Huse & R.P. Gunaga. 2019. Indian Kino Tree (*Pterocarpus marsupium* Roxb.): Biography of excellent timber tree species. *MFP News* 29(1): 4–8.

Mahnaz, M., S. Swapnil, G. Devasagayam & S. Saroj. 2010.Genotoxic assessment of *Pterocarpus marsupium* extract. *Journal of Complementary and Integrative Medicine* 7(1):1–11.

Maneesha, T., M. Sharma & H.N. Khare. 2015. Chemical constituents and medicinal uses of *Pterocarpus marsupium* Roxb. *Flora and Fauna* 21(1): 55–59.

Manisckam, M.M., M.A. Ramanthan, J.P. Jahromi, J. Chasouria & A.B. Ray. 1997. Antihyperglycemic activity of phenolics from *Pterocarpus marsupium. Journal of Natural Products* 60: 609–610.

Mankani, K.L., V. Krishna, B.K. Manjunatha, S.M. Vidya, S.J. Singh, Y.N. Manohara, A.U. Raheman & K.R. Avinash. 2005. Evaluation of hepatoprotective activity of stem bark of *Pterocarpus marsupium* Roxb. *Indian Journal of Pharmacology* 37(3): 165–168.

Maruthupandian, A. & V.R. Mohan. 2011. GC-MS analysis of some bioactive constituents of *Pterocarpus marsupium* Roxb. *International Journal ofChemTech Research* 3(3): 1652–1657.

Mathew, J., A.V. Subba Rao & N.V. Subba Rao. 1977. Photooxidation of pterostilbene from *Pterocarpus marsupuim* Roxb. *Current Science* 46: 337–338.

Mathew, K.M. 1983. *The Flora of Tamil Nadu Carnatic*. St. Josephs College, Tiruchirapalli, India.

Maurya, R., A.B. Ray, F.K. Duah, D.J. Slatkin & P.L. Schiff. 1984. Constituents of *Pterocarpus marsupium*. *Journal of Natural Products* 47(1): 179–181.

Mitra, J. & T. Joshi. 1983. Isoflavonoids from the heartwood of *Pterocarpus marsupium*. *Phytochemistry* 22(10): 2326–2327.

Mohankumar, S.K., T. O'Shea & J.R. McFarlane. 2012. Insulinotrophic and insulin-like effects of a high molecular weight aqueous extract of *Pterocarpus marsupium* Roxb. hardwood. *Journal of Ethnopharmacology* 141(1): 72–79.

Mohire, N.C., V.R. Salunkhe, S.B. Bhise & A.V. Yadav. 2007. Cardiotonic activity of aqueous extract of heartwood of *Pterocarpus marsupium*. *Indian Journal of Experimental Biology* 45: 532–537.

Nair, R., T. Kalariya & S. Chanda. 2005. Antibacterial activity of some selected Indian medicinal flora. *Turkish Journal of Biology* 29: 41–47.

Narendar, K., M.N. Nayak, M.G. Jamadar, A.M. Patil & S. Anand. 2016. Comparison of the effect of *Pterocarpus marsupium* with pioglitazone in dexamethasone-induced insulin resistance. *Asian Journal of Pharmaceutical and Clinical Research* 9(2): 211–214.

Naziroğlu, M. & M. Cay. 2001. Protective role of intraperitoneally administered vitamin E and selenium on the antioxidative defense mechanisms in rats with diabetes induced by streptozotocin. *Biological Trace Element Research* 79: 149–159.

Pan, M.H., Y.H. Chang, V. Badmaev, K. Nagabhushanam & C.T. Ho. 2007. Pterostilbene induces apoptosis and cell cycle arrest in human gastric carcinoma cells. *Journal of Agricultural and Food Chemistry* 55(19): 7777–7785.

Panda, S.K., R.P. Padhy, G.D. Ray & D.C. Hial. 2015. Phytochemical investigation and anthelmintic activity of various leaf extracts of *Pterocarpus marsupium* Roxb. *Biology*: 639–645.

Pandey, M.C. & P.V. Sharma. 1976. Hypoglycaemic effect of bark of *Pterocarpus marsupium* Roxb. (Bijaka) on alloxan induced diabetes. *Medical Surgical* 16: 9–11.

Pant, D.R., N.D. Pant, D.B. Saru, U.N. Yadav & D.P. Khanal. 2017. Phytochemical screening and study of antioxidant, antimicrobial, antidiabetic, anti-inflammatory and analgesic activities of extracts from stem wood of *Pterocarpus marsupium* Roxb. *Journal of Intercultural Ethnopharmacology* 6(2): 170–176.

Rahman, S. & A. Husen. 2021. Potential role of medicinal plants in the cure of liver and kidney diseases. In *Non-Timber Forest Products* (Eds. Husen A., Bachheti R.K., Bachheti A.). Springer International Publishing, Cham, pp. 229–225. https://doi.org/10.1007/978-3-030-73077-2_10

Rahman, S. & A. Husen. 2022. *Asparagus racemosus* (Shatavari) and *Dioscorea villosa* (Wild yams). In *Herbs, Shrubs, and Trees of Potential Medicinal Benefits* (Ed. Husen A.). CRC Press, Boca Raton, FL, pp. 95–112.

Rahman, M.S., M. Mujahid, M.A. Siddiqui, M.S. Rahman, M. Arif, S. Eram, A. Khan & M. Azeemuddin. 2008. Ethnobotanical uses, phytochemistry and pharmacological activities of *Pterocarpus marsupium*: A review. *Pharmacognosy Journal* 10(6): 1–8.

Ramya, S., M. Kalayansundaram, T. Kalaivani & R. Jayakumararaj. 2008. Phytochemical screening and antibacterial activity of leaf extracts of *Pterocarpus marsupium* Roxb. *Ethnobotanical Leaflets* 12: 1029–1034.

Rizvi, S.I. & M.A. Zaid. 2001. Intracellular reduced glutathione content in normal and type 2 diabetic erythrocytes: Effect of insulin and (-) epicatechin. *Journal of Physiology and Pharmacology* 52(3): 483–488.

Salunkhe, V.R., A.V. Yadav, A.S. Shete, S.R. Kane & A.S. Kulkarni. 2005. Antiinflammatory acitivity of hydrogels of extracts of *Pterocarpus marsupium* and *Coccinia indica*. *Indian Drugs* 42(5): 319–321.

Sapha, B. 1956. Clinical observation of antidiabetic properties of *Pterocarpus marsupium*, *Eugenia jambolona*. *Journal of Indian Medical Association* 27: 388–390.

Sashikanth, Y.V., P. Aravindkumar & C. Swarupa. 2012. Two way relation of diabetes mellitus and periodontitis: A review. *Annals and Essences of Dentistry* 4(1): 96–98.

Seshadri, T.R. 1972. Polyphenols of *Pterocarpus* and *Dalbergia* wood. *Phytochemistry* 11(3): 881–898.

Shah, D.S. 1967. A preliminary study of the hypoglycemic action of heartwood of *Pterocarpus marsupium* Roxb. *Indian Journal of Medical Research* 55(2): 166–168.

Sikdar, A., A. Biswas, S. Bhattacharya & M. Biswas. 2013. Assessment of analgesic activity of *Pterocarpus marsupium* leaf extracts in Swiss albino mice. *Journal of Advanced Pharmacy Education and Research* 3(1): 42–45.

Singh, M.V. & M. Ali. 2016. Phytochemical and pharmacological profile of *Pterocarpus marsupium*: A review. *The Pharma Innovation Journal* 5(4): 31–39.

Singh, N.P. & S. Karthikeyan. 2000. *Flora of Maharashtra State, Dicotyledons*. Vol. 1. Botanical Survey of India, Kolkata.

Singh, P.K., D. Baxi, S. Banerjee & A.V. Ramachandran. 2012. Therapy with methanolic extract of *Pterocarpus marsupium* Roxb. and *Ocimum sanctum* Linn reverses dyslipidemia and oxidative stress in alloxan induced type I diabetic rat model. *Experimental and Toxicologic Pathology* 64(5): 441–448.

Singh, U., A.M. Wadhawani & B.M. Johri. 1965. *Dictionary of Economic Plants in India*. Indian Council of Agriculture and Research, New Delhi.

Sonkar, S., A.K. Singh & A. Husen. 2022. *Convolvulus pluricaulis* (Shankhpushpi) and *Erythroxylum coca* (Coca plant). In *Herbs, Shrubs, and Trees of Potential Medicinal Benefits* (Ed. Husen A.). CRC Press, Boca Raton, FL, pp. 83–93.

SubbaRao, V. & J. Mathew. 1982. Marsupol: A novel isoflavonoid glycol from *Pterocarpus marsupium*. *Phytochemistry* 21: 1837–1838.

Takehara, K.O., K.I. Tashima & K.O. Takeuchi. 1997. Alterations in duodenal bicarbonate secretion and mucosal susceptibility to acid in diabetic rats. *Gastroenterology* 112(2): 418–428.

Tchamadeu, M.C., P.D.D. Dzeufiet, P. Nana, C.C.N. Kouambou, F.N. Tsofack, J. Allard, N. Blaes, R. Siagat, L. Zapfack, J.P. Girolami, I.Tack, P. Kamtchouing & T. Dimo. 2011. Acute and sub-chronic oral toxicity studies of an aqueous stem bark extract of *Pterocarpus soyauxii* Taub (Papilionaceae) in rodents. *Journal of Ethnopharmacology* 133(2): 329–335.

Tippani, R., M. Porika, V. Allenki, R.Y. Narsimha & T. Christopher. 2010. Antioxidant and analgesic activities of *Pterocarpus marsupium* Roxb. *Journal of Herbs, Spices and Medicinal Plants* 16(1): 63–68.

Upadhyay, R. 2013. Ethnomedicinal uses of tree barks by tribals of hoshangabad, Madhya Pradesh, India. *International Journal of Biotechnology and Bioengineering Research* 4(7): 671–676.

Valko, M., D. Leibfritz, J. Moncol, M.T. Cronin, M. Mazur & J. Telser. 2007. Free radicals and antioxidants in normal physiological functions and human disease. *The International Journal of Biochemistry & Cell Biology* 39: 44–84.

Vats, V., J.K. Grover & S.S. Rathi. 2002. Evaluation of anti-hyperglycemic and hypoglycemic effect of *Trigonella foenum-graecum* Linn, *Ocimum sanctum* Linn and *Pterocarpus marsupium* Roxb in normal and alloxanized diabetic rats. *Journal of Ethnopharmacology* 79(1): 95–100.

Vats, V., S.P. Yadav, N.R. Biswas & J.K. Grover. 2004. Anti-cataract activity of *Pterocarpus marsupium* bark and *Trigonella foenum-graecum* seeds extract in alloxan diabetic rats. *Journal of Ethnopharmacology* 93(2&3): 289–294.

WHO. 1980. *Second Report of the WHO Expert Committee on Diabetes Mellitus*. World Health Organization, Geneva.

31 Bioactive Compounds and Antidiabetic and Other Health Benefits of *Eucalyptus globulus*

Balahbib Abdelaali, Douae Taha, Zerkani Hannou, El Hamri Ayoub, Naoual El Menyiy, Abdelhakim Bouyahya, and Mohammad S. Mubarak

CONTENTS

31.1 INTRODUCTION

Medicinal plants are a precious heritage for humanity, especially for poor communities in developing countries who depend on them for their daily needs (Husen, 2021, 2022; Husen et al., 2021). Although dating from a time more distant than that of the great civilizations known to human history, they continue to attract great interest. The return to the "natural label" and complementary medicines strongly marked in the last decades in the developed countries, as well as the almost perpetual recourse of the populations of developing countries to plants to assure their health needs reinforce the interest in medicinal plants (Kala et al., 2006; Srivastava & Kumar, 2013).

Traditional medicine has long used medicinal plants as herbal medicines that can be used whole or in part for therapeutic purposes. Currently, nearly 25% of prescriptions worldwide are herbal and, according to WHO estimates in 2002, more than 80% of the African population still use medicinal plants to meet their healthcare needs (Gunjan et al., 2011; Kong et al., 2003). Within this context, *E. globulus* is a medicinal plant of the Myrtaceae family, genus *Eucalyptus*. It is native to Australia, especially Tasmania, and was introduced in the subtropics of Asia and the Mediterranean basin (Boulekbache-Makhlouf et al., 2013). *E. globulus* is a medicinal plant widely used in traditional medicine. In addition, several studies have shown that this plant exhibits several pharmacological activities due to its richness in bioactive molecules (Assaggaf et al., 2022b; Naceiri Mrabti, Rajab, Attar, Hamed, et al., 2022; Bachir & Benali, 2008, 2012; Tan et al., 2008; Vázquez et al., 2008).

DOI: 10.1201/b23347-31

Chemically this plant is rich in phenolic compounds and tannins, and it is recognized as one of the main families that produce flavonoids and aromatic oils. Many species belonging to this family are a source of therapeutic agents; these compounds are called secondary metabolites, which are responsible for the protection against pathogens and for biological activities such as antioxidant, anti-inflammatory, anticancer, antimicrobial, and analgesic (Abdelaali et al., 2021; Al-Mijalli et al., 2022; Assaggaf et al., 2022a; Naceiri Mrabti, Rajab, Attar, Alyamani, et al., 2022; Bouyahya, Bakri, et al., 2017; Bouyahya, Dakka, et al., 2017; Bouyahya et al., 2021). In light of the preceding discussion, this chapter focuses on the pharmacological studies, traditional use, toxicological evaluation, safety issues, cultivation practices, and future remarks related to *E. globulus*.

31.2 BOTANICAL DESCRIPTION

According to evidence obtained in Europe, the broadleaf evergreen plant *E. globulus* can reach a height of 70 meters at its greatest height (Cerasoli, 2016). Although there are more than 700 different species of this plant, *E. globulus* is the most common in East Bay (Paine & Hanlon, 2010). It is an aromatic plant with a straight trunk, a thick crown, and a tap root system that extends down more than ten feet. On new shoots, its leaves are grouped in opposition. On older ones, leaves are arranged alternately with solitary flowers (Fernandes et al., 2022). Eucalyptus species' unique flowers and fruit are their most distinct aspects (capsules or gum nuts). Numerous fluffy stamens, which might be white, cream, yellow, pink, or red in color, are present in flowers. In the bud, the flowers are protected by an operculum, a cap made of fused sepals, petals, or both. Flowers lack petals and instead adorn themselves with many showy stamens. One of the characteristics that unites the genus is the operculum being driven off and separating away from the bowl base of the flower as the filaments expand (Hernández et al., 2022).

The age of the plant, the method the bark sheds, the length of the fibers, the degree of furrowing, the thickness, the hardness, and the color all affect how the eucalyptus bark looks. All mature trees produce an annual layer of bark, thus increasing the stem's diameter. The bark is divided into many distinct flakes, has long fibers that may be pulled off in long lengths, and is hard, rough, and deeply furrowed. Because of these characteristics, the bark can be taken off in long, thin pieces while remaining distantly related in other areas. The tree bark is tough, coarse, and deeply furrowed. Its intense crimson or black coloring is caused by the saturation of dried sap secreted by the tree. The fruit, which is unusual for the genus and bears fruit in the autumn and winter, resembles cone-shaped woody capsules known as "gum nuts." *E. globulus* seeds have a highly varied appearance. Strongly inherited characteristics such as size, color, shape, and surface ornamentation indicate taxonomic groups (Fernandes et al., 2022). Depicted in Figure 31.1 are the flowers and leaves of *E. globulus*.

FIGURE 31.1 (a) Flowers and (b) leaves of *Eucalyptus globulus*.

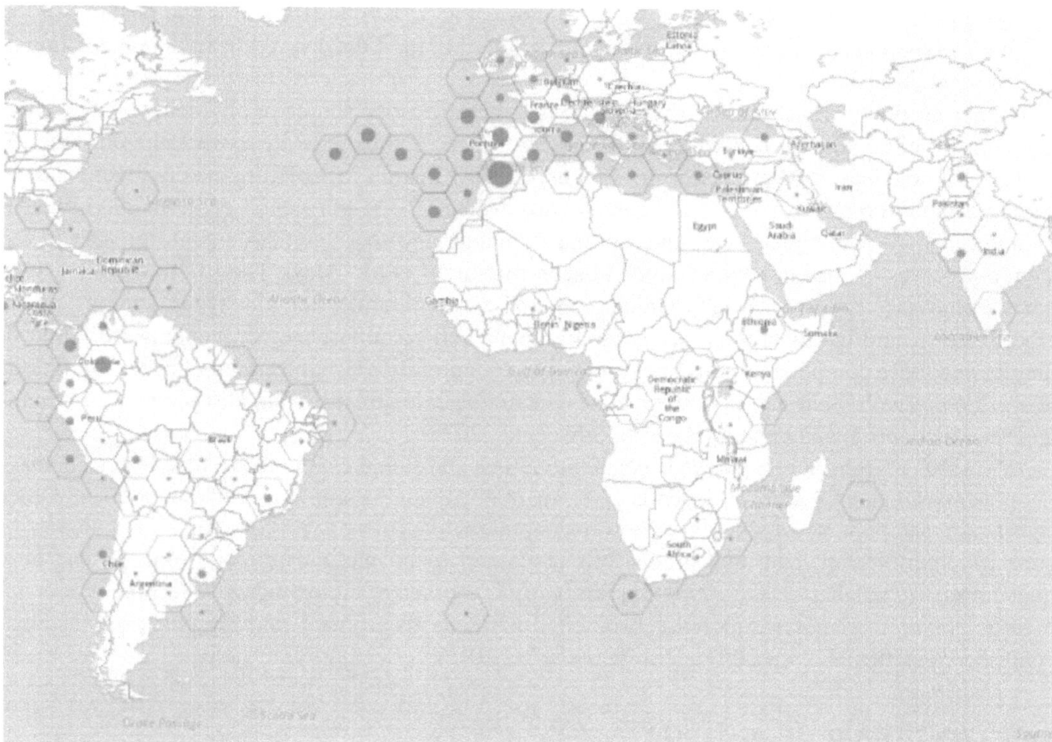

FIGURE 31.2 Geographic distribution of *Eucalyptus globulus*.

31.3 DISTRIBUTION

Although it cannot survive at temperatures below –5 °C, *E. globulus* may grow in mild, warm, and tropical regions with mean annual temperatures ranging from 3–22 to 21–40 °C and mean annual rainfall ranging from 250 to 2500 mm. *E. globulus* is grown in the Mediterranean region and can reach 350 meters above sea level. Young plants are typically planted in the spring or at the end of the summer. *E. globulus* should be planted in an environment with high humidity to prevent leaf border burning. It may thrive in a variety of soil types and with little water. Furthermore, deep, silty, or loamy soils with a clay base and a reachable water table are ideal for the soil type growth. It is one of the species that can handle acidic soils and soils with an optimal pH range of 5.5 to 6.5. Shown in Figure 31.2 is the geographical distribution of *E. globulus*; location is 10.2572°N latitude, 78.8861°E longitude, 216 meters above sea level, typical temperatures ranging from 33.5 to 42.2 °C, and yearly rainfall of 1043.31 mm in India. Red soil comprises the research area's soil type (Muller, 2009); (Adame et al., 2022).

31.4 PHYTOCHEMICAL CONSTITUENTS

Eucalyptus leaves were obtained using water distillation, and the essential oil produced from it is utilized in medicine. When properly constructed, this fluid is translucent, tasteless, and odorless, and it dissolves in its weight of alcohol. Eucalyptol, which makes up 70% of the volume of eucalyptus, is one of its main constituents. Eucalyptol (5.01%–51.13%) is present in the oil found in the leaves, depending on the age and the location of the collection site. In contrast, EOs secreted by the fruit, bud, and branch contain 0.00%, 12.1%, and trace amounts of α-thujene, respectively. In addition, they contain eucalyptol (16.11%, 37.01%, and 55.86%, respectively), and aromadendrene

(24.11%, 15.90%, and 38.01%, respectively) (Moreira et al., 2022). Furthermore, the majority of the oxygenated monoterpenes and sesquiterpenes are found in the EOs derived from the plant (Flores-Macías et al., 2021).

Numerous publications showed that eucalyptus trees are a rich source of phytochemicals including tannins, alkaloids, propanoids, and flavonoids. These compounds are extracted from the tree roots, stem, and leaves. Many researchers focused their research on isolating the phytoconstituents from the plant's organs, particularly eucalyptol aromadendrene, α-gurjunene, epiglobulol, β-pinene, monoterpene ketone, alpha, beta, and gamma terpineol, and these compounds are present in the stem and leaves of the eucalyptus (Boulekbache-Makhlouf et al., 2010). The EO found in various parts comprises various volatile organic compounds, including alcohol, aldehydes, ketones, acids, ethers, esters, and hydrocarbons. Many monoterpene and sesquiterpene components are found in nature, and these components consist of two or more isoprene (C_5H_8) units (Boulekbache-Makhlouf et al., 2010). The EO has a variety of elements, including calcium, nitrogen, phosphorus, iron, manganese, zinc, boron, and copper (R. C. da Silva et al., 2022).

In various regions across the continents, numerous additional elements are discovered in varying concentrations; 1,8-cineole depends upon the maturity and origin of their collection site. Other significant components of the leaf oils were α-pinene, *p*-cymene, cryptone, and spathulenol. In contrast, Spain's fruit, bud, and branch oils contained α-thujene and trace,1,8-cineole, and aromadendrene (Chalchat et al., 1995). In Ethiopia, 1, 8-cineole, cis-ocimene, α-terpineol acetate, α-terpineol, aromadendrene, globulol, β-pinen, β-myrcene, 4-terpineol, and camphene are the most common constituents of eucalyptus (Shiferaw et al., 2019).

31.5 ANTIDIABETIC ACTIVITY

Diabetes is defined as chronic hyperglycemia linked to a disorder of insulin secretion, a condition of insulin action, or both (Raccah, 2004). Type 1 diabetes (T1D) accounts for 5% to 10% of all cases and is characterized by autoimmune or idiopathic destruction of the β-cells of the islets of Langerhans in the pancreas. In contrast, type 2 accounts for 90%–95% of the total number of diabetes cases and can range from predominant insulin resistance with relative insulin deficiency to predominantly secretory defect with or without insulin resistance (Kouchadé et al., 2017). The number of people with diabetes mellitus (DM) is increasing due to population growth, low birth weight, aging, urbanization, sedentary lifestyles with physical inactivity, overworked diets, smoking, psychological stress, and increasing prevalence of obesity (Association, 2004; Rituparna & Agrawal, 2007). Insulin therapy is the mainstay of patients with T1DM, while dietary and lifestyle modifications are the mainstay for the treatment and management of type 2 diabetes mellitus (T2DM). Insulin is also necessary in T2DM when blood sugar levels cannot be controlled by diet, weight loss, exercise, and oral medications (Tripathi & Srivastava, 2006).

Despite the presence of known antidiabetic drugs in the pharmaceutical market, herbal remedies are successfully used to treat this disease (Kooti et al., 2016). Many traditional diabetes treatments are used around the world. Herbal medicines and formulations are often considered less toxic with fewer side effects (Annapurna et al., 2001). For millennia, herbal medicines have been a valuable source of therapeutic agents, and many of today's medicines are natural products obtained from plants or their derivatives (Atanasov et al., 2015).

Currently, more than 400 traditional diabetes treatments have been reported, of which only a small number have received scientific and medical evaluation to assess their effectiveness (Gunjan et al., 2011). The hypoglycemic effect of certain plant extracts has been confirmed in human and animal models of T2D (Preethi, 2013). Indeed, several studies have investigated the antidiabetic activity of medicinal plants (Ahmed et al., 2010; Bhushan, 2010; Omari et al., 2019; Sabu and Kuttan, 2002). In this context, *E. globulus* has traditionally been used to control blood glucose levels. This traditional knowledge has been scientifically authenticated through *in vivo* and *in vitro* studies. The antidiabetic potential of the plant was evaluated using *in vitro* α-amylase and α-glucosidase inhibitory assays. Results showed that *E. globulus* significantly inhibits the activity of α-amylase and α-glucosidase (Usman et al., 2020, 2022). In a study by Capetti et al., 2020, focusing on the α-amylase inhibition assays of 62 EOs from medicinal plants, including *E. globulus*, an interesting α-amylase inhibition (34%) was marked.

In addition, Kamble and coworkers (Kamble et al., 2021) confirmed the antidiabetic potential of *E. globulus* extract in diabetic rats (induced by alloxan). These researchers observed that this extract reduces blood glucose level, with significant oral glucose tolerance. Moreover, the leaf extract of *E. globulus* (200 mg/kg b.w.) prevented hyperglycemia in normal rats (Ajilore et al., 2021). Repeated oral administration of *E. globulus* extract in diabetic rats significantly increased basal plasma insulin concentration (Jouad et al., 2004). Indeed, the leaf extract of *E. globulus* has significant blood-glucose lowering potential in glucose-loaded rats (Houacine et al., 2012). The main active constituents derived from *E. globulus* and other medicinal plants affect various metabolic cascades, directly or indirectly affecting the blood sugar level in the human body (Prabhakar & Doble, 2008). The antidiabetic activity of plants may depend on several mechanisms such as reduction of insulin resistance, stimulation of insulin secretion from β-cells, or/and inhibition of the insulin degradation process, regeneration and repair of damaged pancreatic β-cells, and inhibition of renal glucose reabsorption (Abou Khalil et al., 2016; Jarald et al., 2008; Singh, 2011).

31.6 ANTI-INFLAMMATORY ACTIVITY

Natural products are considered an essential source of bioactive molecules with anti-inflammatory activity (Bouyahya et al., 2021). In this context, eucalyptol contributes, in particular, to the reduction of inflammatory states usually treated with steroids. However, eucalyptol does not have the side effects of long-term steroids (Alaba et al., 2022; Zonfrillo et al., 2022). Several studies reported the anti-inflammatory effects of *E. globulus*. It showed analgesic and anti-inflammatory effects in mice and rats. Research findings indicated that the cineole contained in eucalyptus administered at a dose of 200 mg three times a day inhibited the production of the pro-inflammatory mediators' leukotrienes B4 and prostaglandins E2 (PGE_2) produced by monocytes cultured *ex vivo* in patients suffering from bronchial asthma and in healthy patients (Juergens et al., 1998). Indeed, the cineole contained in eucalyptus reduced the inflammation of rat paws induced by carrageenan and dextran. It exhibited central and peripheral analgesic effects (J. Silva et al., 2003) associated with the inhibition of nitrogen oxide (NO) production and antiradical effects in murine macrophages cultured *in vitro* (Vigo et al., 2004). It exhibits central and peripheral analgesic effects (J. Silva et al., 2003) associated with inhibition of NO production and antiradical effects in murine macrophages cultured *in vitro* (Vigo et al., 2004).

In addition, it also showed immunomodulatory effects. In particular, it stimulated phagocytosis in a dose-dependent manner of human macrophages *in vitro* (Serafino et al., 2008). Cineole inhibits the production of pro-inflammatory cytokines such as tumor necrosis factor-α (TNF-α) and interleukin-1β (IL-1β) produced by lipopolysaccharide (LPS)-stimulated lymphocytes and monocytes *in vitro* (Juergens et al., 2004). In another study, the activity of eucalyptol was evaluated in severe asthmatic patients, suggesting the effectiveness of this molecule for its use as a mucolytic agent in diseases of the upper and lower respiratory tracts (Juergens et al., 2003).

31.7 TRADITIONAL USE

Eucalyptus trees have been employed for centuries as a traditional herbal remedy. Their leaf extracts have been used daily due to their various medicinal properties. Several studies have indicated that *E. globulus* is a plant widely used in traditional medicine against multiple diseases. In fact, *E. globulus* leaves are used in treating conditions affecting the circulatory system; this plant species is highly recommended in treating fever and typhoid. The plant's association with other species, such as cloves, lavandula flowers, and lemon juice, can be used as a compress on the forehead and temples to reduce the temperature and release a large amount of sweat. In addition, this plant is widely used in the care of the respiratory system and is considered an antibiotic for the respiratory system (Lahsissene & Kahouadji, 2010). Using the decoction, the mixture of *E. globulus* leaves with the leafy stems of *Marrubium vulgare*, *Ammi visnaga* Lamk, *Eugenia caryophllata* Thunb., *Brassica rapa* leaves, *Allium cepa*, *Rubus ulmifolius*, and *Vitex agnus-castus* is recommended in cases of diabetes. An infusion of *E. globulus* leaves immediately after harvest is indicated against hyperglycemia (Benkhnigue et al., 2014).

Steaming *E. globulus* leaves, leafy stems of *Marrubium vulgare* L., *Ammi visnaga* Lamk., *Eugenia caryophyllata* Thunb., *Brassica* L., and *Allium cepa* L. are recommended in the case of typhoid fever. In addition, it is recommended to soak young leaves (a handful of leaves per half liter of water; one glass in the evening until recovery) to treat bronchitis and flu-like conditions. It is advisable to use the leaves as compresses (for ten days) in case of rheumatic pains (Salhi et al., 2010). The leaves by infusion and/or inhalation of *E. globulus* are used to relieve stomach pain, liver disorders, headaches, bronchitis, influenza, colds, coughs, asthma, fever, diabetes, immune deficiency, and infections (Mikou et al., 2015). Additionally, the leaves of *E. globulus*, in infusion and/or poultice, have been used by the Kenitra (Morocco) population to treat asthma (El Hachlafi et al., 2020). Stem and leaf vapor is used against influenza (Bouyahya, Abrini, et al., 2017). It is also worth mentioning that the EO from the leaves of *E. globulus* is widely used to treat cold, pulmonary tuberculosis, diabetes, nasal congestion, bronchial diseases, and asthma. It is also used as an antiseptic, antioxidant, and antiseptic agent, especially in treating upper respiratory tract infections (Song et al., 2009). The EO is also used externally to relieve rheumatism, as an anti-irritant, and for specific skin conditions (Salehi et al., 2019; Song et al., 2009). So far, there is no known cure for the novel coronavirus, but many traditional remedies, mostly herbs, have been taken by some people to prevent and treat COVID-19. Indeed, an ethnopharmacological study was conducted on medicinal plants used for prevention purposes during the COVID-19 epidemic in Morocco (EL ALAMI et al., 2020), and to treat or reduce its symptoms. In addition, Wannes and Tounsi (2020) reported that patients infected with this virus used *E. globulus* with honey, sweet orange juice, and tea as a home remedy for COVID-19.

31.8 ANTIBACTERIAL ACTIVITY

The human body, constantly exposed to various microorganisms (bacteria, parasites, fungi, viruses), has a complex defense system that allows it to encounter or harbor these microorganisms without harming its tissues. However, the infection can lead to serious infectious diseases under certain conditions. Thus, during their development, people can be subjected to several microbial infections requiring antimicrobials (antibiotics and antifungals) to regain health. The frequent use of these microbial substances can lead to the selection of multi-resistant strains, hence the importance of directing research toward discovering new pathways, which would constitute a source of inspiration for new drugs. The study of plants, especially medicinal plants, is an excellent avenue to explore (Billing & Sherman, 1998) and develop natural antimicrobials capable of inhibiting and killing microbes. Numerous studies have reported the antimicrobial effect of EO and extracts of *E. globulus* against microbial strains. Ait-Ouazzou and colleagues (Ait-Ouazzou et al., 2012) found that the EO of *E. globulus* has inhibition diameters of 14.0 ± 4.9 and 12.5 ± 0.1 mm against *Staphylococcus aureus* and *Listeria monocytogenes* (EGD-e), respectively.

In addition, Ghaffar et al. (2015) detected the significant antibacterial potential of EO from *E. globulus* against *S. aureus* (28 ± 0.835mm), *B. subtilis* (17 ± 0.833mm), and *E. coli* (13 ± 0.83mm). Thus, the study (Ghalem & Mohamed, 2008) showed that at 10 µL, *E. globulus* EOs from Algeria cause inhibitions of 2.9 cm for *S. aureus* and *E. coli*, respectively. Similarly, the study by Derwich et al. (2009) reported that the EO extracted from the leaves of *E. globulus* collected in Boulemane-Morocco exerts a strong antibacterial effect against *E. coli* (48.15 mm), *S. aureus* (13.50 mm), and *S. intermedius* (10.26 mm). Published research by Damjanović-Vratnica and colleagues (Damjanović-Vratnica et al., 2011) highlighted an interesting antibacterial effect of EO extracted from leaves of *E. globulus* against *S. aureus*, *S. aureus* ATCC 25923, *E. coli*, *E. coli* ATCC 25922, *K. pneumoniae*, and *A. baumannii*, indicating zones of inhibition ranging from 21 to 51 mm. Additionally, Bencheikh et al. (2021) showed that the aqueous extracts of *E. globulus* are active against two gram+ strains with an inhibition diameter varying between 19 and 23 mm for *S. aureus* and between 16.5 to 21.5 mm for *B. subtilis*. On the other hand, previous studies revealed the low antibacterial activity of *E. globulus* EO. In this respect, the study of Miguel et al. (2018) revealed a weak antibacterial activity of the EO from the leaves of *E. globulus* of Portuguese origin, against the bacterial strains *S. pneumoniae* (D39 and TIGR4) and *H. influenza* DSM 9999.

31.9 SAFETY ISSUES

A substance's efficacy in pharmacology is not sufficient to justify its use in therapy. A toxin is a substance capable of disrupting, immediately or in the long term, temporarily or permanently, the normal functioning of a living organism, which can lead to its complete suppression and death (Craig & Stitzel, 2004). It is, therefore, necessary to define the risk-benefit ratio in the therapeutic indication of each substance (Jothy et al., 2011). However, although a group of studies showed that spherical eucalyptus has promising and effective abilities in the treatment of certain diseases, some studies highlighted its toxic effects and the safety of its use. In this context, the histological study of hepatic and renal tissues from rats that received a dose of 2000 mg/kg of body weight showed the presence of structural abnormalities in the tissues (Ajilore et al., 2021). According to Hilpipre (2014), at high or repeated doses eucalyptol could activate drug degradation enzymes (induction of hepatic cytochromes), which could modify their efficacy, particularly in the case of drugs with a low therapeutic margin (Hilpipre, 2014). According to Vigan and Besançon (Vigan & Besançon, 2009), the EO of *E. globulus* should not be used in children under 7 years of age. Additionally, it is recommended not to incorporate camphor, eucalyptol, and menthol in cosmetic products intended for children under 3 years of age. This recommendation does not apply to menthol in oral hygiene products (Vigan & Besançon, 2009).

31.10 CULTIVATION PRACTICES

Planting exotic trees in land benefits poverty reduction and adaptation to climate change. In this context, *E. globulus* has shown an important socioeconomic interest (Morales et al., 2017). Kebebew et al. (Kebebew, 2002) reported that *E. globulus* serves as a cash crop for smallholder farmers and has contributed significantly to farmers' livelihoods (Kebebew, 2002). In another study, Lemenih et al. (2005) reported high clay content under planting of *E. globulus* compared to other soil (Lemenih et al., 2005). Among the beneficial ones are the low density of *E. globulus*, while the soil under *E. globulus* had the lowest bulk density, indicating that the soil is less compacted (Hazelton & Murphy, 2016). In this regard, Yitaferu et al. (2013) reported lower bulk density (1.07 g cm^{-3}) under eucalyptus compared to cropland (1.11 g cm^{-3}) in West Gojam Amhara Regional State, Ethiopia. According to Alemayehu and Melka (Alemayehu & Melka, 2022), eucalyptus offers a much better return on investment than alternative land uses. In addition, eucalyptus contributes significantly to rural development and poverty alleviation. It also promotes biodiversity conservation through the cultivation of eucalyptus (Alemayehu & Melka, 2022). Eucalyptus wood can be used to make floors, barrels, containers, and hard fiberboard (Baso, 1999; Bolaños, 1946).

31.11 CONCLUSION AND PERSPECTIVES

In this review, we focused on the phytochemistry, ethnomedicinal use, and pharmacology investigations of *E. globulus*. This plant, widely grown in several countries, is used by different populations to treat certain diseases, including diabetes, immune failure, infections, and asthma. Phytochemical investigation showed that *E. globulus* contains numerous chemical compounds belonging to several chemical classes, such as terpenoids, flavonoids, and phenolic acids. The variability between these components depends on the plant part used and the origin of the plant. The pharmacological examination of *E. globulus* conducted by researchers showed that the extracts of this species possess potent *in vitro* and *in vivo* biological effects, including antibacterial, antidiabetic, anti-inflammatory, and cytotoxic activities. These biological properties mainly depend on the chemical compounds in *E. globulus* that showed important pharmacological actions on different targets. In this way, further molecular investigations to determine the pharmacodynamic and pharmacokinetic characteristics studies are needed to ensure the bioavailability of these bioactive molecules. Moreover, its main compounds should also be tested for their toxicity in order to complete the toxicity profile. It would also be interesting to make formulations with other plants to study and search for a synergistic or potentiation effect that will be useful for the preparation of a food supplement or a phytomedicine. However, this orientation must be accompanied by rigorous management in terms of the environment and the renewal of species. In view of the lack of regulations regarding aromatherapy, it would be important to take into consideration in the future changes concerning the supervision and regulation of EOs to limit any risk of misuse.

REFERENCES

Abdelaali, B., El Menyiy, N., El Omari, N., Benali, T., Guaouguaou, F.-E., Salhi, N., Naceiri Mrabti, H., & Bouyahya, A. (2021). Phytochemistry, toxicology, and pharmacological properties of Origanum elongatum. *Evidence-Based Complementary and Alternative Medicine, 2021.*

Abou Khalil, N. S., Abou-Elhamd, A. S., Wasfy, S. I., El Mileegy, I. M., Hamed, M. Y., & Ageely, H. M. (2016). Antidiabetic and antioxidant impacts of desert date (Balanites aegyptiaca) and parsley (Petroselinum sativum) aqueous extracts: Lessons from experimental rats. *Journal of Diabetes Research, 2016.*

Adame, P., Alberdi, I., Cañellas, I., Hernández, L., Aguirre, A., Ruano, A., Moreno-Fernández, D., Isabel González, A., Torres, M. B., & Montes, F. (2022). Drivers and spread of non-native pests in forests: The case of Gonipterus platensis in Spanish Eucalyptus plantations. *Forest Ecology and Management, 510*, 120104. https://doi.org/10.1016/j.foreco.2022.120104

Ahmed, M. F., Kazim, S. M., Ghori, S. S., Mehjabeen, S. S., Ahmed, S. R., Ali, S. M., & Ibrahim, M. (2010). Antidiabetic activity of *Vinca rosea* extracts in alloxan-induced diabetic rats. *International Journal of Endocrinology, 2010*, 1–6. https://doi.org/10.1155/2010/841090

Ait-Ouazzou, A., Lorán, S., Arakrak, A., Laglaoui, A., Rota, C., Herrera, A., Pagán, R., & Conchello, P. (2012). Evaluation of the chemical composition and antimicrobial activity of Mentha pulegium, Juniperus phoenicea, and Cyperus longus essential oils from Morocco. *Food Research International, 45*(1), 313–319. https://doi.org/10.1016/j.foodres.2011.09.004

Ajilore, B. S., Oluwadairo, T. O., Olorunnisola, O. S., Fadahunsi, O. S., & Adegbola, P. I. (2021). GC-MS analysis, toxicological and oral glucose tolerance assessments of methanolic leaf extract of Eucalyptus globulus. *Future Journal of Pharmaceutical Sciences, 7*(1), 162. https://doi.org/10.1186/s43094-021-00312-5

Alaba, F. B. J., Avestruz, P. R. S. D., Cordero, R. R. A., Erum, E. L., Ng, M. R. S., Sumampong, M. M. Q., & Abiso-Padilla, J. (2022). Essential oils of allium sativum and eucalyptus globulus as antagonists for sars-cov-2 infection: A review. *Journal Homepage*: www.ijrpr.com ISSN 2582, 7421.

Alemayehu, A., & Melka, Y. (2022). Small scale eucalyptus cultivation and its socioeconomic impacts in Ethiopia: A review of practices and conditions. *Trees, Forests and People, 8*, 100269. https://doi.org/10.1016/j.tfp.2022.100269

Al-Mijalli, S. H., Assaggaf, H., Qasem, A., El-Shemi, A. G., Abdallah, E. M., Mrabti, H. N., & Bouyahya, A. (2022). Antioxidant, antidiabetic, and antibacterial potentials and chemical composition of salvia officinalis and mentha suaveolens grown wild in Morocco. *Advances in Pharmacological and Pharmaceutical Sciences, 2022.*

Annapurna, A., Mahalakshmi, D. K., & Krishna, K. M. (2001). Antidiabetic activity of a poly herbal preparation (tincture of panchparna) in normal and diabetic rats. *Indian Journal of Experimental Biology, 39*(5), 500–502.

Assaggaf, H. M., Naceiri Mrabti, H., Rajab, B. S., Attar, A. A., Alyamani, R. A., Hamed, M., El Omari, N., El Menyiy, N., Hazzoumi, Z., Benali, T., Al-Mijalli, S. H., Zengin, G., AlDhaheri, Y., Eid, A. H., & Bouyahya, A. (2022a). Chemical analysis and investigation of biological effects of salvia officinalis essential oils at three phenological stages. *Molecules*, *27*(16), 5157. https://doi.org/10.3390/molecules27165157

Assaggaf, H. M., Naceiri Mrabti, H., Rajab, B. S., Attar, A. A., Hamed, M., Sheikh, R. A., Omari, N. E., Menyiy, N. E., Belmehdi, O., & Mahmud, S. (2022b). Singular and combined effects of essential oil and honey of eucalyptus globulus on anti-inflammatory, antioxidant, dermatoprotective, and antimicrobial properties: In vitro and in vivo findings. *Molecules*, *27*(16), 5121.

Association, A. D. (2004). Global prevalence of diabetes. *Diabetes Care*, *27*(2), 1047–1053.

Atanasov, A. G., Waltenberger, B., Pferschy-Wenzig, E.-M., Linder, T., Wawrosch, C., Uhrin, P., Temml, V., Wang, L., Schwaiger, S., & Heiss, E. H. (2015). Discovery and resupply of pharmacologically active plant-derived natural products: A review. *Biotechnology Advances*, *33*(8), 1582–1614.

Bachir, R. G., & Benali, M. (2008). Antibacterial activity of leaf essential oils of Eucalyptus globulus and Eucalyptus camaldulensis. *African Journal of Pharmacy and Pharmacology*, *2*(10), 211–215.

Bachir, R. G., & Benali, M. (2012). Antibacterial activity of the essential oils from the leaves of Eucalyptus globulus against Escherichia coli and Staphylococcus aureus. *Asian Pacific Journal of Tropical Biomedicine*, *2*(9), 739–742.

Baso, C. (1999). Usos alternativos de la Madera de eucalipto. *Perspectivas de Utilización En Carpintería y Mobiliario. En "Boletín de Información Técnica AITIM*, *198*, 19–22.

Bencheikh, D., Gueddah, A., Soualat, K., Ben-aissi, H., Benslama, A., Harrar, A., & Khennouf, S. (2021). Polyphenolic contents, antioxidant and antibacterial activities of aqueous extracts of. *Journal of Applied Biological Sciences*, *15*(1), 53–63.

Benkhnigue, O., Akka, F. B., Salhi, S., Fadli, M., & Zidane, L. (2014). Catalogue des plantes médicinales utilisées dans le traitement du diabète dans la région d'Al Haouz-Rhamna (Maroc). *JAPS*, *23*(1), 3539–3568.

Bhushan, M. S. (2010). An analytical review of plants for anti diabetic activity with their phytoconstituent & mechanism of action. *IJPSR*, *1*(1), 18.

Billing, J., & Sherman, P. W. (1998). Antimicrobial functions of spices: Why some like it hot. *The Quarterly Review of Biology*, *73*(1), 3–49. https://doi.org/10.1086/420058

Bolaños, M. M. (1946). *Impresiones comentadas sobre los eucaliptos de Sierra Cabello*. https://biblioteca-digital.infor.cl

Boulekbache-Makhlouf, L., Meudec, E., Chibane, M., Mazauric, J.-P., Slimani, S., Henry, M., Cheynier, V., & Madani, K. (2010). Analysis by high-performance liquid chromatography diode array detection mass spectrometry of phenolic compounds in fruit of eucalyptus globulus cultivated in Algeria. *Journal of Agricultural and Food Chemistry*, *58*(24), 12615–12624. https://doi.org/10.1021/jf1029509

Boulekbache-Makhlouf, L., Slimani, S., & Madani, K. (2013). Total phenolic content, antioxidant and antibacterial activities of fruits of Eucalyptus globulus cultivated in Algeria. *Industrial Crops and Products*, *41*, 85–89. https://doi.org/10.1016/j.indcrop.2012.04.019

Bouyahya, A., Abrini, J., Et-Touys, A., Bakri, Y., & Dakka, N. (2017). Indigenous knowledge of the use of medicinal plants in the North-West of Morocco and their biological activities. *European Journal of Integrative Medicine*, *13*, 9–25. https://doi.org/10.1016/j.eujim.2017.06.004

Bouyahya, A., Bakri, Y., Khay, E. O., Edaoudi, F., Talbaoui, A., Et-Touys, A., Abrini, J., & Dakka, N. (2017). Antibacterial, antioxidant and antitumor properties of Moroccan medicinal plants: A review. *Asian Pacific Journal of Tropical Disease*, *7*(1), 57–64.

Bouyahya, A., Dakka, N., Et-Touys, A., Abrini, J., & Bakri, Y. (2017). Medicinal plant products targeting quorum sensing for combating bacterial infections. *Asian Pacific Journal of Tropical Medicine*, *10*(8), 729–743. https://doi.org/10.1016/j.apjtm.2017.07.021

Bouyahya, A., Guaouguaou, F.-E., El Omari, N., El Menyiy, N., Balahbib, A., El-Shazly, M., & Bakri, Y. (2021). Anti-inflammatory and analgesic properties of Moroccan medicinal plants: Phytochemistry, in vitro and in vivo investigations, mechanism insights, clinical evidences and perspectives. *Journal of Pharmaceutical Analysis*, *12*(1), 35–57.

Capetti, F., Cagliero, C., Marengo, A., Bicchi, C., Rubiolo, P., & Sgorbini, B. (2020). Bio-Guided fractionation driven by in vitro α-amylase inhibition assays of essential oils bearing specialized metabolites with potential hypoglycemic activity. *Plants*, *9*(1242), 1–17. https://doi.org/doi:10.3390/plants9091242

Cerasoli, S. (2016). Usage and threats Eucalyptus globulus and other eucalypts. In *European atlas of forest tree species*.

Chalchat, J. C., Chabard, J. L., Gorunovic, M. S., Djermanovic, V., & Bulatovic, V. (1995). Chemical composition of eucalyptus globulus oils from the montenegro coast and east coast of Spain. *Journal of Essential Oil Research*, *7*(2), 147–152. https://doi.org/10.1080/10412905.1995.9698489

Craig, C. R., & Stitzel, R. E. (2004). *Modern pharmacology with clinical applications*. Lippincott Williams & Wilkins.

Damjanović-Vratnica, B., Đakov, T., Šuković, D., & Damjanović, J. (2011). Antimicrobial effect of essential oil isolated from Eucalyptus globulus Labill. From Montenegro. *Czech Journal of Food Sciences*, *29*(3), 277–284. https://doi.org/10.17221/114/2009-CJFS

Derwich, E., Benziane, Z., & Boukir, A. (2009). GC/MS analysis of volatile constituents and antibacterial activity of the essential oil of the leaves of Eucalyptus globulus in atlas median from Morocco. *Advances in Natural and Applied Sciences*, *3*(3), 305+. Gale Academic OneFile. https://link.gale.com/apps/doc/A235407226/AONE?u=anon~8d422f05&sid=googleScholar&xid=f896bcdc

El Alami, A., Fattah, A., & Chait, A. (2020). Medicinal plants used for the prevention purposes during the Covid-19 pandemic in Morocco. *Journal of Analytical Sciences and Applied Biotechnology*, *2*(1), 4–11. https://doi.org/10.48402/IMIST.PRSM/JASAB-V2I1.21056

El Hachlafi, N., Chebat, A., Soulaymani Bencheikh, R., & Fikri-Benbrahim, K. (2020). Ethnopharmacological study of medicinal plants used for chronic diseases treatment in Rabat-Sale-Kenitra region (Morocco). *Ethnobotany Research and Applications*, *20*, 1–23. https://doi.org/10.32859/era.20.2.1-23

Fernandes, A., Cruz-Lopes, L., Dulyanska, Y., Domingos, I., Ferreira, J., Evtuguin, D., & Esteves, B. (2022). Eco valorization of eucalyptus globulus bark and branches through liquefaction. *Applied Sciences*, *12*(8), 3775. https://doi.org/10.3390/app12083775

Flores-Macías, A., Reyes-Zarate, G. G., Gomes da Camara, C. A., Reyes-López, R., Campos Guillén, J., & Ramos-López, M. Á. (2021). Chemical composition and phytotoxic potential of Eucalyptus globulus essential oil against Lactuca sativa and two herbicide-resistant weeds: Avena fatua and Amaranthus hybridus. *TIP Revista Especializada En Ciencias Químico-Biológicas*, *24*, 1–8. https://doi.org/10.22201/fesz.23958723e.2021.303

Ghaffar, A., Yameen, M., Kiran, S., Kamal, S., Jalal, F., Munir, B., Saleem, S., Rafiq, N., Ahmad, A., Saba, I., & Jabbar, A. (2015). Chemical composition and in-vitro evaluation of the antimicrobial and antioxidant activities of essential oils extracted from seven eucalyptus species. *Molecules*, *20*(11), 20487–20498. https://doi.org/10.3390/molecules201119706

Ghalem, B. R., & Mohamed, B. (2008). Antibacterial activity of leaf essential oils of Eucalyptus globulus and Eucalyptus camaldulensis. *African Journal of Pharmacy and Pharmacology*, *2*(10), 211–215.

Gunjan, M., Ravindran, M., & Jana, G. K. (2011). A review on some potential traditional phytomedicine with antidiabetic properties. *International Journal of Phytomedicine*, *3*(4), 448.

Hazelton, P., & Murphy, B. (2016). *Interpreting soil test results: What do all the numbers mean?* CSIRO Publishing.

Hernández, M. A., Butler, J. B., Ammitzboll, H., Weller, J. L., Vaillancourt, R. E., & Potts, B. M. (2022). Genetic control of the operculum and capsule morphology of Eucalyptus globulus. *Annals of Botany*, *130*(1), 97–108. https://doi.org/10.1093/aob/mcac072

Hilpipre, C. (2014). *Huile essentielle d'eucalyptus globulus: Utilisation et bienfaits*. Passeport Sante. www.passeportsante.net/fr/Solutions/HuilesEssentielles/Fiche.aspx?doc=huile-essentielle-eucalyptus-globulus

Houacine, C., Elkhawad, A. O., & Ayoub, S. M. H. (2012). A comparative study on the anti-diabetic activity of extracts of some Algerian and Sudanese plants. *Journal of Diabetes and Endocrinology*, *3*(3), 25–28. https://doi.org/10.5897/JDE12.002

Husen, A. (2021). *Traditional herbal therapy for the human immune system*. CRC Press, Boca Raton, FL. https://doi.org/10.1201/9781003137955

Husen, A. (2022). *Herbs, shrubs and trees of potential medicinal benefits*. Boca Raton and London: CRC Press. https://doi.org/10.1201/9781003205067

Husen, A., Bachheti, R. K., & Bachheti, A. (2021). *Non-timber forest products: Food, healthcare and industrial applications*. Springer, Cham, Switzerland. https://doi.org/10.1007/978-3-030-73077-2

Jarald, E., Joshi, S. B., & Jain, D. (2008). Diabetes and herbal medicines. *Iranian Journal of Pharmacology & Therapeutics*, *7*(1), 97–106.

Jothy, S. L., Zakaria, Z., Chen, Y., Lau, Y. L., Latha, L. Y., & Sasidharan, S. (2011). Acute oral toxicity of methanolic seed extract of Cassia fistula in mice. *Molecules*, *16*(6), 5268–5282.

Jouad, H., Maghrani, M., Hassani, R. A. E., & Eddouks, M. (2004). Hypoglycemic activity of aqueous extract of *Eucalyptus globulus* in normal and streptozotocin-induced diabetic rats. *Journal of Herbs, Spices & Medicinal Plants*, *10*(4), 19–28. https://doi.org/10.1300/J044v10n04_03

Juergens, U. R., Dethlefsen, U., Steinkamp, G., Gillissen, A., Repges, R., & Vetter, H. (2003). Anti-inflammatory activity of 1.8-cineol (eucalyptol) in bronchial asthma: A double-blind placebo-controlled trial. *Respiratory Medicine*, *97*(3), 250–256.

Juergens, U. R., Engelen, T., Racké, K., Stöber, M., Gillissen, A., & Vetter, H. (2004). Inhibitory activity of 1, 8-cineol (eucalyptol) on cytokine production in cultured human lymphocytes and monocytes. *Pulmonary Pharmacology & Therapeutics*, *17*(5), 281–287.

Juergens, U. R., Stöber, M., & Vetter, H. (1998). The anti-inflammatory activity of L-menthol compared to mint oil in human monocytes in vitro: A novel perspective for its therapeutic use in inflammatory diseases. *European Journal of Medical Research*, *3*(12), 539–545.

Kala, C. P., Dhyani, P. P., & Sajwan, B. S. (2006). Developing the medicinal plants sector in northern India: Challenges and opportunities. *Journal of Ethnobiology and Ethnomedicine*, *2*(1), 1–15.

Kamble, U. B., Chaware, V. J., & Redasani, V. K. (2021). Evaluation of antidiabetic potential of eucalyptus globulus plant extract in alloxan-induced diabetic rats. *Asian Journal of Pharmaceutical Research and Development*, *9*(5), 17–28. http://doi.org/10.22270/ajprd.v9i51021

Kebebew, Z. (2002). *Profitability and household income contribution of growing Eucalyptus globulus (Labill.) to smallholder farmers: The case of the central highlands of Oromia, Ethiopia*. Swedish University of Agricultural Sciences.

Kong, J.-M., Goh, N.-K., Chia, L.-S., & Chia, T.-F. (2003). Recent advances in traditional plant drugs and orchids. *Acta Pharmacologica Sinica*, *24*(1), 7–21.

Kooti, W., Farokhipour, M., Asadzadeh, Z., Ashtary-Larky, D., & Asadi-Samani, M. (2016). The role of medicinal plants in the treatment of diabetes: A systematic review. *Electronic Physician*, *8*(1), 1832.

Kouchadé, S. A., Adjatin, A. R., Adomou, A. C., Dassou, H. G., & Akoègninou, A. (2017). Phytochimiques des plantes médicinales utilisées dans la prise en charge des maladies infantiles au Sud-Bénin. *European Scientific Journal*, *13*(3), 471–488.

Lahsissene, H., & Kahouadji, A. (2010). Usages thérapeutiques traditionnels des plantes médicinales dans le Maroc occidental: Cas de la région de Zaër. *Phytothérapie*, *8*(4), 210–217. https://doi.org/10.1007/s10298-010-0566-5

Lemenih, M., Karltun, E., & Olsson, M. (2005). Assessing soil chemical and physical property responses to deforestation and subsequent cultivation in smallholders farming system in Ethiopia. *Agriculture, Ecosystems & Environment*, *105*(1–2), 373–386.

Miguel, M., Gago, C., Antunes, M., Lagoas, S., Faleiro, M., Megías, C., Cortés-Giraldo, I., Vioque, J., & Figueiredo, A. (2018). Antibacterial, antioxidant, and antiproliferative activities of corymbia citriodora and the essential oils of eight eucalyptus species. *Medicines*, *5*(3), 61. https://doi.org/10.3390/medicines5030061

Mikou, K., Rachiq, S., & Jarrar Oulidi, A. (2015). Étude ethnobotanique des plantes médicinales et aromatiques utilisées dans la ville de Fès au Maroc. *Phytothérapie*, 1–9. https://doi.org/10.1007/s10298-015-0965-8

Morales, M., Quintero, J., & Aroca, G. (2017). Environmental assessment of the production and addition of bioethanol produced from Eucalyptus globulus to gasoline in Chile. *The International Journal of Life Cycle Assessment*, *22*(4), 525–536. https://doi.org/10.1007/s11367-016-1119-4

Moreira, P., Sousa, F. J., Matos, P., Brites, G. S., Gonçalves, M. J., Cavaleiro, C., Figueirinha, A., Salgueiro, L., Batista, M. T., Branco, P. C., Cruz, M. T., & Pereira, C. F. (2022). Chemical composition and effect against skin alterations of bioactive extracts obtained by the hydrodistillation of eucalyptus globulus leaves. *Pharmaceutics*, *14*(3), 561. https://doi.org/10.3390/pharmaceutics14030561

Muller, C. H. (2009). The allelopathic effects of eucalyptus camaldulensis. *The American Midland Naturalist*, *83*(1), 254–282.

Omari, N. E., Sayah, K., Fettach, S., Blidi, O. E., & Bouyahya, A. (2019). Evaluation of in vitro antioxidant and antidiabetic activities of. *Evidence-Based Complementary and Alternative Medicine*, *2019*, 10. https://doi.org/10.1155/2019/7384735

Paine, T. D., & Hanlon, C. C. (2010). Integration of tactics for management of Eucalyptus herbivores: Influence of moisture and nitrogen fertilization on red gum lerp psyllid colonization. *Entomologia Experimentalis et Applicata*, *137*(3), 290–295. https://doi.org/10.1111/j.1570-7458.2010.01064.x

Prabhakar, P. K., & Doble, M. (2008). A target based therapeutic approach towards diabetes mellitus using medicinal plants. *Current Diabetes Reviews*, *4*(4), 291–308.

Preethi, P. J. (2013). Herbal medicine for diabetes mellitus: A review. *International Journal of Phytopharmacy*, *3*(1), 1–22.

Raccah, D. (2004). Epidémiologie et physiopathologie des complications dégénératives du diabète sucré. *EMC-Endocrinologie*, *1*(1), 29–42.

Rituparna, M., & Agrawal, N.K. (2007). Practitioners section-Atherosclerosis in diabetes mellitus: Role of inflammation. *Indian Journal of Medical Sciences*, *61*(5), 292–306.

Sabu, M. C., & Kuttan, R. (2002). Anti-diabetic activity of medicinal plants and its relationship with their antioxidant property. *Journal of Ethnopharmacology*, *6*.

Salehi, B., Sharifi-Rad, J., Quispe, C., Llaique, H., Villalobos, M., Smeriglio, A., Trombetta, D., Ezzat, S. M., Salem, M. A., Zayed, A., Salgado Castillo, C. M., Yazdi, S. E., Sen, S., Acharya, K., Sharopov, F., & Martins, N. (2019). Insights into Eucalyptus genus chemical constituents, biological activities and health-promoting effects. *Trends in Food Science & Technology*, *91*, 609–624. https://doi.org/10.1016/j.tifs.2019.08.003

Salhi, S., Fadli, M., Zidane, L., & Douira, A. (2010). Etudes floristique et ethnobotanique des plantes médicinales de la ville de Kénitra (Maroc). *Lazaroa*, *31*, 133–143. https://doi.org/10.5209/rev_LAZA.2010.v31.9

Serafino, A., Vallebona, P. S., Andreola, F., Zonfrillo, M., Mercuri, L., Federici, M., Rasi, G., Garaci, E., & Pierimarchi, P. (2008). Stimulatory effect of Eucalyptus essential oil on innate cell-mediated immune response. *BMC Immunology*, *9*(1), 1–16.

Shiferaw, Y., Kassahun, A., Tedla, A., Feleke, G., & Abebe, A. A. (2019). Investigation of essential oil composition variation with age of eucalyptus globulus growing in ethiopia. *Natural Products Chemistry & Research*, *7*(2). https://doi.org/10.35248/2329-6836.19.7.360

Silva, J., Abebe, W., Sousa, S. M., Duarte, V. G., Machado, M. I. L., & Matos, F. J. A. (2003). Analgesic and anti-inflammatory effects of essential oils of Eucalyptus. *Journal of Ethnopharmacology*, *89*(2–3), 277–283.

Silva, R. C. da, Beneton, K., Ieda, J. J. C., Azevedo, A. C. de, & Pereira, M. G. (2022). Use of agromineral as substrate for growth of eucalyptus seedlings. *Floresta*, *52*(1), 25. https://doi.org/10.5380/rf.v52i1.73752

Singh, L. W. (2011). Traditional medicinal plants of Manipur as anti-diabetics. *Journal of Medicinal Plants Research*, *5*(5), 677–687.

Song, A., Wang, Y., & Liu, Y. (2009). Study on the chemical constituents of the essential oil of the leaves of Eucalyptus globulus Labill from China. *Asian Journal of Traditional Medicines*, *4*(4), 134–140.

Srivastava, M., & Kumar, S. (2013). Fatty acid compositional studies of lesser known *Acacia tortilis* seed oil. *Food and Nutrition Sciences*, *4*(9), 59–62. https://doi.org/10.4236/fns.2013.49A1010

Tan, M., Zhou, L., Huang, Y., Wang, Y., Hao, X., & Wang, J. (2008). Antimicrobial activity of globulol isolated from the fruits of Eucalyptus globulus Labill. *Natural Product Research*, *22*(7), 569–575.

Tripathi, B. K., & Srivastava, A. K. (2006). Diabetes mellitus: Complications and therapeutics. *Medical Science Monitor*, *12*(7), 130–147.

Usman, L. A., Oguntoye, O. S., & Ismaeel, R. O. (2020). Effect of seasonal variation on chemical composition, antidiabetic and antioxidant potentials of leaf essential oil of *Eucalyptus globulus* L. *Journal of Essential Oil Bearing Plants*, *23*(6), 1314–1323. https://doi.org/10.1080/0972060X.2020.1862710

Usman, L. A., Oguntoye, S. O., & Ismaeel, R. O. (2022). Effect of collection time on the chemical composition, antioxidant and antidiabetic potentials of leaf essential oil of eucalyptus globulus L. growing in North-Central, Nigeria. *Chemistry Africa*, *5*(2), 257–267. https://doi.org/10.1007/s42250-022-00325-4

Vázquez, G., Fontenla, E., Santos, J., Freire, M. S., González-Álvarez, J., & Antorrena, G. (2008). Antioxidant activity and phenolic content of chestnut (Castanea sativa) shell and eucalyptus (Eucalyptus globulus) bark extracts. *Industrial Crops and Products*, *28*(3), 279–285.

Vigan, M., & Besançon, C. H. U. (2009). Les huiles essentielles: Leurs retour et leur toxicité. *Progres En Dermato Allergologie Bordeau*, 123–136.

Vigo, E., Cepeda, A., Perez-Fernandez, R., & Gualillo, O. (2004). In-vitro anti-inflammatory effect of eucalyptus globulus and thymus vulgaris: Nitric oxide inhibition in J774A. 1 murine macrophages. *Journal of Pharmacy and Pharmacology*, *56*(2), 257–263.

Wannes, W. A., & Tounsi, M. S. (2020). Can medicinal plants contribute to the cure of Tunisian COVID-19 patients? *Journal of Medicinal Plants Studies*, *8*(5), 218–226. https://doi.org/10.22271/plants.2020.v8.i5c.1218

Yitaferu, B., Abewa, A., & Amare, T. (2013). Expansion of eucalyptus woodlots in the fertile soils of the highlands of Ethiopia: Could it be a treat on future cropland use? *Journal of Agricultural Science*, *5*(8), 97.

Zonfrillo, M., Andreola, F., Krasnowska, E. K., Sferrazza, G., Pierimarchi, P., & Serafino, A. (2022). Essential oil from eucalyptus globulus (Labill.) activates complement receptor-mediated phagocytosis and stimulates podosome formation in human monocyte-derived macrophages. *Molecules*, *27*(11), 3488.

32 Antidiabetic Potential of Orchid Tree [*Bauhinia variegata* (L.) Benth.]

H. D. T. Madhuranga and D. N. A. W. Samarakoon

CONTENTS

32.1 INTRODUCTION

The genus *Bauhinia* belongs to the family Fabaceae, Leguminosae, consisting of approximately 300 species. It is commonly known as "cow's paw" or "cow's hoof" because of the shape of its leaves. Most of the plants are widely native and distributed in most tropical countries, including Africa, Asia, and South America (Filho, 2009). Their leaves, bark, and stems have been used mostly in folk medicine as a remedy for different kinds of diseases, particularly diabetes, infections, pain, and inflammatory process (Filho, 2009). However, recent interest in these plants has increased considerably worldwide since experimental studies showed their therapeutic properties.

Bauhinia variegata is one of the main species from the genus *Bauhinia* and is commonly known as "orchid tree" in English and also as "camel's foot tree." It is a small to medium-sized tree growing to 10–12 m tall, deciduous in the dry season (Shahana et al., 2017). *Bauhinia variegata* is native to southeastern Asia and grows throughout India and China (Deswal & Arora, 2015). It is usually cultivated in India, ascending altitudes up to 1800 m in the Himalayas (Sharma et al., 2021). The plant's freshly acquired bark is grayish-brown on the outside and cream-colored on the inside

DOI: 10.1201/b23347-32

FIGURE 32.1 *Bauhinia variegata.*

(Mazumder et al., 2022). The fast-growing tree grows 20 to 40 feet in height with a 25- to 35-foot spread; the slender trunks are topped with arching branches clothed in large, two-lobed deciduous leaves, which are 4–6 in. across and rounded with lobed ends and heart-shaped bases. The leaves are shaped a little like a cow's hoof (Deswal & Arora, 2015). More commonly the beautiful white variety of the more commonly seen orchid tree has purple variegated flowers. The flowers have four pure-white petals and one variegated deep purple. Orchid tree is closely related to the peacock flower and is staggeringly beautiful when in bloom, and it blooms for several months.

32.1.1 Botanical Description

In dry forests, *Bauhinia variegata* grows to be small to medium-sized evergreen trees with short trunks and broad crowns that can reach heights of up to 15 meters and a radius of 50 centimeters (Golwala & Patel, 2009) (Figure 32.1).

32.1.2 Bark

The bark shows a light brownish-gray color, smooth to slightly fissured and scaly. The inner bark is pinkish, fibrous, and bitter. The twigs are slender and zigzag; when young, they are light green, slightly hairy, and angled, becoming brownish gray (Pandey & Agrawal, 2009).

32.1.3 Leaves

Leaves consist of minute stipules 1–2 mm, early caducous; petiole puberulous to glabrous, 3–4 cm; lamina broadly ovate to circular, often broader than long, 6–16 cm diameter; 11–13 nerved; tips of lobes rounded widely, base cordate; top surface glabrous, bottom glaucous but glabrous when fully developed (Pandey & Agrawal, 2009).

32.1.4 Flowers

At the tips of twigs, flower bunches (racemes) are unbranched. The few flowers have a stalk-like, green, thin basal tube and short, robust stalks (hypanthium). Five petals that are slightly out of balance have wavy margins and are narrowed at the base; five curved stamens; and a very slender, stalked, curving pistil with a narrow, green, one-celled ovary, style, and dot-like stigma. The light green, moderately hairy calyx forms a pointed, five-angled bud and splits open on one side while staying connected (Yadava & Reddy, 2003).

32.1.5 Seeds

Pods are dehiscent, strap-shaped, obliquely striate, 20–30 by 2–25 cm; long, hard, flat with 10–15 seeds in each; seeds brown, flat, and nearly circular with coriaceus testa (Deswal & Arora, 2015).

32.2 PHYTOCHEMICAL CONSTITUENTS OF *BAUHINIA VARIEGATA*

The different parts of *Bauhinia variegata* have a wide range of phytochemicals (quercitroside, rutoside, isoquercitroside, phosphorus, calcium, 5,7 dimethoxy flavanone-4-o- L rhamnopyrosyl-β-D-glycopyranoside, etc.) which possess different types of biological activities such as anticarcinogenic, antimicrobial, antidiabetic, antiulcer, hematinic, anti-inflammatory, hemagglutinating, immunomodulatory, hepatoprotective, antitumor, antibacterial, antioxidant, hypolipidemic, and wound healing, as reported by various researchers (Sharma et al., 2021) (Table 32.1).

TABLE 32.1
Phytochemical Constituents and Therapeutic Properties of Various Parts of *Bauhinia variegata*

Part of plant	Chemical constituents	Biological activity	References
Flowers	malvidin-3 diglucoside, cynidin-3-glucoside, peonidin-3-diglucoside, malvidin-3 glucoside, quercitroside, rutoside, isoquercitroside, taxifolinerhamnoside, kaempferol-3-glucoside, myricetrol, ascorbic acid, aspartic acid, glutamic acid, keto acids, octadecanoic acid, amino acids, apigenin, tannins	antidiarrheal, antidiabetic, antioxidant, antihyperlipidemic activity	(Dugasani et al., 2010; Gilman et al., 2018; Ahmed et al., 2012)
Leaves	crude protein, phosphorus, calcium, lupeol, carbohydrates, vitamin C, reducing sugars, saponins, fibers, quercetin, quercitrin, β-sitosterol, terpenoids, kaempferol-3-glucoside, tannin, rutin, heptatriacontane-12,13-diol 7 dotetracont-15- en-9-ol ellagic acid, catechol, sterols, tannins, oil, alkaloids, fats, lignin, glycoside, phenolics, apigenin-7-o-glycoside amides	antifungal, antimicrobial, antidiabetic, hypoglycemic, molluscicidal effect, anticancerous activity	(Singh et al., 2006; Shamran et al., 2020)
Stem bark	lupeol, kaempferol-3-glucoside, β-isosterol, 5,7 dimethoxy flavanone- 4-o-L, rhamnopyrosyl-β D-glycopyranoside, hentriacontane, stigmasterol, octacosanol, reducing sugars, nitrogenous substances	antitumor, antiulcer, immunomodulatory, hematinic, antimicrobial, hepatoprotective, antioxidant, antibacterial, anticarcinogenic	(Yadava & Reddy, 2003; Dhale, 2011; Parekh et al., 2009; Singh et al., 2019)
Root Bark	flavanone (2S)-5, dihydrodibenzoxepin, 7-dimethoxy-3,4methylene dioxyflavonone 5,6b dihydroo-1,7-dihydro-1,7 dihydroxy-3,4-dimethoxy-2-methyldibenz oxepin	antioxidant, antiobesity effect	(Yadava & Reddy, 2003) (Maldonado et al., 2003)

(*Continued*)

TABLE 32.1
(Continued)

Part of plant	Chemical constituents	Biological activity	References
Stem	β-sitosterol, naringenin 5,7 dimethyl ether 4- rhamnoglucoside, lupeol	antiulcer	(Zhao et al., 2005) (Pani et al., 2011)
Roots	flavonol glycosides 5,7,3,4 tetrahydroxy-3-methoxy-7-o-α-L rhamnopyranosyl (1–3)-o-β-D-galactopyranoside	anti-inflammatory, wound healing and nephroprotective effect, antimutagenic and antioxidant activity	(Gunalan et al., 2011)
Seeds	Oleic acid, palmitic acid, linoleic acid, stearic acid, proteins.	hemagglutinating	(Shahana et al., 2017; Sharma et al., 2021)

The buds of the plant consist of a high composition of total carbohydrates, crude proteins, fibers, and so on. Other parts of the plant such as flowers and leaves also contain a sufficient amount of nutrients. The leaves of *B. variegata* are very rich in reducing sugars, polyphenolics, and minerals, as well as containing some amount of vitamin C (Krishnamoorthy et al., 2014). The flowers contain digestible carbohydrates, a significant number of proteins, fat content, and so on and are also consistent with high energy value. Further, various phototherapeutic constituents are also present in flowers (Sharma et al., 2021).

32.2.1 Roots

The qualitative chemical analysis of the *B. variegata* root powder revealed the presence of proteins, gums, mucilages, glycosides, flavonoids, tannins, and phenolic substances.

32.2.2 Bark

The bark yields fiber and tannins. In the continuing search for novel agents, seven flavonoids, namely kaempferol (1), ombuin (2), kaempferol 7,4′- dimethyl ether 3-O-β-D-glucopyranoside (3), kaempferol 3-O-β-D-glucopyranoside (4), isorhamnetin 3-O-β-D-glucopyranoside (5) and hesperidin (6), together with one triterpene caffeate, 3β-trans-(3,4-dihydro xycinnamoyloxy) olean-12-en-28-oic acid (7), were isolated from the non-woody aerial parts of *Bauhinia variegata*. Phytochemical analysis of the root bark of *Bauhinia variegata* Linn yielded a new flavanone, (2S)-5,7-dimethoxy-30,40- methylenedioxyflavanone (1), and a new dihydrodibenzoxepin, 5,6-dihydro-1,7-dihydroxy-3,4-dimethoxy-2- methyldibenz [b,f]oxepin (2), together with three known flavonoids (3–5). The structures of the new compounds were determined based on spectral studies (Zaka et al., 1983).

The seeds of *Bauhinia variegata* contain rich phytoconstituents including carbohydrates, proteins, moisture, fats, and fibers, and oils obtained from the seeds of *Bauhinia variegata* contain various fatty acids – linolic acid, stearic acid, oleic acid, and palmitic acid, whereas margaric acid and more content such as arachidic acid, behenic acid, eicosapentaenoic, and nervonic acid are the major fatty acids in crude seed oil. Among all the fatty acids, linoleic acid predominates and amounts to 42.1% of the fatty acids and is known to be the precursor of metabolic regulatory compounds, and it is also an important constituent of the plasma membrane (Arain et al., 2012). Once the oil is removed from the seed, the remaining seed cake can be utilized for poultry and as animal feed. According to Kartik (Sharma et al., 2021), the composition of various parts of *Bauhinia variegata* can be introduced in Figure 32.2 (Sharma et al., 2021).

Constituents	Flowers	Dried leaves	Buds	Seeds
Carbohydrates (%)	16.01	66.82	6.4	28.4±1.6
Proteins (%)	3.24–5.0	15.19	3.70	41.9±1.6
Fats (%)	0.15–2.5	4.15	2.44	0.6±0.1 (FFA)
Moisture (%)	77.80	8.83	84.51	6.7±0.46
Ash (%)	2.81	4.9	4.33	4.8±0.1
Fibers (%)	8.66	4.26	6.8	6.9±0.8
Total oils (%)	–	–	–	18.0±0.1

FIGURE 32.2 Composition of various parts of Bauhinia variegate. Source: (Sharma et al., 2021).

32.3 PHARMACOLOGICAL PROPERTIES OF *BAUHINIA VARIEGATA*

Every part of the plant contains different kinds of bioactive compounds, as mentioned previously, and has health-promoting effects in one or the other form (Kumar et al., 2019). Traditionally, *Bauhinia variegata* is utilized as food, as feed, and for medicinal purposes. Its use as a medicine is for treating obesity, glandular inflammation (Kumar Sharma, 2011), dysentery, menorrhagia (Singh et al., 2002), headache, material fever, dyspepsia, and stomachache (Kumar Sharma, 2011) and for reducing gastric secretion level (Rajkapoor et al., 2003). However, recently researchers have gotten attached to its utilization in food as well as in the pharmaceutical industry (Sharma et al., 2021). The crude extract of various parts and pure isolates of the plant were shown to have antibacterial, antitumor, hypoglycemic, and anti-inflammatory activity effects. This chapter mainly focuses on the antidiabetic property of *B. variegata*. Scientific evidence can be found for the activity of this plant.

32.3.1 Antioxidant Activity

The literature revealed significant antioxidant activity was observed in all the methods, ($P < 0.01$) reducing power and ($P < 0.001$) for scavenging DPPH, superoxide, nitric oxide, and hydrogen peroxide radicals (Rajani & Ashok, 2009). The extract showed a significant reduction ($P < 0.01$) in cholesterol at 6 and 24 h and ($P < 0.05$) at 48 h. There was a significant reduction ($P < 0.01$) in triglyceride levels at 6, 24, and 48 hours. The VLDL level also significantly ($P < 0.05$) reduced from 24 h and maximum reduction ($P < 0.01$) was observed at 48 hours. There was a significant increase ($P < 0.01$) in HDL at 6, 24, and 48 hours (Rajani & Ashok, 2009). Evidence in the literature suggested alcoholic and aqueous extracts of *B. variegata* can effectively reduce plasma cholesterol, triglyceride, LDL, and VLDL and increase plasma HDL levels. Other than that, it showed antioxidant activity. By the virtue of its antioxidant activity, the plant may show antihyperlipidemic activity (Deswal & Arora, 2015).

32.3.2 Antihyperglycemic Activity

Aqueous and ethanolic plant extract in normal and streptozotocin-induced diabetic rats shows antihyperglycemic activity. A study showed a significant reduction in total cholesterol, LDL cholesterol, and VLDL cholesterol, and an improvement in HDL cholesterol in diabetic rats. These kinds of evidence indicate that *B. variegata* possesses a hypoglycemic effect (Reddy et al., 2003).

32.3.3 Antibacterial Activity

Defatted acetone and methanol extract from the bark of the plant showed the most antibacterial activity. Aqueous and methanolic extracts of *B. variegata* bark showed the best antibacterial activity (Parekh et al., 2009).

32.3.4 Immunomodulatory Activity

Ethanolic extract of the stem bark of *B. variegata* showed significant enhancement in the primary and secondary humoral antibody response and was identified and documented as an immunomodulatory agent, probably through stimulation of both specific and nonspecific immunity (Deswal & Arora, 2015).

Other than this pharmacological activity, *Bauhinia variegata* showed antitumor, chemoprotective, anti-inflammatory, antinociceptive, and antiarthritic activity (Deswal & Arora, 2015; Kaur et al., 2019).

32.4 ANTIDIABETIC POTENTIAL OF *BAUHINIA VARIEGATA*

32.4.1 Diabetes Mellitus

Diabetes mellitus is a chronic metabolic disorder of insulin deficiency or ineffectiveness, constituting a global public health burden. Predictions estimate that India, China, and the United States will have the largest number of diabetic people by the year 2030 (Kumar et al., 2012). Diabetes mellitus (DM) has grown to be a major, quickly spreading health issue and a growing financial burden that is impeding the social and economic progress of many nations. A series of metabolic illnesses known as diabetes mellitus is characterized by hyperglycemia brought on by deficiencies in insulin secretion, insulin action, or both. The chronic form of hyperglycemia of diabetes is associated with long-term damage, dysfunction, and failure of various organs, especially the eyes, kidneys, nerves, heart, and blood vessels (Zhang & Tan, 2000). Type 2 diabetes, which frequently affects adults, arises when the body either stops generating enough insulin or develops a resistance to it. Type 2 diabetes prevalence has sharply increased during the past three decades in nations of all income levels. Lack of insulin is the cause of type 1 diabetes. About 422 million people worldwide have diabetes, the majority living in low- and middle-income countries, and 1.6 million deaths occur each year due to diabetes (World Health Organization, 2022).

According to International Diabetes Federation (IDF) data, an estimated 463 million adults aged 20–79 years are currently living with diabetes. This represents 9.3% of the world's population in this age group. The total number is predicted to rise to 578 million (10.2%) by 2030 and to 700 million (10.9%) by 2045 (International Diabeles Federation, 2019). As well as these, the estimated number of adults between 20 and 79 years with impaired glucose tolerance is 374 million (7.5% of the world group). This is predicted to rise to 454 million (8.0%) by 2030 and 548 million (8.6%) by 2045. And an estimated 1.1 million children and adolescents (aged under 20 years) have type 1 diabetes. However, the data estimation is limited among children and adolescents diagnosed with type 2 diabetes (International Diabeles Federation, 2019). An estimated 15.8% (20.4 million) of live births are affected by hyperglycemia in pregnancy in 2019. IDF detailed the number of deaths resulting from diabetes and its complications in 2019 to be estimated at around 4.2 million (International Diabeles Federation, 2019).

Annual global health expenditure on diabetes is estimated to be USD 760 billion. It is projected that expenditure will reach USD 825 billion by 2030 and USD 845 billion by 2045 (International Diabeles Federation, 2019).

Therefore, the search for plant-derived products for the control of diabetes mellitus is continuous, and the World Health Organization (WHO) long ago recommended herbal treatment of diabetes mellitus (Kumar et al., 2012). However, in many developing countries, because the high cost of proprietary medicines is coupled with extremes of poverty that prohibit their regular use, native and traditional medicines are used extensively to treat diseases. Accordingly, botanicals can provide affordable treatment for diabetes worldwide. Plants in different genera have been effectively used in ethnomedicine for the treatment of diabetes. Plants within the genus *Bauhinia* are used in Brazil, Africa, and India to treat diabetes in native populations (Frankish et al., 2010).

Given that background, *Bauhinia variegata* (family Fabaceae), called orchid tree in English and kobolila in Sinhala, is an herbaceous medicinal plant found throughout tropical countries. The leaves of the many *Bauhinia* species are used in antidiabetic treatments by many populations of the world (Jang et al., 2008). The stem bark of. *B. variegata* in the Ayurvedic system of medicine in India is used as an antidiabetic medication (Kumar et al., 2012). In recent *in vivo* investigations, the ethanolic extract of the plant and its major constituent, rose oxide, demonstrated enhanced insulin release from the beta cell lines INS-1 (Frankish et al., 2010). This work studied the influence of the stem bark of *B. variegata* on alloxan or streptozotocin-induced hyperglycemia in rats and men (Frankish et al., 2010; Kumar et al., 2012).

32.4.2 *In Vitro* Antidiabetic Activity of *Bauhinia variegata*

Bauhinia variegata is used in Egyptian folk medicine for its hyperglycemic effects. Assays for the inhibition of α-glucosidase and *in vitro* antioxidant activity were performed. The IC_{50} of leaf extract in α-glucosidase inhibition assay was found to be 0.139 mg/ml compared to acarbose standard (0.789 mg/ml), revealing a strong α-glucosidase inhibitory effect (Hago et al., 2021). The measurement of insulin levels and the biomarkers for both liver and kidney functions were evaluated as well. They were not limited to the hypoglycemic effects but included preventing liver and kidney tissue damage that is associated with diabetes (Hago et al., 2021). Azevedo et al. (2006) showed the hypoglycemic effects of *Bauhinia variegata* leaf extract to a chloroplast protein with a partial amino acid sequence identical to that of bovine insulin, while rose oxide, a major secondary metabolite in the ethanolic leaf extract, demonstrated insulinotropic activity toward pancreatic β-cells in an *in vitro* assay (Azevedo et al., 2006; Wolfe et al., 2003). This literature suggested *Bauhinia variegata* leaves are rich in phenolic compounds, especially flavonoids, which are known for their antioxidant potential and can be especially beneficial in many diseases since oxidative stress is a contributing factor as in diabetes (Hago et al., 2021).

32.4.3 *In Vivo* Antidiabetic Activity of *Bauhinia variegata*

The standardized ethanolic extract showed a significant decrease in fasting blood glucose level at two different doses (250 and 500 mg/kg) in the streptozocin-induced diabetic rats model (Hago et al., 2021). *Bauhinia variegata* in the treatment of diabetes has been practiced for centuries, and the administration of gliclazide (15 mg/kg BW), group 3, and *Bauhinia variegata* leaf extract at two different doses, group 4 through 7, resulted in a significant reduction of fasting blood glucose level when compared to group 2 of diabetic animals (Hago et al., 2021).

Another *in vivo* study was conducted using ethanolic extract of *Bauhinia variegata* and was administered orally to streptozotocin-induced diabetic rats once daily for 21 days. Blood glucose levels were estimated at days 0, 7, 14, and 21 by glucometer (one touch), and lipid profile and histopathological examination of isolated organs (kidney, liver, and pancreas) were also estimated in 21 days. Flower extract of the orchid tree showed a reduction in blood glucose level (90 mg/dl) at the highest dose 400 mg/kg when compared with diabetic control (224.50 mg/dL) and normal control groups (74 mg/dL). However, the reference group's reduced activity was slightly greater than the test groups, e.g. 83.16 mg/dL (Tripathi et al., 2019).

32.4.4 Antioxidant Activity of *Bauhinia variegata*

The body generates free radicals from metabolic processes, while antioxidant systems are present in the body to disarm them. This homeostasis in the body is disturbed due to excess free radical production, depletion of antioxidants, or both. Therefore, when the body's antioxidant system is inadequate, cells are exposed to high levels of free radicals. This condition is called oxidative stress (Corrochano et al., 2018) and causes cell injury, such as protein and lipid peroxidation, DNA fragmentation, racemization or decarboxylation of amino acids, enzyme dysfunction, and breakdown of carbohydrates, and aggravates various chronic ailments such as diabetes, cancer, rheumatoid arthritis, and heart conditions (Li et al., 2015).

Excess concentration of glucose in the blood is one of the most important causes of diabetes and secondary disorders like angiopathy, cataracts, neuropathy, deficiency in the antioxidant defense system, and lipid profile disorders (Ahangarpour et al., 2019). Evidence shows there is a strong relationship between oxidative stress-induced hyperglycemia and the progression of diabetic complications in patients with diabetes mellitus (Tripathi et al., 2019). Natural antioxidants obtained from plants play an important role in protecting the body against reactive oxygen species. Several studies reported that the genus *Bauhinia* contains antioxidant and antidiabetic activities due to the presence of polyphenol and flavonoid contents (Menezes et al., 2007).

Bauhinia variegata leaf extract showed a free radical scavenging activity using DPPH radical scavenging assay, with an IC_{50} of 1.7, while the EC_{50} of the extract relative to ascorbic acid was measured at 1.02 (EC50 0.28). Therefore, further literature revealed antioxidant activity may be related to the high concentration of flavonoids and other phenolic compounds (Hago et al., 2021). The antioxidant activity of plant extract was evaluated by performing 1,1- diphenylpicrylhydrazyl (DPPH) and hydrogen peroxide scavenging (H_2O_2) assays. In addition, the percentage inhibition of *Bauhinia variegata* was 86.6% and 68.47% at 100 μg/mL for DPPH and H_2O_2 radicals respectively, which was near to standard BHT e.g. 91.63% (DPPH) and 73.42% (H_2O_2). These findings concluded that *Bauhinia variegata* possesses significant antioxidant potential (Tripathi et al., 2019).

32.5 CONCLUSION

The genus *Bauhinia* belongs to the family Fabaceae, Leguminosae, consisting of approximately 300 species. It is commonly known as "cow's paw" or "cow's hoof" because of the shape of its leaves. Most of the plants are widely native and distributed in most tropical countries, including Africa, Asia, and South America (Jang et al., 2008). *Bauhinia variegata* is one of the main species from the genus *Bauhinia* and is commonly known as "orchid tree" in English and also as "camel's foot tree." It is a small to medium-sized tree growing to 10–12 m tall, deciduous in the dry season (Shahana et al., 2017). *Bauhinia variegata* is native to southeastern Asia and grows throughout India and China (Deswal & Arora, 2015). Flowers, stems, seeds, roots, bark, and leaves contain chemical constituents such as flavonoids, β-sitosterol, kaempferol, cardiac glycosides, tannins, terpenoids, quercitrin, and apigenin, which play a vital role in promoting human health. As a pharmacological activity, antidiabetic, anti-infective, and antioxidant properties and more are documented. *B. variegata* stems and bark are used as antidiabetics in Ayurvedic medicine in India. This chapter mainly focuses on the antidiabetic property of *B. variegata*. Scientific evidence can be found for the activity of this plant. *In vivo* study conducted using ethanolic and aqueous extracts of *B. variegata* in normal and streptozotocin-induced diabetic rats showed antihyperglycemic activity, reduction in total, LDL, and VLDL cholesterol, and an increase in HDL cholesterol, and ethanolic extract showed a significant reduction of fasting blood sugar levels at two different doses (250 and 500mg/kg) in streptozotocin-induced diabetic rat models it validated for controlling diabetes-related complications such as oxidative stress and diabetes wounds. Significant improvement of diabetic wound healing by topical application of hydrogel preparation of leaf and bark of *B. variegata* on Wister rats was found. *In vitro*, α-glucosidase enzyme inhibition assays were conducted, measuring the insulin levels and the biomarkers for both liver and kidney functions in the treated animals. They were found not to be limited to hypoglycemic effect, but also prevented liver and kidney tissue damage. *In vitro* study showed plant extract enhanced insulin release from the beta cell lines. Hyperglycemia promotes the auto-oxidation of glucose to form free radicals, therefore antioxidants are important in reducing diabetic complications. *B. variegata* has proven activity for antioxidant activity using

DPPH (percentage inhibition 91.63%) assay and H_2O_2 assays (percentage inhibition 73.42%), beta carotene bleaching assay. In conclusion, the plant *B. variegata* is found to be effective against diabetes through *in vivo* and *in vitro* testing; further, the plant has shown the ability to control unwanted side effects through antioxidants and other biochemical investigations.

REFERENCES

Ahangarpour, A., Sayahi, M., & Sayahi, M. (2019). The antidiabetic and antioxidant properties of some phenolic phytochemicals: A review study. *Diabetes and Metabolic Syndrome: Clinical Research and Reviews*, *13*(1), 854–857. https://doi.org/10.1016/j.dsx.2018.11.051

Ahmed, A. S., Elgorashi, E. E., Moodley, N., McGaw, L. J., Naidoo, V., & Eloff, J. N. (2012). The antimicrobial, antioxidative, anti-inflammatory activity and cytotoxicity of different fractions of four South African Bauhinia species used traditionally to treat diarrhoea. *Journal of Ethnopharmacology*, *143*(3), 826–839. https://doi.org/10.1016/j.jep.2012.08.004

Arain, S., Memon, N., Rajput, M. T., Sherazi, S. T. H., Bhanger, M. I., & Mahesar, S. A. (2012). Physicochemical characteristics of oil and seed residues of Bauhinia variegata and Bauhinia linnaei. *Pakistan Journal of Analytical & Environmental Chemistry*, *13*(1), 16–21. https://www.academia.edu/2694226/Physico_chemical_Characteristics_of_Oil_and_Seed_Residues_of_Bauhinia_variegata_and_Bauhinia_linnaei

Azevedo, C. R., Maciel, F. M., Silva, L. B., Ferreira, A. T. S., da Cunha, M., Machado, O. L. T., Fernandes, K. V. S., Oliveira, A. E. A., & Xavier-Filho, J. (2006). Isolation and intracellular localization of insulin-like proteins from leaves of Bauhinia variegata. *Brazilian Journal of Medical and Biological Research*, *39*(11), 1435–1444. https://doi.org/10.1590/S0100-879X2006001100007

Corrochano, A. R., Buckin, V., Kelly, P. M., & Giblin, L. (2018). Invited review: Whey proteins as antioxidants and promoters of cellular antioxidant pathways. *Journal of Dairy Science*, *101*(6), 4747–4761. https://doi.org/10.3168/jds.2017-13618

Deswal, G., & Arora, K. (2015). Ethnobotany and phytopharmacology of Bauhinia variegata. *Ijpda.Com*, *3*(9), 261–263. https://www.neliti.com/id/publications/409245/

Dhale, D. A. (2011). Phytochemical screening and antimicrobial activity of Bauhinia variegata Linn. *Journal of Ecobiotechnology*, *3*(9), 4–7. https://updatepublishing.com/journal/index.php/jebt/article/view/194

Dugasani, S., Balijepalli, M., Tandra, S., & Pichika, M. (2010). Antimicrobial activity of Bauhinia tomentosa and Bauhinia vahlii roots. *Pharmacognosy Magazine*, *6*(23), 204–207. https://doi.org/10.4103/0973-1296.66937

Filho, V. C. (2009). Chemical composition and biological potential of plants from the genus Bauhinia. *Phytotherapy Research: An International Journal Devoted to Pharmacological and Toxicological Evaluation of Natural Product Derivatives*, *23*(10), 1347–1354.

Frankish, N., De Sousa Menezes, F., Mills, C., & Sheridan, H. (2010). Enhancement of insulin release from the β-cell line INS-1 by an ethanolic extract of Bauhinia variegata and its major constituent roseoside. *Planta Medica*, *76*(10), 995–997. https://doi.org/10.1055/s-0029-1240868

Gilman, E. F., Watson, D. G., Klein, R. W., Koeser, A. K., Hilbert, D. R., & Mclean, D. C. (2018). *Bauhinia variegata: Orchid Tree*, 2–4. https://edis.ifas.ufl.edu/publication/ST092

Golwala, D. K., & Patel, L. D. (2009). Pharmacognostical studies of Bauhinia variegata linn root. *Journal of Young Pharmacists*, *1*(1), 36.

Gunalan, G., A. Saraswathy, & Krishnamurthy, V. (2011). Antimicrobial activity of medicinal plant Bauhinia variegata Linn. *International Journal of Pharmacy and Biological Sciences*, *1*(4), 400–408. http://www.ijpbs.com/ijpbsadmin/upload/ijpbs_50c8512db3420.pdf

Hago, S., Mahrous, E. A., Moawad, M., Abdel-Wahab, S., & Abdel-Sattar, E. (2021). Evaluation of antidiabetic activity of Morus nigra L. and Bauhinia variegata L. leaves as Egyptian remedies used for the treatment of diabetes. *Natural Product Research*, *35*(5), 829–835. https://doi.org/10.1080/14786419.2019.1601094

International Diabeles Federation. (2019). *IDF Diabetes Atlas*. 9th edition. International Diabetes Federation. https://diabetesatlas.org/atlas/ninth-edition/

Jang, M. H., Piao, X. L., Kim, J. M., Kwon, S. W., & Park, J. H. (2008). Inhibition of cholinesterase and amyloid-&bgr; aggregation by resveratrol oligomers from Vitis amurensis. *Phytotherapy Research*, *22*(4), 544–549. https://doi.org/10.1002/ptr

Kaur, S., Sharma, N., Vyas, M., Mahajan, R., Satija, S., Mehta, M., & Khurana, N. (2019). A review on pharmacological activities of betaine. *Plant Archives, 19*(2), 1021–1034.

Krishnamoorthy, V., Nagappan, P., Sereen, A. K., & Rajendran, R. (2014). Preliminary phytochemical screening of the fruit rind of Garcinia cambogia and leaves of Bauhinia variegata- A comparative study. *International Journal of Current Microbiology and Applied Sciences, 3*(5), 479–486.

Kumar, P., Baraiya, S., Gaidhani, S. N., Gupta, M. D., & Wanjari, M. M. (2012). Antidiabetic activity of stem bark of Bauhinia variegata in alloxan-induced hyperglycemic rats. *Journal of Pharmacology and Pharmacotherapeutics, 3*(1), 64–66. https://doi.org/10.4103/0976-500X.92518

Kumar, P., Shaunak, I., & Verma, M. L. (2019). Biotechnological application of health promising bioactive molecules. In *Biotechnological Production of Bioactive Compounds* (pp. 165–189). Elsevier. https://doi.org/10.1016/B978-0-444-64323-0.00006-0

Kumar Sharma, R. (2011). In-vitro antioxidant and protective effect of bauhinia variegata linn on gentamicin-induced nephrotoxicity in rats. *International Journal of Research in Ayurveda & Pharmacy, 2*(1), 312–318. www.ijrap.net

Li, S., Tan, H. Y., Wang, N., Zhang, Z. J., Lao, L., Wong, C. W., & Feng, Y. (2015). The role of oxidative stress and antioxidants in liver diseases. *International Journal of Molecular Sciences, 16*(11), 26087–26124. https://doi.org/10.3390/ijms161125942

Maldonado, P. D., Barrera, D., Rivero, I., Mata, R., Medina-Campos, O. N., Hernández-Pando, R., & Pedraza-Chaverrí, J. (2003). Antioxidant S-allylcysteine prevents gentamicin-induced oxidative stress and renal damage. *Free Radical Biology and Medicine, 35*(3), 317–324. https://doi.org/10.1016/S0891-5849(03)00312-5

Mazumder, A., Bharadwaj, P., & Das, D. (2022). Improvement of diabetic wound healing by topical application of hydrogel preparations of leaf and stem bark of Bauhinia Variegata on Wistar rats. *JDDST-D-22-00725.*

Menezes, F. D. S., Minto, A. B. M., Ruela, H. S., Kuster, R. M., Sheridan, H., & Frankish, N. (2007). Hypoglycemic activity of two Brazilian Bauhinia species: Bauhinia forficata L. and Bauhinia monandra Kurz. *Revista Brasileira de Farmacognosia, 17*(1), 8–13. https://doi.org/10.1590/S0102-695X2007000100003

Pandey, S., & Agrawal, R. C. (2009). Effect of Bauhinia variegate bark extract on DMBA-induced mouse skin carcinogenesis: A preliminary study. *Global Journal of Pharmacology, 3*(3), 158–162.

Pani, S., Mishra, S., Sahoo, S., & Panda, P. (2011). Nephroprotective effect of Bauhinia variegata (linn.) whole stem extract against cisplatin-induced nephropathy in rats. *Indian Journal of Pharmacology, 43*(2), 200–202. https://doi.org/10.4103/0253-7613.77370

Parekh, J., Karathia, N., & Chanda, S. (2009). Evaluation of antibacterial activity and phytochemical analysis of Bauhinia variegata L. bark. *African Journal of Biomedical Research, 9*(1), 53–56. https://doi.org/10.4314/ajbr.v9i1.48773

Rajani, G., & Ashok, P. (2009). In vitro antioxidant and antihyperlipidemic activities of Bauhinia variegata Linn. *Indian Journal of Pharmacology, 41*(5), 227–232. https://doi.org/10.4103/0253-7613.58513

Rajkapoor, B., Jayakar, B., & Murugesh, N. (2003). Antitumour activity of Bauhinia variegata on Dalton's ascitic lymphoma. *Journal of Ethnopharmacology, 89*(1), 107–109. https://doi.org/10.1016/S0378-8741(03)00264-2

Reddy, M. V. B., Reddy, M. K., Gunasekar, D., Caux, C., & Bodo, B. (2003). A flavanone and a dihydrodibenzoxepin from Bauhinia variegata. *Phytochemistry, 64*(4), 879–882. https://doi.org/10.1016/S0031-9422(03)00416-3

Shahana, S., Pratima, A., & Nikalje, G. (2017). Brief review on Bauhinia variegata: Phytochemistry, antidiabetic and antioxidant potential. *American Journal of PharmTech Research, 7*(1). https://ajptr.com/archive/volume-7/february-2017-issue-1

Shamran, D. J., Fadel, E., Al-Jumaili, A., & Tawfeeq, A. T. (2020). Cytotoxicity effect of glucokinin isolated from Bauhinia variegata against several cancer cell lines. *Jige.Uobaghdad.Edu.Iq, 19*(1), 69–74. https://jige.uobaghdad.edu.iq/index.php/IJB/article/view/392

Sharma, K., Kumar, V., Kumar, S., Sharma, R., & Mehta, C. M. (2021). Bauhinia variegata: a comprehensive review on bioactive compounds, health benefits and utilization. *Advances in Traditional Medicine, 21*(4), 645–653. https://doi.org/10.1007/s13596-020-00472-4

Singh, A., Singh, N., & Pabla, D. (2019). A review on medicinal uses of Bauhinia variegata Linn. *Pharma Tutor, 7*(6), 12–17. https://doi.org/10.29161/PT.v7.i6.2019.12

Singh, A. K., Raghubanshi, A. S., & Singh, J. S. (2002). Medical ethnobotany of the tribals of Sonaghati of Sonbhadra district, Uttar Pradesh, India. *Journal of Ethnopharmacology, 81*(1), 31–41. https://doi.org/10.1016/S0378-8741(02)00028-4

Singh, R. S., Pandey, H. S., & Ghanshyam. (2006). Two new long chain compounds from Bauhinia variegata Linn. *Indian Journal of Chemistry - Section B Organic and Medicinal Chemistry*, *45*(9), 2151–2153.

Tripathi, A. K., Gupta, P. S., & Singh, S. K. (2019). Antidiabetic, anti-hyperlipidemic and antioxidant activities of Bauhinia variegata flower extract. *Biocatalysis and Agricultural Biotechnology*, *19*, 101142. https://doi.org/10.1016/j.bcab.2019.101142

Wolfe, K., Wu, X., & Liu, R. H. (2003). Antioxidant activity of apple peels. *Journal of Agricultural and Food Chemistry*, *51*(3), 609–614. https://doi.org/10.1021/jf020782a

World Health Organization. (2022). *Diabetes*. World Health Organization. https://www.who.int/news-room/fact-sheets/detail/diabetes

Yadava, R. N., & Reddy, V. M. S. (2003). Anti-inflammatory activity of a novel flavonol glycoside from the Bauhinia variegata Linn. *Natural Product Research*, *17*(3), 165–169. https://doi.org/10.1080/1478641031000104127

Zaka, S., Saleem, M., Shakir, N., & Khan, S. A. (1983). Fatty acid composition of Bauhinia variegata and Bauhinia malabarica seed oils - Comparison of their physico-chemical properties. *Fette, Seifen, Anstrichmittel*, *85*(4), 169–170. https://doi.org/10.1002/lipi.19830850406

Zhang, X. F., & Tan, B. K. H. (2000). Effects of an ethanollic extract of Gynura procumbens on serum glucose, cholesterol and triglyceride levels in normal and streptozotocin-induced diabetic rats. *Singapore Medical Journal*, *41*(1), 9–13.

Zhao, Y. Y., Cui, C. B., Cai, B., Han, B., & Sun, Q. S. (2005). A new phenanthraquinone from the stems of Bauhinia variegata L. *Journal of Asian Natural Products Research*, *7*(6), 835–838. https://doi.org/10.1080/10286020410001721140

Index

T

U

V

Y